Multiple Messengers and Challenges in Astroparticle Physics

Roberto Aloisio · Eugenio Coccia
Francesco Vissani
Editors

Multiple Messengers and Challenges in Astroparticle Physics

GRAN SASSO SCIENCE INSTITUTE
SCHOOL OF ADVANCED STUDIES
Scuola Universitaria Superiore

Springer

Editors
Roberto Aloisio
Gran Sasso Science Institute
L'Aquila
Italy

Francesco Vissani
Gran Sasso Science Institute
L'Aquila
Italy

Eugenio Coccia
Gran Sasso Science Institute
L'Aquila
Italy

ISBN 978-3-030-09739-4 ISBN 978-3-319-65425-6 (eBook)
https://doi.org/10.1007/978-3-319-65425-6

Printed on acid-free paper

This Springer imprint is published by Springer Nature
The registered company is Springer International Publishing AG
The registered company address is: Gewerbestrasse 11, 6330 Cham, Switzerland

Preface

Astroparticle physics is a lively discipline arisen from the observation of many different "signal carriers" from the universe and from the understanding that the laws of physics studied on Earth can be used to make sense of the whole cosmos. In the past, similar observations and studies helped to figure out new laws of physics and it is plausible that this will continue to happen in future.

Astroparticle Physics evolved, thanks to the collaboration of scientists with diverse interests, from particle and nuclear physics to astrophysics, astronomy, plasma physics, and cosmology. In Astroparticle Physics, new exciting discoveries are all but scarce and experiments are blossoming in the hope of solving some of the mysteries of the universe, such as the nature of dark matter and dark energy, the nature of neutrino, the sources of gravitational waves, the origin of cosmic rays, the sources of gamma rays and astrophysical neutrinos, and the study of the final stages of stellar evolution.

Astroparticle Physics is one of the four pillars of the Gran Sasso Science Institute (GSSI) together with Mathematics, Computer Science, and Social Sciences. The GSSI is an internationally renowned center for advanced studies created after the devastating earthquake that hit the city of L'Aquila in 2009. Initially activated by INFN and now established as a new Italian University, the GSSI is a concrete and visible effort to help the city with the reconstruction process, contributing to make it a modern "city of knowledge". Over the years, fulfilling its original mission, the GSSI—and its talented doctoral students—became the symbol that it is possible to start anew by investing in human capital.

The present book "Multiple Messengers and Challenges in Astroparticle Physics" offers an overview of some of the most important topics in the field. The selection of the topics was guided by the intention to emphasize the current observational and theoretical challenges and to identify the most promising lines of research. World-leading experts were asked to contribute to the volume by reviewing the state-of-the-art and the most important open problems in their areas. Each chapter is devoted to a different topic, which include: cosmic rays, gamma rays (observed both from ground and space), astrophysical neutrinos, gravitation, and cosmology. It has been interesting to realize, as a sign of the extreme vitality of this field of research, that each review required multiple revisions to stay up-to-date with the numerous news coming from both experimental and theoretical research in the meanwhile.

The reviews collected in this volume also play a second—and equally important—role, i.e. provide a solid pedagogical basis for all the young researchers entering the field of Astroparticle Physics. Since its inception in 2013, the GSSI welcomed about 200 Ph.D. students, 50 of which in the area of Astroparticle Physics, and many more will come in the future. This book will hopefully become a must-read for those young researchers to help them find their own research agenda.

We are grateful to all the scientists who traveled from all over the world to participate in the activities of the GSSI giving lectures, seminars and contributing to the research activity of the Institute. The present book is also the outcome of the conducive atmosphere established at the GSSI in these years that resulted in many scientific contributions, important achievements, and new collaborations.

L'Aquila, Italy

Roberto Aloisio
Eugenio Coccia
Francesco Vissani

Contents

Selected Topics in Cosmic Ray Physics

Roberto Aloisio, Pasquale Blasi, Ivan De Mitri and Sergio Petrera

Abstract The search for the origin of cosmic rays is as active as ever, mainly driven by new insights provided by recent pieces of observation. Much effort is being channelled in putting the so-called supernova paradigm for the origin of galactic cosmic rays on firmer grounds, while at the highest energies we are trying to understand the observed cosmic-ray spectra and mass composition and relating them to potential sources of extragalactic cosmic rays. Interestingly, a topic that has acquired a dignity of its own is the investigation of the transition region between the galactic and extragalactic components, once associated with the ankle and now increasingly thought to be taking place at somewhat lower energies. Here, we summarize recent developments in the observation and understanding of galactic and extragalactic cosmic rays and we discuss the implications of such findings for the modelling of the transition between the two.

R. Aloisio (✉) · P. Blasi · I. De Mitri · S. Petrera
Gran Sasso Science Institute, Viale F. Crispi 7, L'Aquila, Italy
e-mail: roberto.aloisio@gssi.it

P. Blasi
e-mail: pasquale.blasi@gssi.it

I. De Mitri
e-mail: ivan.demitri@gssi.it

S. Petrera
e-mail: sergio.petrera@aquila.infn.it

R. Aloisio · P. Blasi · I. De Mitri · S. Petrera
Laboratori Nazionali Gran Sasso INFN, Assergi (L'Aquila), Italy

S. Petrera
Dipartimento di Scienze Fisiche e Chimiche, Università Dell'Aquila, L'Aquila, Italy

R. Aloisio et al. (eds.), *Multiple Messengers and Challenges in Astroparticle Physics*, https://doi.org/10.1007/978-3-319-65425-6_1

Introduction

There are different levels of understanding of the origin of cosmic rays (CRs), but a reasonable starting point is to establish some separation between the CRs that can potentially be accelerated inside the Galaxy and the ones that are thought to be produced outside the Milky Way. This separation is somewhat arbitrary and hides our fundamental ignorance of the actual conditions required to define CRs as extra-galactic. Nevertheless, the physical problems associated with galactic CRs appear to be qualitatively different from the ones involved in ultra-high energy CRs, hence we will adopt this separation in this review as well, paying special attention to the underlying assumptions and possibly their failure. While there is a substantial consensus that galactic CRs are somehow related to one or more types of supernova (SN) explosions and that acceleration is mainly due to diffusive transport in the proximity of strong shocks formed as a consequence of these explosions, less consensus exists on whether all or a subset of SNe can actually reach the knee energy. At a few PeV, there is some evidence that chemical composition changes, thereby leading to the formation of the knee in the all-particle spectrum [1], although the details of how this takes place are not well understood: some observations suggest that the knee is made by light elements [1], while others [2] find that light elements disappear at lower energies and the knee gets dominated by elements with intermediate mass. This type of problem is to be considered essentially of experimental nature at this time.

The transport of CRs inside the accelerators and throughout the Galaxy is described by models based on the same physical ingredients: spatial diffusion induced by resonant scattering of charged particles of plasma waves. Such waves are likely to be, at least partially, generated by the same particles during transport, due to instabilities induced by local streaming. This apparently simple picture is in fact deceiving, in that it hides the essentially non-linear nature of the transport phenomenon: the large-scale behaviour of CRs is determined by the superposition of microphysical particle–wave interactions. In this sense, the transport of CRs has become an instance of the so-called inner space—outer space conundrum, well known in the field of cosmology. In much the same way that the laws of particle physics shape the evolution of the universe, the laws of plasma physics on small-scales shape the behaviour of CRs on large scales.

Several instances of self-regulation have been found in such systems, ranging from particle acceleration at supernova shocks to propagation in the Galaxy in a background of self-generated turbulence. The complexity of these situations is often overwhelming and one resolves to adopt effective approaches that, while retaining the main underlying physical aspects, may still allow us to describe nature in a satisfactory way.

The ever increasing quality of observations reveals aspects of Nature that force us to improve the quality of the effective models that we adopt to describe it. This trend is a fair description of the history of CRs in the last few decades: the simple energetic argument that led to propose supernova remnants (SNRs) as sources of the bulk of galactic CRs and the diffusive paradigm for the transport of CRs in the

Galaxy explain, by themselves, the main aspects of the origin of CRs. On the other hand, this simple picture fails to describe many other pieces of observation that have come about with time.

The standard SNR paradigm predicts that the spectrum of CRs accelerated at strong shocks is very close to $\sim E^{-2}$ [3] but in the few cases in which gamma-ray emission can be unequivocally attributed to hadronic interactions, the inferred CR spectrum appears to be steeper than E^{-2}. Moreover, since the spectrum observed at the Earth is $\sim E^{-2.7}$, the SNR paradigm would naively suggest that CR transport is described by a diffusion coefficient $D(E) \propto E^{0.7}$, which however is in contradiction with the measured large scale anisotropy [4, 5]. In addition, such a diffusion coefficient leads to an energy dependence of the B/C ratio, proportional to the grammage traversed by CRs, that is not consistent, at high energy, with measurements from the PAMELA experiment [6] and AMS-02 collaboration [7].

If used to estimate the maximum energy E_{max} of CRs accelerated at SNR shocks, the same diffusion coefficient would lead to expect that $E_{max} \leq 1$ GeV, quite at odds with observations. This fact alone is a signature that the process of particle acceleration at SNR shocks works in a much more complex manner than the basic paradigm would suggest. The recent detection of narrow rims of X-ray emission from virtually all young SNRs [8] provided indirect confirmation that the magnetic field in the shock proximity is amplified by a factor $\sim$10–100. Although the nature of the amplification process is not clear as yet, streaming instability excited by CRs themselves provides the correct order of magnitude to explain the observed rims as a result of synchrotron emission from very high energy electrons. At the same time, the inferred magnetic field would make the acceleration of CRs up to 100–1000 TeV at SNR shocks plausible [9–12]. Interestingly, in order to explain CR energetics on galactic scales, SNRs are required to accelerate CRs with a $\sim$10% efficiency, which is also required for magnetic field amplification, which in turn leads to high values of E_{max}: particle acceleration at a SNR shock is a typical example of a self-regulated non-linear system, in which well-known plasma physics laws combine to provide a complex outcome.

It is likely, though less clear, that a similar chain of processes also works for CR transport through the Galaxy. At present, propagation of CRs is described as diffusive with a diffusion coefficient that is tailored to fit observations. Advection with a wind is treated as an option in most propagation codes (e.g. GALPROP, DRAGON, PICARD and Usine [13–16]).

The spectrum of different elements in CRs has been recently measured at the Earth location by PAMELA [17] and AMS-02 [18, 19] and found to be characterized by small spectral breaks at a few hundred GV rigidity, quite at odds with the standard view of power law injection and diffusion. This phenomenon might be the manifestation of several effects: for instance, it might result from the stochastic overlap of discrete sources around the Sun [20], from re-acceleration in weak SN shocks [21], or it might result from a spatially inhomogeneous diffusion coefficient [22]. Finally, it might show that non-linear production of waves and pre-existing waves are both responsible for CR diffusion, each one of them being important at different energies [23, 24].

Traditionally, the ratios $e^+/(e^- + e^+)$ and $\bar{p}/p$ have also been used to infer the propagation properties of galactic CRs. However this is possible only if both positrons and antiprotons are solely generated as secondary products of CR interactions in the Galaxy, and in this case one expects both these ratios to be monotonically decreasing functions of energy above ~10 GeV. The PAMELA experiment measured the positron ratio and found that it grows with energy [25] at least up to ~100 GeV. This result was later confirmed and extended to higher energies by AMS-02 [26]. The absence of an increasing trend with energy in the $\bar{p}/p$ ratio [27] leads to the conclusion that the positron excess is likely due to new sources of astrophysical positrons that do not produce antiprotons. The same concept also imposes strong constraints on possible Dark Matter (DM) related models of the positrons excess (see [28] for a comprehensive review). The measurement of the spectra of electrons and positrons separately [29, 30] allowed us to conclude that the positron excess stems out of an extra contribution to the positron flux rather than a deficit of the electron flux, namely sources of positrons (but not antiprotons) are required to explain data. It has been speculated that such sources might be old SNRs [31, 32] or pulsar wind nebulae [33, 34].

On the other hand, the recent extension of the measurement of the antiproton flux and the $\bar{p}/p$ ratio to higher energies by AMS-02 [35] has stimulated an exciting discussion on a radically new view of the anomalies in secondary-to-primary ratios: it has been pointed out [35, 36] that the energy spectra of positrons and antiprotons have very similar slopes and such slope is, in turn, very close to that of the proton spectrum at high energies. A similar consideration was put forward earlier in Refs. [37, 38] based on the positron and proton spectra alone. This simple consideration is used by the authors to suggest that both positrons and antiprotons are purely secondary products of CR interactions. Clearly these scenarios are not problem free: for instance, an alternative explanation of the B/C ratio should be sought [36] since no apparent anomaly has been measured in this quantity.

A general consequence of the SNR paradigm outlined above is that the flux of galactic CRs should end with an iron dominated composition at energies ~26 times larger than the knee in the proton spectrum. If such knee is indeed at PeV energies, as KASCADE data suggest, then galactic CRs should end below $\sim 10^{17}$ eV, well below the ankle.

The measurements carried out by the Pierre Auger Observatory [39] have shown that the mass composition of CRs, from prevalently light at $\sim 10^{18}$ eV, becomes increasingly heavier towards higher energies. Several independent calculations [40–42] showed that the observed spectrum and composition can be well explained only if sources of ultra-high energy CRs (UHECRs) provide very hard spectra and a maximum rigidity $\sim 5 \times 10^{18}$ V. One should appreciate the change of paradigm that these recent observations forced us towards: 10 years ago, the general consensus was that UHECRs are protons and that sources should accelerate them to $> 10^{20}$ eV, something that would not be consistent with current Auger data. On the other hand, the Telescope Array (TA), operational in the northern hemisphere, collected data that suggest a somewhat different scenario [43], where the mass composition is compatible with being light for energies above 10^{18} eV, with no apparent transition to a heavier mass composition. A joint working group made of members of both collaborations has recently concluded that the results of the two experiments are not in conflict once systematic and statistical uncertainties have been taken into account. This conclusion, though encouraging on one hand, casts serious doubts on the possibility of reliably measuring the mass composition at the highest energies, unless some new piece of information becomes available. It should be noted that the spectra measured by the two experiments, though being in general agreement, differ beyond the systematic error at the highest energies: the TA spectrum shows a suppression that is consistent with the Greisen–Zatsepin–Kuzmin cutoff, while the shape of the spectrum measured by Auger appears to be in better agreement with propagation of nuclei.

On the other hand, the results of the two experiments in terms of spectra and mass composition show good agreement around 10^{18} eV, where CRs are found to be light. The fact that between 10^{17} and 10^{18} eV the mass composition changes from heavy to light is suggestive of a possible transition from galactic to extragalactic CRs in the same region, well below the ankle.

This paper is structured as follows: in section "Cosmic Rays Observations" we discuss the status of observations. The transport of galactic CRs is discussed in section "Transport of CRs in the Galaxy", while the status of investigation on CR acceleration is summarized in section "Acceleration of Galactic CRs". The transport of ultra-high energy cosmic rays is discussed in section "Transport of Extragalactic CRs" while some considerations about the sources are reported in section "Astrophysical Sources". The possibility to infer useful information on exotic physics (such as top-down models and violations of Lorentz invariance) are discussed in section "Exotic Models". In section "Transition Between Galactic and Extragalactic Cosmic Rays" we summarize different models of the transition from galactic to extragalactic cosmic rays. We conclude in section "Conclusions".

Cosmic Rays Observations

In this section, a short review of experimental results on some selected topics on CR physics will be given. For each topic, a discussion on new and future projects has also been added in order to focus on the key issues that, from the experimental side, could bring to more and better information for the understanding of the relevant physics phenomena. After a section dedicated to the observation of electrons/positrons and antiprotons, the measurements on protons and nuclei will be discussed starting from ballon and spaceborne experiments up to the highest energies, currently covered with giant ground arrays.

Observations of Electrons, Positrons and Antiprotons

Even if the electron/positron component, i.e. $(e^- + e^+)$, accounts for approximatively 1% only of the total CR flux, it is deeply studied in order to infer important information on propagation processes. In the standard scenario, secondary electrons and positrons are (equally) produced via interactions of primary CRs with the InterStellar Medium (ISM), therefore the observed overabundance of electrons on positrons is a clear indication that most of the electrons have a primary origin. Because of the low mass, this component suffers significant energy losses during propagation in the Galaxy. At high energies, such losses produce a steeper energy spectrum compared to that of protons and actually place upper limits on the age and distance (at about 10^5 yr and 1 kpc, respectively) of the astrophysical sources of TeV electrons. Since the number of such nearby objects is limited, the electron energy spectrum above 1 TeV is then expected to exhibit spectral features, and a sizeable anisotropy in the arrival directions is also foreseen at very high energies [44].

Measurements of the CR electron/positron fluxes have been pursued since many years by balloon-borne and space-based experiments. Because of the low intensity of the signals and the large proton-induced background, the main requirements for these instruments are a large exposure time and a sufficient e/p separation capability. Calorimeters can be used to measure the inclusive so-called all-electron, i.e. $(e^- + e^+)$, spectrum, while separating electrons form positrons obviously requires the determination of the sign of the charge through a magnetic spectrometer, that puts anyway severe limits to the highest possible detectable energy, this being limited by the Maximum Detectable Rigidity (MDR). Important progress was made in the field in the last years, due to the use of magnetic spectrometers in space. The positron fraction was shown to grow with energy by the PAMELA experiment [25] at least up to $\sim$100 GeV, this result being confirmed with precision measurements by AMS-02 [26], that also extended the covered range up to about 500 GeV.

These findings were also confirmed, even though with larger systematic uncertainties by the Fermi-LAT experiment [45], which is not equipped with a magnetic spectrometer but used the Earth magnetic field as a charge sign separation tool.

Experimental results show evidence for an excess of the positron fraction with respect to the standard production mechanism (i.e. primary CR interaction with the ISM), in the form of an increase with energy above approximatively 10 GeV (see Fig. 1). The latest precision measurements of the AMS-02 experiment [30] ascribe the positron fraction excess to a hardening of the positron flux, showing a spectral index above 50 GeV compatible with that of primary protons. Moreover, as Fermi-LAT recently showed [47], no anisotropy signal has been detected in the inclusive electron spectrum with current sensitivities.

Understanding the origin of this excess of positrons in the cosmic radiation requires measurements up to the highest possible energies, where both spectral features and/or anisotropies might be detected. In this context, the multi-TeV, largely unexplored, region is very interesting because of the high potential for studying local sources. Indirect measurements made by imaging atmospheric Čerenkov telescopes

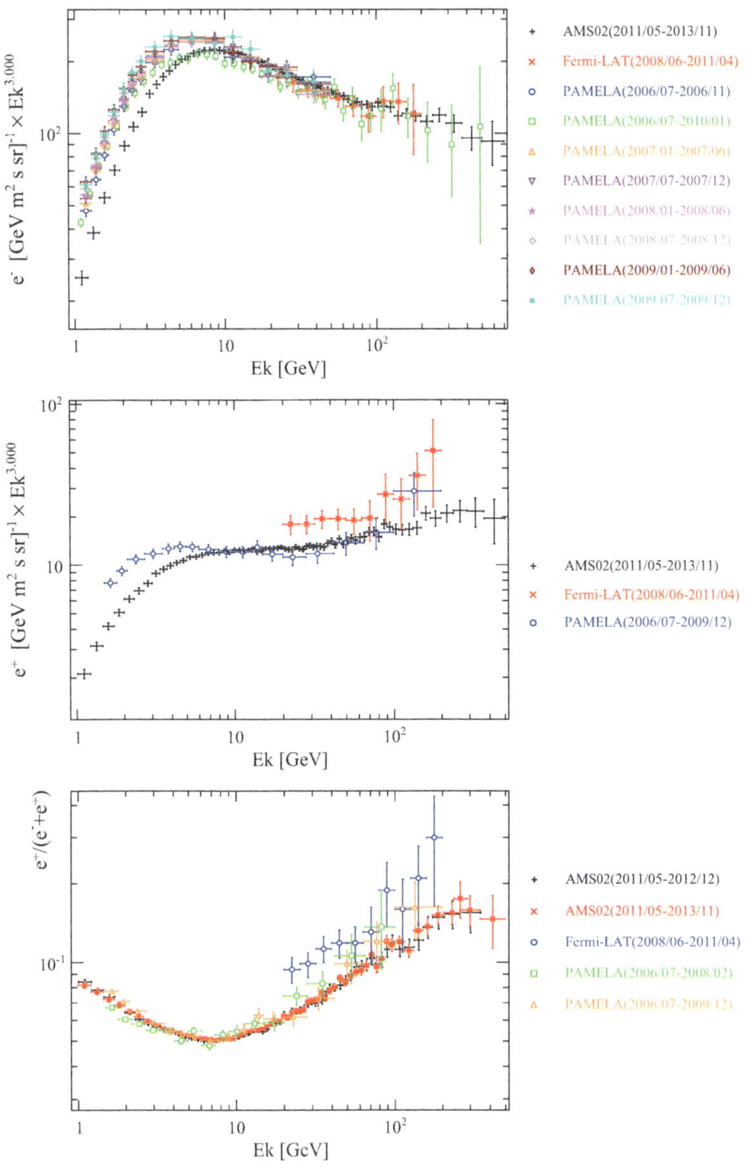

Fig. 1 Fluxes of cosmic electrons and positrons (*upper* and *central panel*, respectively) as measured by AMS-02, PAMELA, and Fermi-LAT experiments. The positron fraction is shown in the *lower panel*. The dates in the experiment labels refer to the analyzed data sample [46]

suggest, even though with large uncertainties, an exponential cutoff at about 2 TeV [48, 49]. The analysis of 7 years Fermi-LAT data [50] recently extended the spectral measurements up to 2 TeV. Fermi-LAT data alone exclude an exponential cutoff below 1.8 TeV at 95% C.L., while a combined fit of Fermi-LAT and HESS data would lead to a cutoff at energies larger than 2.1 TeV. The exploration of the high energy part of the spectrum with high precision direct measurements is then mandatory.

New technologies might extend current MDR values up to few TeV for future missions, while deep homogeneous calorimeters in space, with large geometric factors, will reach even higher energies, but obviously without matter/antimatter separation. The recently launched CALET and DAMPE detectors might return interesting results on the high energy all-electron component.

CALET (CALorimetric Electron Telescope) is a space mission led by the Japanese Space Agency (JAXA) with the participation of the Italian Space Agency (ASI) and NASA. The payload was launched on 19 August 2015 and installed on the Japanese Experiment Module Exposure Facility (JEM-EF) of the International Space Station (ISS) on August 24. The mission is foreseen to last for 2 years, with a possible first extension to 5 years [52]. The main scientific goal is to search for possible nearby sources of high energy electrons or signatures of DM, by measuring accurately the all-electron spectrum from 1 GeV up to several TeV. It will also measure the energy spectra and elemental composition of CR nuclei from H to Fe up to hundreds of TeV (see below). The instrument consists of two layers of segmented plastic scintillators (for particle charge determination), a thin tungsten-scintillating fiber imaging calorimeter providing accurate particle tracking and identification by multiple dE/dx sampling, and a thick PWO crystal calorimeter to measure the energy of CRs with excellent resolution and electron/hadron separation up to the multi-TeV scale. The total thickness is equivalent to 30 radiation lengths and 1.3 proton interaction lengths with a geometric factor of about 0.1 m^2sr. An extensive campaign of beam tests for calibration was carried out at GSI and CERN [53].

The DAMPE (DArk Matter Particle Explorer) satellite was launched on 17 December 2015 and is in smooth data taking since few days after. It was designed in order to properly work for at least 3 years and, thanks to its large geometric factor (about 0.3 m^2sr for protons and nuclei and even larger for electrons), it already integrated one of the largest exposures for galactic CR studies in space. The detector, built and operated by a collaboration of Chinese, Italian and Swiss institutions, is made by 12 layers of Si-W tracker followed by a $32X_0$ BGO calorimeter. A plastic scintillator detector on top and a neutron detector on bottom, for ion charge and shower neutron content measurements, respectively, actually complete the setup. As also resulted from a large set of beam test measurements with a full-scale detector prototype at CERN, the BGO calorimeter actually provides an energy resolution for electrons at the level of 5% at 1 GeV and better than 2% above 10 GeV. The information from the various sub-detectors (e.g. ion charge measurement, precision tracking, shower topology) allows an efficient identification of the electron signal over the large (mainly proton induced) background. As a result, the all-electron spectrum will be measured with excellent resolution form few GeV up to few TeV [51]. The DAMPE contribution to the measurement of the all-electron energy spectrum,

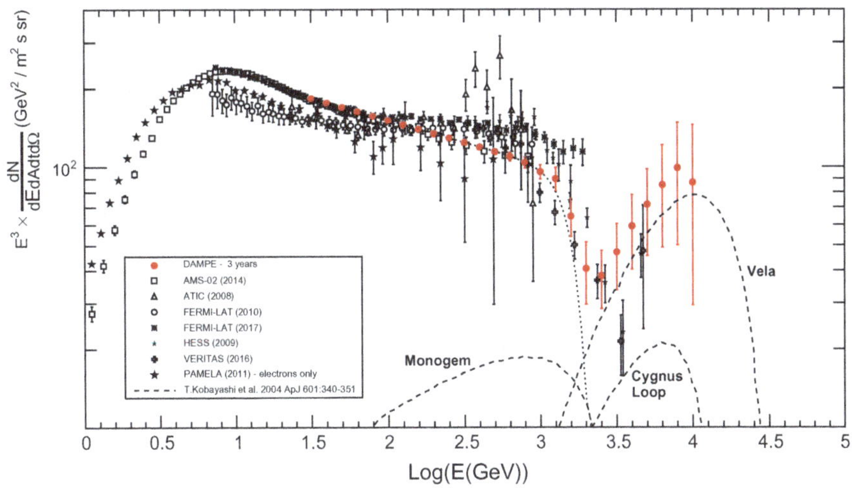

Fig. 2 Inclusive all-electron energy spectrum as measured by several experiments. The possible contribution of the DAMPE mission after 3 years of operation is also shown (see *text* for details). Plot taken from [51]

after 3 years of operation, is shown in Fig. 2. The DAMPE spectrum was simulated by assuming a power law with a spectral index as given by AMS-02 [54] data above 30 GeV, a cutoff at about 1.5 TeV as suggested by HESS and VERITAS in [48, 49], and then a possible contribution of three nearby sources as parametrized in [44]. In the figure, the result is compared with existing measurements and with the model given in [44].

As can be seen, this will allow a direct and precise detection of a possible cutoff at about 1–2 TeV. Moreover further structures/excesses due to nearby sources will be clearly identified below few TeV, together with possible indirect evidence for a DM-induced excess.

In the case of the HERD mission (see below), a larger acceptance and an even deeper calorimeter would provide a unique tool to investigate all the spectral features also above the TeV region. In particular, the contribution of nearby sources could be clearly identified and studied. In the case of sizeable contribution of nearby sources, a large anisotropy is expected at high energy, which could be easily detected by HERD, giving important clues to the understanding of diffusion processes in the Galaxy.

The CR antiproton component can only be identified by using magnetic spectrometers together with sufficient MDR and particle identification capability. Many balloon and space born experiments contributed to this field with important progress due to the PAMELA [27], AMS-02 [35], and (at low energies) BESS [55] experiments (see Fig. 3). The measurements, currently carried out up to few hundreds GeV, are in fair agreement with secondary production due to primary CR interactions with the interstellar medium. Interestingly enough, the same spectral index as the one for protons is suggested by data, for both positrons and antiprotons (see

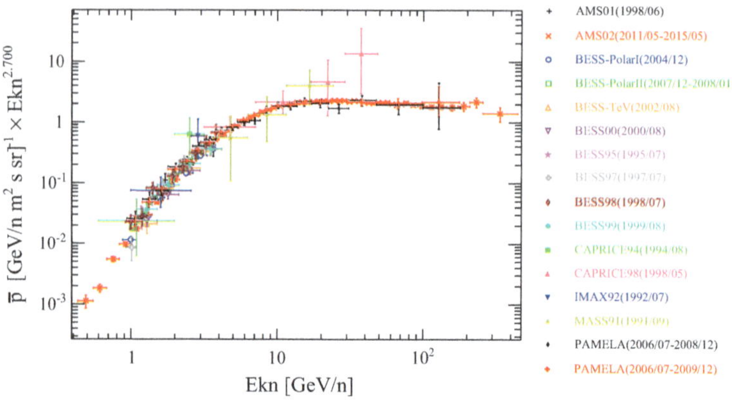

Fig. 3 Fluxes of antiprotons as measured by several experiments. The dates in the experiment labels refer to the analyzed data sample [46]

section "Positrons and Antiprotons" for a discussion). New important inputs on this topic might be provided by the search for antinuclei (e.g. anti-deuterons) in the CR flux by both current, e.g. AMS-02, and future experiments, such as GAPS [56].

Observations of Protons and Nuclei up to Hundreds PeV

From Low Energies up to 100 TeV: Flux Hardenings and Secondary-to-Primary Abundances

Recent direct measurements of primary protons and nuclei shed new light on acceleration and propagation mechanisms. The paradigm of a unique power law energy spectrum below the knee, down to the region where solar modulation effects become sizeable, might have been invalidated. In 2010 the CREAM (Cosmic Ray Energetics And Mass) experiment showed evidence for a hardening in the spectra of protons and nuclei with different ('discrepant') spectral index changes. This is summarized in Fig. 4 where CREAM data, also fitted by (broken) power laws, are shown together with other measurements [57]. Even with large error bars (mainly at high energy and/or heavy primaries), a change of spectral index is suggested at about 200 GeV/n. Both the energy ranges and the flux uncertainties prevented anyway a clear claim for a break in the proton and helium spectra.

This became possible with the analysis of PAMELA results [17], later confirmed by AMS-02 [18, 19]. As can be seen in Fig. 5 a clear change of spectral index is shown by data, even though different experiments return slightly different slopes at energies above the breaks.

Recent results of the analysis of CREAM-III [58] and NUCLEON [59] data did confirm the scenario up to about 100 TeV but with large uncertainties. More data are

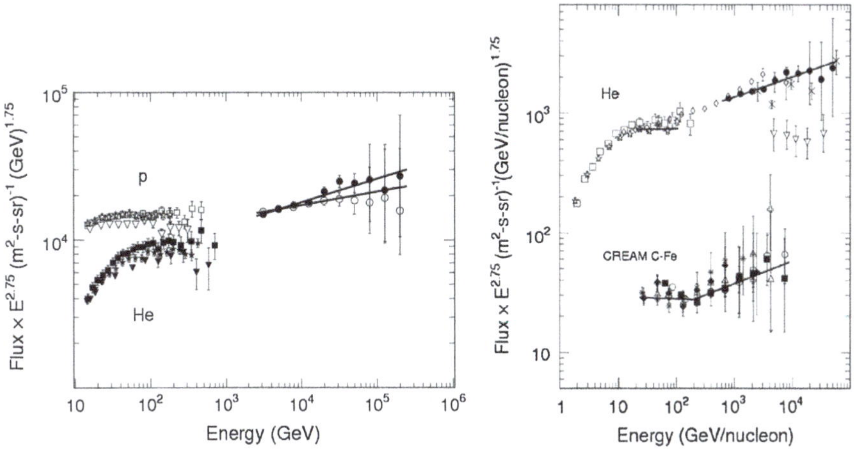

Fig. 4 Measurements of proton, helium and nuclei fluxes as for year 2010: first evidence for discrepant hardenings by the CREAM experiment (see *text*). Plots taken from [57]

Fig. 5 Recent measurements on proton and helium fluxes (*upper* and *lower panel*, respectively). Plots taken from [18, 19]

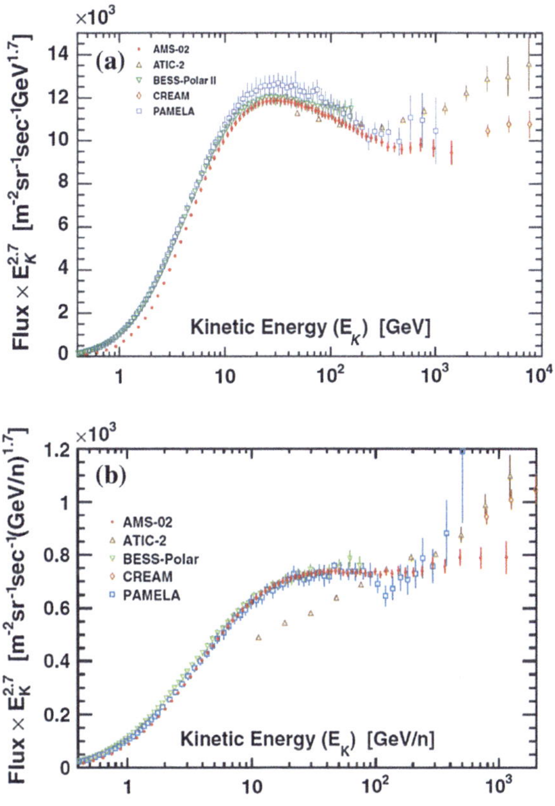

then needed at high energy in order to measure, with a single experiment, both the region across the breaks and the high energy one, with sufficiently small uncertainties. Current missions like CALET and DAMPE have the size and the needed resolution in order to check the break region and uniquely determine the spectral behaviour up to more than 100 TeV.

Even though primarily optimized for the study of electrons and gamma rays (see above), the DAMPE detector provides good tracking and calorimetric performances also in the case of protons and nuclei, together with the possibility of ion identification through charge measurements in the top scintillator layer, in the tracker, and in the calorimeter itself. This allows precise measurements of proton and nuclei energy spectra from tens of GeV up to about 100 TeV, the high energy limit being essentially determined by the overall geometric factor and the calorimeter's dynamic range.

In particular, the energy region above about 50 GeV will be explored with higher precision compared to previous experiments [60]. Spectral indexes for individual species could then be well measured and evidence for the observed hardenings could be checked and better quantified. This would be very important for a comparison with state-of-the-art models of galactic CR acceleration/propagation mechanisms, and to assess the contribution of nearby sources. Moreover measurements of important quantities like the boron-to-carbon ratio will be improved and extended to higher energies. Similar contributions are expected from the CALET mission, even if the smaller geometric factor (by about a factor three) would result in larger uncertainties. Further extensions in energy, towards the all-particle knee, are expected for the ISS-CREAM and HERD projects (see below).

In the low energy range, important information can be provided by the study of the production rate of secondary CRs. Recent results from PAMELA and BESS-Polar [61] on the isotopic abundance ratios $^2H/^1H$ and $^3He/^4He$ in the range 0.1–2 GeV/n provide essential information to better understand the history of cosmic-ray propagation in the Galaxy. On the other hand, as discussed in section "Transport of CRs in the Galaxy", elemental secondary-to-primary ratios (such as Boron/Carbon or subFe/Fe) can be employed, to infer information on the nature and size of the cosmic-ray confinement region and on the propagation properties of CRs in the Galaxy. Current measurements of the B/C ratio are shown in Fig. 6. While measurements performed by balloon-born experiments suffer from small statistics and large systematic errors (due to short exposure times and to the effects of the residual overburden atmosphere, respectively), data from space spectrometers, like PAMELA and AMS-02, can provide an accurate spectral measurement up to about 1 TeV/n [6, 62]. Also in this case, from the experimental point of view, the challenge is then to extend the energy range to the multi-TeV/n region by using large geometric factor instruments in space.

Approaching the Knee(s) with Direct Measurements

Since the first experimental evidence in 1958 [63], the energy region around the knee in the all-particle cosmic ray spectrum, at about 3 PeV, has been investigated by many experiments with different approaches [64].

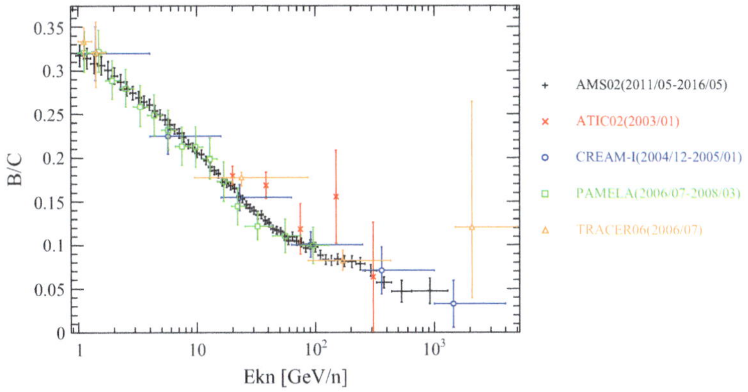

Fig. 6 Measurements B/C ratio made by several experiments. The dates in the experiment labels refer to the analyzed data sample [46]

Several theoretical explanations have been proposed exploiting different hypotheses on source properties/populations, acceleration/propagation mechanisms and particle physics issues at high energies [65]. After the first results at the Large Hadron Collider (LHC), the 'particle physics' origin of the knee seems to be disfavoured, confirming that it is a genuine property of the CR spectrum itself [66]. It remains still unsolved whether the (dominant) origin of the knee is due to the reaching of the maximum energy achievable at the sources or to diffusion processes in the Galaxy. In both cases, a rigidity dependent sequence of knees in individual elemental spectra is the most likely scenario [65].

For the analysis of the CR flux, direct measurements carried on space or stratospheric balloons actually provide the best performance in terms of both energy resolution and charge identification. However, due to their limited acceptance and the steeply falling fluxes, they could hardly reach, up until now, energies of hundreds of TeV and then did not yet provide clear information on the steepening of the spectrum of various elements nor on the knee of each species or of the all-particle spectrum itself [60, 67].

As shown in the previous section, current data suggest a hardening of the spectra above about 0.2 TeV/nucleon and spectral indexes γ (above that energy) of about -2.6 for all considered elements but for protons, that show a softer spectrum with $\gamma \simeq -2.7$ [57, 68]. Moreover, the chemical composition is shown to evolve towards heavier nuclei, with helium becoming more abundant than hydrogen at energies of about 10–20 TeV [60]. It is then mandatory to explore the sub PeV region with high precision direct measurements in order to study the energy spectra of each nuclear species, to measure the various spectral indexes, to detect any possible hardening and to establish mass composition below the knee of the all-particle spectrum. The measurement of the spectrum of individual elements and an understanding of the nature of the knee in the all-particle spectrum would represent a result of unprecedented importance in CR physics, and would provide a crucial insight into unveiling the transition between galactic and extragalactic CRs.

The ISS-CREAM detector has been built by transforming the CREAM payload, flown in several flights over Antarctica, for accommodation on the ISS for a 3-year mission [69]. The exposure will be increased by about one order of magnitude allowing to extend the measurements on nuclei up to hundreds of TeV. The detector includes four layers of Si pixels to measure the particle charge; a carbon target to induce the inelastic interaction of the incoming nuclei; a sampling calorimeter made of 20 layers of alternating tungsten plates and scintillating fibres, providing energy measurement, particle tracking and trigger; top and bottom plastic scintillator counters and a boronated scintillator detector for e/p separation. The Si charge detector and the calorimeter were already used in CREAM, while the last two detectors have been newly developed for the space mission, in order to add sensitivity also to CR electrons.

The HERD (High Energy Radiation Detector) experiment [70] is being proposed by an international collaboration as a space mission for the study of the high energy cosmic radiation with a detector characterized by an unprecedented geometric factor, to be installed onboard the CSS (Chinese Space Station) around 2023. Current detector design includes a cubic calorimeter (about $55X_0$ and 3Λ in depth) made by ($3\,\text{cm} \times 3\,\text{cm} \times 3\,\text{cm}$) LYSO crystals, readout individually, a high precision Si-W tracker covering 5 out of the 6 calorimeter faces and an array of plastic scintillator for the charge measurements. This innovative setup allows a jump in the geometric factor, with respect to previous experiments, of more than one order of magnitude for an estimated value of the geometric factor $\sim 3\,\text{m}^2\,\text{sr}$. Together with the unprecedented depth of the calorimeter and the high-resolution tracker, this will allow the extension of high precision measurements on proton and nuclei spectra up to PeV energies. Moreover, a clear identification of each nuclear species will be possible through the charge measurements made by the plastic scintillators, the Si-W tracker and by the calorimeter itself. Energy resolution for the electromagnetic and hadronic showers will be at the 1% and 30% level, respectively.

Simulated HERD results for the measurement of protons and nuclei, after 3 years of operation, obtained by assuming flux parametrization as given in [71], are shown in Fig. 7. HERD points (in colour) are compared with measurements from PAMELA, CREAM, AMS-02 and ATIC experiments (shown in black) and theoretical models [71, 72].

As can be seen, features like single element spectral indexes and spectral hardenings/steepenings could be carefully studied from hundreds GeV/n up to hundreds TeV/n. In particular, the proton and helium component could be measured up to PeV energies, thus providing a test of the origin of the knee in the spectrum of the light CR component, possibly associated with the maximum energy reached by CR at the source (see next section).

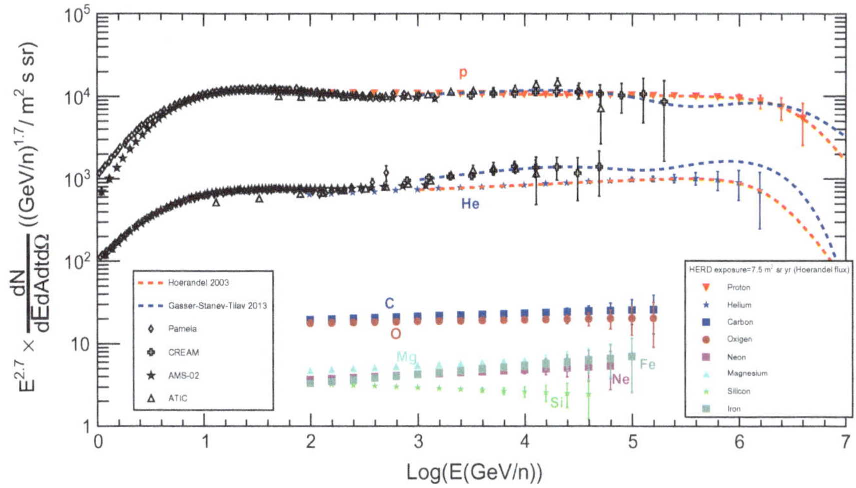

Fig. 7 Contributions of the HERD mission after 3 years of operation: individual proton and nuclei energy spectra (see *text* for details)

Ground-Based CR Observations up to Hundreds PeV

Indirect measurements fully explore the energy region above 0.5–1 PeV (or even few TeV, if located at high altitude) through the detection of Extensive Air Showers (EAS) in the atmosphere. Data show a general agreement, within the systematic uncertainties, on the all-particle spectrum, also suggesting evidence of a second knee at about 100 PeV [73–76]. However systematic uncertainties related to the experimental procedure itself and intrinsic in the assumptions adopted for the hadronic interaction models do not allow an easy and straightforward estimate of the mass composition nor of the single species (or mass group) energy spectra [64, 77].

One or more EAS observables (e.g. the lateral distribution of particles at the ground, the longitudinal development in the atmosphere, the muon content of the shower, etc.) are measured in order to estimate, by adopting a given assumption on the primary interaction and the shower development in the atmosphere, the CR composition. Results are often given in terms of the energy dependence of the mean logarithmic mass, defined as $\langle lnA \rangle = \sum_i \eta_i lnA_i$, where η_i is the fraction of nuclei of mass A_i in the CR beam. A compilation of $\langle lnA \rangle$ measurements can be found in [64] and [77], with a comprehensive discussion on the results and their uncertainties. As can be seen in Fig. 8, data show large uncertainties, mainly coming from the systematics associated to the adopted interaction models. Moreover, a somewhat different trend with energy might be identified by dividing experiments in two large classes: the ones measuring charged particles at the ground and those detecting the Čerenkov or fluorescence light emissions in the atmosphere [71]. Even with these uncertainties, data collected across the knee region show an evolution towards heavier mass groups as expected from acceleration/propagation models. A tendency towards

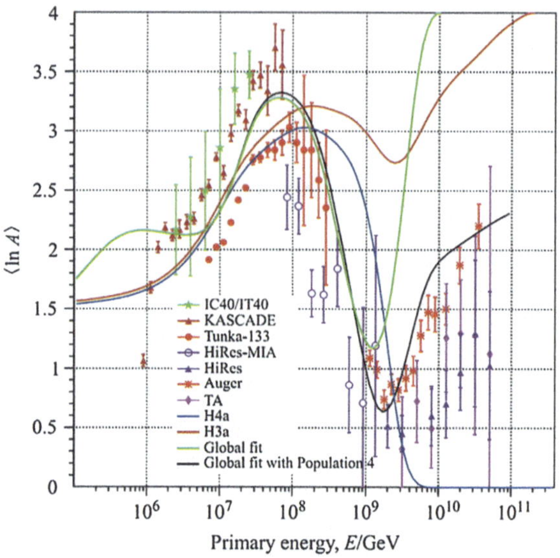

Fig. 8 Average value of the estimated CR logarithmic mass, $\langle lnA \rangle$, from several ground-based experiments. Superimposed are the *lines* corresponding to mixed composition resulting from muti-population models as given in [72]. Plot taken from [72]

lighter elements is then observed starting at energies compatible with the position of the second knee, while a new trend towards the medium mass group can be recognized above the ankle region (see next section).

The energy spectra of individual elements (or mass groups) are even more difficult to be measured. Results from the KASCADE experiment, even with sizeable systematic uncertainties on the individual fluxes mainly coming from the dependence on the hadronic interaction model, imply an average composition at the knee that is dominated by light elements, and the knee itself is interpreted as the steepening of the p and He spectra [1]. The KASCADE-Grande experiment returned results consistent, at higher energies, with this scenario [78, 79], and ascribed the second knee to the steepening of the heavy component [73].

Several different experimental results suggested a somewhat heavier composition at energies around the knee. For instance, a hybrid measurement was carried out by the EAS/TOP and MACRO experiments (by detecting, in coincidence, EAS Čerenkov light at 2000 m a.s.l. and underground muons below about 3000 m of water equivalent depth, respectively). The result implied a decreasing proton contribution to the primary flux at energies well below the observed knee in the primary spectrum [80].

The same indication was previously obtained through the analysis of the underground muon component alone in the MACRO experiment [81]. In addition, the results of the Tibet-ASγ experiment, located at 4300 m a.s.l, do favour a heavier composition because the proton component is no longer dominant at the knee [82]. In particular, the fraction of the light component (i.e. protons and helium nuclei) is shown to be of 50% at about 500 TeV and decreasing with energy [82], this also being consistent with later measurements of the upgraded Tibet array showing a steepening of the proton spectrum above few hundred TeV [83].

This is also in agreement with results from the CASA-MIA experiment, showing a decreasing proton content at about 600 TeV [84] and with a series of measurements on Mount Chacaltaya (about 5200 m a.s.l.) giving a steady increase of the average mass number of primary CRs with energy above $10^{14.5}$ eV [85, 86].

Furthermore, indications for a substantial fraction of nuclei heavier than helium at 1 PeV have also been obtained in measurements with delayed hadrons [87]. Finally, the compilation of measurements of the energy spectrum of the so-called CNO group (i.e. Carbon-Nitrogen-Oxygen) shows a knee at energies not larger than about 7 PeV (see for instance [64]). In a scenario with a rigidity dependent knee position, this is not consistent with a position of the proton knee at about 3 PeV.

Similar conclusions have been reached by the combined analysis of data coming from the ARGO-YBJ experiment and a wide field of view Čerenkov telescope (a prototype of the future LHAASO experiment [2]): the measured energy spectrum of the proton + helium component shows a break at $(700 \pm 230 \pm 70)$ TeV [2]. Preliminary results from two independent analyses of the ARGO-YBJ data alone do confirm this picture, within the quoted uncertainties [88, 89].

An overall picture of indirect measurements of the all-particle spectrum, below 10^{18} eV, is shown in Fig. 9. In the same plot, the measurements of the so-called light component (i.e. proton + helium) are also given, showing a clear bending at energies below the knee of the all-particle spectrum. For comparison, the combination of indirect measurements is shown at lower energies, while the results for the light and heavy component as identified by KASCADE and KASCADE-Grande are shown

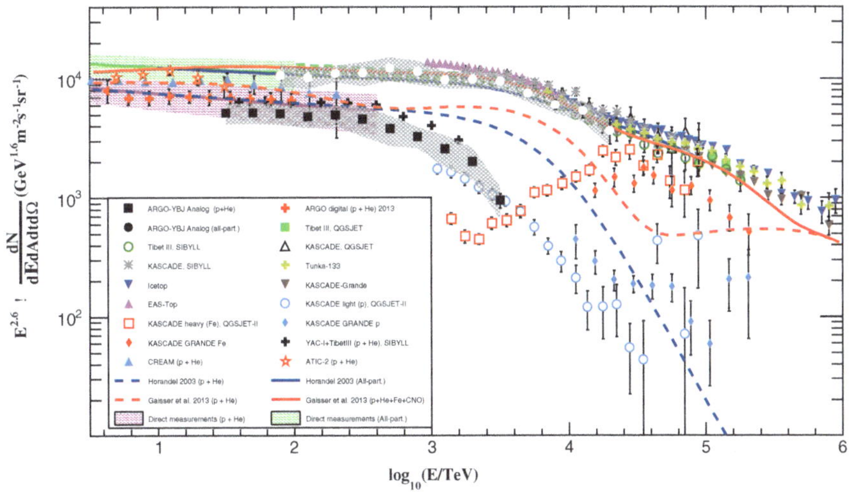

Fig. 9 Indirect measurements of the all-particle CR energy spectrum below 10^{18} eV. Also shown are the combination of high energy direct measurements, and the energy spectrum for the light (i.e. proton + helium) component. Plot taken from [88]

for higher energies ([1, 78] and references therein). Besides a knee-like behaviour in the heavy elements at about 10^{17} eV (consistent with the second knee), KASCADE-Grande data also suggest an ankle-like structure in the light elements at the same energies [78]. The uncertainties on both energy spectra and mass composition are reduced at higher energies, due to the possibility to detect fluorescence light emission along the whole shower development in the atmosphere (see next session).

Flux Anisotropies

A complementary approach to the study of CR sources and propagation, with respect to the analysis of energy spectra and composition, is provided by the measurement of anisotropy signals. This also possibly leads to some information on the galactic magnetic field, which is mainly responsible for the highly isotropic CR flux. Even though the first evidences for anisotropies (resulting from the CR intensity variations with sidereal time) dates back to Hess and Steinmaurer in 1932 [91], in recent years the huge event statistics collected by several experiments with good pointing accuracy allowed a detailed analysis of two dimensional arrival direction distribution maps (right ascension and declination) and their evolution with time. As a consequence, anisotropy signals at the level of 10^{4}–10^{-3} were found at different angular scales in both hemispheres (see for instance [90] and refs. therein).

A so-called Large Scale Anisotropy (LSA) has been measured by several experiments (e.g. Tibet-ASγ [92], Milagro [93], ARGO-YBJ [94], IceCube [95]) showing an approximate dipole-like feature with an excess region between 40°–90° in right ascension (around the heliospheric tail) and a deficit between 150°–240° (in the direction of the galactic north pole), referred to as *tail-in* and *loss cone* regions, respectively.

These observations are likely to reflect the combination of several effects, namely the relative motion of the solar system with respect to the frame in which CRs are isotropic (Compton–Getting effect [96]), the orientation of the local magnetic field [97] and the overall gradient in the CR local density (see for instance [5, 98, 99]).

As can be seen in Fig. 10, the amplitude of the observed signal is of the order of 10^{-4}–10^{-3} with a wide maximum in the multi-TeV region and a stable phase. An increase in the amplitude and a dramatic change of phase (pointing to the opposite direction) are then suggested by data above 100 TeV up to 5 PeV (see [90] and ref. therein).

Recently an additional anisotropy signal has been found in the few TeV energy region, with excesses at angular scales of about 10° (the so-called Medium/Small-Scale Anisotropy, MSA) by Tibet-ASγ [100], Milagro [101], ARGO-YBJ [102], IceCube [95] and HAWC [103]. Such signals have a quite large statistical significance (up to 15 standard deviations) and a nice matching between observations from both hemispheres. Moreover, there are hints for a harder energy spectrum in the excess regions with respect to the isotropic CR background. For both LSA and MSA signals, most observations suggest time stability over several years time scales, which would exclude a correlation with the solar activity.

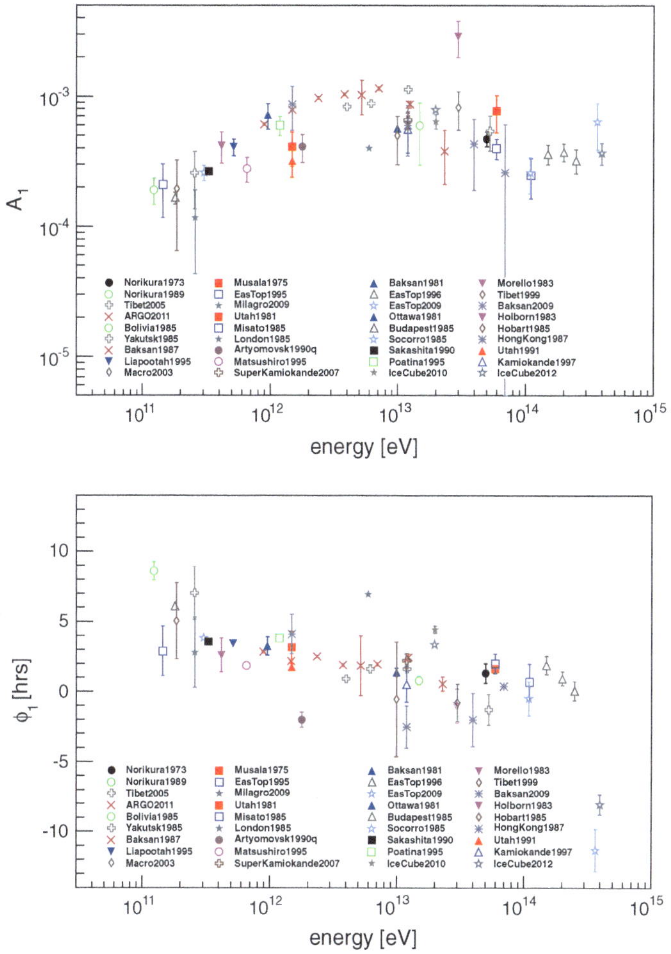

Fig. 10 Amplitude and phase (*upper* and *lower panel*, respectively) of the sidereal CR flux daily variation (first harmonic) as measured by several (under)ground experiments. Plot taken from [90]

The discovery of anisotropy on small angular scales was rather surprising in that the basic expectation of the theory of CR diffusive transport is that only a dipole anisotropy should be expected. On the other hand, it has been noted by several authors that small-scale anisotropies may develop because of the local configuration of the magnetic field, within, say, a few pc from Earth. For instance, in Ref. [104] the author describes the propagation of CRs arriving in the neighborhood of the solar system in several realizations of the local magnetic field and small-scale anisotropies are in fact found, mainly as a result of the fact that fluctuations in the deflections are not averaged to zero. In other words, the transport is not fully in the diffusive limit. An elegant derivation of the same result was found by [105], in which the author shows that these small-scale fluctuations naturally arise as a consequence of the Liouvilles theorem.

A better knowledge of CR physics up to the ankle will need measurements of energy spectra and anisotropy maps of individual species (or at least mass groups) with better resolution and larger statistics. This will also depend on future experiments trying to use new observables (e.g. radio emission [106]) and/or to combine several techniques to be used at the same time (e.g. the LHAASO project [107]).

Observations of Ultra-High Energy Cosmic Rays

Above 10^{18} eV (UHE), the two largest and most precise detectors to date are the Pierre Auger Observatory in Argentina (Mendoza) and the Telescope Array in the USA (Utah). Both detectors exploit the hybrid concept, combining an array of surface detectors to sample extensive air showers when they reach the ground and telescopes, overlooking the surface array, to collect the fluorescence light of the excited atmospheric nitrogen. The advent of the hybrid approach has been a major breakthrough in the detection of UHECRs since the method allows to have the same energy scale in the surface detectors and the fluorescence telescopes. In fact the absence of an energy scale common to both detection methods had led to the puzzle about the existence of the flux suppression around 5×10^{19} eV, which was observed by HiRes [108] but not present in AGASA data [109], whose energy calibration was based on Monte Carlo simulations. The first hybrid measurements were done in HiRes/MIA [110] with a detector array of limited size; the Auger project, for the first time, adopted the hybrid approach [111] as the basis of the detector design to definitely attack the suppression puzzle.

Pierre Auger Observatory and Telescope Array

The two detectors are very similar, but the different sizes and operation times make them differ sizably in the collected data sets and exposures. They are both located at similar average elevation, about 1,400 m a.s.l., and roughly similar longitudes, Auger in the southern hemisphere and TA in the northern. A detailed description of the experiments can be found in [112, 113]; the main features of the basic detectors are summarized in Table 1 [114].

The most remarkable difference lies in their surface detectors (SD) which are based on different detection methods. The particle detectors in the Auger SD are cylindrical tanks of 10 m^2 surface and 1.2 m height, filled with purified water, with three photomultiplier tubes (PMT) to detect the Čerenkov light of particles in the shower front. In the TA they consist of two 3 m^2 slabs of plastic scintillator on top of each other which give light pulses also read by PMTs. The water tanks are relatively much more sensitive to shower muons which usually traverse the tank from wall to wall while the counts in the TA detectors are dominated by electrons and positrons in

Table 1 Comparison of characteristics of the Pierre Auger Observatory and the Telescope Array. The low energy extensions for each observatory, HEAT and TALE, are not included

			Auger	TA
SD	Average latitude		35.3° S	39.4° N
	Average altitude		1,400 m	1,400 m
	Surface area		3,000 km^2	700 km^2
	Lattice		1.5 km hexagon	1.2 km square
	Detector	Type	water-Čerenkov	Plastic scintillator
		Size	10 m^2 × 1.2 m	(2×) 3 m^2 × 1.2 cm
		Sampling	25 ns	20 ns
FD	Sites		4	3
	Telescopes	Number	24	36
		Size	13 m^2	6.8 m^2/3 m^2
		Field of view	28.5° × 30°	16° × 14°/18° × 15°
		Pixels	440	256

the shower front. Furthermore, because of their height, the Auger detectors are well suited to detect highly inclined showers, thereby increasing the exposure and the sky coverage. Inclined showers are also used for neutrino searches and to establish the muon content of the showers. In Auger, an array of radio antennas (AERA) complements the data with the detection of the shower radiation in the hundred MHz region.

The fluorescence telescopes are located on the boundary of the two observatories to overlook the whole atmospheric volume just above the surface arrays and are based on similar detector components. The Pierre Auger Observatory contains a smaller area of 23.5 km^2 with stations separated by 750 m (*infill* array) which can be combined with three additional telescopes pointing at higher elevations (HEAT) for lower energy measurements. Similarly, TA has two sub-arrays of 46 and 35 stations separated by about 600 and 400 m over a surface of 20 km^2, together with ten telescopes covering from 31° to 59° of elevation (TALE).

Event Classes and Energy Calibration

A hybrid experiment collects shower events of different classes. The separation into classes is a natural consequence of the different on-time (generally called *duty cycle*) of the two detector components: the surface array is able to collect showers at any time, whereas the fluorescence detectors can operate only during clear moonless nights (≈15% duty cycle). Taking into account geometry and quality cuts applied at the event reconstruction level, the common dataset is only few percent. Therefore only a small part of the SD showers are actually reconstructed by the FD. Nonetheless, this sub-sample (the *hybrid* dataset) is very valuable, including events having both the footprint of the shower at ground and the longitudinal profile measured. The advantage of this approach is twofold:

- the energy estimator used in the surface detector can be compared on an event-by-event basis with the shower energy reconstructed by the FD. The latter measurement is based on the total amount of light emitted along the shower, which is in turn proportional to the energy deposited by the shower particles. Apart from the missing energy carried by neutrinos and energetic muons, the FD performs a calorimetric measurement of the shower energy. The SD estimator is given by the particle density at 1000 m (800 m) from the shower core in Auger (TA), corrected for the shower attenuation in the atmosphere depending on the zenith angle. The correlation between the FD energy and the SD energy estimator provides the energy calibration that is used for the whole SD data set [114, 115].
- The hybrid events are higher quality showers, because the availability of the longitudinal profiles allows to access the most prominent information about the primary mass (the maximum of the shower depth). These events have also a superior definition of the shower geometry, even if the SD data are coming from a single surface detector [111]. Therefore the hybrid dataset, though being reduced in size, constitutes a selection of well-reconstructed events and a reference for all methods, based on SD data, aiming to obtain mass discriminating parameters.

The SD energy calibration through the hybrid dataset is a technique adopted by both collaborations. The Pierre Auger collaboration calculates the correlation between the energy estimators (for the three classes: standard, inclined and infill) and the FD energy [116]. TA has found that the SD energy derived from the energy estimator via Monte Carlo simulation has to be multiplied by a constant factor of 1.27 to ensure a good matching to the measured FD energy [117]. The latter result is a remarkable evidence of the inadequacy of energy calibrations based on Monte Carlo methods as done in the past.

It has finally to be noted that the relation between the longitudinal profile density and the measured light is provided by the combination of the fluorescence yield and the measured light transmission. The former has been established experimentally, the latter is obtained from the atmospheric monitoring data system operated at each site. Unfortunately, the collaborations use different parameterizations of the fluorescence yield. Including all that the quoted systematic uncertainty in the energy scale is 14% for Auger and 20% for TA.

The Energy Spectrum

The energy spectra measured at the two observatories are shown in Fig. 11. A more comprehensive review of spectrum data, including other experiments, e.g. IceCube and Yakutzk, can be found in [115]. Yet, especially for energies above 10^{18} eV, the

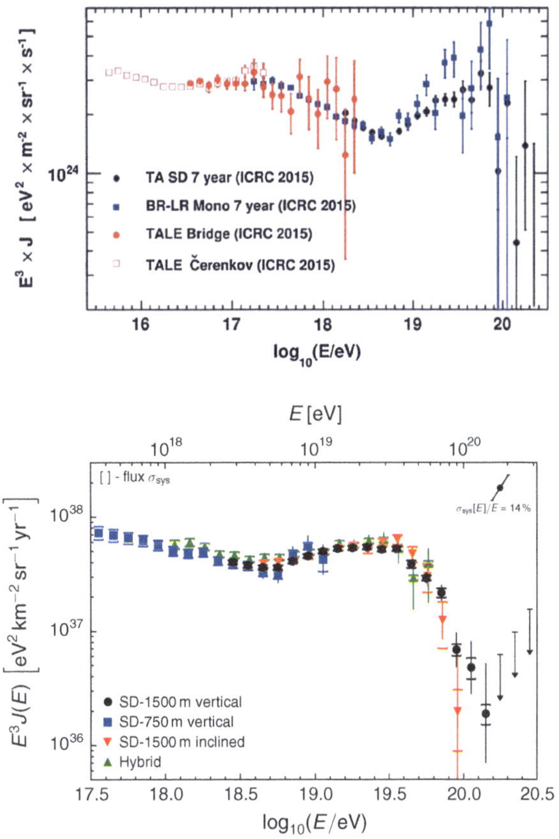

Fig. 11 Energy spectra presented at ICRC 2015 by the Telescope Array (*upper panel*) and Auger (*lower panel*) collaborations. The data from the different sub-detectors are shown separately

bulk of the data comes from Auger and TA. The two panels show the spectra as originating from different detector components for TA (upper panel) and Auger (lower panel).

The most prominent features appear similar in the common energy interval with a break (the *ankle*) at around $10^{18.7}$ eV and a flux suppression, quite evident (at several standard deviations for both experiment) in both cases, but exhibiting somewhat different shapes. It has to be noted that for both experiments the data above the ankle are dominated by the respective ground arrays.

Both the collaborations exploit procedures to combine the different spectrum components into a unique spectrum. For a better comparison, the combined energy spectra are superimposed in Fig. 12, which provides also the values of the main spectral features [115]. The corresponding exposures are about 6,300 km^2 sr yr for TA and 50,000 km^2 sr yr for Auger. Comparing the values of the ankle energy (E_{ankle}) and of the cutoff ($E_{1/2}$) (the energy at which the integral flux drops to half of what is expected in the absence of a cutoff) one finds that the ankle energies are consistent within the systematic uncertainties in the energy scale, but the discrepancy between the cutoff energies is not explained by systematics.

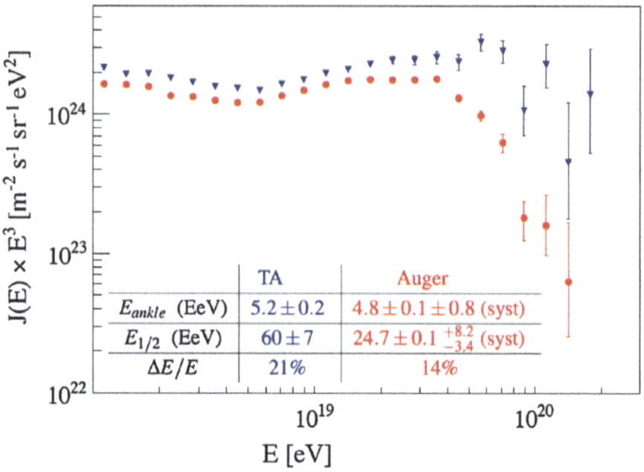

Fig. 12 Comparison between TA (*blue*) and Auger (*red*) combined energy spectra

The different behaviour in the cutoff region is apparently beyond the expectations of the current knowledge of systematics. A possible contribution to this difference in terms of declination dependence of the flux has been investigated by Auger [118]. No significant variation in the flux measured with the SD in four declination bands has been found that could account for the discrepancy between spectra measured from different hemispheres. The differences found between the measurements in two separate declination bands are instead compatible with the variations expected from a dipolar modulation (see below) of the flux [119].

Mass Composition

Composition is addressed using the depth of the position of the maximum in the number of shower particles, $X_{\max}$, which is measured by the FD. In a simplistic picture, the sensitivity of $X_{\max}$ to mass composition relies on the fact that showers from heavier (lighter) nuclei develop higher (deeper) in the atmosphere and their profiles fluctuate less (more).

The measurements by Auger are the most robust for both the data selection and the quality of the $X_{\max}$ distributions that are obtained. For the limited field of view of the telescopes, depending on the zenith angle and impact point of the shower, a fluorescence detector views a different range of $X_{\max}$. The Auger analysis adopts event selection and quality cuts that allow to get rid of this bias and thus obtain unbiased $X_{\max}$ distributions. Correcting for detector resolution and acceptance, the first two moments of the distributions (mean and standard deviations) can be directly compared to air shower simulations. The Auger collaboration has published $X_{\max}$ measurements for hybrid showers having energies above $10^{17.8}$ eV [39] and recently reported preliminary results extending these measurements down to 10^{17} eV [118]. Figure 13 shows the published data.

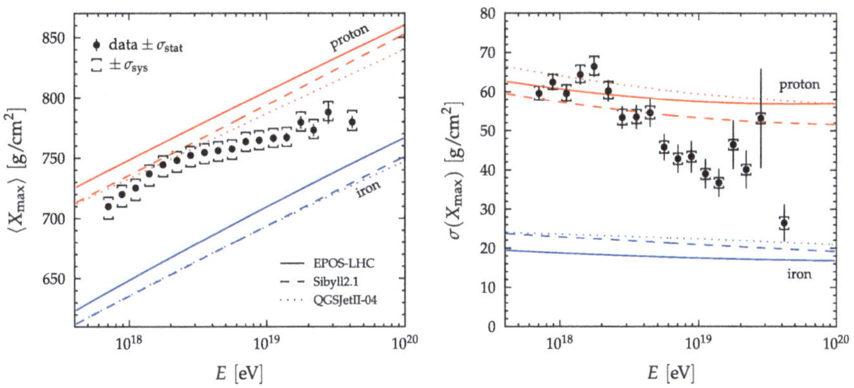

Fig. 13 The mean (*left*) and the standard deviation (*right*) of the X_{max} distributions measured by Auger, as a function of energy compared to air-shower simulations for protons and iron primaries

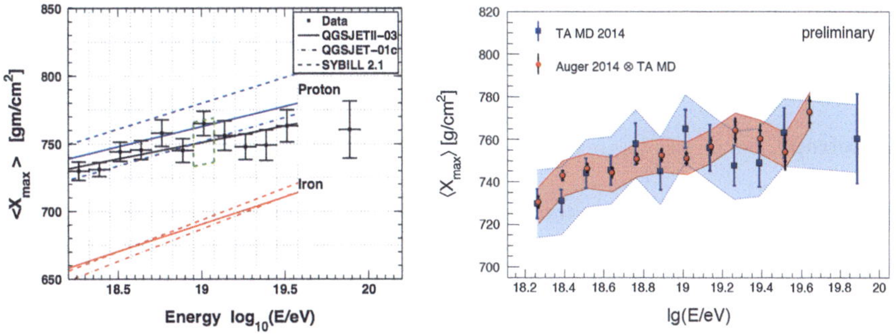

Fig. 14 (*Left panel*) The TA Middle Drum hybrid composition result using geometry and pattern recognition cuts. The *solid black line* is a fit to the data. *Coloured lines* are fits to MC, for the used hadronic models. The *green hashed box* indicates the total systematic error on ⟨X_{max}⟩. (*Right panel*) Comparison of ⟨X_{max}⟩ as measured with the MD of TA (*blue squares*) and the ⟨X_{max}⟩ of the Auger data folded with the MD acceptance. The *coloured bands* show the systematic uncertainties of the X_{max} scales of each experiment

Telescope Array has reported data for different data selections: monocular, stereo and hybrid data. The measurements of the mean depth for hybrid events observed from the Middle Drum fluorescence detector [43] are shown in Fig. 14 (left panel).

It has to be noted that, contrary to Auger, the TA X_{max} distributions are folded with the detector effects such as the selection efficiency and acceptance, and therefore biased by experimental effects. This approach is not only due to an analysis choice, but is primarily determined by the limited size of the data that prevents from applying cuts so selective as the ones adopted by Auger. The interpretation of TA data is then performed using Monte Carlo predictions folded with the same detector efficiency and resolution.

The different approaches of the two collaborations imply that the $\langle X_{max}\rangle$ values obtained cannot be directly compared to one another. Also, the MC predictions for the measurements are based on different hadronic interaction models. A joint working group has been setup for comparing the two approaches. The primary abundances which best describe the Auger data have been simulated and analyzed by the TA collaboration using the same procedure as applied to their data. The result, which is a simulated version of the Auger data as it would be observed by TA, is shown in Fig. 14 (right panel). The average difference between the two data sets was found to be (2.9 $\pm$ 2.7 (stat.) $\pm$ 18 (syst.)) g/cm^2.

The Auger X_{max} data (moments and distributions) enable a step further in the interpretation of mass composition. In fact, the mean log mass can be derived from the measurement of $\langle X_{max}\rangle$, on the basis of the superposition model or a simple parametric extension of the same model [120]. Yet from the mean X_{max} alone it is not possible to retrieve information about the relative weights of primaries contributing to the actual value of $\langle \ln A\rangle$. Using their data the Auger collaboration has obtained the evolution with energy of the first two moments of $\ln A$ [120] and of the fractions of four mass groups (H, He, N and Fe) from the fit of the X_{max} distributions [121].

Anisotropy

The search for anisotropies in the arrival directions of cosmic rays aims at spotting their sources or a global inhomogeneity in the source distribution on scales comparable with the loss length at given energy. At the highest energies, finding small angular scale anisotropies would represent the access gate to the beginning of charged particle astronomy. This possibility is however tightly related to the composition of the CRs at such high energies, since only for protons the deflection angles induced by the (poorly known) magnetic fields are expected to be small. Both the Auger and TA collaborations have intense programmes to search for anisotropies. These include several tools like auto-correlation, correlation with source catalogues, search for flux excesses (*hotspots*) in the visible sky and correlation with other experiments.

At present, none of the tests show statistically significant evidence of anisotropy [122] [123]. Yet remarkable flux excesses are observed at intermediate scales in the North (South) hemisphere by TA (Auger). Using cosmic ray events with energy $E > 57$ EeV, TA have observed a cluster of events, centred at R.A. = 146.7°, Dec.= 43.2° [123], of about 20° radius and with a calculated probability of appearing by chance in an isotropic cosmic-ray sky of 3.7×10^{-4} (3.4σ). In Auger, the strongest departures from isotropy (post-trial probability $\sim$1.4%) are obtained for cosmic rays with $E > 58$ EeV around the direction of Cen A (15° radius) [122]. In any case, it will be interesting to follow the evolution of these excesses with future data.

Another recent correlation study is based on the attempt to exploit IceCube neutrino observations to identify the sources of UHECRs [124]. The study is a common effort by the IceCube, Auger and TA collaborations and is based on the fact that neutrinos can be related to charged particles both at astrophysical sources, through their interaction with the ambient matter and radiation, and along their propagation through the cosmic background radiation. The UHECRs comprise about 300 events above 50 EeV collected by Auger and TA. The neutrino events used are the most energetic (30 TeV to 2 PeV), which provided evidence for a neutrino flux of astrophysical origin. These comprise 39 *cascades* (signatures of charged-current ν_e interactions as well as neutral-current interactions of all flavors) and 16 high energy *tracks* (signatures of charged-current ν_μ interactions). Another correlation study involves the so-called IceCube point-source sample of about 400,000 tracks with a sub-degree angular resolution. No indications of correlations at discovery level are obtained for any of the searches performed. The smallest of the p-values comes from the search for correlation between UHECRs with IceCube high energy cascades, a result that should continue to be monitored.

Large-scale anisotropies are not suited to correlate directly cosmic rays to sources, but can be used to infer aspects of the global distribution of sources. Some large-scale anisotropies (below the percent level) are expected as well because of the relative motion of cosmic rays with respect to the rest frame of background radiation [125]. Furthermore, a large-scale analysis as a function of energy is important since it could provide information about the transition from galactic to extragalactic dominance in the cosmic ray flux.

The Auger collaboration has published evidence [126] of a smooth phase transition of the first harmonic modulation in right ascension distribution from 270° to 100° around 1 EeV. This fact might correspond to a transition from dipole direction pointing to the galactic centre, at low energies, to one which is rotated by about 180° above a few EeV. This study is based on both Rayleigh and East-West analyses of the counting rates. This analysis has been recently updated and a prescribed test is running to confirm it [118]. Figure 15 shows the results from the latest data. The amplitude is still given as upper limits, but for two energy bins, between 1 and 2 EeV and above 8 EeV, the statistical significance is high enough to provide measured amplitudes.

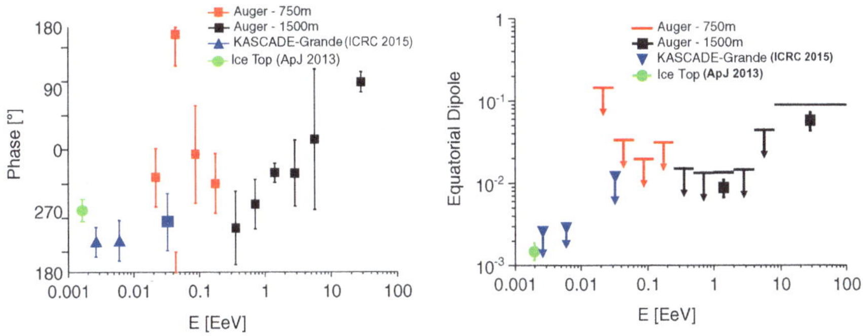

Fig. 15 Auger first harmonic analysis. *Left* Measured phases of the first harmonic modulation in RA. *Right* Upper limits of the dipole equatorial component. Amplitudes are also reported in the two energy bins when the corresponding p-value expected from isotropy is below 10^{-3}

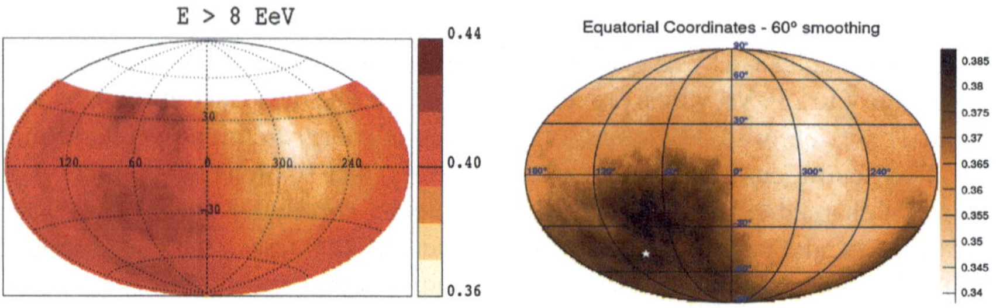

Fig. 16 Sky maps in equatorial coordinates of flux, smoothed in angular windows of 45° (60°) radius, for Auger (Auger and Telescope Array) events with $E > 8$ EeV (10 EeV), *left* (*right*) *panel*

The large-scale distribution of arrival directions has been also studied by combining the data of Auger and Telescope Array. Thanks to the full-sky coverage, the measurement of the angular power spectrum does not rely on any assumption on the underlying flux of cosmic rays. The study, carried out using a spherical harmonic analysis for cosmic rays above 10^{19} eV, has been published in [127] and updated at ICRC 2015 [128]. No deviation at discovery level from isotropy is found at any multipoles. The largest deviation, with a p-value of 5×10^{-3}, occurs for the dipole, with an amplitude of $(6.5 \pm 1.9)\%$, pointing to $93° \pm 24°$ in right ascension and $-46° \pm 18°$ in declination (see Fig. 16, right). It is worth noting that this result agrees with the one found with Auger-only data [119], assuming pure dipolar or dipolar-quadrupolar distributions (see Fig. 16, left).

Future Developments

UHECR data provide other interesting outcomes on several aspects of cosmic ray physics (e.g. muon component in the shower events, photon and neutrino limits) and particle physics (e.g. proton cross section, hadronic interactions at energies higher than LHC) that are not included in this section and will be partly considered in the following. Considering all data, it appears difficult to build a consistent picture of the origin of UHECRs in the presence of so many unknowns about source distribution, composition, galactic and extragalactic magnetic fields, etc. To make further progress in this direction more accurate and extended information on the nature of the primaries is required: mass composition is currently unavailable above 40 EeV due to the intrinsic duty cycle of the FD and the scarce accuracy of the composition sensitive methods based on the surface array data.

The AugerPrime [129] upgrade of the Pierre Auger Observatory has been specifically designed to improve mass composition in the whole energy range. Along the line of a hybrid design, each SD will be equipped with a top scintillator layer. Shower particles will be sampled by two detectors (scintillators and water-Čerenkov stations) having different responses to the muonic and electromagnetic components, thus allowing to reconstruct each of them separately. The muonic component will be derived in each station by subtracting the signal observed in the scintillator from that seen in the water-Čerenkov tank. The upgraded array will provide data with no duty cycle limitation and then the access to the highest energies will be made possible.

The Telescope Array collaboration plan to extend the SD array by a factor four, TAx4 [130], by adding 500 surface detectors on a square grid of about 2 km spacing. With this new design the overall area will be approximately 3000 km^2. With enhanced statistics, they expect to improve the investigation of the observed hotspot and possibly the correlation with other sources. TA is also designing new muon detectors [131] to enable the simultaneous detection of electromagnetic and muonic components in shower events.

A possible jump in integrated statistics, even though with limited resolution on $X_{\rm max}$, would be provided by the observation of EAS fluorescence light from a space-based detector looking towards Earth's surface. The Extreme Universe Space Observatory on board the Japanese Experiment Module of the International Space Station, JEM-EUSO, is being designed for such a mission. It is based on a wide field of view (60°) near-UV telescope with a diameter of 2.5 m, orbiting at an altitude of about 400 km, that would provide an annual exposure larger than 50,000 km^2 sr yr/yr, above 6×10^{19} eV [132, 133].

Transport of CRs in the Galaxy

The transport of CRs in the Galaxy has been subject of active investigation for quite some time. The main handle we have on the propagation of galactic CRs comes from measurements of secondary-to-primary ratios, such as Boron/Carbon (B/C), which provide us with an estimate of the grammage that CRs traverse during propagation. In fact, this indicator is the main reason why we describe the CR transport as mainly diffusive in nature: a ballistic propagation of CRs would make CRs escape the Galaxy in times which are much shorter than the one necessary to explain the observed B/C ratio.

The B/C ratio is related to the grammage traversed by CRs, $X(E) = \bar{n}\mu v \tau_{esc}(E)$, where $\bar{n}$ is the mean gas density in the confinement volume of the Galaxy (disc plus halo), μ is the mean mass of the gas, v is the speed of particles. For particles with energy per nucleon of 10 GeV/n the measured B/C corresponds to $X \sim 10\,\mathrm{g\,cm^{-2}}$. If the sources are located in the thin disc of the Galaxy with half thickness $h = 150$ pc and the halo extends to a height H, the mean density can be estimated as $\bar{n} = n_{disc}h/H = 5 \times 10^{-2} \left(\frac{n_{disc}}{1cm^{-3}}\right) \left(\frac{H}{3kpc}\right)^{-1}$ cm^{-3}. For a standard chemical composition of the ISM ($n_{He} \approx 0.15 n_H$) the mean mass is $\mu = (n_H + 4n_{He})/(n_H + n_{He}) \approx 1.4 m_p$. It follows that for a proton with energy $E_* = 10$ GeV the typical escape time is

$$\tau_* \sim \frac{X(E_*)}{\bar{n}\mu c} = 90\left(\frac{H}{3kpc}\right)\text{Myr}, \qquad (1)$$

which exceeds the ballistic propagation time scale by at least three orders of magnitude. This remains the strongest evidence so far for diffusive motion of CRs in the Galaxy. A diffusion coefficient can be introduced as $\tau_{esc}(E) = H^2/D(E) = \tau_*(E/E_*)^{-\delta}$, so that at 10 GeV $D(E) \simeq 3 \times 10^{28}\left(\frac{H}{3kpc}\right)\text{cm}^2\text{s}^{-1}$. The grammage (and therefore the escape time) decreases with energy (or rather with rigidity) as inferred from the B/C ratio, illustrated in Fig. 6, which shows a collection of data points on the ratio of fluxes of boron and carbon. Figure 6 illustrates the level of uncertainty in the determination of the slope of the B/C ratio at high energies, which reflects on the uncertainty in the high energy behaviour of the diffusion coefficient. At low energies, the uncertainty is due to the effects of solar modulation which suppresses CR fluxes in a different way during different phases of the solar activity (see [134] for a recent review). The effect of modulation is more pronounced on the spectra of individual elements than on the B/C ratio. The high rigidity behaviour of the B/C ratio is compatible with a power law grammage $X(R) \propto R^{-\delta}$ with $\delta = 0.3$–0.6.

While at high energy CR transport is most likely diffusive, at low energies other processes may become important or even dominant. For instance, particles can be advected with a galactic wind. This phenomenon leads to CR spectra at the position of the Sun that are much harder than the ones observed at high energies, a phenomenon that is apparent in the recent Voyager I data [135]. At energies below $\sim GeV$, energy losses due to ionization also become important and result in spectral hardenings as compared with the high energy trend. For nuclei, spallation energy losses also become important and may harden nuclear spectra, as compared with the proton spectrum.

Much can be learned in favour and against current models of CR propagation from the understanding of the microphysics of CR transport: spatial diffusion results from pitch angle diffusion of charged particles propagating in a background of Alfvén waves with random phases and a given power spectrum. This process requires resonance between the gyration radius of the particle in the background magnetic field and the wavelength of the relevant Alfvén waves. If the waves are assumed to propagate in both directions along the magnetic field, as it is usually implicitly assumed in standard propagation calculations, then CRs are scattered to reach diffusion, and gain energy through second order Fermi acceleration (diffusion in momentum space).

The debate on whether the background Alfvén waves are the result of environmental processes (for instance SN explosions or other types of stirring of the ISM) or rather produced non-linearly by CRs themselves remains a hot topic in the field. Below we discuss the implications of some recent ideas concerning this point and some possible observational consequences.

Self-generation of Waves

It was first shown by [136] that the super-Alfvenic streaming of charged particles may result in the excitation of a streaming instability, namely generation of weakly modified Alfvén waves with wavenumber $k \sim 1/r_L(p)$, where $r_L(p)$ is the Larmor radius of particles with momentum p that are responsible for the instability. It is important to realize that these waves have the correct wavenumber k to be effective for particle pitch angle scattering. Hence the question arises of whether it is possible to think of particle diffusion in self-generated waves. The problem of particle transport, that is usually solved in a test-particle approximation (the diffusion coefficient is pre-assigned and independent upon the particles that are being propagated), becomes non-linear when self-generation is included. The growth of the instability is limited by different mechanisms of wave damping. In the Galaxy, the two main damping mechanisms are ion-neutral damping [137] and non-linear Landau damping (NLLD) [138, 139], which are important in partially ionized and in totally ionized plasmas, respectively.

As recognized by [140, 141], the effect of ion-neutral damping in the ISM is expected to be so strong that the growth of Alfvén waves through streaming instability is strongly hindered. In these conditions, the wave excitation is possible only far from the galactic disc, where the density of neutral hydrogen drops to very low values. On the other hand, while the average value of such density along a line of sight is well determined by observations [142], it is possible that neutrals may be spatially segregated, so that dense regions of high density may be surrounded by vast regions where the gas is mostly ionized and ion-neutral damping is not very important. In this case, Alfvén waves can be excited by CRs and the growth of the instability occurs at a rate

$$\Gamma_{\rm cr} = \frac{16\pi^2}{3} \frac{v_A}{\mathscr{F}(k) B_0^2} \left[p^4 v(p) \frac{\partial f}{\partial z} \right]_{p=eB_0/kc}, \tag{2}$$

where the gradient in the CR distribution has been assumed to be only along the z direction perpendicular to the galactic plane. Here $\mathscr{F}(z, k)$ is the fractional power $\delta B^2(k)/B_0^2$ per unit logarithmic interval of k, which also determines the diffusion coefficient through

$$D(z, p) = \frac{1}{3} r_L(p) v(p) \frac{1}{\mathscr{F}(k)|_{k=r_L(p)}}. \tag{3}$$

Equation 3 returns Bohm diffusion coefficient only if there is the same power on all scales and $\delta B/B_0 = 1$. In general, the real diffusion coefficient is larger than Bohm diffusion, as one can see from Eq. 3, since $\mathscr{F}(z, k) \ll 1$ (Eq. 3 is strictly valid only in this limit). The rate of NLLD can be written as [143]:

$$\Gamma_{\rm nlld} = (2c_k)^{-3/2} k v_A \mathscr{F}(k)^{1/2}, \tag{4}$$

with $c_k = 3.6$.

Imposing that growth and damping balance each other locally, one can determine the power spectrum $\mathscr{F}(z,k)$. At high energy, where transport is dominated by diffusion, one can see that $\partial f/\partial z = f_0/H$, where $f_0(p) \sim p^{-\gamma}$ is the CR spectrum in the disc and H is the size of the galactic halo. Hence $\mathscr{F}(z,k) \sim k^{-\frac{2}{3}(5-\gamma)}$, and consequently $D(p) \propto p^{-\frac{7}{3}+\frac{2}{3}\gamma}$. For $\gamma \approx 4.7$, one easily infers $D(p) \sim p^{0.8}$. The fast momentum dependence of the diffusion coefficient implies that the CR confinement due to self-generated waves, at least in purely diffusive models, is bound to be effective only at relatively low energies (typically below ~ few 100 GeV).

After the discovery of the spectral breaks by PAMELA [6] and AMS-02 [18, 19] (see discussion in section "Observations of Protons and Nuclei up to Hundreds PeV"), models of self-generation have been reconsidered [23, 24, 144]: this work showed that the combination of self-generated waves and waves produced by cascading of large scale pre-existing turbulence may explain the spectral breaking: at high energies turbulence injected at large scale (possibly due to SN explosions) and cascading towards smaller scales dominates CR scattering; at low energies the self-generated waves provide the necessary scattering. This change of regime naturally leads to a spectral break at a few hundred GV rigidity. Moreover [144] pointed out that, since the self-generated waves all move on average away from the disc and along the CR gradient, advection of such CRs with the waves may become important at low energies (below 10 GeV/n) and be responsible for the spectral hardening at such energies that seems to fit well the recent Voyager I data [135]. These phenomena are illustrated in Fig. 17 (from [144]): the flux of protons as measured by PAMELA, AMS-02, CREAM and Voyager is shown, together with the prediction [144] with (red solid line) and without (blue dotted line) the effect of solar modulation. A combination of advection with the self-generated waves and proton energy losses seems to provide a good description of the Voyager data in the ISM.

The calculations of [24, 144] were extended to the spectra of all nuclei and provided a good description of their spectra, as illustrated in Fig. 18.

Interestingly, despite the good agreement with the spectra of primary nuclei, the B/C ratio as calculated in Ref. [144], shows a slight excess at high energies (black

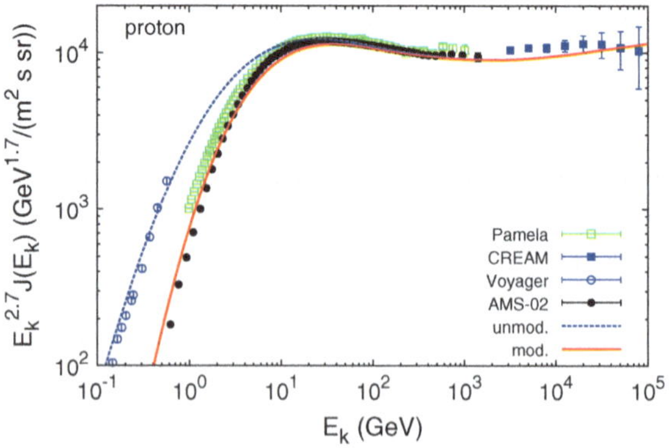

Fig. 17 Proton spectrum as measured by some experiments (see *labels*) compared with the flux as predicted by [144], with and without accounting for solar modulation (*solid* and *dashed lines*, respectively). The flux without modulation fits well the recent Voyager data [135]

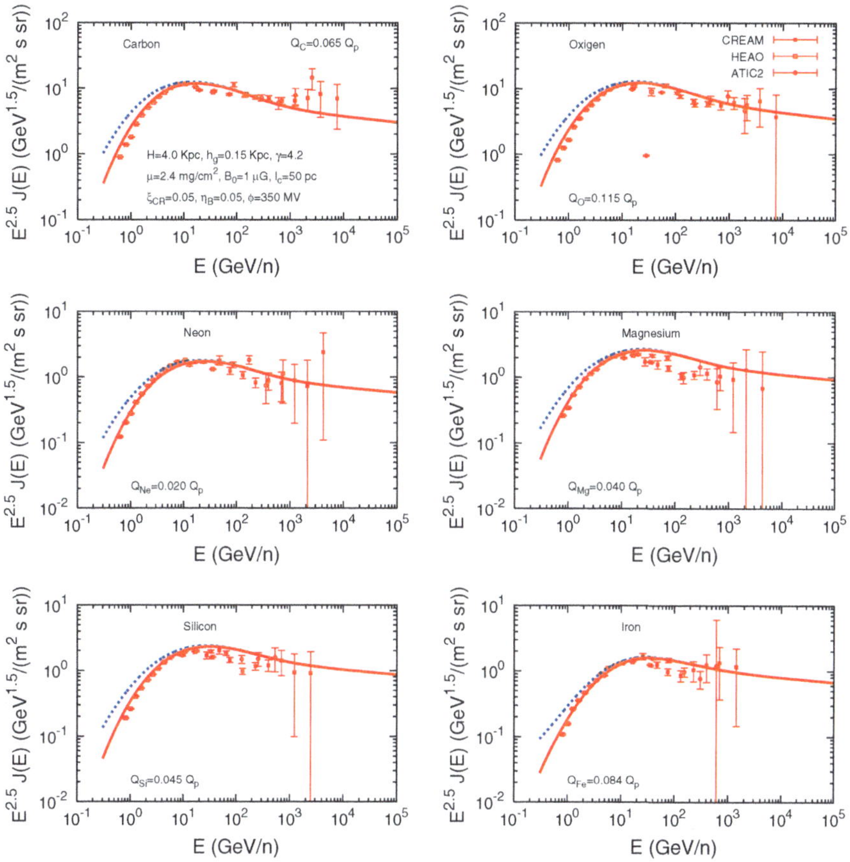

Fig. 18 Spectra of nuclei [24] in a model with self-generated waves and pre-existing turbulence, compared with available data

solid line in Fig. 19). The red solid line in Fig. 19 shows the B/C ratio obtained by accounting for a grammage of $\sim 0.17\,\mathrm{g\,cm}^{-2}$ accumulated by CRs while being downstream of a supernova shock for $\sim 10^4$ years. This irreducible contribution seems to be required if to improve the fit to the observed B/C data (this conclusion, that was found by [144] using preliminary AMS-02 data, seems to apply also to the recently AMS-02 data [7] on the B/C ratio).

A spectral breaking at ∼200 GV similar to the one discussed above may also be induced by a spatial dependence of the diffusion coefficient in the halo, as discussed by [22]: if the diffusion coefficient is $D_1(p) \propto p^{\delta_1}$ closer to the disc, and $D_2(p) \propto p^{\delta_2}$ farther away, then the spectrum at the Earth is $\propto p^{-\alpha-\delta_2}$ at low energies, and $\propto p^{-\alpha-\delta_1}$ at high energy, if injection is $\propto p^{-\alpha}$.

The propagation of CRs in the Galaxy leaves an imprint in measurable quantities other than the B/C ratio. The galactic emission of gamma rays reflects the density of

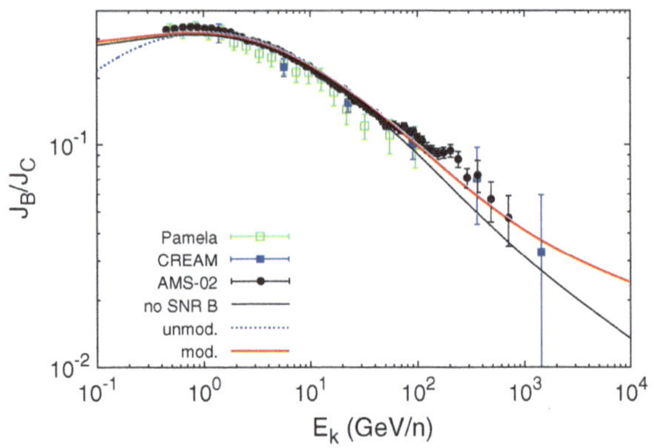

Fig. 19 B/C ratio (from [144]) in a model with self-generated waves and pre-existing turbulence (*black solid line*), compared with available data. The *red line* is obtained by adding an energy-independent grammage of $\sim$0.17 gcm^{-2} accumulated by CRs while being downstream of a supernova shock for $\sim 10^4$ years

hadronic and leptonic CRs and the density of gas and photons in the environment. In the disc of the Galaxy, most gamma-ray emission is due to pp inelastic collisions, which result in neutral pion production and decays. If the density of gas is reliably traced, the detection of gamma radiation from a given line of sight results in a measurement of the local density of CRs. This analysis has been recently carried out by the *Fermi*-LATcollaboration [145] and revealed an interesting trend: (1) the density of CRs in the inner Galaxy (within a few kpc from the galactic centre) is rather peaked where the density of SNRs is also observed to be peaked, while the CR density decreases very slowly with Galactocentric distance in the outer Galaxy; (2) the spectrum of CRs with energies $\leq$100 GeV in the inner Galaxy is somewhat harder than in the outer regions. These findings were qualitatively confirmed by an independent analysis of the *Fermi*-LATdata [146].

The weak dependence of the CR density on Galactocentric distance R for $R > 5$ kpc is the well known CR gradient problem [147–149]: the CR density drops much slower than proportional to the density of sources. These findings are difficult to reconcile with the standard approach to CR propagation, which is based upon solving the transport equation under the assumption that the diffusive properties are the same in the whole propagation volume [150]. Within the context of this approach, several proposals have been put forward to explain the radial gradient problem. Among them: (a) assuming a larger halo size or (b) a flatter distribution of sources in the outer Galaxy [148]; (c) accounting for advection effects due to the presence of a galactic wind [151]; (d) assuming a sharp rise of the CO-to-H_2 ratio in the external Galaxy [152]; (e) speculating on a possible radial dependence of the injected spectrum [153]. None of these ideas, taken individually, can simultaneously account for both the spatial gradient and the spectral behaviour of CR protons. Moreover, many of them have issues in accounting for other observables [154].

A different class of solutions invokes the breakdown of the hypothesis of a spatially constant diffusion coefficient. For instance, [154] proposed a correlation between the diffusion coefficient parallel to the galactic plane and the source density in order to account for both the CR density gradient and the small observed anisotropy of

CR arrival directions. Ref. [155] followed the same lines of thought and showed that a phenomenological scenario where the transport properties (both diffusion and convection) are position-dependent can account for the observed gradient in the CR density. It is however unsatisfactory that these approaches do not provide a convincing physical motivation for the assumed space properties of the transport parameters.

In the context of models of CR transport with self-generated diffusion and advection the CR accumulation and the harder spectra in the inner Galaxy find a relatively simple explanation [156]: waves are excited more easily where there is more injection, so that the diffusion coefficient is correspondingly smaller and CRs are accumulated there for longer times resulting in a higher CR density. At the same time, advection with self-generated waves remains dominant up to higher energies, thereby implying harder CR spectra. One should keep in mind that so far the evidence for harder spectra in the inner Galaxy derives from low energy gamma-ray observations. It is not clear if observations will show that this phenomenon holds even at higher energies. In that case, self-generated waves alone would not provide the full explanation of observations.

The CR density and spectral slope as found by [156] in the context of self-generated diffusion are shown in Fig. 20, together with the findings of the analysis of the *Fermi*-LATdata. Both the density and slope are well fitted if an exponential cutoff is assumed in the spatial distribution of the background magnetic field at distances beyond 10 kpc from the galactic centre. In the absence of this magnetic field drop at large galactocentric distances, the CR gradient problem becomes even more severe in non-linear modes than it is in the standard model, because the density of CRs at large R drops even faster as a result of the smaller density of sources and larger values of the diffusion coefficient.

Finally, we stress that, as discussed above, if the effect of an R-dependent slope is confirmed to exist at higher energies as well then an alternative explanation of the CR gradient problem should be sought.

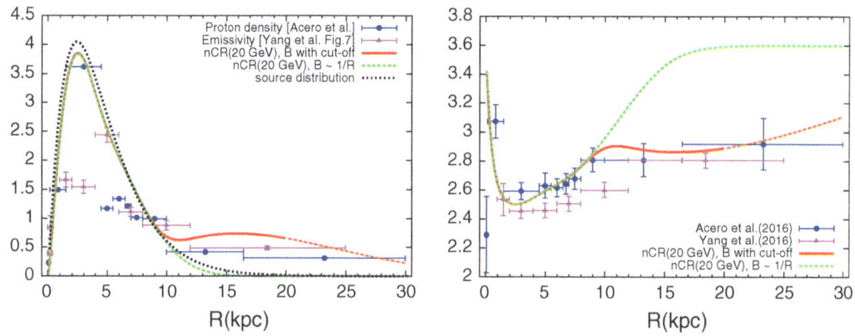

Fig. 20 Density (*left panel*) and spectral slope (*right panel*) of CRs with energy ~20 GeV as a function of the Galactocentric distance. The data points are from [145] and [146], while the curves refer to the results of [156]

CR Driven Galactic Winds

The possibility that a galaxy may launch winds has attracted attention for many different reasons. For instance, star formation is regulated by the amount of gas available, and winds modify the availability of such gas. In fact, galactic models that do not include feedback processes suffer from over-predicting the amount of baryons and star formation rates [157, 158]. Winds also pollute galactic halos with hot dilute plasma that may provide an important contribution to the number of baryons in the Universe [159–161]. Such gas might in fact have already been detected [162] in the form of X-ray emitting plasma with temperature of several million degrees, and possibly associated with a galactic wind [163, 164] (see also [165]). Finally, winds can affect the transport of cosmic rays (CRs) in a galaxy, by advecting them away from their sources.

Galactic winds may be thermally driven, namely powered by core-collapse SNe [166] or momentum driven, powered by starburst radiation [167, 168]. These two mechanisms of wind launching are thought to be at work in starburst galaxies and galaxies with active nuclei [169]. On the other hand in a galaxy like the Milky Way, winds are unlikely to be due to such processes because thermal and radiation pressure gradients are expected to be too small. A possible exception is the innermost part of the galactic Centre region where the recently discovered Fermi Bubbles may originate from direct bursting activity of Sgr A* [170, 171] or past starburst activities [172]. On the other hand, CRs can play an important role in launching winds because of the gradient that their pressure develops as a consequence of the gradual escape of CRs from the Galaxy. The force $-\nabla P_{\mathrm{CR}}$ associated with such gradient is directed opposite to the gravitational force, and in certain conditions the plasma above and below the disc can be lifted off to form a CR driven wind. Notice that the gravitational force may be dominated by the dark matter component or the baryonic (gas and stars) components depending on the location. The force exerted by CRs depends in a complicated manner on the density of sources of CRs but also on non-linear processes of excitation of Alfvén waves through streaming instability (see discussion above). Both the force induced by CRs on the background plasma and the streaming instability induced by CRs depend on the gradient in CR density. In turn, the distribution function of CRs is affected by their transport: diffusion is self-regulated through the production of Alfvén waves, and advection is determined by the velocity of the wind, if any is launched, and by the Alfvén waves' velocity, directed away from the sources of CRs. This complex interplay makes the problem non-linear. The first pioneering attempt to describe the hydrodynamics of a CR driven wind was described in Ref. [173], where the author used a spherically symmetric model of the Galaxy and considered only baryons and stars for the calculation of the gravitational potential. Later [174] presented an extensive discussion of the hydrodynamics of CR driven winds: dark matter was included and a realistic geometry of the wind was considered, in which the launch takes place at some distance from the galactic disc and proceeds in a roughly cylindrical symmetry out to a distance of about ~15 kpc, where the flow opens up into a spherical shape. The calculations of [174] treated CRs as a fluid, hence no information on the spectrum of CRs was retained. The important role of wave damping in the wind region was also discussed in [174], although only in the simplified case of a spherical outflow. As mentioned above, [174] assumed that the wind is launched some distance away from the disc of the Galaxy and this assumption raises the issue of what happens in the region between the disc and the base of the wind, a problem of both mathematical and physical importance [175].

The dynamical role of CRs in launching winds was also studied via purely hydrodynamical simulations [176–178] and through MHD simulations [179–181]. These simulations, with their progressive level of sophistication, demonstrated that CRs play an important role in wind launching. Nevertheless, all this bulk of work treated CRs as a fluid, thereby not providing any information on the CR spectrum.

At present, the only two attempts at calculating the spectrum of galactic CRs in the presence of CR driven winds in a self-consistent manner were made by [182] and [183]. The analytical approach put forward by [182] is illuminating in terms of understanding the basic physical aspects of CR driven winds and CR transport in such words, although the conclusions may be rather different in more realistic winds, as discussed by [183].

In the cases studied in Ref. [182], the Alfvén velocity at the base of the wind is larger than the wind launching speed v_w, and increases linearly with the distance z from the galactic disc. There is a distance $s_*(p)$ where the time scale for advection and diffusion are equal:

$$\frac{s_*^2}{D(p)} = \frac{s_*(p)}{v_A + u_w} \approx \frac{s_*(p)}{v_A} \propto \sim \text{constant}, \tag{5}$$

since $v_A \propto z \sim s_*$. This implies that the critical distance $s_*(p)$, as a function of the momentum p scales as $s_*(p) \propto D^{1/2}(p)$. In CR-driven wind models of the origin of CRs, the distance s_* plays the role of the halo size H in the standard leaky box model. By analogy, the spectrum of CRs in the disc can be written as $f(p) \propto Q(p)s_*(p)/D(p) \propto Q(p)/D^{1/2}(p)$, which is quite different from the standard result $f(p) \propto Q(p)/D(p)$.

Although these scalings are very useful to pin down the essential physical ingredients of the problem, they do not fully reflect the complexity of the CR transport in winds: as recently discussed in [183], most hydrodynamical wind solutions lead to CR spectra at the Earth which are quite unlike the observed ones. The main reason for this result is that the advection velocity at the base of the wind is usually found to be around ~100 km/s, which leads to advection dominated transport even at relatively high energies, say in the TeV range, at odds with the observed CR spectra. Moreover, standard wind solutions are characterized by a spectral softening at energies above ~TeV, rather than the observed hardening found by PAMELA and AMS-02. On the other hand, [183] pointed out that the region between the base of the wind and the disc, where the sources are assumed to be localized, is crucial for the determination of the spectrum: for instance, assuming that the wind is actually launched at $z_0 \sim 1$ kpc from the disc and that for $|z| < z_0$ the diffusion coefficient is fixed (not self-generated) with a Kolmogorov-like shape, the overall structure of the CR spectrum at the Earth may be recovered: a low energy hardening due to advection is visible, a steep spectrum at energies $20 \leq E \leq 1000$ GeV is produced by the self-generation of waves in the wind region, and finally at high energy a hardening is produced as due to the dominance of the Kolmogorov spectrum upon the spectrum of self-generated waves.

Positrons and Antiprotons

The ratios $e^+/(e^- + e^+)$ and $\bar{p}/p$ are often used to infer the propagation properties of galactic CRs. If positrons and antiprotons are solely produced in hadronic interactions of primary CRs with ISM gas, it is easy to demonstrate that both ratios should be decreasing functions of energy, at least for energies high enough that radiative energy losses of electrons dominate their propagation and solar modulation can be neglected. These conditions are typically satisfied for energies above $\sim$10 GeV. For both ratios, it is expected that they decrease with energy as $\sim 1/D(E)$, where $D(E)$ is the diffusion coefficient of particles with energy E (see, for instance, Ref. [28] and [184] for a recent review), although the uncertainties in the cross section of antiproton production [185] may affect such conclusion as far as the $\bar{p}/p$ ratio is concerned.

The PAMELA experiment measured the positron ratio and found that it grows with energy for $E > 10$ GeV [25], at least up to $\sim$100 GeV. A similar trend was also obtained by analyzing lepton fluxes from the Fermi-LAT telescope [45]. These results were later confirmed and extended to higher energies by AMS-02 [26]. The PAMELA [29, 186] and AMS-02 [187] measurements of the separate fluxes of electrons and positrons showed that the increasing trend in the ratio $e^+/(e^- + e^+)$ is due to an excess of positrons rather than a deficit of electrons, at odds with the simplest interpretation of positrons as secondary products of hadronic interactions.

As discussed in section "Cosmic Rays Observations" (see also [184]), the $\bar{p}/p$ does not show any rise with energy, thereby suggesting that whatever the sources of positrons, they should not produce appreciable amounts of antiprotons. However, it has been suggested that the recent AMS-02 measurement of the flux of antiprotons [35], which extends the previous measurements to higher energies, indicate that the spectrum of $\bar{p}$ is harder than expected based on the standard model of CR transport. This conclusion is, at present, questionable, in that the uncertainties in the cross sections of $\bar{p}$ production and in the parameters of CR transport to be adopted do not allow to reach a definite conclusion in this matter. Nevertheless, several authors noticed the intriguing similarity between the spectrum of antiprotons and those of positrons and protons (see for instance [36] and references therein) and postulated that both positrons and antiprotons might still be the result of CR interactions provided the model of CR transport is changed in a suitable way. We will comment further on these models below.

The discovery of an increasing positron ratio stimulated much interest in dark matter annihilation as the source of the *excess* positrons. Such phenomenon typically leads to copious production of pions and finally gamma rays, electrons and positrons and, to a lesser extent, hadrons. The explanation of the positron excess in terms of dark matter annihilation requires the dark matter candidate to have peculiar properties: it should be leptophylic (otherwise the $\bar{p}/p$ would be affected), and should be characterized by Sommerfeld enhancement of the cross section (in order to explain the normalization of the positron spectrum). In addition, the contribution of clumps of dark matter in the Milky Way's halo should be prominent, in order to account for the observed flux of positrons at the Earth. As discussed in detail in Ref. [28], these conditions appear to be rather *ad hoc*, and an explanation of the positron excess in terms of dark matter annihilation is, at present, considered as disfavored, at least by these authors.

Several astrophysical explanations of the excess of CR positrons have also been put forward. It has been suggested that the positron excess may be due to old SNRs [31]: the idea is that electrons and positrons are also produced as secondary products of hadronic interactions inside accelerators, such as SNR shocks. The peculiarity of these electrons and positrons is that, since they are produced inside the acceleration region, they also take part in the acceleration process which leads to particularly hard spectra, required to explain the observed $e^+/(e^- + e^+)$ ratio. It was soon realized that the same fate would occur to antiprotons [32] and to nuclear secondaries, such as Boron [188]: in both the $\bar{p}/p$ ratio and B/C ratio one would expect a rising trend at sufficiently high energies. The measurements of the B/C ratio and of the $\bar{p}/p$ ratio by PAMELA and AMS-02 experiments have shown no sign of such rise, thereby providing strong constraints on the applicability of such model.

Electron-positron pairs are also copiously produced in pulsar magnetospheres. In the vicinity of pulsars, the electromagnetic fields are so intense that pair cascades develop with very high multiplicity: each electron extracted from the star surface typically produces $\sim 10^4$–10^6 e^+-e^- pairs. Aside from being theoretically predicted, direct evidence of this phenomenon is provided by multi-wavelength observations of Pulsar Wind Nebulae (PWNe), bright synchrotron and IC nebulae surrounding many young pulsars (see e.g. [189] for a review). Since these particles will have to be released in the ISM at some point, their contribution to CR leptons is unavoidable and must be taken into account in any model aimed at explaining the positron excess. In fact, it has been argued [33, 34] that PWNe could be the most important sources of the excess positrons observed by PAMELA and AMS-02 (the possibility that pulsars could be sources of CR positrons had been put forward long before the recent developments in the field [190]).

The e^+-e^- pairs created in the pulsar magnetosphere become part of the relativistic wind into which pulsars convert most of their rotational energy. The interaction between the wind and the surrounding medium, the SNR during early stages and the ISM later on, is what makes the PWN shine: a shock develops from this interaction and propagates towards the pulsar down to a distance that guarantees pressure equilibrium between the unshocked wind and the downstream nebula. Extremely efficient particle acceleration occurs at this shock: long power law spectra extending from about 1 GeV to even 1 PeV are formed and the radiation of these particles in the ambient magnetic field directly reveals the accelerated particle spectrum. The e^+-e^- pairs are seen to be described by a flat spectrum ($N(E) \propto E^{-\gamma}$ with $1 < \gamma < 2$), at low energies, which then steepens to $\gamma > 2$ beyond a few hundred GeV. The hard lepton spectrum at low energies is extremely appealing in terms of explaining the CR positron spectrum if, as expected, at some point in the history of the PWN these particles are released into the ISM. What needs to be assessed is the effective rate of release of the pairs and their potential contribution to the CR spectrum.

Most of the positrons produced in a PWN are likely to be confined in the nebula for long times and perhaps lose their energy there, but pairs produced after the pulsar escapes the parent remnant and forms a bow shock nebula could leak into the ISM and account for the positron excess [191]. In this model, no antiprotons are expected to accompany the pairs, hence the absence of a rise in the $\bar{p}/p$ ratio is easily accounted for. The pulsar explanation of the positron excess leads to expecting several interesting and potentially observable effects, that are currently being investigated (see for instance [192]).

As anticipated above, the approximate similarity of the high energy spectra of positrons, antiprotons and protons have stimulated some speculations on the possibility that e^+ and $\bar{p}$ could be solely secondary products of CR interactions [36–38], by invoking modifications of the standard model of CR transport.

In Ref. [36], it is noticed that the observed e^+ and $\bar{p}$ production rates are in perfect accord with those calculated from standard CR interactions in the ISM and that their spectral shape is consistent with being the same as that of protons at energies above ~300 GeV, thereby supporting the idea that the observed flux of antiprotons and positrons may be interpreted in terms of CR interactions. On the other hand, in the standard model of CR transport, where the grammage is inferred from the B/C ratio, the propagation of leptons (electrons and positrons) is dominated by radiative losses for energies above ~10 GeV, which steepen the spectrum of leptons and lead to a spectral difference with antiprotons. It follows that the conclusion of Ref. [36] can only be considered as potentially viable if energy losses are negligible, which requires a residence time in the Galaxy much smaller than usually assumed. This also implies that in this model the boron production must be decoupled from the production of other secondaries, which is not very appealing from the theoretical point of view, although one implementation of this scenario is already present in the literature, the so-called nested leaky box model [193, 194] (NLB).

The basic assumption of the NLB model is that CRs accumulate most grammage in *cocoons* around the sources, while lesser grammage is accumulated during propagation throughout the Galaxy. The former grammage is assumed to be energy dependent while the latter is assumed to be energy independent. By construction, the two values of the grammage become comparable around few hundred GeV/n. Since secondary Boron nuclei are produced by primary CRs (mainly Carbon and Oxygen) at the same energy per nucleon, the B/C ratio reflects the energy dependence of the grammage at the same energy, hence one should expect that the B/C should be decreasing with energy (reflecting the near source grammage) below 100 GeV/n and become energy independent at higher energies. Since the e^+ and $\bar{p}$ production is characterized by a large inelasticity ($E_{e^+,\bar{p}} \sim 0.1 E_p$ with E_p the proton energy), the production of e^+ and $\bar{p}$ at energy ~10–100 GeV reflects the grammage traversed by CR at 0.1–1 TeV, where the grammage is assumed to be flat in the context of the NLB model. One should keep in mind that the recent paper by AMS-02 on the measurement of the B/C ratio up to ~1 TeV/n did not find evidence for a flattening of the ratio at the highest energies. The idea of cocoons around sources, presented in Ref. [193] as a speculative possibility, might find a theoretical justification in the mechanism of CR self-confinement near sources discussed in Ref. [195].

Acceleration of Galactic CRs

There are several aspects of the problem of the origin of CRs that are tightly linked to the physical processes responsible for particle energization in astrophysical sources: (1) why are there non-thermal particles in the first place? (2) what are the physical mechanisms through which nature energizes a small fraction of the particles in a plasma to non-thermal energies? (3) What is the spectrum of the accelerated particles? (4) What is the maximum energy that particles can be accelerated to?

In terms of energetics of CR injection in the Galaxy, once the confinement time of CRs has been normalized to the B/C ratio, the only class of sources that are left as plausible sources of CRs are supernovae, with typical efficiency ~5–20%. The question of which types of SNe contribute the most is very complex, and it is probably dependent upon the energy of CRs we are interested in.

The mechanism of particle acceleration that is expected to account for the required acceleration efficiency is diffusive shock acceleration (DSA) at the forward shock that accompanies the supersonic motion of the plasma associated with the SN explosions. The theory of DSA was initially developed in [3, 196–199] in the so-called test-particle regime.

Let us consider a shock front characterized by a Mach number M_s. The compression factor at the shock is $r = u_1/u_2$ where u_1 and u_2 are the plasma velocities upstream and downstream of the shock, respectively. The compression factor can be expressed in terms of the Mach number using conservation of mass, momentum and energy at the shock:

$$r = \frac{4M_s^2}{M_s^2+3}, \tag{6}$$

which tends to 4 in the limit of strong shocks, $M_s \to \infty$. A test-particle diffusing in the upstream or downstream plasma alone does not gain or lose energy (although the second order Fermi process discussed above may be at work).

For a stationary parallel shock, namely a shock for which the normal to the shock is parallel to the orientation of the background magnetic field, the transport of particles is described by the diffusion–convection equation (see for instance [197]), which in the shock frame reads:

$$u\frac{\partial f}{\partial z} = \frac{\partial}{\partial z}\left[D\frac{\partial f}{\partial z}\right] + \frac{1}{3}\frac{du}{dz}p\frac{\partial f}{\partial p} + Q, \tag{7}$$

where $f(z, p)$ is the distribution function of accelerated particles, normalized in a way that the number of particles with momentum p at location z is $\int dp 4\pi p^2 f(p, z)$. In Eq. 7 the LHS is the convection term, the first term of the RHS is the spatial diffusion term. The second term on the RHS describes the effect of fluid compression on the accelerated particles, while $Q(x, p)$ is the injection term.

A few comments on Eq. 7 are in order: (1) the shock will appear in this equation only in terms of a boundary condition at $z = 0$, and the shock is assumed to have infinitely small size along z. This implies that this equation cannot properly describe the thermal particles in the fluid. The distribution function of accelerated particles is continuous across the shock. (2) In a self-consistent treatment in which the acceleration process is an integral part of the processes that lead to the formation of the shock one would not need to specify an injection term. Injection would result from the microphysics of the particle motions at the shock.

For the purpose of the present discussion we assume that injection only takes place at the shock surface, immediately downstream of the shock, and that it only consists of particles with given momentum p_{inj}:

$$Q(p, x) = \frac{\eta n_1 u_1}{4\pi p_{inj}^2} \delta(p - p_{inj}) \delta(z) = q_0 \delta(z), \tag{8}$$

where n_1 is the fluid density upstream of the shock and η is the acceleration efficiency, defined here as the fraction of the incoming number flux across the shock surface that takes part in the acceleration process.

The compression term vanishes everywhere but at the shock since $du/dz = (u_2 - u_1)\delta(z)$. Integration of Eq. 7 around the shock surface (between $z = 0^-$ and $z = 0^+$) leads to:

$$\left[D\frac{\partial f}{\partial z}\right]_2 - \left[D\frac{\partial f}{\partial z}\right]_1 + \frac{1}{3}(u_2 - u_1) p \frac{df_0}{dp} + q_0(p) = 0, \tag{9}$$

where $f_0(p)$ is now the distribution function of accelerated particles at the shock surface. Particle scattering downstream leads to a homogeneous distribution of particles, at least for the case of a parallel shock, so that $\left[\partial f / \partial z\right]_2 = 0$. In the upstream region, where $du/dz = 0$ the transport equation reduces to:

$$\frac{\partial}{\partial z}\left[uf - D\frac{\partial f}{\partial z}\right] = 0, \tag{10}$$

and since the quantity in parenthesis vanishes at upstream infinity, it follows that

$$\left[D\frac{\partial f}{\partial z}\right]_1 = u_1 f_0. \tag{11}$$

Using this result in Eq. 9, we obtain an equation for $f_0(p)$

$$u_1 f_0 = \frac{1}{3}(u_2 - u_1) p \frac{df_0}{dp} + \frac{\eta n_1 u_1}{4\pi p_{inj}^2} \delta(p - p_{inj}), \tag{12}$$

which is easily solved to give:

$$f_0(p) = \frac{3r}{r-1} \frac{\eta n_1}{4\pi p_{inj}^2} \left(\frac{p}{p_{inj}} \right)^{-\frac{3r}{r-1}}. \tag{13}$$

The spectrum of accelerated particles is a power law in momentum (not in kinetic energy) with a slope α that only depends on the compression ratio r:

$$\alpha = \frac{3r}{r-1}. \tag{14}$$

The slope tends asymptotically to $\alpha = 4$ in the limit $M_s \to \infty$ of an infinitely strong shock front. The number of particles with energy ε is $n(\varepsilon)d\varepsilon = 4\pi p^2 f_0(p)$ $(dp/d\varepsilon)d\varepsilon$, therefore $n(\varepsilon) \propto \varepsilon^{-\alpha}$ for relativistic particles and $n(\varepsilon) \propto \varepsilon^{(1-\alpha)/2}$ for non-relativistic particles. In the limit of strong shocks, $n(\varepsilon) \propto \varepsilon^{-2}$ ($n(\varepsilon) \propto \varepsilon^{-3/2}$) in the relativistic (non-relativistic) regime.

Even in the pioneering work of Refs. [3, 199], it was already recognized that for acceleration efficiency ~10% non-linear effects would become important. A full non-linear theory of DSA was developed later, first in the hydrodynamical limit (two fluid models) and then in kinetic models (see [200] for a complete review of these approaches).

The main effects of this non-linearity can be summarized as follows:

(1) *Dynamical reaction of accelerated particles*
For typical efficiencies of CR acceleration at a SN shock, ~10%, the pressure exerted by accelerated particles on the plasma around the shock affects the shock dynamics as well as the acceleration process. The dynamical reaction that accelerated particles exert on the shock is due to two different effects: (a) the pressure in accelerated particles slows down the incoming upstream plasma as seen in the shock reference frame, thereby creating a precursor. In terms of dynamics of the plasma, this leads to a compression factor that depends on the location upstream of the shock [201, 202]. (b) The escape of the highest energy particles from the shock region makes the shock *radiative-like* [203], thereby inducing an increase of the compression factor between upstream infinity and downstream. Both these effects result in a modification of the spectrum of accelerated particles, which turns out to be no longer a perfect power law [203–207]. Moreover, the fact that a sizeable fraction of the ram pressure at the shock is converted to accelerated particles implies that the temperature of the shocked gas is lower than in the absence of particle acceleration.

(2) *Plasma instabilities induced by accelerated particles*
As discussed above, SNRs can be the source of the bulk of CRs in the Galaxy, up to rigidities of order ~10^6 GV only if substantial magnetic field amplification

takes place at the shock surface. Since this process must take place upstream of the shock in order to reduce the acceleration time, it is likely that it is driven by the same accelerated particles, which would therefore determine the diffusion coefficient that describes their motion. The existence of magnetic field amplification is also the most likely explanation of the observed bright, narrow X-ray rims of non-thermal emission observed in virtually all young SNRs (see [8, 208] for recent reviews). The non-linearity here reflects in the fact that the diffusion coefficient becomes dependent upon the distribution function of accelerated particles, which is in turn determined by the diffusion coefficient in the acceleration region.

(3) *Dynamical reaction of the amplified magnetic field*
The magnetic fields required to explain the X-ray filaments are of order 100–1000 μG. The magnetic pressure is therefore still a fraction of order 10^{-2}–10^{-3} of the ram pressure ρv_s^2 for typical values of the parameters. However, the magnetic pressure may easily become larger than the upstream thermal pressure of the incoming plasma, so as to affect the compression factor at the shock. A change in the compression factor affects the spectrum of accelerated particles which in turn determines the level of magnetic field amplification, another non-linear aspect of DSA.

The dynamical reaction of accelerated particles leads to concave spectra, and above $\sim$10–100 GeV are harder than p^{-4}, the standard prediction of DSA in the test-particle approximation. This simple finding refers to the instantaneous spectrum at a given time in the evolution of a SN shock. The time integrated spectrum may be somewhat different, but in general it is not easy to make it steeper than p^{-4}. In addition, one should notice that the CR spectrum released into the ISM by an individual SNR, integrated over the injection history, is made of two contributions: the spectrum of particles accelerated at the shock and advected downstream, and eventually released after the remnant ends its evolution, and the spectrum of particles that escape at the maximum momentum at any given time. This point was discussed in detail in [209].

The hardness of the spectra of injected CRs into the ISM is somewhat of an issue for the SNR paradigm: in fact the rule of thumb that the spectrum observed at the Earth should be $\sim E^{-\gamma-\delta}$, where δ identifies the energy dependence of the diffusion coefficient, would imply that in order to have, at Earth, a slope $\sim$2.6, a diffusion coefficient $\propto E^{0.7-0.8}$ would be required. In turn, this would, most likely cause problems with anisotropy [5], although some solutions may be devised to mitigate such a problem [210].

Observations of gamma-ray emission from young SNRs also suggests that the spectra of accelerated particles are somewhat steeper than predicted by diffusive shock acceleration [211, 212]. However, it should be stressed that such spectra may also reflect a complex morphology of the emission region (see [213] for an implementation of this idea to the case of the Tycho SNR) or the presence of neutral material in the acceleration region [214, 215]. Finally, this phenomenon might be due to subtle aspects of DSA, such as the role played by the velocity of scattering centres in the acceleration process, as discussed in Refs. [209, 216] and in Ref. [217] for the case of the Tycho SNR.

The Quest for PeVatrons

The name *PeVatron* is used to indicate an astrophysical source able to accelerate protons to energies of order $\sim$1 PeV (nuclei with charge Z would therefore be accelerated to Z times larger energies). The need for the existence of PeVatrons mainly derives from the empirical fact that at the knee the mass composition of CRs is dominated by light nuclei. As discussed in section "Cosmic Rays Observations", this piece of observation has recently been questioned, in that some experiments show evidence of a knee-like structure in the spectrum of light nuclei at $\sim$700 TeV, well below the knee. This finding is at odds with the KASCADE measurement of the spectrum in the same energy region, and the nature of this difference is not clear at present.

As discussed by many authors, the assumption that CR protons may accelerate protons up to a few PeV may lead to a satisfactory description of both the knee and the transition region between galactic and extragalactic CRs. Most members of the community would agree that at least a class of sources must accelerate CRs up to PeV energies, while it is less clear whether such sources are also the ones that provide the main contribution to the CR spectrum below the knee. These issues were discussed, for instance, in Refs. [11, 12, 218].

What are the conditions needed for particle acceleration to $\sim$PeV energies? As discussed early on [219], even in case of magnetic field amplification $\delta B/B \sim 1$ and Bohm diffusion, the maximum energy in a typical SNR is unlikely to be higher than $\sim$100 TeV, thereby failing to reach the knee by more than one order of magnitude.

The situation evolved quite abruptly in the last decade or so, mainly because of two findings: (1) X-ray observations of young SNRs led to the discovery that virtually all of them are characterized by thin filaments of non-thermal emission, to be interpreted as the result of synchrotron emission of high energy electrons [8]. In the assumption of Bohm diffusion, the thickness of such filaments can be used to infer the strength of the magnetic field in the emission region, and in all cases this estimate returns $B \sim$ 100–1000 μG, about $\sim$100 times larger than the typical magnetic field in the ISM. Such finding can be only explained as a result of efficient magnetic field amplification, which may be due to hydrodynamical instabilities [220] and/or to CR-induced instabilities. (2) Investigation of the streaming instability induced by CRs at a SNR shock led [221] to find a non-resonant branch of quasi-purely growing modes with a growth rate much higher than the resonant modes studied in [219]. These modes could explain the large magnetic fields at SNR shocks, and possibly lead to higher values of the maximum energy of accelerated particles.

There are however some subtle features concerning these fastly growing modes that need to be taken into account if to assess their importance for particle acceleration [11, 218, 221]. From the physical point of view, the instability is due to the return current generated by electrons in the background plasma in order to compensate the CR current J_{CR} upstream of the shock. Hence the scales that grow the fastest are the ones on very small spatial scales (large wavenumber k). The growth rate of the fastest modes can be written as:

$$\gamma_M = k_M v_A, \tag{15}$$

where v_A is the Alfvén speed in the unperturbed magnetic field B_0. The wavenumber where the growth is the fastest can also be easily estimated using the condition

$$k_M B_0 \cong \frac{4\pi}{c} j_{CR}, \tag{16}$$

which corresponds to balance between current and magnetic tension. For a spectrum of particles accelerated at the shock $\propto p^{-4}$, the above expression can be rewritten as:

$$k_M r_L = \xi_{CR} \frac{1}{\Lambda} \left(\frac{v_{sh}}{V_A}\right)^2 \left(\frac{v_{sh}}{c}\right) \gg 1, \tag{17}$$

where r_L is the gyration radius of the particles dominating the current, $\Lambda = \ln\left(\frac{p_{max}}{m_p c}\right)$ and ξ_{CR} is the CR acceleration efficiency. Equation 17 illustrates well the fact that the fast growing modes are the ones that grow on scales much smaller than the Larmor radius of the particles. Since the resonance condition $kr_L^{-1} = 1$ cannot be achieved, scattering of particles off these waves is not effective, hence it might seem at first sight that the excitation of these modes should not appreciably impact the scattering of particles near the shock, and not help achieving higher values of the maximum momentum p_{max}. However, this conclusion only applies to the linear growth of the modes, while at later stages the situation becomes more interesting.

An element of background plasma is subject to a force $\sim \frac{1}{c} j_{CR} \delta B$ due to the exponentially growing magnetic field δB, so that within the growth time γ_M^{-1} the fluid element is displaced by

$$\delta x \approx \frac{J_{CR} \delta B}{c \rho \gamma_M^2}. \tag{18}$$

When the loops of magnetic fields get stretched by $\delta x \sim r_L$ (where now the Larmor radius is calculated in the amplified magnetic field), the spatial scale in the magnetic field becomes sufficient to cause particle scattering and the current gets destroyed. Hence, one can envision the condition $\delta x \sim r_L$ as the saturation condition for the instability. The corresponding value of the magnetic field is given by

$$\frac{\delta B^2}{4\pi} \approx \frac{\xi_{CR}}{\Lambda} \rho v_s^2 \frac{v_s}{c}. \tag{19}$$

The right end side of Eq. 19 represents the energy density of accelerated particles escaping the accelerator from upstream, so that Eq. 19 suggests that the saturation level is reached when equipartition between magnetic energy density (LHS) and energy density of escaping particles (RHS) is reached. The implications of this finding are potentially very impressive: particles at a given maximum energy E_M escape the accelerator because there are no scattering centres able to scatter them back to the shock; however, the growth of the waves that they excite leads to particles of the same energy at a later time to be confined to the shock region, thereby reaching higher energy.

The maximum energy that can be reached at a SNR shock as due to the mechanism above has been calculated by several authors [11, 12, 218]. For a type Ia SN with ejecta mass M_{ej} and total kinetic energy E_{SN}, exploding in the ISM with density n_{ISM}, the maximum energy can be written as

$$E_M \cong 130 \left(\frac{\xi_{CR}}{0.1}\right)\left(\frac{M_{ej}}{M_\odot}\right)^{-\frac{2}{3}}\left(\frac{E_{SN}}{10^{51}\text{erg}}\right)\left(\frac{n_{ISM}}{\text{cm}^{-3}}\right)^{\frac{1}{6}} \text{TeV}. \tag{20}$$

Despite the efficient magnetic field amplification due to CR streaming, the maximum energy falls short of the knee by about one order of magnitude, thereby leading to the conclusion that type Ia SNe is unlikely to act as PeVatrons.

Most core-collapse SNe, on the other hand, explode in the wind produced by their pre-SN red giant progenitor. For a reference value of the mass loss rate $\dot{M} \sim 10^{-5} M_\odot \text{yr}^{-1}$ and wind speed $V_w \sim 10$ km/s, the maximum energy can be estimated as

$$E_M \approx 1 \left(\frac{\xi_{CR}}{0.1}\right)\left(\frac{M_{ej}}{M_\odot}\right)^{-1}\left(\frac{E_{SN}}{10^{51}\text{erg}}\right)\left(\frac{\dot{M}}{10^{-5}M_\odot\text{yr}^{-1}}\right)^{\frac{1}{2}}\left(\frac{V_w}{10\,\text{km s}^{-1}}\right)^{-\frac{1}{2}} \text{PeV}. \tag{21}$$

In principle, SNe of this type can accelerate particles to PeV energies, although one should notice that the values of the parameters have been rather optimized and while it is easy to lower this estimate, it is not that easy to raise it to higher values. In this sense, acceleration to PeV energies remains challenging although for the first time, at least theoretically, it appears possible to explain how SNe can accelerate particles to these very high energies. In both cases of type Ia and type II SNe, the spectrum of accelerated particles is expected to steepen at E_M, reached at the beginning of the Sedov–Taylor phase. Notice however that while such stage is reached a few hundred years after the explosion for type Ia SNe, a type II SN enters its adiabatic stage a few tens of years after the explosion. This implies that catching a PeVatron of this type in our Galaxy in the act of accelerating to the highest energies, for instance using gamma-ray observations, is and will remain in the near future rather challenging. Notice that in the case of particle acceleration in type II SNe exploding in the wind of the pre-SN red giant, the maximum energy E_M is defined as the highest energy of accelerated particles at the beginning of the Sedov–Taylor phase. However, even higher energies can be reached at earlier times, when the mass processed by the shock is appreciably smaller than the ejecta mass. This phenomenon reflects in a steepening in the CR spectrum at energy E_M that is not exponential, but rather a steeper power law. This finding has important implication for the description of the transition from galactic to extragalactic CRs [12].

The investigation of the effects of streaming instability in the regime discussed in Ref. [221] has recently received a strong boost thanks to the adoption of hybrid and Particle-in-cell (PIC) codes to describe particle acceleration at non-relativistic shocks (see [222] for a recent review).

These simulations aim at describing the basic physics of the formation of a collisionless shock wave: particle acceleration seems to be a by-product of the formation of such shocks. Hybrid simulations of diffusive shock acceleration at parallel shocks [223] have shown that the mechanism works and that CR acceleration efficiency of 10–20% can be achieved. The spectrum of accelerated particles is confirmed to be consistent with the standard prediction, $\sim p^{-4}$. On the other hand, the temperature of the downstream plasma is measured to be lower by the amount predicted by the conservation relations at the shock and accounted for in non-linear theories of DSA. This is the first time, to our knowledge, that non-linear effects induced by particle acceleration are detected in simulations based on basic physical principles of the formation of collisionless shocks. These simulations also show magnetic field amplification [224], especially through the non-resonant channel proposed in [221]. The instability is seen to grow on small scales as predicted by quasi-linear treatments [221, 225]. On the other hand, the saturation level of the magnetic field is found to depend upon the strength of the pre-existing magnetic field B_0, contrary to what one would expect based on Eq. 19, and to be somewhat lower than the value returned by Eq. 19.

One of the concepts that are most central to the investigation of DSA at collisionless shocks is that of injection, namely of establishing what are the physical criteria that differentiate thermal from non-thermal particles. A substantial advancement was recently achieved in this direction in Ref. [226], where injection was studied as a function of the orientation of the pre-existing magnetic field and a semi-analytical theory of injection, supported by hybrid simulations, was proposed. This study revealed that ions are effectively injected for quasi-parallel shocks (inclination angle of $\lesssim 45°$ of the magnetic field with respect to the shock normal) with the help of shock drift acceleration that serves as a pre-energisation process. For more inclined shocks, [226] found a strong suppression of injection, namely the acceleration process does not get bootstrapped. It is not clear at present whether this conclusion may change by assuming the presence of some level of pre-existing turbulence on small scales, independent of CR-induced instabilities.

Recently, PIC simulations have been used to investigate the problem of electron acceleration at collisionless shocks [227]. For the first time, these simulations have demonstrated that at quasi-parallel shocks both ions and electrons can be accelerated through DSA, with a ratio of fluxes at given momentum $K_{ep} \sim 10^{-3}$–10^{-2}, that compares well with the ratio observed at the Earth between the electron and proton flux in the GeV range, where both protons and electrons lose a negligible fraction of their energy during propagation.

Finally, it is worth recalling that recently gamma-ray observations carried out with the HESS telescope provided the very first evidence of a PeVatron in our Galaxy [228]. The source is spatially located at the galactic centre, coincident with the position of Sagittarius A^*. At the present time, it is not clear as yet whether the black hole at the galactic centre or some other type of source could be responsible for this emission.

DSA in the Presence of Neutral Hydrogen

The presence of neutral atomic hydrogen in the acceleration region may potentially lead to modifications in the spectrum of accelerated particles [214] for shock velocities below ~3000 km/s. The theory of DSA in the presence of neutrals was recently developed in [214] in the test-particle regime and in [229] in its non-linear version. The spectrum of accelerated particles is modified because of the so-called neutral return flux: neutral hydrogen atoms crossing the (collisionless) shock towards downstream can suffer a charge exchange reaction with a hot ion downstream of the shock. There is a finite probability that the resulting neutral atom moves in the direction of the upstream of the shock, where it can damp its energy and momentum onto the plasma, thereby heating it and slowing it down. This phenomenon leads to a reduction of the Mach number and to a correspondingly steeper spectrum of accelerated particles for momenta such that the diffusion length $D(p)/v_s < \lambda_n$, where λ_n is the pathlength for reactions of charge exchange and ionization. For typical values of the parameters, the effect of neutrals on the spectrum of accelerated particles is limited to energies below ~1–10 TeV. The neutral return flux requires that charge exchange reactions downstream occur faster than ionization, a condition that restricts the importance of this phenomenon to shocks moving with velocity below $\sim(3\text{–}4) \times 10^3$ km/s.

The presence of neutrals in the acceleration region also leads to Balmer line emission, which represents a powerful diagnostic tool for testing particle acceleration. In fact the presence of CRs at the shock leads to an enhanced width of the narrow Balmer line (due to charge exchange reactions in the CR induced precursor discussed above) and to a reduced width of the broad Balmer line (due to the lower gas temperature downstream of the shock that follows from non-linear DSA). These effects are described quantitatively in [229–231], where the appearance of a Balmer line component with intermediate width ~ few hundred km/s was also discussed, as resulting from the phenomenon of neutral return flux.

In principle, the simultaneous measurement of the width of the narrow and broad component of the Balmer line at a SNR shock may allow us to measure the CR acceleration efficiency and to infer the existence of a CR induced precursor, due to the pressure of accelerated particles slowing down the plasma incoming into the shock from upstream. The phenomenon of neutral return flux discussed above also gives rise to an intermediate component of the Balmer line, with typical width of 100–300 km/s. The cases of SNRs RCW 86 (G315.4-2.3) and SNR 0509-67.5 were recently studied in [232, 233], respectively. Unfortunately, at the present time it is difficult to extract unambiguous information about CRs from these observations, due to uncertainties in the level of thermalization between electrons and ions in the system and poor knowledge of the shock velocity (see [234] for a recent review of these issues). On the other hand, if this information becomes available, the observation of Balmer line emission has the potential to provide precious insights in the acceleration process at SNR shocks.

Transport of Extragalactic CRs

The extragalactic origin of UHECRs, at least at energies above the ankle $E > 10^{19}$ eV, is widely accepted [235]. The propagation of UHECRs across intergalactic space is conditioned primarily by astrophysical photons backgrounds and, if any, by the presence of extragalactic magnetic fields.

The astrophysical photon backgrounds relevant for the propagation of UHECRs are the Cosmic Microwave Background (CMB) and the Extragalactic Background Light (EBL). The former background, with the highest density, is the well-known relic radiation from the big bang, while the latter is composed of infrared, optical and ultraviolet photons produced and reprocessed by astrophysical sources at present and past cosmological epochs. While the cosmological evolution of the CMB is analytically known, the case of the EBL is model dependent. In the past years, several models for the cosmological evolution of the EBL have been proposed [236–239]. These models show sizeable differences only at high red-shift ($z > 4$), not actually relevant in the propagation of UHECRs but affecting the production of secondary neutrinos [40, 240], as we discuss in section "Secondary Cosmogenic Messengers".

In the forthcoming sections "Interactions of Ultra-High Energy Cosmic Rays", "Propagated Spectra", "Secondary Cosmogenic Messengers" and "Intergalactic Magnetic Fields" we discuss the details of UHECRs propagation, through astrophysical backgrounds and intergalactic magnetic fields, and the production of secondary particles.

Interactions of Ultra-High Energy Cosmic Rays

Ultra-high energy protons[1] or nuclei propagating in the intergalactic space interact with CMB and EBL photons through the processes of pair production, photo-pion production and, only in the case of nuclei heavier than protons, photo-disintegration. Given the distribution of background photons and the energies involved, the propagation of protons is substantially affected only by the CMB radiation field, while in

[1] Here we do not consider the case of neutrons because their decay time is much shorter than all other scales involved in the propagation of UHECR [241, 242].

the case of nuclei, and only for the photo-disintegration process, also the EBL field is important [241, 242]. As we discuss below, the effect of the EBL on proton propagation plays a role only for the production of secondary particles, but it negligibly affects the expected proton flux.

These mechanisms of energy losses and their relevance for the propagation of UHECRs were discussed soon after the discovery of the CMB radiation field [243]. Greisen [244] and, independently, Zatsepin and Kuzmin [245] realized that, due to photo-pion production, the interaction of UHE protons with CMB would produce a sharp suppression in the expected UHECR spectrum at $E \geq 10^{20}$ eV, the so-called GZK feature.[2] In the same period, Hillas [246] and Blumenthal [247] studied the effect of pair production of UHE protons with $E > 10^{18}$ eV on the CMB radiation field and Berezinsky and Zatsepin [248–251] realized that the propagation of UHE protons would give rise to the production of secondary cosmogenic particles, generated by the decay of photo-produced mesons, such as neutrinos and gamma rays.

The interaction rate associated with processes involving UHECRs can be written in a very general form as [241, 252]:

$$\frac{1}{\tau} = \frac{c}{\Gamma^2} \int_{\varepsilon'_{\min}}^{+\infty} \varepsilon' \sigma(\varepsilon') \int_{\varepsilon'/2\Gamma}^{+\infty} \frac{n_\gamma(\varepsilon)}{2\varepsilon^2} d\varepsilon \, d\varepsilon', \tag{22}$$

where Γ is the Lorentz factor of the particle, $\sigma(\varepsilon')$ is the total cross section associated with the particle interactions, ε' is the background photon energy in the particle rest frame, $\varepsilon'_{\min}$ is the lowest value of ε' above which the interaction is kinematically possible (threshold), and $n_\gamma(\varepsilon)\, d\varepsilon$ is the number per unit volume of background photons with energy between ε and $\varepsilon + d\varepsilon$ in the laboratory reference frame. The photon energy in the particle rest frame is related to that in the laboratory frame by $\varepsilon' = \Gamma\varepsilon(1 - \cos\theta)$, where θ is the angle between the particle and photon momenta $(0 \leq \varepsilon' \leq 2\Gamma\varepsilon)$.

Nucleons (N), whether free or bound in nuclei, with Lorentz factor larger than $\Gamma \geq m_\pi/2\varepsilon(1 - \cos\theta) \simeq 10^{10}$ interacting with the CMB photons give rise to the photo-pion production process:

$$N + \gamma \rightarrow N + \pi^0 \qquad N + \gamma \rightarrow N + \pi^\pm. \tag{23}$$

At lower energies ($\Gamma < 10^{10}$) the same processes can occur on the EBL field [240], although with a lower probability.

The photo-pion production process involves a sizeable energy loss for protons resulting in the GZK feature [244, 245] which arises at the threshold for photo-pion production, which in the nucleon rest frame reads $\varepsilon'_{\min} = m_\pi + m_\pi^2/2m_N \approx$ 145 MeV. The photo-pion production cross section has a complex behaviour with a number of peaks corresponding to different hadronic resonances, the largest one being the Δ resonance placed at $\varepsilon' = \varepsilon_\Delta \approx 340$ MeV [253]. At energies much larger

[2] Also referred as GZK cutoff or suppression.

than ε_Δ the cross section is approximately constant [253]. The photo-pion production process holds also for nucleons bound within UHE nuclei, being the interacting nucleon ejected from the parent nucleus. This process is subdominant if compared with photo-disintegration except at extremely high energies [40] and, as we discuss later, has some relevance only in the case of production of secondary cosmogenic particles.

UHE nuclei propagating through astrophysical backgrounds can be stripped of one or more nucleons through the interactions with the CMB and EBL photons, a process named photo-disintegration:

$$(A, Z) + \gamma \rightarrow (A - n, Z - n') + nN \tag{24}$$

being A and Z the atomic mass number and atomic number of the nucleus, n (n') the number of stripped nucleons. In the nucleus rest frame the energy involved in such processes is usually much less than the rest mass of the nucleus itself, hence in the laboratory frame all fragments approximately inherit the same Lorentz factor of the parent nucleus, i.e. we can neglect nucleus recoil [241, 242].

For photon energies close to the threshold (8 MeV $\approx \varepsilon'_{\rm min} < \varepsilon' \leq 30$ MeV), the cross section is dominated by a smooth peak, the giant dipole resonance, that corresponds to the extraction of one nucleon and it is the dominating process in UHE nuclei propagation [241, 242, 254, 255]. At larger energies $\varepsilon' > 30$ MeV the quasi-deuteron process dominates with the extraction of two or more nucleons. This regime corresponds to an almost constant cross section and has a small impact on the propagation of UHE nuclei [241, 242, 254, 255].

The process of photo-disintegration is responsible for the production of secondary, hadronic, particles that will compose a sizeable fraction of the flux of UHECRs observed at the Earth.

The interaction rate associated with the processes of photo-pion production and photo-disintegration can be written using Eq. (22) specifying the cross section, the background photon density and all relevant kinematical thresholds of the process. It is interesting that, while the first process changes the particle's Lorentz factor, leaving the nature of the particle unchanged, the second process changes the particle's nature leaving the Lorentz factor unchanged.

Protons and nuclei with Lorentz factor $\Gamma \geq 2m_e/\varepsilon(1 - \cos\theta) \simeq 10^9$ can undergo the process of pair production $p + \gamma \rightarrow p + e^+ + e^-$. The mean free path associated with pair production is relatively short compared with all other length scales of UHECR propagation, with a very small amount of energy lost by the propagating particle in each interaction [253] so that frequently this process is treated in the approximation of continuum energy losses. In this case, the rate of energy losses due to pair production is given by Eq. (22) substituting $\sigma \rightarrow \sigma f$ being f the inelasticity of the process, i.e. the average fraction of energy lost by the particle in one interaction [150]. In the case of nuclei, the rate of pair production energy losses can be computed starting from that of protons and taking into account that $f^A = f^p/A$ and $\sigma^A = Z^2\sigma^p$.

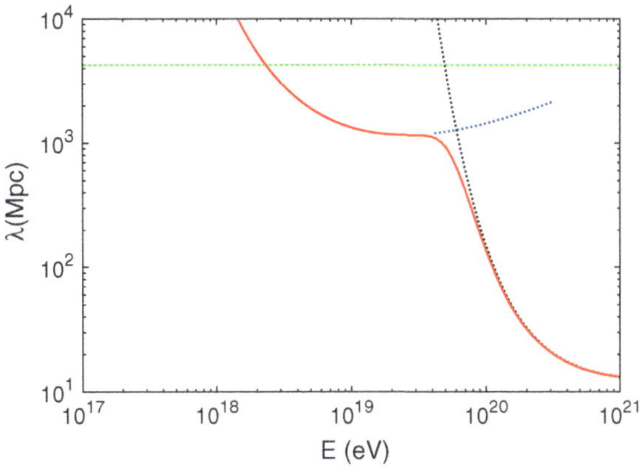

Fig. 21 Loss length of UHE protons (*red solid line* total energy losses). Photo-pion production *dotted black line*, pair production *dotted blue line*. The size of the visible universe is seen by the *dashed green line*

UHECRs can propagate through cosmological distances, in this case the expansion of the universe produces adiabatic energy losses. Assuming standard ΛCDM cosmology, we can write the energy lost per unit time by UHECRs (protons or nuclei) as

$$\left(-\frac{1}{\Gamma}\frac{d\Gamma}{dt}\right)_{\rm ad} = H(z) = H_0\sqrt{(1+z)^3\Omega_{\rm m} + \Omega_\Lambda}, \tag{25}$$

where z is the red-shift at time t, $H_0 \simeq 70$ km/s/Mpc is the Hubble constant, $\Omega_{\rm m} \simeq 0.26$ is the matter density, and $\Omega_\Lambda \simeq 0.74$ is the dark energy density [256].

In Figs. 21 and 22, we plot the interaction path length of protons and nuclei computed at zero red-shift as a function of energy.

In the case of protons (Fig. 21), at low energies, up to few 10^{18} eV, energy losses are dominated by the expansion of the universe. The pair production process starts to be relevant at the energy 2×10^{18} eV; pion photo-production becomes important at 5×10^{19} eV where the loss length drops to very low values and the large scale universe becomes opaque to UHECRs.

In Fig. 22, the loss length of nuclei is plotted as function of the Lorentz factor Γ, assuming the EBL model presented in [238]. Two sharp drops can be seen in the pathlength of nuclei: the first drop, at energy $A \times 10^{18}$ eV ($\Gamma \simeq 10^9$), is due to the combined effect of photo-disintegration on far infrared photons (low energy EBL) and pair production on the CMB; the second drop, even more pronounced, is due to photo-disintegration on CMB photons and arises at energies $A \times 4 \times 10^{18}$ eV ($\Gamma \simeq 4 \times 10^9$). In particular, the position of this last drop is less model dependent, being related to photo-disintegration on CMB photons only, and fixes the highest energy behaviour of the fluxes of nuclei expected at the Earth.

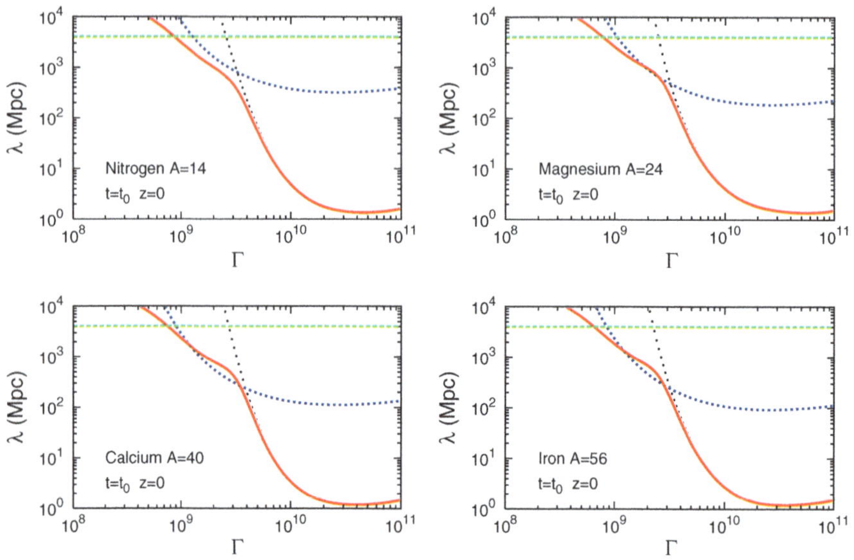

Fig. 22 Loss length of UHE nuclei. *Red solid line* photo-disintegration and *blue dotted line* pair production. The effect of EBL on photo-disintegration is seen by the *black dotted line* that shows photo-disintegration due to the sole CMB field. The size of the visible universe is seen by the *dashed green line*

Propagated Spectra

As discussed in section "Observations of Ultra-High Energy Cosmic Rays", the main spectral features of UHECRs observed at the Earth are: (i) the ankle, a flattening of the spectrum at energy around $\simeq 5 \times 10^{18}$ eV, observed since 1960s (Volcano Ranch experiment [257]) and confirmed by all observations [118, 258]; (ii) a sharp suppression of the spectrum at the highest energies. The energy of such suppression is not clearly identified, due to some tension between the observations of Auger and TA (see Fig. 12 and section "Observations of Ultra-High Energy Cosmic Rays" for a detailed discussion).

Once energy losses of UHECRs are specified one can determine the spectra expected at the Earth by assuming an injection spectrum at the sources and their cosmological evolution.[3] In the following, we will assume that the energy per unit volume injected by the sources in the form of UHECRs only depends on red-shift, that sources share the same injection spectral index γ_g (with a power law injection $\propto E^{-\gamma_g}$) and the same maximum energy E_{max} at the sources.

As discussed in section "Observations of Ultra-High Energy Cosmic Rays", Auger and TA are currently not providing an unambiguous measurement of the mass

[3]We will not discuss here the case of the possible presence of extragalactic magnetic fields, we will come back to this point in the forthcoming section "Intergalactic Magnetic Fields".

composition. Hence, in what follows, we discuss separately the two cases of a pure proton composition (according to TA data) and a mixed composition with heavy nuclei contributing to the UHECR flux (according to Auger data).

Protons and the Dip Model

In the case of a pure proton composition, the only relevant astrophysical background is the CMB [241, 242]. This fact makes the propagation of UHE protons free from the uncertainties related to the background, being the CMB exactly known as a pure black body spectrum that evolves with red-shift through its temperature. In this case, therefore, any signature of the propagation in the observed spectrum will depend only on the assumptions made at the source and the details of the interactions suffered by propagating protons.

In order to isolate the effects of energy losses in the propagated proton spectrum it is useful to use the so-called modification factor $\eta(E)$ defined as the ratio:

$$\eta(E) = \frac{J_p(E)}{J_{unm}(E),} \tag{26}$$

where J_p is the proton spectrum, computed with all energy losses taken into account, and $J_{unm}(E)$ is the unmodified spectrum computed taking into account only adiabatic energy losses due to the expansion of the universe.

Assuming an injection spectrum at the source with a power law form $\propto E^{-\gamma_g}$, in Fig. 23 we plot the modification factor $\eta(E)$ as computed for different values of the injection power law γ_g as labelled. From this figure, it is evident that $\eta(E)$ clearly shows the signatures of energy losses suffered by protons, being almost independent

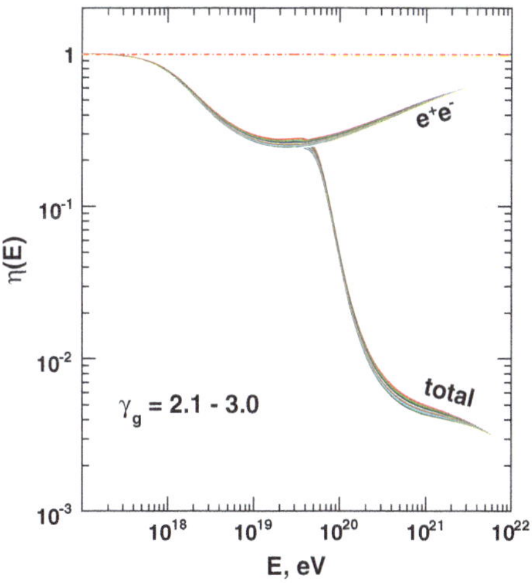

Fig. 23 Theoretical modification factor computed for different values of the injection power law index as labelled

of the injection. At low energy, losses are dominated by the adiabatic expansion of the universe, the spectrum at the Earth keeps the injection shape, and the modification factor is $\eta(E) = 1$; above the pair production threshold (around 2×10^{18} eV) pair production energy losses become important and the propagated spectrum hardens (the curves labelled with e^+e^- take into account only pair production); above the photo-pion production threshold (around 5×10^{19} eV) the propagated spectrum experiences an abrupt steepening which corresponds to the GZK feature (curves labelled with *total* take into account all relevant energy losses).

Particularly relevant is the feature associated with the pair production energy losses, named 'dip' [253, 264], that reproduces quite well the ankle observed in the UHECR spectrum, provided that the injection power law at the source is around $\gamma_g = 2.6$–2.7. In Fig. 24 we plot the theoretical modification factor together with the experimental data of several detectors as labelled, which all claim a pure proton

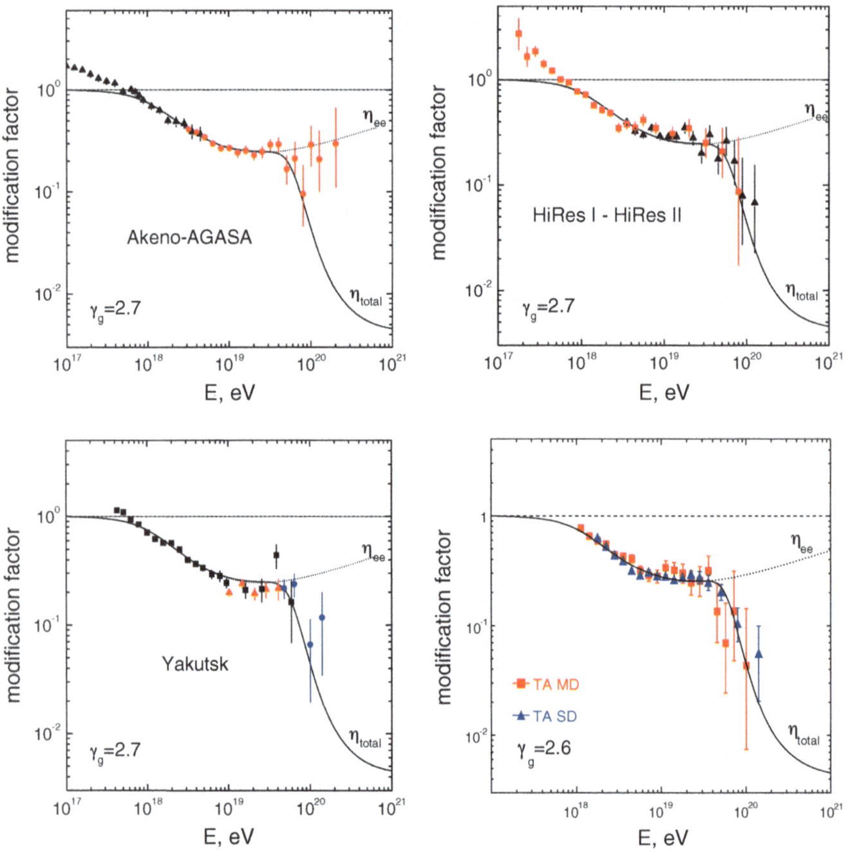

Fig. 24 Comparison of the modification factor with data [259–263] of the UHECR experiments favouring a proton dominated composition

composition [259–263]. From this figure it is evident that the behaviour of the pair production dip reproduces quite well the observations.

The results presented in Fig. 24 refer to the case without a cosmological evolution of the sources, i.e. density and luminosity of sources are independent of red-shift. Assuming a cosmological evolution, which typically gives a larger weight to distant sources, the transition between adiabatic and pair production energy losses arises at lower energy and the pair production dip will be deeper and slightly shifted to lower energies. As a consequence of this fact the spectral index at the sources needed to reproduce the observations will be lowered with respect to the best fit value (see Fig. 24) obtained without evolution [253, 264].

Remarkably, the dip model explains the observed flux with only one extragalactic component of pure protons, directly linking the flux behaviour to energy losses. Hence, in the case of the dip model, the transition between galactic and extragalactic cosmic rays occurs at energies below the pair production threshold, i.e. $E_{tr} < 2 \times 10^{18}$ eV [235, 253, 264–267].

The source parameters that can be fitted by the comparison of Fig. 24 are basically only two: the injection power law index γ_g and the source emissivity $\mathscr{L}_S$, i.e. the energy emitted (in the form of UHECRs) per unit time and volume by sources.[4] The value of the required emissivity depends on the power law index, with a value that ranges from $\gamma_g = 2.5$ (for strong cosmological evolution) up to $\gamma_g = 2.7$ (without evolution) [253, 264]. Using these values of γ_g and assuming a single power law injection down to the lowest energies (GeV) results in a prohibitive energy budget for any astrophysical source. To avoid this problem, in the original papers introducing the dip model, a change in the spectral index at injection was assumed [253]: at energies below 10^{18} eV, $\gamma_g = 2.0$ while at larger energies γ_g takes the best fit values quoted above. Under this assumption the required emissivity (at red-shift $z = 0$) necessary to reproduce UHECR data is around $\mathscr{L}_S = 10^{45}$–10^{46} erg/Mpc^3/yr [240, 253, 264].

In Fig. 24 the same maximum acceleration energy $E_{max} = 10^{21}$ eV for all sources is adopted. Releasing this hypothesis and taking into account that sources can be distributed over different values of the maximum energy we can assume an injection power law index as $\gamma_g = 2.0$ for all sources [264, 268]. At energies above the minimal maximum energy, a softer spectral index arises from the convolution of the distribution function over E_{max} ($dN(E_{max})/dE_{max} \propto E_{max}^{-\alpha}$ with $\alpha = 1.6$–1.7) leading to an 'effective' spectral index $\gamma_g^{eff} = \gamma_g + \alpha - 1$ at the highest energies [264, 268].

As discussed above, we stress that in the framework of the dip model, a suppression of the flux at low energy ($<10^{18}$ eV) is needed not only to avoid too high luminosity of the sources but also not to overshoot the observed flux at energies below 10^{18} eV. As we discuss later (see section "Intergalactic Magnetic Fields"), magnetic horizon effects can also be invoked to reduce the proton fraction at $E \leq 10^{18}$ [269] even if such effects would leave the energy budget of the sources unaffected.

[4] Given a distribution of sources with a number density n_S each with the same luminosity L_S, the energy emitted per unit time and volume (emissivity) is given by $\mathscr{L}_S = n_S L_S$.

Mixed Composition Model

The discussion in the previous section was centred around the hypothesis of a pure proton composition of UHECRs. However, as discussed in section "Observations of Ultra-High Energy Cosmic Rays", a somewhat different picture arises from the Auger observations that claim a mixed composition of UHECRs characterized by light nuclei at low energies ($\leq 5 \times 10^{18}$ eV) and heavier nuclei at the highest. The qualitative new finding that mass composition of UHECRs might be mixed has served as a stimulus to build models that can potentially explain the phenomenology of Auger data. These models all show that the Auger spectrum and mass composition at $E \geq 5 \times 10^{18}$ eV can be fitted at the same time only at the price of requiring very hard injection spectra for all nuclei ($\propto E^{-\gamma_g}$ with $\gamma_g = 1$–1.6) and a maximum acceleration energy of $E_{max} \leq 5Z \times 10^{18}$ eV [42, 270–272].

The need for hard spectra can be understood taking into account that the low energy tail of the flux of UHECRs reproduces the injection power law. Therefore, taking $\gamma \geq 2$ causes the low energy part of the spectrum to be polluted by heavy nuclei thereby producing a disagreement with the light composition observed at low energy.

One should appreciate here the change of paradigm that these findings imply: while in the case of a pure proton composition it is needed to find sources and acceleration mechanisms able to energize CR protons up to energies larger than 10^{20} eV with steep injection ($\gamma_g \simeq 2.5$–2.7), the Auger data require that the highest energy part of the spectrum ($E > 5 \times 10^{18}$ eV) has a flat injection ($\gamma_g \simeq 1.0$–1.6) being dominated by heavy nuclei with protons' maximum energy not exceeding few $\times 10^{18}$ eV.

By accepting the new paradigm, it follows that the Auger spectrum at energies below 5×10^{18} eV requires an additional component, with a steep injection spectrum ($\gamma_g \simeq 2.5$–2.7), composed of protons and helium nuclei that in principle could be both of galactic or extragalactic origin [42, 271, 273]. However, the anisotropy expected for a galactic light component extending up to 10^{18} eV exceeds by more than one order of magnitude the upper limit measured by Auger [274]. This observation, just restricting the analysis to Auger data, would constrain the transition between galactic and extragalactic CRs to energies below 10^{18} eV [235, 275].

The Auger data can be modelled essentially in two ways: (i) assuming the presence of two classes of sources: one injecting heavy nuclei with a hard injection and the other only protons and helium nuclei with a soft spectrum [42, 271] or (ii) identifying a peculiar class of sources that could provide at the same time a steep light component and a flat heavy one [273, 276, 277]. These findings were recently confirmed by the Auger collaboration through a combined fit of the flux and mass composition data at energies $E > 5 \times 10^{18}$ eV [272].

In Figs. 25, 26 and 27, we plot the comparison of Auger data on flux and chemical composition with the theoretical expectation in the case of two classes of generic sources discussed in [240] (left panel Figs. 25 and 26) and in the case of a single class of sources (internal shocks in Gamma Ray Bursts, GRBs) discussed in [276] (right panel of Figs. 25 and 27). In the left panel of Fig. 25 we also plot the spectra

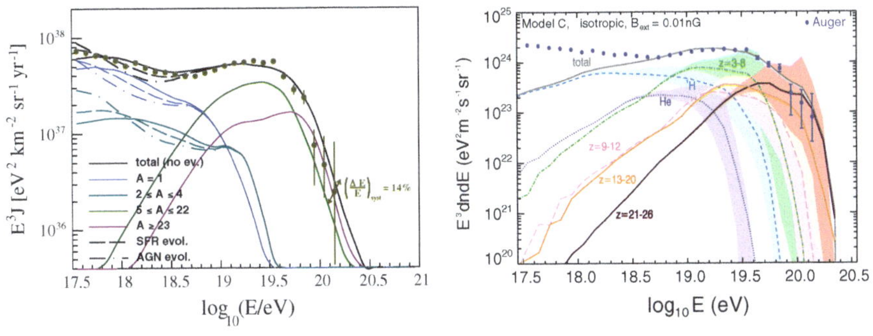

Fig. 25 Comparison of the Auger spectrum with theoretical expectations in the case of models with mixed composition. (*Left panel*) Model with two classes of sources as in [240]. Continuous, *dashed* and *dot-dashed lines* correspond, respectively, to the cases of: no cosmological evolution of sources, evolution as the star formation rate and as active galactic nuclei. (*Right panel*) Model with UHECR production in the internal shock of GRB as discussed in [276]

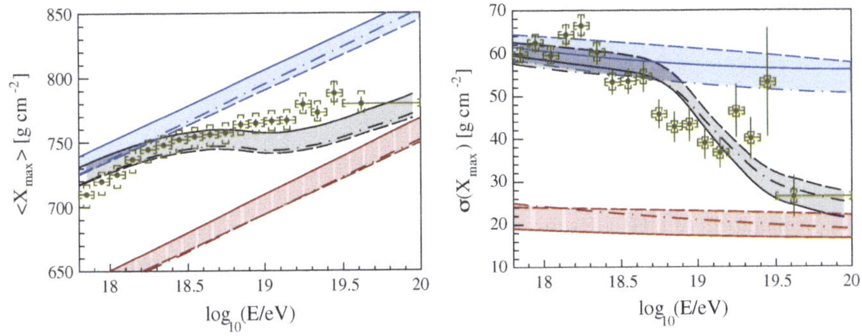

Fig. 26 Comparison of the elongation rate and its root mean square computed assuming the model with two classes of sources as discussed in [240]

computed with different assumptions on the cosmological evolution of sources as discussed in [240]: no cosmological evolution, evolution following the star formation rate [278–280] and of Active Galactic Nuclei (AGN) [278, 281, 282].

In Figs. 26 and 27, chemical composition is inferred from the mean value of the depth of shower maximum $\langle X_{\max} \rangle$ and its dispersion (RMS) $\sigma(X_{\max})$. As discussed in section "Observations of Ultra-High Energy Cosmic Rays" (see also [77, 120, 266]), the combined analysis of $\langle X_{\max} \rangle$ and $\sigma(X_{\max})$ allows one to obtain less model dependent information on the mass composition of UHECRs. The main uncertainties in such a procedure are introduced by the dependence of $\langle X_{\max} \rangle$ and its fluctuations on the hadronic interaction model used to describe the shower development. Most of such models fit low energy accelerator data while providing somewhat different results when extrapolated to the energies of relevance for UHECRs (for a review see [283] and references therein). In Fig. 26, to highlight the uncertainties in the atmospheric shower development, four different models of hadronic interaction were

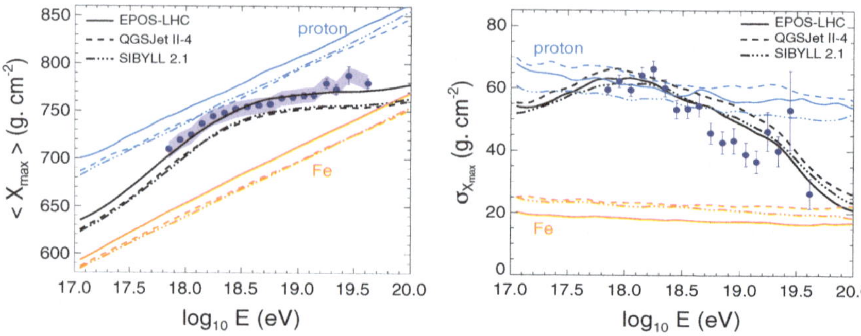

Fig. 27 Comparison of the elongation rate and its root mean square in the case of UHECR production in the internal shock of GRB as discussed in [276]

included in the coloured bands, namely EPOS 1.99 [284], Sibyll 2.1 [285], QGSJet 01 [286] and QGSJet 02 [287]. In Fig. 27, different lines correspond to different interaction models as labelled.

Secondary Cosmogenic Messengers

The propagation of UHECRs through intergalactic space gives rise to the production of several unstable particles, produced by photo-hadronic interactions with CMB and EBL photons, that in turn produce high energy photons, electrons and neutrinos. The possible detection of these signal carriers, as realized soon after the proposal of the existence of the GZK cutoff [248, 249, 251, 252], is extremely important to constrain models for UHECR sources, mass composition and the details of propagation [240, 282, 288–299].

Neutrinos

There are two processes that lead to neutrino production in the propagation of UHECRs: (i) the decay of charged pions, produced by photo-pion production, $\pi^{\pm} \to \mu^{\pm} + \nu_{\mu}(\bar{\nu}_{\mu})$ and the subsequent muon decay $\mu^{\pm} \to e^{\pm} + \bar{\nu}_{\mu}(\nu_{\mu}) + \nu_e(\bar{\nu}_e)$; (ii) the beta-decay of neutrons and nuclei produced by photo-disintegration: $n \to p + e^- + \bar{\nu}_e$, $(A, Z) \to (A, Z-1) + e^+ + \nu_e$, or $(A, Z) \to (A, Z+1) + e^- + \bar{\nu}_e$. These processes produce neutrinos in different energy ranges: in the former the energy of each neutrino is around a few percent of that of the parent nucleon, whereas in the latter it is less than one part per thousand (in the case of neutron decay, larger for certain unstable nuclei). This means that in the interactions with CMB photons, which have a threshold Lorentz factor around $\Gamma \geq 10^{10}$, neutrinos are produced with energies of the order of 10^{18} eV and 10^{16} eV, respectively. Interactions with

EBL photons contribute to a much lower probability than CMB photons, affecting a small fraction of the propagating protons and nuclei. Neutrinos produced through interactions with EBL, characterized by lower thresholds, have energies of the order of 10^{15} eV in the case of photo-pion production and 10^{14} eV in the case of neutron decay (see [240] and references therein).

As discussed in the previous sections, theoretical models aiming at describing UHECR observations can be distinguished in two general scenarios depending on the mass composition: the dip model, based on a pure proton composition, and mixed composition models, with protons and heavy nuclei contributing to the flux of UHECRs. The flux of secondary neutrinos can be a powerful tool to investigate the actual composition of UHECRs. In Fig. 28 we plot the flux of cosmogenic neutrinos expected in the case of the dip model (left panel) and in the case of mixed composition (right panel). Comparing the two panels of Fig. 28 it is evident the huge impact of the composition on the expected neutrino flux: heavy nuclei provide a reduced flux of neutrinos because the photo-pion production process in this case is subdominant.

The production of cosmogenic neutrinos during the propagation of UHECRs is not sensitive to variations in the distribution of sources, the whole universe contributes to the flux of neutrinos up to the maximum red-shift of astrophysical structures able to energize UHECRs. This red-shift is typically placed around $z_{max} \simeq 10$, which is the red-shift of the first stars (pop III) [303].

Once produced at cosmological distances neutrinos travel toward the observer almost freely, except for the adiabatic energy losses and flavour oscillations, being the opacity of the universe to neutrinos relevant only at $z \gg 10$ [304, 305]. Hence, cosmogenic neutrinos are also a viable probe of the cosmological evolution of sources while UHE protons and nuclei, given their energy losses, can be observed only if produced at red-shifts $z < 3$–4.

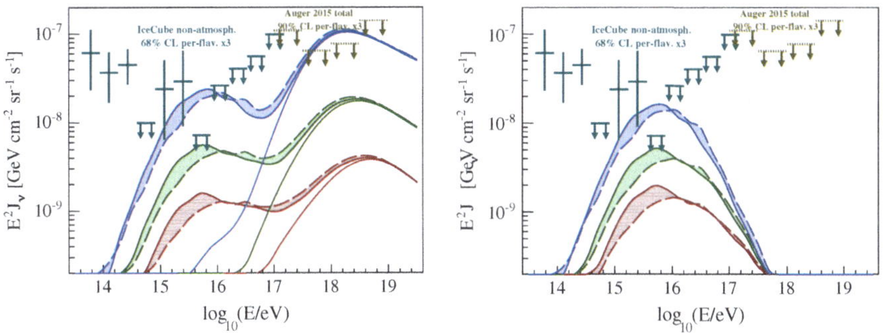

Fig. 28 (*Left panel*) Fluxes of neutrinos in the case of the dip model. The three different fluxes correspond to different assumptions on the cosmological evolution of sources (from *bottom* to *top*): no evolution (*red*), SFR (*green*) and AGN (*blue*), *coloured bands* show the uncertainties due to the EBL model considered [237–239]. Thin *solid lines* are neutrino fluxes obtained taking into account the sole CMB field. (*Right panel*) Neutrino fluxes in the case of mixed composition (as shown in Fig. 25 left panel) with the same colour code of *left panel*. Experimental points are the observation of IceCube on extraterrestrial neutrinos [300, 301] and the Auger limits on neutrino fluxes [302]

In Fig. 28, three different hypotheses on the cosmological evolution of sources are taken into account, following the same assumptions used in Fig. 25 (left) [278, 281, 282].

There is a solid consensus about the light composition of UHECRs in the low energy part of the observed spectrum. This assures a flux of cosmogenic neutrinos in the PeV energy region produced by the protons' photo-pion production process on the EBL photons. Coloured bands in Fig. 28 show the uncertainties connected with the EBL background [237–239].

Gamma Rays

While neutrinos reach the observer without being absorbed, high energy photons and electrons colliding with astrophysical photon backgrounds (CMB and EBL) produce electromagnetic cascades (EMC) through the processes of pair production (PP, $\gamma + \gamma_{CMB,EBL} \to e^+ + e^-$) and Inverse Compton Scattering (ICS, $e + \gamma_{CMB,EBL} \to \gamma + e$). While PP is characterized by a kinematic threshold, the ICS occurs at all energies. From this simple observation follows that once a cascade is started by a primary photon/electron/positron it develops for as long as the energy of photons produced through ICS is above the PP threshold. The final output of the cascade, namely what is left behind when the cascade is completely developed, is a flux of low energy photons all with energies below the PP threshold.

The two astrophysical backgrounds, CMB and EBL, against which the EMC develops are characterized by typical energies $\varepsilon_{CMB} \simeq 10^{-3}$ eV and $\varepsilon_{EBL} \simeq 1$ eV. Hence, the typical threshold energy scale for pair production will be, respectively, $\mathcal{E}_{CMB} = m_e^2/\varepsilon_{CMB} = 2.5 \times 10^{14}$ eV and $\mathcal{E}_{EBL} = m_e^2/\varepsilon_{EBL} = 2.5 \times 10^{11}$ eV; the radiation left behind by the cascade will be restricted to energies below $\mathcal{E}_{EBL}$. Clearly, numerical values quoted here should be intended as reference values being background photons distributed over energy and not monochromatic.

The cascade development has a universal nature independent of the spectrum of the initial photon/pair. It can be proven[5] that the spectrum of photons produced in the cascade, those left behind with energy below threshold, is always of the type:

$$n_\gamma(E_\gamma) \propto \begin{cases} E_\gamma^{-3/2} & E_\gamma < \mathcal{E}_X \\ E_\gamma^{-2} & \mathcal{E}_X \le E_\gamma \le \mathcal{E}_{EBL} \end{cases} \tag{27}$$

being $\mathcal{E}_X = (1/3)\mathcal{E}_{EBL}\varepsilon_{CMB}/\varepsilon_{EBL}$ the (average) minimum energy of a photon produced through the ICS mechanism by an electron with the minimum allowed energy $\mathcal{E}_{EBL}/2$ [150, 247, 306]. The normalization of the spectrum (27) can be easily determined imposing energy conservation, i.e. the total energy of cascading photons should correspond to the energy of the photon/pair that started the cascade.

[5] For a recent detailed discussion of EMC development on CMB and EBL see [299] and references therein.

The propagation of UHECRs produces EMCs started by pairs and photons produced in the processes of pair production and photo-pion production. These cascades transform the energy lost by UHECRs into low energy gamma-ray photons, with the characteristics discussed above, that in turn contribute to the diffuse gamma-ray background [248, 251, 278, 297, 299, 307, 310, 311]. Hence, the observation of a diffuse extragalactic gamma-ray background by the Fermi-LAT satellite [308, 312] can be used to constrain models of UHECRs.

A pure proton composition maximizes the production of secondary gamma rays. The fast decrease in energy ($\propto E^{-2.4}$) of the diffuse background as observed by the Fermi-LAT satellite already constrains versions of the dip model (see section "Protons and the Dip Model") with strong red-shift evolution of sources [297]. These constraints could become even more severe taking into account a recent analysis of the Fermi-LAT collaboration [309] that shows how the 'true' diffuse extragalactic gamma-ray background could be lower than what considered in the past if the contribution of unresolved sources is taken into account [307, 313]. In Fig. 29 we plot the expected gamma-ray background in the case of the dip model without cosmological evolution of sources in comparison with the experimental data of Fermi-LAT in two cases of models for unresolved sources as discussed in [309].

At extreme energies ($>10^{19}$ eV), given the behaviour of PP cross section, the universe becomes more and more transparent to photons [150, 247, 306], so that very high energy gamma rays can propagate few tens of Mpc without being absorbed. This high energy gamma radiation was discussed in the literature either as a signature of exotic models for the production of UHECRs [314–319] (see section "Exotic Models") or as a probe of the astrophysical acceleration of UHECRs in the local universe [278, 320, 321].

The detection of point-like gamma-ray sources at GeV–TeV energies could also be a promising way to reveal powerful astrophysical accelerators of UHECRs [310, 322, 323]. This possibility critically depends on the magnetic field in the intergalactic

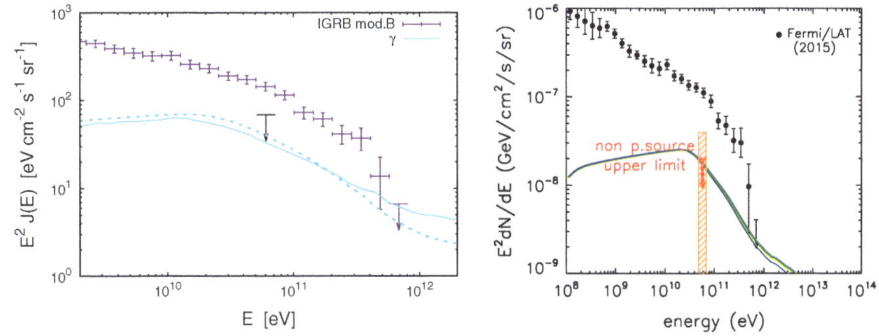

Fig. 29 Spectra of cosmogenic gamma rays obtained in the case of pure proton composition of UHECR without cosmological evolution of sources, as computed in [297] (*left panel*) and in [307] (*right panel*), together with the Fermi-LAT data on diffuse gamma-ray background, as in model-B (*left panel*) and model-A (*right panel*) of the analysis presented in [308, 309]

space as it could modify the spatial development of the cascade. Hence, to reveal a point-like source of this kind it is needed to confine the cascade within small angular size around the source line of sight and the corresponding magnetic field should be quite low ($B < 10^{-14}$ G) [310, 322, 323]. The detection of such effect and its firm correlation with a source of UHECRs is anyway very difficult as the electromagnetic cascading gamma rays have a universal spectrum independent of the primary particle type and energy (provided it is large enough, see above).

Let us conclude stressing the importance of the magnetic field in the physics of EMC. Cascades can be sustained only if the process of ICS dominates over electron interactions, nevertheless increasing the magnetic field synchrotron interaction becomes more and more important with the net result of producing low energy ($\leq$MeV) photons thus damping the cascade development.

Intergalactic Magnetic Fields

The propagation of UHECRs can be heavily affected by the presence of intergalactic magnetic fields (IMF). Here we will not discuss the effect of the galactic magnetic field which affects only the arrival distribution of particles leaving unchanged spectrum and mass composition [324–329].

Our experimental knowledge of intergalactic magnetic fields is still poor and fragmented, even if several important constraining observations were achieved (see [330–336] and references therein). In certain environments of the universe such as galaxy clusters, which could harbour sources of UHECR, the magnetic field is better known with typical observed values in the range of 1 μG [330–336].

It is outside clusters, far the largest space traversed by extragalactic CR, in filaments and voids, that the value of the magnetic field is not known and, until now, no convincing mechanism to produce strong fields over very large (supra-cluster) scales have been clearly found. The most reliable observations of the IMF are those of synchrotron emission, its polarization and Faraday rotation at radio frequencies (0.1–10 GHz) [330–336]. These measurements imply an upper limit for the IMF that depends on the coherence length of the field itself. For instance, according to [337], in the case of an inhomogeneous universe, $B < 4$ nG with a coherence scale of about $l_c = 50$ Mpc.

The detection of magnetic fields in voids or, more generally, along paths with very low field intensities can be achieved by observing cascading propagation of TeV gamma rays coming from a point-like source [338, 339]. As discussed in section "Gamma Rays", for low enough magnetic fields secondary pairs produced in the cascade will be weakly deflected producing a narrow emission of ICS gamma rays around the source line of sight. The variation with energy of the emission size can be detected allowing for a measure of the magnetic field strength in the range 10^{-16}–10^{-12} G [338, 339].

Apart from observations, the IMF can be predicted, in principle, implementing Magneto-Hydrodynamics (MHD) evolution of magnetic fields in numerical simulation of Large-Scale Structures (LSS) formation [340–344]. The main ambiguities in these simulations are related to the assumed seed magnetic fields, to the mechanism invoked in their growth and to the capability of reproducing the local density velocity field (constrained [340, 341] and unconstrained simulations [342]). Unfortunately, because of these uncertainties, MHD simulations are not completely conclusive. The volume filling factor for strong fields, of the order of 1 nG, vary by several orders of magnitude from one simulation another. The predicted magnetic field in voids (filaments) vary from 10^{-3} nG (10–1 nG) [341, 343, 344] up to 10^{-1} nG (10 nG) [342].

In order to discuss the effects of magnetic fields on the propagation of UHECRs, let us consider the ideal configuration of a homogeneous turbulent magnetic field with strength B on the coherence scale l_c. A charged particle with energy E in a magnetic field B has a Larmor radius given by:

$$r_L(E) \simeq 1 \left(\frac{E_{EeV}}{ZB_{nG}} \right) \text{Mpc} \tag{28}$$

being E_{EeV} the energy in units of 10^{18} eV, B_{nG} the magnetic field in units of nano-Gauss and Z the charge of the particle.

Depending on their energy, particles can feel the effect of the magnetic field in different ways: if the Larmor radius of the particles is smaller than l_c, then the particles can resonantly scatter against turbulence at the appropriate wavelength, a phenomenon that naturally leads to diffusion. This happens for

$$E < E_c \simeq 10^{18} ZB_{nG}(l_c/Mpc) \text{ eV}.$$

The diffusion length l_D, i.e. the distance that corresponds to a typical particle deflection of 1 rad, depends on the turbulent power encountered by particles at the Larmor radius scale, therefore it depends on the turbulent spectrum of the magnetic field. It can be easily shown that $l_D \simeq l_c(E/E_c)^\alpha$ [345–348] being α related to the turbulent spectrum ($\alpha = 1/3$ Kolmogorov, $\alpha = 1/2$ Kraichnan and $\alpha = 1$, i.e. $l_D = r_L$, in the case of Bohm diffusion). As discussed earlier in this review, the quasi-linear expression for the diffusion coefficient reads

$$D = \frac{1}{3} c l_D = \frac{1}{3} c l_c \left(\frac{E}{E_c} \right)^\alpha .$$

At higher energies, namely for $r_L > l_c$ ($E > E_c$), the scattering becomes non-resonant and particles' propagation results in a series of small deflections $\delta\theta \simeq l_c/r_L$ in each coherence length l_c. The diffusion length l_D can be easily evaluated as the space traversed by a particle to suffer a deflection $\Delta\theta \simeq 1$ rad, one has $l_D \simeq l_c(E/E_c)^2$ with a diffusion length that strongly increases with energy [345, 347, 348]. A formal derivation of the overall transport of particles in a spectrum of turbulence in the different regimes was recently presented in [349].

In this case, the propagation of particles can be described as diffusive or rectilinear depending on the distribution of sources, namely on the average distance between sources r_s. If $l_D < r_s$ the propagation is still diffusive, with a diffusion coefficient

$$D = \frac{1}{3} c l_D = \frac{1}{3} c l_c \left(\frac{E}{E_c} \right)^2 ,$$

while at the highest energies when $l_D > r_S$ the rectilinear propagation regime is recovered and the magnetic field has no effects on propagation.

On general grounds, the effect of the magnetic field on spectrum and mass composition of UHECRs is closely tied to the assumptions made about density and luminosity of sources. In other words, even a very strong field would have no effect on UHECRs if the mean separation between sources is smaller than all other propagation length scales, i.e. diffusion and energy losses lengths [345, 347].

Typical density expected for UHECR sources is in the range $n_s = 10^{-4}$–10^{-6} Mpc^{-3}, with a rather large uncertainty, as can be estimated based on detection (or non-detection) of small-scale clustering in the arrival directions [350–353]. The typical separation distance between sources will be $r_s = (n_s)^{-1/3} \simeq 10$–$10^2$ Mpc.

As discussed above, assuming a non-resonant diffusion regime ($r_L > l_c$), cosmic rays from a source at distance r_s will diffuse whenever $l_D \leq r_s$, or in terms of energy:

$$E \leq E_D \simeq 10^{18} Z B_{nG} \sqrt{\frac{r_s l_c}{\text{Mpc}^2}} \text{eV}. \tag{29}$$

Particles with energies below E_D, travelling for longer times with respect to rectilinear propagation, lose more energy. The net effect of this kind of transport is to reduce the expected flux at energies below E_D.

As was first realized in [269, 346], the (IMF induced) flux suppression (in the energy range 10^{17}–10^{18} eV) is of paramount importance in tagging the transition between galactic and extragalactic CR. Such scenario can also be invoked to reduce the flux of protons below 10^{18} eV in the case of the dip model (see section "Protons and the Dip Model") or to allow for softer injection spectra in the case of mixed composition (see section "Mixed Composition Model"). As follows from Eq. (29), the viability of these scenarios clearly depends on the assumptions made about the magnetic field configuration that should be at the nG level. It is worth recalling here that, as follows from equipartition, a nG field is an absolute upper limit of the expected magnetic field strength in voids.

Considering more realistic configurations with inhomogeneous magnetic fields, i.e. taking into account different field intensities and coherence scales in voids and filaments, the suppression effect on the low energy flux will be less pronounced and shifted to lower energies [354–356].

Let us conclude by stressing that the simple description given above of the role of magnetic fields for UHECR propagation is likely to be an oversimplification, mainly because it fails to catch the easiest ingredients of the interaction between

UHECRs and magnetic fields. For instance, within several Mpc from any source of UHECRs, the current induced by the escape of the accelerated particles from the source is likely to induce instabilities that are bound to affect both the local magnetic fields and the transport of UHECRs in that region. As discussed in [357], these phenomena may well induce a natural confinement of the lowest energy end of UHECRs close to sources because of the excitation of plasma instabilities. At present the phenomenological implications of these phenomena have not yet been investigated.

Astrophysical Sources

We do not know what the sources of UHECRs are. Hence, it is important to define general requirements that a hypothetical accelerator should fulfil in order to reach the extreme energies observed. If the size of the accelerator is R, a necessary condition to reach the energy E is that particles with this energy would remain trapped inside the source, hence the Larmor radius of the particle should be: $r_L(E) < R$. This condition fixes a relation, at the base of the so-called Hillas plot, between the size of the accelerator R and the magnetic field B in the acceleration region. In Fig. 30 we show the Hillas plot [358] with the curves relative to the condition $r_L < R$ (see Eq. (28)) in the case of protons and iron nuclei, the corresponding energy of accelerated particles, and several astrophysical objects that match this condition. The acceleration of nuclei, thanks to larger electric charge, is less demanding and can be easier achieved with respect to the case of protons.

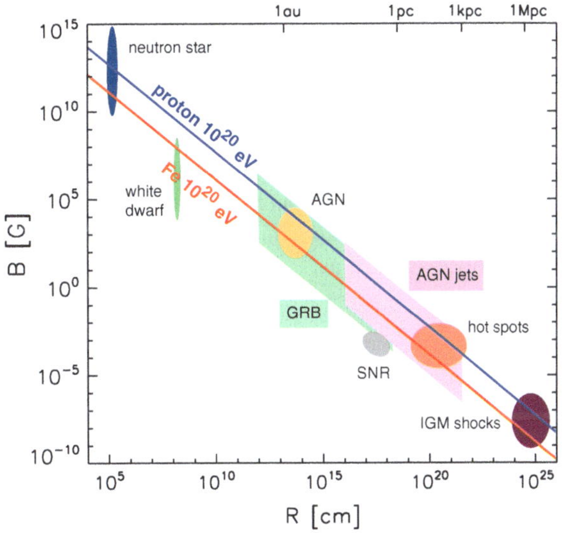

Fig. 30 Hillas plot [358] obtained by imposing the condition $r_L(E) < R$, *dashed line* corresponds to protons with energy 10^{20} eV = 100 EeV while the two *solid lines* correspond to protons with $E = 10^{21}$ eV = 1 ZeV and iron nuclei with $E = 10^{20}$ eV as labelled. Also shown are several astrophysical objects that meet the requirements of size/magnetic field needed for the acceleration process. The figure is taken from [359]

Following [360, 361], the general idea at the base of the Hillas plot can be further refined by assuming that the accelerator moves (as for shocks) with either relativistic or non-relativistic velocity. Let us consider first the non-relativistic case. The condition $r_L(E) < R$ on the magnetic field can be easily transformed in a condition on the magnetic energy density $\varepsilon_B = B^2/4\pi$. This quantity should be lower that the total ram pressure $\varepsilon_B < \rho V^2$ and this fixes a lower limit on the luminosity of the source:

$$L = 4\pi R^2 V \frac{\rho V^2}{2} > 2\pi R^2 V \varepsilon_B \simeq 1.6 \times 10^{45} Z^{-2} \left(\frac{E}{10^{20}\,\text{eV}}\right)^2 \beta \text{ erg/s} \qquad (30)$$

being Z the electric charge of the particle and $\beta = V/c$ the velocity of the accelerator. As discussed in [361], the condition (30) is somewhat uncertain in the scaling with β, as it depends on the details of particles transport in the acceleration region. What is more relevant in Eq. (30) is the strong dependence on the electric charge of the particle Z^{-2} that softens the limit in the case of nuclei.

In the case of a relativistic motion of the acceleration region, i.e. with a Lorentz factor $\Gamma \gg 1$, it is useful to introduce the co-moving reference frame, all quantities in this frame will be indicated with a prime. The condition discussed above for acceleration in the co-moving frame becomes: $r_L(E') < R' = R/\Gamma$, using the Lorentz contraction of length. Moreover, since the energy density transforms as Γ^2, we can rewrite the condition on luminosity in the case of a relativistic motion of the accelerator as:

$$L > 4\pi R^2 c \Gamma^2 \varepsilon_{B'} \simeq 10^{47} \Gamma^2 Z^{-2} \left(\frac{E}{10^{20}\,\text{eV}}\right)^2 \text{ erg/s.} \qquad (31)$$

The two conditions, Eqs. (30) and (31), are very stringent in the case of protons, while in the case of nuclei with high electric charge, as for iron $Z = 26$, the two conditions become less constraining.

It is also important to keep in mind that the constraints obtained above only apply to stationary sources, or sources that can be considered as stationary on time scales relevant for the propagation of UHECRs. The constraints do not apply in a straightforward manner to bursting sources.

Another quantity that can be used to constrain sources of UHECRs is the energy input per unit volume and time, i.e. the sources emissivity. The flux of UHECRs can be roughly estimated as:

$$J(E) \simeq \frac{c}{4\pi} \dot{n}(E) \tau_{loss}(E) \qquad (32)$$

where $\dot{n}$ is the injection rate per unit volume and τ_{loss} the time scale of energy losses. Comparing Eq. (32) with the observed flux we can immediately deduce, at fixed energy, the expected emissivity of the sources: at the level of 3×10^{45} erg/Mpc^3/yr at 10^{19} eV. This quantity can be compared with the known emissivity of well-defined sources. For instance, AGN typically shows bolometric luminosities in gamma rays in the range between $L_{bol} \simeq 10^{43}$ erg/s, for Seyfert galaxies and radio-quite quasars, and $L_{bol} \simeq 10^{47}$ erg/s, for radio-loud quasars, with typical number density in the range 10^{-5}–10^{-4} Mpc^{-3} [362, 363]. Hence AGN would meet the energy requirements if they emit a fraction in the range 10^{-4}–10^{-3} of their bolometric luminosity in the form of UHECRs.

In the following, we will address three general categories of possible astrophysical sources distinguishing among the acceleration mechanisms at work: relativistic and non-relativistic shocks and strong electric fields, as those produced by rotating magnetized stars.

Non-relativistic Shocks and Large Scale Structures

There are not many examples of non-relativistic shocks able to accelerate particles at the extreme energies of UHECRs. The most noteworthy case is certainly represented by shocks produced during the formation of clusters of galaxies [364–366]. Generally speaking, the formation of large-scale structures in the universe naturally leads to supersonic motion of plasma that, fueled by gravitational forces, gives rise to shock waves. These shocks can be formed either during cluster mergers or during accretion of cold gas on an isolated cluster.

In filaments, that develop between clusters during large scale structure formation, typical flow velocities are around 10^3–10^4 km/s with a background temperature at the level of 10^5 K. The corresponding shock waves are usually relatively strong, with typical Mach numbers of order $M_s \simeq 10$–100. On the other hand, shocks produced during cluster mergers are weaker ($M_s \ll 10$) because formed in the intra-cluster medium with a typical temperature at the level of 10^8 K, being clusters already virialized structures [367]. In both cases of accretion and merging, the available luminosity in the form of accelerated particles can reach $L \simeq 10^{45}$–10^{46} erg/s, compatible with the estimates discussed above.

These kinds of accelerators operate for a very long time, of the order of the age of the universe, and the accelerated particles are typically confined inside the cluster volume [368]. The maximum energy attainable with these mechanisms is constrained by the magnetic field at the shock, which fixes the acceleration time, and by the energy losses of the accelerated particles. Particularly relevant are losses due to photo-pion production and photo-disintegration that limit the maximum energy at the level of $\sim 5 \times 10^{19}$ eV [364–366]. Hence, shocks in clusters of galaxies, also called cosmological shocks, may represent an option for UHECR acceleration only in the case of no substantial flux observed at energies higher than few $\times\, 10^{19}$ eV. It is unlikely that a substantial fraction of heavy nuclei may be accelerated at these pristine shocks, because the intergalactic medium in the outskirts of clusters is probably not reach enough in such heavy elements. Finally, the spectrum of the accelerated particles produced by cosmological shocks can be determined applying the general theory of particle acceleration at non-relativistic shocks which is relatively well understood and discussed in section "Acceleration of Galactic CRs".

Relativistic Shocks

As discussed in section "Acceleration of Galactic CRs", particle acceleration at shock fronts requires multiple shock crossings, which in turn depend upon the return probability from downstream being sizeable. While this does not, in general, represent a problem in the case of newtonian shocks, it becomes critical in the case of relativistic shocks. In fact, for such shocks both the particles and the front move at speeds very close to the speed of light and this results in a large anisotropy of the particle distribution at the shock. The first point to notice is that a relativistic shock is superluminal for all orientations of the background magnetic field that form an angle $>1/\Gamma$ with the shock normal. For large values of the shock Lorentz factor Γ, it becomes hard to avoid this condition. The consequences are quite important for shock acceleration: for $\Gamma \gg 1$, the shock velocity in the frame comoving with the downstream plasma is $\sim c/3$. On average a particle takes a time $\tau = 2\pi r_L/c$ to cover one Larmor rotation. In this time the shock moves by $\tau c/3 = (2\pi/3)r_L > r_L$, namely the particle is trapped downstream and its probability of returning upstream is greatly reduced. This fact leads to expect steeper spectra for acceleration at relativistic shocks, as discussed in [369].

The first time that a particle traverses the shock from upstream to downstream and back its energy can increase by a large factor of the order of $\simeq 4\Gamma^2$. For large values of Γ, as in GRBs that can achieve $\Gamma \sim 300$, the energy gain can be remarkable with particles that acquire energies at the level of 10^5–10^6 GeV, note that this energy scale will be a low energy cutoff in the spectrum of accelerated particles. After the first shock crossing particles distribution will be beamed within an angle of the order of $1/\Gamma$ around the shock normal. The highly anisotropic distribution of particles implies a much lower energy gain at any subsequent shock crossing, if any, of the order of $\Delta E/E \simeq 2$. Note that the simple picture discussed here applies to planar shocks, it might become somewhat different (and more complicated) for non-planar relativistic shocks as, for instance, in the case of relativistically moving plasmoids as those observed in AGN jets.

Based on the simple argument illustrated above, one can infer that reaching ultra-high energies at a relativistic shock requires efficient scattering downstream, on scales smaller than or comparable with the Larmor radius of particles. The presence of such turbulence is in fact implicitly present in all models of particle acceleration at relativistic shocks [370–375], which show a general consensus on the accelerated spectrum being $N(E) \propto E^{-2.3}$ following from the assumption of small pitch angle scattering downstream. However, the nature of the magnetic turbulence in the shock region determines whether the maximum energy achieved by accelerated particles is of interest for UHECRs or not. If the power spectrum of the turbulence upstream of the shock is peaked at scales larger than the Larmor radius of the particles, then the compressed field downstream of the shock front is again quasi-perpendicular and the same argument above holds, namely the return probability is small and the spectrum of accelerated particles is steep, thereby making the process of poor interest for the application to UHECRs. On the other hand, the turbulence upstream may be on scales much smaller than the Larmor radius, as would be the case for the field generated through Bell-like instabilities. Such phenomena appear to be needed to explain typical strengths of the field of 10–100 μG [376, 377] as inferred from X-ray observations. In this case, the return probability from downstream may not be small, but the scattering is weak, thereby making the acceleration time longer and the maximum energy correspondingly lower. These arguments, that here we described in a rather qualitative manner, have been formalized and investigated in detail in the context of PIC simulations in [378–380].

Recently, it has also been proposed that UHECRs may in fact result from re-acceleration of lower energy CRs (around 10^{17} eV): CRs accelerated to such energies in the host galaxy of an AGN may penetrate the jet of the AGN sideways, thereby receiving a one-shot boost in energy by a factor Γ^2 that allows the particles to reach ultra-high energies [381].

Acceleration Inspired by Unipolar Induction

The rotation of a magnetized star implies potentially large induced electric fields that, in turn, could accelerate particles to ultra-high energies [150]. Several astrophysical objects show strong magnetization and may be suitable to accelerate UHECRs: noticeable examples are black hole magnetospheres and fastly spinning Neutron Stars (NSs). NSs were proposed as sources of UHECRs some time ago [382–384] while such models were recently updated mainly to accommodate the new Auger findings [41, 385, 386], with special attention for the mixed composition.

Magnetized, fast spinning NSs present important advantages as sources of UHECRs with respect to more sources more traditionally associated to UHECRs, such as AGN and gamma-ray bursts: first, the energy budget is favourable with a NS population density $\dot{n}_s = 3 \times 10^{-3}$ Mpc^{-3}yr^{-1} [387] and a very large reservoir of rotational energy, at the level of

$$E_{rot} \simeq 2 \times 10^{52} \left(\frac{I}{10^{45}\text{g cm}^2} \right) \left(\frac{P}{10^{-3}\text{s}} \right)^{-2} \text{ erg} \tag{33}$$

with I the moment of inertia and P the rotation period of the star [387]. Comparing these numbers with the emissivity expected from UHECR sources (see Eq. (32)), one finds that only a tiny fraction of NSs, $\leq 10^{-4}$, is expected to contribute to the observed flux of UHECRs.

Another important point in favour of the NS hypothesis is represented by the NS surface which is naturally rich in heavy elements that may potentially be accelerated thereby providing a possible explanation of the composition observed by Auger [386], while in the case of other kind of sources, such as GRB, the acceleration of heavy nuclei seems challenging [388–390].

The crust of a NS is thought to be made of condensed matter tightly bound in long molecular chains oriented along the magnetic field lines [391, 392]. These chains are thought to be made of iron nuclei ordered in a one-dimensional lattice with an outer sheath of electrons. The binding energy of iron nuclei can be estimated as ~14 keV and the lattice spacing $d \simeq 10^{-9}$ cm [393]. Hence, the electric field needed to extract an iron nucleus is $\mathscr{E}_0 = 14\,\mathrm{keV}/(\mathrm{Zea}) \simeq 1.4 \times 10^{13}/(Zd_{-9})$ V/cm, being Z the electric charge of the extracted nucleus ($Z = 26$ for iron) and d_{-9} the lattice spacing in units of 10^{-9} cm.

Extraction of nuclei can be achieved by the electric field generated at the NS surface by the star rotation, estimated to be [384]

$$\mathscr{E} = \frac{2\pi B R_s}{Pc} \simeq 6.3 \times 10^{14} \left(\frac{B}{10^{13}\mathrm{G}}\right)\left(\frac{R_s}{10^{6}\mathrm{cm}}\right)\left(\frac{P}{10^{-3}\mathrm{s}}\right)^{-1} \frac{\mathrm{V}}{\mathrm{cm}}, \tag{34}$$

where B is the strength of the magnetic field of the star at its surface, R_s is the radius of the NS and P the star rotation period.

The main effect of the electric field in Eq. (34) is to extract electrons from the NS crust electrons, since they are much less bounded than nuclei. Such electrons suffer curvature losses in the strong dipolar magnetic field of the star. Photons produced as curvature radiation by electrons can in turn give rise to pairs by scattering on the virtual photons of the magnetic field, pairs will in turn generate other curvature photons, giving rise to an electromagnetic cascade. This chain of events leads to a multiplication of the number of electron-positron pairs that eventually fill the magnetosphere of the star. The number of pairs generated by a single extracted electron is in the range 10–10^4 depending on local conditions. Pairs in the magnetosphere have a screening effect, which reduces the effective electric field available for particle acceleration.

At least in principle, the total potential drop available for particle acceleration in the magnetosphere is [384]:

$$\Phi = \frac{2\pi^2 B R_s^3}{P^2 c^2} \simeq 7 \times 10^{19} \left(\frac{B}{10^{13}\mathrm{G}}\right)\left(\frac{R_s}{10^{6}\mathrm{cm}}\right)^3\left(\frac{P}{10^{-3}\mathrm{s}}\right)^{-2} \mathrm{V}, \tag{35}$$

that would correspond to a maximum particles' Lorentz factor $\gamma_\Phi = Ze\Phi/(Am_pc^2)$, namely an energy exceeding 10^{20} eV for iron nuclei. In fact, the maximum energy of nuclei accelerated in the NS magnetosphere is limited by curvature losses. Assuming that the total potential drop Φ is available over a gap of length ξR_L, being R_L the radius of the light cylinder[6] of the star $R_L = cP/2\pi$, and equating curvature energy losses with energy gain one gets an upper bound to the acceleration Lorentz factor as [386]

[6]Distance at which the peripheral velocity of the star reaches the speed of light $2\pi R_L/P = c$ and the magnetic field lines spiral outwards along the azimuth.

$$\gamma_{curv} = \left(\frac{3\pi B R_s^3}{2ZeP\xi c}\right)^{1/4} \simeq 10^8 \left(\frac{\xi Z}{26}\right)^{-1/4} \left(\frac{B}{10^{13}\text{G}}\right)^{1/4} \left(\frac{P}{10^{-3}\text{s}}\right)^{-1/4} \left(\frac{R_s}{10^6\text{cm}}\right)^{3/4}, \quad (36)$$

hence the actual maximum energy that particles can attain within the co-rotating magnetosphere will be set by $\gamma_{max} = min(\gamma_{cur}, \gamma_\Phi)$. The parameter ξ takes into account the screening effect of pair creation and it can be estimated at the level of $O(1)$ [386], signalling that the gap cannot be too far from the star surface.

If the NS wind has a Lorentz factor larger than γ_{max}, particles that end up in the wind will be advected with it at the Lorentz factor of the wind irrespective of the energy reached in the magnetosphere. This mechanism discussed in [386] provides a way of increasing UHECRs energies independently of the curvature losses.

The discussion above focuses on the most classical scenario of particles acceleration nearby the stellar surface. There are in literature other scenarios in which acceleration happens at the light cylinder or further out [394] that are not discussed here.

The spectrum of UHECRs accelerated by a NS is determined by the evolution of the rotational frequency: as the star spins down the energy of the accelerated particles decreases (see Eq. (35)). In general, a NS is powered by the rotational kinetic energy and loses energy by accelerating particle winds and by emitting electromagnetic radiation. Because of this, the rotation frequency of the star decreases with time following the relation $\dot{\nu} = -K\nu^n$ where n is the braking index ($n = 3$ for a pure magnetic dipole) [391] and K is a positive constant, which depends on the moment of inertia and on the magnetic dipole moment of the star [391]. The spectrum of UHECRs accelerated by the NS integrated over the history of the star, is found to be $N(E) \propto E^{\frac{1-n}{2}}$ [383], which can be as hard as $N(E) \propto E^{-1}$ in the reference case $n = 3$ and even harder for braking index $n < 3$, which are actually the norm.

Once nuclei are extracted from the stellar crust and accelerated by the potential gap Φ they are advected with the NS wind and interact with the environment of the star, suffering mainly photo-hadronic interactions [41, 385, 386]. As a result, both the energy of accelerated particles and their mass composition change in time.

The nature of the nuclei that manage to escape the wind region at later times is fully determined by how effective photo-disintegration of nuclei on the thermal photons coming from the star's surface is in breaking heavy nuclei into lighter ones. As discussed in [386, 395], for reasonable values of the NS surface temperature ($T < 10^7$ °K), nuclei are not completely destroyed, hence the mass composition of UHECR nuclei from NSs is naturally expected to be mixed [41, 385, 386].

Exotic Models

The extreme energies of UHECRs, as high as 10^{11} GeV, eleven orders of magnitude above the proton mass and 'only' eight below the Planck mass, are a unique workbench to probe new ideas, models and theories which show their effects at energies much larger than those ever obtained, or obtainable in the future, in accelerator experiments.

There are two general classes of exotic theories that can be tested using UHECRs: top-down models for the production of these extremely energetic particles and models that imply extensions and/or violations of Lorentz invariance, as in certain theories of quantum gravity.

Top-Down Models and Super-Heavy Dark Matter

In top-down models, UHECRs are not accelerated particles but rather result from either the decay or the annihilation of particles with very large masses, produced as relics of early universe phenomena. The idea of generating UHECRs in this way arose in the 90s in the aftermath of the AGASA claim of absence of the GZK feature in the data [396].

The two main classes of top-down models are associated with topological defects and super-heavy relics (see [315] for a discussion of these two classes). The former are usually associated with symmetry breaking of some type; known instances of topological defects are monopoles, cosmic strings and necklaces. The latter, super-heavy relic particles, may form as a consequence of quantum processes during inflation and have been widely discussed as candidates for dark matter, the so-called super-heavy dark matter (SHDM) to distinguish it from the more familiar WIMP candidates. In fact, the connection of super-heavy relics to dark matter is, nowadays, the main scientific motivation for pursuing this type of investigation.

The existence of SHDM has been postulated based on the possibility of particle production due to a non-adiabatic expansion of the background space-time acting on vacuum quantum fluctuations. In quantum theories the possibility of producing particle pairs through the effect of a strong (classical) external field is well known; consider, for instance, the case of $e^{\pm}$ pair creation by strong electromagnetic fields. The idea to apply such a mechanism to the cosmological context using external strong gravitational fields dates back to E. Schrödinger in 1939 [397].

The theory of particle creation during the expansion of the Universe has been developed through the last 40 years, starting with the pioneering work discussed in Refs. [398–402]. More recently, in the framework of inflationary cosmologies, it has been shown that particle creation is bound to be a common phenomenon, independent of the specific cosmological scenario. Moreover, the resulting SHDM particles (labelled as X) have been shown to potentially account for most dark matter, $\Omega_X(t_0) \simeq 1$ [318, 319, 403–408]. This conclusion can be drawn under three general hypotheses: (i) SHDM in the early Universe never reaches local thermal equilibrium (LTE); (ii) SHDM particles have mass M_X of the order of the inflation mass M_ϕ; and (iii) SHDM particles are long living with a lifetime exceeding the age of the Universe, $\tau_X \gg t_0$.

Precision measurements of CMB polarization and observations of UHECRs up to energies $\simeq 10^{20}$ eV enable a direct experimental test of the three hypotheses listed above. As discussed in [319], the first two hypotheses can be probed through the measurement of CMB polarization. The third hypothesis, particle lifetime longer than the age of the Universe, is a general requirement of any DM model based on the existence of new particles. As in the case of WIMPs, discrete gauge symmetries protecting particles from fast decays need to be introduced (see [317, 318] and references therein).

The best way to test the existence of SHDM is through the indirect detection of its annihilation and/or decay products (direct detection is unattainable). In general, since the annihilation cross section of a (point) particle is bound by unitarity, $\sigma_{ann} \propto 1/M_X^2 \sim 1/M_\phi^2$, the annihilation process results in a small annihilation rate [318], although alternative theoretical models can be constructed (see for instance [410]) in which this limitation is overcome. Below we will specialize our discussion to decay channels.

If SHDM particles decay, under general assumptions on the underlying theory (see [317, 318] and references therein), we can determine the composition and spectra of the standard model particles produced. Typical decay products are neutrinos, gamma rays and nucleons with a relatively hard spectrum, that at the relevant energies can be approximated as $dN/dE \propto E^{-1.9}$, independently of the particle type, with a photon/nucleon ratio of about $\gamma/N \simeq 2$–3 and a neutrino-nucleon ratio $\nu/N \simeq 3$–4, rather independent of the energy range [317]. The most constraining limits on SHDM lifetime are those coming from the (non) detection of UHE photons and neutrinos.

Auger observations provide very stringent limits on the flux of photons at energies above 10^{19} eV: at the level of 2% [411]. This fact already constrains the SHDM lifetime to values $\tau_X \geq 10^{21}$–10^{22} yr, depending on the underlying inflationary potential, and implies that SHDM models can be effectively probed only by the next generation of UHECR experiments, those designed to maximise statistics at the highest energies [133, 412, 413], together with new and more refined observations of the CMB polarization pattern.

In the left panel of Fig. 31, as discussed in [319], we plot the flux of UHECRs coming from the decay of SHDM in a specific model of inflation with $M_X = 4.5 \times 10^{13}$ GeV and $\tau_X = 2.2 \times 10^{22}$ yr (solid lines); we also show the expected sensitivities of the proposed JEM-EUSO space mission (thick red solid line) [133] and, for UHE neutrinos, of the upcoming ARA observatory (thick blue solid line) [414, 415]. In the right panel, we show the corresponding fractions of fluxes of protons, photons and neutrinos produced by the decay of SHDM.

Tests of Lorentz Invariance

Lately, a general consensus has emerged that UHECR observations can be used as a powerful tool to put under experimental scrutiny the validity of Lorentz Invariance (LI). The possibility of testing LI at scales not probed so far is interesting per se, as any new experimental scrutiny of theoretical foundations. Moreover, the need for

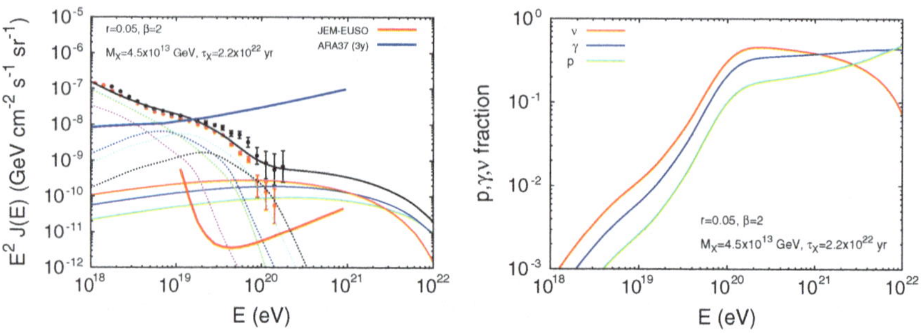

Fig. 31 (*Left panel*) Flux of UHECR from the decay of SHDM (*thin solid lines*) with parameters as labelled together with the flux expected in the framework of the mixed composition model of [240]. Also shown is the sensitivity to SHDM decay products: of the proposed JEM-EUSO space mission (*thick red solid line*) and, for UHE neutrinos, the upcoming ARA observatory (*thick blue solid line*). Experimental data are those of Auger (*red points*) [118] and TA (*black points*) [409]. (*Right panel*) Fraction over the total flux of protons, photons and neutrinos by SHDM decay as follows from fluxes in *right panel*. Both figures are taken from [319]

extensions and/or violations of LI can be connected to the long-standing problem of the construction of a quantum theory of gravity (QG).

Our universe is very well described by quantum mechanics at small scales and general relativity at large scales, but a unified theory of QG is still out of reach. While all other fundamental interactions propagate through space-time, gravity turns out to be a property of space-time itself. This simple statement, at the base of General Relativity, has important implications for the construction of a quantum theory of gravity, as it implies that the structure of space-time itself has quantum fluctuations. In other words, at the scales where quantum effects of gravity arise, space-time could have a granular (or foamy) structure where the size of space-time cells fluctuates stochastically causing an intrinsic uncertainty in the measurements of space-time lengths and, indirectly, of the energy and momentum of particles. The typical scale at which quantum gravitational effects are supposed to become relevant is the so-called Planck length, i.e. the length scale given by $l_p = \sqrt{\hbar G/c^3} \simeq 1.6 \times 10^{-33}$ cm. It is generally argued that measurements of distances (times) smaller than the Planck length (time) are conceptually unfeasible, since the process of measurement collects in a Planck size cell an energy exceeding the Planck mass ($M_P = \sqrt{\hbar c/G} \simeq 1.2 \times 10^{19}$ GeV) hence forming a black hole in which information is lost.

As was immediately realized after the proposal of the GZK suppression [416], in the reference frame in which astrophysical photon backgrounds are isotropic, an UHE nucleon only needs a fractional gain in energy at the level of 10^{-22}–10^{-21} to perform the transition to its final state (photo-pion production or photo-disintegration). LI guarantees that this is exactly the same to what happens in the reference frame in which the nucleon is at rest and the photon has 10–100 MeV energy. But this also shows that even very tiny violations of LI are bound to give, in some selected reactions at least, observable effects. The kind of reactions typically very sensitive to LI violations are those characterized by a particle production threshold [417–420].

In recent times LI violating models have been investigated in depth and their implications compared with available experimental data [421, 422]. Particularly interesting is the approach of Effective Field Theories (EFT) in which LI or CPT symmetry (and renormalizability) is no longer a guide; in this kind of theories the number of possible terms violating LI is very high. Those that can be tested experimentally (several hundreds) are described in [422] and can be generally modelled through modifications of the dispersion relation of particles [423] (which in the EFT approach corresponds to modifications of kinetic terms in the Lagrangian density), such as:

$$E^2 - p^2 = \mu^2(E, p, M_P) \simeq m^2 + fE^{2+n}/M_P^n, \tag{37}$$

where μ is a generalized 'mass' that can be always approximated as the mass of the particle m plus terms that violate LI at the strength fixed by f ($f = 0$ corresponds to the standard invariant relation).

The firm experimental evidence of the suppression in the spectrum of UHECRs around few $\times\, 10^{19}$ eV implies very stringent limits on the possible violations of LI. Using the parameterization introduced in Eq. (37), the case $f > 0$ is strongly excluded by observations, because in this case thresholds for particles' production move to lower energies and new exotic processes are allowed, such as vacuum Čerenkov $p \to p\gamma$ for which very strong bounds exists [424]. As soon as f moves toward negative values, thresholds for particle's production slightly increase up to the point where the process becomes kinematically forbidden. In this case, limits of LI violations obtained from the observed spectral steepening are reported in literature [425, 426]. These limits, however, depend crucially on the assumption that the steepening in the flux is originated by the propagation of UHECRs. As discussed in section "Observations of Ultra-High Energy Cosmic Rays", Auger data can be very well accommodated in models in which the flux suppression is connected with low values (see section "Mixed Composition Model") of the maximum acceleration energy at the source. In this case, no relevant limit on LI violations can be placed using the observed flux of UHECRs [427].

Violations of LI can also produce important effects in the development of showers produced by the interaction of UHECRs with nuclei of the Earth's atmosphere. These effects typically reduce the kinematical phase space for certain processes modifying the particles content of the cascading shower. The most important process in the physics of cascades is the neutral pion decay $\pi^0 \to \gamma\gamma$, which has a reduced kinematic phase space in the case of LI violations ($f < 0$) with stable neutral pions at energies larger than $E > (M_P^n m_\pi^2/|f|)^{\frac{1}{2+n}}$ [427]. This modification of particles' cascade has the net effect to move the shower maximum to higher altitudes as the electromagnetic part of the shower consumes faster. Moreover, it produces an increased number of high energy muons in the shower due to the interaction of 'non-decaying' neutral pions. As of today, observations of the shower development in the atmosphere are not able to exclude LI violations effects, that are however much weaker and difficult to tag than in the case of UHECRs propagation.

Transition Between Galactic and Extragalactic Cosmic Rays

In sections "Transport of CRs in the Galaxy", "Acceleration of Galactic CRs" and "Transition Between Galactic and Extragalactic Cosmic Rays" we discussed the transport of galactic CRs, their acceleration, mainly in SNR shocks, and the transport of extragalactic CRs, respectively. We left on purpose the definition of galactic and extragalactic CRs somewhat vague. In this section, we discuss the transition between the two in the different scenarios that we introduced earlier in this review paper.

Historically, the transition from galactic CRs to extragalactic CRs has been assumed to take place at the ankle: CR iron nuclei were assumed to be accelerated up to energies in excess of $\sim 10^{19}$ eV, where they would leave room to extragalactic CR protons, injected with spectrum $\sim E^{-2}$. This picture is typically invoked in models of GRBs as sources of UHECRs [428, 429].

This picture remained virtually untouched until it was realized that an ankle-like feature would naturally appear in the spectrum of extragalactic CRs due to the combination of the expansion of the universe and pair production of protons propagating in the CMB photons, the so-called dip scenario [253, 264], discussed in section "Protons and the Dip Model". In this model the galactic CR spectrum is required to end with a heavy composition at much lower energies, $\sim 10^{17}$ eV, two orders of magnitude below the energy of the ankle. This picture is roughly consistent with the idea that galactic SNRs may accelerate protons up to the knee.

As discussed in section "The Quest for PeVatrons", acceleration of CRs to PV rigidity at SNR shocks is also all but trivial: only in core collapse SNe exploding in the wind of their red giant companions there seem to be the conditions to reach maximum energies around $E_{max} \sim 10^{15}$ eV for protons [11, 12, 218], reached at the beginning of the Sedov–Taylor phase, that occurs about $\sim$30 years after the explosion for these supernovae. In fact, it was pointed out that particles could be accelerated to somewhat higher energies at earlier times, but the number of particles processed is smaller, so that the transition between the ejecta dominated and the Sedov phase leads to a broken power law rather than an exponential cutoff at E_{max}.

This finding was used in Ref. [12] to calculate the shape of CRs between the end of the galactic component and the beginning of the extragalactic one, assuming that in the energy region between 10^{17} and 10^{18} eV the flux of light CRs as measured by KASCADE-Grande [73, 78] is of extragalactic origin.

A careful investigation of the transition region in the context of the dip and ankle models was carried out in Refs. [264, 266], assuming an exponential cutoff in the galactic CR component. The authors concluded that the ankle model is basically ruled out by measurements of the depth of shower maximum, $X_{\max}(E)$, and its rms fluctuations.

As mentioned above, a turning point in the investigation of the transition region is represented by the recent measurement of the spectra of the light and heavy components of CRs in the energy region 10^{17}–10^{18} eV by KASCADE-Grande [73, 78]. These measurements found evidence for an ankle-like feature in the light component at $\sim 10^{17}$ eV and a knee-like feature in the heavy component at roughly the same energy. The former was interpreted by many as a signature of the transition to a light extragalactic CR composition, while the latter was interpreted as the end of the galactic heavy CR component.

The main reason to believe that the light KASCADE-Grande component is of extragalactic origin is the low level of anisotropy observed by Auger at 10^{18} eV, that seems to be at odds with a galactic origin of protons at those energies [275]. This view has been recently questioned in Ref. [430], where the authors proposed that CRs might all be generated in occasional galactic GRBs, at all energies, and advocate that the low observed anisotropy can still be compatible with this picture [431] if the transport of CRs in the galactic magnetic field is properly taken into account. Further investigation is probably needed to identify the reasons for the different conclusions reached by different numerical simulations of CR propagation.

It is worth stressing that if indeed the light CR component measured by KASCADE-Grande is of extragalactic origin, the transition region becomes weakly dependent upon whether extragalactic CRs are all light (dip model) or have mixed mass composition, since in the energy region 10^{17}–10^{18} eV the expected composition is light in both scenarios (see discussion in section "Mixed Composition Model"). What is less clear is whether this light component reflects a class of sources which are different from those that produce nuclei (with a hard injection spectrum) or rather the light component is due to interaction processes inside the sources, as advocated in Ref. [273].

Conclusions

While writing this review article on the origin of the cosmic radiation, we often found ourselves in need of editing the manuscript to update it with the most recent ideas and/or measurements appearing in the literature. This is a clear evidence that one century after the discovery of cosmic rays, this field of research remains extremely lively, mainly driven by new exciting experimental findings.

This wealth of data pouring in the field has the effect of continuously challenging models, even those that in time appear to be well established. In this context, the main role has been played by measurements of spectra and mass composition. Although almost all spectra in CR science have a power law-like shape, it is the deviations from such power laws that typically provide clues to the underlying physics. For instance, the knee, probably the most prominent and best studied feature in the all-particle CR spectrum, is by most considered as emerging from the combination of two factors: (1) the maximum energy of accelerated protons, and (2) the rigidity dependence of the acceleration process, so that a knee results from the overlap of the spectra of elements with different atomic number Z. This implies that the mass composition at the knee should be predominantly light (H and He), a picture that is supported by the KASCADE data. Recent data by ARGO-YBJ and by CREAM cast a doubt on this picture, showing some evidence of a flux reduction in the light component

(H + He) at energies well below the knee in the all-particle spectrum, which would then correspond to intermediate mass nuclei.

At first sight it may seem surprising that decades after the experimental discovery of the knee there are still uncertainties on its interpretation, but the truth is that this is the first time that we are actually probing this region with direct measurements on one side, and the first time that we studied EAS very close to the shower maximum (high altitude), and its core, with full coverage arrays. This is the crucial point: understanding the physics of the knee requires a credible measurement of the mass composition in this energy region. The fact that available data return somewhat conflicting outcomes provide us with an estimate of the systematic uncertainties intrinsic in experimental approaches, mainly coming from indirect measurements. As long as these systematic uncertainties are not understood the issue of the physical origin of the knee will remain open. In this aspect, the extension to larger energies of direct (space based) detections is a key issue.

On the other hand, whenever the mass composition has been measured accurately the improvement of our knowledge has been astonishing: a clear instance is the measurement of the spectra of protons and helium nuclei at energies $\leq$1 TeV by PAMELA and AMS-02, that showed the existence of a spectral break at rigidity $\sim$300 GV and systematically harder spectrum of helium nuclei. This finding is leading to the development of alternative models of CR transport and the investigation of the implications for secondary nuclei is only starting now with the most recent measurements of the B/C ratio by AMS-02.

A similarly problematic situation exists in the ultra-high energy region: while the all-particle spectrum has been reliably measured and the existence of features such as the ankle and the so-called GZK suppression has been confirmed, their interpretation is still subject of an active debate, mainly triggered by an experimental assessment of the mass composition that is far from being completely understood. The Auger Observatory measured $X_{\max}$ and its fluctuations and concluded that the mass composition is bound to be mixed, being predominantly light around 10^{18} eV and gradually heavier at higher energies, though iron seems to be basically absent. In this picture, the GZK feature is not associated with energy losses but mainly to an intrinsically low value of the maximum energy at the sources. The Telescope Array, with a smaller surface and different systematics, also measures the GZK suppression but such feature seems to be compatible with a light composition. Again, as for the case of the knee, the physical picture remains unclear and can be possibly outlined only when the observational situation will be better assessed. The common effort of the Auger and TA collaborations made setting up joint working groups to compare data and analysis tools will certainly help to assess the experimental situation. This situation also affects the related problem of transition from galactic to extragalactic CRs.

Therefore, the reliable measurement of the mass composition of CRs appears to be central to the investigation of the origin of CRs throughout the spectrum. It is certainly positive that the issue is mainly experimental in nature: one should aim at building detectors with large exposure and accurate mass discrimination in the knee region in order to make sure that this region is well understood. ISS-CREAM and HERD are appropriate responses to this requirement. In the transition region the infill and low energy enhancements of Auger and TA are definitely a move in the right direction.

At the highest energies the problem is somewhat more complex, this being mainly a consequence of the limited exposure of fluorescence data, that at present constitute the bulk of composition measurements. The AugerPrime upgrade of the Pierre Auger Observatory has been specifically designed to extend the energy range of these measurements. Yet, the accurate measurement of the mass does not only rely on an increased statistics of events, but is also strongly dependent upon the understanding of shower development. Although much progress has been taking place in this field, especially in the aftermath of LHC data collection, there are still aspects of the shower development that need improvement to be reliably applied to mass discrimination. Finally, new experimental approaches to the extreme energies, i.e. UHECR induced fluorescence light observation from space, are currently in R&D phase and, if the sufficient mass resolution will be reached, would help understanding the mass composition providing an important jump in exposure.

Acknowledgements This review paper sprang out of the workshop 'Multiple Messengers and Challenges in Astroparticle Physics', held at GSSI from 6 to 17 October 2014. The authors acknowledge all participants for making many discussions possible during and after the workshop. The authors are also grateful to their colleagues at the GSSI, at the Arcetri Astrophysical Observatory, at the LNGS, at the University of Salento and the University of L'Aquila for stimulating discussions on the topics illustrated in the review.

References

1. W.D. Apel et al., Astropart. Phys. **31**, 86 (2009). https://doi.org/10.1016/j.astropartphys.2008.11.008
2. B. Bartoli et al., Phys. Rev. D **92**(9), 092005 (2015). https://doi.org/10.1103/PhysRevD.92.092005
3. A.R. Bell, Mon. Not. R. Astron. Soc. **182**, 147 (1978)
4. V. Ptuskin, J. Phys. Conf. Ser. **47**, 113 (2006). https://doi.org/10.1088/1742-6596/47/1/014
5. P. Blasi, E. Amato, JCAP **1201**, 011 (2012). https://doi.org/10.1088/1475-7516/2012/01/011
6. O. Adriani et al., Astrophys. J. **791**(2), 93 (2014). https://doi.org/10.1088/0004-637X/791/2/93
7. M. Aguilar, L. Ali Cavasonza, G. Ambrosi, L. Arruda, N. Attig, S. Aupetit, P. Azzarello, A. Bachlechner, F. Barao, A. Barrau et al., Phys. Rev. Lett. **117**(23), 231102 (2016). https://doi.org/10.1103/PhysRevLett.117.231102
8. J. Vink, Astron. Astrophys. Rev. **20**, 1 (2012). https://doi.org/10.1007/s00159-011-0049-1
9. V.N. Zirakashvili, V.S. Ptuskin, H.J. Volk, Astrophys. J. **678**, 255 (2008). https://doi.org/10.1086/529579
10. H.J. Volk, E.G. Berezhko, L.T. Ksenofontov, Astron. Astrophys. **433**, 229 (2005). https://doi.org/10.1051/0004-6361:20042015
11. A. Bell, K. Schure, B. Reville, G. Giacinti, Mon. Not. R. Astron. Soc. **431**, 415 (2013). https://doi.org/10.1093/mnras/stt179
12. M. Cardillo, E. Amato, P. Blasi, Astropart. Phys. **69**, 1 (2015). https://doi.org/10.1016/j.astropartphys.2015.03.002
13. A.W. Strong, I.V. Moskalenko, Ap. J. **509**, 212 (1998). https://doi.org/10.1086/306470

14. C. Evoli, D. Gaggero, D. Grasso, L. Maccione, JCAP **10**, 018 (2008). https://doi.org/10.1088/1475-7516/2008/10/018
15. R. Kissmann, Astropart. Phys. **55**, 37 (2014). https://doi.org/10.1016/j.astropartphys.2014.02.002
16. D. Maurin, F. Donato, R. Taillet, P. Salati, Ap. J. **555**, 585 (2001). https://doi.org/10.1086/321496
17. O. Adriani et al., Science **332**, 69 (2011). https://doi.org/10.1126/science.1199172
18. M. Aguilar et al., Phys. Rev. Lett. **114**, 171103 (2015). https://doi.org/10.1103/PhysRevLett.114.171103
19. M. Aguilar et al., Phys. Rev. Lett. **115**(21), 211101 (2015). https://doi.org/10.1103/PhysRevLett.115.211101
20. S. Thoudam, J.R. Hrandel, Mon. Not. R. Astron. Soc. **435**, 2532 (2013). https://doi.org/10.1093/mnras/stt1464
21. S. Thoudam, J.R. Hrandel, Astron. Astrophys. **567**, A33 (2014). https://doi.org/10.1051/0004-6361/201322996
22. N. Tomassetti, Astrophys. J. **752**, L13 (2012). https://doi.org/10.1088/2041-8205/752/1/L13
23. P. Blasi, E. Amato, P.D. Serpico, Phys. Rev. Lett. **109**, 061101 (2012). https://doi.org/10.1103/PhysRevLett.109.061101
24. R. Aloisio, P. Blasi, JCAP **1307**, 001 (2013). https://doi.org/10.1088/1475-7516/2013/07/001
25. O. Adriani et al., Nature **458**, 607 (2009). https://doi.org/10.1038/nature07942
26. M. Aguilar et al., Phys. Rev. Lett. **110**, 141102 (2013). https://doi.org/10.1103/PhysRevLett.110.141102
27. O. Adriani et al., Phys. Rev. Lett. **102**, 051101 (2009). https://doi.org/10.1103/PhysRevLett.102.051101
28. P.D. Serpico, Astropart. Phys. **39–40**, 2 (2012). https://doi.org/10.1016/j.astropartphys.2011.08.007
29. O. Adriani et al., Phys. Rev. Lett. **111**, 081102 (2013). https://doi.org/10.1103/PhysRevLett.111.081102
30. M. Aguilar et al., Phys. Rev. Lett. **113**, 121102 (2014). https://doi.org/10.1103/PhysRevLett.113.121102
31. P. Blasi, Phys. Rev. Lett. **103**, 051104 (2009). https://doi.org/10.1103/PhysRevLett.103.051104
32. P. Blasi, P.D. Serpico, Phys. Rev. Lett. **103**, 081103 (2009). https://doi.org/10.1103/PhysRevLett.103.081103
33. D. Hooper, P. Blasi, P.D. Serpico, JCAP **0901**, 025 (2009). https://doi.org/10.1088/1475-7516/2009/01/025
34. S. Profumo, Cent. Eur. J. Phys. **10**, 1 (2012). https://doi.org/10.2478/s11534-011-0099-z
35. M. Aguilar, L. Ali Cavasonza, B. Alpat, G. Ambrosi, L. Arruda, N. Attig, S. Aupetit, P. Azzarello, A. Bachlechner, F. Barao et al., Phys. Rev. Lett. **117**(9), 091103 (2016). https://doi.org/10.1103/PhysRevLett.117.091103
36. P. Lipari, Phys. Rev. D **95**(6), 063009 (2017). https://doi.org/10.1103/PhysRevD.95.063009
37. B. Katz, K. Blum, J. Morag, E. Waxman, MNRAS **405**, 1458 (2010). https://doi.org/10.1111/j.1365-2966.2010.16568.x
38. K. Blum, B. Katz, E. Waxman, Phys. Rev. Lett. **111**(21), 211101 (2013). https://doi.org/10.1103/PhysRevLett.111.211101
39. A. Aab et al., Phys. Rev. D **90**(12), 122005 (2014). https://doi.org/10.1103/PhysRevD.90.122005
40. D. Allard, Astropart. Phys. **39–40**, 33 (2012). https://doi.org/10.1016/j.astropartphys.2011.10.011
41. K. Fang, K. Kotera, A.V. Olinto, Astrophys. J. **750**, 118 (2012).https://doi.org/10.1088/0004-637X/750/2/118
42. R. Aloisio, V. Berezinsky, P. Blasi, JCAP **1410**(10), 020 (2014).https://doi.org/10.1088/1475-7516/2014/10/020

43. R.U. Abbasi et al., Astropart. Phys. **64**, 49 (2015). https://doi.org/10.1016/j.astropartphys.2014.11.004
44. T. Kobayashi, Y. Komori, K. Yoshida, J. Nishimura, Astrophys. J. **601**, 340 (2004). https://doi.org/10.1086/380431
45. M. Ackermann et al., Phys. Rev. Lett. **108**, 081104 (2012). https://doi.org/10.1103/PhysRevLett.108.011103
46. C.R. Data Base (2017), http://lpsc.in2p3.fr/crdb
47. S. Abdollahi et al., Phys. Rev. Lett. **118**(9), 091103 (2017). https://doi.org/10.1103/PhysRevLett.118.091103
48. F. Aharonian et al., Astron. Astrophys. **508**, 561 (2009). https://doi.org/10.1051/0004-6361/200913323
49. D. Staszak, in *PoS ICRC2015*, vol. 411 (2016)
50. S. Abdollahi et al., Phys. Rev. D **95**(8), 082007 (2017). https://doi.org/10.1103/PhysRevD.95.082007
51. I. De Mitri et al., EPJ Web Conf. **136**, 02010 (2017). https://doi.org/10.1051/epjconf/201713602010
52. P.S. Marrocchesi, Nucl. Instrum. Meth. **A692**, 240 (2012). https://doi.org/10.1016/j.nima.2011.12.108
53. Y. Asaoka et al., Astropart. Phys. **91**, 1 (2017). https://doi.org/10.1016/j.astropartphys.2017.03.002
54. M. Aguilar et al., Phys. Rev. Lett. **113**, 221102 (2014). https://doi.org/10.1103/PhysRevLett.113.221102
55. K. Sakai, et al., in *PoS ICRC2015*, vol. 436 (2016)
56. T. Aramaki, C.J. Hailey, S.E. Boggs, P. von Doetinchem, H. Fuke, S.I. Mognet, R.A. Ong, K. Perez, J. Zweerink, Astropart. Phys. **74**, 6 (2016). https://doi.org/10.1016/j.astropartphys.2015.09.001
57. H.S. Ahn et al., Astrophys. J. **714**, L89 (2010). https://doi.org/10.1088/2041-8205/714/1/L89
58. Y.S. Yoon et al., Astrophys. J. **839**(1), 5 (2017). https://doi.org/10.3847/1538-4357/aa68e4
59. E. Atkin et al., Astropart. Phys. **90**, 69 (2017). https://doi.org/10.1016/j.astropartphys.2017.02.006
60. E.S. Seo, Astropart. Phys. **39–40**, 76 (2012). https://doi.org/10.1016/j.astropartphys.2012.04.002
61. N. Picot-Clemente, et al., in *PoS ICRC2015*, vol. 425 (2015)
62. A. Oliva, et al., in *PoS ICRC2015*, vol. 425 (2015)
63. G. Kulikov, G. Khristiansen, J. Exp. Theor. Phys. **35**, 635 (1958)
64. J. Blumer, R. Engel, J.R. Horandel, Prog. Part. Nucl. Phys. **63**, 293 (2009). https://doi.org/10.1016/j.ppnp.2009.05.002
65. J.R. Hoerandel, Astropart. Phys. **21**, 241 (2004). https://doi.org/10.1016/j.astropartphys.2004.01.004
66. D. d'Enterria, R. Engel, T. Pierog, S. Ostapchenko, K. Werner, Astropart. Phys. **35**, 98 (2011). https://doi.org/10.1016/j.astropartphys.2011.05.002
67. M. Boezio, E. Mocchiutti, Astropart. Phys. **39–40**, 95 (2012). https://doi.org/10.1016/j.astropartphys.2012.05.007
68. Y.S. Yoon et al., Astrophys. J. **728**, 122 (2011). https://doi.org/10.1088/0004-637X/728/2/122
69. E. Seo, in *PoS ICRC2015*, vol. 574 (2015)
70. M. Xu, Nucl. Part. Phys. Proc. **279–281**, 161 (2016). https://doi.org/10.1016/j.nuclphysbps.2016.10.023
71. J.R. Hoerandel, Astropart. Phys. **19**, 193 (2003). https://doi.org/10.1016/S0927-6505(02)00198-6
72. T.K. Gaisser, T. Stanev, S. Tilav, Front. Phys. (Beijing) **8**, 748 (2013). https://doi.org/10.1007/s11467-013-0319-7
73. W.D. Apel et al., Phys. Rev. Lett. **107**, 171104 (2011). https://doi.org/10.1103/PhysRevLett.107.171104

74. N. Budnev et al., Astropart. Phys. **50–52**, 18 (2013). https://doi.org/10.1016/j.astropartphys.2013.09.006
75. V.V. Prosin et al., Nucl. Instrum. Meth. **A756**, 94 (2014). https://doi.org/10.1016/j.nima.2013.09.018
76. M.G. Aartsen et al., Phys. Rev. D **88**(4), 042004 (2013). https://doi.org/10.1103/PhysRevD.88.042004
77. K.H. Kampert, M. Unger, Astropart. Phys. **35**, 660 (2012). https://doi.org/10.1016/j.astropartphys.2012.02.004
78. W.D. Apel et al., Astropart. Phys. **47**, 54 (2013). https://doi.org/10.1016/j.astropartphys.2013.06.004
79. W.D. Apel et al., Adv. Space Res. **53**, 1456 (2014). https://doi.org/10.1016/j.asr.2013.05.008
80. M. Aglietta et al., Astropart. Phys. **21**, 223 (2004). https://doi.org/10.1016/j.astropartphys.2004.01.005
81. M. Ambrosio et al., Phys. Rev. D **56**, 1418 (1997). https://doi.org/10.1103/PhysRevD.56.1418
82. M. Amenomori et al., Phys. Lett. B **632**, 58 (2006). https://doi.org/10.1016/j.physletb.2005.10.048
83. M. Amenomori et al., Adv. Space Res. **47**, 629 (2011). https://doi.org/10.1016/j.asr.2010.08.029
84. M.A.K. Glasmacher et al., Astropart. Phys. **12**, 1 (1999). https://doi.org/10.1016/S0927-6505(99)00076-6
85. S. Ogio et al., Astrophys. J. **612**, 268 (2004). https://doi.org/10.1086/422510
86. H. Tokuno et al., Astropart. Phys. **29**, 453 (2008). https://doi.org/10.1016/j.astropartphys.2008.05.001
87. H.T. Freudenreich, A.I. Mincer, D. Berley, J.A. Goodman, S. Tonwar, A. Wrotniak, G.B. Yodh, R.W. Ellsworth, Phys. Rev. D **41**, 2732 (1990). https://doi.org/10.1103/PhysRevD.41.2732
88. I. De Mitri, A. D'Amone, L. Perrone, A. Surdo, in *PoS ICRC2015*, vol. 366 (2016)
89. S.M. Mari, P. Montini, EPJ Web Conf. **121**, 03008 (2016). https://doi.org/10.1051/epjconf/201612103008
90. G. Di Sciascio, R. Iuppa, *in Homage to the Discovery of Cosmic Rays* (Nova Science Publishers, New York, 2013). arXiv:1407.2144
91. O. Wollan, Rev. Mod. Phys. **11**, 11 (1939)
92. M. Amenomori et al., Astropart. Phys. **36**, 237 (2012). https://doi.org/10.1016/j.astropartphys.2012.06.005
93. A.A. Abdo et al., Astrophys. J. **698**, 2121 (2009). https://doi.org/10.1088/0004-637X/698/2/2121
94. B. Bartoli et al., Astrophys. J. **809**(1), 90 (2015). https://doi.org/10.1088/0004-637X/809/1/90
95. M.G. Aartsen et al., Astrophys. J. **826**(2), 220 (2016). https://doi.org/10.3847/0004-637X/826/2/220
96. A.H. Compton, I.A. Getting, Phys. Rev. **47**(11), 817 (1935). https://doi.org/10.1103/PhysRev.47.817
97. P. Mertsch, S. Funk, Phys. Rev. Lett. **114**(2), 021101 (2015). https://doi.org/10.1103/PhysRevLett.114.021101
98. A.D. Erlykin, A.W. Wolfendale, Astropart. Phys. **25**, 183 (2006). https://doi.org/10.1016/j.astropartphys.2006.01.003
99. V.S. Ptuskin, F.C. Jones, E.S. Seo, R. Sina, Adv. Space Res. **37**, 1909 (2006). https://doi.org/10.1016/j.asr.2005.08.036
100. M. Amenomori, Science **314**, 439 (2006). https://doi.org/10.1126/SCIENCE.1131702
101. A.A. Abdo et al., Phys. Rev. Lett. **101**, 221101 (2008). https://doi.org/10.1103/PhysRevLett.101.221101
102. B. Bartoli et al., Phys. Rev. D **88**(8), 082001 (2013). https://doi.org/10.1103/PhysRevD.88.082001

103. A.U. Abeysekara et al., Astrophys. J. **796**(2), 108 (2014). https://doi.org/10.1088/0004-637X/796/2/108
104. G. Giacinti, G. Sigl, Phys. Rev. Lett. **109**(7), 071101 (2012). https://doi.org/10.1103/PhysRevLett.109.071101
105. M. Ahlers, Phys. Rev. Lett. **112**(2), 021101 (2014). https://doi.org/10.1103/PhysRevLett.112.021101
106. A. Aab et al., Phys. Rev. Lett. **116**(24), 241101 (2016). https://doi.org/10.1103/PhysRevLett.116.241101
107. S. Vernetto, J. Phys. Conf. Ser. **718**(5), 052043 (2016). https://doi.org/10.1088/1742-6596/718/5/052043
108. R.U. Abbasi et al., Phys. Rev. Lett. **100**, 101101 (2008). https://doi.org/10.1103/PhysRevLett.100.101101
109. M. Takeda et al., Astropart. Phys. **19**, 447 (2003). https://doi.org/10.1016/S0927-6505(02)00243-8
110. T. Abu-Zayyad et al., Phys. Rev. Lett. **84**, 4276 (2000). https://doi.org/10.1103/PhysRevLett.84.4276
111. P. Abreu et al., Astropart. Phys. **34**, 368 (2011). https://doi.org/10.1016/j.astropartphys.2010.10.001
112. A. Aab et al., Nucl. Instrum. Meth. **A798**, 172 (2015). https://doi.org/10.1016/j.nima.2015.06.058
113. T. Abu-Zayyad et al., Nucl. Instrum. Meth. **A689**, 87 (2012). https://doi.org/10.1016/j.nima.2012.05.079
114. E. Zas, J. Phys. Conf. Ser. **632**(1), 012104 (2015). https://doi.org/10.1088/1742-6596/632/1/012104
115. V. Verzi, Rapporteur report at ICRC 2015 (2015)
116. J. Abraham et al., Phys. Rev. Lett. **101**, 061101 (2008). https://doi.org/10.1103/PhysRevLett.101.061101
117. T. Abu-Zayyad et al., Astrophys. J. **768**, L1 (2013). https://doi.org/10.1088/2041-8205/768/1/L1
118. *The Pierre Auger Observatory: Contributions to the 34th International Cosmic Ray Conference (ICRC 2015)*
119. A. Aab et al., Astrophys. J. **802**(2), 111 (2015). https://doi.org/10.1088/0004-637X/802/2/111
120. P. Abreu et al., JCAP **1302**, 026 (2013). https://doi.org/10.1088/1475-7516/2013/02/026
121. A. Aab et al., Phys. Rev. D **90**(12), 122006 (2014). https://doi.org/10.1103/PhysRevD.90.122006
122. A. Aab et al., Astrophys. J. **804**(1), 15 (2015). https://doi.org/10.1088/0004-637X/804/1/15
123. R.U. Abbasi et al., Astrophys. J. **790**, L21 (2014). https://doi.org/10.1088/2041-8205/790/2/L21
124. M.G. Aartsen et al., JCAP **1601**(01), 037 (2016). https://doi.org/10.1088/1475-7516/2016/01/037
125. M. Kachelriess, P.D. Serpico, Phys. Lett. B **640**, 225 (2006). https://doi.org/10.1016/j.physletb.2006.08.006
126. P. Abreu et al., Astropart. Phys. **34**, 627 (2011). https://doi.org/10.1016/j.astropartphys.2010.12.007
127. A. Aab et al., Astrophys. J. **794**(2), 172 (2014). https://doi.org/10.1088/0004-637X/794/2/172
128. R.U. Abbasi, et al., Pierre Auger and Telescope Array Collaborations, in *Joint Contributions to the 34th International Cosmic Ray Conference (ICRC 2015)* (2015), arXiv:1511.02103
129. A. Aab, et al. (2016), arXiv:1604.03637
130. H. Sagawa, JPS Conf. Proc. **9**, 010014 (2016). https://doi.org/10.7566/JPSCP.9.010014
131. T. Nonaka, M. Takamura, K. Honda, J.N. Matthews, S. Ogio, N. Sakurai, H. Sagawa, B.T. Stokes, M. Tsujimoto, K. Yashiro, JPS Conf. Proc. **9**, 010013 (2016). https://doi.org/10.7566/JPSCP.9.010013

132. F. Fenu (2017), arXiv:1703.01875
133. T. Ebisuzaki, G. Medina-Tanco, A. Santangelo, Adv. Space Res. **53**, 1499 (2014). https://doi.org/10.1016/j.asr.2013.11.056
134. M. Potgieter, Living Rev. Solar Phys. **10**, 3 (2013). https://doi.org/10.12942/lrsp-2013-3
135. E.C. Stone, A.C. Cummings, F.B. McDonald, B.C. Heikkila, N. Lal, W.R. Webber, Science **341**, 150 (2013). https://doi.org/10.1126/science.1236408
136. J. Skilling, MNRAS **173**, 255 (1975). https://doi.org/10.1093/mnras/173.2.255
137. R.M. Kulsrud, C.J. Cesarsky, ApJ Lett. **8**, 189 (1971)
138. Y. Zhou, W.H. Matthaeus, J. Geoph. Res. **95**, 14881 (1990). https://doi.org/10.1029/JA095iA09p14881
139. Y. Zhou, W.H. Matthaeus, JGR **95**, 10291 (1990). https://doi.org/10.1029/JA095iA07p10291
140. J. Skilling, Astrophys. J. **170**, 265 (1971). https://doi.org/10.1086/151210
141. J.A. Holmes, MNRAS **170**, 251 (1975). https://doi.org/10.1093/mnras/170.2.251
142. K.M. Ferriere, Rev. Mod. Phys. **73**, 1031 (2001). https://doi.org/10.1103/RevModPhys.73.1031
143. V.S. Ptuskin, V.N. Zirakashvili, A&A **403**, 1 (2003). https://doi.org/10.1051/0004-6361:20030323
144. R. Aloisio, P. Blasi, P. Serpico, Astron. Astrophys. **583**, A95 (2015). https://doi.org/10.1051/0004-6361/201526877
145. F. Acero et al., Astrophys. J. Suppl. **223**(2), 26 (2016). https://doi.org/10.3847/0067-0049/223/2/26
146. R.z. Yang, F. Aharonian, C. Evoli (2016), arXiv:1602.04710
147. F.W. Stecker, F.C. Jones, Astrophys. J. **217**, 843 (1977). https://doi.org/10.1086/155631
148. M. Ackermann et al., Astrophys. J. **726**, 81 (2011). https://doi.org/10.1088/0004-637X/726/2/81
149. M. Ackermann et al., Astron. Astrophys. **538**, A71 (2012). https://doi.org/10.1051/0004-6361/201117539
150. V.S. Berezinsky, S.V. Bulanov, V.A. Dogiel, V.L. Ginzburg, V.S. Ptuskin, *Astrophysics of Cosmic Rays* (Amsterdam, 1990) 534 p
151. J.B.G.M. Bloemen, V.A. Dogiel, V.L. Dorman, V.S. Ptuskin, A&A **267**, 372 (1993)
152. A.W. Strong, I.V. Moskalenko, O. Reimer, S. Digel, R. Diehl, Astron. Astrophys. **422**, L47 (2004). https://doi.org/10.1051/0004-6361:20040172
153. A. Erlykin, A. Wolfendale, V. Dogiel, Adv. Space Res. **57**, 519 (2016). https://doi.org/10.1016/j.asr.2015.10.010
154. C. Evoli, D. Gaggero, D. Grasso, L. Maccione, Phys. Rev. Lett. **108**, 211102 (2012). https://doi.org/10.1103/PhysRevLett.108.211102
155. D. Gaggero, A. Urbano, M. Valli, P. Ullio, Phys. Rev. D **91**(8), 083012 (2015). https://doi.org/10.1103/PhysRevD.91.083012
156. S. Recchia, P. Blasi, G. Morlino (2015), arXiv:1604.07682
157. R.A. Crain, V.R. Eke, C.S. Frenk, A. Jenkins, I.G. McCarthy, J.F. Navarro, F.R. Pearce, Mon. Not. R. Astron. Soc. **377**, 41 (2007). https://doi.org/10.1111/j.1365-2966.2007.11598.x
158. G. Stinson, C. Brook, A.V. Maccio, J. Wadsley, T.R. Quinn, H.M.P. Couchman, Mon. Not. R. Astron. Soc. **428**, 129 (2013). https://doi.org/10.1093/mnras/sts028
159. P.M.W. Kalberla, G. Westphalen, U. Mebold, D. Hartmann, W.B. Burton, A&A **332**, L61 (1998)
160. P.M.W. Kalberla, L. Dedes, Astron. Astrophys. **487**, 951 (2008). https://doi.org/10.1051/0004-6361:20079240
161. M.J. Miller, J.N. Bregman, Astrophys. J. **770**, 118 (2013). https://doi.org/10.1088/0004-637X/770/2/118
162. S.L. Snowden, M.J. Freyberg, P.P. Plucinsky, J.H.M.M. Schmitt, J. Truemper, W. Voges, R.J. Edgar, D. McCammon, W.T. Sanders, ApJ **454**, 643 (1995). https://doi.org/10.1086/176517
163. D. Breitschwerdt, T. Schmutzler, Nature **371**, 774 (1994). https://doi.org/10.1038/371774a0
164. D. Breitschwerdt, T. Schmutzler, Astron. Astrophys. **347**, 650 (1999)

165. J.E. Everett, E.G. Zweibel, R.A. Benjamin, D. McCammon, L. Rocks, J.S. Gallagher, Astrophys. J. **674**, 258 (2008). https://doi.org/10.1086/524766
166. R.A. Chevalier, A.W. Clegg, Nature **317**, 44 (1985). https://doi.org/10.1038/317044a0
167. N. Scoville, J. Korean Astronom. Soc. **36**, 167 (2003). https://doi.org/10.5303/JKAS.2003.36.3.167
168. N. Murray, E. Quataert, T.A. Thompson, Astrophys. J. **618**, 569 (2005). https://doi.org/10.1086/426067
169. S. Veilleux, G. Cecil, J. Bland-Hawthorn, Ann. Rev. Astron. Astrophys. **43**, 769 (2005). https://doi.org/10.1146/annurev.astro.43.072103.150610
170. K.S. Cheng, D.O. Chernyshov, V.A. Dogiel, C.M. Ko, W.H. Ip, Astrophys. J. **731**, L17 (2011). https://doi.org/10.1088/2041-8205/731/1/L17
171. K. Zubovas, A.R. King, S. Nayakshin, Mon. Not. R. Astron. Soc. **415**, 21 (2011). https://doi.org/10.1111/j.1745-3933.2011.01070.x
172. B.C. Lacki, Mon. Not. R. Astron. Soc. (2013). https://doi.org/10.1093/mnrasl/slu107
173. F.M. Ipavich, Astrophys. J. **196**, 107 (1975). https://doi.org/10.1086/153397
174. D. Breitschwerdt, J.F. McKenzie, H.J. Voelk, A&A **245**, 79 (1991)
175. D. Breitschwerdt, J.F. McKenzie, H.J. Voelk, Astron. Astrophys. (ISSN 0004-6361) **269**, 54 (1993)
176. M. Uhlig, C. Pfrommer, M. Sharma, B.B. Nath, T.A. Ensslin, V. Springel, Mon. Not. R. Astron. Soc. **423**, 2374 (2012). https://doi.org/10.1111/j.1365-2966.2012.21045.x
177. C.M. Booth, O. Agertz, A.V. Kravtsov, N.Y. Gnedin, Astrophys. J. **777**, L16 (2013). https://doi.org/10.1088/2041-8205/777/1/L16
178. M. Salem, G.L. Bryan, Mon. Not. R. Astron. Soc. **437**(4), 3312 (2014). https://doi.org/10.1093/mnras/stt2121
179. P. Girichidis, T. Naab, S. Walch, M. Hanasz, M.M. Mac Low, J.P. Ostriker, A. Gatto, T. Peters, R. Wünsch, S.C.O. Glover, R.S. Klessen, P.C. Clark, C. Baczynski, ApJ Lett. **816**, L19 (2016). https://doi.org/10.3847/2041-8205/816/2/L19
180. T. Peters, P. Girichidis, A. Gatto, T. Naab, S. Walch, R. Wnsch, S.C.O. Glover, P.C. Clark, R.S. Klessen, C. Baczynski. Astrophys. J. **813**(2), L27 (2015). https://doi.org/10.1088/2041-8205/813/2/L27
181. M. Ruszkowski, H.Y.K. Yang, E. Zweibel (2016), arXiv:1602.04856
182. V.S. Ptuskin, H.J. Voelk, V.N. Zirakashvili, D. Breitschwerdt, A&A **321**, 434 (1997)
183. S. Recchia, P. Blasi, G. Morlino, MNRAS **462**, 4227 (2016). https://doi.org/10.1093/mnras/stw1966
184. P.D. Serpico, in *PoS ICRC2015*, vol. 009 (2016)
185. M. di Mauro, F. Donato, A. Goudelis, P.D. Serpico, PRD **90**(8), 085017 (2014). https://doi.org/10.1103/PhysRevD.90.085017
186. O. Adriani et al., Phys. Rev. Lett. **106**, 201101 (2011). https://doi.org/10.1103/PhysRevLett.106.201101
187. M. Aguilar, D. Aisa, A. Alvino, G. Ambrosi, K. Andeen, L. Arruda, N. Attig, P. Azzarello, A. Bachlechner, F. Barao et al., Phys. Rev. Lett. **113**(12), 121102 (2014). https://doi.org/10.1103/PhysRevLett.113.121102
188. P. Mertsch, S. Sarkar, Phys. Rev. Lett. **103**, 081104 (2009). https://doi.org/10.1103/PhysRevLett.103.081104
189. E. Amato, Int. J. Mod. Phys. Conf. Ser. **28**, 1460160 (2014). https://doi.org/10.1142/S2010194514601604
190. F.A. Aharonian, A.M. Atoyan, H.J. Voelk, A&A **294**, L41 (1995)
191. P. Blasi, E. Amato, Astrophys. Space Sci. Proc. 623–641 (2011). https://doi.org/10.1007/978-3-642-17251-9_50
192. T. Linden, S. Profumo, Astrophys. J. **772**, 18 (2013). https://doi.org/10.1088/0004-637X/772/1/18
193. R. Cowsik, B. Burch, PRD **82**(2), 023009 (2010). https://doi.org/10.1103/PhysRevD.82.023009

194. R. Cowsik, T. Madziwa-Nussinov, Ap. J. **827**, 119 (2016). https://doi.org/10.3847/0004-637X/827/2/119
195. M. D'Angelo, P. Blasi, E. Amato, PRD **94**(8), 083003 (2016). https://doi.org/10.1103/PhysRevD.94.083003
196. G.F. Krymskii, Akademiia Nauk SSSR Doklady **234**, 1306 (1977)
197. R.D. Blandford, J.P. Ostriker, Astrophys. J. **221**, L29 (1978). https://doi.org/10.1086/182658
198. W.I. Axford, E. Leer, G. Skadron, Int. Cosm. Ray Conf. **11**, 132 (1977)
199. A.R. Bell, Mon. Not. R. Astron. Soc. **182**, 443 (1978)
200. M. Malkov, L.O. Drury, Rep. Prog. Phys. **64**(4), 429 (2001). https://doi.org/10.1088/0034-4885/64/4/201
201. L.O. Drury, J.H. Voelk, ApJ **248**, 344 (1981). https://doi.org/10.1086/159159
202. W.I. Axford, E. Leer, J.F. McKenzie, A&A **111**, 317 (1982)
203. E.G. Berezhko, D.C. Ellison, ApJ **526**, 385 (1999). https://doi.org/10.1086/307993
204. E.G. Berezhko, V.K. Elshin, L.T. Ksenofontov, Astropart. Phys. **2**, 215 (1994). https://doi.org/10.1016/0927-6505(94)90043-4
205. E.G. Berezhko, H.J. Volk, Astropart. Phys. **7**, 183 (1997). https://doi.org/10.1016/S0927-6505(97)00016-9
206. M.A. Malkov, Astrophys. J. **511**, L53 (1999). https://doi.org/10.1086/311825
207. P. Blasi, Astropart. Phys. **16**, 429 (2002). https://doi.org/10.1016/S0927-6505(01)00127-X
208. J. Ballet, Adv. Space Res. **37**, 1902 (2006). https://doi.org/10.1016/j.asr.2005.03.047
209. D. Caprioli, E. Amato, P. Blasi, Astropart. Phys. **33**, 160 (2010). https://doi.org/10.1016/j.astropartphys.2010.01.002
210. P. Mertsch, S. Funk, Phys. Rev. Lett. **114**(2), 021101 (2015). https://doi.org/10.1103/PhysRevLett.114.021101
211. D. Caprioli, JCAP **1105**, 026 (2011). https://doi.org/10.1088/1475-7516/2011/05/026
212. F. Acero et al., Astrophys. J. Suppl. **224**(1), 8 (2016). https://doi.org/10.3847/0067-0049/224/1/8
213. E.G. Berezhko, L.T. Ksenofontov, H.J. Vlk, Astrophys. J. **763**, 14 (2013). https://doi.org/10.1088/0004-637X/763/1/14
214. P. Blasi, G. Morlino, R. Bandiera, E. Amato, D. Caprioli, Astrophys. J. **755**, 121 (2012). https://doi.org/10.1088/0004-637X/755/2/121
215. G. Morlino, P. Blasi, Astron. Astrophys. **589**, A7 (2016). https://doi.org/10.1051/0004-6361/201527761
216. D. Caprioli, JCAP **1207**, 038 (2012). https://doi.org/10.1088/1475-7516/2012/07/038
217. G. Morlino, D. Caprioli, Astron. Astrophys. **538**, A81 (2012). https://doi.org/10.1051/0004-6361/201117855
218. K.M. Schure, A.R. Bell, Mon. Not. R. Astron. Soc. **435**, 1174 (2013). https://doi.org/10.1093/mnras/stt1371
219. P.O. Lagage, C.J. Cesarsky, Astron. Astrophys. **118**, 223 (1983)
220. J. Giacalone, J.R. Jokipii, Astrophys. J. **663**, L41 (2007). https://doi.org/10.1086/519994
221. A.R. Bell, Mon. Not. R. Astron. Soc. **353**, 550 (2004). https://doi.org/10.1111/j.1365-2966.2004.08097.x
222. D. Caprioli, in *PoS ICRC2015*, vol. 008 (2015)
223. D. Caprioli, A. Spitkovsky, Astrophys. J. **783**, 91 (2014). https://doi.org/10.1088/0004-637X/783/2/91
224. D. Caprioli, A. Spitkovsky, Astrophys. J. **794**(1), 46 (2014). https://doi.org/10.1088/0004-637X/794/1/46
225. E. Amato, P. Blasi, Mon. Not. R. Astron. Soc. **392**, 1591 (2009). https://doi.org/10.1111/j.1365-2966.2008.14200.x
226. D. Caprioli, A.R. Pop, A. Spitkovsky, Astrophys. J. **798**(2), L28 (2015). https://doi.org/10.1088/2041-8205/798/2/L28
227. J. Park, D. Caprioli, A. Spitkovsky, Phys. Rev. Lett. **114**(8), 085003 (2015). https://doi.org/10.1103/PhysRevLett.114.085003
228. A. Abramowski et al., Nature **531**, 476 (2016). https://doi.org/10.1038/nature17147

229. G. Morlino, P. Blasi, R. Bandiera, E. Amato, D. Caprioli, Astrophys. J. **768**, 148 (2013). https://doi.org/10.1088/0004-637X/768/2/148
230. G. Morlino, R. Bandiera, P. Blasi, E. Amato, Astrophys. J. **760**, 137 (2012). https://doi.org/10.1088/0004-637X/760/2/137
231. G. Morlino, P. Blasi, R. Bandiera, E. Amato, Astron. Astrophys. **558**, A25 (2013). https://doi.org/10.1051/0004-6361/201322006
232. G. Morlino, P. Blasi, R. Bandiera, E. Amato, Astron. Astrophys. **562**, A141 (2014). https://doi.org/10.1051/0004-6361/201322986
233. G. Morlino, P. Blasi, R. Bandiera, E. Amato, Astron. Astrophys. **557**, A142 (2013). https://doi.org/10.1051/0004-6361/201322161
234. P. Ghavamian, S.J. Schwartz, J. Mitchell, A. Masters, J.M. Laming, Space Sci. Rev. **178**, 633 (2013). https://doi.org/10.1007/s11214-013-9999-0
235. R. Aloisio, V. Berezinsky, A. Gazizov, Astropart. Phys. **39–40**, 129 (2012). https://doi.org/10.1016/j.astropartphys.2012.09.007
236. A. Franceschini, G. Rodighiero, M. Vaccari, Astron. Astrophys. **487**, 837 (2008). https://doi.org/10.1051/0004-6361:200809691
237. F.W. Stecker, M. Malkan, S. Scully, Astrophys. J. **658**, 1392 (2007). https://doi.org/10.1086/511738
238. F.W. Stecker, M. Malkan, S. Scully, Astrophys. J. **648**, 774 (2006). https://doi.org/10.1086/506188
239. T.M. Kneiske, T. Bretz, K. Mannheim, D. Hartmann, Astron. Astrophys. **413**, 807 (2004). https://doi.org/10.1051/0004-6361:20031542
240. R. Aloisio, D. Boncioli, A. di Matteo, A.F. Grillo, S. Petrera, F. Salamida, JCAP **1510**(10), 006 (2015). https://doi.org/10.1088/1475-7516/2015/10/006
241. R. Aloisio, V. Berezinsky, S. Grigorieva, Astropart. Phys. **41**, 73 (2013). https://doi.org/10.1016/j.astropartphys.2012.07.010
242. R. Aloisio, V. Berezinsky, S. Grigorieva, Astropart. Phys. **41**, 94 (2013). https://doi.org/10.1016/j.astropartphys.2012.06.003
243. A.A. Penzias, R.W. Wilson, Astrophys. J. **142**, 419 (1965). https://doi.org/10.1086/148307
244. K. Greisen, Phys. Rev. Lett. **16**, 748 (1966). https://doi.org/10.1103/PhysRevLett.16.748
245. G. Zatsepin, V. Kuzmin, JETP Lett. **4**, 78 (1966)
246. A.M. Hillas, Phys. Lett. A **24**, 677 (1967)
247. G.R. Blumenthal, Phys. Rev. D **1**, 1596 (1970). https://doi.org/10.1103/PhysRevD.1.1596
248. V. Berezinsky, G. Zatsepin, Phys. Lett. B **28**, 423 (1969). https://doi.org/10.1016/0370-2693(69)90341-4
249. F. Stecker, Astrophys. Space Sci. **20**, 47 (1973). https://doi.org/10.1007/BF00645585
250. A. Strong, A. Wolfendale, J. Wdowczy, Nature **241**, 109 (1973)
251. V. Berezinsky, A.Y. Smirnov, Astrophys. Space Sci. **32**, 461 (1975). https://doi.org/10.1007/BF00643157
252. F.W. Stecker, Phys. Rev. Lett. **21**, 1016 (1968). https://doi.org/10.1103/PhysRevLett.21.1016
253. V. Berezinsky, A. Gazizov, S. Grigorieva, Phys. Rev. D **74**, 043005 (2006). https://doi.org/10.1103/PhysRevD.74.043005
254. J. Puget, F. Stecker, J. Bredekamp, Astrophys. J. **205**, 638 (1976). https://doi.org/10.1086/154321
255. D. Allard, E. Parizot, E. Khan, S. Goriely, A. Olinto, Astron. Astrophys. **443**, L29 (2005). https://doi.org/10.1051/0004-6361:200500199
256. P.A.R. Ade, et al., Planck collaboration, eprint 1502.01589 (2015)
257. J. Linsley, in *Proceedings of 8th ICRC (Jaipur 1963)*, vol. 4, p. 77 (1963)
258. R.U. Abbasi et al., Astropart. Phys. **68**, 27 (2015). https://doi.org/10.1016/j.astropartphys.2015.02.008
259. R.U. Abbasi et al., Phys. Rev. Lett. **92**, 151101 (2004). https://doi.org/10.1103/PhysRevLett.92.151101
260. V. Egorova et al., [Yakutsk Collaboration] Nucl. Phys. Proc. Suppl. **3**, 136 (2004)
261. K. Shinozaki et al., [AGASA Collaboration] Nucl. Phys. Proc. Suppl. **3**, 151 (2006)

262. M. Honda et al., [Akeno Collaboration] Phys. Rev. D **70**, 525 (1993)
263. C.C.H. Jui, J. Phys. Conf. Ser. **404**, 012037 (2012). https://doi.org/10.1088/1742-6596/404/1/012037
264. R. Aloisio, V. Berezinsky, P. Blasi, A. Gazizov, S. Grigorieva, B. Hnatyk, Astropart. Phys. **27**, 76 (2007). https://doi.org/10.1016/j.astropartphys.2006.09.004
265. V. Berezinsky, A.Z. Gazizov, S.I. Grigorieva, Phys. Lett. B **612**, 147 (2005). https://doi.org/10.1016/j.physletb.2005.02.058
266. R. Aloisio, V. Berezinsky, P. Blasi, S. Ostapchenko, Phys. Rev. D **77**, 025007 (2008). https://doi.org/10.1103/PhysRevD.77.025007
267. D. De Marco, T. Stanev, Phys. Rev. D **72**, 081301 (2005). https://doi.org/10.1103/PhysRevD.72.081301
268. M. Kachelriess, D.V. Semikoz, Phys. Lett. B **634**, 143 (2006). https://doi.org/10.1016/j.physletb.2006.01.009
269. R. Aloisio, V.S. Berezinsky, Astrophys. J. **625**, 249 (2005). https://doi.org/10.1086/429615
270. R. Aloisio, V. Berezinsky, A. Gazizov, Astropart. Phys. **34**, 620 (2011). https://doi.org/10.1016/j.astropartphys.2010.12.008
271. A.M. Taylor, Astropart. Phys. **54**, 48 (2014). https://doi.org/10.1016/j.astropartphys.2013.11.006
272. A. Aab et al., JCAP **1704**(04), 038 (2017). https://doi.org/10.1088/1475-7516/2017/04/038
273. N. Globus, D. Allard, E. Parizot, Phys. Rev. D **92**(2), 021302 (2015). https://doi.org/10.1103/PhysRevD.92.021302
274. P. Abreu et al., Astrophys. J. **762**, L13 (2012). https://doi.org/10.1088/2041-8205/762/1/L13
275. G. Giacinti, M. Kachelriess, D.V. Semikoz, G. Sigl, JCAP **1207**, 031 (2012). https://doi.org/10.1088/1475-7516/2012/07/031
276. N. Globus, D. Allard, R. Mochkovitch, E. Parizot, Mon. Not. R. Astron. Soc. **451**(1), 751 (2015). https://doi.org/10.1093/mnras/stv893
277. M. Unger, G.R. Farrar, L.A. Anchordoqui, Phys. Rev. D **92**(12), 123001 (2015). https://doi.org/10.1103/PhysRevD.92.123001
278. G.B. Gelmini, O. Kalashev, D.V. Semikoz, JCAP **1201**, 044 (2012). https://doi.org/10.1088/1475-7516/2012/01/044
279. H. Yuksel, M.D. Kistler, J.F. Beacom, A.M. Hopkins, Astrophys. J. **683**, L5 (2008). https://doi.org/10.1086/591449
280. X.Y. Wang, R.Y. Liu, F. Aharnonian, Astrophys. J. **736**, 112 (2011). https://doi.org/10.1088/0004-637X/736/2/112
281. G. Hasinger, T. Miyaji, M. Schmidt, Astron. Astrophys. **441**, 417 (2005). https://doi.org/10.1051/0004-6361:20042134
282. M. Ahlers, L.A. Anchordoqui, S. Sarkar, Phys. Rev. D **79**, 083009 (2009). https://doi.org/10.1103/PhysRevD.79.083009
283. R. Engel, D. Heck, T. Pierog, Ann. Rev. Nucl. Part. Sci. **61**, 467 (2011). https://doi.org/10.1146/annurev.nucl.012809.104544
284. T. Pierog, K. Werner, Phys. Rev. Lett. **101**, 171101 (2008). https://doi.org/10.1103/PhysRevLett.101.171101
285. E.J. Ahn, R. Engel, T.K. Gaisser, P. Lipari, T. Stanev, Phys. Rev. D **80**, 094003 (2009). https://doi.org/10.1103/PhysRevD.80.094003
286. N. Kalmykov, S. Ostapchenko, A. Pavlov, Nucl. Phys. Proc. Suppl. **52B**, 17 (1997)
287. S. Ostapchenko, Phys. Rev. D **74**, 014026 (2006). https://doi.org/10.1103/PhysRevD.74.014026
288. R. Engel, D. Seckel, T. Stanev, Phys. Rev. D **64**, 093010 (2001). https://doi.org/10.1103/PhysRevD.64.093010
289. O.E. Kalashev, V.A. Kuzmin, D.V. Semikoz, G. Sigl, Phys. Rev. D **66**, 063004 (2002). https://doi.org/10.1103/PhysRevD.66.063004
290. D. Hooper, A. Taylor, S. Sarkar, Astropart. Phys. **23**, 11 (2005). https://doi.org/10.1016/j.astropartphys.2004.11.002

291. D. Seckel, T. Stanev, Phys. Rev. Lett. **95**, 141101 (2005). https://doi.org/10.1103/PhysRevLett.95.141101
292. D. De Marco, T. Stanev, F.W. Stecker, Phys. Rev. D **73**, 043003 (2006). https://doi.org/10.1103/PhysRevD.73.043003
293. D. Allard, M. Ave, N. Busca, M. Malkan, A. Olinto et al., JCAP **0609**, 005 (2006). https://doi.org/10.1088/1475-7516/2006/09/005
294. L.A. Anchordoqui, H. Goldberg, D. Hooper, S. Sarkar, A.M. Taylor, Phys. Rev. D **76**, 123008 (2007). https://doi.org/10.1103/PhysRevD.76.123008
295. H. Takami, K. Murase, S. Nagataki, K. Sato, Astropart. Phys. **31**, 201 (2009). https://doi.org/10.1016/j.astropartphys.2009.01.006
296. K. Kotera, D. Allard, A.V. Olinto, JCAP **1010**, 013 (2010). https://doi.org/10.1088/1475-7516/2010/10/013
297. V. Berezinsky, A. Gazizov, M. Kachelriess, S. Ostapchenko, Phys. Lett. B **695**, 13 (2011). https://doi.org/10.1016/j.physletb.2010.11.019
298. T. Stanev, Comptes Rendus Physique **15**, 349 (2014). https://doi.org/10.1016/j.crhy.2014.02.013
299. V. Berezinsky, O. Kalashev (2016), arXiv:1603.03989
300. M. Aartsen et al., Science **342**, 1242856 (2013). https://doi.org/10.1126/science.1242856
301. M. Aartsen et al., Phys. Rev. Lett. **111**, 021103 (2013). https://doi.org/10.1103/PhysRevLett.111.021103
302. P. Abreu et al., Adv. High Energy Phys. **2013**, 708680 (2013). https://doi.org/10.1155/2013/708680
303. V. Berezinsky, P. Blasi, Phys. Rev. D **85**, 123003 (2012). https://doi.org/10.1103/PhysRevD.85.123003
304. T.J. Weiler, Phys. Rev. Lett. **49**, 234 (1982). https://doi.org/10.1103/PhysRevLett.49.234
305. P. Gondolo, G. Gelmini, S. Sarkar, Nucl. Phys. B **392**, 111 (1993). https://doi.org/10.1016/0550-3213(93)90199-Y
306. G.R. Blumenthal, R.J. Gould, Rev. Mod. Phys. **42**, 237 (1970). https://doi.org/10.1103/RevModPhys.42.237
307. R.Y. Liu, A.M. Taylor, X.Y. Wang, F.A. Aharonian (2016), arXiv:1603.03223
308. M. Ackermann et al., Astrophys. J. **799**, 86 (2015). https://doi.org/10.1088/0004-637X/799/1/86
309. M. Ackermann et al., Phys. Rev. Lett. **116**(15), 151105 (2016). https://doi.org/10.1103/PhysRevLett.116.151105
310. C. Ferrigno, P. Blasi, D. De Marco, Astropart. Phys. **23**, 211 (2005). https://doi.org/10.1016/j.astropartphys.2004.04.013
311. M. Ahlers, L.A. Anchordoqui, M.C. Gonzalez-Garcia, F. Halzen, S. Sarkar, Astropart. Phys. **34**, 106 (2010). https://doi.org/10.1016/j.astropartphys.2010.06.003
312. A.A. Abdo et al., Phys. Rev. Lett. **104**, 101101 (2010). https://doi.org/10.1103/PhysRevLett.104.101101
313. V. Berezinsky, A. Gazizov, O. Kalashev, Astropart. Phys. **84**, 52 (2016). https://doi.org/10.1016/j.astropartphys.2016.08.007
314. V. Berezinsky, M. Kachelriess, A. Vilenkin, Phys. Rev. Lett. **79**, 4302 (1997). https://doi.org/10.1103/PhysRevLett.79.4302
315. V. Berezinsky, P. Blasi, A. Vilenkin, Phys. Rev. D **58**, 103515 (1998). https://doi.org/10.1103/PhysRevD.58.103515
316. G. Sigl, S. Lee, P. Bhattacharjee, S. Yoshida, Phys. Rev. D **59**, 043504 (1999). https://doi.org/10.1103/PhysRevD.59.043504
317. R. Aloisio, V. Berezinsky, M. Kachelriess, Phys. Rev. D **69**, 094023 (2004). https://doi.org/10.1103/PhysRevD.69.094023
318. R. Aloisio, V. Berezinsky, M. Kachelriess, Phys. Rev. D **74**, 023516 (2006). https://doi.org/10.1103/PhysRevD.74.023516
319. R. Aloisio, S. Matarrese, A.V. Olinto, JCAP **1508**(08), 024 (2015). https://doi.org/10.1088/1475-7516/2015/08/024

320. S. Yoshida, M. Teshima, Prog. Theoret. Phys. **89**, 833 (1993). https://doi.org/10.1143/PTP.89.833
321. A.M. Taylor, J.A. Hinton, P. Blasi, M. Ave, Phys. Rev. Lett. **103**, 051102 (2009). https://doi.org/10.1103/PhysRevLett.103.051102
322. W. Essey, O.E. Kalashev, A. Kusenko, J.F. Beacom, Phys. Rev. Lett. **104**, 141102 (2010). https://doi.org/10.1103/PhysRevLett.104.141102
323. W. Essey, O. Kalashev, A. Kusenko, J.F. Beacom, Astrophys. J. **731**, 51 (2011). https://doi.org/10.1088/0004-637X/731/1/51
324. T. Stanev, Astrophys. J. **479**, 290 (1997). https://doi.org/10.1086/303866
325. D. Harari, S. Mollerach, E. Roulet, JHEP **02**, 035 (2000). https://doi.org/10.1088/1126-6708/2000/02/035
326. M. Prouza, R. Smida, Astron. Astrophys. **410**, 1 (2003). https://doi.org/10.1051/0004-6361:20031281
327. H. Yoshiguchi, S. Nagataki, K. Sato, Astrophys. J. **596**, 1044 (2003). https://doi.org/10.1086/378201
328. M. Kachelriess, P.D. Serpico, M. Teshima, Astropart. Phys. **26**, 378 (2006). https://doi.org/10.1016/j.astropartphys.2006.08.004
329. H. Takami, H. Yoshiguchi, K. Sato, Astrophys. J. **639**, 803 (2006). https://doi.org/10.1086/499420. [Erratum: Astrophys. J. 653, 1584 (2006)]
330. P.P. Kronberg, Rept. Prog. Phys. **57**, 325 (1994). https://doi.org/10.1088/0034-4885/57/4/001
331. D. Grasso, H.R. Rubinstein, Phys. Rept. **348**, 163 (2001). https://doi.org/10.1016/S0370-1573(00)00110-1
332. C.L. Carilli, G.B. Taylor, Ann. Rev. Astron. Astrophys. **40**, 319 (2002). https://doi.org/10.1146/annurev.astro.40.060401.093852
333. R.M. Kulsrud, E.G. Zweibel, Rep. Prog. Phys. **71**, 0046091 (2008). https://doi.org/10.1088/0034-4885/71/4/046901
334. R. Beck, A.I.P. Conf, Proc. **1381**, 117 (2011). https://doi.org/10.1063/1.3635828
335. D. Ryu, D.R.G. Schleicher, R.A. Treumann, C.G. Tsagas, L.M. Widrow, Space Sci. Rev. **166**, 1 (2012). https://doi.org/10.1007/s11214-011-9839-z
336. R. Durrer, A. Neronov, Astron. Astrophys. Rev. **21**, 62 (2013). https://doi.org/10.1007/s00159-013-0062-7
337. P. Blasi, S. Burles, A.V. Olinto, Astrophys. J. **514**, L79 (1999). https://doi.org/10.1086/311958
338. A. Elyiv, A. Neronov, D.V. Semikoz, Phys. Rev. D **80**, 023010 (2009). https://doi.org/10.1103/PhysRevD.80.023010
339. A. Neronov, D.V. Semikoz, Phys. Rev. D **80**, 123012 (2009). https://doi.org/10.1103/PhysRevD.80.123012
340. K. Dolag, M. Bartelmann, H. Lesch, Astron. Astrophys. **387**, 383 (2002). https://doi.org/10.1051/0004-6361:20020241
341. K. Dolag, D. Grasso, V. Springel, I. Tkachev, JCAP **0501**, 009 (2005). https://doi.org/10.1088/1475-7516/2005/01/009
342. G. Sigl, F. Miniati, T.A. Ensslin, Phys. Rev. D **70**, 043007 (2004). https://doi.org/10.1103/PhysRevD.70.043007
343. J. Donnert, K. Dolag, H. Lesch, E. Muller, Mon. Not. R. Astron. Soc. **392**, 1008 (2009). https://doi.org/10.1111/j.1365-2966.2008.14132.x
344. D. Ryu, H. Kang, J. Cho, S. Das, Science **320**, 909 (2008). https://doi.org/10.1126/science.1154923
345. R. Aloisio, V. Berezinsky, Astrophys. J. **612**, 900 (2004). https://doi.org/10.1086/421869
346. M. Lemoine, Phys. Rev. D **71**, 083007 (2005). https://doi.org/10.1103/PhysRevD.71.083007
347. R. Aloisio, V. Berezinsky, A. Gazizov, Astrophys. J. **693**, 1275 (2009). https://doi.org/10.1088/0004-637X/693/2/1275
348. S. Mollerach, E. Roulet, JCAP **1310**, 013 (2013). https://doi.org/10.1088/1475-7516/2013/10/013

349. P. Subedi, W. Sonsrettee, P. Blasi, D. Ruffolo, W.H. Matthaeus, D. Montgomery, P. Chuychai, P. Dmitruk, M. Wan, T.N. Parashar, R. Chhiber, Ap. J. **837**, 140 (2017). https://doi.org/10.3847/1538-4357/aa603a
350. S.L. Dubovsky, P.G. Tinyakov, I.I. Tkachev, Phys. Rev. Lett. **85**, 1154 (2000). https://doi.org/10.1103/PhysRevLett.85.1154
351. Z. Fodor, S.D. Katz, Phys. Rev. D **63**, 023002 (2001). https://doi.org/10.1103/PhysRevD.63.023002
352. P. Blasi, D. De Marco, Astropart. Phys. **20**, 559 (2004). https://doi.org/10.1016/j.astropartphys.2003.07.002
353. M. Kachelriess, D. Semikoz, Astropart. Phys. **23**, 486 (2005). https://doi.org/10.1016/j.astropartphys.2005.03.004
354. K. Kotera, M. Lemoine, Phys. Rev. D **77**, 023005 (2008). https://doi.org/10.1103/PhysRevD.77.023005
355. K. Kotera, M. Lemoine, Phys. Rev. D **77**, 123003 (2008). https://doi.org/10.1103/PhysRevD.77.123003
356. S. Kalli, M. Lemoine, K. Kotera, Astron. Astrophys. **528**, A109 (2011). https://doi.org/10.1051/0004-6361/201015688
357. P. Blasi, E. Amato, M. D'Angelo, Phys. Rev. Lett. **115**(12), 121101 (2015). https://doi.org/10.1103/PhysRevLett.115.121101
358. A.M. Hillas, Ann. Rev. Astron. Astrophys. **22**, 425 (1984). https://doi.org/10.1146/annurev.aa.22.090184.002233
359. K. Kotera, A.V. Olinto, Ann. Rev. Astron. Astrophys. **49**, 119 (2011). https://doi.org/10.1146/annurev-astro-081710-102620
360. E. Waxman, Phys. Scr. **T121**, 147 (2005). https://doi.org/10.1088/0031-8949/2005/T121/022
361. P. Blasi, EPJ Web Conf. **53**, 01002 (2013). https://doi.org/10.1051/epjconf/20135301002
362. J.H. Woo, C.M. Urry, Astrophys. J. **579**, 530 (2002). https://doi.org/10.1086/342878
363. L.H. Jiang, X.H. Fan, Z. Ivezic, G.T. Richards, D.P. Schneider, M.A. Strauss, B.C. Kelly, Astrophys. J. **656**, 680 (2007). https://doi.org/10.1086/510831
364. H. Kang, D. Ryu, T.W. Jones, Astrophys. J. **456**, 422 (1996). https://doi.org/10.1086/176666
365. H. Kang, D. Ryu, Astrophys. J. **764**, 95 (2013). https://doi.org/10.1088/0004-637X/764/1/95
366. S.E. Hong, D. Ryu, H. Kang, R. Cen, Astrophys. J. **785**, 133 (2014). https://doi.org/10.1088/0004-637X/785/2/133
367. S. Gabici, P. Blasi, Astrophys. J. **583**, 695 (2003). https://doi.org/10.1086/345429
368. V.S. Berezinsky, P. Blasi, V.S. Ptuskin, Astrophys. J. **487**, 529 (1997). https://doi.org/10.1086/304622
369. M. Lemoine, G. Pelletier, B. Revenu, Astrophys. J. **645**, L129 (2006). https://doi.org/10.1086/506322
370. P. Schneider, J.G. Kirk, Astrophys. J. Lett. **323**, L87 (1987). https://doi.org/10.1086/185063
371. D.C. Ellison, S.P. Reynolds, F.C. Jones, Astrophys. J. **360**, 702 (1990). https://doi.org/10.1086/169156
372. J.G. Kirk, A.W. Guthmann, Y.A. Gallant, A. Achterberg, Astrophys. J. **542**, 235 (2000). https://doi.org/10.1086/309533
373. M. Vietri, Astrophys. J. **591**, 954 (2003). https://doi.org/10.1086/375534
374. M. Vietri, D. De Marco, D. Guetta, Astrophys. J. **592**, 378 (2003). https://doi.org/10.1086/375719
375. P. Blasi, M. Vietri, Astrophys. J. **626**, 877 (2005). https://doi.org/10.1086/430164
376. Z. Li, E. Waxman, Astrophys. J. **651**, 328 (2006). https://doi.org/10.1086/507413
377. Z. Li, JCAP **1105**, 008 (2011). https://doi.org/10.1088/1475-7516/2011/05/008
378. L. Sironi, U. Keshet, M. Lemoine, Space Sci. Rev. **191**(1–4), 519 (2015). https://doi.org/10.1007/s11214-015-0181-8
379. L. Sironi, A. Spitkovsky, J. Arons, Astrophys. J. **771**, 54 (2013). https://doi.org/10.1088/0004-637X/771/1/54

380. A. Marcowith et al., Rept. Prog. Phys. **79**, 046901 (2016). https://doi.org/10.1088/0034-4885/79/4/046901
381. D. Caprioli, Astrophys. J. **811**(2), L38 (2015). https://doi.org/10.1088/2041-8205/811/2/L38
382. A. Venkatesan, M.C. Miller, A.V. Olinto, Astrophys. J. **484**, 323 (1997). https://doi.org/10.1086/304304
383. P. Blasi, R.I. Epstein, A.V. Olinto, Astrophys. J. **533**, L123 (2000). https://doi.org/10.1086/312626
384. J. Arons, Astrophys. J. **589**, 871 (2003). https://doi.org/10.1086/374776
385. K. Fang, K. Kotera, A.V. Olinto, JCAP **1303**, 010 (2013). https://doi.org/10.1088/1475-7516/2013/03/010
386. K. Kotera, E. Amato, P. Blasi, JCAP **1508**(08), 026 (2015). https://doi.org/10.1088/1475-7516/2015/08/026
387. D.R. Lorimer, Living Rev. Rel. **11**, 8 (2008). https://doi.org/10.12942/lrr-2008-8
388. M. Lemoine, Astron. Astrophys. **390**, L31 (2002). https://doi.org/10.1051/0004-6361:20020939
389. J. Pruet, S. Guiles, G.M. Fuller, Astrophys. J. **580**, 368 (2002). https://doi.org/10.1086/342838
390. S. Horiuchi, K. Murase, K. Ioka, P. Meszaros, Astrophys. J. **753**, 69 (2012). https://doi.org/10.1088/0004-637X/753/1/69
391. M. Ruderman, Ann. Rev. Astron. Astrophys. **10**, 427 (1972). https://doi.org/10.1146/annurev.aa.10.090172.002235
392. H.H. Chen, M.A. Ruderman, P.G. Sutherland, Astrophys. J. **191**, 473 (1974). https://doi.org/10.1146/annurev.aa.10.090172.002235
393. M.A. Ruderman, P.G. Sutherland, Astrophys. J. **196**, 51 (1975). https://doi.org/10.1086/153393
394. K.S. Cheng, C. Ho, M.A. Ruderman, Astrophys. J. **300**, 500 (1986). https://doi.org/10.1086/163829
395. R.J. Protheroe, W. Bednarek, Q. Luo, Astropart. Phys. **9**, 1 (1998). https://doi.org/10.1016/S0927-6505(98)00014-0
396. M. Takeda et al., Phys. Rev. Lett. **81**, 1163 (1998). https://doi.org/10.1103/PhysRevLett.81.1163
397. E. Schrodinger, Physica **6**, 399 (1939)
398. N.A. Chernikov, E.A. Tagirov, Ann. Inst. H. Poincare Phys. Theor. **A9**, 109 (1968)
399. L. Parker, Phys. Rev. Lett. **21**, 562 (1968). https://doi.org/10.1103/PhysRevLett.21.562
400. A. Grib, S. Mamayev, Sov. J. Nucl. Phys. **10**, 722 (1970)
401. Ya.B. Zeldovich, A.A. Starobinsky, Sov. Phys. JETP **34**, 1159 (1972). [Zh. Eksp. Teor. Fiz. 61, 2161 (1971)]
402. L.P. Grishchuk, Sov. Phys. JETP **40**, 409 (1975). [Zh. Eksp. Teor. Fiz. 67, 825 (1974)]
403. D.J.H. Chung, E.W. Kolb, A. Riotto, Phys. Rev. D **59**, 023501 (1999). https://doi.org/10.1103/PhysRevD.59.023501
404. V. Kuzmin, I. Tkachev, JETP Lett. **68**, 271 (1998). https://doi.org/10.1134/1.567858. [Pisma Zh. Eksp. Teor. Fiz. 68, 255 (1998)]
405. D.J.H. Chung, E.W. Kolb, A. Riotto, I.I. Tkachev, Phys. Rev. D **62**, 043508 (2000). https://doi.org/10.1103/PhysRevD.62.043508
406. D.J.H. Chung, P. Crotty, E.W. Kolb, A. Riotto, Phys. Rev. D **64**, 043503 (2001). https://doi.org/10.1103/PhysRevD.64.043503
407. E.W. Kolb, A.A. Starobinsky, I.I. Tkachev, JCAP **0707**, 005 (2007). https://doi.org/10.1088/1475-7516/2007/07/005
408. M.A. Fedderke, E.W. Kolb, M. Wyman, Phys. Rev. D **91**(6), 063505 (2015). https://doi.org/10.1103/PhysRevD.91.063505
409. D. Ivanov, in *PoS ICRC2015*, vol. 349 (2016)
410. P. Blasi, R. Dick, E.W. Kolb, Astropart. Phys. **18**, 57 (2002). https://doi.org/10.1016/S0927-6505(02)00113-5
411. J. Abraham et al., Astropart. Phys. **29**, 243 (2008). https://doi.org/10.1016/j.astropartphys.2008.01.003

412. D.B. Cline, F.W. Stecker, in *OWL/AW Neutrino Workshop on Observing Ultrahigh Energy Neutrinos*, Los Angeles, California, November 1–3, 1999
413. A. Petrolini, Nucl. Instrum. Meth. **A630**, 131 (2011). https://doi.org/10.1016/j.nima.2010.06.044
414. P. Allison, J. Auffenberg, R. Bard, J. Beatty, D. Besson et al., Astropart. Phys. **35**, 457 (2012). https://doi.org/10.1016/j.astropartphys.2011.11.010
415. P. Allison et al., Astropart. Phys. **70**, 62 (2015). https://doi.org/10.1016/j.astropartphys.2015.04.006
416. D.A. Kirzhnits, V.A. Chechin, Yad. Fiz. **15**, 1051 (1972)
417. R. Aloisio, P. Blasi, P.L. Ghia, A.F. Grillo, Phys. Rev. D **62**, 053010 (2000). https://doi.org/10.1103/PhysRevD.62.053010
418. R. Aloisio, P. Blasi, A. Galante, P.L. Ghia, A.F. Grillo, Astropart. Phys. **19**, 127 (2003). https://doi.org/10.1016/S0927-6505(02)00196-2
419. R. Aloisio, P. Blasi, A. Galante, A.F. Grillo, Lect. Notes Phys. **669**, 1 (2005). https://doi.org/10.1007/11377306_1
420. R. Aloisio, A. Galante, A.F. Grillo, S. Liberati, E. Luzio, F. Mendez, Phys. Rev. D **74**, 085017 (2006). https://doi.org/10.1103/PhysRevD.74.085017
421. S. Liberati, Class. Quantum Grav. **30**, 133001 (2013). https://doi.org/10.1088/0264-9381/30/13/133001
422. V.A. Kostelecky, N. Russell, Rev. Mod. Phys. **83**, 11 (2011). https://doi.org/10.1103/RevModPhys.83.11
423. J.M. Carmona, J.L. Cortes, F. Mercati, Phys. Rev. D **86**, 084032 (2012). https://doi.org/10.1103/PhysRevD.86.084032
424. F.R. Klinkhamer, M. Risse, Phys. Rev. D **77**, 117901 (2008). https://doi.org/10.1103/PhysRevD.77.117901
425. T. Jacobson, S. Liberati, D. Mattingly, Ann. Phys. **321**, 150 (2006). https://doi.org/10.1016/j.aop.2005.06.004
426. A. Saveliev, L. Maccione, G. Sigl, JCAP **1103**, 046 (2011). https://doi.org/10.1088/1475-7516/2011/03/046
427. R. Aloisio, D. Boncioli, A. di Matteo, P.L. Ghia, A.F. Grillo, S. Petrera, F. Salamida, Frascati Phys. Ser. **58**, 274 (2014)
428. E. Waxman, Phys. Rev. Lett. **75**, 386 (1995). https://doi.org/10.1103/PhysRevLett.75.386
429. M. Vietri, Ap. J. **453**, 883 (1995). https://doi.org/10.1086/176448
430. D. Eichler, N. Globus, R. Kumar, E. Gavish, Astrophys. J. **821**(2), L24 (2016). https://doi.org/10.3847/2041-8205/821/2/L24
431. R. Kumar, D. Eichler, Astrophys. J. **781**(1), 47 (2014). https://doi.org/10.1088/0004-637X/781/1/47

Selected Topics in Gamma-Ray Astronomy: Very High Energy Gamma-Rays as Tracers of Galactic Cosmic-Rays

Felix Aharonian and Sabrina Casanova

Introduction

Photons from the highest energy end of the spectrum of cosmic electromagnetic radiation, dubbed gamma-rays, are messengers carrying crucial information about the most energetic and violent phenomena in the Universe. The respective research area, *gamma-ray astronomy*, is a part of the modern interdisciplinary field called *astroparticle physics*, a discipline in its own right. It addresses an impressively broad range of topics related to nonthermal processes in a diverse variety of astrophysical environments: from compact relativistic objects like pulsars and accreting black holes to large-scale cosmological structures such as clusters of galaxies.

Except for a range of specific processes related to the decays of long-lived radioactive nuclei (leftovers from powerful cosmic events like Supernovae explosions), as well as the decay or annihilation of Dark Matter, cosmic γ-rays are produced in

F. Aharonian (✉) · S. Casanova
Max-Planck-Institut Für Kernphysik, Saupfercheckweg 1,
69117 Heidelberg, Germany
e-mail: aharon@mpi-hd.mpg.de

F. Aharonian
Dublin Institute for Advanced Studies, 31 Fitzwilliam Place,
Dublin 2, Ireland

F. Aharonian
Gran Sasso Science Institute, Viale Francesco Crispi, 7,
67100 L'Aquila, Italy

F. Aharonian
National Research Nuclear University MEPhI, Kashirskoje Shosse, 31,
115409 Moscow, Russia

S. Casanova
Instytut Fizyki Jądrowej PAN, ul. Radzikowskiego 152,
31-342 Kraków, Poland
e-mail: sabrinacasanova@gmail.com

R. Aloisio et al. (eds.), *Multiple Messengers and Challenges in Astroparticle Physics*, https://doi.org/10.1007/978-3-319-65425-6_2

interactions of suprathermal (accelerated) electrons, protons, and nuclei with the ambient medium. In this way, γ-rays carry information about the cosmic factories of relativistic particles—electrons (positrons), protons, and nuclei. These nonthermal particles can be considered as the fourth form of substance (after matter, radiation, and magnetic fields) of which the visible Universe consists. In certain environments, e.g., in the Interstellar Medium (ISM), the pressure in relativistic particles can be comparable or even exceed the pressure of the thermal gas and magnetic fields. These particles may constitute not only a significant fraction of the overall energy budget, but also play an important role in the dynamics and evolution of astrophysical source populations. Historically, the locally measured relativistic particles are called Cosmic-Rays (CRs). However, presently, high energy astrophysicists use the term "Cosmic-Rays" in a broader context, implying also the relativistic particles which do not reach us, but are traced and studied indirectly, through the neutral and stable secondaries—photons and neutrinos.

The γ-ray production often proceeds under extreme physical conditions, in the presence of huge gravitational, magnetic and electric fields, relativistic bulk motions and shock waves, highly excited (turbulent) plasma, etc. Consequently, any coherent description and interpretation of phenomena related to gamma-rays requires deep knowledge of many disciplines of physics including nuclear and particle physics, quantum and classical electrodynamics, special and general relativity, plasma physics, magnetohydrodynamics.

The energy range covered by cosmic γ-rays spans from MeV to EeV energies.[1] While the lower bound is associated with the domain of nuclear reactions, the upper bound is linked to the interactions of the highest energy particles observed in cosmic-rays. Because of the essentially different detection methods and approaches applicable to various energy bands, the enormous energy domain of cosmic electromagnetic radiation is covered inhomogeneously. In particular, so far cosmic γ-rays are detected in the "low" (LE or MeV), "high" (HE or GeV) and very high (VHE or TeV) energy bands.

Observational Gamma-Ray Astronomy: Status

(i) Low Energy Band: 0.1–100 MeV

At low energies, the sky remains a poorly explored frontier. A combination of several factors—limited detection efficiency, the poor accuracy of reconstruction of arrival directions of MeV γ-rays, the presence of "heavy" backgrounds of different origin—substantially constrain the potential of low energy gamma-ray astronomy. Meanwhile, this important energy interval is uniquely linked to a number of phenomena, for example, to Gamma Ray Bursts, Solar flares, the nucleosynthesis of heavy elements, heating and ionization of the interstellar gas by sub-relativistic CRs, etc.

[1]Below we will use the following energy units common in high energy physics and astrophysics: 1 MeV=10^6 eV, 1 GeV=10^9 eV, 1 TeV=10^{12} eV, 1 PeV=10^{15} eV, 1 EeV=10^{18} eV.

The great challenge of MeV gamma-ray astronomy remains the design and construction of detectors with a sensitivity compatible to the conservative flux predictions. More specifically, the minimum detectable energy fluxes at MeV energies should achieve a level as low as $10^{-12}\,\mathrm{erg/cm^2\,s}$, in order to address effectively a number of long-standing astrophysical issues.

This concerns, for example, the probes of sub-relativistic ($E \leq 100\,\mathrm{MeV}$) CRs in ISM through the prompt *de-excitation* γ-ray lines (see, e.g., Ref. [66]). Also, the γ-ray continuum at MeV energies produced via *bremsstrahlung* of electrons (positrons), as well as at the *positron annihilation in flight*, contains unique information about relativistic electrons in the same (below 100 MeV) energy range [22]. Note that in typical astrophysical environments with low magnetic field, in particular in the ISM ($B \leq 10\,\mu\mathrm{G}$), such information is not accessible via synchrotron radio emission because the latter appears at non-visible frequencies below 1 MHz. The information about nonthermal electrons and protons (nuclei) is important for derivation of the energy balance between different forms of matter, magnetic fields, and cosmic-rays. In the galactic disk, such measurements have some other astrophysical implications. For example, they provide direct estimates of the ionization and heating rates of the interstellar gas by low energy cosmic-ray protons and electrons. This channel of information is unique and very important for understanding the star forming processes in our Galaxy.

Another important implication of MeV gamma-ray emission is related to studies of mildly relativistic *thermal* plasmas formed in the vicinity of compact relativistic objects such as neutron stars and black holes. The characteristic radiation dominated by Comptonized free-free emission and by electron-positron annihilation contains direct information about the electron temperature and density (see e.g., [84]). Generally, because of fast radiative cooling of electrons, the proton temperature significantly exceeds the electron temperature. The formation of hot two-temperature plasmas, $T_\mathrm{i} \gg T_\mathrm{e}$, in strong shock waves or in accretion flows can be studied by detection of characteristic gamma-ray line emission produced through the chain of spallation and excitation reactions [62]. A clear signature of radiation at the final stage of such hot plasmas when all nuclei are destroyed and the nucleonic component of plasma basically consists of protons and neutrons (with a small fraction of deuterium in equilibrium), is the continuum due to the proton-neutron bremsstrahlung and the broadened and "blue-shifted" deuterium line [27]. In accreting solar-mass black holes, this radiation is released, depending on T_i, typically between 1 and 30 MeV.

One of the most promising directions in MeV astronomy is believed to be linked to the studies of the (e^+, e^-) annihilation. Prolific production of positrons in various astrophysical environments, and their subsequent annihilation with electrons from the ambient thermal gas create favorable conditions for formation of the annihilation line around 0.5 MeV. The profile of this spectral feature (the gravity center and the shape of the line) carries unique information about the environment, in particular about the temperature, density and the bulk motion speed of fast-moving gas flows, on the gravitational potential of compact objects, e.g., at the surface of neutron starts, etc. The detection of the narrow annihilation line at 0.511 MeV from the direction of the Galactic Centre (see, e.g., the recent review articles [44, 82]) is undoubtedly one of the highlight results of observational gamma-ray astronomy. It is expected that the future MeV gamma-ray telescopes could result in detections of the e^+, e^- annihilation radiation from other astrophysical objects.

(ii) High Energy Band: 0.1–100 GeV

Before the launch of the *Fermi Gamma-ray Space Telescope* in 2008, only two source populations—the active galactic nuclei and pulsars—had been clearly identified as high energy gamma-ray emitters. With the *Fermi* LAT (Large Area Telescope) the HE gamma-ray astronomy entered a new era. This instrument with angular resolution of about 0.5° at 1 GeV, flux sensitivity better than 10^{-12} erg/cm^2 s), and an effective field of view of 2 steradian [32], is a powerful tool for continuous monitoring of a significant fraction of the sky.

The high energy γ-ray sky as seen by *Fermi* LAT (see Fig. 1) is impressive! These results confirm the most optimistic prelaunch expectations concerning, in particular, the dramatic increase of the number of γ-ray emitting pulsars and AGN. On the other hand, the discovery of new classes of *compact/variable* and *extended* γ-ray emitters, revealed the richness of γ-ray phenomena on both galactic and extragalactic scales. Among these discoveries are results of exceptional importance for the solution of a number problems of high energy astrophysics. Just to mention a few examples: the discovery of short flares from the Crab, the detection of multi-GeV components of Gamma Ray Bursts (GRBs), the discovery of the so-called *Fermi Bubbles*, etc. Currently, the list of GeV γ-ray emitters consists of more 2000 galactic and extra-galactic sources. While more than half of these objects have firmly identified or likely counterparts representing different source populations, the origin of a very

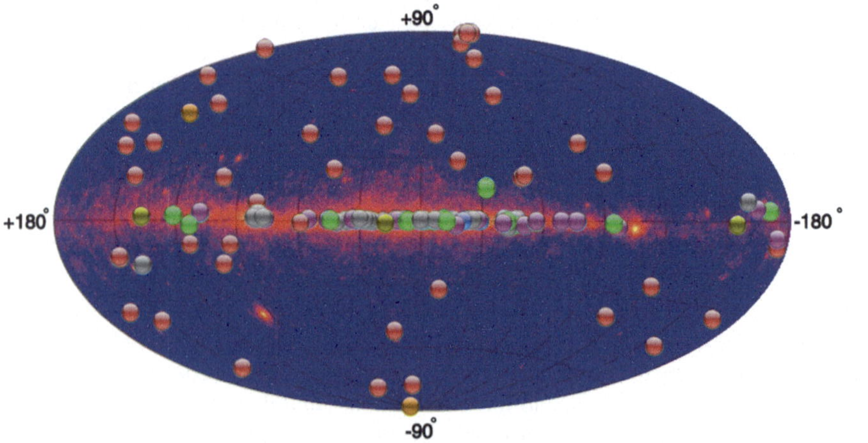

Fig. 1 The overall MeV/GeV gamma-ray sky (the *blue* to *red color background*) as seen by *Fermi* [76] and the positions of discrete TeV gamma-ray sources detected with ground-based instruments; the regularly updated version of this figure can be found on the "TeVCat" webpage: http://tevcat.uchicago.edu. The SNRs (in association with GMCs or without) are shown with a *green symbol*, the pulsar wind nebulae are shown in *violet*, the binary systems—in *yellow*, starburst galaxies—in *brown*, AGN (all types)—in *red*. The so-called dark sources without any reliable association with the well-known objects are shown with a *grey symbol*

large fraction of these objects remains an open issue. This concerns, first of all, the extended sources located in the galactic plane, e.g., SNRs and PWNe, for which the confusion with the structured diffuse emission of the galactic disk is especially high.

Because of the limited angular resolution, the most reliable approach for identification of GeV gamma-ray sources is based on temporal studies. It is quite natural that the best "astronomical clocks"—the Pulsars—constitute the largest population of firmly identified GeV gamma-ray sources. The periodic character of gamma-ray emission of the galactic binary sources or the sporadic flares of AGN provide another approach for identification of variable gamma-ray sources based on simultaneous observations in different energy bands. In general, the multiwavelength studies represent a key component in the attempts of identification of gamma-ray emitters. And, of course, it is difficult to overestimate the significance of multiwavelength studies for a deeper understanding of the nature of these objects.

The *Fermi* observations greatly enhanced our knowledge of the diffuse γ-ray backgrounds of different origin. In particular, *Fermi* has extended the range of observations of the diffuse emission of the galactic disk and the isotropic (extragalactic) gamma-ray background to several hundreds of GeV and helped to clarify some controversial issues related to the contributions of different source populations and the relevant radiation mechanism. The results are of prime interest for understanding of the origin of both the galactic and extragalactic components of CRs.

A number of important results, especially at energies below 1 GeV, have been reported also by the Italian gamma-ray satellite AGILE [79]. In general, the observations by *Fermi* and *Agile* support many phenomenological concepts and theoretical models in different areas of astrophysics. At the same time, these observations resulted in a number of "unplanned" discoveries and revealed some puzzling phenomena like flares of the Crab (Nebula?) or existence of multi-kpc scale nonthermal structures—giant reservoirs of relativistic particles centered on the core of the Galaxy.

Concerning the next generation space-based gamma-ray detectors, it is likely that in the foreseeable future one cannot expect significant developments beyond the level achieved by *Fermi* LAT, except perhaps for the energy band below 1 GeV down to several tens of MeV. The design and construction of a space-based instrument in this energy interval with an affordable effective detection area of order of 1 m^2 and angular resolution of about 1°, would increase significantly the detection rate (photon statistics) and improve the flux sensitivity of *Fermi* LAT below 1 GeV potentially by an order of magnitude. The optimization of the pair-conversion tracking detection technique with a focus on energies around 100 MeV would be an attractive and promising strategy given the number of outstanding astrophysical questions relevant to nearly all source populations, but not fully addressed by *Fermi LAT*.

(iii) Very High Energy Band: 0.1–100 TeV

One of the remarkable achievements of recent years in astrophysics is related to the great results obtained with ground-based detectors in the Very High Energy (VHE) domain. As a result, VHE gamma-ray astronomy quickly emerged as a truly astronomical discipline. The observations conducted by HESS (Fig. 3), MAGIC (Fig. 4), and VERITAS atmospheric Cherenkov telescope arrays resulted in many discoveries of topical importance. The number of reported γ-ray emitters currently exceeds

150 galactic and extragalactic objects (see Fig. 1). These sources represent almost all major *nonthermal* astrophysical source populations, including shell type Supernova Remnants, Pulsar Wind Nebulae, Star Forming Regions, Giant Molecular Clouds, X-ray Binary Systems, Blazars, Radiogalaxies, Starburst Galaxies (for a review see, e.g., [10, 59]). To a large extent, this success was a big surprise, especially given the rather difficult and controversial history of the field over the last four decades (see, e.g., [18, 57, 92]). Therefore, a question arises concerning the reasons which made possible this success. The answer to this question can be formulated as a fortunate combination of two independent factors (Fig. 2):

(i) the practical realization of the great potential of stereoscopic arrays of Imaging Atmospheric Cherenkov Telescopes (IACTs) as effective multifunctional tools for spectral, temporal, and morphological studies of VHE gamma-ray sources;

(ii) the existence of a large variety of perfectly designed machines—*TeVatrons* and *PeVatrons* "factories" of relativistic matter where effective particle acceleration is accompanied by efficient radiation processes.

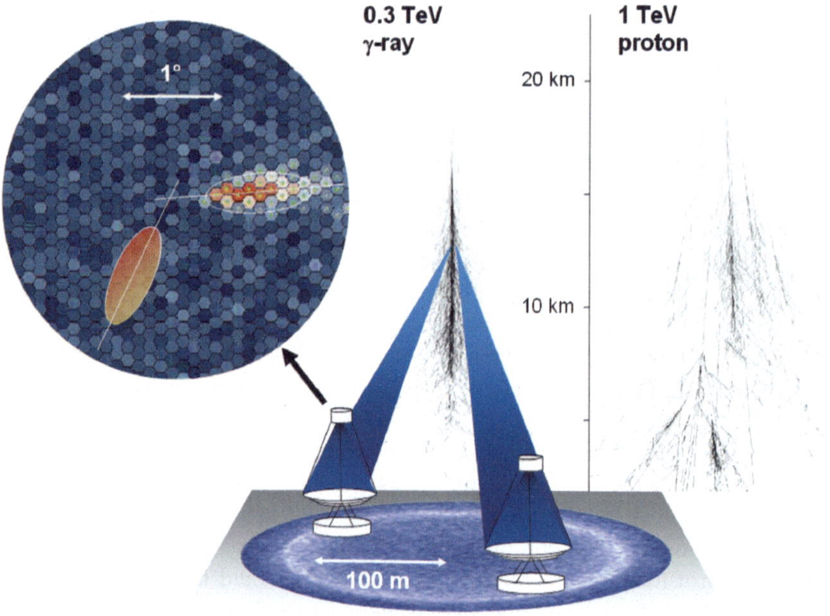

Fig. 2 The operation of the imaging atmospheric Cherenkov telescope technique (from Ref. [59])

Performance of Ground-Based Detectors

The Earth's atmosphere is not transparent to γ-rays. Therefore, in order to avoid any significant absorption, the γ-ray detectors should be installed out of the atmosphere, i.e., on space platforms. However, as long as the satellites cannot offer detection areas significantly exceeding 1 m^2, effective studies of the tiny fluxes of cosmic γ-rays will be limited to energies ≤ 100 GeV. Fortunately, at higher energies, an alternative method of registration of the so-called extensive atmospheric showers (EAS) initiated by interactions of γ-ray with the atmosphere, can be used. This can be realized through direct detection of the main components of EAS (electrons, photons, muons, hadrons) or through the Cherenkov radiation of relativistic electrons. The detectors are respectively called EAS arrays and Atmospheric Cherenkov Telescopes (ACTs).

ACT Technique

The faint Cherenkov radiation from relativistic electrons of an air shower lasts only several nanoseconds, but it can be detected by large optical reflectors equipped with fast multi-pixel cameras. With a telescope consisting of an optical reflector of diameter $D \approx 10$ m, as well as a camera with pixel size $0.1°$–$0.2°$ and field of view $\Theta \geq 3°$, primary gamma-rays of energy ≥ 100 GeV can be collected from distances as large as 100 m. This provides huge detection areas, $A \geq 3 \times 10^4$ m^2, which compensate the weak gamma-ray fluxes at these energies. The total number of photons in the registered Cherenkov light image is a measure of energy, the orientation of the image correlates with the arrival direction of the gamma-ray, and the shape of the image contains information about the origin of the primary particle (a proton or photon) (see for a review Ref. [57]). The basic principles of operation of the IACT technique are illustrated in Fig. 2 from Ref. [59].

The stereoscopic observations of air showers with two or more 10 m diameter telescopes located at distances of about 100 m from each other provide a quite low energy threshold around 100 GeV, effective (by a factor of 100) rejection of hadronic showers, and good angular ($\approx 0.1°$) and energy ($\approx 15\%$) resolutions (see, e.g., [20]). At energies around 1 TeV, the performance achieves the best sensitivity, a minimum detectable energy flux of 10^{-13} erg/cm^2 (see Fig. 5). This is a quite impressive achievement even in the standards of advanced branches of astrophysics. In particular, it is much better than in any other γ-ray domain, including the GeV energy band. Moreover, the very large collection area provides large γ-ray photon statistics even from relatively modest TeV γ-ray sources. In combination with good energy and angular resolutions, the rich photon statistics allows effective morphological, spectral, and temporal studies. This makes the IACT arrays perfect multifunctional and multipurpose astronomical tools for exploration of a broad range of nonthermal objects and phenomena. Currently, three major IACT arrays HESS (High Energy Stereoscopic System), MAGIC (Major Atmospheric Imaging Cherenkov), and VERITAS (Very Energetic Radiation Imaging Telescope Array System)—located both in the northern (MAGIC, VERITAS) and southern (HESS) hemispheres, determine the status of VHE gamma-ray astronomy. The HESS and MAGIC arrays are shown in Figs. 3 and 4.

Regarding the previous generation instruments, one should mention two detectors, the 10 m diameter single dish of the Whipple Observatory (south Arizona) and the HEGRA array of five relatively modest (4 m diameter) Cherenkov telescope (La Palma, Canary Islands). These instruments can be considered as prototypes of the current IACT arrays. They played a crucial role in the development of ground-based gamma-ray astronomy. While the Whipple collaboration pioneered the implementation and successful realization of the *imaging* atmospheric Cherenkov technique, the HEGRA collaboration convincingly demonstrated the power of the *stereoscope* approach.

Fig. 3 The HESS system of four 13 m diameter Cherenkov telescopes. While this system has been operating in Namibia since 2003, the 28 m diameter telescope has been operating since 2012

Fig. 4 The system of two 17 m diameter MAGIC gamma-ray telescopes at the Roque de los Muchachos observatory (La Palma, Canary Islands)

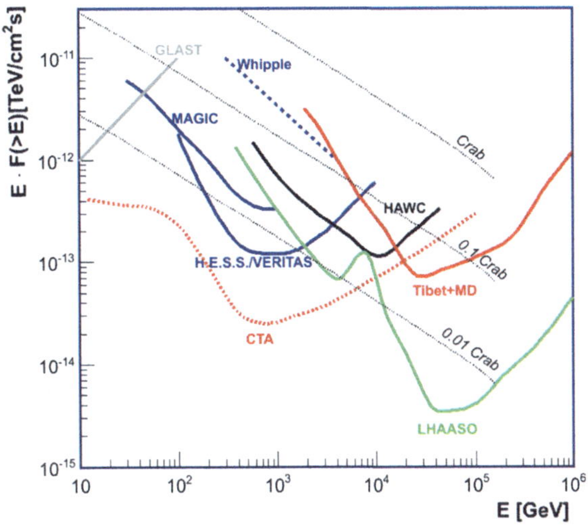

Fig. 5 The energy flux sensitivities of the current and future ground-based detectors—the IACT and EAS arrays in the energy range 10^{10}–10^{16} eV

The IACT arrays are designed to observe point-like or moderately extended (with angular size 1° or less) objects with known celestial coordinates. However, the high sensitivity and relatively large (up to ~10°) field of view of IACT arrays allow effective all-sky surveys. On the other hand, the potential of IACT arrays is limited for the search for very extended structures (like Fermi Bubbles and giant Radio Lobes of the radiogalaxy Centaurus A), as well as for solitary or transient γ-ray phenomena. In this regard, a detection technique based on direct registration of particles that comprise the extensive air showers (EAS), is a complementary approach to the IACT technique.

Future IACT Arrays

Among the major motivations of the next generation of Imaging Atmospheric Cherenkov Telescope (IACT) arrays two objectives are of particular relevance: (i) dramatic, by an order of magnitude, improvement of the flux sensitivity in the standard 0.1–10 TeV interval, and (ii) aggressive expansion of the energy domain down to 10 GeV and up to 300 TeV.

TeV regime: 0.1–10 TeV

The IACT technique achieves the best performance in this energy regime, and still the potential is not saturated. The three fundamental factors, (i) high efficiency of detection and identification of electromagnetic showers, (ii) good accuracy of reconstruction of the direction and the energy of the primary γ-ray photon, and (iii) large γ-ray photon statistics, allow angular resolution $\delta\theta \approx 1-2$ arcminutes, and flux sensitivity at the level of 10^{-14} erg/cm^2 s,

Such a nice performance can be achieved by a stereoscopic array consisting of tens of 10 m diameter class telescopes. The flux sensitivity 10^{-14} erg/cm^2 s at TeV energies would be a great achievement even in the standards of the most advanced branches of observational astronomy. This should allow us to probe the γ-ray luminosities of potential TeV emitters at the levels of $10^{32}\,(d/10\,\mathrm{kpc})^2$ erg/s. For an extended source, e.g., of an angular size $\Psi \sim 1^\circ$, the minimum detectable energy flux will be by an order of magnitude higher. Yet it would be better than the energy flux sensitivities of the best current X-ray satellites, *Chandra* and *XMM*-Newton. This should allow the deepest probes of nonthermal processes, in particular, in shell type Supernova Remnants (SNRs), Pulsar Driven Nebulae (Plerions), Giant Radio Lobes of Radiogalaxies, Clusters of Galaxies, etc. Such a system of 10–12 m diameter class IACTs with a field of view (FoV) of $6-8^\circ$, constitutes the core of the Cherenkov Telescope Array (CTA)—the major ground-based gamma-ray detector (see Fig. 6) [6].

Sub-PeV regime

The general tendency of reduction of γ-ray fluxes with energy becomes dramatic above 10 TeV, in particular, because of the limited efficiency of acceleration of parent particles, the escape of the highest energy particles from the production region, the external and internal absorption of γ-rays, etc. Ideally, effective studies of γ-rays of energy larger than 10 TeV require detection areas significantly larger than 1km^2. This can be realized by an array consisting of tens of small (a few meter in diameter) reflectors, separated from each other by 200–300 m. The pixel size of cameras could be rather modest, 0.25° or so, however, they should have large (up to 10°) FoV for detection of showers from distances significantly exceeding 100 m [81]. A sub-array consisting of several tens of such telescopes is included in the concept of CTA [6]. It will serve as a powerful tool for searches of galactic cosmic-ray "PeVatrons," as well as multi-TeV γ-rays from nearby ($R \ll 100$ Mpc) radiogalaxies and starburst galaxies.

Fig. 6 CTA

Energy domain below 100 GeV

The energy threshold of IACTs is defined as a characteristic energy at which the detection rate for the γ-ray power-law spectrum with a photon index between 2 and 3 achieves its maximum. However, the best performance, in particular, the flux sensitivity, is achieved at energies exceeding several times the threshold energy. Therefore, for optimization of gamma-ray detection around 100 GeV, the energy threshold should be reduced to $\varepsilon_{\rm th} \leq 30$ GeV. On the other hand, the reduction of the detection threshold to 30 GeV is an important scientific issue in its own right; the intermediate interval between 30 and 300 GeV is a crucial energy regime for certain classes of galactic and extragalactic gamma-ray source populations. This can be done with very large, more than 20 m diameter reflectors. A sub-array of several large 23 m diameter telescopes foreseen in CTA [6] (see Fig. 6) will substantially broaden the scientific objectives of CTA.

The three sub-arrays of CTA are characterized by corresponding energy intervals in which the best energy flux sensitivity is achieved. Each of these arrays covers at least 2 decades in energy, thus the energy domains of the sub-arrays significantly overlap. Since these sub-arrays contain the same basic elements, regarding the technical realization of telescopes, and similar, to a large extent, scientific motivations, an ideal arrangement of the telescopes would be the combination of these sub-arrays in a single facility which would have a homogeneous coverage throughout the energy region from approximately 30 Ge–300 TeV. The conceptual design of CTA is motivated, to a large extent, by this objective [6]. The high detection rates, together with good angular and energy resolutions over four energy decades will make CTA a powerful multifunctional gamma-ray observatory with a great capability for spectrometric, morphological and temporal studies of a diverse range of persistent and transient high energy phenomena in the Universe.

Perspectives in the Multi-GeV domain

Despite the remarkable results achieved by the AGILE and *Fermi* LAT telescopes, further progress in space-based gamma-ray astronomy at high (GeV) energies seems rather uncertain, given the objective difficulties regarding the improvement of the performance of satellite-borne detectors at a level which would justify any initiative towards new projects. This concerns the low angular resolution, especially at energies below 1 GeV, and the poor photon statistics above 10 GeV. While there is room for improvement of the angular resolution at low energies, e.g., reducing it to 1° around 100 MeV, the enhancement of photon statistics at high energies, where the angular resolution is significantly improved, is more problematic. It would be possible only via dramatic increase of the detector size.

One should note, in this regard, that the flux sensitivity of *Fermi* LAT above 1 GeV of about 10^{-12} erg/cm^2 s can be achieved only after one year all-sky survey. While for persistent gamma-ray sources this seems to be an adequate sensitivity (given that a huge number of sources are simultaneously monitored within the large, almost $\sim 2\pi$ steradian homogeneous FoV), the small, ≈ 1 m^2 detection area limits significantly the potential for detailed studies of the temporal and spectral characteristics of highly variable sources. Apparently, the sensitivity achieved by *Fermi* at high energies will not be easy to improve significantly in any future space-based mission, unless the Moon could be used in the (far) future as a possible platform for installation of very large ($\gg 10\,\mathrm{m}^2$) area pair-conversion tracking detectors. Thus, it is clear that the space-based resources of GeV gamma-ray astronomy have achieved a point where any further progress would appear extremely difficult and very expensive.

On the other hand, the principal possibility of extension of the Imaging Atmospheric Cherenkov Telescope (IACT) technique towards 10 GeV (or even below) promises a new breakthrough in gamma-ray astronomy [10]. The (relatively) large gamma-ray fluxes in this energy interval, together with the huge detection areas offered by the IACT technique, can provide the best γ-ray photon statistics compared to any other energy band of cosmic gamma-radiation. In the case of realization of 10 GeV threshold IACT arrays, the presently almost unexplored interval between 10 and 100 GeV could become one of the most advanced domains of gamma-ray astronomy, especially for studies of highly variable phenomena.

The reduction of the energy threshold down to 10 GeV or even less is in principle possible within the basic concept of the IACT technique, but it requires an extreme approach—25 m diameter class telescopes equipped with high ($\simeq$40%) quantum efficiency focal plane imagers, operating in a robotic regime at very high (5 km or) mountain altitudes [25].

The energy range from several GeV to 30 GeV has very specific astrophysical and cosmological objectives: exploration of highly variable nonthermal phenomena, in particular, in the remote universe at redshifts up to $z = 5$ and beyond, as well as in compact galactic objects like pulsars and microquasars.

On the Potential of EAS Arrays

The traditional EAS technique, based on scintillators or water Cherenkov detectors spread over large areas, works quite effectively for detection of cosmic-rays at ultra-high energies, $E \geq 100$ TeV. In order to make this technique more adequate to purposes of gamma-ray astronomy, the detection energy threshold should be reduced by two orders of magnitude. This can be achieved using dense particle arrays located at very high altitudes. The feasibility of both approaches has recently been successfully demonstrated by the ARGO and Milagro groups.

The prospects of continuous monitoring of a significant part of the sky might lead to exciting discoveries of yet unknown VHE transient phenomena in the Universe. This motivates, to a large extent, the proposals of construction of high altitude EAS detectors (see for a review [10]) like HAWC, a High Altitude Water Cherenkov detector completed on a site close to Sierra Negra, Mexico and the ambitious LHAASO (Large High Altitude Air Shower Observatory) detector facility under construction at Yangbajing, Tibet. Presently some more projects of very high altitude (around 5 Km above sea level) detectors are under intensive discussions. The sensitivity of these instruments at energies between 1–10 TeV is expected to be comparable to the sensitivity of *Fermi* around 1 GeV. In this regard, they will be complementary to *Fermi* for continuous monitoring of more than 1 steradian of the sky at TeV energies. At the same time, these instruments can serve as powerful tools for hunting solitary γ-ray events like GRBs and for the study of transient phenomena like AGN flares at energies down to 100 GeV.

Radiation Process

One of the key aspects of interpretation of astrophysical observations is the correct identification of relevant radiation mechanism(s). Each γ-ray domain has its specifics and is characterized by several radiation processes. With some exceptions, all essential gamma-ray production and absorption mechanisms have been studied in great detail using the methods and tools of modern experimental and/or theoretical physics.

Basic definitions

To calculate the γ-ray emissivity, one should be aware of two functions—the differential cross section of the process, $\frac{d\sigma}{dEd\Omega}$, and the energy distribution of parent particles, $f(E)$. Generally, for the given energy of the primary particle E, the γ-ray production cross sections are well-known (calculated or measured) functions of the energy and the emission angle of secondary products. On the other hand, the distributions of parent particles, electrons, protons, and nuclei depend not only on the specifics of the acceleration mechanism, but also on the character of propagation of these particles before they reach the target where γ-rays are produced. This is in contrast to the case of emission of thermal gas (plasma). The energy distribution of thermal particles is always Maxwellian and, therefore, is characterized by a single parameter, the temperature.

The general kinetic equation that describes the evolution of the energy distribution of charged particles, $f(E, \boldsymbol{r}, t)$, can be written in the form

$$\frac{\partial f}{\partial t} = \nabla \cdot (D_r \nabla f) - \nabla \cdot (\mathbf{u}_r f) + \frac{\partial}{\partial E}(P_r f) - \frac{\partial}{\partial E}(b_r f) + \frac{\partial^2}{\partial E^2}(d_r f), \quad (1)$$

which includes the terms responsible for the diffusion, acceleration, and energy losses of particles (see Ref. [51] for derivation of this equation, as well for its approximate solutions under specific assumptions).

Strictly speaking, this equation should be invoked in any detailed treatment of γ-ray production. This implies that one has to make an assumption about the acceleration mechanism which in many cases is highly unknown. Moreover, in the formation of the energy distribution of radiating relativistic particles, the propagation effects dominate over the role of the acceleration. Thus, the inclusion of the acceleration aspect in the overall treatment of the problem not only introduces additional uncertainties, but in some cases could be superficial. In such cases, it would be reasonable to assume an injection of particles, protons and/or electrons, into the γ-ray production rate with a given time history of the injection spectrum, $Q(E, t)$. Such an approach could be appropriate, for example, when the accelerator of particles and the target (the γ-ray production region) are separated. Then, Eq. (1) integrated over the volume of the radiation region, is reduced to

$$\frac{\partial N}{\partial t} = \frac{\partial}{\partial E}(PN) - \frac{N}{\tau_{\rm esc}} + Q(E, t) \tag{2}$$

where

$$N(E, t) \equiv \int f\, d^3r. \tag{3}$$

Here $P(E, t) = -\partial E/\partial t$ is the energy loss rate for a particle with energy E, and $\tau_{\rm esc}(E, t)$ is the characteristic escape time of particles from the source, due to either diffusion or convection, $\tau_{\rm esc} = (1/\tau_{\rm diff} + 1/\tau_{\rm conv})^{-1}$.

The injection spectrum generally is assumed to be a power-law with a high energy cutoff, given in the form $Q(E, t) \propto E^{-\alpha} \exp\left[-(E/E_0)^{\beta}\right]$; the parameter β characterizes the sharpness of the spectrum in the cutoff region. The characteristic energy E_0 is estimated by equating the acceleration rate to the cooling rate, $t_{\rm cool} = t_{\rm acc}$. Due to energy losses, the particle spectrum established in the radiation region can deviate significantly from the original injection spectrum. The steady-state particle distribution is obtained by setting $\partial N/\partial t = 0$ in Eq. (2). Assuming that the term proportional to $\tau_{\rm esc}^{-1}$ can be neglected compared to the term containing the energy loss rate, one finds the following simple solution:

$$N(E) = \left|\frac{dE}{dt}\right|^{-1} \int_E^{\infty} Q(E)\, dE. \tag{4}$$

Several specific realizations for energy losses are of particular interest. For example, if the energy losses are proportional to the particle energy, $dE/dt \propto E$, which is the case of relativistic Bremsstrahlung, proton-proton inelastic collisions and adiabatic losses, the initial (injection) power-law spectrum of particles remains unchanged, $N(E) \propto E^{-\alpha}$. The steady-state distribution is steeper if electrons are cooled through the synchrotron radiation or inverse Compton scattering (IC) in the Thomson regime, when $dE/dt \propto E^2$. For such losses, $N(E) \propto E^{-(\alpha+1)}$. On the other hand, the spectrum becomes harder if dE/dt is constant (the case of ionization losses) or $dE/dt \propto E^{-1}$ (the case of IC scattering in the Klein-Nishina limit): $N(E) \propto E^{-(\alpha-1)}$ and $N(E) \propto E^{-(\alpha-2)}$, respectively. These features are reflected in the steady-state γ-ray spectra.

The γ-ray luminosity is characterized by the cooling time of parent particles, $t_{\rm cool} = E/\left|dE/dt\right|$, namely,

$$L_{\gamma} \approx W_{e(p)} t_{\rm cool}^{-1}, \tag{5}$$

where $W_{e(p)}$ is the total energy contained in relativistic particles. For the given volume V of radiation region,

$$W_{e(p)} = \int_V \int_{E_{\min}}^{E_{\max}} E\, N_{e(p)}(E)\, dE\, d^3x. \tag{6}$$

Here $E_{\min}$ and $E_{\max}$ are the minimum and maximum energy of particles.

The energy spectrum of γ-rays can be obtained, often with a very good accuracy, using the so-called δ-function approximation. This method assumes that the same fixed fraction of energy of the parent particle transfers to a secondary γ-ray photon. However, the δ-function formalism should be applied with care, especially near sharp spectral features like a cutoff or a break.

Major elementary processes

In standard astrophysical environments, the most important γ-ray production mechanisms are

- *electron bremsstrahlung*
- *decay of neutral π^0-mesons produced in p-p and p-A interactions*
- *inverse Compton (IC) scattering*

The first two mechanisms operate efficiently in relatively dense and extended gaseous environments. Typically, the electron bremsstrahlung contributes to relatively small (sub-TeV) energies, while the π^0-decay γ-rays are important practically in all energy bands above 100 MeV. The IC scattering is the most universal γ-ray emission mechanism regarding both the energy domain and the type of objects. It is a very effective channel for production of TeV γ-rays, but could be a dominant radiation process also at lower energies. It works effectively almost everywhere, from compact objects like pulsars and AGN to extended sources like supernova remnants and clusters of galaxies.

In environments with dense radiation fields, ultra-relativistic electrons release a significant fraction of their energy through the IC channel, unless the energy density of the magnetic field significantly exceeds the energy density of radiation. Therefore, from the point of view of production of high energy γ rays, the synchrotron cooling generally plays a negative (dissipative) role. However, in the so-called *extreme accelerators*, where the particle acceleration proceeds at the maximum possible rate allowed by classical electrodynamics [23], the γ-ray production can be dominated by

- *synchrotron radiation*
  In this regime, the synchrotron radiation of both electrons and protons is released mainly in the γ-ray band. Indeed, the position of the synchrotron cutoff is determined by two fundamental constants, the mass of the charged particle, m and the fine structure constant $\alpha_f = 1/137$: $E_0 = \frac{9}{4}\alpha_f^{-1} mc^2$, approximately 160 MeV for

the electron synchrotron radiation, and 300 GeV for the proton synchrotron radiation [17]. Note that in this extreme regime of particle acceleration and radiation, E_0 does not depend on the magnetic field and the energy of the emitting particle. In a small-scale turbulent medium, the radiation of charged particles due to the magnetic bremsstrahlung can be significantly shifted (compared to the nominal synchrotron radiation) toward higher energies. This happens when the nonrelativistic Larmor radius $R_L = mc^2/eB$ exceeds the correlation length Λ of the magnetic field. Under such conditions, the magnetic bremsstrahlung of charged relativistic particles proceeds to the so-called

- *jitter radiation*
  regime [64, 72, 88]. For the given average magnetic field, the electron cooling timescale is identical to the synchrotron cooling time, thus this radiation regime can be produced with very high efficiency in different astrophysical sources characterized by high turbulence and large average magnetic field. The jitter radiation has distinct spectral features. In particular, compared to synchrotron radiation, its maximum is shifted toward much higher energies [64]. This makes the jitter mechanism an attractive broad-band γ-ray production channel, which, in highly magnetized and turbulent environments, can compete or even dominate over the other major radiation mechanisms.
  In strong magnetic fields, when particles move along the magnetic field lines, they radiate through the so-called
- *curvature radiation*
  This mechanism (see, e.g., [65, 77]) is generally believed to be the most important process responsible for gamma-radiation of pulsars, but it can effectively work also in other astronomical objects, in particular, in the vicinity of rotating supermassive black holes [23, 69]. Thus, in compact and highly magnetized objects, high energy gamma-rays can be effectively produced through *magnetobremsstrahlung of electrons and protons* under which one should understand the radiation of electrons in the magnetic fields in the most general sense. The emission of relativistic particles in a curved magnetic field is often treated in either synchrotron or curvature radiation modes. However, in strong curved magnetic fields, a significant fraction of the energy of relativistic electrons can be radiated away in the intermediate, so-called *synchro-curvature* regime [65]. Because of the persistent change of the trajectory curvature, the radiation varies with the frequency of particle gyration. While this effect can be ignored in the synchrotron and curvature regimes, the variability plays a key role in the formation of the synchro-curvature radiation. The transition from the synchrotron to curvature radiation regimes is determined by the relation between the drift velocity and the component of the particle velocity perpendicular to the drift trajectory. The detailed numerical calculations demonstrate that even very small pitch angles lead to significant deviations from the energy spectrum of the "standard" curvature radiation [65].
  In the presence of dense radiation fields, and effective confinement of relativistic particles in their acceleration sites, γ-rays can be effectively produced at hadronic interactions of very high energy protons with low energy photons, through the so-called

- *photomeson reactions:* $p + \gamma \rightarrow \pi + N...$
  Gamma-rays are produced indirectly, at decays of π^0-mesons, or through the IC scattering and synchrotron radiation of secondary electrons and positrons, the products of $\pi^{\pm}$-meson decays. Another important source for secondary electrons and positrons is the process of
- *Bethe-Heitler pair production:* $p + \gamma \rightarrow e^+e^-$
  In the case of interactions of relativistic nuclei with low-frequency photons, γ-rays can appear also through the *de-excitation* of secondary nuclei produced in the *photo-disintegration interactions*
- $A + \gamma \rightarrow A'^* + X;\ A^* \rightarrow A' + \gamma$
  However, the efficiency of this channel is rather low. Despite some optimistic claims (see, e.g., Ref. [29, 75]), the efficiency of the conversion of energy of accelerated nuclei to γ-rays in optically thin objects appears very small, $\eta \ll 1$. This robust limit [16] renders the role of this mechanism negligible in almost all realistic γ-ray production scenarios. In any case, the photo-disintegration reaction rates are significantly below the rates of the photomeson and Bethe-Heitler pair production (BHPP) processes.
  Because of small cross sections, the interactions of ultra-relativistic protons and nuclei with photon fields generally are "slow" processes. Formally, the small cross sections can be compensated by assuming very dense radiation fields. This could be the case, for example, in compact relativistic objects. However, the γ-ray production rates cannot be arbitrary increased, otherwise the source would become optically thick regarding the process of
- *photon-photon pair production*: $\gamma + \gamma \rightarrow e^+e^-$
  The cross section of this process exceeds, by orders of magnitude, the cross sections of the photomeson and pair production cross sections. Therefore, the condition $\tau_{\gamma\gamma} \leq 1$ implies a very low efficiency of γ-ray production through the photomeson and BHPP processes, unless the protons are very effectively confined in the source. And, vice versa, the high efficiency of the photomeson and BHPP processes implies a large optical depth for the source. In this case, the products of these reactions, electrons, positrons, and gamma-ray, initiate electromagnetic cascades in the low energy photon gas, which wash out the signatures of primary particles. The energy spectrum of γ-rays that emerge from the source is basically determined by the specifics of the cascade development, rather than the initial spectra of primary particles.
  Gamma-rays can also be absorbed due to interactions with the gas and magnetic fields. While the first process generally has little importance in the most astrophysical environments, the pair production of high energy gamma-rays is one of the major absorption processes in objects with very strong magnetic fields.

Relativistic electron-photon cascades

Any γ-ray production process is effective when the corresponding cooling time does not exceed the *source age*, the cooling time of the *competing radiation mechanisms*, and the time of *non-radiative losses* (e.g., due to the particle escape or adiabatic expansion of the source). A prolific γ-ray production typically requires dense targets. However, if the charged particles are well-confined, these conditions could be fulfilled even in environments with relatively low gas or photon densities. Note that the γ-ray production efficiency could be close to 1 even when $t_{\rm rad} \gg R/c$ (R is the characteristic linear size of the production region, c is the speed of light), provided that $t_{\rm rad}$ is comparable to or shorter than the confinement time.

The major γ-ray production mechanisms associated with relativistic electrons have their "counterpart" *absorption* processes—the electron-positron pair production in the matter (the counterpart of bremsstrahlung), in radiation (the counterpart of inverse Compton scattering), and magnetic (the counterpart of synchrotron radiation) fields. The γ-ray production and absorption counterparts have comparable cross sections, therefore, the condition $t_{\rm rad} \geq R/c$ implies an optically thin source; for the corresponding γ-ray absorption process, the optical depth $\tau_{\rm abs} \leq 1$.

In optically thick sources with $\tau_{\rm abs} \geq 1$, both the absorption and radiation processes proceed with high efficiency. Therefore, in such sources, the problem cannot be reduced merely to the absorption effect. The secondary electrons produce a new generation of high energy γ-rays, the annihilation of these photons leads to electron-positron pairs, so an electromagnetic cascade develops.

The characteristics of electromagnetic cascades in the matter have been comprehensively studied in accelerator experiments as well as for interactions of cosmic-rays with the Earth's atmosphere. These results can be applied to a broad class of the so-called hidden sources—the high energy particle accelerators surrounded by dense shells of gas: young pulsars, compact X-ray binaries, supermassive black holes, etc. (see, e.g., Ref. [37]). If the column density of the surrounding gas exceeds $100\,{\rm g/cm^2}$, the protons accelerated in the central source initiate electron-photon cascades. The "hidden sources" are considered as potential high energy neutrino emitters. Note that the γ-ray emission in these objects is not entirely suppressed. The recycled radiation with spectral features determined by the thickness ("grammage") of the gas shell, should be seen in γ-rays, albeit at lower energies.

For typical astrophysical environments, the electromagnetic cascades are more often triggered in magnetic and radiation fields. Formally, as long as the study can be limited by a one-dimensional treatment of the cascade development, the integro-differential equations derived by Landau and Rumer [68] for cascades in the matter, can also be applied, after some modifications (to take into account the specifics of cross sections of relevant radiation and absorption processes), to the cascades in the radiation and magnetic fields. However, for detailed description of the spectral and spatial distributions of electrons and photons, more sophisticated numerical methods of modeling the cascade development are required.

The cascades in the magnetic and radiation fields can be evolved in a broad range of astronomical sources, from compact objects like pulsars and AGN, to the entire Universe (the intergalactic medium).

The development of a cascade in a radiation dominated source is essentially supported by two processes: γ-γ pair production and inverse Compton scattering. It is described by the so-called compactness parameter [52] $l = L\sigma_{\rm T}/Rm_{\rm e}c^3$, where L is target photon luminosity and R is the radius of the source. For l less than 10, the cascade evolves in the linear regime, when the soft radiation produced by electrons does not have a significant feedback effect on the cascade development [26].

A pair cascade in the magnetic fields is effectively triggered when the product of the particle (photon or electron) energy and the strength of the B-field approaches the "quantum threshold" of about $EB \geq B_{\rm crit} m_{\rm e} c^2 \simeq 2 \times 10^7$ TeV Gauss. The cascades initiated in magnetic fields have been best studied for pulsar magnetospheres (see, e.g., Refs. [53, 87]. Such cascades can be developed in other sites as well, e.g., in the Earth's geomagnetic field [80], in blazars [34], etc.

In general, the development of cascades in "pure" environment, i.e., in a medium dominated by matter, radiation or the magnetic field) can be considered as an appropriate and fair approximation. However, under certain conditions, the interference of the processes associated with interactions of cascade electrons and γ-rays with both the ambient photon gas and magnetic field (or matter) can significantly change the character of the cascade development. The impact of such interference, in particular, on the formation of γ-ray spectra, is quite complex and rather sensitive to the choice of the model parameters. Therefore, each specific case is subject to independent studies

In the following sections, the recent results obtained in the VHE γ-ray domain will be discussed with an emphasis on the topical areas related to the origin of galactic cosmic-rays.

Gamma-Ray Signatures of Galactic Cosmic-Rays

Supernova Remnants

General remarks

The energy spectrum of locally measured CRs (electrons,positrons, protons, and nuclei) spans over eleven energy decades extending up to 10^{20} eV. There is a consensus in the CR community that the fraction of CRs below the distinct spectral feature around 1 PeV, the so-called knee, is contributed by relativistic particles of galactic origin. Concerning the possible factories of CRs, the Supernova Remnants (SNRs) are believed to be the major suppliers to the galactic component of CRs. This conviction is based on two arguments. The first one has a phenomenological reasoning. Namely, the CR production rate in the Galaxy, estimated within $\dot{W}_{\rm CR} \approx (0.3-1) \times 10^{41}$ erg/s, can be supported by SNRs if approximately 10% of the kinetic energy of supernova explosions in our Galaxy is released in CRs (for a review, see, e.g., Ref. [37]). The second argument has a more *theoretical* background. It is based on the effective operation of the so-called Diffusive Shock Acceleration (DSA) mechanism under conditions which can be created in young SNRs. DSA not only provides hard power-law spectra of CRs with a required spectral index $p \sim 2$, but also allows effective conversion of the kinetic energy of bulk motion in SNRs to relativistic particles (for a review see Ref. [71]).

Yet, despite the recent progress in observations and theoretical studies, the SNR paradigm of the origin of galactic CRs is not yet firmly confirmed. The direct measurements of CRs are,of course, important, but they cannot address the principal issue related to the localisation and identification of the particle acceleration sites.

The ultimate solution to this long-standing problem can be found only by astronomical means. Namely, the acceleration of CRs in SNRs and their subsequent interactions with the ambient matter make these objects potentially detectable sources of γ-rays and neutrinos. It has been predicted long ago that the detection of hadronic TeV γ-rays—directly from young remnants [46] and/or from nearby dense gas clouds overtaken by the expanding shells [24]—could provide the key test of particle acceleration in SNRs. The optimism of these early works was based on the theoretical estimates of sufficiently high γ-ray fluxes from young SNRs, and on the perception of the great potential of stereoscopic IACT arrays for morphological and spectroscopic studies of TeV γ-rays from moderately extended ($\leq 1°$) sources [46]. Since then, the search for γ-rays from SNRs has been among the top priorities of gamma-ray astronomy. The reports on detection of VHE γ-rays from several prominent SNRs, in particular, from Cas A, Tycho, SN 1006, RX J1713.7-3946, RX J0852-4622 (Vela Junior), RCW 86, clearly demonstrate that young SNRs are effective TeV γ-ray emitters (see, e.g., [19]). Interestingly, the SNR G353.6-0.7 has been first discovered (serendipitously) in TeV γ-rays, and only later confirmed by radio and X-ray observations. This remarkable fact indicates that we might expect a significantly larger number of γ-ray emitting SNRs compared to the conservative estimates based on the current catalogues SNRs. Such an optimistic view is also supported (independently) by another interesting finding of recent years—the discovery of SNRs with exceptionally luminous TeV γ-ray emission: G338.3-0.0 (HESS 1640-465) in the galactic plane at a distance of 10 kpc [4], and SNR N 132D in the Large Magellanic Cloud, a nearby satellite galaxy of the Milky Way. The γ-ray luminosities of both objects in the TeV band are close to 10^{35} erg/s. Given that the total energy budget in accelerated CRs in a typical SNR cannot significantly exceed 10^{50} erg, these large γ-ray luminosities imply presence of very dense gas targets in these objects. Independently from the origin of these targets (which it not yet identified), if this is the case also for some other young SNRs, the chances of detection of SNRs would be dramatically increased. In particular, CTA should be able to reveal all young SNRs in our Galaxy.

While the discovery of γ-rays from SNRs is undoubtedly one of the major achievements of recent years in high energy astrophysics, the fact of the detection of VHE γ-rays itself does not yet prove the dominant role of SNRs in the production of galactic CRs. VHE γ-rays from young SNRs demonstrate the effective acceleration of CRs to energies to 100 TeV, but it is not yet clear that the detected γ-rays are of hadronic origin.

Hadronic or leptonic?

In SNRs, in addition to the γ-ray production in interactions of CR protons and nuclei with the surrounding gas, an equally important γ-ray production process is the IC scattering of ultra-relativistic electrons on the CMB photons. The ratio of contributions of these two channels is sensitive to a few parameters, first of all to the gas density ($\propto n$) and the magnetic field ($\propto 1/B^2$).

If the target for the IC scattering is dominated by CMB photons, for production of a γ-ray photon of energy 1 TeV, an electron of energy 20 TeV is required. The characteristic IC cooling time of these electrons is $\approx 5 \times 10^4$ year. This is longer than the age of young SNRs (from several hundred to several thousand years) by a factor of 10–100 implying that the efficiency of conversion of energy of ultra-relativistic electrons to TeV gamma-rays is quite high, from 1 to 10%. The efficiency of production of hadronic γ-rays is significantly lower. In hadronic interactions, 1 TeV γ-rays are produced by $\approx$20 TeV protons for a characteristic cooling time $5 \times 10^7 (n/1\,\mathrm{cm}^{-3})^{-1}$ year. Thus, at 1 TeV the ratio of production rates of "IC" and "π^0"-decay γ-rays, is $10^3 (W_e/W_p)(n/1\,\mathrm{cm}^{-3})^{-1}$, where W_e and W_p are the total energies in 20 TeV electrons and protons, respectively. One can conclude that for a typical gas density in SNRs $n \leq 1\,\mathrm{cm}^{-3}$, even for a very small electron-to-proton ratio, $e/p = 10^{-3}$, the contribution of the IC component would dominate over the π^0 decay channel, unless the γ-ray production takes place in a dense environment, $n \gg 1\,\mathrm{cm}^{-3}$. Alternatively, the p/e ratio of accelerated particles should exceed 10^3. Given the uncertainty in one of the key issues of the DSA paradigm related to the so-called injection problem (see Ref. [71]), such a large p/e ratio during the process of acceleration cannot be excluded.

Moreover, the relativistic electrons suffer severe radiative losses. Typically, at very high energies the radiative cooling time of electrons, t_{rad}, is shorter than the age of the accelerator, t_0, and their confinement time (time of their escape from the source), t_{conf}. Therefore, the total energy of electrons contained *presently* in the source is determined as $W_e = \dot{W}_e \times t_{\mathrm{rad}}$, where $\dot{W}_e$ is the (quasi) continuous acceleration rate of electrons. On the other, the radiative cooling time of protons typically is much longer than their confinement time in the remnant, therefore, the total energy of protons is $W_p = \dot{W}_p \times t_{\mathrm{conf}}$. Correspondingly, $W_p/W_e = (\dot{W}_p/\dot{W}_e) \times (t_{\mathrm{conf}}/t_{\mathrm{rad}})$. Thus, the ratio W_p/W_e can be larger than the proton-to-electron ratio at the acceleration stage, $\dot{W}_p/\dot{W}_e$, especially at very high energies given that $t_{\mathrm{rad}} \propto 1/E$. In any case, the ratio $p/e \sim 100$, as detected in local CRs, cannot be taken as a reference point; inside the CR sources, it can be much larger than 100.

Since the radiative cooling of electrons is due to the losses through the synchrotron radiation and the IC scattering of electrons, $t_{\mathrm{rad}} = (1/t_{\mathrm{synch}} + 1/t_{\mathrm{IC}})^{-1}$, the condition $t_{\mathrm{rad}} \leq t_0$ does not yet guarantee a high efficiency of γ-ray production which proceeds through the IC scattering. In the case of target photons from the CMB, the γ-ray production efficiency remains high until the magnetic field exceeds a few μG. In an environment with the magnetic field exceeding $10\,\mu$G, relativistic electrons are cooled faster through synchrotron radiation, thus only a small fraction of the energy, $w_{MBR}/w_{\mathrm{B}} \approx 0.1(B/10\,\mu\mathrm{G})^{-2}$, is released in IC γ-rays.

On the other hand, for acceleration of particles to energies beyond 100 TeV by shocks in SNRs, the magnetic field should be much larger than $10\,\mu$G [67]. In other words, a SNR can operate, at a certain stage of its evolution, as a PeVatron, i.e., accelerate particles to energies of 1 PeV and beyond, only if the magnetic field is enhanced, compared to the interstellar medium, at least by an order of magnitude.

This makes the problem of formation of strong magnetic fields in SNRs, e.g., due to the instabilities driven by CRs in the region upstream the shock [35], one of the central issues in the SNR paradigm of the origin of Galactic CRs.

Because of the dramatic suppression of production of IC γ-rays in SNRs with strongly enhanced magnetic fields, the VHE γ-radiation in PeVatrons should be dominated by the hadronic component. However, so far we do not have a substantial evidence of extension of hard γ-ray spectra, without a break or a cutoff, well beyond 10 TeV, which would be an indication of operation of SNRs as PeVatrons [19].

Meanwhile, the origin of γ-radiation detected from more than ten SNRs is still under intense debates. The interpretations within both leptonic and hadronic models have their *pros* and *contras*, and both can formally, within the uncertainties of model parameters, fit the broad-band γ-ray spectra of SNRs. In particular, this is the case of the RX J1713.7-3946, one of the strongest VHE γ-ray sources on the sky.

RX J1713.7-3946 is a young SNR with a clear shell type structure seen in VHE γ-rays (see Fig. 7). At the same time, this γ-ray source has quite unusual features. While the strong nonthermal radio emission and thermal X-rays are two distinct radiation components of shell type SNRs in general, RX J1713.7-3946 shows weak radio emission, and no thermal X-radiation at all.

On the other hand, this object is a powerful nonthermal X-ray and TeV gamma-ray emitter.

The broad-band energy spectrum of RX J1713.7-3946 over 5 decades in energy is shown in Fig. 8. The measurements of *Fermi* LAT [2] and HESS [12] are presented together with the model calculations within the leptonic (left panel) and hadronic (right panel) scenarios of γ-ray production. Since the γ-rays from the IC scattering

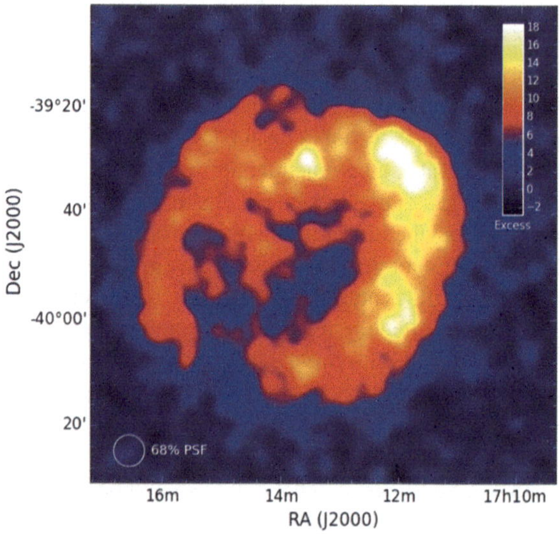

Fig. 7 The VHE γ-ray image of RX J1713.7-3946 above 250 GeV obtained by the H.E.S.S. telescope [13]

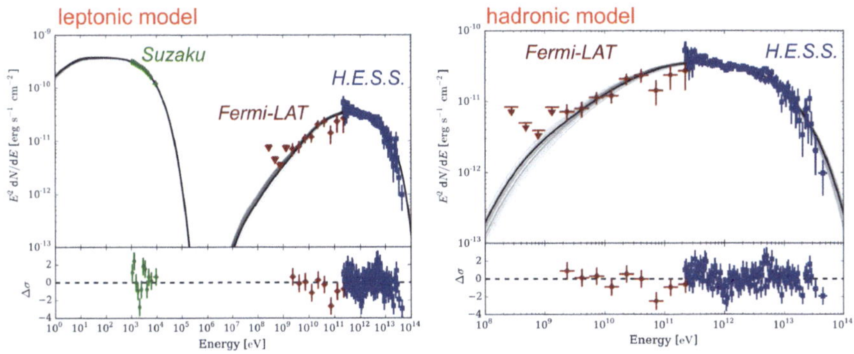

Fig. 8 *Top panel* The spectral energy distribution of gamma-ray emission of RX J1713.7−3946. The *Fermi* and *HESS* γ-ray data are shown together with the theoretical spectra calculated within the "hadronic" and "leptonic" scenarios of γ-ray production (V. Zabalza, private communication)

of electrons, are tightly linked to synchrotron X-ray emission of the same electrons, in the left panel of Fig. 8 are shown also the Susaku [85] X-ray data. The theoretical curves are obtained under the simplifying assumption that the GeV and TeV γ-ray production regions fully overlap and represent a homogeneous emitter within the shell. In both models, the γ-ray points at low energies require hard distributions of the parent particles, namely, power-law indices $\Gamma_p = 1.5$ and $\Gamma_e = 1.7$ for the relativistic protons and electrons, respectively. At higher energies one should require a significant steepening of the electron and proton spectra with a break around 1 TeV for protons and 2.5 TeV for electrons; the power-law indices in this region should be close to $\Gamma_p = 1.9$ and $\Gamma_p = 3$. And, finally, both the proton and electron spectra should have a sharp, exponential type cutoff at energies below 100 TeV to match the highest energy γ-ray points.

These features can be clearly seen in the energy distributions of the parent electrons and protons directly derived from the γ-ray data using the *naima* package [93]. Remarkably, these distributions can be derived model-independently under the assumption that γ-rays are fully contributed either by the IC component of radiation (the "leptonic" scenario) or by the inelastic *pp* interactions (the "hadronic" scenario). Moreover, if the target photons for the IC scatterings are provided by photons of 2.7 K CMB, the detected γ-ray flux also allows precise measurement of the total energy contained in relativistic electrons in the relevant energy band without any additional model assumption. In the case of hadronic ("π^0-decay") γ-rays, the estimate of the total energy in relativistic protons requires an information about the gas density ($W_p \propto 1/n$). The accuracy of these estimates is determined only by the uncertainty in the distance to the source (W_p, $W_e \propto d^2$).

A remarkable (and quite unexpected!) feature seen in Fig. 9 is the break in the distributions of both electrons and protons around 1 TeV. It could have different implications for the leptonic and hadronic models of γ-ray production.

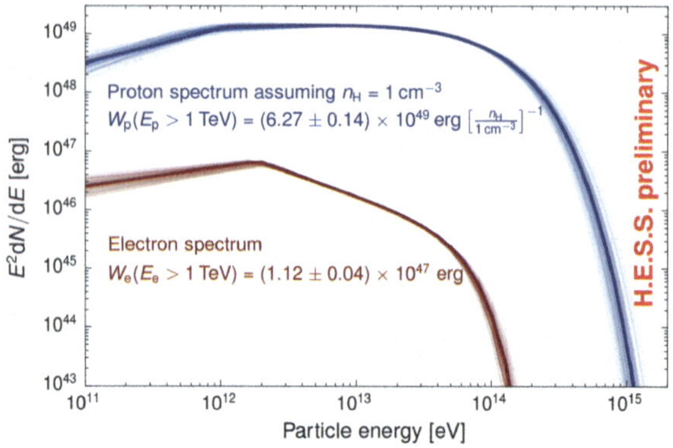

Fig. 9 The energy distributions of the proton and electron populations of parent relativistic particles in RX J1713.7-3946, derived the *Fermi* and *HESS* γ-ray observations from (see Fig. 8), under the assumptions that the detected γ-rays from sub-GeV to multi-TeV energies are dominated either the "π^0-decay" (hadronic) or "IC" (leptonic) components of radiation (V. Zabalza, private communication)

In the leptonic models, the change of the spectral index of electrons from 1.7 to 3 cannot formally be referred to the radiative (synchrotron + IC) cooling break since the latter predicts a change in the slope of the electron spectrum by exactly 1. Nevertheless, the uncertainties in the derived electron spectrum at low energies do not exclude the explanation of observations by the cooling break. However, for the given age of the remnant RX J1713.7-3946 (less than 2000 years), this interpretation requires a very strong magnetic field, $B \geq 100\,\mu$G which is in contradiction with the X-ray measurements. Indeed, assuming that IC γ-rays and synchrotron X-rays are produced by the *same* population of electrons and in the *same* region of the remnant, one should require a quite modest magnetic field of order 15 μG [85]. An alternative assumption that the magnetic field is small but the source is much older than 10^3 year, could, in principle, reproduce the cooling break in the electron spectrum around 1 TeV. But, such an assumption is not supported by the multiwavelength data either. It should be noted, however, that the constraints on the strength of the magnetic field are less robust, if the IC and synchrotron components of radiation are formed in different zones [21]. Such a scenario in young SNRs is possible. In particular, it can be realized in the forward and reverse shocks, in which the magnetic fields are essentially different [94].

In the hadronic models, the spectral break in the proton population cannot be explained by the radiative cooling effects. At low energies, the proton spectrum is significantly harder than the "nominal" E^{-2} type acceleration spectrum predicted by the "standard" models applied to this source [36, 47, 74, 94]. However, such hard spectra in SNRs cannot be excluded, at least formally, from first principles (see, e.g., Ref. [71]). Moreover, the proton spectrum derived from the γ-ray data is representative not for the entire proton population in the nebula, but for the part confined currently in the region where from the γ-rays are observed. Therefore, in the case of inhomogeneous distribution of the gas, the proton spectrum in the densest regions, where the major fraction of γ-rays is produced, can significantly deviate, due to the propagation effects, from the acceleration spectrum. In this regard, the spectral break in the proton population can be explained if γ-rays are produced mostly in dense gas condensations. Because of the slow diffusion, the penetration of low energy particles into the dense cores of these condensations can take longer than the age of the SNR. Correspondingly, the low energy gamma-ray emission can be suppressed [49, 50, 61, 94]. Figure 11 demonstrates the potential of this model [50] to explain the broad-band γ-ray spectrum of RX J1713.7-3946. Interestingly, the concept of the presence of high-density gas targets in SNRs for production of γ-rays is supported by detection of VHE γ-rays from two supernova remnants exceptionally luminous in TeV γ-rays like HESS J1640-465 [4] located in our Galaxy and N132D in the Large Magelanic Cloud [54]. Moreover, it could be a clue for understanding of another puzzle related to RX J1713.7-3946, namely the lack of thermal X-ray emission from the shell. It implies a very low gas density, as low as 0.1 cm^{-3}, which excludes the realization of the *standard* hadronic scenarios [47, 63, 94]. On the other hand, even at a very low gas density of the shell, the contribution of hadronic γ-rays can be dramatically enhanced, if the accelerated protons interact with the dense cores of molecular clouds embedded in the shell. Indeed, the corresponding γ-ray flux may exceed the flux from the shell by a factor equal to the ratio of the total mass of the dense condensations to the mass swept up by the forward shock, provided that all particles freely enter the dense clouds (Fig. 10).

The production of γ-rays in *pp* interactions in *dense* gas condensations embedded in the *low-density* shell seems to be a viable scenario with several distinct spectral and morphological signatures. In this regard, CTA with its significantly improved, compared to the current detectors, flux sensitivity, and angular resolution, has a great potential to reveal and study these features. In particular, this concerns the separation of contributions of individual bright clumps and the diffuse emission of

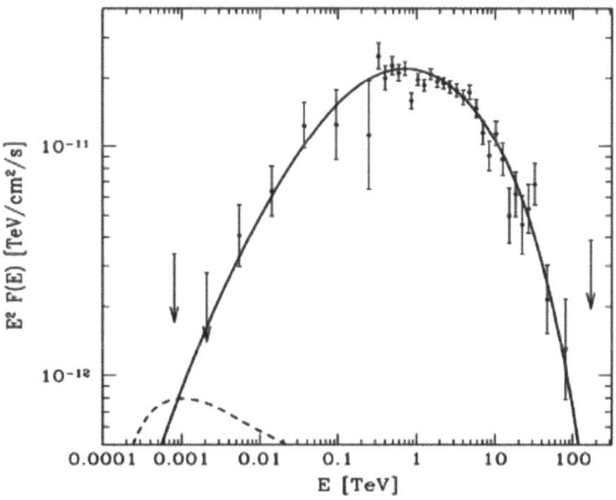

Fig. 10 Gamma-rays from SNR RX J1713.7-3946. Data points from the *Fermi* and HESS observations are shown together with the fluxes calculated under the assumption that *dense* gas condensations are embedded in the *low-density* shell. See Ref. [50] for the details regarding the acceleration of propagation of protons as well as the gas distribution in the shell

the shell between the clumps. The detailed spectral measurements of the clumps in a broad energy band, from 100 GeV to 100 TeV, especially around 1 TeV, would give a unique information about the propagation of relativistic protons in the shell.

Of course, the detailed spectral and morphological studies of young SNRs are of prime interest, far beyond the inspection of a specific hadronic model with the γ-ray production in hypothetical clumps. Such studies provide definitive information about the radiation mechanism in a more general sense. For example, the low magnetic field, which is a key element of any IC model, allows the multi-TeV electrons to propagate to large distances, and thus to fill a large volume. Since the target photon fields are homogeneously distributed, the spatial extension of the secondary IC γ-rays appears quite broad. Because of the strongly amplified magnetic field, as well as the enhanced gas distribution in the compressed shell, the hadronic models predict narrower and sharper spatial distribution of γ-rays as is seen in Fig. 11. However, because of the limited angular resolution of current instruments, it is not an easy task to distinguish between the radial distributions predicted by the two models. This is demonstrated in Fig. 11 where the radial profiles are smoothed with a typical, for the current Cherenkov telescope arrays, point spread function of $\delta\psi = 3$ arcmin. For decisive conclusions about the sharpness of the shell emission, and, therefore, on the nature of parent particles, the detector's angular resolution should be around 1–2 arcmin. This seems to be a difficult but feasible task for the CTA.

Do SNRs Operate as PeVatrons?

The question "leptonic or hadronic?" is one of the important aspects of the current studies of γ-ray emitting SNRs, but it still does not address the most fundamental

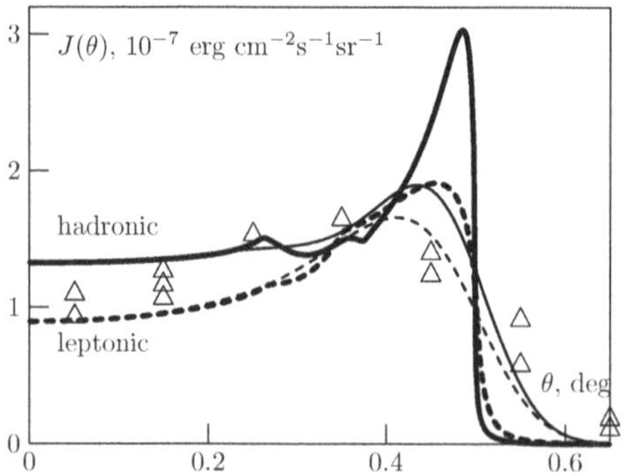

Fig. 11 Radial profiles of 1 TeV γ-rays in RXJ1713.7-3946 calculated in Ref. [94] for the hadronic scenario in the uniform medium (*solid*) and for the leptonic scenario with the unmodified forward shock (*dashed*). The profiles smoothed with a Gaussian point spread function with $\sigma = 3$ arcminute are shown by thin lines. The *triangles* correspond to the azimuthally averaged TeV γ-ray radial profile as observed by HESS [12]

question—whether SNRs are the major contributors to the Galactic Cosmic-Rays? Strictly speaking, this issue should be split into two questions:
(i) whether SNRs can convert the available energy from the SN explosion to relativistic protons and nuclei with an efficiency which would be sufficient for explanation of the locally detected flux of CRs at GeV and TeV energies?
(ii) if yes, whether they can be responsible for the locally observed CR flux up to the "knee" around the energy of 1 PeV?

The proof of the hadronic origin of γ-radiation would be a rather positive answer to the first question. Indeed, almost all of the pronounced young SNRs have been detected as GeV/TeV γ-ray emitters. Moreover, the total energy budget in CRs $W_{\rm p}$ derived from γ-ray data in these objects is close to 10^{50} erg, as required to explain the flux of Galactic CRs. One should note, however, that the estimates of $W_{\rm p}$ depend strongly on the ambient gas density ($\propto 1/n$) and, therefore, contain significant uncertainties.

Concerning the second question, so far no observational evidence is registered which would indicate particle acceleration in SNRs up to PeV energies.

The key energy domain for the straightforward search for galactic PeVatrons is the interval between 10 and 100 TeV. In SNRs with shock speeds of several thousand km/s, the acceleration rate is relatively low, so the synchrotron losses prevent acceleration of electrons to energies more than 100 TeV. Moreover, because of the Klein-Nishina effect, the IC scattering of electrons at such high energies is suppressed. Therefore, the contribution of the IC component of γ-rays above 10 TeV is gradually faded out. On the other hand, the term "PeVatron" implies, by definition, an object accelerating protons with a hard energy (in the case of SNRs, E^{-2} type) spectrum without a break to energies over $E_0 \sim 1$ PeV. The spectrum of secondary γ-rays mimic the spectrum of the parent protons but is shifted toward low energies by a factor of 20–30. Thus, a detection of γ-rays with a hard power-law energy spectrum extending several tens of TeV would imply an unambiguous detection of a PeVatron.

Figure 12 shows the SED of the broad-band emission of a PeVatron initiated by the interaction of accelerated protons with the ambient gas. The radiation consists of two components: γ-rays from the decays of π^0-mesons, and synchrotron radiation of secondary electrons, the products of $\pi^{\pm}$-decays. The lifetime of electrons producing X-rays, $t_{\rm synch} \simeq 1.5 B_{\rm mG}^{-3/2} (E_{\rm X}/1\,{\rm keV})^{-1/2}$ year, is very short ($\leq$50 year) compared to the typical age of young SNRs. Therefore, X-rays could be treated as a "prompt" radiation emitted simultaneously with γ-rays.

The γ-ray luminosity of the source depends on the present total energy of accelerated protons accumulated in the source, and on the density of the ambient matter. The synchrotron luminosity depends also on the magnetic field, but in the X-ray band the dependence is quite weak since the radiation proceeds in the saturation regime when the synchrotron cooling time is shorter than the age of the source. The ratio of the γ-ray to X-ray luminosity depends on the proton spectrum as well as on the particle injection history; typically, for the E^{-2} type proton spectrum, it is between a factor of 3–5.

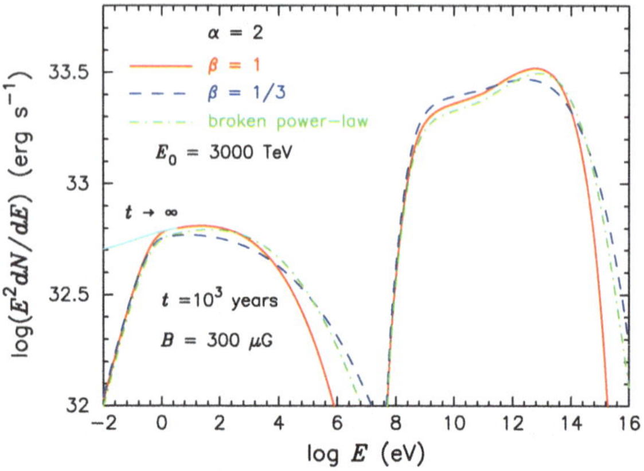

Fig. 12 The luminosities of electromagnetic radiation of a PeVatron produced in interactions of protons with the ambient gas. The calculations are performed for different energy distributions of protons: **a** "power-law with an exponential cutoff," $E^{-\alpha}\exp(-E/E_0)^{\beta}$ with $\alpha = 2$, $E_0 = 3$ PeV for two values of the parameter β - $\beta = 1$ (*solid curve*), $\beta = 1/3$ (*dashed curve*), and **b** "broken power-law" when the spectral index is changed at $E = 1$ PeV from $\alpha = 2$ to $\alpha = 3$. The gas density $n = 1\ \text{cm}^{-3}$, magnetic field $B = 300\ \mu$G, and the age of the source $t = 10^3$ years

The spectrum of highest energy gamma-rays contains important information about the shape of the proton spectrum around the cutoff E_0 which is crucial for identification of acceleration mechanisms in SNRs, as well as for the understanding of the role of different processes responsible for the formation of the knee in the CR spectrum.

The luminosities in Fig. 12 are calculated for a constant injection rate of protons resulting in their current content in the source $W_p = L_p \cdot T \simeq 3 \times 10^{49}$ erg. Thus, to estimate the energy fluxes (in units of erg/cm^{2}s) from an arbitrary PeVatron, one should multiply the luminosities in Fig. 12 by the normalization factor of $A \approx 10^{-44}(\text{nW}_p/3 \times 10^{49}\ \text{erg/cm}^3)(d/1\ \text{kpc})^{-2}$. Remarkably, all galactic PeVatrons up to distances of 10 kpc and with $\text{nW}_p \geq 10^{49}$ erg/cm^3, can be probed by CTA, the sensitivity of which above 10 TeV is expected to be better than 10^{-13} erg/cm^2 s [8].

The extension of the proton spectrum to 1 PeV is crucial for the efficient neutrino production in the 10–100 TeV range—the most optimal energy interval for the high energy neutrino detectors [90]. However, the sensitivity of neutrino detectors, like IceCube or KM3NeT, is limited; only the brightest SNRs in γ-rays are expected to be marginally detected by these instruments. Also, the potential of the neutrino detectors, because of their rather modest angular and energy resolutions, is quite limited for detailed morphological and spectroscopic studies. The strength of neutrino studies is in the absolute assurance of their hadronic origin which makes them unique among other messengers of information of high energy processes in general, and for the search for PeVatrons, in particular.

Finally, the synchrotron X-radiation of secondary ($\pi^{\pm}$-decay) electrons has been proposed as a complementary (to multi-TeV γ-rays and neutrinos) tool for the search of cosmic PeVatrons [18]. In principle, the current *Chandra* and *XMM-Newton* X-ray satellites have adequate sensitivities to perform such studies. However, the energy range of these detectors below 10 keV is not optimal for subtraction of this component of X-ray emission from the synchrotron radiation of directly accelerated electrons as well as from the thermal component of X-rays.

The energy range of hard X-rays between 10 and 100 keV is more suitable for such studies. The synchrotron emission of secondary electrons for typical parameters of SNRs operating as PeVatrons effectively extends well beyond 10 keV. Meanwhile, at these energies, both the X-ray flux of the thermal plasma and the synchrotron emission of directly accelerated electrons is suppressed given that in young SNRs, the SED of synchrotron radiation of directly accelerated electrons peaks, independent of the magnetic field, at $h\nu \sim 1$ keV [21]. Remarkably, the spectrum of secondary synchrotron radiation in the most informative (from the point of view of the mechanism of particle acceleration) cutoff region is smoother and broader than the γ-ray spectrum around the cutoff (see Fig. 12). This opens unique possibilities, given the superior energy and angular resolutions of the current (NewStar) and the planned hard X-ray imagers, for exploration of the physics of cosmic PeVatrons.

The cosmic *PeVatrons* should not necessarily be linked to SNRs. Some other objects like stellar winds, OB associations, superbubbles (see, e.g., Ref. [56]) or remnants of γ-ray bursts in our Galaxy [30] have been proposed as effective accelerators of particles to PeV energies and as potential contributors to galactic cosmic-rays. One can anticipate that the galactic plane survey by CTA might result in an exciting discovery of a new class of $E \gg 10$ TeV gamma-ray sources of unknown origin. The outcome could be a dramatic revision of the concept of the origin of galactic CRs without a major contribution from SNRs.

Star Forming Regions

The collision of stellar winds and the multiple shocks initiated by supernovae explosions in clusters of massive stars have long been suggested as possible alternative to individual SNRs as the major "factories" of the Galactic CRs (see, e.g., Refs. [39–41, 78]. The massive stars produced at the collapse of giant molecular clouds typically form compact groups and remain close to each other during their entire life. The core-collapse supernovae and colliding stellar winds in such dense stellar clusters can drive giant *superbubbles* filled with a hot and highly turbulent plasma. The effects related to the stellar winds of massive stars, and the collective supernova explosions result in huge energy release in the form of multiple (interacting with each other) shock waves inside the superbubbles and thus create favorable conditions for particle acceleration. An attractive aspect of particle acceleration in supperbubbles is the maximum achievable energy of accelerated particles. Namely, in superbubbles, the acceleration of protons to PeV energies can be realized more efficiently than in the shocks of individual SNRs. This makes the associations of massive OB stars

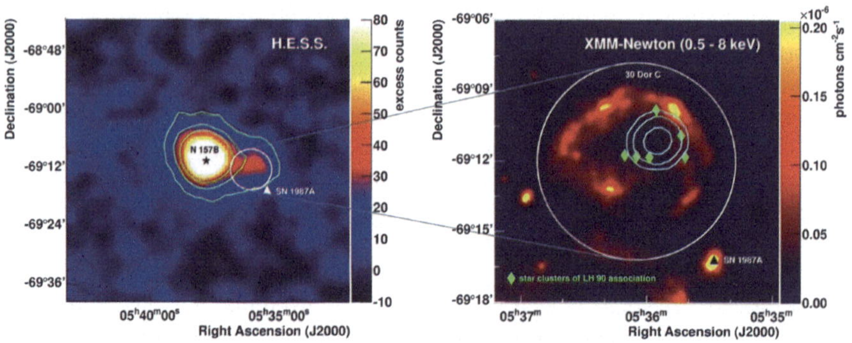

Fig. 13 (*Left panel*) The VHE γ-ray image of a part of the Large Magelanic Cloud observed by the HESS array [54]. It contains the star formation region 30 Dor C. The *green lines* represent contours of 5, 10, and 15 σ statistical significance of the γ-ray signal. (*Right panel*) X-ray image of the region of 30 Dor C. The superimposed *cyan lines* represent contours of 68, 95, and 99% confidence levels of the position of the gamma-ray source. *Diamonds* denote the positions of star clusters of the LH 90 association

and the superbubbles potential PeVatrons and likely sites for the production of the highest energy galactic CRs around the "knee."

The TeV γ-ray observations of massive stellar clusters, such as Westerlund 1 [3] and the Cygnus region [1, 9, 28] in our Galaxy, as well as 30 Dor C in the Large Magelanic Cloud (LMC), a satellite galaxy of the Milky Way located at a distance of 50 kpc [54], support, to a large extent, the hypothesis that star forming regions are sites of high energy particle acceleration (Fig. 13).

VHE γ-ray emission from 30 Dor C, the largest starburst region in LMC, has been serendipitously discovered during the survey of LMC by the HESS array of telescopes [54]. The γ-ray luminosity in the 1–10 TeV interval is estimated $(0.9 \pm 0.2) \times 10^{35}\,\mathrm{erg\,s^{-1}}$. The TeV explanation of this emission by interactions of CR protons with the surrounding gas requires $W_\mathrm{p} = (0.7 - 25) \times 10^{52}\,(n/1\,\mathrm{cm}^{-3})^{-1}$ erg. On the other hand, the plasma density based on the thermal X-ray emission is estimated within $n \approx 0.1$–$0.4\ \mathrm{cm}^{-3}$ [33]. This implies an unrealistically large total energy in relativistic protons, $W_\mathrm{p} \sim 10^{52}$ erg. In particular, if the acceleration of protons is initiated by supernovae events, the number of explosions experienced by 30 Dor C should exceed 100! However, the requirement to the total energy can be dramatically reduced assuming that γ-rays are produced in the dense outer shell or in compact dense clumps inside the rarified interior of the bubble [50]

For the leptonic scenario of the TeV emission, assuming that the synchrotron X-rays and TeV gamma-rays are produced in the same region by the same population of electrons, the broad-band SED requires an average magnetic field of $15\,\mu$G and a total energy energy in accelerated multi-TeV electrons of $\sim 4 \times 10^{48}$ erg (see Fig. 14). Although these seem quite reasonable numbers, the leptonic model requires an extremely rarified bubble with a density as low as $n \approx 5 \times 10^{-4}\ \mathrm{cm}^{-3}$.

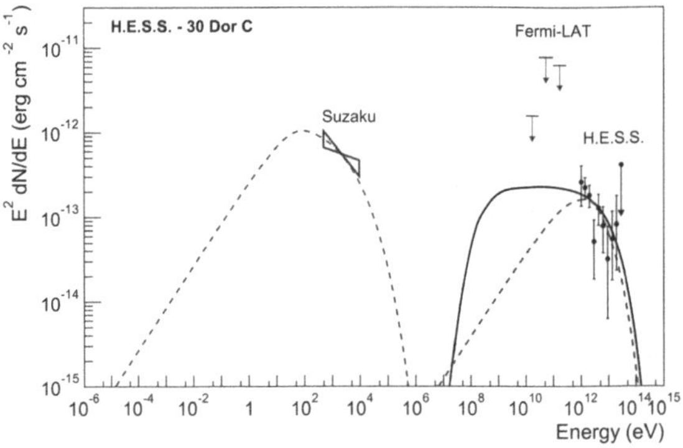

Fig. 14 The spectral energy distribution of 30 Dor C (from Ref. [54]). The X-ray and γ-ray data are shown together with the theoretical curves corresponding to the hadronic (*solid lines*) and leptonic (*dashed lines*) models

The X-ray and γ-ray observations tell us that the particle acceleration takes place, most likely, in a very large and fast expanding shell. Together with the highly excited plasma with turbulent magnetic field, this may provide the confinement of particles for million of years, and thus create the conditions required for acceleration of protons to energies exceeding 1 PeV. In this regard, the future observations of this source with CTA, with adequate sensitivity for $\geq$10 TeV γ-rays may prove that this object operates as a Pevatron.

TeV γ-rays have also been reported from the direction of Westerlund 1 [3], the most massive ($M \geq 10^5_\odot$) stellar cluster in our Galaxy. The source dubbed as HESS J1646-458 is unique amongst the other sources reported by HESS in the sense of its angular size ($\approx$2°) and very hard power-law spectrum with a photon index 2.2, extending from 0.5 TeV up to 20 TeV without any indication of a steepening or a cutoff (see Fig. 15). The large angular size implies a huge extension of the source over 100 pc or more. This almost excludes an association of the source with an individual pulsar wind nebula or a supernova remnant. This also makes quite unlikely the leptonic (IC) origin of γ-rays because the electrons of the energy of $\sim$100 TeV cannot propagate over such large distance. Also, the hard power-law energy spectrum of γ-rays without an indication of characteristic hardening at low energies, does not agree with typical spectra of IC γ-rays produced either in a pulsar wind nebula or a supernova remnant.

The association of HESS J1646-458 with Westerlund 1 is likely, although the γ-ray image of HESS J1646-458 is larger, by an order of magnitude than the size of Westerlund 1. This can be explained by the fact the parent protons are accelerated in Westerlund 1, then escape and later interact with the surrounding material and a nearby HII region. This scenario is further supported by the morphology of the HI gas in this region, which shows a bubble-like structure where the dense gas partly overlaps with the regions of strong TeV emission. The total energy released during the age of Westerlund 1 by stellar winds and supernovae explosions is estimated

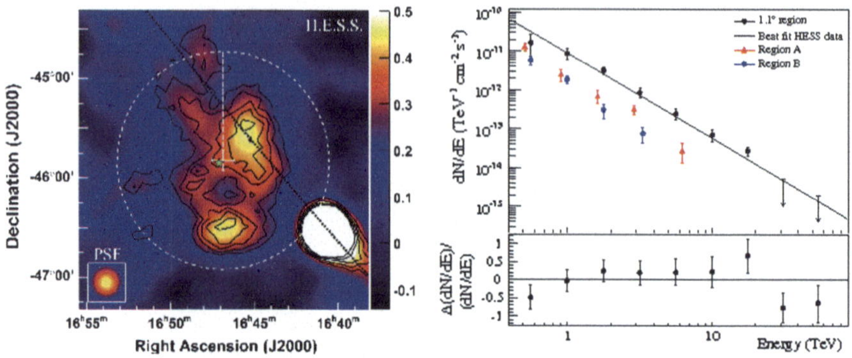

Fig. 15 The image and energy spectrum of the γ-ray source HESS J1646-458 associated with Westerlund 1. *Left panel* The image of HESS J1646-458 surrounding Westerlund 1 (indicated with a *green cross*). The *white cross* corresponds to the best-fit position of the VHE gamma-ray emission and the *black dashed line* indicates the Galactic plane. The *green lines* represent contours of 5, 10, and 15 σ statistical significance of the γ-ray signal. *Right Panel* The differential VHE *gamma*-ray energy spectrum of HESS J1646-458. *Arrows* indicate the 95% upper limits for spectral bins which are compatible with a zero flux within 1 σ. Also shown are spectra as obtained for region A and B (see [3] for details). Figure from [3]

as $W_{\rm tot} \approx 3 \times 10^{53}$ erg, while the energy contained in the parent relativistic protons should be $W_{\rm p} \approx 3 \times 10^{51}(n/1\ {\rm cm}^{-3})^{-1}$ erg in order to explain the detected γ-ray flux [3].

The extension of the hard power-law γ-ray spectrum of HESS J1646-458 beyond 10 TeV and its association with Westerlund 1 makes the latter a possible detectable neutrino source. It has been argued that the supernovae shocks in Westerlund 1 should be able to accelerate protons to $\geq$40 PeV. Their escape from the accelerator sites and penetration into the surrounding gas clouds results in the production of neutrinos with fluxes sufficient for interpretation of a fraction of the events detected by IceCube from the inner Galaxy [38].

The Cygnus region, located at a distance of 1.5 kpc, hosts some of the most remarkable star forming regions in the Milky Way. The total mass in the form of molecular gas is estimated of a few million solar masses, and the total mechanical stellar wind energy input is about 10^{39} erg s^{-1}.

At TeV energies, the Cygnus region shows two distinct features, one possibly connected to the complex Cygnus X, another to the so-called Fermi Cocoon discovered at GeV energies [7].

Within the Cygnus X complex, HEGRA collaboration serendipitously discovered the first TeV γ-ray source which does not have an obvious counterpart at other wavelengths, TeV J2032+4130 [9]. TeV J2032+4130 was also the first extended source to be discovered in the VHE range, with a spectrum fitted as a pure hard power-law spectrum, with no evidence for a cutoff up to 20 TeV. The HEGRA result of this source was later confirmed by the MAGIC and VERITAS telescopes, as well as by observations with the air shower particle detectors Milagro and ARGO.

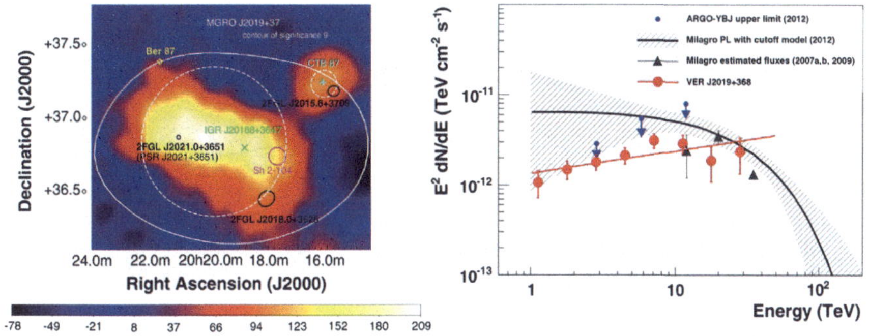

Fig. 16 *Left panel* VHE γ-ray excess map of MGRO J2019+37 region as observed by VERITAS above 600 GeV. The locations of possible counterparts are marked using different colors. *Right panel* The spectral energy distribution of MGRO J2019+37/VER J2019+368 based on the Milagro and VERITAS measurements

In the Cyg OB1 region, the Milagro collaboration discovered another extended multi-TeV source, MGRO J2019+37 [1]. Later, the observations with the VERITAS array of IACTs resolved the Milagro source into two sources [28] (see Fig. 16). One of them, VER J2019+378, is a bright 1° extended source that is likely powered by the nearby star formation region Sh 2-104. It has a very hard power-law spectrum with a photon index extending to at least 30 TeV.

Galactic Centre

The Galactic Centre (GC) is an extraordinary site in our Galaxy. It can be used as a "local" laboratory for the study of the complex processes and phenomena in the central cores of other galaxies. Although the central supermassive black hole (SMBH) in our Galaxy currently operates in a quite modest regime, in particular, compared to the Active Galactic Nuclei (AGN), it is believed that it was more active in the past. Formally, the mass of the central SMBH allows much higher luminosities in various energy bands supported by the energy release through the accretion of the gas onto SMBH. The energy release can proceed in different forms, in particular, through acceleration of relativistic particles either in the vicinity of the SMBH (close to the "event horizon") or at large distances from the central engine, e.g., at the termination of relativistic outflows (a jet or a wind) that originate close to the black hole and carry a substantial fraction of energy extracted in the accretion disk. Because of diffusion in the interstellar magnetic fields, the relativistic particles, unlike the electromagnetic radiation, "remember" the activity in GC in the past. Thus, they can be used as a unique tool for exploring the time-evolution of the GC.

The nonthermal phenomena that take place in the galactic centre may have a direct link to the origin of galactic CRs. The SMBH itself can operate as a prolific "factory"

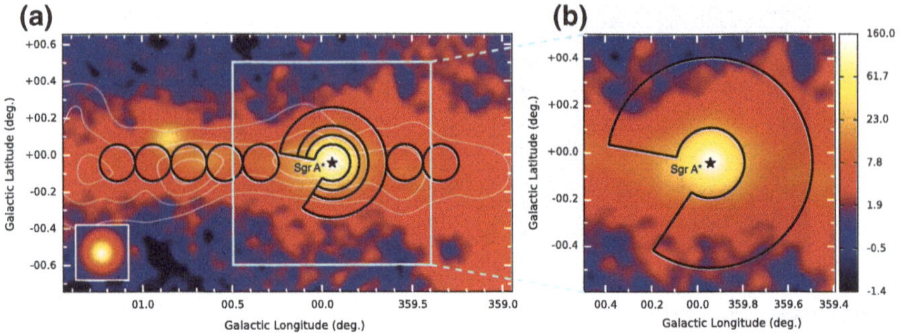

Fig. 17 VHE γ-ray image of the Galactic Centre region (from Ref. [55]). *Left panel* The *black lines* outline the regions used to extract the energy density of CRs. White contours indicate the density distribution of molecular gas. *Right panel* Zoomed view of the inner ~70 pc and the contour of the region used to extract the spectrum of the diffuse emission

of relativistic particles and contribute substantially to the flux of the locally observed CRs. Moreover, the galactic centre hosts, besides the SMBH, some other remarkable objects which also can be effective particle accelerators. In this regard, the galactic centre is considered as one of the highest priority targets in gamma-ray astronomy.

The galactic centre harbors a bright, point-like γ-ray source surrounded (HESS J1745-290), by a diffuse component of VHE γ-ray emission [55] (see Fig. 17). The central source positionally coincides with the location of the compact radio object Sgr A*, a suspected SMBH, at the dynamical center of the Galaxy. Unlike the radio and X-ray bands, no variability has been observed so far in γ-rays. This disfavors but still cannot exclude Sgr A* as a possible counterpart of HESS J1745-290.

On the other hand, the upper limit on the angular size of a few arcminutes is still large to exclude any sources located within the central $\leq$10 pc region. It could be, in particular, an extended source with a size of several pc. Moreover, it could be not an independent source, but an unresolved core of the diffuse component. In this interpretation, it peaks toward the direction of Sgr A* because of the higher concentration of both the gas and relativistic particles [15, 43]. However, the central source and the diffuse component have different energy spectra. While the spectrum of the diffuse component extends to several tens of TeV without any indication of a cutoff, the energy spectrum of the central region shows an apparent steepening that starts before 10 TeV (see Fig. 18). This excludes the common origin of the central source and the diffuse component unless one provides an explanation of the cutoff in the spectrum of the central source (see below).

Gamma-Rays from the central molecular zone

Only two radiation mechanisms can be responsible for the diffuse VHE γ-ray emission of CMZ—interactions of relativistic protons and electrons with the ambient gas and the radiation fields, respectively. However, in the CMZ, the γ-ray production proceeds under conditions which exclude the leptonic origin of γ-rays. For an explanation of the hard γ-ray spectrum, as shown in Fig. 18, the power-law spectra of

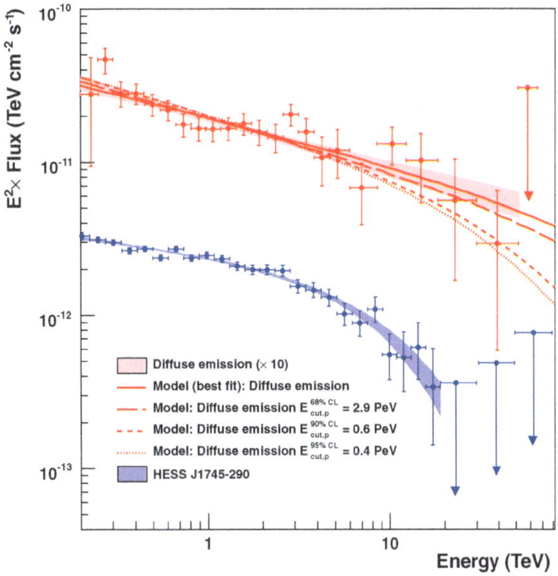

Fig. 18 VHE γ-ray spectra of the diffuse emission and the central source (HESS J1745-290) (from Ref. [55]). The 1σ confidence bands of the best-fit spectra of the diffuse and HESS J1745-290 are shown in *red* and *blue shaded areas*, respectively. The *red lines* show the numerical computations assuming that γ-rays result from the decay of neutral pions produced by proton-proton interactions. The fluxes of the diffuse emission spectrum and models are multiplied by 10

protons and electrons should extend to ~1 PeV and ~100 TeV, respectively. Apparently, these particles should effectively propagate and fill the entire region of CMZ. In the case of electrons, the acceleration to the energy of 100 TeV seems almost unrealistic given the severe radiative (IC and synchrotron) losses in the GC. For the same reason, they hardly can escape the sites of their acceleration and propagate over distances of tens of parsecs. The correlation of the brightness distribution of VHE γ-rays with the location of massive gas complexes of the CMZ is an independent argument which points toward a hadronic nature of the diffuse emission.

At the same time, the spatial distribution of the gas alone is not the only factor shaping the observed morphology of the CMZ seen in VHE γ-rays. The second factor is the spatial distribution of parent particle—the accelerated protons and nuclei. The latter depends on the location of the accelerator(s), the history of (time-dependent) injections of these particles, and the character of their propagation. In general, one should not expect homogeneous distributions of these particles in space and time, as it often assumed in the treatments of calculations of the diffuse γ-ray component. Therefore, there is no reason to expect a linear correlation between the brightness distribution of γ-radiation and the gas column density.

Indeed, the comparison of the observed flux of the VHE emission throughout the CMZ, combined with the information on the gas spatial distribution in this region, reveals an order of magnitude enhancement of the CR density in the CMZ compared to the "sea" of CRs that fill the Galactic disk (see Fig. 19). Moreover, the derived radial distribution of CRs appears close to $1/r$, where r is the distance to the GC. The radial distribution of the CR density implies that (i) the accelerator has to be centrally located within CMZ (in our case, it implies anywhere within the unresolvable

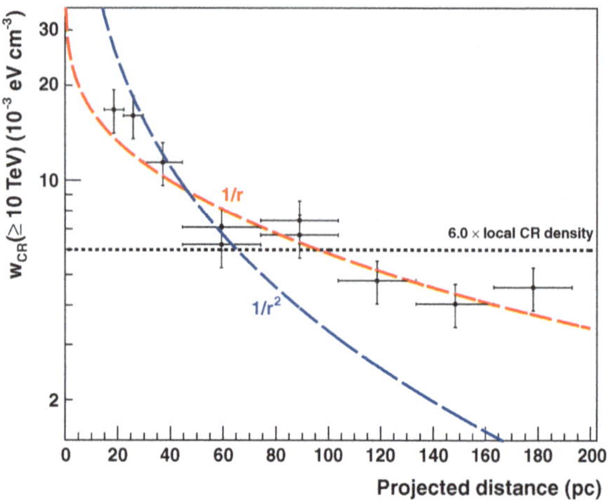

Fig. 19 Spatial distribution of the CR density versus projected distance from Sgr A* (from [55]). The best fit of a $1/r^{\alpha}$ profile to the data is found for $\alpha = 1.10 \pm 0.12$

inner 10 pc region of GC); (ii) the accelerator should supply accelerated particles at a (quasi)continuous rate on time scales determined by the diffusion of $\geq$10 TeV protons in the CMZ (10^4 years, if the diffusion coefficient in the CMZ does not significantly deviate from the one in the Galactic Disk). In a complex environment, such as the GC region, one may expect some deviation from the strict $1/r$ dependence, in particular, due to a radial variation of the diffusion coefficient. Nevertheless, the apparent increase in the CR density toward the GC excludes the origin of CRs as a result of a single burst-like event, which would imply a rather flat radial distribution of CRs. The HESS data also disfavor $1/r^2$ type distribution of CRs, which would be realized in the case of advection of CRs in a wind.

It should be noted that already in the first observations of GC by HESS, a lack of γ-ray emission has been found from a massive gaseous complex at $\sim 1.3°$ angular (or $\sim$200 pc projected) distance from the GC. The "deficit" of emission is apparent when comparing with the γ-ray flux detected from other giant clouds in CMZ, in particular, from the complex Sgr B2 [11]. A solution suggested that time was a burst-type scenario of particle acceleration. Namely, assuming that the accelerator in the past was more active than at the present epoch, the low rate of γ-ray production could be explained by the time-delay of arrival of accelerated particles to that region located in outskirts of CMZ [11]. However, this scenario now seems disfavoured. A solution that would satisfy both requirements, namely the $1/r$ decrease of the CR density within $r \lesssim 100$ pc region, and its significant drop before $\sim$200 pc, is possible if the so-called diffusion radius (see, e.g., Ref. [18]) is located somewhere between 100 and 200 pc, i.e., $r_{\rm diff} \sim 150$ pc. For a given diffusion coefficient, the duration of the recent activity of the accelerator can be estimated as $t = r_{\rm diff}^2/4D \approx 2 \times 10^3$ year, where a diffusion coefficient $D(10{\rm TeV}) \approx 10^{30}$ cm^2 s^{-1} close to the value that characterizes the diffusion of multi-TeV CRs in the Galaxy, has been adopted. Unfortunately, the poorly known diffusion coefficient in the CMZ introduces rather large uncertainty in this estimate.

Figure 18 shows the VHE γ-ray spectrum of the diffuse emission, extracted from the central ~100 pc region excluding the source HESS J17145-290 [55]. Remarkably, the spectral fit is well described by a pure power-law of index $\simeq$2.3 extending up to ~50 TeV, without any indication of an energy cutoff, and a luminosity of $L_\gamma(> 1\text{TeV}) \sim 5.7 \times 10^{34}$ erg/s. This is the highest energy ever reached for a hard γ-ray source following a pure power-law spectrum. The hadronic nature of the diffuse VHE emission implies that the γ-ray emission results from the decay of neutral pions produced by relativistic protons interacting with the interstellar gas. Thus, an injection proton spectrum is fitted to the γ-ray data assuming either a pure power-law or a power-law with an exponential cutoff. No exponential cutoff is found in the injection proton spectrum, and a lower-limit on the cutoff energy, for the power-law index of 2.4, can be placed at 2.9 PeV, 0.6 PeV, and 0.4 PeV at 68%, 90%, and 95% confidence level, respectively (see Fig. 18). This is the first indication of the existence of a hadronic accelerator capable of providing CRs at PeV energies, i.e., a cosmic *PeVatron*. Using the CMZ mass distribution from CS maps [89], and assuming that the CRs are produced at the GC and then propagate diffusively, the Pevatron injection power of $\geq$10 TeV protons is estimated at $L_p \sim 2 \times 10^{38}(D_{10\text{TeV}}/10^{30}\,\text{cm}^2\,\text{s}^{-1})$ erg/s, where $D_{10\text{TeV}}$ is the diffusion coefficient of 10 TeV protons, and $10^{30}\,\text{cm}^2\,\text{s}^{-1}$ is its typical value in the Galaxy. Thus, we can conclude that, given the pronounced gradient of the CR density toward the GC and the rather large injection power derived above, a potential source in this region capable to supply adequate "fuel" for operation of such a powerful proton accelerator, could be Sgr A* itself.

Possible links between the PeVatron, Sgr A* and HESS J1745-290

The spectrum of the central source, HESS J1745-290 is shown in Fig. 18. The spectrum is well described by a power-law of index $\simeq$ 2.15 with an exponential cutoff at $\simeq$10.7 TeV. Of course, HESS J1745-290 and the Pevatron should not necessarily be connected. There are several other possible counterparts for HESS J1745-290 in the central 10 pc zone, e.g., G359.95-0.04, a pulsar wind nebula not far from Sgr A* [58, 91], a hypothetic "BH nebula" around Sgr A* [31], or Sgr A* itself (the inner parts of the accretion flow close to the gravitational radius of the SMBH) [14]).

However, the location of both sources in the inner 10 pc makes a compelling case for their intrinsic connection. Interestingly, on scales $r \leq \lambda$, where $\lambda = 3D(E)/c$ is the CR scattering mean free path and c is the speed of light, the particle propagation has a ballistic rather than diffusive character [43]. For the characteristic CR diffusion coefficient in the Galactic Disk, the mean free path for 10 TeV protons would be $\lambda \sim 30$ pc. In this case, the luminosity of γ-rays at $E \geq 1$ TeV is determined by the injection power of $\geq$ 10 TeV protons via the simple relation $L_p \approx L_\gamma(t_{\text{pp}\rightarrow\text{pion}^0}/t_\text{b}) \approx L_\gamma(l\ n/1.5 \times 10^{26}\ \text{cm}^{-2})^{-1}$, where $t_\text{b} = l/c$ is the propagation time of protons in a ballistic motion through a γ-ray production region of size l and a gas density n. Moreover, γ-rays resulting from interactions of protons with gas are produced almost parallel to the proton propagation direction. As a consequence, although the γ-rays are produced in an extended region, they will be detected, independent of the gas distribution, as photons from a *point-like* source at the location of the particle accelerator [15]. This could explain not only the lack of variability of the γ-ray signal from the central point-like source, but also the proximity of its gravity center to the position of Sgr A* [5]. A favorable site for γ-ray production within a few pc of this source could be the circumnuclear ring, a dense ($n = 8 \times 10^4$ cm^{-3}) region confined between 1.2 and 3 pc [70]. However, because of the large inclination of the circumnuclear ring [48], most of the ballistically moving protons will just pass next to it without interacting with its gas. The γ-rays are, therefore, most likely produced within the central cavity, a less dense environment ($n \sim 2 \times 10^3$ cm^{-3}) with a radial extension of $\lesssim$3 pc and an average column density $\lesssim 2 \times 10^{22}$ cm^{-2}. Thus, the injection power of $\geq$ 10 TeV protons should exceed 4×10^{38} erg s^{-1} in order to explain the γ-ray luminosity of the central source. This estimate is rather close to that derived above from the γ-ray flux of the diffuse emission. Notably, the two independent estimates refer to quite different timescales characterizing the CR propagation in the central cavity and the CMZ—less than 10 year and more than 10^3 year, respectively.

The results presented in Fig. 18 demonstrate that one can explain, for a reasonable value of the diffusion coefficient in the CMZ and using the measured gas distribution within the central several pc region [48, 89], the absolute fluxes of γ-rays of both the core and the extended diffuse emissions. Moreover, the detailed numerical studies [43, 70] show that, indeed, the "runaway CR" scenario can explain the broad-band γ-ray emission of the GC, including three distinct features—the flat spectral energy distribution at GeV energies, the steepening in the GeV-TeV transition region, and the recovery of a flat spectral energy distribution at TeV energies (see Fig. 20).

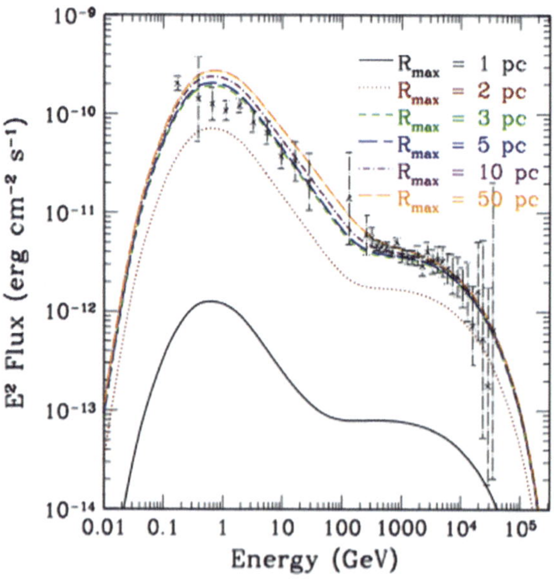

Fig. 20 Energy spectra of gamma-ray emission from GC (from Ref. [70]). The *Fermi* and HESS data are shown together with calculations of gamma-rays from *pp* interactions within radial cones of various size up to 50 pc. The flux falls off rapidly after 3 pc because the main contribution comes from the 1.2–3 pc circumnuclear ring

In the previous studies of the concept of "runaway protons," it has been suggested that the cutoff in the γ-ray spectrum of the central source at $\approx$10 TeV would require a corresponding cutoff in the spectrum of parent protons at $\approx$100 TeV [43, 70]. This, however, is not supported by the recent HESS results regarding the energy spectrum from the extended region. Thus, the attractive idea of explaining the entire γ-ray emission of the GC by runaway protons from the *same* centrally located accelerator would not work unless one finds an alternative explanation for both the cutoff at 10 TeV in the spectrum of the central source, and the lack of it in the spectrum of the extended diffuse component. The simplest possible solution of this problem would be an assumption that the central TeV source has different (from the diffuse component) origin, for example, is a result of the inverse Compton scattering of ultra-relativistic electrons in a pulsar wind nebula. However, the problem should not be exaggerated, given the essentially different "ages" of the parent protons responsible for γ-radiation of the central point-like source and the surrounding diffuse emission. This effect could thus be caused by the change of acceleration efficiency of the highest energy particles. Alternatively, the accelerator could have been fully switched off over the last decades, and while the main fraction of accelerated particles are still confined in the source, the highest energy ($E \geq 100$ TeV) cosmic-ray particles might have already left the source. Apparently, both scenarios can readily explain the cutoff in the γ-ray spectrum of the central source. The search for a variability of the central source on decade timescales, in particular, in the cutoff region of the spectrum, seems a straightforward approach for inspecting this possibility. Another possibility is that the spectral cutoff in the central source can be caused by absorption due to photon-photon pair production in the ambient infrared radiation field. The spectral energy distribution of the infrared emission from the inner several parsecs of the GC region peaks at a wavelength $\lambda_{IR} \sim 20\,\mu$m, thus most significant absorption of γ-rays should happen at energies of 10–20 TeV, given that the cross section of the process achieves its maximum when $\lambda_{IR} \sim (E/1\,\text{TeV})\,\mu$m (see, e.g., [18]). To explain the cutoff in the γ-ray spectrum of the central source by photon-photon interactions, the optical depth at 20 TeV should be close to $\tau = 1$. The conservative calculations show, however, that the diffuse IR radiation observed from the 10 pc region of GC cannot provide such a large optical depth.

A possible way to gain the missing factor (from 3 to 10, depending on the chosen geometry) in the optical depth, we can assume that both the observed IR and γ-ray photons originate from dense clumps. Indeed, the optical depth of photon-photon pair production is proportional to the ratio $\tau \propto L_{\rm IR}/R_{\rm IR}$, where $L_{\rm IR}$ is the luminosity of the infrared radiation of a spherical source of radius $R_{\rm IR}$. Therefore, assuming that the gas is concentrated in N dense clumps where both the γ-rays and IR radiation are produced, and assuming that the average size of these clumps does not exceed $1/10N$ fraction of the radius of the entire region, the average ratio L/R of each clump could be increased by a factor of 10, and correspondingly the optical depth of

20 TeV γ-rays would increase by the same factor. These simple estimates show that the several-pc core region can indeed be optically thick for $\geq$10 TeV γ-rays, and thus explain the appearance of the cutoff in the spectrum of the central source contrary to the diffuse component of radiation, which is produced in more extended regions and, therefore, does not suffer from noticeable absorption.

Multimessanger Implications of the PeVatron in GC

The PeVatron in the Galactic Centre has unique "multiwavelength" and "multimessenger" signatures carried out by the neutral secondary products resulting from hadronic interactions of highest energy ($E \geq 0.1$ PeV) parent protons. The secondaries produced at low energies, namely the MeV/GeV γ-rays and the radio synchrotron emission of primary and secondary electrons, are also important, but, strictly speaking, their information is rather circumstantial as long as it concerns the exploration of the nature of the PeVatron.

The neutral and stable products of interactions of relativistic protons—gamma-rays and neutrinos—generated through the production and decays of π^0, π^+ and π^- mesons, receive approximately 10% of the energy of primary protons. Therefore, the multi-TeV secondary neutrals carry the most straightforward and model-independent information about the primary PeV protons. This concerns, first of all, $\geq$10 TeV γ-rays because of the adequate performance of IACT arrays for deep spectroscopic and morphological studies.

Although at such high energies the non-leptonic nature of γ-rays hardly could be debated, the most straightforward inspection of the hadronic origin of diffuse emission of CMZ would be the detection of multi-TeV neutrinos spatially correlated with γ-rays. The predictions regarding the detectability of high energy neutrinos by the current or planned telescopes are robust (i.e., less sensitive to the spectral shape) when normalized at the neutrino flux at energy of $\approx$20 TeV. Namely, neutrinos can be detected by a km^3 volume detector if the differential flux of accompanying γ-rays at 20 TeV exceeds 10^{-12} TeV cm^{-2} s^{-1} [90]. Since the γ-ray fluxes in Fig. 18 are quite close to this value, a km^3-scale detector located in the Northern Hemisphere should be able to detect, after several years of exposure, a positive signal from GC.

Other messengers of information about the PeVatron in GC are relativistic neutrons which are produced in the reaction $pp \rightarrow pn\pi^+$. However, because of the neutron's lifetime at rest $\tau \simeq 10^3$ s, only particles of energy exceeding $E_{\rm n} = m_{\rm n}d/(c\tau) \simeq 10^{18}$ eV could reach us during the 8.5 kpc journey from the Galactic Centre. Thus, this channel of information can present a practical interest only if the spectrum of the PeVatron in GC extends beyond 10^{18} eV.

The γ-ray fluxes from the GC region are almost insensitive (for any reasonable CR diffusion coefficient) to the history of operation of the proton Pevatron on large timescales, $t \geq 10^4$ year. Such information, however, could be memorized in the

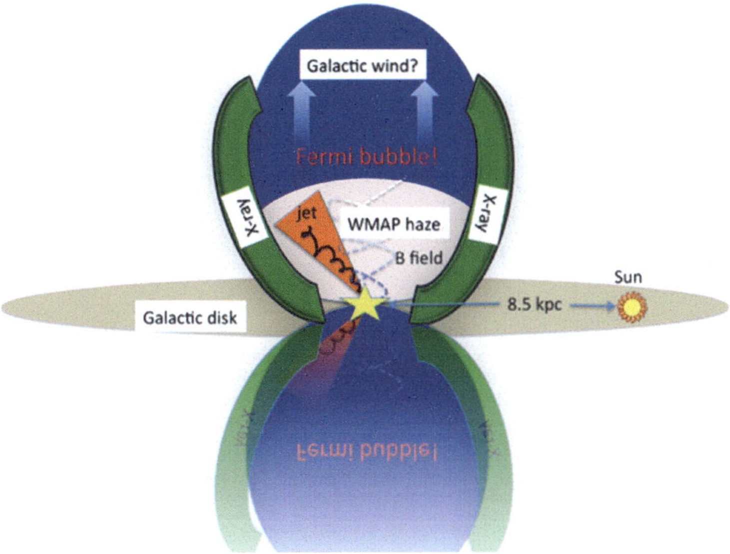

Fig. 21 The location and geometry of the *Fermi* Bubbles shown schematically together with the *WPAP* haze and the X-ray emission structure detected by ROSAT (from Ref. [83]). The gamma-ray structures could be initiated by cosmic-rays related to the PeVatron in GC

radiation of "relic" protons which already escaped the GC, and presently interact with the low-density plasma beyond the Galactic Disk. Interesting sites, in this context, are the so-called Fermi Bubbles (see Fig. 21) or even larger structures of the Galactic Halo. The detection of γ-rays with *Fermi* LAT has been reported for energies up to 100 GeV [83]. The detection of higher energy γ-rays is hard because of the limited detection area of *Fermi LAT* and the small field of view of the Cherenkov telescopes. Nevertheless, one of the possible scenarios initiating nonthermal processes in Fermi Bubbles could be related to multi-TeV protons (see Fig. 22). If so, and if the proton spectrum in Fermi Bubbles extends to 100 TeV and beyond, we should expect (marginally) detectable fluxes of high energy neutrinos as well. Interestingly, the recently reported detection of $\geq$30 TeV neutrinos [60] in principle can be linked to the same Pevatron in the GC. This would be possible if the latter has been accelerating protons over the age of the Galaxy ($\sim$10^{10}year) with an average power of $\sim$10^{39} ergs^{-1} [86], i.e., comparable to the current acceleration power derived from the luminosity of the central point-like γ-ray source. Remarkably, such a power would be sufficient to account for most of the flux of galactic CRs up to the *knee* around 1 PeV.

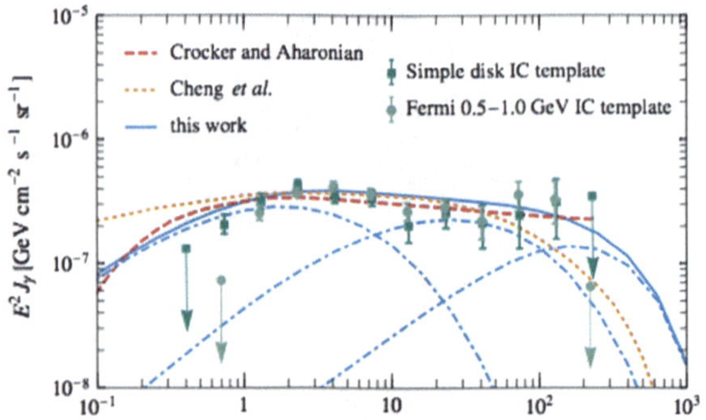

Fig. 22 The spectral energy distribution of gamma-rays from the *Fermi* bubbles compared to the theoretical predictions. (i) IC model of Ref. [73] (*solid line*) assuming stochastic acceleration of electrons in the bubbles (the contributions from the scattering on the CMB, FIR, and optical/UV backgrounds are shown separately); (ii) IC model of Ref. [42] (*dotted line*) assuming diffusive shock acceleration of electrons; (iii) hadronic model of Ref. [45] (*dashed line*). The figure is from Ref. [73]

References

1. A.A. Abdo, U. Abeysekara, B.T. Allen, T. Aune, D. Berley, E. Bonamente, G.E. Christopher, T. DeYoung, B.L. Dingus, R.W. Ellsworth, J.G. Galbraith-Frew, M.M. Gonzalez, J.A. Goodman, C.M. Hoffman, P.H. Hüntemeyer, C.M. Hui, B.E. Kolterman, J.T. Linnemann, J.E. McEnery, A.I. Mincer, T. Morgan, P. Nemethy, J. Pretz, J.M. Ryan, P.M. Saz Parkinson, A. Shoup, G. Sinnis, A.J. Smith, V. Vasileiou, G.P. Walker, D.A. Williams, G.B. Yodh, Spectrum and morphology of the two brightest milagro sources in the cygnus region: MGRO J2019+37 and MGRO J2031+41. ApJ **753**, 159 (2012). https://doi.org/10.1088/0004-637X/753/2/159
2. A.A. Abdo et al., (Fermi collaboration): Observations of the young supernova remnant RX J1713.7-3946 with the fermi large area telescope. ApJ **734**, 28 (2011)
3. A. Abramowski, F. Acero, F. Aharonian, A.G. Akhperjanian, G. Anton, A. Balzer, A. Barnacka, U. Barres de Almeida, Y. Becherini, J. Becker, B. Behera, K. Bernlöhr, E. Birsin, J. Biteau, A. Bochow, C. Boisson, J. Bolmont, P. Bordas, J. Brucker, F. Brun, P. Brun, T. Bulik, I. Büsching, S. Carrigan, S. Casanova, M. Cerruti, P.M. Chadwick, A. Charbonnier, R.C.G. Chaves, A. Cheesebrough, L.M. Chounet, A.C. Clapson, G. Coignet, G. Cologna, J. Conrad, M. Dalton, M.K. Daniel, I.D. Davids, B. Degrange, C. Deil, H.J. Dickinson, A. Djannati-Ataï, W. Domainko, L. O'C. Drury, F. Dubois, G. Dubus, K. Dutson, J. Dyks, M. Dyrda, K. Egberts, P. Eger, P. Espigat, L. Fallon, C. Farnier, S. Fegan, F. Feinstein, M.V. Fernandes, A. Fiasson, G. Fontaine, A. Förster, M. Füßling, Y.A. Gallant, H. Gast, L. Gérard, D. Gerbig, B. Giebels, J.F. Glicenstein, B. Glück, P. Goret, D. Göring, S. Häffner, J.D. Hague, D. Hampf, M. Hauser, S. Heinz, G. Heinzelmann, G. Henri, G. Hermann, J.A. Hinton, A. Hoffmann, W. Hofmann, P. Hofverberg, M. Holler, D. Horns, A. Jacholkowska, O.C. de Jager, C. Jahn, M. Jamrozy, I. Jung, M.A. Kastendieck, K. Katarzyński, U. Katz, S. Kaufmann, D. Keogh, D. Khangulyan, B. Khélifi, D. Klochkov, W. Kluźniak, T. Kneiske, N. Komin, K. Kosack, R. Kossakowski, H. Laffon, G. Lamanna, D. Lennarz, T. Lohse, A. Lopatin, C.C. Lu, V. Marandon, A. Marcowith, J. Masbou, D. Maurin, N. Maxted, M. Mayer, T.J.L. McComb, M.C. Medina, J. Méhault, R. Moderski, E. Moulin, C.L. Naumann, M. Naumann-Godo, M. de Naurois, D. Nedbal, D. Nekrassov, N. Nguyen, B. Nicholas, J. Niemiec, S.J. Nolan, S. Ohm, E. de Oña Wilhelmi, B. Opitz, M. Ostrowski, I. Oya, M. Panter, M. Paz Arribas, G. Pedaletti, G. Pelletier, P.O. Petrucci, S. Pita, G. Pühlhofer, M. Punch, A. Quirrenbach, M. Raue, S.M. Rayner, A. Reimer, O. Reimer,

M. Renaud, R. de Los Reyes, F. Rieger, J. Ripken, L. Rob, S. Rosier-Lees, G. Rowell, B. Rudak, C.B. Rulten, J. Ruppel, V. Sahakian, D. Sanchez, A. Santangelo, R. Schlickeiser, F.M. Schöck, A. Schulz, U. Schwanke, S. Schwarzburg, S. Schwemmer, F. Sheidaei, M. Sikora, J.L. Skilton, H. Sol, G. Spengler, Ł. Stawarz, R. Steenkamp, C. Stegmann, F. Stinzing, K. Stycz, I. Sushch, A. Szostek, J.P. Tavernet, R. Terrier, M. Tluczykont, K. Valerius, C. van Eldik, G. Vasileiadis, C. Venter, J.P. Vialle, A. Viana, P. Vincent, H.J. Völk, F. Volpe, S. Vorobiov, M. Vorster, S.J. Wagner, M. Ward, R. White, A. Wierzcholska, M. Zacharias, A. Zajczyk, A.A. Zdziarski, A. Zech, H.S. Zechlin, Discovery of extended VHE γ-ray emission from the vicinity of the young massive stellar cluster Westerlund 1. A&A **537**, A114 (2012). https://doi.org/10.1051/0004-6361/201117928

4. A. Abramowski, F. Aharonian, F.A. Benkhali, A.G. Akhperjanian, E. Angüner, G. Anton, S. Balenderan, A. Balzer, A. Barnacka, Y. Becherini et al., HESS J1640–465 - an exceptionally luminous TeV γ-ray supernova remnant. MNRAS **439**, 2828–2836 (2014). https://doi.org/10.1093/mnras/stu139
5. F. Acero et al., (HESS Collaboration): localizing the VHE γ-ray source at the galactic Centre. MNRAS **402**, 1877–1882 (2010)
6. B.S. Acharya, M. Actis, T. Aghajani, G. Agnetta, J. Aguilar, F. Aharonian, M. Ajello, A. Akhperjanian, M. Alcubierre, J. Aleksić et al., Introducing the CTA concept. Astropart. Phys. **43**, 3–18 (2013). https://doi.org/10.1016/j.astropartphys.2013.01.007
7. M. Ackermann, M. Ajello, A. Allafort, L. Baldini, J. Ballet, G. Barbiellini, D. Bastieri, A. Belfiore, R. Bellazzini, B. Berenji, R.D. Blandford, E.D. Bloom, E. Bonamente, A.W. Borgland, E. Bottacini, M. Brigida, P. Bruel, R. Buehler, S. Buson, G.A. Caliandro, R.A. Cameron, P.A. Caraveo, J.M. Casandjian, C. Cecchi, A. Chekhtman, C.C. Cheung, J. Chiang, S. Ciprini, R. Claus, J. Cohen-Tanugi, A. de Angelis, F. de Palma, C.D. Dermer, E. do Couto e Silva, P.S. Drell, D. Dumora, C. Favuzzi, S.J. Fegan, W.B. Focke, P. Fortin, Y. Fukazawa, P. Fusco, F. Gargano, S. Germani, N. Giglietto, F. Giordano, M. Giroletti, T. Glanzman, G. Godfrey, I.A. Grenier, L. Guillemot, S. Guiriec, D. Hadasch, Y. Hanabata, A.K. Harding, M. Hayashida, K. Hayashi, E. Hays, G. Jóhannesson, A.S. Johnson, T. Kamae, H. Katagiri, J. Kataoka, M. Kerr, J. Knödlseder, M. Kuss, J. Lande, L. Latronico, S.H. Lee, F. Longo, F. Loparco, B. Lott, M.N. Lovellette, P. Lubrano, P. Martin, M.N. Mazziotta, J.E. McEnery, J. Mehault, P.F. Michelson, W. Mitthumsiri, T. Mizuno, C. Monte, M.E. Monzani, A. Morselli, I.V. Moskalenko, S. Murgia, M. Naumann-Godo, P.L. Nolan, J.P. Norris, E. Nuss, T. Ohsugi, A. Okumura, E. Orlando, J.F. Ormes, M. M. Ozaki, D. Paneque, D. Parent, M. Pesce-Rollins, M. Pierbattista, F. Piron, M. Pohl, D. Prokhorov, S. Rainò, R. Rando, M. Razzano, T. Reposeur, S. Ritz, P.M.S. Parkinson, C. Sgrò, E.J. Siskind, P.D. Smith, P. Spinelli, A.W. Strong, H. Takahashi, T. Tanaka, J.G. Thayer, J.B. Thayer, D.J. Thompson, L. Tibaldo, D.F. Torres, G. Tosti, A. Tramacere, E. Troja, Y. Uchiyama, J. Vandenbroucke, V. Vasileiou, G. Vianello, V. Vitale, A.P. Waite, P. Wang, B.L. Winer, K.S. Wood, Z. Yang, Z. Zimmer, S. Bontemps, A cocoon of freshly accelerated cosmic rays detected by Fermi in the cygnus superbubble. Science **334**, 1103 (2011). https://doi.org/10.1126/science.1210311
8. M. Actis et al., (CTA consortium): design concepts for the cherenkov telescope array CTA: an advanced facility for ground-based high-energy gamma-ray astronomy. Exp. Astron. **32**, 193–316 (2011)
9. F. Aharonian, A. Akhperjanian, M. Beilicke, K. Bernlöhr, H. Börst, H. Bojahr, O. Bolz, T. Coarasa, J. Contreras, J. Cortina, S. Denninghoff, V. Fonseca, M. Girma, N. Götting, G. Heinzelmann, G. Hermann, A. Heusler, W. Hofmann, D. Horns, I. Jung, R. Kankanyan, M. Kestel, J. Kettler, A. Kohnle, A. Konopelko, H. Kornmeyer, D. Kranich, H. Krawczynski, H. Lampeitl, M. Lopez, E. Lorenz, F. Lucarelli, N. Magnussen, O. Mang, H. Meyer, M. Milite, R. Mirzoyan, A. Moralejo, E. Ona, M. Panter, A. Plyasheshnikov, J. Prahl, G. Pühlhofer, G. Rauterberg, R. Reyes, W. Rhode, J. Ripken, A. Röhring, G.P. Rowell, V. Sahakian, M. Samorski, M. Schilling, F. Schröder, M. Siems, D. Sobzynska, W. Stamm, M. Tluczykont, H.J. Völk, C.A. Wiedner, W. Wittek, Y. Uchiyama, T. Takahashi, HEGRA collaboration: an unidentified tev source in the vicinity of cygnus OB2. A&A **393**, L37–L40 (2002). https://doi.org/10.1051/0004-6361:20021171

10. F. Aharonian, J. Buckley, T. Kifune, G. Sinnis, High energy astrophysics with ground-based gamma ray detectors. Rep. Prog. Phys. **71**(9), 096,901 (2008)
11. F. Aharonian et al., (HESS collaboration): discovery of very-high-energy γ-rays from the galactic centre ridge. Nature **439**, 695–698 (2006)
12. F. Aharonian et al. (HESS collaboration): primary particle acceleration above 100 TeV in the shell-type supernova remnant RX J1713.7-394 with deep HESS observations. A&A **464**, 235–243 (2007)
13. F. Aharonian et al., (HESS collaboration): primary particle acceleration above 100 TeV in the shell-type supernova remnant RX J1713.7-3946 with deep HESS observations. A&A **464**, 235–243 (2007)
14. F. Aharonian, A. Neronov, High-energy gamma rays from the massive black hole in the galactic center. ApJ **619**, 306–313 (2005)
15. F. Aharonian, A. Neronov, TeV gamma rays from the galactic center direct and indirect links to the massive black hole in Sgr A. Ap&SS **300**, 255–265 (2005)
16. F. Aharonian, A.M. Taylor, Limitations on the photo-disintegration process as a source of VHE photons. Astropart. Phys. **34**, 258–266 (2010)
17. F.A. Aharonian, TeV gamma rays from BL Lac objects due to synchrotron radiation of extremely high energy protons. New Astr. **5**, 377–395 (2000)
18. F.A. Aharonian, *Very High Energy Cosmic Gamma Radiation : A Crucial Window on the Extreme Universe* (World Scientific Publishing, River Edge, NJ, 2004)
19. F.A. Aharonian, Gamma rays from supernova remnants. Astropart. Phys. **43**, 71–80 (2013). https://doi.org/10.1016/j.astropartphys.2012.08.007
20. F.A. Aharonian, C.W. Akerlof, Gamma-ray astronomy with imaging atmospheric cerenkov telescopes. Annu. Rev. Nucl. Part. Sci. **47**, 273–314 (1997)
21. F.A. Aharonian, A.M. Atoyan, On the origin of TeV radiation of SN 1006. A&A **351**, 330–340 (1999)
22. F.A. Aharonian, A.M. Atoyan, Broad-band diffuse gamma ray emission of the galactic disk. A&A **362**, 937–952 (2000)
23. F.A. Aharonian, A.A. Belyanin, E.V. Derishev, V.V. Kocharovsky, V.V. Kocharovsky, Constraints on the extremely high-energy cosmic ray accelerators from classical electrodynamics. Phys. Rev. D **66**(2), 023005 (2002)
24. F.A. Aharonian, L.O. Drury, H.J. Voelk, GeV/TeV gamma-ray emission from dense molecular clouds overtaken by supernova shells. A&A **285**, 645–647 (1994)
25. F.A. Aharonian, A.K. Konopelko, H.J. Völk, H. Quintana, 5@5—a 5 GeV energy threshold array of imaging atmospheric Cherenkov telescopes at 5 km altitude. Astropart. Phys. **15**, 335–356 (2001)
26. F.A. Aharonian, V.G. Kririllov-Ugriumov, V.V. Vardanian, Formation of relativistic electron-photon showers in compact X-ray sources. Ap&SS **115**, 201–225 (1985)
27. F.A. Aharonian, R.A. Sunyaev, Gamma-ray line emission, nuclear destruction and neutron production in hot astrophysical plasmas—the deuterium boiler as a gamma-ray source. MNRAS **210**, 257–277 (1984)
28. E. Aliu, T. Aune, B. Behera, M. Beilicke, W. Benbow, K. Berger, R. Bird, A. Bouvier, J.H. Buckley, V. Bugaev, M. Cerruti, X. Chen, L. Ciupik, M.P. Connolly, W. Cui, J. Dumm, V.V. Dwarkadas, M. Errando, A. Falcone, S. Federici, Q. Feng, J.P. Finley, H. Fleischhack, P. Fortin, L. Fortson, A. Furniss, N. Galante, G.H. Gillanders, E.V. Gotthelf, S. Griffin, S.T. Griffiths, J. Grube, G. Gyuk, D. Hanna, J. Holder, G. Hughes, T.B. Humensky, C.A. Johnson, P. Kaaret, O. Kargaltsev, M. Kertzman, Y. Khassen, D. Kieda, F. Krennrich, M.J. Lang, A.S. Madhavan, G. Maier, S. McArthur, A. McCann, J. Millis, P. Moriarty, R. Mukherjee, D. Nieto, A. O'Faoláin de Bhróithe, R.A. Ong, A.N. Otte, D. Pandel, N. Park, M. Pohl, A. Popkow, H. Prokoph, J. Quinn, K. Ragan, J. Rajotte, L.C. Reyes, P.T. Reynolds, G.T. Richards, E. Roache, M. Roberts, G.H. Sembroski, K. Shahinyan, A.W. Smith, D. Staszak, I. Telezhinsky, J.V. Tucci, J. Tyler, S. Vincent, S.P. Wakely, A. Weinstein, R. Welsing, A. Wilhelm, D.A. Williams, B. Zitzer, Spatially resolving the very high energy emission from MGRO J2019+37 with VERITAS. ApJ **788**, 78 (2014). https://doi.org/10.1088/0004-637X/788/1/78

29. L.A. Anchordoqui, J.F. Beacom, H. Goldberg, S. Palomares-Ruiz, T.J. Weiler, TeV γ rays and neutrinos from photodisintegration of nuclei in cygnus OB2. Phys. Rev. D **75**(6), 063001 (2007)
30. A. Atoyan, J. Buckley, H. Krawczynski, A gamma-ray burst remnant in our galaxy: HESS J1303–631. ApJL **642**, L153–L156 (2006)
31. A. Atoyan, C.D. Dermer, TeV emission from the galactic center black hole plerion. ApJL **617**, L123–L126 (2004)
32. W.B. Atwood, A.A. Abdo, M. Ackermann, W. Althouse, B. Anderson, M. Axelsson, L. Baldini, J. Ballet, D.L. Band, G. Barbiellini et al., The large area telescope on the fermi gamma-ray space telescope mission. ApJ **697**, 1071–1102 (2009)
33. A. Bamba, M. Ueno, H. Nakajima, K. Koyama, Thermal and nonthermal x-rays from the large magellanic cloud superbubble 30 doradus c. ApJ **602**, 257–263 (2004). https://doi.org/10.1086/380957
34. W. Bednarek, Gamma-rays from synchrotron pair cascades in blazars? MNRAS **285**, 69–81 (1997)
35. A.R. Bell, Turbulent amplification of magnetic field and diffusive shock acceleration of cosmic rays. MNRAS **353**, 550–558 (2004)
36. E.G. Berezhko, H.J. Völk, Hadronic versus leptonic origin of the gamma-ray emission from supernova remnant RX J1713.7-3946. A&A **492**, 695–701 (2008)
37. V.S. Berezinskii, S.V. Bulanov, V.A. Dogiel, V.S. Ptuskin, *Astrophysics of Cosmic Rays* (North-Holland, Amsterdam, 1990)
38. A.M. Bykov, D.C. Ellison, P.E. Gladilin, S.M. Osipov, Ultrahard spectra of PeV neutrinos from supernovae in compact star clusters. MNRAS **453**, 113–121 (2015). https://doi.org/10.1093/mnras/stv1606
39. A.M. Bykov, I.N. Toptygin, A model of particle acceleration to high energies by multiple supernova explosions in OB associations. Astron. Lett. **27**, 625–633 (2001). https://doi.org/10.1134/1.1404456
40. M. Casse, J.A. Paul, Local gamma rays and cosmic-ray acceleration by supersonic stellar winds. ApJ **237**, 236–243 (1980). https://doi.org/10.1086/157863
41. C.J. Cesarsky, T. Montmerle, Gamma rays from active regions in the galaxy—the possible contribution of stellar winds. Sp. Sci. Rev. **36**, 173–193 (1983). https://doi.org/10.1007/BF00167503
42. K.S. Cheng, D.O. Chernyshov, V.A. Dogiel, C.M. Ko, W.H. Ip, Origin of the fermi bubble. ApJL **731**, L17 (2011)
43. M. Chernyakova, D. Malyshev, F.A. Aharonian, R.M. Crocker, D.I. Jones, The high-energy. Arcminute-scale galactic center gamma-ray source. ApJ **726**, 60 (2011). https://doi.org/10.1088/0004-637X/726/2/60
44. E. Churazov, S. Sazonov, S. Tsygankov, R. Sunyaev, D. Varshalovich, Positron annihilation spectrum from the galactic centre region observed by SPI/INTEGRAL revisited: annihilation in a cooling ISM? MNRAS **411**, 1727–1743 (2011). https://doi.org/10.1111/j.1365-2966.2010.17804.x
45. R.M. Crocker, F. Aharonian, Fermi bubbles: giant, multibillion-year-old reservoirs of galactic center cosmic rays. Phys. Rev. Lett. **106**(10), 101102 (2011). https://doi.org/10.1103/PhysRevLett.106.101102
46. L.O. Drury, F.A. Aharonian, H.J. Voelk, The gamma-ray visibility of supernova remnants. A test of cosmic ray origin. A&A **287**, 959–971 (1994)
47. D.C. Ellison, D.J. Patnaude, P. Slane, J. Raymond, Efficient cosmic ray acceleration, hydrodynamics, and self-consistent thermal x-ray emission applied to supernova remnant RX J1713.7-3946. ApJ **712**, 287–293 (2010)
48. K. Ferrière, Interstellar gas within ~10 pc of sagittarius A*. A&A **540**, A50 (2012). https://doi.org/10.1051/0004-6361/201117181
49. Y. Fukui, H. Sano, J. Sato et al., A Detailed Study of the Molecular and Atomic Gas Toward the gamma-ray SNR RX J1713.7-3946: Spatial TeV gamma-ray and ISM Gas Correspondence. ArXiv e-prints (2011)

50. S. Gabici, F.A. Aharonian, Hadronic gamma-rays from RX J1713.7-3946? MNRAS **445**, L70–L73 (2014). https://doi.org/10.1093/mnrasl/slu132
51. V.L. Ginzburg, S.I. Syrovatskii, *The Origin of Cosmic Rays* (Macmillan, New York, 1964)
52. P.W. Guilbert, A.C. Fabian, M.J. Rees, Spectral and variability constraints on compact sources. MNRAS **205**, 593–603 (1983)
53. A.K. Harding, A.G. Muslimov, Pulsar pair cascades in a distorted magnetic dipole field. ApJL **726**, L10 (2011). https://doi.org/10.1088/2041-8205/726/1/L10
54. H.E.S.S. Collaboration, A. Abramowski, F. Aharonian, F. Ait Benkhali, A.G. Akhperjanian, E.O. Angüner, M. Backes, S. Balenderan, A. Balzer, A. Barnacka et al., The exceptionally powerful TeV γ-ray emitters in the large magellanic cloud. Science **347**, 406–412 (2015). doi:https://doi.org/10.1126/science.1261313
55. H.E.S.S. Collaboration, A. Abramowski, F. Aharonian, F.A. Benkhali, A.G. Akhperjanian, E.O. Angüner, M. Backes, A. Balzer, Y. Becherini, J.B. Tjus et al., Acceleration of petaelectronvolt protons in the galactic centre. Nature **531**, 476–479 (2016). doi:https://doi.org/10.1038/nature17147
56. J.C. Higdon, R.E. Lingenfelter, OB associations, supernova-generated superbubbles, and the source of cosmic rays. ApJ **628**, 738–749 (2005)
57. A.M. Hillas, Evolution of ground-based gamma-ray astronomy from the early days to the Cherenkov Telescope Arrays. Astropart. Phys. **43**, 19–43 (2013). https://doi.org/10.1016/j.astropartphys.2012.06.002
58. J.A. Hinton, F.A. Aharonian, Inverse compton scenarios for the TeV gamma-ray emission of the galactic center. ApJ **657**, 302–307 (2007)
59. J.A. Hinton, W. Hofmann, **47**, 523–565 (2009)
60. IceCube Collaboration, Evidence for High-Energy Extraterrestrial Neutrinos at the IceCube Detector. Science **342**, 1242856 (2013). https://doi.org/10.1126/science.1242856
61. T. Inoue, R. Yamazaki, S.I. Inutsuka, Y. Fukui, Toward understanding the cosmic-ray acceleration at young supernova remnants interacting with interstellar clouds: possible applications to RX J1713.7-3946. ApJ **744**, 71 (2012)
62. E. Kafexhiu, F. Aharonian, G.S. Vila, Nuclear reactions in hot astrophysical plasmas with $T \geq 10^{10}$ K. Int. J. Mod. Phys. D **21**, 1250009-1-1250009-39 (2012). https://doi.org/10.1142/S0218271812500095
63. B. Katz, E. Waxman, In which shell-type SNRs should we look for gamma-rays and neutrinos from P-P collisions? JCAP **1**, 18 (2008)
64. S.R. Kelner, F.A. Aharonian, D. Khangulyan, On the jitter radiation. ApJ **774**, 61 (2013). https://doi.org/10.1088/0004-637X/774/1/61
65. S.R. Kelner, A.Y. Prosekin, F.A. Aharonian, Synchro-curvature radiation of charged particles in the strong curved magnetic fields. AJ **149**, 33 (2015). https://doi.org/10.1088/0004-6256/149/1/33
66. B. Kozlovsky, R.J. Murphy, R. Ramaty, Nuclear deexcitation gamma-ray lines from accelerated particle interactions. ApJS **141**, 523–541 (2002)
67. P.O. Lagage, C.J. Cesarsky, The maximum energy of cosmic rays accelerated by supernova shocks. A&A **125**, 249–257 (1983)
68. L. Landau, G. Rumer, The cascade theory of electronic showers. Roy. Soc. Lond. Proc. Ser. A **166**, 213–228 (1938)
69. A. Levinson, Particle acceleration and curvature TeV emission by rotating. Supermassive black holes. Phys. Rev. Lett. **85**, 912–915 (2000)
70. T. Linden, E. Lovegrove, S. Profumo, The morphology of hadronic emission models for the gamma-ray source at the galactic center. ApJ **753**, 41 (2012). https://doi.org/10.1088/0004-637X/753/1/41
71. M.A. Malkov, L. O'C Drury, Nonlinear theory of diffusive acceleration of particles by shock waves. Rep. Prog. Phys. **64**, 429–481 (2001)
72. M.V. Medvedev, Theory of "Jitter" radiation from small-scale random magnetic fields and prompt emission from gamma-ray burst shocks. ApJ **540**, 704–714 (2000). https://doi.org/10.1086/309374

73. P. Mertsch, S. Sarkar, Fermi gamma-ray "bubbles" from stochastic acceleration of electrons. Phys. Rev. Lett. **107**(9), 091101 (2011). https://doi.org/10.1103/PhysRevLett.107.091101
74. G. Morlino, E. Amato, P. Blasi, Gamma-ray emission from SNR RX J1713.7-3946 and the origin of galactic cosmic rays. MNRAS **392**, 240–250 (2009)
75. K. Murase, J.F. Beacom, Very-high-energy gamma-ray signal from nuclear photodisintegration as a probe of extragalactic sources of ultrahigh-energy nuclei. Phys. Rev. D **82**(4), 043008 (2010)
76. P.L. Nolan, A.A. Abdo, M. Ackermann, M. Ajello, A. Allafort, E. Antolini, W.B. Atwood, M. Axelsson, L. Baldini, J. Ballet et al., Fermi large area telescope second source catalog. ApJS **199**, 31 (2012)
77. I.P. Ochelkov, V.V. Usov, Curvature radiation of relativistic particles in the magnetosphere of pulsars I—theory. Ap&SS **69**, 439–460 (1980)
78. E. Parizot, A. Marcowith, E. van der Swaluw, A.M. Bykov, V. Tatischeff, Superbubbles and energetic particles in the Galaxy. I. Collective effects of particle acceleration. A&A **424**, 747–760 (2004)
79. C. Pittori et al., (AGILE collaboration): first AGILE catalog of high-confidence gamma-ray sources. A&A **506**, 1563–1574 (2009)
80. A.V. Plyasheshnikov, F.A. Aharonian, Characteristics of air showers produced by extremely high energy gamma-rays. J. Phys. G Nucl. Phys. **28**, 267–288 (2002)
81. A.V. Plyasheshnikov, F.A. Aharonian, H.J. Völk, On the potential of the imaging atmospheric Cherenkov technique for study of cosmic gamma rays in the energy range 10–100 TeV. J. Phys. G Nucl. Phys. **26**, 183–201 (2000)
82. N. Prantzos, C. Boehm, A.M. Bykov, R. Diehl, K. Ferrière, N. Guessoum, P. Jean, J. Knoedlseder, A. Marcowith, I.V. Moskalenko, A. Strong, G. Weidenspointner, The 511 keV emission from positron annihilation in the galaxy. Rev. Mod. Phys. **83**, 1001–1056 (2011)
83. M. Su, T.R. Slatyer, D.P. Finkbeiner, Giant gamma-ray bubbles from fermi-LAT: active galactic nucleus activity or bipolar galactic wind? ApJ **724**, 1044–1082 (2010)
84. R. Svensson, Steady mildly relativistic thermal plasmas—processes and properties. MNRAS **209**, 175–208 (1984)
85. T. Tanaka, Y. Uchiyama, F.A. Aharonian, T. Takahashi, et al., Study of nonthermal emission from SNR RX J1713.7-3946 with suzaku. ApJ **685**, 988–1004 (2008)
86. A.M. Taylor, S. Gabici, F. Aharonian, Galactic halo origin of the neutrinos detected by IceCube. Phys. Rev. D **89**(10), 103003 (2014). https://doi.org/10.1103/PhysRevD.89.103003
87. A.N. Timokhin, Time-dependent pair cascades in magnetospheres of neutron stars—I. Dynamics of the polar cap cascade with no particle supply from the neutron star surface. MNRAS **408**, 2092–2114 (2010)
88. I.N. Toptygin, G.D. Fleishman, A role of cosmic rays in generation of radio and optical radiation by plasma mechanics. Ap&SS **132**, 213–248 (1987). https://doi.org/10.1007/BF00641755
89. M. Tsuboi, T. Handa, N. Ukita, Dense molecular clouds in the galactic center regionI. Observations and data. ApJS **120**, 1–39 (1999). https://doi.org/10.1086/313165
90. F. Vissani, F. Aharonian, N. Sahakyan, On the detectability of high-energy galactic neutrino sources. Astropart. Phys. **34**, 778–783 (2011)
91. Q.D. Wang, F.J. Lu, E.V. Gotthelf, G359.95-0.04: an energetic pulsar candidate near Sgr A*. MNRAS **367**, 937–944 (2006)
92. T.C. Weekes, *Very High Energy Gamma-Ray Astronomy*. The Institute of Physics Publishing (2003)
93. V. Zabalza, Naima: a Python package for inference of relativistic particle energy distributions from observed nonthermal spectra. ArXiv e-prints (2015)
94. V.N. Zirakashvili, F.A. Aharonian, Nonthermal Radiation of Young Supernova Remnants: the Case of RX J1713.7-3946. ApJ **708**, 965–980 (2010)

Space-Based Gamma-Ray Astrophysics

Marco Tavani

Abstract Cosmic gamma-rays of energies above the electron mass are messengers of the most powerful phenomena in our Universe. In this review we focus on gamma-ray astrophysics in the energy range 30 MeV–100 GeV which requires sophisticated instruments on board of satellites operating above the Earth atmosphere. Gamma-ray sources in this energy range involve systems of great power for which gravitational, electromagnetic, and particle interactions contribute to generate very energetic emissions. During the last decade our view of the gamma-ray Universe accessible from space observations has been revolutionized by the advent of the AGILE and *Fermi* space missions. We summarize here the main discoveries made during the last years and the most pressing theoretical challenges and surprises. Gamma-rays are detected from our Sun, from systems in our Galaxy (pulsars, compact object binaries, Supernova Remnants, the Galactic Center) in addition to the Galactic diffuse emission. Furthermore, a variety of far-away extragalactic sources shine in gamma-rays all in time-variable fashion, most notably relativistic jets of Active Galactic Nuclei of the "blazar" class and Gamma-Ray Burst sources. Space-based gamma-ray astrophysics is at the forefront of research and stimulates the most advanced theoretical investigations involving relativistic plasmas and particle acceleration in extreme environments.

Introduction

Cosmic gamma-rays above X-ray energies up to hundreds of GeV can be detected only in space. A very broad range of cosmic physical phenomena can be studied in the energy range 0.5 MeV–hundreds of GeV, and the drive for detecting gamma-ray radiation from space has been one of the most challenging tasks of the astrophysics

M. Tavani (✉)
INAF-IAPS, via Fosso del Cavaliere 100, I-00133 Rome, Italy
e-mail: marco.tavani@inaf.it

M. Tavani
Dipartimento di Fisica, Università Roma Tor Vergata,
via O. Raimondo 18, I-00173 Rome, Italy

R. Aloisio et al. (eds.), *Multiple Messengers and Challenges in Astroparticle Physics*, https://doi.org/10.1007/978-3-319-65425-6_3

community since the late 1960s. As compared to X-ray space astrophysics, gamma-ray astrophysics deals with definitely nonthermal distributions of radiation sources and particle acceleration processes that can be very extreme. The number of photons that can be reasonably collected by current instruments operating at gamma-ray energies are definitely less than those obtainable by X-ray detectors. Detection techniques are also completely different: X-ray detectors evolved toward focusing optics and CDD detectors, whereas the gamma-ray instruments visualize individual photons in "cube-like" devices that have to be capable of tracking all particles (electrons and positrons) knocked-off and/or produced by the impinging gamma-ray interacting with the instrument material. Instrumental backgrounds in space are also completely different for X rays and gamma-rays: the latter has to sustain a rate 10^4 times larger of particle background compared to the gamma-ray signal. No wonder then for different development routes of space high-energy community that ended up specializing in two different subfields. In this report, we will focus on the gamma-ray part of the work.

A very large number of results and papers have been produced over the past 10 years; it is impossible to provide a detailed account of all the important developments of gamma-ray astrophysics in a concise review. We necessarily focus here on only some of the most relevant results, apologizing for the unavoidable omissions.

A Brief History

During the 1960s many detectors in the hard X-ray soft gamma-ray range were launched in space (for a review see [1]). The realization that cosmic X-ray sources could be detected in large numbers justified an effort to enlarge the energy range to the highest accessible energies. Technology development in event classification and photon selection was crucial to make advancements in the astrophysics context. A number of detectors were developed to study the MeV range in the Compton regime [1]. In the following, we focus on the gamma-ray detection above 20 MeV well in the pair-producing regime.

Early Experiments

The first gamma-ray telescopes launched in space were scintillator-Cherenkov detectors surrounded by an efficient anticoincidence dome to reject charged particles [2, 3]. The first experiment was launched on the Explorer 11 satellite in 1961 and detected a few tens of events reconstructed as cosmic gamma-rays [2]. A few years later, a first systematic detection of cosmic gamma-rays above 20 MeV was provided by an improved version of the same instrument flown on the NASA OSO-3 satellite [4]. This detector could obtain only a preliminary and sort of "primitive" glimpse of the gamma-ray sky with a clear indication of an enhancement of emission in the Galactic plane. No individual sources could be detected. Other similar detectors were flown on the Cosmos 208 mission [5], on OSO-1 [6], and on OSO-3 [7]. The path toward the observation of the gamma-ray sky started at that time. Clearly, improvements were necessary to gain information on photon energies and directions, and to improve sensitivity and particle background rejection.

SAS-2

Spark chambers provided in the 1970s the technological boost to gamma-ray detection. They provided a high-discrimination and a picture-type device that made a big advance in detection. Different types of imaging spark chambers were developed at the end of the 1960s. The first space mission carrying a spark-chamber instrument devoted to gamma-ray astrophysics was the NASA small mission SAS-2 launched in an equatorial orbit from the San Marco Space Center in Malindi (Kenya) on November 15, 1972, and operating through June 1973 [8]. Sixteen spark chamber modules interleaved with tungsten absorbers of 0.03 radiation length each provided an efficient gamma-ray imaging camera. An anticoincidence dome provided the veto system for the charged particle background. Each SAS-2 module had a 25×25 cm^2 active area with planes separated by 3.6 mm. The effective area was $\sim$500 cm^2 for a quite large field of view (FoV) $\sim$1/4 of the whole sky. Photon energy and directional information for individual photons could be obtained in the range 35–200 MeV. Despite the limited time of operations, SAS-2 was a very important precursor for space-based gamma-ray astrophysics [9, 10]. Several gamma-ray sources could be detected, including the Crab and Vela pulsars as well as a clear enhancement of emission from the Galactic plane.

COS-B

The European Space Agency mission COS-B was launched on an elliptical orbit in August 1975 carrying a spark-chamber instrument similar to SAS-2 but enhanced by an additional "calorimetric" detector positioned at the bottom of the instrument [11]. This configuration allowed a more precise determination of the photon energy and extension of the energy range. In addition, despite the high particle background, the COS-B elliptical orbit allowed prolonged observations without Earth occultations. COS-B exposed mostly the Galactic plane with a few pointings at high latitudes. The Second COS-B Catalogue lists about 30 entries [12], the famous "COS-B sources" that will be sources of puzzlement and the focus of many investigations by gamma-ray astronomers in the following years. Among them, we recall: 2CG 078+01 (that will turn out to be the SNR Gamma-Cygni and its pulsar), 2CG 135+01 (the famous "135+1" that is associated with a compact object, most likely a pulsar, orbiting around the Be star LS I 61 303), 2CG 195+04 (at that time, the mysterious "Geminga"), and then the only extragalactic gamma-ray source detected in the 1980s, 2CG 289+64 (that is, the quasar 3C 273).

The Compton Gamma-Ray Observatory

The Compton Gamma-Ray Observatory (CGRO) has been a major NASA space mission launched in April 2991 and operating in the period 1991–2000. Four instruments covered the spectrum from hard X-rays and intermediate gamma-rays up to pair-producing gamma-rays (for a review, see [1]). OSSE, a non-imaging detector operating in the hard X-ray/MeV range with good line sensitivity; BATSE, an instrument devoted to the detection of gamma-ray bursts (GRBs) in the energy range of 50 keV–a few MeV; COMPTEL, a Compton telescope sensitive in the range 1–30 MeV; EGRET an imaging gamma-ray instrument sensitive in the range 30 MeV–30 GeV [13]. The EGRET instrument had an upper spark chamber with 28 elements interleaved with 0.02 radiation length plates, and a calorimeter made of NaI(Tl) scintillator crystals for a total 8-radiation length. It constituted a definite advance with respect to its predecessors, both in terms of sensitivity and angular resolution [14]. The EGRET field of view (FOV) of 0.5 sr was adequate to monitor multiple sources. For the first time, a coverage of the entire sky above 100 MeV was obtained by EGRET during the first phase of operations lasting since 1991 until 1996.

Among the many achievements by EGRET we can mention here: the discovery that a class of AGNs (blazars) are intense gamma-ray sources, the detection of several gamma-ray pulsars, the unveiling of some mysterious sources (e.g., Geminga as a pulsed gamma-ray source), the detection of high-energy solar flares, the first study of the Galactic diffuse emission, first attempts to study and model the extragalactic gamma-ray background. EGRET legacy relies on its 300 gamma-ray sources whose positioning in the sky constituted a sort of "gamma-ray constellations" for the next generation of instruments.

AGILE

A quite long hiatus separates EGRET's end of operations (1996) and the advent of a new generation of space instruments devoted to gamma-ray astrophysics. The AGILE satellite was launched in space from the Shriharikota base (India) on April 23, 2007, in a quasi-equatorial orbit at 530 km of altitude (see Fig. 1). It is a mission of the Italian Space Agency (ASI) and the first space experiment based on solid-state silicon detectors [15]. Its design is based on a pair-conversion silicon tracker (made of 12 planes of silicon microstrip detectors of width of 121 μm with analog readout interleaved with 0.07 radiation length layers of tungsten, W), a calorimeter (Cesium Iodine bars sensitive in the 0.4–100 MeV range), and a light coded-mask instrument sensitive in the

Fig. 1 The AGILE satellite during qualification testing at the IABG facility (Munich), 2006 [15]

20–60 keV band (called Super-AGILE). Despite its relatively small size (weight ~100 kg, total power ~60 W) the AGILE instrument represented the first new-technology experiment with no expendable elements, and therefore of potentially long duration in orbit. Relevant parameters are: the FoV (2.5 sr), the effective area (500 cm^2 at 100 MeV and 200 cm^2 at 1 GeV on-axis) the angular resolution (0.7° at 1 GeV), the energy resolution (40% on-axis at 100 MeV). The AGILE FoV at GeV energies (2.5 sr) was unprecedented at the time of the launch; together with an excellent PSF (that turns out to be similar to *Fermi*-LAT [16]) the mission is quite competitive for fast detection of bright transients and sources especially in the range 100 MeV–1 GeV with unique co-axial imaging capabilities in the 20–60 keV. Excellent timing and a special trigger supports the operation of the calorimeter. During the period 2007–2009 the satellite operated in pointing mode. Since 2010 the satellite is in "spinning mode," rotating every 7 min around the satellite-Sun axis and sweeping about 80% of the entire sky for each revolution. The AGILE mission is currently (July 2017) operating nominally in orbit, 10 years after the launch.

The Fermi Mission

The *Fermi* mission is the NASA continuation of the program in gamma-ray astrophysics which started with the earliest experiments and continued with SAS-2 and CGRO. The mission (of the Probe class) was launched by NASA on June 11, 2008, on a Delta II Heavy Launch vehicle in an orbit with 25° inclination. It is focused to gamma-ray astrophysics in the range 30 MeV–300 GeV (using the LAT instrument [17]) as well as GRB detection in the hard X-ray energy range (by the GBM instrument [18]). The gamma-ray imager is a pair-conversion instrument (the Large Area Telescope, LAT). It is a 4×4 array of 16 modules based on solid state technology with silicon detectors. Each module is made of 18 detection planes (each approximately $40 \times 40\,\text{cm}^2$) with bidimensional tracking capability. The silicon microstrips have 228 μm pitch and digital readout. The first 12 planes ("front converter") are interleaved with W absorbers 0.03 radiation length each. The "back converter" is made of 4 additional planes with W absorbers of 0.18 radiation length each. At the bottom of the silicon tracker is positioned a calorimeter module with 96 CsI(Tl) crystals, arranged in an 8-layer hodoscopic configuration for a total depth of 8.6 radiation lengths. Important characteristics are: the FoV (2.5 sr), the effective area (1500 cm^2 at 100 MeV and 10000 cm^2 at 10 GeV on-axis) the angular resolution (0.7° at 1 GeV), the energy resolution (10% on-axis at 1 GeV). The total LAT weight is 2789 kg, and total power 650 W. The gamma-ray burst monitor (GBM) is the second *Fermi* instrument [18]. It is devoted to GRB detection and other fast transients. The GBM includes two sets of detectors: twelve sodium iodide (NaI) scintillators sensitive from a few keV to about 1 MeV, and two cylindrical bismuth germanate (BGO) scintillators covering the energy range 150 keV–30 MeV. Figure 2 shows the satellite during ground testing. The *Fermi* mission is operating nominally (July 2017) in its ninth year in orbit (see Fig. 3 for the gamma-ray sky detected by the Fermi-LAT).

Fig. 2 The *Fermi* satellite during pre-launch testing [19]. The LAT instrument is visible on *top* of the spacecraft with the GBM detectors positioned at the *lateral sides*

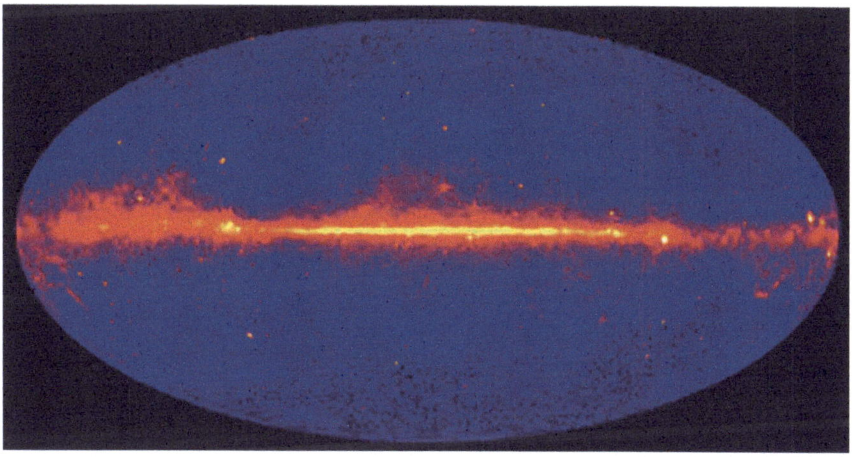

Fig. 3 The gamma-ray sky above 100 MeV as detected by the *Fermi* -LAT during the first 60 months of operations [19]

The Gamma-Ray Sky

Gamma-Ray Surprises

A number of unexpected discoveries and surprises in the gamma-ray energy range filled the last decade with exciting challenges.

- the flaring Crab Nebula, indicating the existence of a new and still poorly understood particle acceleration mechanism;
- the Fermi bubbles, showing that the Galactic Center is the source of particles emanating over long timescales at very high latitudes;
- the detection of neutral pion decay emission from Supernova Remnants that contributed to resolving some crucial issues on the origin of cosmic rays;
- transient gamma-ray emission from the microquasar Cygnus X-3 simultaneously with relativistic jet ejections;
- the puzzling radiation properties of jets in massive black hole indicating that magnetic instabilities and reconnection power the gamma-ray emission from active galactic nuclei (AGNs).

In addition to "positive" surprises, a number of "negative" surprises challenge us both observationally and theoretically, as they were expected to be addressed and/or resolved by the current generation of instruments:

- the lack of a convincing gamma-ray signal from dark matter particles;
- the complexity of the Galactic Center region and of the inner Galaxy that makes difficult detailing the interplay between diffuse and point-like source gamma-rays

and star forming region, and the possibility of detecting cosmic-ray anisotropy throughout the disk;
- the missing evidence for Galactic sources able to accelerate cosmic rays up to PeV energies and beyond;
- the difficulty of distinguishing hadronic versus leptonic signatures in relativistic jets;
- the challenge of intense gamma-ray emission decoupled from optical and X-ray radiation in relativistic jets of blazars;
- the lack of gamma-ray emission from clusters of galaxies.

As we summarize in this report, space-based gamma-ray astrophysics in the GeV range has made great progress in a broad number of issues. It is projected into the future with great possibilities. Also worth considering are the prospects for the 1–100 MeV range, a largely unexplored range that might be accessible with more sensitive instruments in the next decade.

A Solar Intermezzo

Before venturing outside our Solar system, it is important to report the important contributions to the physics of solar flares provided by gamma-ray instruments. CGRO observed several solar flares by detecting their hard X-ray-MeV continuum, line emission, and occasional GeV emission [20]. In particular, the June 11, 1991, solar flare was very intense; it was detected with a remarkable "delayed gamma-ray emission" up to GeV energies for a duration of about 10 h [21]. *Fermi*-LAT also detected several solar flares, the most prominent one being the event on March 7, 2012. This flare produced a gigantic MeV–GeV outburst lasting many hours, by far the most intense gamma-ray source ever detected (at its peak, approximately 100 times the GeV flux of the Vela pulsar!) [22]. The location of the gamma-ray flare is coincident with that of X-ray emission (Fig. 4, left panel). Gamma-rays are most likely produced from the decay of pions produced by accelerated protons and ions with a power-law spectrum of index 3 (see the peak emission spectrum, Fig. 4, right panel). Both impulsive and "delayed" phases of gamma-ray emission are clearly detected. They can be modelled by continuous acceleration of protons in the solar corona that penetrate the lower atmosphere producing pions and then gamma-rays. Figure 5 shows the X-ray, particle and gamma-ray light curves of this remarkable manifestation of solar activity.

Low-energy cosmic rays modulated by solar activity can be studied through the effects of their bombardment of the lunar surface. The Moon indeed is producing a feeble but detectable "gamma-ray glowing" as observed by EGRET [23] and *Fermi*-LAT [24]. Figure 6 shows the spectrum produced by cosmic-ray bombardment of our Moon, a quite interesting result [24].

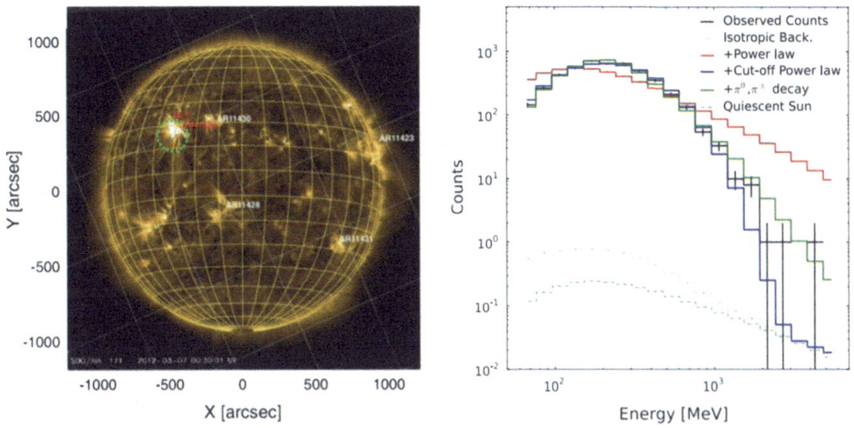

Fig. 4 *Left panel* Location of the gamma-ray emission produced by the solar flare of March 7, 2012, as detected by *Fermi*-LAT [22]. *Right panel* gamma-ray spectrum at the peak of the flare emission [22]. It is the spectrum of the most intense source of gamma-rays ever detected

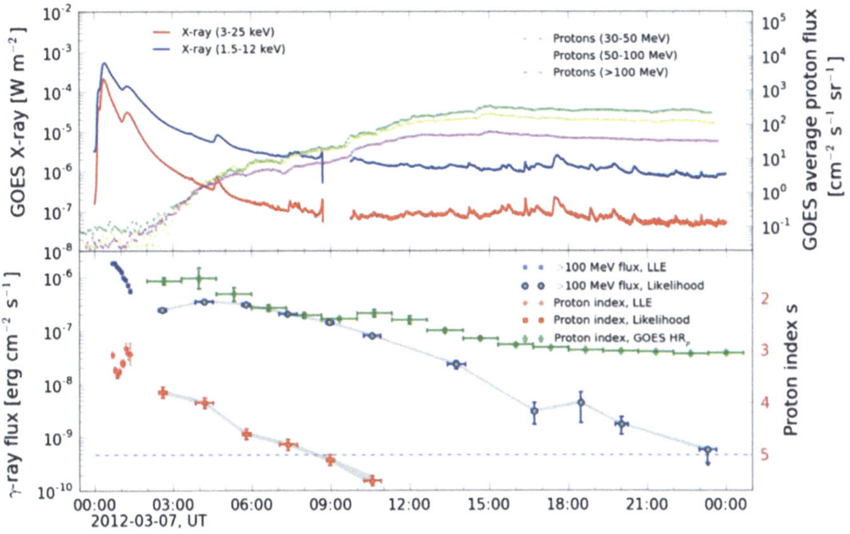

Fig. 5 Time development of the high-energy photon and particle emission the March 7, 2012, solar flare [22]. *Top panel* X-ray and particle data from the GOES experiment.) *Bottom panel* Gamma-ray emission above 50 MeV as detected by *Fermi*-LAT

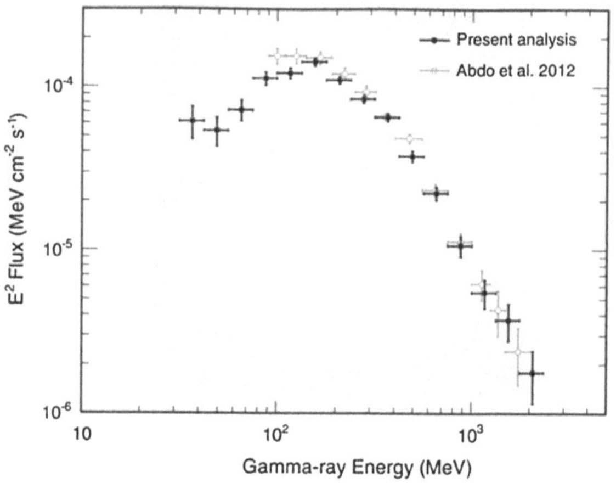

Fig. 6 The gamma-ray spectrum detected by *Fermi*-LAT from the surface of the Moon as a consequence of cosmic-ray bombardment of an atmosphere-less planetary environment [24]

Galactic Sources

Gamma-Ray Pulsars and Millisecond Pulsars

EGRET detected a handful of gamma-ray pulsars, i.e., rotationally powered neutron stars with pulsed gamma-ray emission. The burden and the excitement of discovering more pulsars passed then to the next generation of instruments. As a reminder of the power of the new generation of instruments today, all three main gamma-ray pulsars (the *Grande Triade* consisting of the Crab, Vela and Geminga pulsars) could be detected in a *single* pointing covering 120° of diameter by AGILE and *Fermi*-LAT (see Fig. 7 for an early detection by an AGILE pointing in 2007). A success that preludes an unprecedented exposure for hundreds of candidate gamma-ray sources for which pulsations could be looked for.

It is then natural to expect that a large fraction of EGRET unidentified sources in the Galactic plane are today identified or strongly suspected to be gamma-ray pulsars. From the very early AGILE detection of PSR J2021+365 in the Cygnus region [25] to the many pulsars detected by *Fermi* [26] the path of discovery has been exciting. As of today (end of July 2017) *Fermi*-LAT detected more than 200 gamma-ray pulsars, of which 7 are a confirmation of those detected by EGRET, more than 50 are young radio-selected PSRs, more than 60 are young X-ray or gamma-ray-selected, and about 100 are millisecond pulsars (MSPs). Many of these objects are coincident with the EGRET and *Fermi*-LAT unidentified sources. Fig. 8 shows the P-$\dot{P}$ diagram of all 2,000 radio pulsars currently known with measured $\dot{P}$ (black dots) with the gamma-ray pulsars superimposed in color.

In general, the gamma-ray pulse profile is non-trivially connected with the pattern of radio emission. Gamma-ray peaks are detected either lagging the radio peaks, being aligned, or leading the radio peaks [26]. Photon number spectra can be described as $dN/d\varepsilon = N_o\,\varepsilon^{-\alpha}\,e^{-(\varepsilon/\varepsilon_c)^b}$, with spectral indices ranging from 0.7 to 2, and

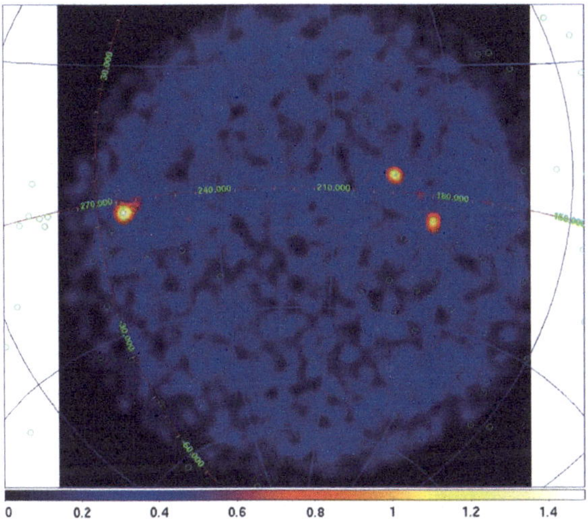

Fig. 7 The *Grande Triade* of gamma-ray pulsars detected by AGILE in a 1-day integration, July 2007 (from [15]). The gamma-ray intensity map above 100 MeV shows all three main gamma-ray pulsars detected in a single pointing (from *left* to *right*: Vela, Geminga and Crab). For the first time, a very large field of view (FoV) of about 2.5 sr (120° diameter) was achieved. The *Fermi*-LAT FoV above 100 MeV is similar [17]

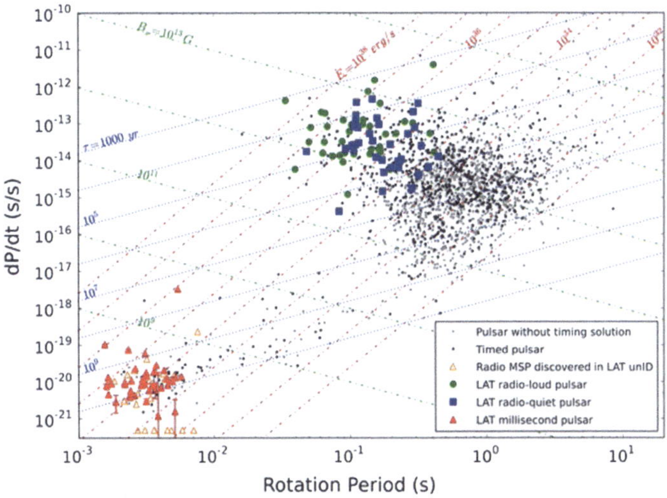

Fig. 8 The P-$\dot{P}$ diagram of pulsars showing in color the pulsars detected above 100 MeV [26]. Radio-quiet pulsars detected in gamma-rays are marked with *blue squares*; millisecond pulsars are marked by *red triangles*

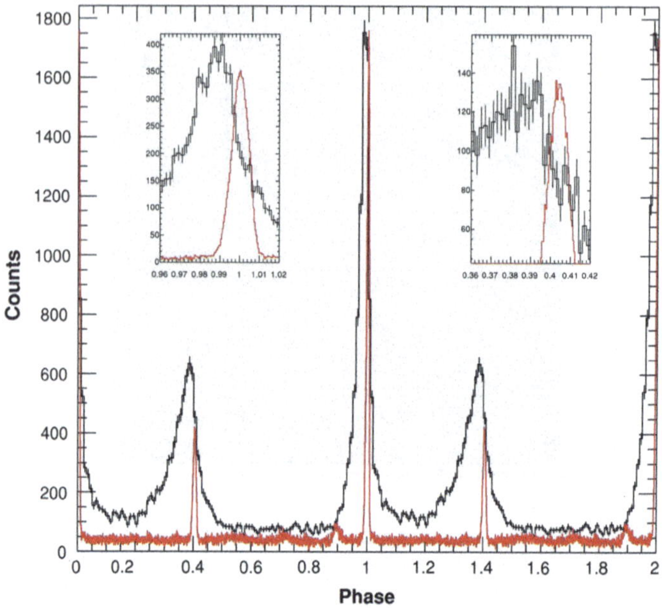

Fig. 9 The gamma-ray light curve above 100 MeV of the Crab pulsar [27] (*black curve*). The radio light curve is overlaid in *red color*

sub-exponential cutoffs (with index b smaller than unity, suggesting emission not too close to the neutron star) of energies ε_c in the range of a few GeV (generally peaking as the gamma-ray flux peaks). As an example, Fig. 9 shows the gamma-ray and radio light curve of the Crab PSR [27]. Figure 10 provides a schematic view of the main radiation models and sites suggested for particle acceleration and radiation.

Pulsars are great accelerators. They are known to emit, in addition to electromagnetic waves, a highly relativistic wind made of electrons, positrons, and possibly ions. The manifestation of pulsar winds (PWs) depends on the environment and conditions of the surrounding medium. Pulsar wind nebulae are indeed powered by the kinetic and radiative dissipation of PWs (see next section). If the pulsar wind encounters the surface of a companion star in a binary system, several interesting phenomena occur. Irradiated stars can lift a mass outflow that interacts with the source pulsar in a complex way. Under special conditions, the irradiation-driven mass outflow from the surface of a low-mass star can eventually lead to its complete or almost total "evaporation" as predicted in 1988 [29, 30] and later verified with the first prototypical "black widow" binary system, PSR B1957+20 [31]. Today, several compact binaries containing MSPs and low-mass and very-low-mass companion stars are described by two different classes: "redbacks" and "black widows," respectively (e.g., [32–37]. In addition, "transitional" binary MSPs are discovered for which periods of pulsar emission phases alternate with phases of radiopulsar quenching and "disk" emission [38–40]. Gamma-ray emission is of crucial importance for all of these

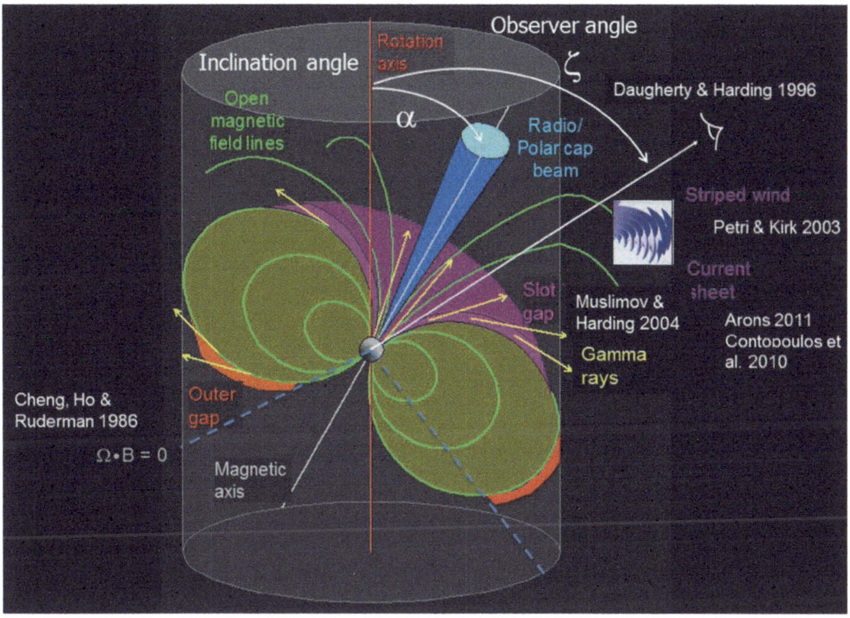

Fig. 10 A schematic view of the sites and mechanisms of particle acceleration and radiation based on current theoretical models of pulsar emission (from [28])

binary MSPs, as expected in systems that can switch between radiopulsar emission, intermediate phases with a partial or totally surrounding "disk," and eventually accretion. "Hidden" pulsars [41], predicted after the discovery of the first eclipsing MSP PSR B1957+20 [31], naturally produce gamma-rays because of enhanced IC and of pulsar wind termination shock acceleration and radiation.

PSRs are known for their stability. Despite timing noise and glitches, the average flux properties of PSRs from radio to gamma-rays are usually quite constant in time. It is then interesting to point at exceptions of this rule. A special case is the unusually bright gamma-ray pulsar PSR J2021+4026 located in the SNR Gamma-Cygni. Since the beginning of the AGILE and *Fermi*-LAT observations of the Cygnus region, the intense gamma-ray source associated with Gamma-Cygni has been a source of puzzlement. Its nature has been finally unveiled with the discovery of pulsed gamma-ray emission with a spin period of 0.245 s [43]. This pulsar turns out to be radio-quiet and can be studied only in gamma-rays [26]. Surprisingly, long timescale gamma-ray monitoring revealed a nonconstant flux subject to sudden "jumps" [44]. For example, a quite sudden transition occurred on MJD 55850 with a gamma-ray flux decrease of the flux by 20% and an increase of the spin-down rate by 4%. The pulsar stayed in this state for a couple of years, and then apparently went back to its previous state [42]. Figure 11 shows the gamma-ray flux history and the two state transitions. The mechanism for producing these variations is probably related to a change of external torque induced by the magnetospheric coupling with the pulsar

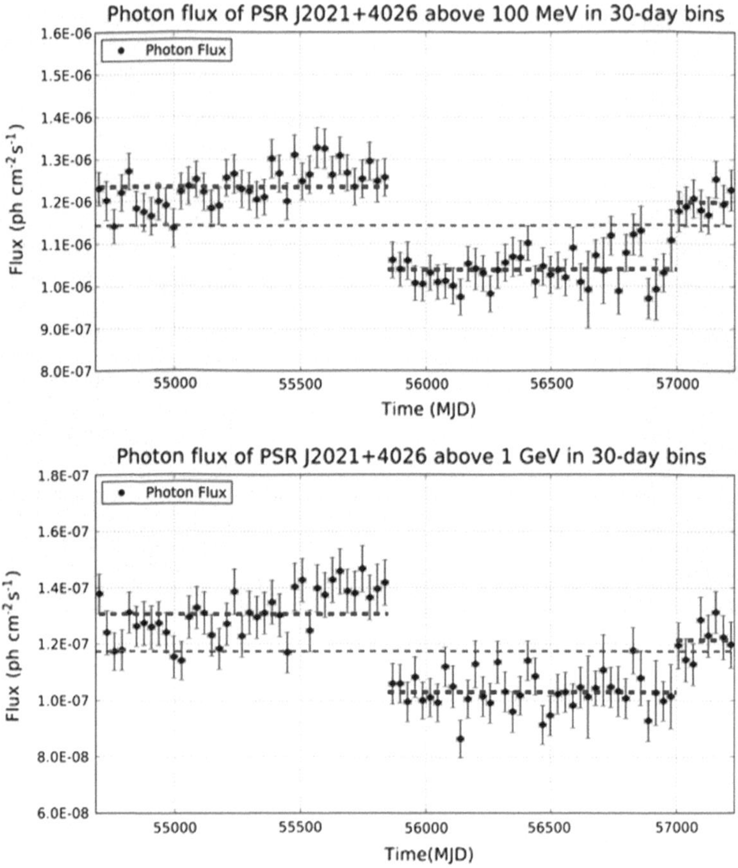

Fig. 11 The gamma-ray ligth curve of the pulsar PSR J2021+4026 showing two clear flux transitions (from [42])

wind. PSR spin-down rates can change and recovery was recently noticed also in several "intermittent" radio pulsars [45].

A challenging question arises: can pulsars have their wind properties change so significantly to affect the torque on the neutron star and eventually the far-away features of the outflow (see also the discussion on the Crab Nebula)?

Pulsar Wind Nebulae

Pulsar Wind Nebulae (PWNe) are produced by relativistic pulsar winds interacting with the ISM or with their surrounding SNR as in the case of the young Crab, Geminga and Vela PWNe (for recent reviews, see [46, 47]). Their emissions range from radio to optical and X-rays up to GeV and TeV gamma-rays. In addition to the prototypical Crab Nebula, about a dozen PWNe have been detected in the GeV range [48, 49]. Gamma-rays are produced by relativistic e-/e+ pairs emitting inverse Compton (IC) radiation in the environment of soft photons provided by the CMB and interstellar field. Gamma-ray observations provide then a unique way to sample the energetic particle distributions of pulsar winds, and the likely re-acceleration at the pulsar wind termination shock either within the SNR or in the ISM.

As an example, we discuss here the case of Vela-X, the PWN associated with the 0.89 ms Vela pulsar which shows interesting properties at GeV energies [50, 52]. Figure 12 shows the AGILE 100 MeV–GeV map with the pulsar contribution subtracted out (left panel, from [50]) compared with the HESS TeV map (middle panel, from [51]). The spectral peaks at GeV and TeV energies (Fig. 12) clearly point to different particle populations. Yet, the spatial agreement between the AGILE and HESS maps shows that these two populations spatially coexist. Furthermore, AGILE spectral data are inconsistent with the fluxes expected from the electron population simultaneously fitting the synchrotron X-rays (peaking at 1 keV) and the IC TeV emission. The additional electron population can be related with the radio synchrotron emitting electrons observed in the Vela-X structure. These electrons may be responsible for the IC bump in the GeV band arising from scattering on the CMB and Galactic and starlight photon fields, as predicted in [53] and modelled in [50]. The fact that the Vela-X PWN shows two accelerated particle components with well-defined characteristics (one producing the radio and GeV emissions, the other one radiating the X-ray and TeV emissions, see also [54]) can be considered a cornerstone in our understanding of the interplay between synchrotron and inverse Compton emission from different populations in a "nebular" environment.

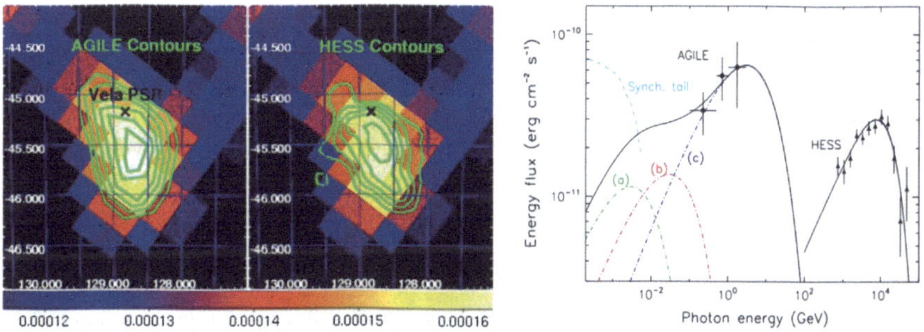

Fig. 12 The gamma-ray emission detected by AGILE from the Vela-X PWN [50]. *Left panel* the AGILE detection above 100 MeV. *Central panel* the AGILE detection (in *background*) compared with the TeV emission detected by HESS (*green contours*, from [51]). *Right panel* Gamma-ray spectrum showing AGILE and HESS results (from [50])

The Flaring Crab Nebula

The Crab Nebula, the famous "Messier 1" source, can be considered the "Queen of Nebulae;" it is an invaluable source of information for high-energy astrophysics (e.g., [55]). Exploded in 1054 AD, as recorded in the Chinese astronomical annals, the Supernova Remnant is expanding in a relatively dilute external environment. Its ejecta are wonderfully depicted by the Hubble image (Fig. 13). At the center of the Nebula, a 33 ms pulsar is powering the inner parts of the remnants with a relativistic outflow of particles and electromagnetic fields of spin-down luminosity $L_{sd} \simeq 5 \times 10^{38}$ erg s^{-1}. It is well known from the high-resolution X-ray imaging by Chandra (probing the synchrotron radiating electrons and positrons) that emission in the inner Nebula is highly anisotropic. An inner ring at the distance of about 3×10^{17} cm appears to be stable and surrounded by wisps and striations that form a disk-like structure in (what turns out to be) the pulsar equatorial plane; "plumes" of X-ray (and optical) emissions emanate perpendicularly, in what seem to be "jet"-like structures (see Fig. 13). Decades of HST and Chandra observations uncovered local variations of optical and X-ray brightness that appear to be localized in "spots" erratically flickering on timescales of days/weeks/months. Despite its local flickering, the Crab Nebula overall X-ray and gamma-ray flux (once the pulsed emission from the PSR is properly subtracted out) was measured to be apparently stable over many decades. The Crab (in first approximation summed as PSR plus Nebula contributions) has been therefore considered the ideal high-energy standard candle to be used for calibration. In the X-ray range (1–10 keV) dominated by nebular emission, the Crab flux became the standard "Crab unit" used to characterize X-ray sources. In the GeV range, the Crab is also particularly important for calibration purposes because of its

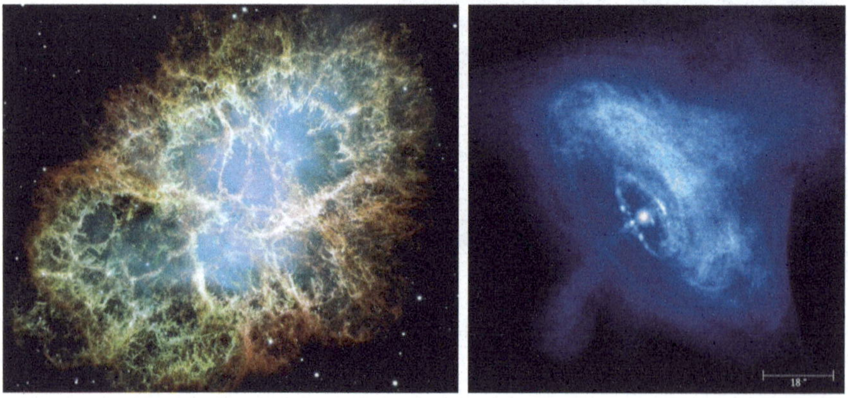

Fig. 13 The Crab Nebula as imaged in the optical band by the Hubble Space Telescope (*left panel*), and the inner Nebula in the 1–10 keV range as observed by Chandra (*right panel*). The (X-ray emitting) pulsar is clearly visible at the center of the image on the right

E^{-2} photon energy spectrum (dominated by the PSR emission) that turns out to be close to the spectra of the majority of Galactic gamma-ray sources.

Before moving on, let us pause for a second, and consider the nebular emission from radio to TeV [55]. The Nebula is dominated by a low-energy population of electrons/positrons emitting synchrotron radiation in the radio band, and by another (maybe more) component(s) at larger particle energies that are synchrotron radiating in the optical, X-ray and gamma-ray energies up to about 100 MeV (e.g., [56–58]). Emission above 10–100 GeV up to TeV energies is interpreted as IC radiation in the remnant and CMB environment (e.g., [56]).

The standard (and supposed to be stable) nebular spectrum in the GeV spectral range has been determined by *Fermi*-LAT that improved previous investigations by EGRET in this energy range (Fig. 14). The spectral cutoff observed near 100 MeV is in agreement with the idea of a "synchrotron burn-off." Photons radiated by synchrotron emission from a population of electrons and positrons have their energies limited in the magnetic field which provides the seed of acceleration by electromagnetic waves through a relativistic cyclotron mechanism [57]. In ideal MHD, the electric field of an e.m. wave is limited by the magnetic field of the perturbation itself $E \leq B$. For an acceleration rate induced by cyclotron resonance (as is usually assumed in a nebular environment) proportional to Ω_c/γ (with $\Omega_c = eB/mc$, and γ the particle relativistic Lorentz factor) we can establish a limiting particle energy by equating acceleration and synchrotron cooling by a transverse $B_\perp$. The consequence is that the maximum emitted photon energy by synchrotron radiation $\varepsilon_{s,m}$ is formally independent of the local magnetic field, and can be written as

$$\varepsilon_{s,m} = \frac{9\, m_e c^2}{4\, \alpha_F} \frac{E}{B} \simeq 150\,\mathrm{MeV} \qquad (1)$$

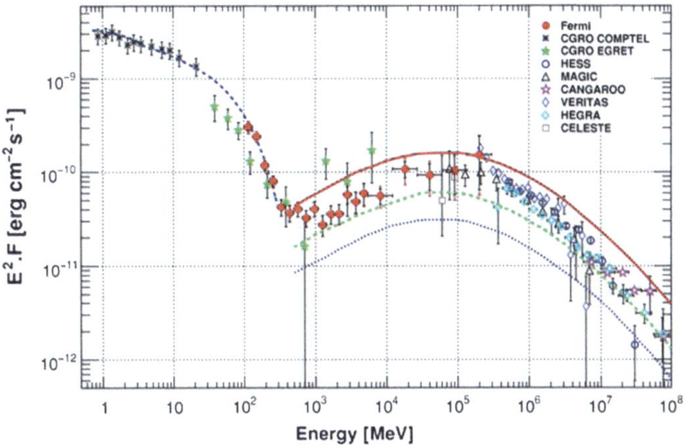

Fig. 14 The gamma-ray spectrum above 1 MeV of the Crab Nebula [27]. Pulsed emission from the Crab pulsar has been subtracted

where m_e is the electron mass, c the speed of light, and α_F the fine structure constant $\alpha_F = e^2/\hbar c \simeq 1/137$. Obviously, in ideal MHD, $E/B \leq 1$. This approach explains well the observed spectral cutoff (see Fig. 14), and that would have settled it, if not for a big surprise.

On September 22nd, 2010 a very unusual announcement was made by the AGILE Team [59]. An enhanced gamma-ray flux was detected from the direction of the Crab; it reached its peak on 19–21 September 2010 with a 2-day flux of $F_{\gamma,p1} = (7.2 \pm 1.4) \times 10^{-6}\,\mathrm{ph\,cm^{-2}\,s^{-1}}$ ($\alpha'' = 2.03 \pm 0.18$). This gamma-ray "flare" was confirmed within 1 day by *Fermi*-LAT [60]. The flux subsequently decayed within 2–3 days to normal average values. At first, it was not obvious whether the pulsar or the Nebula were producing the flare. However, two facts were immediately clear: (1) that the Crab (pulsar or Nebula) is not a stable gamma-ray source, and (2) that Eq. 1 is violated because of the detection of ~GeV photons during the event [61, 62]. The Crab was flaring in gamma-rays with a flux increase of about three times the standard value in a timescale of 1–2 days! The initial announcements were immediately followed by a wealth of multifrequency observations of the Crab Nebula, ranging from radio, optical, X-rays up to TeV energies (see [63] for a summary of follow-up observations). A more refined analysis of gamma-ray data [61, 62] together with radio, optical and X-ray data established that the enhanced gamma-ray emission was unpulsed and that the Crab pulsar did not show any appreciable change or glitch, therefore excluding a pulsar origin of the flare. Also, a random coincidence with a field source was excluded with high confidence [61]. Furthermore, no other nebular variations at wavelengths different from GeV photons could be related to the gamma-ray flares [62, 64]. As of today (July 2017) several major gamma-ray flares have been detected from the Crab Nebula, confirming the absence of an obvious relation with pulsar activity, or the presence of a field transient source [65]. Gamma-ray flares are then attributed to the Nebula, and occur, on the average, once per year [66].

In order to appreciate the relevance of this phenomenon, let us consider the brightest of all gamma-ray flare detected to date, that one of April 2011. The flare, lasting about 2 weeks, was observed by AGILE [67] and *Fermi*-LAT [68] in great detail with good optical (Hubble) and X-ray (Chandra) coverage. Figure 15 shows the gamma-ray light curve reaching the record peak flux of $F = (22 \pm 0.85) \times 10^{-6}\mathrm{ph\,cm^{-2}\,s^{-1}}$ on a 12-h timescale [67, 68] (i.e., twice the value of the brightest steady gamma-ray source in the sky, the Vela pulsar). Furthermore, the peak spectrum of Fig. 16 shows a clear violation of Eq. 1. Since that equation is derived from very general principles of ideal magnetohydrodynamics and classical radiation processes believed to apply to nebular environments and relativistic outflows (e.g., [57]), the crisis could not be more dramatic. The issue is made worse by the absence of positive clues about the flare localization in the Nebula, despite many optical (HST, Keck) and X-ray (Chandra) observations, some of them just at the gamma-ray flaring peak (as for the April, 2011 flare [64]; see also [69]).

What is needed is a sort of "super-acceleration" (in the spirit of [66]) taking place on a sufficiently short timescale to explain the observations (hours), and that impulsively produces either a quasi mono-energetic distribution of particle energies (well approximated by a relativistic Maxwellian of the type $f_M(\gamma) = dn_M/d\gamma =$

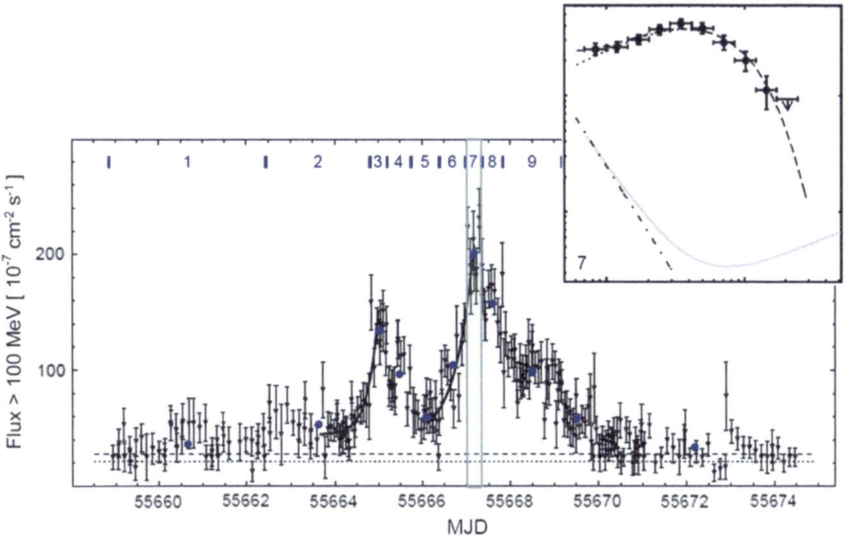

Fig. 15 The gamma-ray light curve above 100 MeV of the Crab during the major gamma-ray flare in April 2011 [68]. The *inset* shows the gamma-ray spectrum at the peak of the emission as detected by *Fermi*-LAT

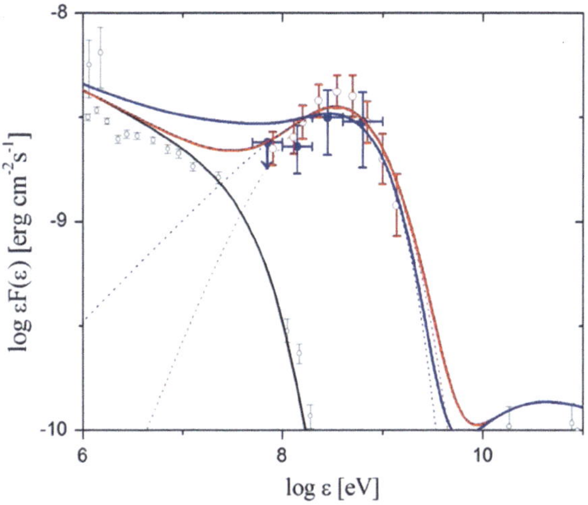

Fig. 16 The gamma-ray spectrum of the Crab during the major gamma-ray flare in April 2011 as determined by *Fermi*-LAT (data marked in *red* [68]) and AGILE (data in *blue* [67]). Data in *black color* represent the spectrum during the low-state of emission. The *red dotted curve* is a power-law model for the particle distribution of index 2.1 and maximum energy $\gamma_m = 5 \times 10^9$ producing emission possibly linking flaring gamma-rays with optical and X-ray simultaneous enhanced emission. The *blue dotted curve* represents a model based on an electron/positron relativistic Maxwellian distribution of average Lorentz factor $\gamma = 4 \times 10^9$ and local magnetic field $B = 1$ mG. Modelling from [67]

$K_M \gamma \exp(-\gamma/\gamma^*)$, with n the local density of particles, K_M a constant, and γ^* a typical value of the particle Lorentz factor), or a power-law (PL) distribution ($f_{pl}(\gamma) = dn_{pl}/d\gamma = K_{pl}\,\gamma^{-p_1}$, up to a maximum energy γ_{max}, with K_{pl} a constant, and p_1 an index). Super-accelerated particles (with "parallel" acceleration initially not bound by synchrotron cooling and most likely limited only by radiation reaction, see below) then evolve by pitch angle scattering developing a perpendicular component of their impulse and then radiating by synchrotron emission. Whether parallel electric field acceleration in the Crab Nebula can be efficiently produced by magnetic reconnection is a crucial theoretical issue today. Radiation reaction ultimately limits the maximum particle energy not only in the perpendicular (synchrotron) but also in the parallel direction. It is tantalizing to note that current observations of the Crab strongly favor a quasi "mono-energetic" particle distribution with energies that violate Eq. 1 unless the effective ratio $E/B > 1$. Understanding the Crab gamma-ray flares is an important topic of current theoretical research (e.g., [70]) with crucial applications in many sectors of relativistic astrophysics (from particle acceleration in jets to plasma instabilities in blazars and GRB sources). The Crab is more relevant than ever in studying extreme acceleration processes in our cosmos and many mysteries remain to be solved.

Microquasars: Cygnus X-3 and Others

Microquasars are Galactic binary systems (either persistent or transient X-ray sources) hosting a compact object which can be a confirmed (by mass) black hole (such as Cygnus X-1 and GRS 1915+10) or a peculiar compact object of uncertain nature (e.g., Cygnus X-3). Microquasars are special in their manifestations. They are outbursting X-ray binaries, in most cases with massive companion stars, sporadically (or steadily as in the case of SS 433) producing gigantic ejections of relativistic jets. Many decades of X-ray observations of these sources lead to a classification based on different spectral "states" presumably related with conditions at the inner edge of the accretion disk and of its surrounding "corona" of relativistic plasma.

Determining the existence of (transient) gamma-ray emission from microquasars is relevant for understanding the physical conditions corresponding to different spectral states. COMPTEL observations of Cygnus X-1 led to identify a possible episode of gamma-ray emission up to 5–10 MeV in the so-called soft state [71]. For many years, Cygnus X-1 remained the only X-ray source with a detection above a few MeV. Given the lack of continuous monitoring up to the 1990s, it was difficult to determine how common is this transient emission, and the connection with other radio/X-ray states.

The situation was clarified with the advent of quasi-continuous monitoring obtained by AGILE and *Fermi*-LAT. For the first time, practically all Galactic microquasars could be monitored over extended periods of time, and the relation between X-ray states and gamma-ray emission (if any) could be determined by simultaneous observations. It is interesting to note that the AGILE co-axial arrangement of a hard X-ray imager and a gamma-ray detector is particularly suited to detect X-ray outburst and gamma-ray enhancements. Surprisingly, the majority of Galactic microquasars were not detected simultaneously in hard X-rays and gamma-rays above 10 MeV. On the contrary, for a few cases of gamma-ray detections of the most active source (Cygnus X-3), the hard X-rays and gamma-rays have been observed to be anticorrelated.

Instead of Cygnus X-1, the most prolific source of gamma-rays in our Galaxy turned out to be Cygnus X-3. AGILE [72, 73] and *Fermi*-LAT [74, 75] detected several gamma-ray outbursts from this erratic microquasar. A more careful analysis of these detections above 100 MeV (lasting from one to a few days) clearly shows that gamma-rays are indeed emitted by Cygnus X-3 in coincidence with significant (either decreasing or increasing) changes of the hard X-ray flux. Remarkably, the episodes of gamma-ray flaring are all in coincidence with major instabilities of the inner accretion disk that make the system change from "radio-quenched" with soft X-ray states back to normal hard X-ray states in coincidence with relativistic jet ejections. The connection of particle acceleration producing gamma-rays in coincidence with major disk readjustments" leading to relativistic ejections is of great importance for modelling microquasars and in general accreting compact objects producing jets. Figure 17 shows one such episode detected from Cygnus X-3 in April 2008 showing a clear spectral transition in coincidence with a major relativistic plasmoid ejection [73]. The bottom panel of Fig. 17 shows how the hard X-ray flux goes back to normal (presumably connected with the re-establishment of a Comptonizing corona) within a few hours. The whole thing happened in coincidence with a major relativistic ejection and detectable gamma-ray emission. This pattern remarkably repeats itself in several other detections of Cygnus X-3 [72, 75]. Many issues are open for future investigations, including the origin of the emission (hadronic or leptonic), the nature of the disk instability, the details of the acceleration and magnetic dissipation between the disk and the corona (see, e.g., [76, 77]).

Other microquasars are much less active in gamma-rays. Sporadic emission has been claimed from Cygnus X-1 [78, 79], and a recent weak detection of V404 Cygni has been reported by *Fermi*-LAT [80] and AGILE [81] with characteristics similar to those of Cygnus X-3. Clearly, we need improved monitoring and better sensitivities in the MeV range to better understand the nature of the emission and the disk-jet interplay in microquasars. This effort, a crucial item in planning a new gamma-ray mission, may be rewarding not only for Galactic microquasars but also for understanding, in general, the compact object/jet relation.

Gamma-Ray Binaries

Among the many hundreds X-ray binaries in our Galaxy, only very few ones are observed to emit gamma-rays above 30 MeV. Here we comment on some of the most relevant gamma-ray binaries.

PSR 1259-63—This remarkable binary system hosts a 47.76 ms pulsar orbiting in a very eccentric orbit (with orbital period $P_{orb} = 3.4$ years) around a Be star

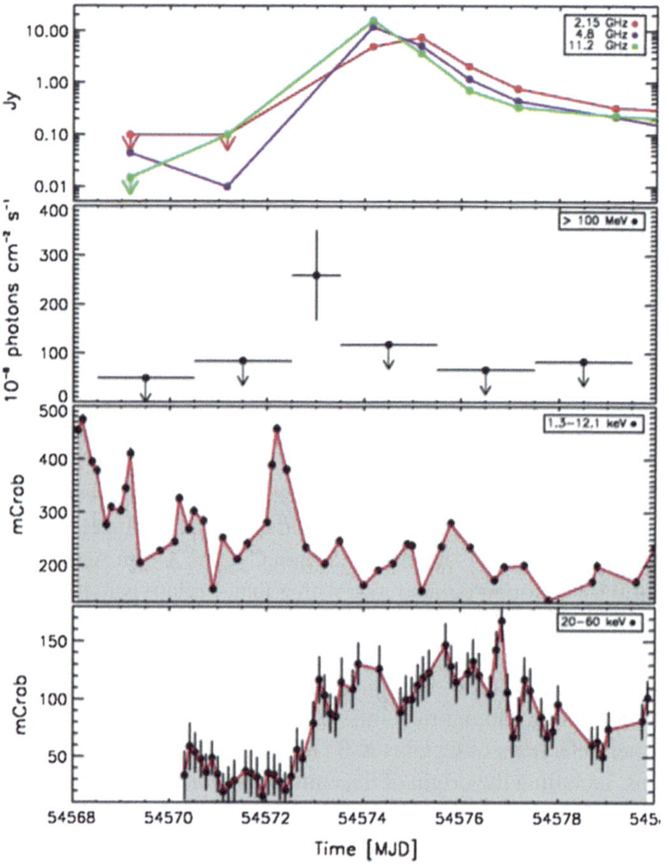

Fig. 17 Multifrequency data of the major outburst from Cygnus X-3 during the period 13–27 April 2008 [73]. *Top panel* radio monitoring from the RATAN observatory. *Second panel* gamma-ray lightcurve above 100 MeV detected by AGILE. *Third panel* soft X-ray lightcurve detected by the ASM on board of XTE. *Bottom panel* hard X-ray lightcurve detected by Super-AGILE

companion. This binary is a fantastic laboratory to study the interaction of a pulsar wind with a variable gaseous environment provided by the equatorial slow-mass outflow from the Be star in addition to a faster isotropic outflow component (e.g., [82]). The fact that the pulsar orbit is misaligned with the Be star equatorial outflow makes the whole process even more interesting. Gamma-rays are detected as the pulsar approaches the denser regions of the mass outflow and the periastron area. A variable pulsar wind termination shock leads to variable synchrotron emission and probably to additional particle acceleration near periastron. The source has been detected near periastron passages with a complex pattern by *Fermi*-LAT [83] and at TeV energies [84].

LS I +61 303—This somewhat mysterious binary hosts a compact star completely enshrouded in the gaseous wind from a massive companion. This system is probably similar to PSR 1259-63 except that in this case the orbit is tighter ($P_{orb} = 26.5$ d) and the radio pulsar is invisible because of wave dispersion in the dense mass outflow from the companion star. It can be considered an ideal case of a "hidden pulsar" [41]. Gamma-rays are now clearly detected and modulated with the binary period [85]. The broad hump dominating the spectral shape and the different GeV and TeV components indicate that the pulsar wind/outflow interaction is complex with multiple populations of accelerated particles.

LS 5039 can be considered similar to LS I +61 303 from the point of view of gamma-ray modulation and lack of evidence for radio pulsations from a compact source. The spectrum is broad and most likely due to a shocked pulsar wind enshrouded in the gaseous mass outflow from the companion star orbiting with a period of 3.9 days [86].

1FGL J1018.6-5856 was discovered by *Fermi*-LAT to be modulated with a period of 16.6 days [87]. The spectrum is qualitatively similar to those of LS I +61 303 and LS 5039 and therefore can be modelled in terms of an enshrouded relativistic wind of a hidden pulsar. As for the similar systems, anisotropic Compton scattering can explain the modulation in terms of different conditions of emission as a function of orbital phase. The pattern is quite complex, as shown Fig. 18.

Eta Carinae is one of the most famous binaries for studying colliding winds from massive stars showing strong variability from radio to X-ray energies with a period of 5.54 years. The primary star is believed to be of mass $\sim$100 $M_{\odot}$ and its companion has mass $\sim$30 $M_{\odot}$. Both stars eject powerful mass outflows that power an impressive nebula (e.g., [88]). Both AGILE and *Fermi*-LAT clearly detect the source [89, 90], which might variability on a few day timescale. The analysis of [91] uses almost 9 years of *Fermi*-LAT observations to establish a pattern of gamma-ray variation along the orbit that can be modelled by a combination of hadronic and leptonic emission of accelerated particles. Analysis during a more extended timeline is necessary to confirm these features and their interpretation.

Novae

Discovery of gamma-ray emission from novae came as a surprise. Novae are white dwarfs undergoing thermonuclear explosions on their surfaces fueled by mass accreted from a companion star. Shocked expanding gas from the nova shell produce X-rays and gamma-rays were not expected by standard shock models. *Fermi*-LAT detected transient gamma-ray emission from the symbiotic star V407 Cygni during a major outburst in 2010 [92]. A most likely interpretation of the emission is based on efficiently shock-accelerated hadrons at the nova shell that produce neutral pions and then gamma-rays. A leptonic model cannot be excluded, even though appears less likely. Interesting constraints on the acceleration model can be derived from these

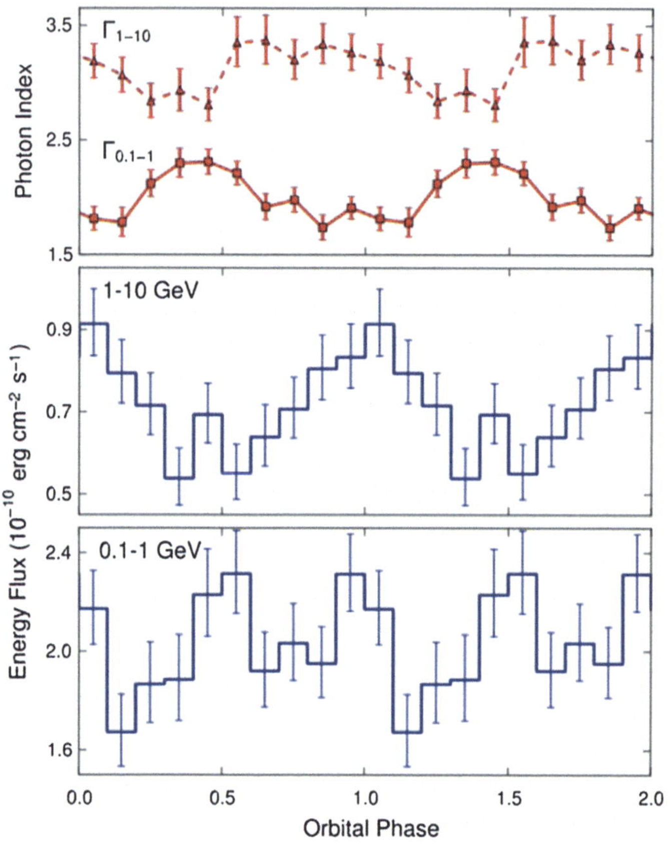

Fig. 18 Gamma-ray modulation of 1FGL J1018.6-5856 as detected by *Fermi*-LAT [87]

observations, and theoretical work is ongoing to interpret them (e.g., [93]). Several novae have been detected by *Fermi*-LAT and the number may be increasing in the near future [94].

Supernova Remnants and the Origin of Cosmic Rays

Cosmic rays are energetic particles (ions and electrons) reaching the Earth with a very broad spectrum (from a few GeV up to $\sim 10^{20}$ eV). The hadronic component of cosmic rays dominates over the leptonic one, and has a distinctive spectrum of index 2.7 for kinetic energies up to about 1 PeV. Above this energy, the observed spectrum changes slope a first time (the "knee") and becomes steeper; then it goes back to a harder value (the "ankle") at energies $\sim 10^{18}$ eV. The debate about the origin of hadronic cosmic rays has been central to the community of theoretical and observational space physics for decades. Clearly, cosmic rays are a manifestation of the most energetic processes in our Universe. But how and where are they produced?

Focussing here on the cosmic rays up to 1 PeV, there is a large consensus about their Galactic origin. However, despite decades of investigations, several important details of the process are not satisfactorily understood yet. The most likely source candidates for the GeV-TeV-PeV cosmic rays are considered to be Supernova Remnants (SNRs) (e.g., [95, 96]). Both energetic and physical reasons make SNRs ideal candidates for energizing hadronic cosmic rays (e.g., [97]), even though other sources could in principle contribute in the Galaxy (pulsars, compact binaries with neutron stars and black holes, interstellar shocks in "super-bubbles," the Galactic center).

Space observations of SNRs are crucial to determine their MeV-GeV spectra, a piece of information that cannot be obtained otherwise. In particular, what we are looking for is a direct "hadronic signature" of accelerated protons/ions in the 100 MeV–1 GeV spectra: the existence of a distinct spectral feature (a low-energy decrement near 100–200 MeV) that is a consequence of the almost-at-rest neutral pion decay $\pi^o \to \gamma\,\gamma$. The problem, that haunted gamma-ray observations for many years, is *distinguishing leptonic from hadronic contributions* to the emission in that energy range. In many cases, SNRs are embedded in a complex environment, and leptonic relativistic Bremsstrahlung contributions may dominate the GeV emission. What we look for are SNRs accelerating hadronic cosmic rays that diffuse out from the moving shock and hit gaseous surroundings (spread out or in dense clouds) producing gamma-rays from neutral pions induced by the reaction $p\,p \to \pi^o + \ldots$. Young and middle-aged SNRs are the most promising sources, but their surroundings have to "cooperate." Are they?

EGRET could establish only in a preliminary way a relation between enhanced gamma-ray emission and a selected number of SNRs in the Galactic plane [98]. The breakthrough was provided by AGILE and Fermi-LAT. After several years of observations and attempts, a clear case for hadronic acceleration was finally provided by the middle-aged SNR W44 [99, 100]. In this case, evidence was obtained for a well-defined synchrotron emitting shell interacting with dense surroundings (as determined by CO measurements) with clearly overlapping gamma-ray emission (see Fig. 19). A first spectral determination by Fermi-LAT was not conclusive because of its lack of sensitivity at energies below 200 MeV [101]. A much improved spectrum was originally obtained by AGILE below 200 MeV [100] (and later refined by Fermi-LAT through an improved processing [99]). The result is an impressive spectrum peaking at a few hundreds of MeV and showing a very distinct decrement at lower energies up to 50 MeV (Fig. 20). Given the spatial superposition between gamma-ray emission and synchrotron radiating electrons, one could unambiguously establish their Bremsstrahlung contribution, and demonstrate that it cannot reproduce the observed gamma-ray spectrum. Direct evidence for neutral pion decay emission was then established for the first time almost after 50 years since the hypothesis was consistently formulated! Furthermore, W44 shows also an interesting spectral decrement at photon energies higher than 1 GeV with photon index $\sim$3, indicating that this class of middle-aged SNRs does not contribute to the bulk of cosmic-ray spectrum above TeV energies (see also [102]). In any case, this breakthrough measurement

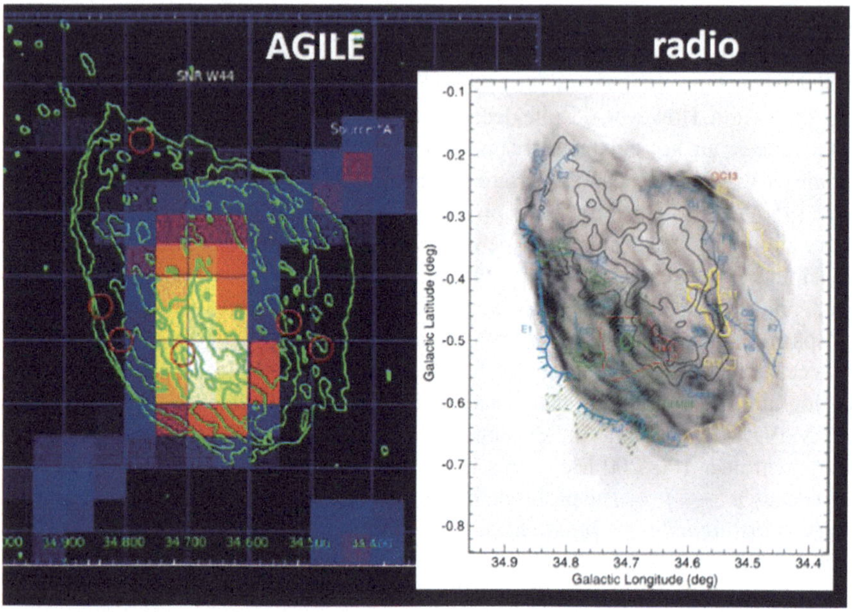

Fig. 19 Gamma-ray imaging of the SNR W44 above 100 MeV obtained by AGILE superimposed with radio data (from [103])

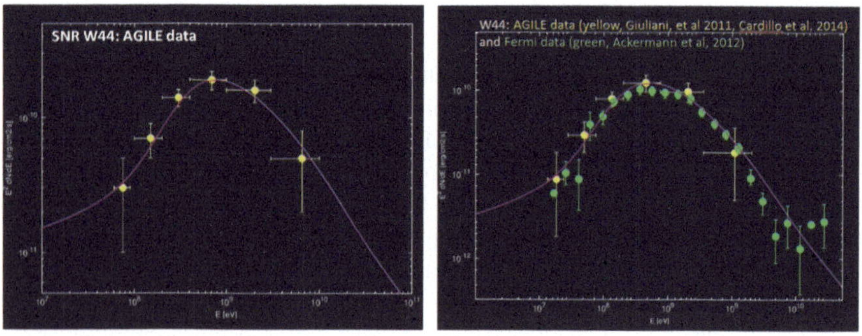

Fig. 20 *Left panel* the gamma-ray spectrum of W44 as detected by AGILE [103]. *Right panel* the concordance between the AGILE (in *yellow*) and *Fermi*-LAT (in *green color*) spectral data of W44 (data from [103] and [99])

demonstrates that hadronic cosmic rays are accelerated by SNRs most likely in sufficiently large numbers to explain the low-energy part of the cosmic-ray spectrum. As of July 2017, evidence for hadronic gamma-ray emission from SNRs with a spectral pion-signature has been obtained for several sources, most notably W44, IC443, and W51C.

The theory of diffusive shock acceleration (DSA) (e.g., [104–107]) with its nonlinear variants (NLDSA, e.g., [108, 109]) has been applied to SNRs in a variety of conditions. These models are based on a modern version of the "Fermi mechanism," a way to accelerate particles at the strong supersonic shocks of SNRs. Particles (hadrons and electrons) diffuse in-and-out the shock front being scattered by electromagnetic waves induced by magnetic turbulence. In doing so, they get a net energy gain at every "cycle," and the process can lead to large particle energies for a sufficient number of "cycles" [104, 105]. Recent literature is devoted to detailing the process and its efficiency. For strong shocks (and "compression ratio" $\mathscr{R} = 4$) DSA "naturally" produces a power-law index of accelerated particles $\delta = 2$, a value similar to what is requested by cosmic ray sources once the effect of Galactic diffusion is taken into account. Pressure from accelerated cosmic rays influences the shock structure, and nonlinear effects may alter the final outcome. A prediction of nonlinear DSA is the production of upward "concave" energy spectra of accelerated particles, an interesting prediction that can be tested by observations.

It is interesting to compare the gamma-ray spectra of different SNRs by combining Fermi-LAT and TeV Cherenkov detectors. Three classes of spectra are clearly shown in Fig. 21, see [110]. Young SNRs appear to be harder (even though they show a cutoff at TeV energies) than middle-aged SNRs. The latter class divides into two categories: SNRs with GeV maxima and neutral pion-signatures, and other SNRs peaking at TeV energies with very low, if any, GeV emission (probably dominated by leptonic inverse Compton radiation, for a discussion see, e.g., [111]).

We note that the upward concave spectra expected from NLDSA models have not been detected yet in any SNRs. Furthermore, most of SNRs show a spectral index in the 1–10 GeV range steeper than 2, indicating that modifications of the models are necessary even in the linear regime.

Young SNRs have the most powerful shocks and are the primary candidates for cosmic-ray acceleration up to PeV energies and beyond. However, it is uncertain today whether their number and duty cycle of their most accelerating phases are sufficient to explain observations up to the "knee." Figure 22 shows the combined Fermi-LAT and VERITAS spectra of two young SNRs, Tycho [112] and Cas-A [113]. The spectra up to 10 GeV energies are overall considerably harder than those of older SNRs (see Fig. 23 for IC443); however, even in these cases, spectral cutoffs

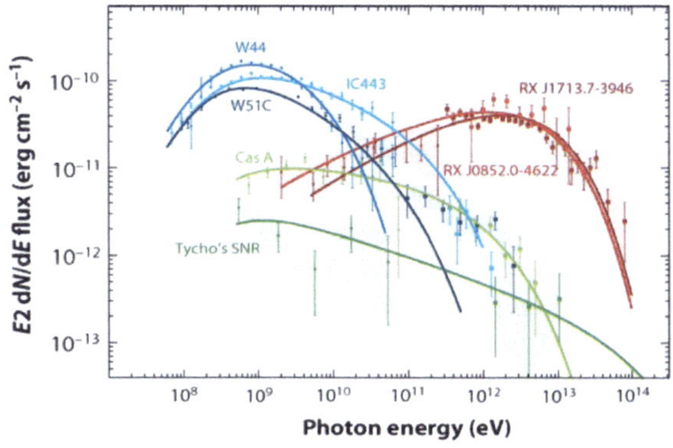

Fig. 21 Gamma-ray spectra above 10 MeV of SNRs detected by Fermi-LAT and TeV Cherenkov telescopes (HESS, VERITAS, and MAGIC), as of 2015, from [110]

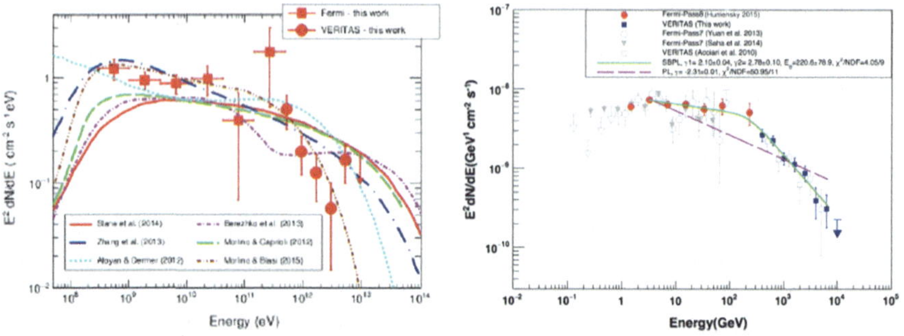

Fig. 22 *Left panel* gamma-ray spectrum of the Tycho SNRs detected by Fermi-LAT and VERITAS [112]. *Right panel* gamma-ray spectrum of the SNR Cas-A detected by *Fermi*-LAT and VERITAS [113]

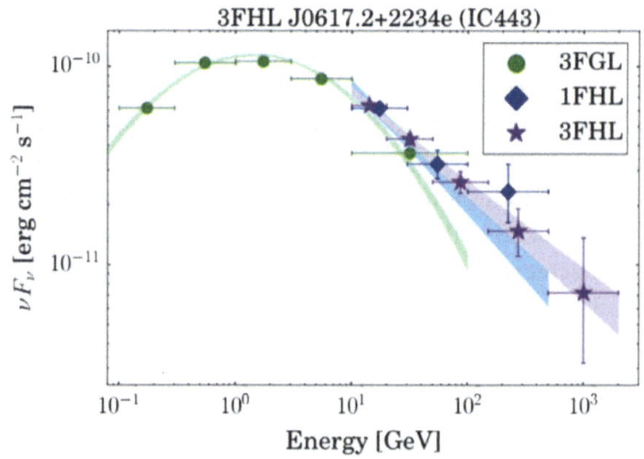

Fig. 23 Gamma-ray spectrum of the SNR IC443 detected by Fermi-LAT up to TeV energies [114]

are evident at higher energies. Most likely, SNRs can accelerate cosmic rays only during their very early phases of expansion (as suggested by models invoking Bell's non resonant instability [115], e.g., [116]). Models SNRs as hadronic Pevatrons still escape us!

Large-Scale Particle Acceleration in Galactic Super-Bubbles

Interstellar shocks and Galactic super-bubbles driven by interacting powerful stellar winds can in principle accelerate cosmic rays (e.g., [117]). The efficiency and the energy spectrum that can be produced by super-bubble acceleration is under debate [118]. Evidence for hadronic particle acceleration in the complex interstellar medium of the Cygnus region was remarkably detected after subtraction of known gamma-ray sources [119]. Figure 24 shows the residual gamma-ray flux compared with IR emission in the region: the existence of enhanced diffuse gamma-ray emission strongly suggests the presence of an accelerated population of particles. If the nature of this

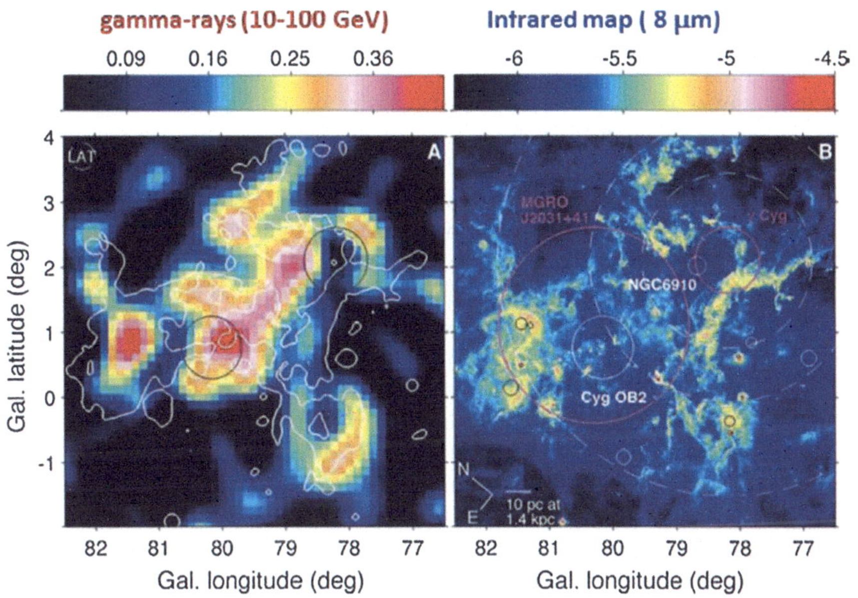

Fig. 24 Diffuse gamma-ray emission above 1 GeV versus infrared emission in the Cygnus region, from [119]. *Left panel* residual gamma-ray emission detected by *Fermi*-LAT once known sources have been subtracted. *Right panel* IR-emission in the 8-μ band

Fig. 25 Combined gamma-ray spectrum of the Cygnus cocoon with *Fermi*-LAT, ARGO, and MILAGRO data, from [120]

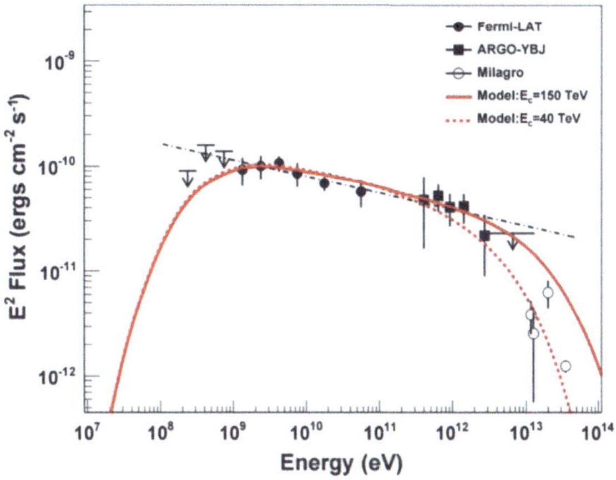

emission (which extends up to TeV energies [120]) turns out to be hadronic, super-bubble acceleration of cosmic rays will have a case to be considered for more detailed studies. Figure 25 shows the spectrum of the Cygnus cocoon extending up to TeV energies. Detecting super-bubble acceleration in the Galaxy is a difficult task, but it is definitely a topic of great interest for future observations.

The Galactic Center

The inner part of our Galaxy is the most complex region for gamma-ray observations. The density of point-like gamma-ray sources (or potential gamma-ray emitters: SNRs, pulsars, binaries) is substantially larger compared to other regions in the bulge and disk. A massive black hole of $\sim 4 \times 10^6 \, M_\odot$ resides at the Galactic Center (GC) in the Sgr A* region, a potential source of high-energy emission. In addition, diffuse gamma-ray emission masks the presence of weak sources and efficient modelling and "subtraction" of the diffuse emission is necessary to unveil underlying sources. It is therefore not surprising that all gamma-ray space missions experienced considerable trouble in data analysis of this region. Angular resolution at GeV energies improved substantially from EGRET to AGILE and Fermi-LAT, but even the current generation of instruments is challenged. In addition to diffuse and source astrophysics, the GC is expected to be the most promising site for detecting radiation produced by dark matter annihilations. The interest in the region is therefore enormous.

EGRET observations showed the existence of a prominent gamma-ray source (2EG J1746-2852) centered at $\sim 0.2°$ from the GC but still consistent with the position of Sgr A* [121] (see also the refined analysis of [122]). Today, we have a much sharper view of the inner Galactic region. Yet, it is difficult to provide a complete understanding of a very crowded region. Several sources within 10° from the GC have been now been catalogued and studied. In principle, a large variety of point-like sources can contribute to gamma-ray emission in addition to the enhanced diffused radiation: regular pulsars and millisecond pulsars, compact binaries, SNRs, molecular clouds, exotic objects. The recently discovered TeV source positionally coincident with Sgr A* is intriguing [123]. Its relation with nearby or positionally coincident GeV sources is unclear and requires additional investigations. It has been suggested to be a source of hadronic cosmic rays that in principle might explain the observed distribution up to the knee. If confirmed, this hypothesis would have far reaching consequences.

Fermi Bubbles

Totally unexpected, the so-called Fermi bubbles were discovered in 2010 in the *Fermi*-LAT data after a few years of operations [124]. Figure 26 gives the original discovery map at GeV energies showing a clear "emanation" from the inner regions of our Galaxy that is capable of radiating gamma-rays at very large distances from the plane. Initially noticed as an additional gamma-ray "haze" in the inner Galaxy [125], the feature was recognized as being produced by a particle outflow from inner Galaxy [124]. Morphological and spectral studies of the gamma-ray emission [126] show its non-thermal nature possibly produced by accelerated and IC radiating electrons or by high-energy diffusing protons (e.g., [127]).

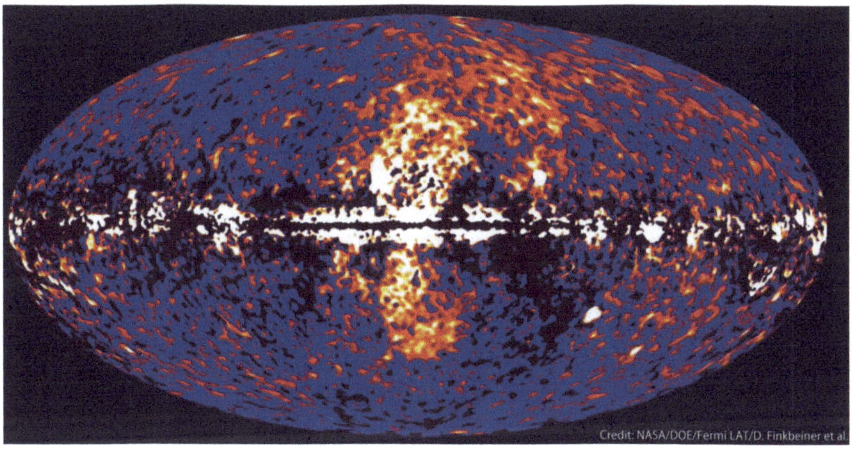

Fig. 26 *Fermi*-LAT gamma-ray map of residual GeV emission showing the existence of "Fermi bubbles" apparently emanating from the inner Galaxy, from [124]

Is the somewhat mysterious black hole at the center of our Galaxy producing this gigantic outflow? Or, is the inner star forming region producing a particle and field emanation reaching large distances [128] as observed also in other radio-galaxies? Is the nature of the emission leptonic or hadronic? The issue is being investigated intensively (e.g., [127]).

Blazars

Blazars are Active Galactic Nuclei (AGNs) hosting a massive black hole whose relativistic jet is pointed toward us (see Fig. 27 for a schematic picture). Since the first serendipitous discovery of 3C273 by COS-B, these enigmatic objects have been sources of amazement and challenge for all gamma-ray space missions. After the first and necessarily limited pointings above the Galactic plane by SAS-2 and COS-B, the first all-sky survey above 100 MeV was performed by EGRET. It produced a first surprising view of blazar emissions. Not only blazars emit gamma-rays with large intensities, but the observed variability clearly indicated the existence of a complex "machine" capable of radiating a large fraction of accretion luminosity into high-energy gamma-rays. The EGRET catalog of gamma-ray sources lists a few hundred blazars that therefore dominate the extragalactic sky. Immediately a number of questions arise especially in connection with the blazar properties observed at other wavelengths. How and when are the gamma-rays produced? What is that triggers the gamma-ray emission? What are the physical properties of blazars' relativistic jets? What can be learned by simultaneous observations in the radio, optical, X-ray and gamma-ray ranges?

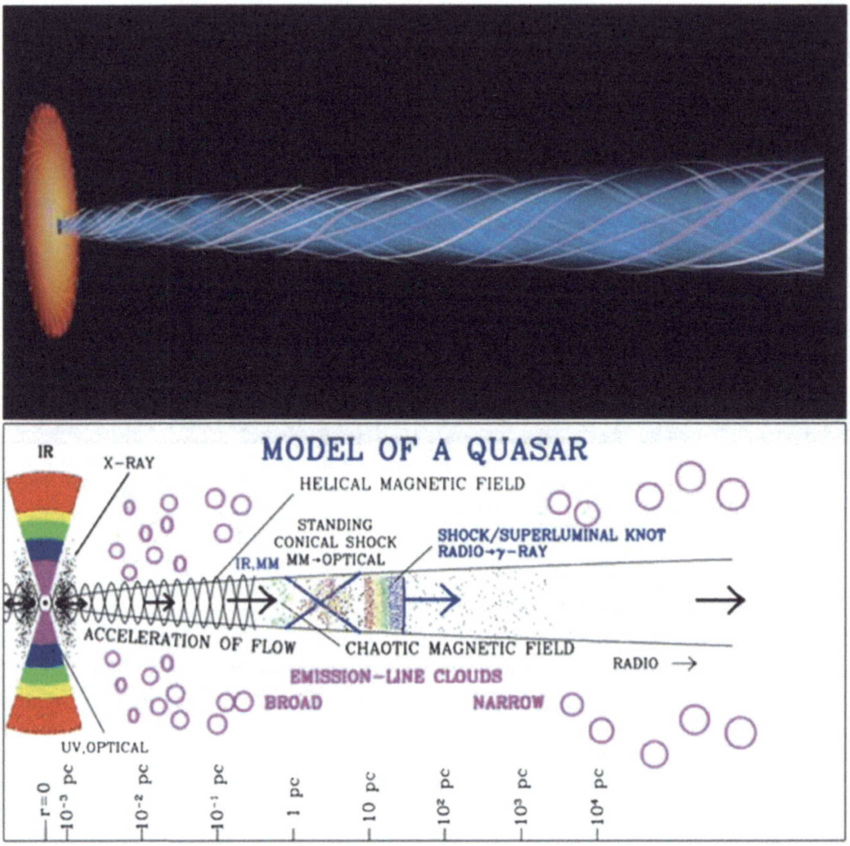

Fig. 27 Schematic representation of a magnetized relativistic jet of an Active Galactic Nucleus of the blazar class

3C 279 turned out to be among the most active blazars during the EGRET active time (1990–1996), and several campaigns involving observers from radio to space satellites were organized for the first time [131–133]. One can say that multifrequency gamma-ray astrophysics started at the time of the EGRET-driven campaigns on 3C279 [129, 134] (see Fig. 27). Today multifrequency high-energy astronomy is given for granted, but during EGRET time it was far from obvious: those who worked hard those days to make it happen can be happy today of their efforts. We will come back to 3C 279 shortly.

Indeed, the simultaneous multifrequency spectra of blazars that could be obtained for the first time confirmed the existence of two major ranges of emission where most of the emitted power is concentrated. The power spectra (obtained by multiplying the photon spectrum by the ε^2 power of photon energies) showed peaks near IR-optical and GeV gamma-rays for blazars of the Flat-Spectrum-Radio-Quasar (FSRQ) type (such as 3C273 and 3C279). It was soon realized that the other class of blazars,

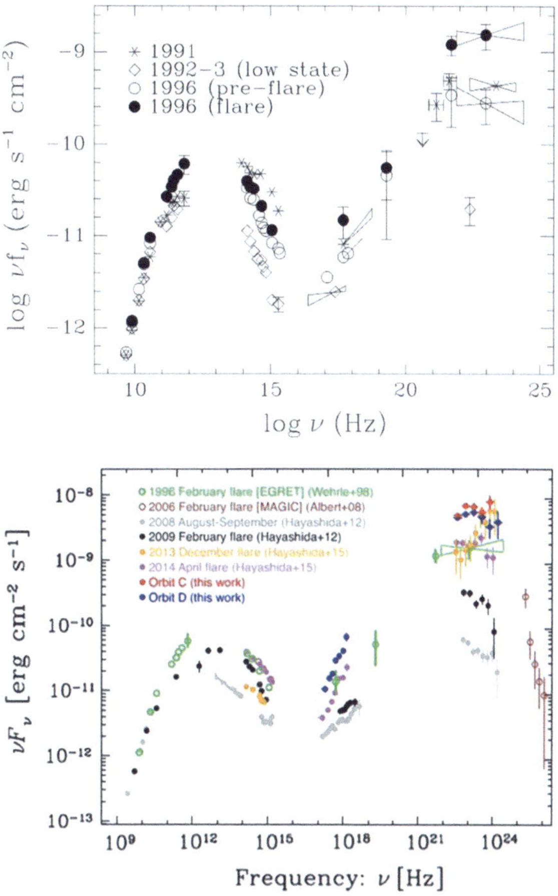

Fig. 28 Old and new gamma-ray detections of the quasar 3C 279. *Top panel* The multifrequency energy power spectrum of the 3C 279 during EGRET time [129]. Except for very low-states of emission, the gamma-ray peak dominates over the synchrotron peak, a phenomenon called "Compton dominance". *Bottom panel* The power spectrum of 3C 279 during the major flare on June 2015 [130]. The gamma-ray peak flux is, in this case, one order of magnitude larger than what detected in the 1990's, and the Compton dominance can reach values remarkably larger than 100

that of BL Lacs objects, had the peaks of the νF_ν spectrum shifted toward higher energies usually with equal peak power. A physical interpretation based on realistic e.m. fields and particle densities led to interpreting the first peak as produced by synchrotron radiation, and the second peak (that at GeV or TeV energies) as caused by inverse Compton (IC) scattering. Figure 28 shows the multifrequency results of several campaigns [129].

The physical mechanism of IC requires a bath of soft photons of large photon densities. Many theoretical studies focused on the inner region of the system called the "broad line region" (BLR) where a sort of "Hohlraum" can be realized. If L_d is the accretion disk luminosity, the photon energy density U_{ph} is calculated to be simply $U_{ph} = L_d/(4\pi c r^2)$, where r is the radial distance from the central black hole. Often the assumption $r = R_{BLR}$ is made, with $R_{BLR} \sim 3 \times 10^{17}$cm the size of the BLR. A blob of particles and fields ejected in the jet and moving relativistically within the BLR with bulk Lorentz factor Γ would see a soft photon bath boosted by a factor of Γ^2 in its comoving frame, that is $U'_{ph} \simeq U_{ph}\,\Gamma^2$. It is ultimately this quantity, together with the number and spectra of radiating particles within the blob, that determines the output in gamma-rays. This emission mechanism, that we can call "within the BLR" has a number of predictions that are verified in the majority (by number) of gamma-ray emission episodes of blazars. One of the consequences is the correlation between the synchrotron radiating particles and their IC emissions that should result in an optical-gamma-ray correlation. This is realized in a large number of blazars' flaring episodes, but as we will see shortly, there are remarkable exceptions. Furthermore, in this mechanism is quite difficult to account for strongly "Compton dominated" spectra, as those of 3C279 in flaring states (Fig. 28).

Among the brightest blazars detected by EGRET also showing rapid variability was PKS 1622-297 [135]. The gamma-ray light curve showed a clear "plateau"-like behavior, rising from nothingness, superimposed on a wild sequence of peaks that reached the world-record flux (for that time), $F_\gamma = (1700 \pm 300)\ 10^{-8}\ \mathrm{cm^{-2}\ s^{-1}}$ for ($\varepsilon > 100$ MeV) [135]. The optical coverage was not optimal and quite sparse; nevertheless, it showed large variations of optical intensity that appeared not obviously correlated with the gamma-rays.

The small EGRET FoV did not allow a systematic monitoring of blazars flaring in gamma-rays. EGRET exposure of extragalactic fields was most of the times randomly scheduled with respect to phases of activity of bright blazars, and only in a few occasions, a multifrequency campaign could be arranged. A number of questions arise: (1) what makes the class of gamma-ray blazars so special compared to the others? (2) what is the duty cycle of gamma-ray emission? (3) is there any difference between FSRQs and BL Lacs? (4) is the inverse Compton model of gamma-ray emission consistent with the multifrequency data? The almost 10-year hiatus between EGRET observations and AGILE first and later *Fermi* observations of blazars passed by with expectations and speculations. The most intense EGRET blazars have been monitored in the radio and optical wavelengths, sometimes with amazing detections of large flux increases (see, e.g., the case of the 2005 optical flaring of 3C454.3 [136]).

Finally, we get to the AGILE/Fermi observations of the gamma-ray sky with large FoV instruments. Since the beginning of the AGILE observations in 2007, the extragalactic sky revealed a few fundamental properties: blazars dominate the extragalactic emission; their gamma-ray emission above 100 MeV is time variable; the emission duty cycle for the most active blazars is about 10–20 % of the whole monitoring period; both FSRQs and BL Lacs contribute to the emission. After almost 10 years of observations, thousands of blazars have been detected by *Fermi*-LAT extending the range of detection up to redshifts $z \sim 3 - 4$. A complete account of blazars observations goes beyond the scope of this concise review. We mention here some of the most relevant issues that are stimulated by *Fermi*-LAT and AGILE observations: (1) the ultimate origin and site of the gamma-ray emission that in several prominent cases appears to be decoupled from synchrotron emission; (2) the nature of particle acceleration that now seems suggesting that prominent magnetic phenomena (reconnection?) are dominating the flow and the radiation of BL Lacs and FSRQ jets; (3) the origin of the very strong Compton dominance observed in 3C 454.3, 3C 279 and other FSRQs; (4) the stability of jets that occasionally show kinked structure as driven by magnetic instabilities; (5) very fast acceleration, with observed radiation timescales of minutes also in the GeV energy range, that dramatically challenge shock diffuse acceleration models. Blazar gamma-ray astrophysics is extremely rich and will continue to be the focus of exciting future theoretical investigations and observations by gamma-ray telescopes.

Sources Above 50 GeV

Space experiments can contribute to clarify important detections obtained on the ground at energies above 100 GeV. Space-based instruments are limited in their size and therefore are not competitive for the short timescale detection of photons at ~TeV energies. Nevertheless, by integrating over timescales of many years, a cosmic population of very high-energy (VHE) emitters can be revealed. It is a recent important result by *Fermi*-LAT to have successfully accomplished this difficult task. The third catalog of hard *Fermi*-LAT sources (3FHL) contains 1556 objects detected during 7 years in the range 10 GeV–2 TeV [114] (see Fig. 29). The vast majority of these sources are BL Lacs (59%), followed in numbers by blazars of uncertain type (25%) and by FSRQs (14%). Galactic sources (pulsars, PWNe, SNRs, binaries) amount to a few tens in total (see Fig. 23 for a beautiful spectrum of a middle-aged SNR extending up to TeV energies). Interestingly, a non-negligible fraction of 3FHL sources are not detected by *Fermi*-LAT at lower energies. Figure 30 shows two examples of spectra of two prominent blazars with emissions extending beyond 100 GeV. These results are of crucial importance for future projects in the TeV range, both ground-based ones and from space.

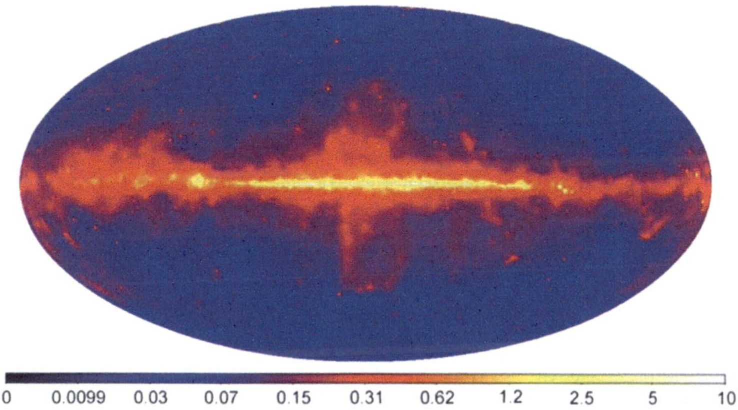

Fig. 29 Fermi-LAT counts map in the 10 GeV–2 TeV band represented in Galactic coordinates and Hammer-Aitoff projection [114]. Color scale is logarithmic and the units are counts per $(0.1°)^2$ pixel

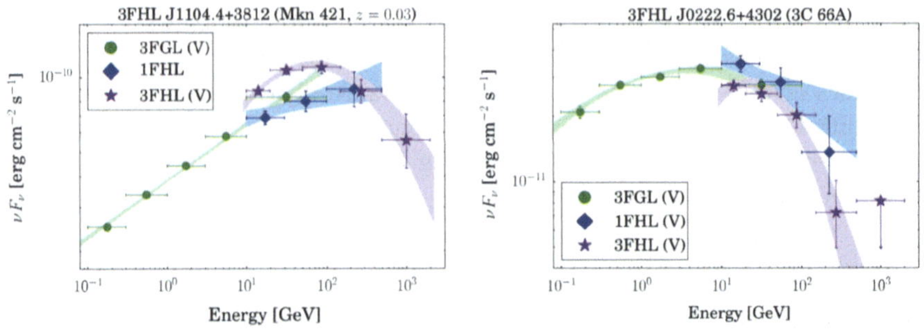

Fig. 30 Average spectra of two prominent extragalactic objects (Mrk 421 and 3C 33A) as determined in the 3FHL catalogue by *Fermi*-LAT [114]

Gamma-Ray Bursts

Cosmic Gamma-Ray Bursts (GRBs) are transient phenomena producing strongly variable emission from radio, optical, X-rays and gamma-rays up to 100 GeV. Discovered in the late 1960s by military satellites aimed at checking agreements under the nuclear nonproliferation treaties between the US and the Soviet Union, GRBs have fascinated astrophysicists ever since. Observationally, several types of GRBs can be distinguished according to their timing and spectral properties. An important distinction is made between "short GRBs" (lasting less than ~1 s, hereafter s-GRBs) and "long GRBs" (in the duration range ~1– hundreds of seconds). Lightcurves are usually "spiky" with strong spectral evolution (usually "hard-to-soft") within the peaks and in between them. In addition to the initial most intense ("prompt") phase, many if not all GRBs show a delayed emission phase ("afterglow") that may last hours-days-weeks with a decaying light curve of X-rays, optical and radio emission. Most of the GRB prompt emission is usually radiated in hard X-rays, and that is the main energy range used for their detection. However, it was clear from EGRET observations of the few GRBs that occurred in its (small) FoV, that a fraction of events could extend their spectra well beyond 100 MeV.

Today, many tens of GRBs have been detected at gamma-ray energies by AGILE and *Fermi*-LAT out of thousands of GRBs detected at X-ray energies. It is now clearly determined that gamma-ray emission above 100 MeV is not always produced by all intense GRBs within the instruments' FoVs (resolving an issue raised by EGRET observations). Gamma-rays belong to a spectral component usually different from the prompt hard X-ray initial peak. In addition to the prompt phase, gamma-rays, if radiated, are also detected in a prolonged "delayed" phase lasting ten-hundred-thousand of seconds after the prompt emission.

A remarkable case of a GRB showing all these characteristics is the short GRB 090510 [137–139]. Figure 31 displays its light curve detected by *Fermi* GBM and LAT showing a prominent prompt phase (with negligible emission above 100 MeV) which is anticipated by a low-intensity "precursor" and then followed by an

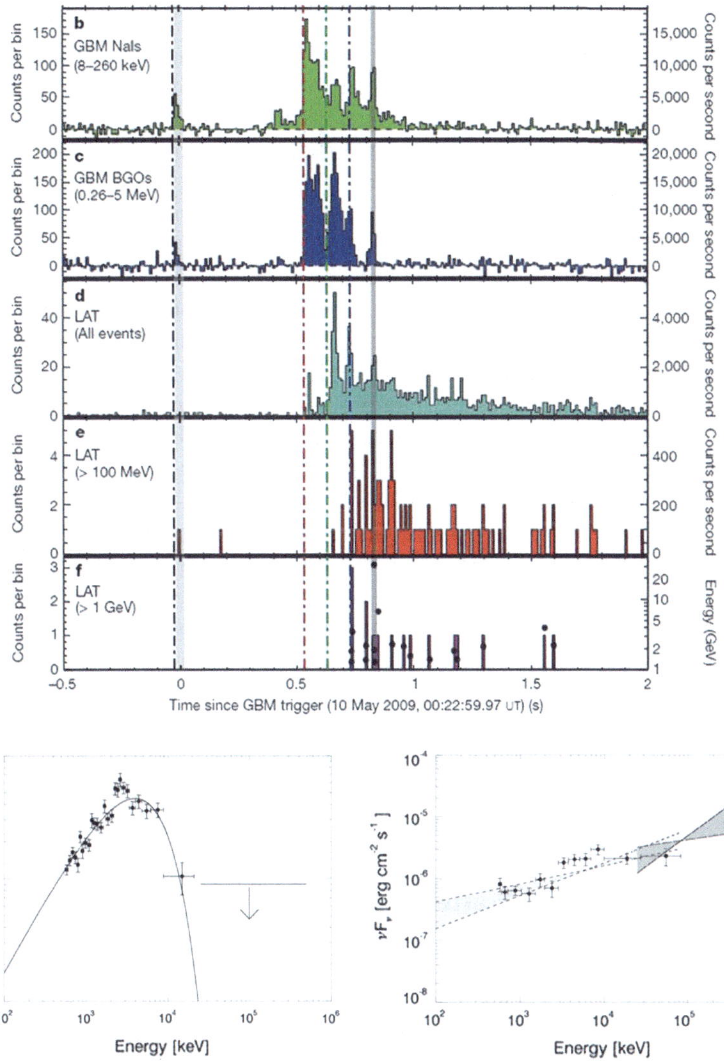

Fig. 31 The remarkable GRB 090510 detected by *Fermi* and AGILE. *Top panel* Lightcurves from hard X-rays to gamma-rays above 1 GeV detected by *Fermi*-LAT from the short GRB 091510 (adapted from [137]). *Bottom panels* Dramatic spectral change in GRB 091510 detected by AGILE (adapted from [138]). *Left panel* emission detected during the prompt phase lasting ~0.3 s. *Right panel* emission detected during the following tens of seconds

intense gamma-ray afterglow lasting hundreds of seconds. The dramatic spectral change between the prompt and gamma-ray delayed phase is shown in Fig. 31 (bottom panel). The spectrum changes from a quasi-thermal distribution with peak energy of a few MeV to a power-law with no apparent cutoff detected by AGILE [138] or *Fermi*-LAT [139]. A very efficient non-thermal acceleration and emission ensues following the prompt phase. The quasi-thermal nature of the prompt phase spectrum is also intriguing, suggesting an optically thick initial phase of relativistic expansion. If the emission is dominated by electron/positron pairs, from the observed spectral peak, we would deduce a bulk Lorentz factor $\Gamma \sim 10$. However, other interpretations are possible and the GRB is the subject of very intense theoretical research.

Another case of a surprising GRB is the gamma-ray bright GRB 130427A [140]. Figure 32 shows the gamma-ray light curve detected by *Fermi*-LAT extending for about 20 h after the prompt phase. Not only we witness here a remarkable manifestation of delayed acceleration "longevity," but also the detection after 244 s of a gamma-ray photon with energy equal to 95 GeV is challenging standard interpretations based on synchrotron emission of afterglow photons [140]. The crisis is related to the discussion made for Eq. (1): the maximum synchrotron photon energy can be written as $\varepsilon_{m,s} \simeq (80\,\mathrm{MeV})\,\Gamma(t)$, with $\Gamma(t)$ the bulk Lorentz factor of the emitting region. For reasonable values of $\Gamma(t)$, it is not possible to reproduce the observed energy in standard modelling. Inverse Compton models of the event suffer from several difficulties (as discussed in [140]). The event is challenging theoretical models, and the conundrum is possibly related with the very fast and "non-ideal" acceleration witnessed in the Crab Nebula. Probably, we are observing a phenomenon driven by magnetic field reconnection of an expanding shock front influenced by non-ideal MHD phenomena. That would be remarkable.

The Search for Dark Matter

Non-baryonic cold dark matter (DM) is modelled to dominate (84%) the matter density of the Universe (e.g., [142]). DM particle candidates, such as weakly interacting massive particles (WIMPs), can annihilate or decay into detectable standard particles and photons (e.g., Bertone et al. 2005; Feng 2010). Depending on the DM particle properties and annihilation cross section or decay rates, detectable gamma-rays can be produced. The expected flux is proportional to an integral of the square of the DM density (the so-called J-factor) integrated over the line of sight. It is natural to expect that a concentration of DM may reside at the Galactic Center (GC) which becomes a primary target for searching gamma-ray excesses. A recent example of analysis is the work of the Fermi team on a possible GeV excess from the GC and unaccounted for by the diffuse and standard source modelling [143]. Figure 33 shows the excess GeV emission as resulting from a 6.5 year data accumulation once known sources of radiation are subtracted (diffuse, point-like sources and SNRs).

The current conclusion by the Fermi team is that overall systematic uncertainties in the Galactic source/diffuse model of GeV emission do not make possible determining

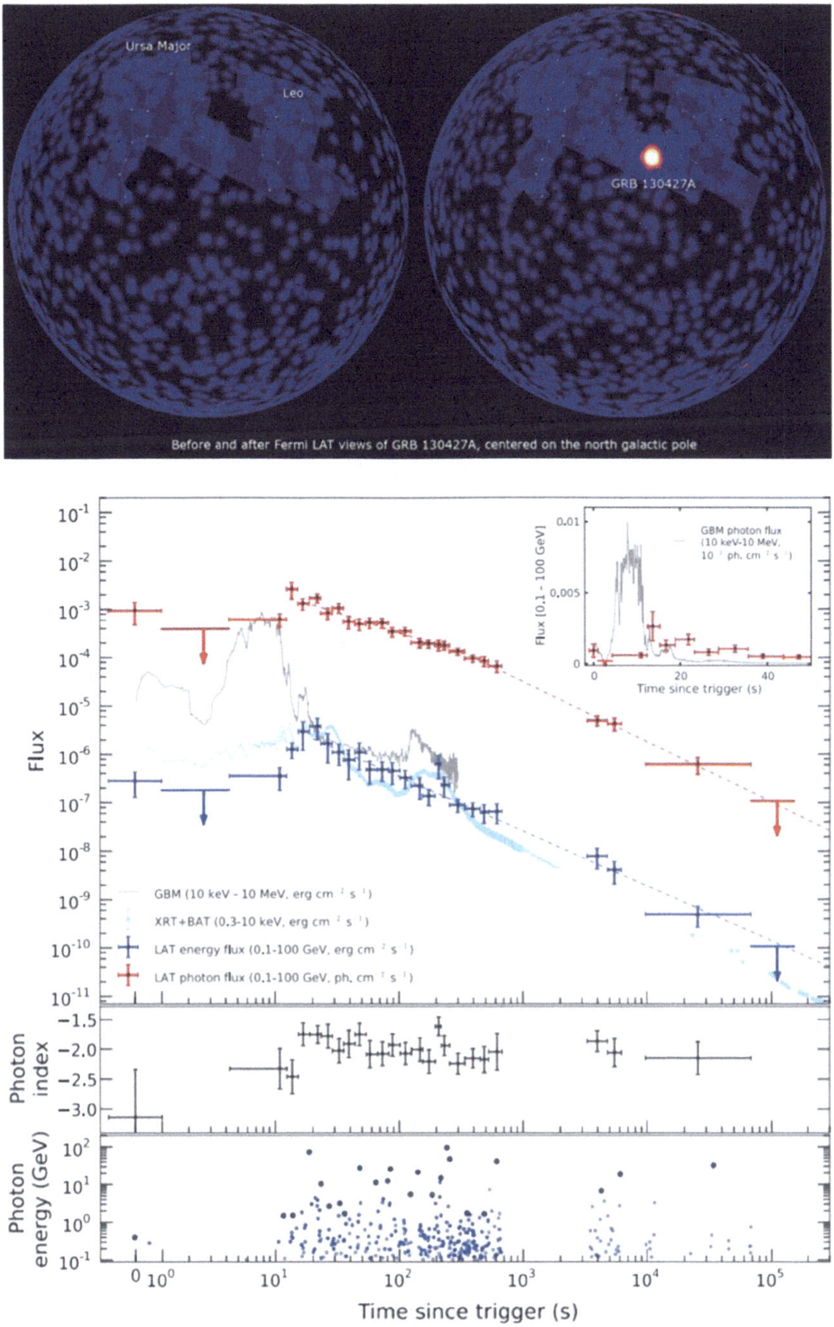

Fig. 32 GRB 130427A as detected by *Fermi*. It is a remarkable GRB, with gamma-ray emission lasting almost a day with the detection of a 95.4 GeV photon that challenges synchrotron models of GRB afterglows *Top panel* gamma-ray map by *Fermi*-LAT above 100 MeV [141]. *Bottom panel* light curve of GRB 130427A showing GBM and LAT data [140]

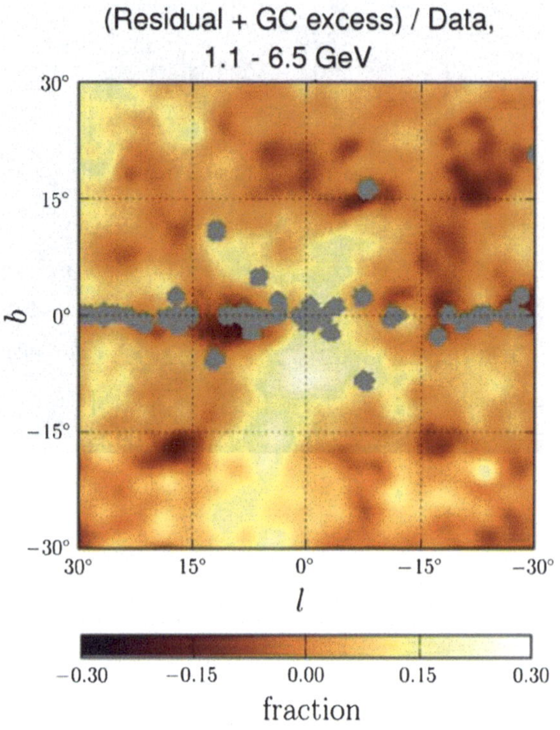

Fig. 33 Gamma-ray map of the residual GeV emission from the inner Galaxy as determined by Fermi-LAT, from [143]. The *gray circles* correspond to the highest flux (>1 GeV) 3FGL sources in the field

the existence of an additional component of emission at GeV energies (possibly to be attributed to DM). It can be noted that other sources of gamma-ray emission exist in the GC or can be reasonably assumed to contribute there in an individual or collective manner. A recent discussion pointed at a population of millisecond pulsars likely to exist at the GC [144, 145]. Current searches at the GC or in the disk or halo (DM clumps) are so far inconclusive.

Since galactic centers can be the sites of action, it is interesting to study and search nearby massive galaxies. A recent study by *Fermi*-LAT interestingly indeed detects gamma-ray emission concentrated in the bulge of the Andromeda (M31) galaxy [146]. Contrary to expectations, the extent of the gamma-ray emission excludes detectable contributions from the main gas ring of M31 and from star formation sites. Again, a population of unresolved gamma-ray sources concentrated in the bulge can explain the emission. Or dark matter.

Venturing outside our Galaxy and nearby massive galaxies, considerable attention has been devoted in searching for a gamma-ray signal from spheroidal galaxies in the surrounding of our Milky Way. Fifteen nearby dark-matter-dominated dwarf spheroidal galaxies (dSphs) have been initially studied by *Fermi*-LAT with no detections [147]. A more recent combined analysis of a sample of about 30 stellar systems (including newly discovered dSph candidates) yields no globally significant excess of gamma-ray emission [148] (see Fig. 34).

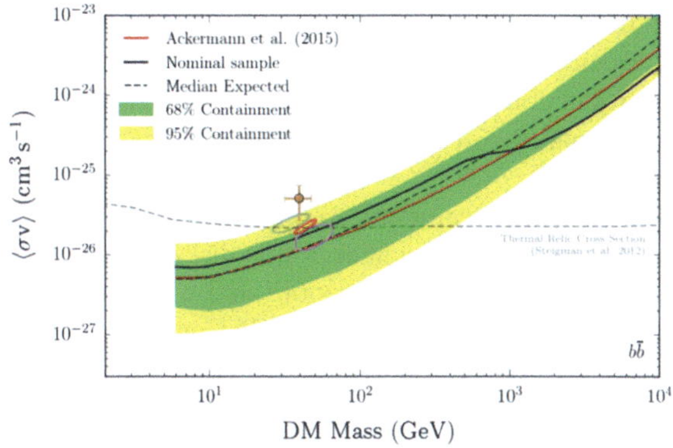

Fig. 34 Upper limits (95% confidence level) on the DM annihilation cross section derived from a combined analysis of the nominal target sample for the bb channel (from [148]). The *dashed line*shows the median expected sensitivity while the bands represent the 68 and 95% quantiles. The *solid black line* shows the observed limit from the combined analysis of 15 dSphs from [147]. The *closed contours* and *marker* show the best-fit regions (at 2σ confidence) in cross section and mass from several DM interpretations of the GC GeV excess, as discussed in [148]. The *dashed gray curve* corresponds to the thermal relic cross section from [149]

Despite the current lack of conclusive evidence, searching for gamma-ray signals linked to dark matter annihilations remains one of the most fundamental research topics for high-energy experiments and satellites. Future experiments in the MeV–GeV range need to improve their sensitivity and angular resolution to confront the difficult problem of detecting a DM signal from our Galaxy and its surroundings.

The New Astronomy of Gravitational Wave Sources

The 2015 first detection of gravitational waves by the LIGO/Virgo Collaboration opened a new field of astronomy with crucial implications for fundamental physics [150]. At the time of writing of this report (early July 2017), three gravitational wave (GW) events have been reported and studied by follow-up multifrequency observations. At the moment, no electromagnetic counterparts have been detected in coincidence with a GW event. The gravitational signals detected so far are interpreted as originating from the coalescence of quite massive black holes (BHs) of several tens of solar masses. In addition to BH-BH binaries, also relevant are compact systems containing BH and neutron stars (NS), and NS-NS binaries. In the framework of current models of formation and evolution of BH-BH binaries, no disk or substantial matter is expected to be surrounding the BHs in their final stage of coalescence. Therefore, standard models do not predict an electromagnetic signal in coincidence with this type of BH binaries. However, the situation can be more complex, and a variety of effects can play a role in making the environment surrounding coalescing

BHs or BH-NS systems interesting for electromagnetic emission. Residual disk matter can be present, and non-trivial electromagnetic fields can surround the BH or NS to make possible the production of a detectable electromagnetic signal before or immediately following the final time of coalescence. This is particularly true of course for NS-NS binaries.

For these reasons, and also waiting for events associated to NS companions (and therefore with the formation of transient accretion disks) the search for electromagnetic signals from GW events is certainly motivated with the highest priority. The current gamma-ray space-based experiments (AGILE and *Fermi*) can, therefore, play a crucial role because of their very large fields of view (2.5 sr) and multifrequency capability from tens of keV to hundreds of GeV. These detection features are unique in the framework of follow-up studies. The satellite response can cover a very large portion of the sky, therefore allowing the exposure of a substantial if not all of the GW event sky localization region (currently quite large). AGILE and Fermi have been reacting to GW events, and several searches were done (e.g., [151, 152]). Important upper limits have been obtained, excluding, so far and as expected, the detection of bright events similar to short-GRBs. However, the electromagnetic signal from BH GW events can be different and substantially weaker than for canonical short-GRBs, and a big effort is ongoing in trying to improve the space instruments' sensitivity to weak and fast events.

The search for an X-ray or gamma-ray signal from GW events constitutes today probably the most exciting frontier of research of gamma-ray astrophysics. If a signal associated with a GW event will be detected by X-ray and gamma-ray satellites (including the more active missions so far in the search: Swift, Integral, AGILE, *Fermi*) a new window will be open with unimaginable consequences for our understanding of fundamental processes. This goal may be within reach in the near future.

The Future

This short review focused on some of the relevant discoveries and issues in high-energy astrophysics as obtained during the last years. The field is very active and projected into future investigations with great impetus. In several cases, modelling of fundamental processes needed to be revised to address the challenges derived from new observations. This is true for both Galactic and extragalactic sources. Compact objects, shocks, relativistic jets, acceleration mechanisms, plasma physics all combine to provide a very active Universe at gamma-ray energies. A glimpse on the near future is provided by the momentous discovery of gravitational waves that recently opened a new chapter of observational (and theoretical) astronomy. We are just beginning to appreciate the relevance of this discovery. Future doors might open soon. Astrophysical sources of neutrinos might be discovered: it is a difficult observational task but it can be reached hopefully within the next decade.

Several ground-based experiments will be operating in the next years in the TeV range (HAWC, CTA, LHAASO) trying to reach not only the highest energies up to

100 TeV, but also the 50–100 GeV range. *Fermi* will continue its operations hopefully for several more years, and a possible overlap with CTA and LHAASO would be highly desirable. There is considerable interest, of course, for a new gamma-ray mission that would improve our knowledge in energy sectors that are currently not explored despite their relevance. Especially important is the MeV range, that is really the "bridge" between X-ray and GeV emission and can be of great importance for our understanding of many challenging and still unexplained phenomena. Mission concepts based on realizing the "MeV-GeV bridge" have been recently proposed to ESA (e-ASTROGAM [153, 154]) and to NASA (AMEGO [155]) which possibly might enter into an advanced study phase toward selection for the next decade.

Gamma-ray sources populate the Galactic and extragalactic sky as messengers of the most extreme particle acceleration processes in our Universe. Let us continue during the next decade the impressive path of discovery demonstrated by space-based gamma-ray experiments. We have to take the challenge and sharpen more and more our detection capabilities. Fast reaction to transients and multifrequency observations should be improved. Detectors should be made more efficient and more manageable. Theoretical modelling should continue to explore the links among gravitationally driven processes, relativistic hydrodynamics of magnetized flows, plasma physics, and radiative processes. The theory of acceleration processes should evolve more and more in exploring non-ideal MHD situations and magnetic field reconnection. The new astronomy of gravitational wave sources today and of neutrino sources possibly discovered within the next decade is waiting for us. If we are brave, we will be rewarded by surprises and gamma-ray wonders!.

Acknowledgements The author wishes to thank the many collaborators who inspired him during the years, especially the colleagues of EGRET, AGILE and *Fermi* teams. Felix Aharonian, Roberto Aloisio, Eugenio Coccia and Francesco Vissani supported the idea of this report and encouraged me with unfailing patience. Martina Cardillo helped in commenting section "Supernova Remnants and the Origin of Cosmic Rays."

References

1. C.E. Fichtel, J.J. Trombka, *Gamma-Ray Astrophysics* (NASA Reference Publication no. 1386, 1997)
2. W.L. Kraushaar, G.W. Clark, Search for primary cosmic gamma rays with the satellite explorer XI. Phys. Rev. Lett. **8**, 106 (1962)
3. W.L. Kraushaar, G.W. Clark, G. Garmire, H. Helmken, P. Higbie, M. Agogino, Explorer XI experiment on cosmic gamma rays. ApJ **141**, 845 (1965)
4. G.W. Clark, G.P. Garmire, W.L. Kraushaar, Observations of high-energy cosmic gamma rays. ApJL **153**, L203 (1968)
5. L.I. Bratolyubova-Tsulukidze, N.L. Grigorov, L.F. Kalinkin et al., Proceedings of the 11th International Cosmic Ray Conference, vol. 1, issue 123 (1969)
6. G.G. Fazio, E.M. Hafner, The OSO 1 high-energy gamma-ray experiment. J. Geophys. Res. **72**, 2452 (1967)

7. D. Valentine, M.F. Kaplon, G. Badhwar, High energy gamma ray observations from OSO-III, in *Proceedings of the 11th International Conference on Cosmic Rays, held in Budapest*, 25 August–4 September, 1969. Edited by A. Somogyi, vol. 1. Acta Physica, Supplement to Volume 29, Origin and Galactic Phenomena, 101 (1969)
8. S.M. Derdeyn, D.H. Ehrmann, C.E. Fichtel et al., SAS-B digitized spark chamber gamma ray telescope. Nucl. Instr. Meth **98**, 557 (1972)
9. G.F. Bignami, C.E. Fichtel, R.C. Hartman, D.J. Thompson, Galaxies and gamma-ray astronomy. ApJ **232**, 649 (1979)
10. C.E. Fichtel, R.C. Hartman, D.A. Kniffen et al., High-energy gamma-ray results from the second small astronomy satellite. ApJ **198**, 163 (1975)
11. G.F. Bignami, G. Boella, J.J. Burger et al., The COS-B experiment for gamma-ray astronomy. Space Sci. Instr. **1**, 245–268 (1975)
12. B.N. Swanenburg, K. Bennett, G.F. Bignami, R. Buccheri, P. Caraveo, W. Hermsen, G. Kanbach, G.G. Lichti, J.L. Masnou, H.A. Mayer-Hasselwander, J.A. Paul, B. Sacco, L. Scarsi, R.D. Wills, Second COS-B catalog of high-energy gamma-ray sources. ApJ **242**, L69 (1981)
13. C.E. Fichtel, R. Hofstadter, K. Pinkau et al., Proposal for a high energy gamma ray telescope on the gamma ray observatory (1978)
14. D.J. Thompson, D.L. Bertsch, C.E. Fichtel et al., Calibration of the energetic gamma-ray experiment telescope (EGRET) for the compton gamma-ray observatory. ApJS **86**, 629 (1993)
15. M. Tavani, G. Barbiellini, A. Argan et al., The AGILE mission, A&A, **502**(3), 995 (2009)
16. S. Sabatini, I. Donnarumma, M. Tavani et al., On the angular resolution of the AGILE gamma-ray imaging detector. ApJ **809**, 60 (2015)
17. W.B. Atwood, A.A. Abdo, M. Ackermann, W. Althouse, B. Anderson, M. Axelsson, L. Baldini, J. Ballet, D.L. Band, G. Barbiellini et al., The large area telescope on the Fermi gamma-ray space telescope mission. ApJ **697**, 1071–1102 (2009)
18. Fermi-GBM website. http://gammaray.msfc.nasa.gov/gbm/
19. Fermi mission website. https://fermi.gsfc.nasa.gov
20. G.H. Share, R.J. Murphy, J. Ryan, Solar and stellar gamma-ray observations with Compton, in *Proceedings of the Fourth Compton Symposium (AIP Conference Proceedings)* no. 410 (1997)
21. D. Bertsch, R. Hartman, S. Hunter et al., Gamma ray observations of the June 1991 solar flares. Bull. Am. Astron. Soc. **29**, 782 (1997)
22. M. Ajello, A. Albert, A. Allafort et al., Impulsive and long duration high-energy gamma-ray emission from the very bright 2012 March 7 solar flares. ApJ **789**, 20A (2014)
23. D.J. Thompson, D.L. Bertsch, D.J. Morris, R. Mukherjee, Energetic gamma ray experiment telescope high-energy gamma ray observations of the Moon and Quiet Sun. J. Geophys. Res. **102**, 14735 (1997)
24. M. Ackermann, M. Ajello, A. Albert et al., Measurement of the high-energy gamma-ray emission from the Moon with the Fermi large area telescope. Phys. Rev. D **93**, 082001 (2016)
25. J.P. Halpern, F. Camilo, A. Giuliani, E.V. Gotthelf, M.A. McLaughlin, R. Mukherjee, A. Pellizzoni, S.M. Ransom, M.S.E. Roberts, M. Tavani, Discovery of high-energy gamma-ray pulsations from PSR J2021+3651 with AGILE. ApJL **688**, L33 (2008)
26. A.A. Abdo, M. Ajello, A. Allafort, L. Baldini, J. Ballet, G. Barbiellini et al., The second fermi large area telescope catalog of gamma-ray pulsars. ApJS **208**, 17 (2013)
27. A.A. Abdo, M. Ackermann, M. Ajello, W.B. Atwood, M. Axelsson, L. Baldini et al., Fermi large area telescope observations of the Crab pulsar and nebula. ApJ **708**, 1254 (2010)
28. A. Harding, Talk given at the Heidelberg *Gamma-2016 Symposium, July 2016*
29. M. Ruderman, J. Shaham, M. Tavani, Accretion turnoff and rapid evaporation of very light secondaries in low-mass X-ray binaries. ApJ **336**, 507 (1989)
30. M. Ruderman, J. Shaham, M. Tavani, D. Eichler, Late evolution of very low mass X-ray binaries sustained by radiation from their primaries. ApJ **343**, 292 (1989)
31. A.S. Fruchter, D.R. Stinebring, J.H. Taylor, A millisecond pulsar in an eclipsing binary. Nature **333**, 237 (1988)

32. H.J. Pletsch et al., Binary millisecond pulsar discovery via gamma-ray pulsations. Science **338**, 1314 (2012)
33. P.S. Ray, A.A. Abdo, D. Parent, D. Bhattacharya, B. Bhattacharyya, F. Camilo, Radio searches of fermi LAT sources and blind search pulsars: *The Fermi Pulsar Search Consortium* (2012). arXiv:1205.3089R
34. M.S.E. Roberts, Surrounded by Spiders! New black widows and redbacks in the galactic field, in neutron stars and pulsars: challenges and opportunities after 80 years, in *Proceedings IAU Symposium* no. 291, 127 (2012)
35. R.W. Romani, A.V. Filippenko, J.M. Silverman, S.B. Cenko, J. Greiner, A. Rau, J. Elliott, H.J. Pletsch, PSR J1311–3430: a heavyweight neutron star with a flyweight helium companion. ApJ **760**, 36 (2012)
36. C.W. Ng, K.S. Cheng, J. Takata, Exploring the extra component in the gamma-ray emission of the new redback candidate 3FGL J2039.6-5618. J. Astron. Space Sci. **33**, 93 (2016)
37. R.W. Romani, A likely millisecond pulsar binary counterpart for Fermi source 2FGL J2039.6-5620, ApJL **812**, L24 (2015)
38. S. Bogdanov, J.P. Halpern, Identification of the high-energy gamma-ray source 3FGL J1544.6-1125 as a transitional millisecond pulsar binary in an accreting state. ApJL **803**, L27 (2015)
39. T.J. Johnson et al., Discovery of gamma-ray pulsations from the transitional redback PSR J1227–4853. ApJ **806**, 91 (2015)
40. B.W. Stappers et al., A state change in the missing link binary pulsar system PSR J1023+0038. ApJ **790**, 39 (2014)
41. M. Tavani, Hidden millisecond pulsars. ApJ **379**, 69 (1991)
42. C.W. Ng, J. Takata, K.S. Cheng, Observation and simulation of the variable gamma-ray emission from PSR J2021+4026. ApJ **825**, 18 (2016)
43. A.A. Abdo et al., Detection of 16 gamma-ray pulsars through blind frequency searches using the Fermi LAT. Science **325**, 840 (2009)
44. A. Allafort, L. Baldini, J. Ballet, G. Barbiellini, M.G. Baring, D. Bastieri, R. Bellazzini et al., PSR J2021+4026 in the gamma Cygni region: the first variable gamma-ray pulsar seen by the fermi LAT. ApJL **777**, 2 (2013)
45. A. Lyne, G. Hobbs, M. Kramer, I. Stairs, B. Stappers, Switched magnetospheric regulation of pulsar spin-down. Science **329**, 408 (2010)
46. A.M. Bykov, E. Amato, A.E. Petrov, A.M. Krassilchtchikov, K.P. Levenfish, Pulsar wind nebulae with bow shocks: non-thermal radiation and cosmic ray leptons. Space Sci. Rev. (2017). arXiv:1705.00950
47. S.P. Reynolds, G.G. Pavlov, O. Kargaltsev, N. Klingler, M. Renaud, S. Mereghetti, Pulsar-wind nebulae and magnetar outflows: observations at radio, X-Ray, and gamma-ray wavelengths. Space Sci. Rev. (2017). arXiv:1705.08897
48. F. Acero, M. Ackermann, M. Ajello, A. Albert, W.B. Atwood et al., Fermi large area telescope third source catalog. ApJS **218**, 23 (2015)
49. M. Ackermann, M. Ajello, W.B. Atwood et al., 2FHL: The second catalog of hard Fermi-LAT sources. ApJS **222**, 5 (2016)
50. A. Pellizzoni, A. Trois, M. Tavani et al., Detection of gamma-ray emission from the Vela pulsar wind nebula with AGILE. Science **327**, 663 (2010)
51. F. Aharonian, A.G. Akhperjanian, A.R. Bazer-Bachi et al., First detection of a VHE gamma-ray spectral maximum from a cosmic source: HESS discovery of the Vela X Nebula. A&A **448**, L43 (2006)
52. A.A. Abdo, M. Ackermann, M. Ajello et al., Fermi large area telescope observations of the Vela-X pulsar wind nebula. ApJ **713**, 146A (2010)
53. O.C. de Jager, P.O. Slane, S. La Massa, Probing the radio to X-Ray connection of the Vela X pulsar wind nebula with Fermi LAT and H.E.S.S. ApJ **689**, L125 (2008)
54. J.A. Hinton, S. Funk, R.D. Parsons, S. Ohm, Escape from Vela X. ApJL **743**, L7 (2011)
55. J.J. Hester, The Crab Nebula: an astrophysical chimera. Ann. Rev. Astron. Astrophys. **46**, 127 (2008)

56. A.M. Atoyan, F.A. Aharonian, On the mechanisms of gamma radiation in the Crab Nebula. MNRAS **278**, 525A (1996)
57. O.C. de Jager, A.K. Harding, P.F. Michelson, H.I. Nel, P.L. Nolan, P. Sreekumar, D.J. Thompson, Gamma-ray observations of the Crab Nebula: a study of the synchro-compton spectrum. ApJ **457**, 253 (1996)
58. M. Meyer, D. Horns, H.-S. Zechlin, The Crab Nebula as a standard candle in very high-energy astrophysics. A&A **523**, 2 (2010)
59. M. Tavani et al., Astron. Telegr. **2855** (2010)
60. R. Buehler et al., Astron. Telegr. **2861** (2010)
61. A.A. Abdo, M. Ackermann, M. Ajello et al., Gamma-ray flares from the Crab Nebula. Science **331**, 739 (2011)
62. M. Tavani, A. Bulgarelli, V. Vittorini et al., Discovery of powerful gamma-ray flares from the Crab Nebula. Science **331**, 736 (2011)
63. E. Bernardini, Astronomy in the time domain. Science **331**, 686 (2011)
64. M.C. Weisskopf, A.F. Tennant, J. Arons, R. Blandford, R. Buehler, P. Caraveo, C.C. Cheung, E. Costa, A. de Luca, C. Ferrigno, H. Fu, S. Funk, M. Habermehl, D. Horns, J.D. Linford, A. Lobanov, C. Max, R. Mignani, S.L. O'Dell, R.W. Romani, E. Striani, M. Tavani, G.B. Taylor, Y. Uchiyama, Y. Yuan, Chandra, Keck, and VLA observations of the Crab Nebula during the 2011-April gamma-ray flare. ApJ **765**, 56 (2013)
65. E. Striani, M. Tavani, V. Vittorini et al., Variable gamma-ray emission from the Crab Nebula: short flares and long "waves". ApJ **765**, 52 (2013)
66. M. Tavani, Super-acceleration in the flaring Crab Nebula. Nucl. Phys. B **243**, 131 (2013)
67. E. Striani, M. Tavani, G. Piano et al., The Crab Nebula super-flare in 2011 April: extremely fast particle acceleration and gamma-ray emission. ApJ **741**, L5 (2011)
68. R. Buehler, J.D. Scargle, R.D. Blandford et al., Gamma-ray activity in the Crab Nebula: the exceptional flare of 2011 April. ApJL **749**, 26 (2012)
69. A. Rudy, D. Horns, A. De Luca et al., Characterization of the inner knot of the Crab: the site of the gamma-ray flares? ApJ **811**, 24 (2015)
70. R. Buehler, R. Blandford, The surprising Crab pulsar and its nebula: a review. Rep. Progress Phys. **77**, 066901 (2014)
71. M.L. McConnell, J.M. Ryan, W. Collmar, V. Schnfelder, H. Steinle, A.W. Strong, H. Bloemen, W. Hermsen, L. Kuiper, K. Bennett, B.F. Phlips, J.C. Ling, A high-sensitivity measurement of the MeV gamma-ray spectrum of Cygnus X-1. ApJ **543**, 928 (2000)
72. G. Piano, M. Tavani, V. Vittorini et al., The AGILE monitoring of Cygnus X-3: transient gamma-ray emission and spectral constraints. A&A **545**, 110 (2012)
73. M. Tavani, A. Bulgarelli, G. Piano, S. Sabatini et al., Extreme particle acceleration in the microquasar Cygnus X-3. Nature **462**, 620 (2009)
74. A.A. Abdo, M. Ackermann, M. Ajello et al., Modulated high-energy gamma-ray emission from the Microquasar Cygnus X-3. Science **326**, 1512 (2009)
75. S. Corbel, G. Dubus, J.A. Tomsick et al., A giant radio flare from Cygnus X-3 with associated gamma-ray emission. MNRAS **421**, 2947 (2012)
76. A.A. Zdziarski, C. Maitra, A. Frankowski, G.K. Skinner, R. Misra, Energy-dependent orbital modulation of X-Rays and constraints on emission of the jet in Cyg X-3. MNRAS **426**, 1031 (2012)
77. A.A. Zdziarski, P. Pjanka, M. Sikora, L. Stawarz, Jet contributions to the broad-band spectrum of Cyg X-1 in the hard state. MNRAS **442**, 3243 (2014)
78. A. Bodaghee, J.A. Tomsick, K. Pottschmidt, J. Rodriguez, J. Wilms, G.G. Pooley, Gamma-ray observations of the Microquasars Cygnus X-1, Cygnus X-3, GRS 1915+105, and GX 339–4 with the fermi large area telescope. ApJ **775**, 98 (2013)
79. S. Sabatini, M. Tavani, P. Coppi et al., Gamma-ray observations of Cygnus X-1 above 100 MeV in the hard and soft states. ApJ **766**, 83 (2013)
80. A. Loh, S. Corbel, G. Dubus, J. Rodriguez, I. Grenier, T. Hovatta, T. Pearson, A. Readhead, R. Fender, K. Mooley, High-energy gamma-ray observations of the accreting black hole V404 Cygni during its 2015 June outburst. MNRAS **462**, 111L (2016)

81. G. Piano, P. Munar-Adrover, F. Verrecchia, M. Tavani, S.A. Trushkin, High-energy gamma-ray activity from V404 Cygni detected by AGILE during the 2015 June outburst. ApJ **839**, 84 (2017)
82. M. Tavani, J. Arons, Theory of high-energy emission from the pulsar/be star system PSR 1259-63: I, radiation mechanisms and interaction geometry, ApJ **477**, 439
83. A.A. Abdo, M. Ackermann, M. Ajello et al., Discovery of high-energy gamma-ray emission from the binary system PSR B1259–63/LS 2883 around periastron with fermi. ApJ **736**, 11L (2011)
84. F. Aharonian, A.G. Akhperjanian, G. Anton et al., Very high energy gamma-ray observations of the binary PSR B1259–63/SS2883 around the 2007 Periastron. A&A **507**, 389 (2009)
85. A.A. Abdo, M. Ackermann, M. Ajello et al., Fermi LAT observations of LS I +61303: first detection of an orbital modulation in GeV gamma rays. ApJL **701**, 123L (2009)
86. A.A. Abdo et al., Fermi LAT observations of LS5039. ApJL **706**, L56 (2009)
87. M. Ackermann, M. Ajello, J. Ballet, Periodic emission from the gamma-ray binary 1FGL J1018.6-5856. Science **335**, 189
88. M.F. Corcoran, K. Hamaguchi, J.M. Pittard, C.M.P. Russell, S.P. Owocki, E.R. Parkin, A. Okazaki, Recent X-ray variability of Eta Carinae: the quick road to recovery. ApJ **725**, 1528 (2010)
89. A.A. Abdo, M. Ackermann, M. Ajello et al., Fermi large area telescope observation of a gamma-ray source at the position of Eta Carinae. ApJ **723**, 649 (2010)
90. M. Tavani, S. Sabatini, E. Pian et al., Detection of gamma-ray emission from the Eta-Carinae region. ApJL **698**, 142L (2009)
91. M. Balbo, R. Walter, Fermi acceleration along the orbit of Eta Carina (2017). arXiv:170502706B
92. A.A. Abdo, M. Ackermann, M. Ajello, Gamma-ray emission concurrent with the nova in the symbiotic binary V407 Cygni. Science **329**, 817 (2010)
93. P. Martin, G. Dubus, Particle acceleration and non-thermal emission during the V407 Cygni Nova outburst. A&A **551**, 37 (2013)
94. C.C. Cheung, P. Jean, S.N. Shore, J.E. Grove, M. Leising et al., Fermi reveals new light on novae in gamma rays, in *Proceedings of the 34th International Cosmic Ray Conference*, 30 July–6 August 2015, The Hague, The Netherlands (2016)
95. W. Baade, F. Zwicky, Remarks on super-novae and cosmic rays. Phys. Rev. **46**, 76 (1934)
96. V.L. Ginzburg, S.I. Syrovatsky, *The Origin of Cosmic Rays* (Pergamon Press, 1964)
97. J. Vink, Supernova Remnants: the X-Ray perspective. Astron. Astrophys. Rev. **20**, 49 (2012)
98. J.A. Esposito, S.D. Hunter, G. Kanbach et al., EGRET observations of radio-bright supernova Remnants. ApJ **461**, 820 (1996)
99. M. Ackermann, M. Ajello, A. Allafort et al., Detection of the characteristic Pion-Deacy signature in supernova Remnants. Science **339**, 807 (2013)
100. A. Giuliani, M. Cardillo, M. Tavani et al., Neutral pion emission from accelerated protons in the supernova Remnant W44. ApJL **742**, 30G (2011)
101. A.A. Abdo, M. Ackermann, M. Ajello et al., Gamma-ray emission from the shell of supernova Remnant W44 revealed by the Fermi LAT. Science **327**, 1103 (2010)
102. M. Cardillo, E. Amato, P. Blasi, Supernova Remnant W44: a case of cosmic-ray reacceleration. A&A **595**, 58C (2016)
103. M. Cardillo, M. Tavani, A. Giuliani et al., The supernova Remnant W44: confirmations and challenges for cosmic-ray acceleration. A&A **565**, 74C (2014)
104. A.R. Bell, The acceleration of cosmic rays in shock fronts: I. MNRAS **182**, 147 (1978)
105. A.R. Bell, The acceleration of cosmic rays in shock fronts: II. MNRAS **182**, 443 (1978)
106. R.D. Bandford, J.P. Ostriker, Particle acceleration by astrophysical shocks. ApJL **221**, L29 (1978)
107. G.F. Krymskii, A regular mechanism for the acceleration of charged particles on the front of a shock wave. Akademiia Nauk SSSR Doklady **234**, 1306 (1977)
108. P. Blasi, The origin of galactic cosmic rays. Astron. & Astrophys. Rev. **21**, 70 (2013)

109. M. Malkov, L. Drury, Nonlinear theory of diffusive acceleration of particles by shock waves. Rep. Progress Phys. **64**, 429 (2001)
110. S. Funk, Ground- and space-based gamma-ray astronomy. Annu. Rev. of Nucl. Part. Sci **65**, 245 (2015)
111. S. Gabici, F.A. Aharonian, Hadronic gamma-rays from RX J1713.7-3946? MNRAS **445**, 70 (2014)
112. S. Archambault, A. Archer, W. Benbow et al., Gamma-ray observations of Tycho's SNR with VERITAS and Fermi. ApJ **836**, 26A (2017)
113. A. Ghiotto et al., A deep observation of gamma-ray emission from Cassiopeia A using VERITAS (2015). arXiv:151100309G
114. M. Ajello, W.B. Atwood, L. Baldini, et al., 3FHL: The third catalog of hard fermi-LAT sources (2017). arXiv:1702.00664v2
115. A.R. Bell, Turbulent amplification of magnetic field and diffusive shock acceleration of cosmic rays. MNRAS **353**, 550 (2004)
116. M. Cardillo, E. Amato, P. Blasi, On the cosmic ray spectrum from type-II supernovae expanding in their red giant pre-supernova wind. Astrop. Phys. **69**, 1 (2015)
117. E. Parizot, A. Marcowith, E. van der Swaluw et al., Superbubbles and energetic particles in the galaxy. A&A **424**, 747 (2004)
118. N. Prantzos, On the origin and composition of galactic cosmic rays. A&A **538**, 80 (2012)
119. M. Ackermann, M. Ajello, A. Allafort et al., A cocoon of freshly accelerated cosmic rays detected by Fermi in the Cygnus Superbubble. Science **334**, 1103A (2011)
120. B. Bartoli, P. Bernardini, X.J. Bi et al., Identification of the TeV gamma-ray source ARGO J2031+4157 with the Cygnus Cocoon. ApJ **790**, 152 (2014)
121. J.R. Mattox et al., The likelihood analysis of EGRET data. ApJ **461**, 396 (1996)
122. D. Hooper, B.L. Dingus, Limits on supersymmetric dark matter from EGRET observations of the galactic center region. Phys. Rev. D **70**, 113007 (2004)
123. A. Abramowski, F. Aharonian, F.A. Benkhali et al., Acceleration of petaelectronvolt protons in the galactic centre. Nature **531**, 476 (2016)
124. M. Su, T.R. Slatyer, D.P. Finkbeiner, Giant gamma-ray bubbles from Fermi-LAT: active galactic nucleus activity or bipolar galactic wind? ApJ **724**, 1044 (2010)
125. G. Dobler, D.P. Finkbeiner, I. Cholis et al., The Fermi haze: a gamma-ray counterpart to the microwave haze. ApJ **717**, 825 (2010)
126. M. Ackermann, A. Albert, W.B. Atwood et al., The spectrum and morphology of the Fermi bubbles. ApJ **793**, 64A (2014)
127. R. Yang, F. Aharonian, R. Crocker, The Fermi bubbles revisited. A&A **567**, 19 (2014)
128. R.M. Crocker, G.V. Bicknell, A. Taylor, E. Carretti, A unified model of the Fermi bubbles. Microwave haze, and polarized radio lobes: reverse shocks in the galactic centers giant outflows, ApJ **808**, 107C (2015)
129. A.E. Wehrle, E. Pian, C.M. Urry, L. Maraschi, I.M. McHardy, A.J. Lawson, G. Ghisellini, R.C. Hartman et al., Multiwavelength observations of a dramatic high-energy flare in the blazar 3C 279. ApJ **497**, 178 (1998)
130. M. Ackermann, R. Anantua, K. Asano et al., Minute-timescale >100 MeV gamma-ray variability during the giant outburst of Quasar 3C 279 observed by fermi-LAT in 2015 June. ApJ **824**, 20A (2016)
131. R.C. Hartman, D.L. Bertsch, C.E. Fichtel, S.D. Hunter, G. Kanbach, D.A. Kniffen, Detection of high-energy gamma radiation from quasar 3C 279 by the EGRET telescope on the Compton gamma ray observatory. ApJL **385**, 1 (1992)
132. D.A. Kniffen, D.L. Bertsch, C.E. Fichtel, R.C. Hartman, S.D. Hunter, G. Kanbach et al., Time variability in the gamma-ray emission of 3C 279. ApJ **411**, 133 (1993)
133. L. Maraschi, P. Grandi, C.M. Urry, A.E. Wehrle, G.M. Madejski, H.H. Fink, G. Ghisellini et al., The 1993 multiwavelength campaign of 3C 279: the radio to gamma-ray energy distribution in low state. ApJL **435**, 91 (1994)
134. R.C. Hartman, J.R. Webb, A.P. Marscher, J.P. Travis, C.D. Dermer, H.D. Aller et al., Simultaneous multiwavelength spectrum and variability of 3C 279 from 10^9 to 10^{24} Hz. ApJ **461**, 698 (1996)

135. J.R. Mattox, S.J. Wagner, M. Malkan, T.A. McGlynn, J.F. Schachter, J.E. Grove, W.N. Johnson, J.D. Kurfess, An intense gamma-ray flare of PKS 1622–297. ApJ **476**, 692 (1997)
136. M. Villata, C.M. Raiteri, T.J. Balonek, M.F. Aller et al., The unprecedented optical outburst of the quasar 3C 454.3, A&A **453**, 817–822 (2006)
137. A.A. Abdo, M. Ackermann, M. Ajello, A limit on the variation of the speed of light arising from quantum gravity effects. Nature **462**, 331A (2009)
138. A. Giuliani, F. Fuschino, G. Vianello et al., AGILE detection of delayed gamma-ray emission from the short gamma-ray burst GRB 090510. ApJ **708L**, 84G (2010)
139. M. Ackermann, K. Asano, W.B. Atwood et al., Fermi observations of GRB 090510: a short-hard gamma-ray burst with an additional, hard power-law component from 10 keV to GeV energies. ApJ **716**, 1178A (2010)
140. M. Ackermann, M. Ajello, K. Asano et al., Fermi-LAT observations of the gamma-ray burst GRB 130427A. Science **343**, 42A (2014)
141. Fermi mission website, GRB research. https://fermi.gsfc.nasa.gov/science/eteu/grbs/
142. Planck Collaboration: P.A.R. Ade, N. Aghanim, M. Arnaud et al., Planck 2015 results. XIII. Cosmological parameters. A&A **594**, A13 (2016)
143. M. Ackermann, M. Ajello, A.W. Albert et al., The Fermi galactic center GeV excess and implications for dark matter. ApJ **840**, 43A (2017)
144. R. Bartels, S. Krishnamurthy, C. Christoph Weniger, Strong support for the millisecond pulsar origin of the galactic center GeV excess. Phys. Rev. Lett. **116**, 051102 (2016)
145. S. Lee, M. Lisanti, B.R. Safdi, T.R. Slatyer, W. Xue, Evidence for unresolved gamma-ray point sources in the inner galaxy. Phys. Rev. Lett. **116**, 051103 (2016)
146. M. Ackermann, M. Ajello, A. Albert et al., Observations of M31 and M33 with the Fermi large area telescope: a galactic center excess in Andromeda? ApJ **836**, 201 (2017)
147. M. Ackermann, A. Albert, B. Anderson et al., Searching for dark matter annihilation from milky way dwarf spheroidal galaxies with six years of Fermi large area telescope data. Phys. Rev. Lett. **115**, 231301 (2015)
148. A. Albert, B. Anderson, K. Bechtol et al., Searching for dark matter annihilation in recently discovered milky way satellites with Fermi-Lat. ApJ **834**, 110 (2017)
149. G. Steigman, B. Dasgupta, J.F. Beacom, Precise relic WIMP abundance and its impact on searches for dark matter annihilation. Phys. Rev. D **86**, 023506 (2012)
150. B.P. Abbott, R. Abbott, T.D. Abbott et al., Observation of gravitational waves from a binary black hole merger. Phys. Rev. D **93**, 122003 (2016)
151. M. Ackermann, M. Ajello, A. Albert, Fermi-LAT observations of the LIGO event GW150914. ApJL **823**, 2 (2016)
152. M. Tavani, C. Pittori, F. Verrecchia, AGILE observations of the gravitational wave event GW150914. ApJL **825**, 4 (2016)
153. A. De Angelis, V. Tatischeff, M. Tavani, U. Oberlack, I. Grenier, L. Hanlon, R. Walter et al., The e-ASTROGAM mission—exploring the extreme universe with gamma rays in the MeV–GeV range, experimental astronomy (2017). https://doi.org/10.1007/s10686-017-9533-6
154. M. Tavani, V. Tatischeff, P. von Ballmoos et al., ASTROGAM: a proposal for the ESA M4 mission programme (2015)
155. J. McEnery et al., AMEGO: a gamma-ray mission (2017)

Neutrino Astronomy

Antonio Capone, Paolo Lipari and Francesco Vissani

Abstract Neutrino astronomy is a lively discipline that has born at the cross road of particle physics, nuclear physics, and astrophysics. Many low-energy neutrino observatories have demonstrated the possibility of investigating the functioning of the Sun, the terrestrial radiactivity and crucial astrophysical phenomena as the gravitational collapse. There are mounting evidences that extraterrestrial high-energy neutrinos are observable. This impacts strongly on our understanding of the high-energy phenomena from the cosmos, and the dataset and the sensitivity continue to improve. In this chapter, the status of the investigations of neutrino radiation is outlined, aiming to provide the reader with a unified description of this discipline that covers its main observational and theoretical aspects. We collect, for a readership at Ph.D. level, the most relevant formulae, expectations, and information concerning neutrino astronomy and astrophysics. Connections with other disciplines and selected applications to particle physics are discussed. Each section begins with an overview of the material included and a brief annotated bibliographical selection to books and review papers, aimed to favor "staged access" into the vast scientific literature.

Basic Considerations

The last years have witnessed the flourishing of new astronomies. The detection of solar and supernova neutrinos permitted us to observe the center of the stars, as recognized in 2002 by a Nobel prize. High-energy neutrino astronomy, dreamt since sixties, is finally becoming a reality and has the potential to shed light on one century-

A. Capone
INFN-Sezione di Roma and Università di Roma La Sapienza, Rome, Italy
e-mail: antonio.capone@roma1.infn.it

P. Lipari
INFN-Sezione di Roma, Rome, Italy
e-mail: paolo.lipari@roma1.infn.it

F. Vissani (✉)
INFN-LNGS and GSSI, L'aquila, Italy
e-mail: francesco.vissani@lngs.infn.it

R. Aloisio et al. (eds.), *Multiple Messengers and Challenges in Astroparticle Physics*, https://doi.org/10.1007/978-3-319-65425-6_4

old unsolved mystery, the origin of the cosmic radiation. Ultra-high-energy neutrino astronomy could ultimately permit us to explore the highest energy processes in the Universe. We are witnessing the birth of a new discipline, with a great potential and many facets: neutrino astronomy.

Here, the Reader is introduced to the aims of neutrino astronomy (section "Aims of Neutrino Astronomy"), to the specific features of this elusive but yet observable radiation (discussed in general theoretical terms in section "Neutrino Interactions and Properties: Generalities"). A more advanced but practical discussion on neutrino properties and interaction is given in section "Neutrino Interactions and Properties: Formulae", and the principles of various types of neutrino telescopes will be examined in section "Neutrino Telescopes".

General references:

There are several excellent books in the scientific literature that concern neutrinos and that can help a student to have a first access or to go very deep into the matter. The choice among them is not only dictated by the content but it is also, to good extent, matter of convenience and personal taste. Here, we present a rather incomplete selection that should be however sufficient to provide a first access to the topics we will deal with. We will not be diffuse on historical aspects here, but, for readers especially interested in this aspect, we would like to begin by pointing out a couple of books [1, 2] and one review [20] written by pioneers of this field.

Aims of Neutrino Astronomy

Neutrinos are interesting particles with curious behaviors. The main goal of neutrino astronomy is to exploit these particles to study and understand the various astronomical environments where they are produced. In order to achieve this goal, neutrino properties should be known sufficiently enough. This is a requirement that is well-evident in any new branch of astronomy, such as neutrino astronomy or gravity wave astronomy, and that it is instead given for granted for the usual astronomy (i.e., we know how to use the light, much more than neutrinos). Another scientific goal, where neutrinos are also involved, is to use observations to learn about neutrino properties. Even if this is more properly a characteristic task of particle physics, rather than an astronomical task, the two goals have been often (and perhaps are still) entangled. However, only a few sources of neutrinos can be observationally investigated.[1] In order to introduce the matter as directly as possible, we proceed to present a list of interesting and important sources from which observations are possible or plausible, leaving a more detailed explanation and discussion to the rest of this chapter.

[1] Sometimes one can probe the effect of these particles (only) indirectly: e.g., in cosmology, it is possible to reveal the imprint of the presence of neutrinos on the distribution of visible matter.

Table 1 We use for illustration a few specific detectors and reactions, discussed later on. Their abbreviations are, (1) BX = Borexino; SK = Super-Kamiokande (22.5 kton); Ice^3 = IceCUBE. (2) ES = elastic scattering; IBD = inverse beta decay; QEL = quasi-elastic nucleon interaction; DIS = deep-inelastic scattering. (3) e = electron; p = proton; N = nucleon. 1 pc = 3×10^{18} cm. Question marks denote purely theoretical predictions. Adapted from [332]

Neutrino source	Flux (1/cm^{2}s)	Target # and type	Time	Cross section (cm^2)
Sun (pp)	3×10^{10}	3×10^{31} e	1 day	10^{-45} (ES)
Earth	3×10^5	10^{31} p	4 year	3×10^{-44} (IBD)
Reactor	3×10^4	10^{31} p	4 year	10^{-42} (IBD)
Relic SN	2.5	10^{33} p	1 year	7×10^{-42} (IBD)
Galactic SN	10^{10}	10^{33} p	10 s	3×10^{-41} (IBD)
Atmosphere	1	10^{34} N	1 year	10^{-38} (QEL+DIS)
Accelerators ν_μ	0.1	6×10^{32} N	1 year	10^{-37} (DIS)
Accelerators ν_τ	0.01	10^{34} N	3 year	10^{-38} (DIS)
Galactic source	10^{-12}	2×10^{39} N	1 year	10^{-35} (DIS)
HE neutrinos	2×10^{-12}	6×10^{38} N	3 year	10^{-34} (DIS)
	Average energy	Distance	Detector	Number of events
Sun (pp)	0.2 MeV	1.5×10^8 km	BX	100
Earth	2 MeV	10^{2-4} km	BX	10
Reactor	4 MeV	1000 km	BX	30
Relic SN	10 MeV	300 kpc	SK	0.6?
Galactic SN	20 MeV	10 kpc	SK	5000
Atmosphere	1 GeV	500 km	SK (all ν)	3000
Accelerators ν_μ	10 GeV	500 km	SK	3000
Accelerators ν_τ	120 GeV	700 km	Opera	6
Galactic source	3 TeV	1 kpc	Ice^3	1?
HE neutrinos	100 TeV	?	Ice^3	10

Number of events

For any source of neutrinos, the most important characteristic for neutrino astronomy is just the number of events in the existing or future neutrino telescopes. In Table 1, various cases (neutrino sources and sources) are considered for illustration. For each of them, we list various features, emphasizing the number of observed (or potentially observable) events. This is given by,

$$\text{events} = \mathscr{N} \times T \times \Phi \times \sigma \tag{1}$$

where the 4 terms in the r.h.s. are the number of targets $\mathscr{N}$, the time of data taking T, the flux Φ and the cross section of the relevant interaction σ. Few experiments are mentioned in Table 1, but making reference only to the number of relevant targets and to the time of data taking. We fixed the value of the total cross section at some relevant energy, and then checked that with a suitable average value of the flux, the product gives the number of events, measured or expected. (It is important that the reader understands that this table is not meant to be used for reference but only to help for a first orientation.)

Fluence, exposure, effective area

Note that in the scientific literature the time T is sometimes included in the *fluence* F, i.e., the time integrated flux

$$F = \int \Phi(t)\, dt = \Phi \times T \tag{2}$$

where Φ in the r.h.s. is the time-averaged flux. Alternatively, the time is included in the *exposure*, namely the product $\mathscr{N} \times T$. Another combination that is commonly used is the *effective area* given by $\mathscr{N} \times \sigma$. Here, we do not take into account explicitly any "efficiency factor"[2] that can be attached to the effective cross section or to the exposure or to the effective area. Note finally that in the table and in our simplified estimations, the flux (and the fluence) are always integrated in the relevant energy range. Let us repeat that the estimations in the table are not supposed to be precise, they should only convey the correct order of magnitude of the number of events.

Intensity and luminosity

If we know the average distance of production of the flux, D, and when the emission is isotropic—to some degree of approximation—we can connect the observed flux with the *intensity* of emission I (i.e., number of neutrinos per second) namely,

$$\Phi = \frac{I}{4\pi D^2} \tag{3}$$

In this case, the total power radiated in neutrinos—in astrophysical parlance, the *luminosity*—will be

$$\mathscr{L} = I \times \langle E \rangle \tag{4}$$

where we introduced the average energy of the neutrinos, $\langle E \rangle$. If the emission is not isotropic, we have to replace 4π with Ω, the solid angle of emission.

Neutrino Interactions and Properties: Generalities

Here, we recall the main aspects of neutrino interactions that are relevant for neutrino astronomy (section "Neutrino Interactions") and also the "oscillation" phenomena,

[2]This is a factor or function smaller than one that describes the deviation of the detector from the ideal performances, and in particular the possibility that some events are lost.

due to neutrino masses that modify neutrino propagation in an observable manner (section "Flavor Tranformations/Oscillations"). The present section discusses only the general aspects and can be used as an introduction to the basic underlying theory. A collection of formulae of practical usage is given later, in section "Neutrino Interactions and Properties: Formulae".

Basic terminology and overview

We know three negative fundamental particles, called the electron, the muon and the tau that are identified by the symbols e, μ, τ (or sometimes e^-, μ^-, τ^-) and that are distinguished by the mass,

$$m_e = 0.5110\ \text{MeV}, m_\mu = 105.7\ \text{MeV}, m_\tau = 1776\ \text{MeV} \tag{5}$$

Each of them is accompanied by its own antiparticle, predicted by relativistic field theory, with the same mass and positive charge, i.e., e^+, μ^+, τ^+.

Moreover, each one is associated by the charged current weak interactions (discussed in detail in the following) to a light neutral particle, named the *neutrino*. The names and the symbols of the three neutrinos are

$$\text{electron neutrino } \nu_e, \text{ muon neutrino } \nu_\mu, \text{ tau neutrino } \nu_\tau. \tag{6}$$

The antineutrinos are the neutral particles associated by the charged current weak interactions with the charged leptons of positive charge and denoted $\bar{\nu}_e$, $\bar{\nu}_\mu$, $\bar{\nu}_\tau$.

The common properties of these charged and neutral particles is to be spin 1/2 fermions and to be without strong interactions with the nuclei. They are collectively called *leptons* and together with the *quarks*, they are the basic constituents of the matter. So e.g., when we speak of a neutral lepton we refer to one of the six particles: ν_e, ν_μ, ν_τ, $\bar{\nu}_e$, $\bar{\nu}_\mu$, and $\bar{\nu}_\tau$; very often one speaks of "neutrinos" to refer to the same concept. The type of lepton (neutral or charged) is often called "flavor". E.g., the electron e and its corresponding neutrino ν_e are told to have the electronic flavor or flavor "e" and similarly for the other two.

The most characteristic properties of neutrinos are two: they have very weak interactions with the matter and very small masses. In the standard model of particles and interactions, their charge and their mass are zero. The fact that they are neutral is deeply connected to a gage symmetry, and it is considered established in a very reliable way. Instead, their masslessness is due to an "accidental" symmetry, and it does not hold anymore if the standard model is modified adding new matter or new scalar fields. Moreover, a series of experiments have provided us with evidences that the mass of neutrinos is small but non-zero, thereby showing that the standard model has to be modified (as recognized by the Nobel in physics of 2015).

We have already discussed that the main goal of neutrino astronomy is to observe and study neutral leptons and to extract the useful information they carry. As we shall see in a while, the existence of neutrino masses imply important observable effects:

in practice, this means that we have to account for these effects in order to interpret correctly (astronomical and terrestrial) observations of neutrinos.

Before proceeding, it is important to remark that there are many more new phenomena concerning neutrinos, on top of the ones that we emphasize here. For instance, some kinematical effects of neutrino masses during their propagation, or phenomena due to a special character of their mass that is likely to differ from the one of the other fermions ("Majorana mass"); neutrinos could feel the presence of other hypothetical light neutrino states, of the supposed existence of a (small) magnetic moment, they could decay, etc. We will discuss these possibilities later where appropriate, but in most of this chapter, we prefer to emphasize only the phenomena that are established and surely relevant for neutrino astronomy.

Low and high energy neutrinos

The definition of what are low and high-energy neutrinos is, just as in other astronomies, linked to the type of telescopes and of the objects that are observed, but also, to some extent, simply matter of convention. A reasonable value that we will keep in mind as the upper boundary of low-energy neutrinos is,

$$E_\nu < E_{\rm low} = 100\ {\rm MeV} \tag{7}$$

One could adopt other reference masses as m_μ, or the pion mass $m_{\pi^\pm} = 139.57$ MeV or $m_{\pi^0} = 134.98$ MeV, or even $\Lambda_{QCD} \sim m_p/3$, but in practical terms, this is not so important. Instead the lower boundary of the high-energy neutrinos will be set (somewhat arbitrarily) to,

$$E_\nu > E_{\rm high} = 100\ {\rm GeV} \tag{8}$$

that is close to the maximum value of artificially produced neutrinos. This could be raised more conservatively to 1 TeV, which is generally considered the minimum energy neutrino that could be attributed to some cosmic sources, without being overwhelmed by the background of the neutrinos produced in the atmosphere of the Earth.

With these choices, all neutrino sources, deriving from nuclear transitions almost at rest (such as those happening in the stars, in the Earth radioactivity, or during big-bang nucleosynthesis) fall in the low-energy regime, whereas the neutrinos that derive from decays of particles in fast motion, as those produced in collisions of cosmic rays (possibly near their sources) fall mostly in the high energy regime instead.

Note incidentally that the region between 100 MeV and 100 GeV encompasses certain sources of neutrinos. E.g., it includes atmospheric neutrinos and the typical neutrinos produced in terrestrial accelerators. Also stellar collapses produce neutrinos of similar energies in their center (even if it not considered plausible that they succeed to emerge becoming observable). However, to date we do not have to date any clear case of a cosmic source of neutrinos (confirmed or supposed) that yields an observable neutrinos in this intermediate range. This remark helps us to delimit the current goals of neutrino astronomy and the scope of this writeup.[3]

Neutrino Interactions

Here, we begin to discuss the general aspects of neutrino interactions, introducing the general terminology and a few key concepts. Important examples of cross sections are given later, section "A Gallery of Relevant Cross Sections".

Fermi coupling and weak interactions

The hamiltonian that causes the weak charged currents transitions at low energies is,

$$H_{cc} = \frac{G_F}{\sqrt{2}} \int d^3x \; (J^{\mu}_{\rm cc})^{\dagger} J_{\rm cc}, {}_{\mu} \text{ where we sum over } \mu = 0, 1, 2, 3 \tag{9}$$

The factor $\sqrt{2}$ is purely conventional. The amplitude of any weak transition at low energy is proportional to the Fermi coupling G_F, thus any decay width or cross section is proportional to its square.[4] The numerical value can be written as,

$$G_F^2 = 5.297 \times 10^{-44} \, \frac{\text{cm}^2}{\text{MeV}^2} \tag{10}$$

where we have used $\hbar c \sim 200$ MeV fm, with 1 fm = 10^{-13} cm. With this convenient choice of units, the typical size of weak cross sections at low energies becomes evident.

It is important to recall that,

- From the point of view of the standard model, the above interaction derives from the low-energy limit of the interaction that describes the exchange of a W boson at the lowest perturbative order: Thus, we get,

$$\frac{G_F}{\sqrt{2}} = \frac{g^2}{8M_W^2} \tag{11}$$

  where the W mass is $M_W \sim 80$ GeV and its coupling to the fermions is $g \sim 0.65$.

[3]Conversely, it does not imply that this intermediate range of energies should be considered uninteresting: In particular, it is hoped that a certain type of dark matter particle could annihilate or decay in celestial bodies (such as the Earth, the Sun or the Galactic center), thereby producing neutrino signals of similar energies and characterized by a well-defined direction.

[4]As we will show later, the description of neutrino oscillations in matter implies effects that are linear in G_F rather than quadratic. The reason is that these effects concern the *amplitude* of oscillation rather than the oscillation probability.

- At high energies, the W boson cannot be treated as a point particle and the full propagator has to be included. This will be discussed in details later.
- Besides the charged currents interactions, the standard models describes also *neutral current* interactions, that are due to Z boson exchange. These treat all three neutrinos in the same manner and have a size similar to those of the charged current interactions.

As we discuss in the following, neutral current interactions are important in several context concerning neutrino astronomy and astrophysics. However in this introductory considerations, we will emphasize a bit more on the charged current interactions that, as a rule, give events that more easily detected.

The nuclear decays caused by weak interactions are called beta decay, since conforming to Rutherford's terminology, the electrons (or positrons) emitted in nuclear transitions are called β-rays. Important nuclear manifestations of this type include,

$$\begin{array}{ll} n \to p + e^- + \bar{\nu}_e & \text{[neutron decay]} \\ (A, Z) \to (A, Z+1) + e^- + \bar{\nu}_e & [\beta^- \text{ decay}] \\ (A, Z) \to (A, Z-1) + e^+ + \nu_e & [\beta^+ \text{ decay}] \\ e^- + (A, Z) \to (A, Z-1) + \nu_e & \text{[electron capture]} \end{array} \tag{12}$$

where (A, Z) indicate a nucleus with mass number A and atomic number Z.

Charged currents and formal definition of neutrino

The weak charged current $J^\mu_{\rm cc}$ is the operator that transforms (along with its hermitian conjugate) neutral in charged leptons and viceversa. It decreases the electric charge of the fermionic state by one unit, resembling somewhat a ladder operator, and describing a wide variety of transitions, e.g., the passage from one electron to an electronic neutrino, or the passage from an electronic antineutrino to an antielectron, or the interaction of an electron and an electronic antineutrino, etc. The current $J^\mu_{\rm cc}$ contains two parts, one leptonic and one hadronic.[5] The first one is,

$$J^\mu_{\rm cc,lept}/2 = \bar{e}\gamma^\mu P_L \nu_e + \bar{\mu}\gamma^\mu P_L \nu_\mu + \bar{\tau}\gamma^\mu P_L \nu_\tau \tag{13}$$

where e, μ, τ, and ν_e, ν_μ, ν_τ are relativistic quantum fields, γ^μ are the 4×4 Dirac matrices, and where the chiral projector $P_L = (1 - \gamma_5)/2$ selects the states with left chirality ($\gamma_5 = i\gamma^0\gamma^1\gamma^2\gamma^3$). This current makes more precise the formal definition of neutrino "flavor" (or "type", or "species"): the electronic neutrino field is the one associated to electronic field, and similarly for the other ones. Thus, by definition, the neutrino emitted in the pion decay $\pi^+ \to \mu^+ + \nu_\mu$ is a muonic neutrino, whereas the one emitted in the beta decay of the neutron $n \to p + e^- + \bar{\nu}_e$ is an electronic antineutrino.

[5] In the standard model language, it could be advantageous to use the term "quarkonic" rather than hadronic. This is not only matter of semantics; indeed the various analogies between quarks and leptons are a key aspect of modern understanding of particles.

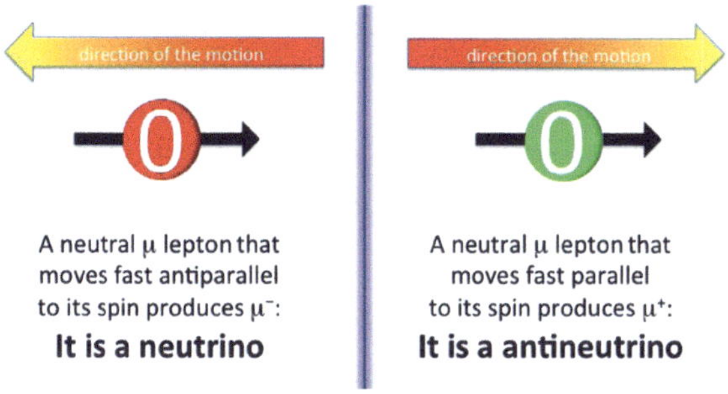

Fig. 1 In the ultra relativistic limit, neutrinos and antineutrinos are well characterized. Neutral leptons with spin opposite to the momentum (=negative helicity) produce negative leptons and thus are called "neutrinos". Neutral leptons with spin like the momentum (=positive helicity) produce positive leptons and thus are called "antineutrinos"

In the ultrarelativistic limit $E_\nu \gg m_\nu$ (that includes most of the cases of practical interest for neutrino astronomy) the mass of the neutrinos m_ν is negligible in comparison to the kinetic energy. The presence of the chiral projector has an interesting implication in this limit: the chirality projector implies that the particles have a well-defined value of the projection of the spin on the momentum, namely, a specific value of their helicity. More precisely, neutrinos have negative helicity and antineutrinos positive helicity: see Fig. 1 for illustration and Appendix 2 for a review. Unfortunately we have observed neutrinos and antineutrinos only in this limit, and we do not know what happens when neutrinos move more slowly or are at rest, as happens (most likely) to the neutrinos that were produced in the big-bang.

The number of leptons of any type (e.g., of electronic type) is unchanged in any reaction described by the standard model, but the type of neutrinos can change during their propagation: This phenomenon is called neutrino oscillations or flavor transformation and will be discussed in the next section. No known transition violates the total number of leptons[6]; in particular we have no evidence that neutrinos can transform into antineutrinos or viceversa. However, it can be argued that something similar happens and could be relevant for non-relativistic neutrinos: This is discussed in the section on big-bang neutrinos.

Flavor Tranformations/Oscillations

Neutrino astronomy has to take into account a very characteristic behavior of this type of radiation; in most occasions, the type of emitted neutrino is not the same of the type of detected radiation. This phenomenon is known as neutrino oscillations

[6]The most sensitive investigation involve the search of the nuclear transition where a nucleus changes its electric charge Z by two units by emitting two electrons $(A, Z) \to (A, Z+2) + 2e^-$, called *neutrinoless double beta decay*.

or flavor transformation, and it is important to account for it, in order to interpret the observations correctly and/or to learn the most from them. Here, we introduce the basic concepts and in a subsequent section (section "Parameters of Neutrino Oscillations and Conditions") we provide the reader with formulae of practical usage. (A reminder of basic concepts on spinors is given in Appendix 2.)

Basic idea

Neutrino oscillations were hypothesized by B. Pontecorvo about 60 years ago [55], just after the discovery of neutral kaon oscillations, and were put on firmer theoretical bases in [56]. A crucial step forward was the hypothesis that the neutrinos produced by weak interactions are not pure states, but mixed states; this was proposed by two independent groups [57–59] after realizing that there is more than one type of neutrinos. These pioneers are recalled together in the acronym PMNS (used in the terminology PMNS mixing matrix, aka leptonic mixing matrix).

Neutrino oscillation is a typical quantum phenomenon, analogous to the transformation between two different polarization states of the photon, induced in media where the velocity of propagation depends upon the direction of polarization. E.g., a photon linearly polarized in the direction of the $x - y$ bisector (moving in the 1st–3rd quadrant) transforms into its orthogonal state (moving in the 2nd–4th quadrant) when the phase of the x component increases by 180° in comparison to the phase of the y component, after propagating in the medium.

Each type of neutrino, e.g., an electronic neutrino, is supposed to be composed by more neutrinos with given mass, and the component waves propagate with different phase velocities. Therefore, the transformations among neutrinos happens also in vacuum. However, it can be influenced by the presence of the media where the neutrino propagate, as understood by L. Wolfenstein [60], S. Mikheyev and A.Yu. Smirnov [61, 62] that are recalled together in the acronym MSW (used in the terminology MSW effect aka matter effect).

Many experiments proved that this phenomenon positively occur and measured its characteristics. An incomplete list includes: Homestake, Kamiokande, Gallex/GNO and SAGE, Super-Kamiokande, MACRO, Soudan, SNO, KamLAND, K2K, MINOS, T2K, Borexino, Daya Bay, RENO, Double Chooz, Antares, IceCube.[7] For this reason, two these experiments, Super-Kamiokande and SNO, were recognized by the 2015 Nobel prize in physics.

To date, there are a few anomalous results that could indicate that neutrino oscillations are even more complex, and include more than 3 light neutrino states. But none of them is confirmed, there are contradictory aspects, and the global picture is not coherent. For these reasons (and few more that we will discuss later) we will focus on the simplest theoretical picture here on, with three light neutrinos only.

See the general references given above and in particular [13, 20, 21, 349] for a deeper access the original literature.

[7] All of them will be discussed in the following.

Neutrino mixing

We know that neutrinos masses are non-zero and very small. Each quantized neutrino field of given mass is given by[8]

$$\hat{\nu}_\ell = U_{\ell i} \hat{\nu}_i \text{ where } \ell = e, \mu, \tau \text{ and } i = 1, 2, 3 \tag{14}$$

The fields with given mass $\hat{\nu}_i$ are operator-valued functions over the ordinary space-time $x = (t, \mathbf{x})$ that can be decomposed in particle and antiparticle oscillators, $\hat{a}$ and $\hat{b}$, respectively, with given momentum $\mathbf{p}$ and helicity λ, as usual,

$$\hat{\nu}_i(x) = \sum_{\mathbf{p}\lambda} \hat{a}_i(\mathbf{p}\lambda)\, \psi_i(\mathbf{p}\lambda, x) + \hat{b}_i^\dagger(\mathbf{p}, \lambda)\, \psi_i^c(\mathbf{p}\lambda, x) \tag{15}$$

where ψ is the spinorial wavefunction and we use the charge conjugation matrix C to define $\psi^c = C\bar{\psi}^t$. We adopt the normalization for the wave functions: $\int d\mathbf{x}\ |\psi(t, \mathbf{x})|^2 = 1$, and for the oscillators: $\hat{a}(\mathbf{p}\lambda)\, \hat{a}^*(\mathbf{p}'\lambda') + \hat{a}^*(\mathbf{p}'\lambda')\, \hat{a}(\mathbf{p}\lambda) = \delta_{\mathbf{pp}'}\delta_{\lambda\lambda'}$. For Dirac neutrinos the oscillators of particles and antiparticles are independent whereas for Majorana neutrinos we should identify $\hat{a}(\mathbf{p}\lambda) = \hat{b}(\mathbf{p}\lambda)$. As we shall see in a while (and as elaborated further in the Appendix 2) this difference is not important when we consider oscillations of ultra relativistic neutrinos, since in this case only two types of oscillators matter, or in other terms, Dirac or Majorana masses produce the same type of oscillation phenomena.

We will assume that the matrix $U_{\ell i}$ is unitary, and the admixture with other hypothetical heavy neutrinos is small. To describe the phenomena in which we are interested (=oscillations) we are free to define the phases of the 6 fields $\hat{\nu}_\ell$ and $\hat{\nu}_i$ in the previous equation: therefore, we remain with and one physical phase and three mixing angles, corresponding to Euler angles of a unitary and real (=orthogonal) matrix. Another way to understand that there are 4 physical parameters is to note that, due to unitarity, there are just 4 independent parameters $|U_{\ell i}^2|$ that evidently do not depend upon the unphysical phases of the fields.

Formalism of oscillations in vacuum

Consider a neutrino, whose wave packet contains only ultra relativistic components

$$E_i = \sqrt{|\mathbf{p}|^2 + m_i^2} \approx |\mathbf{p}| \gg m_i \tag{16}$$

and the momentum width of the packet is much larger than the differences of energies $E_i - E_j = (m_i^2 - m_j^2)/(E_i + E_j) \approx (m_i^2 - m_j^2)/(2|\mathbf{p}|)$. The field of given flavor $\hat{\nu}_\ell$ can be written in terms of oscillators for given flavor, defined by,

$$\hat{a}_\ell(\mathbf{p}) = U_{\ell i}\ \hat{a}_i(\mathbf{p}, -) \text{ and } \hat{b}_\ell(\mathbf{p}) = U_{\ell i}^*\ \hat{b}_i(\mathbf{p}, +) \tag{17}$$

[8] In this section, for maximum clarity, we use the symbol "hat", i.e., $\hat{}$, in order to distinguish the operators. A very essential introduction to neutrino wavefunction can be found in the Appendix 2.

since the chirality projector P_L selects the neutrino oscillators with negative helicity and the antineutrino oscillators with positive helicity, and the wavefunctions are very close to the massless ones in the ultra relativistic limit. Thus, it is a good approximation to assume that the vector states that are produced by charged current weak interactions are,

$$\begin{aligned} |\nu_\ell, \mathbf{p}\rangle &\approx \hat{a}_\ell^\dagger(\mathbf{p})|0\rangle = U_{\ell i}^* \hat{a}_i^\dagger |\mathbf{p}, i\rangle = U_{\ell i}^* |\nu_i, \mathbf{p}\rangle, \\ |\bar{\nu}_\ell, \mathbf{p}\rangle &\approx \hat{b}_\ell^\dagger(\mathbf{p})|0\rangle = U_{\ell i} \hat{b}_i^\dagger |\mathbf{p}, i\rangle = U_{\ell i} |\bar{\nu}_i, \mathbf{p}\rangle \end{aligned} \tag{18}$$

for neutrinos and antineutrinos, respectively.

Suppose that such a neutrino of given flavor is produced in some distant source. In the Schrodinger representation, each momentum component of the wave packet will change in the course of the time as,

$$|\nu_\ell, \mathbf{p}, t\rangle = e^{-iE_i t} U_{\ell i}^* |\nu_i, \mathbf{p}\rangle \tag{19}$$

thereby acquiring a finite amplitude to change flavor

$$A_{\ell\to\ell'}(t) = \langle \nu_{\ell'}, \mathbf{p} | \nu_\ell, \mathbf{p}, t\rangle = \sum_i e^{-iE_i t} U_{\ell' i} U_{\ell i}^* \neq 0 \text{ when } \ell \neq \ell' \tag{20}$$

At the same time, the probability to remain in the same flavor will diminish. We can define a normalized wavefunction on flavor space that changes in the course of the time, simply by letting $\Psi_{\ell'}(t) = A_{\ell\to\ell'}(t)$ such that $\Psi_{\ell'}(0) = \delta_{\ell\ell'}$ that satisfies the Schrodinger like equation,

$$\begin{cases} i\partial_t \Psi(t) = H_0 \Psi \\ \Psi_{\ell'}(0) = \delta_{\ell\ell'} \end{cases} \quad \text{with } H_0 = U \text{diag}(E_i) U^\dagger \tag{21}$$

where H_0 is called the (effective) hamiltonian in vacuum and we use the shorthand $\partial_t \equiv \partial/\partial t$ for the time derivative. The treatment of antineutrinos is absolutely analogous, modulo the replacement $U \to U^*$.

Formalism of oscillations in matter

As we have just seen, the fact that the neutrinos produced in weak interactions do not coincide with the stationary states (i.e., the neutrinos with given mass) produces an effect on the flavor of the neutrino in the course of its propagation. This is formally similar to the change of the polarization of the light, and one is lead to wonder whether neutrino propagation is affected by the presence of the medium in which it propagates. We do not allude to the possibility that individual neutrinos are scattered away or absorbed, but to some refraction effect. Indeed, it can be noted that the ordinary matter is charged under weak interactions. It is sufficient to consider the following term in the hamiltonian,

$$H_{\nu e} = \frac{4G_F}{\sqrt{2}} \int d^3x \; \bar{\nu}_e \gamma^a P_L e \;\; \bar{e} \gamma_a P_L \nu_e = \frac{4G_F}{\sqrt{2}} \int d^3x \; \bar{\nu}_e \gamma^a P_L \nu_e \;\; \bar{e} \gamma_a P_L e \quad (22)$$

we have omitted the "hat" and did not indicate the dependence of the fields from the spacetime to simplify the notation; moreover, we have used the Fierz identity to rewrite the operator in the last, quite useful form. When we consider the average of the last operator over a background of non-relativistic electrons with number density $n_e(x)$, we find that $\langle \bar{e}\gamma_a P_L e \rangle = (1, 0, 0, 0) \times n_e(x)/2$. Therefore, the hamiltonian that dictates the behavior of the neutrinos acquires the new term $\langle H_{\nu e} \rangle = \sqrt{2} G_F n_e(x) \nu_e^\dagger \nu_e$. The Schrodinger like equation that regulates the flavor state gets modified as follows,

$$\begin{cases} i\partial_t \Psi(t) = (H_0 + H_{\text{matter}})\Psi \\ \Psi_{\ell'}(0) = \delta_{\ell\ell'} \end{cases} \quad \text{with } H_{\text{matter}} = \sqrt{2} G_F n_e(x) \text{diag}(1, 0, 0) \quad (23)$$

where the last term affects only the phase of propagation of the electron neutrinos.[9] This is called the matter term or also the MSW term of the hamiltonian. It should be noted that this term can modify oscillations, but it does not produce oscillations on its own (i.e., not without the vacuum term discussed above).

Neutrino Interactions and Properties: Formulae

In this section we discuss the properties and interactions of neutrinos that are relevant for neutrino astronomy. The material is selected emphasizing formulae of practical usage rather than the theoretical aspects. For general references, see the first paragraph of section "Basic Considerations". Note that there are several reference books [3–5] useful to understand neutrino properties; a few review works on the same subject include [21, 22]. Some textbooks that concern neutrino interactions, and more in general weak interactions, are [6–8], and they will be usefully complemented with the review papers [23, 24].

A Gallery of Relevant Cross Sections

Here, we provide the Reader with several examples of cross sections, that allow detecting neutrinos. We do not aim to achieve a complete discussion but rather to offer a general idea of the various possibilities and of the size of these cross sections.

[9] It is common to write $n_e = \rho_e N_A$ where ρ_e is the electron molar density and N_A the Avogadro number. For neutral matter, $\rho_e = \rho_p = Y_p \rho$, where Y_p is the proton fraction and ρ the density.

Other cross sections will be discussed later in this chapter, where appropriate. See [21, 24, 265] for a more complete discussion.

Electron antineutrino scattering on protons (IBD or QEL)

The electron antineutrinos above few MeV have a relatively large cross section with free proton nuclei, as known since the early days and as widely verified. The most common detectors are made of water or hydrocarbon compounds, and therefore are rich of protons. The reaction is called *inverse beta decay* (IBD),

$$\bar{\nu}_e + p \rightarrow e^+ + n \tag{24}$$

and its cross section is reliably known. Precise calculations of this cross section are given in [63, 64].

In the energy region below 200 MeV, of particular interest for neutrino astronomy, it can be usefully parameterized as [64],

$$\sigma = 10^{-43} p_e E_e \; E^{-0.07056+0.02018 \ln E - 0.001953 \ln^3 E} \tag{25}$$

with $E_e = E - 1.293$ MeV and $E_e = \sqrt{p_e^2 + m_e^2}$. In this expression, all energies, momenta and masses are given in MeV units (or if one prefers, are to be divided by 1 MeV) and the resulting cross section is in cm^2. Evidently the correct order of magnitude of the cross section is accounted for by the explicit presence of the square of the Fermi constant, G_F^2, see Eq. 10. The positron energy correlates closely with the electron antineutrino reaction, but the positron is emitted almost isotropically. Other useful expressions, of practical use for effective and precise numerical evaluation at low energies, are given in [37]. At higher energies, instead (say above few 100 MeV) this reaction is commonly considered as one of the *quasi-elastic scattering reactions* with nucleons (QEL); its evaluation requires to know the nucleon form factors that are to some extent uncertain.

Elastic scattering on electrons (ES)

The number of electrons is never less than the one of free protons and in usual detectors it is typically larger. The cross section of interaction is smaller but all neutrinos can interact with electrons; the typical number of expected events is 1/20-1/30 than those of the previous class at low energies. The reaction is

$$\nu + \mathrm{e} \rightarrow \nu + \mathrm{e} \tag{26}$$

In this case, it is convenient to exhibit the analytical formula. The tree level expression is,

$$\frac{d\sigma}{dy} = \frac{2G_F^2}{\pi} m_e E \left(g^2 + g'^2 (1-y)^2 - g g' \frac{m_e}{E} y \right) \text{ with } 0 < y < \frac{1}{1 + \frac{m_e}{2E}} \tag{27}$$

where y is the kinetic energy of the electron over the neutrino energy,

$$y = \frac{T_e}{E} \text{ with } T_e = E_e - m_e \tag{28}$$

The various adimensional constants are,

	g	g'
ν_e	$\frac{1}{2} + \sin^2 \theta_W$	$\sin^2 \theta_W$
$\bar{\nu}_e$	$\sin^2 \theta_W$	$\frac{1}{2} + \sin^2 \theta_W$
ν_μ	$-\frac{1}{2} + \sin^2 \theta_W$	$\sin^2 \theta_W$
$\bar{\nu}_\mu$	$\sin^2 \theta_W$	$-\frac{1}{2} + \sin^2 \theta_W$

where the squared sine of the Weinberg angle is $\sin^2 \theta_W = 0.223 \approx \frac{1}{4}$. Note that the electron neutrino cross section is larger, since it receives contributions from charged currents and neutral currents; instead, muon and tau neutrinos interact only via neutral currents (i.e., via Z exchange).

As a check and as a good mnemonic trick, one can verify that at high energies, the total cross section satisfies

$$\begin{aligned} &\sigma(\nu_e) = 0.9 \times 10^{-44} \text{ cm}^2 \; (E/\text{MeV}) \\ &\sum_{\ell=e,\mu,\tau} [\sigma(\nu_\ell) + \sigma(\bar{\nu}_\ell)] = 1.8 \times 10^{-44} \text{ cm}^2 \; (E/\text{MeV}) \end{aligned} \tag{29}$$

Stated otherwise, the sum of all neutrino cross sections is (about) twice the one due to electron neutrinos alone. Indeed, it is easy to check that this property follows from the approximation $\sin^2 \theta_W \approx \frac{1}{4}$ given above. Note that owing to the neutrino in the final state, a large fraction of the initial energy remains invisible, and should be accounted for to describe detector response properly.

Neutral current reactions below 100 MeV

In the region below 100 MeV, of particular interest for us, the detection cross sections can be parameterized numerically as

$$\frac{1}{2}(\sigma_\nu + \sigma_{\bar{\nu}}) = G_F^2 \, (E_\nu - E_{thr.})^2 \; 10^p \text{ with } p = \sum_n c_{n=0}^N \left(\frac{E_\nu}{100 \text{ MeV}}\right)^n \tag{30}$$

where $E_{thr.}$ is the threshold (minimum energy) and only a small number N of non-zero coefficients are needed. This is true, e.g., in the following two cases, where we have considered two common type of nuclear targets used in neutrino astronomy: free protons and carbon nuclei.

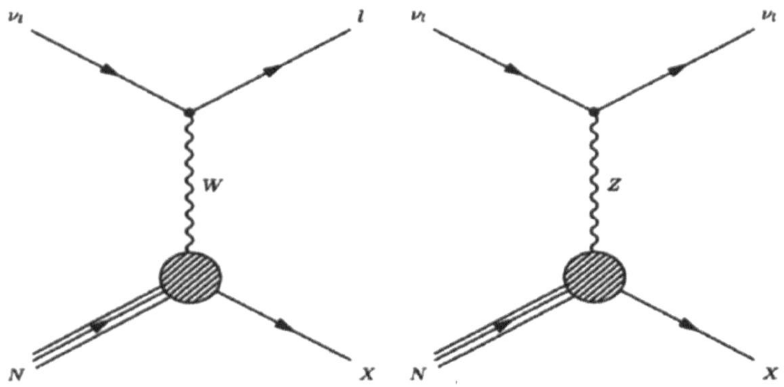

Fig. 2 Feynman diagram for CC (*left*) and NC (*right*) neutrino interaction on Nucleon

- For the reaction
$$\nu + p \rightarrow \nu + p \tag{31}$$
the threshold is zero and we need only 2 non-zero coefficients $c_0 = -0.333$ and $c_1 = -0.16$. These are the value obtained assuming the proton strangeness suggested by the experiment BNL 734 $\eta = 0.12$, note that the cross section scales with $(1+\eta)^2$ and theoretical models suggest $\eta \sim 0$.
- For the reaction
$$\nu + {}^{12}\mathrm{C} \rightarrow \nu + {}^{12}\mathrm{C}^* \text{ followed by } {}^{12}\mathrm{C}^* \rightarrow {}^{12}\mathrm{C} + \gamma(15.1\text{ MeV}) \tag{32}$$
instead we have $E_{thr.} = 15.11$ MeV and $c_0 = -0.645$, $c_1 = -0.184$ $c_2 = -0.884$ and $c_3 = +0.233$. This expression agrees very well with the theoretical expression that however has a large theoretical error about 20%.

Deep-inelastic scattering (DIS)

When neutrino energies $E_\nu \gg$ GeV, their total cross section for interaction with the nucleons can be described in terms of scattering with its constituents, namely quarks ("partonic formalism"). One focusses on the inclusive description of the reaction, ignoring (in first approximation at least) the details of how the quark, hit by the neutrino, gets "hadronized". In Fig. 2 the Feynman diagrams of the neutrino-nucleon interaction is illustrated for both charged current (CC) and neutral current (NC) channels.

Let us discuss in some details the CC case. Consider a charged current interaction of a neutrino ν_ℓ with a nucleon N ($N = p$ or n) that produces a lepton with energy E_ℓ. Consider the scattering with a quark carrying momentum fraction x. Its cross section at the leading order (LO)—namely at the tree level in W exchange and neglecting the corrections due to strong interactions—is,

$$\begin{aligned}
\frac{d\sigma_q}{dy}(\nu_\ell d \to \ell u) &= \frac{d\sigma_q}{dy}(\bar\nu_\ell \bar d \to \bar\ell \bar u) = \frac{G_F^2 s_q}{\pi}\mathscr{P}_W \\
\frac{d\sigma_q}{dy}(\nu_\ell \bar u \to \ell \bar d) &= \frac{d\sigma_q}{dy}(\bar\nu_\ell u \to \bar\ell d) = \frac{G_F^2 s_q}{\pi}(1-y)^2\mathscr{P}_W
\end{aligned} \quad \text{with } \ell = e, \mu, \tau \tag{33}$$

where $0 < x < 1$ and $0 < y < 1$ and we have[10] $s_q = (k + xp)^2 \approx x\, s$ and $y = -t_q/s_q \approx 1 - E_\ell/E_\nu$, neglecting fermion masses. The propagator of the W boson gives

$$\mathscr{P}_W = \frac{1}{(1 + Q^2/M_W^2)^2} \text{ with } Q^2 = 2m_N E_\nu\, x\, y \tag{34}$$

The total cross section is found by multiplying the above quark contributions by the distribution of the momentum of the quarks inside the nucleon, and integrating over x. The distributions are given conventionally for the proton, and one can obtain the corresponding distributions for the neutron applying isospin invariance, namely exchanging $u \leftrightarrow d$. Note that quite often one is interested in "isoscalar" target N, i.e., one takes the average between a proton and a neutron. (For refined treatments of the cross section, one should consider the specific target and include higher orders in QCD and nuclear corrections.)

At low energies, the typical value of Q^2 is, $\langle Q^2\rangle = \langle s\, xy\rangle \approx 2m_N E_\nu \times 1/2 \times 1/3 = M_W^2(E_\nu/20\text{ TeV})$; this explains why at energies below $E_\nu = 10$ TeV we can set $\mathscr{P}_W \approx 1$. In this regime, we find that the total cross section increases linearly with $s \propto E_\nu$. Instead at higher energies the effect of $\mathscr{P}_W$ is conspicuous, and the cross section scales with a milder power of E_ν. Note that at very high energies, values of x and y close to zero are favored, and in this limit the four expressions for the neutrino and the antineutrino cross sections given above coincide in practice.

Similar expressions are valid for the neutral current cross section. For a more complete but introductory discussion see [21].

Parameters of Neutrino Oscillations and Conditions

We begin by resuming what it is known on three flavor neutrino oscillations. Then, we present numerical formulae for the phase of vacuum oscillations that allows one to check and estimate when the oscillation effects are sizeable; we proceed similarly for matter effects. Finally, we offer a number of explicit formulae to describe neutrino oscillations that are useful for the subsequent discussion.

---

[10]We use the Mandelstam variables s, t, u here that in the example of a muon neutrinos scattering onto a nucleon N, namely $\nu_\mu(k) + N(p) \to \mu(k') + X(p')$, are given by $s = (k+p)^2$, $t = (k - k')^2 \equiv -Q^2$, $u = (p - k')^2$. Moreover we adopt the usual definitions $q = k - k'$ (4-momentum transfer) $x = Q^2/(2pq)$ and $y = pq/pk$. Note that $s \approx 2pk = 2m_N E_\nu$, $Q^2 = 2pqx = 2xypk \approx s\,xy$.

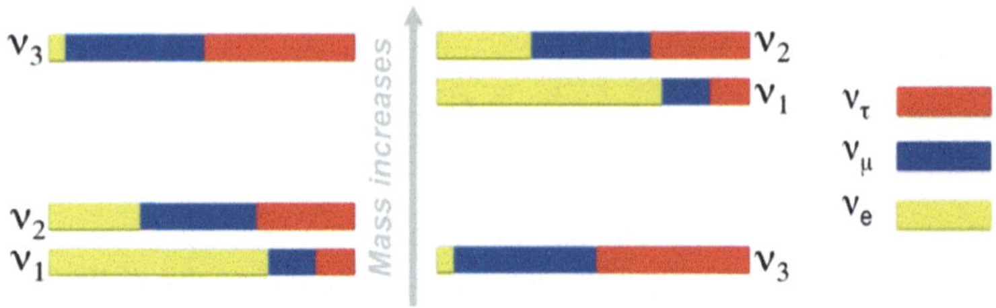

Fig. 3 Illustration of the mass spectra compatible with the data from neutrino oscillations; *left* normal hierarchy; *right* inverted hierarchy. The minimum mass is not probed by oscillations

What we know on 3 flavor neutrino oscillations

Neutrino oscillations (in vacuum or in matter) depend upon the difference of neutrino masses, squared, sometimes indicated as delta-m-squared. The analysis of oscillation data have allowed to discover and measure two different values of delta-m-squared, in a manner that will be recalled shortly here below. The results are illustrated in Fig. 3; note that two different types of neutrino mass hierarchies (or orderings, or spectra) are compatible with the existing data.

Moreover, the three mixing angles named θ_{12}, θ_{13} and θ_{23} and the single phase δ that matters for neutrino oscillations have been determined within their experimental uncertainties. The conventional decomposition of the mixing matrix is,

$$U = \begin{pmatrix} c_{12}c_{13} & s_{12}c_{13} & s_{13}e^{-i\delta} \\ -s_{12}c_{23} - c_{12}s_{13}s_{23}e^{i\delta} & c_{12}c_{23} - s_{12}s_{13}s_{23}e^{i\delta} & c_{13}s_{23} \\ s_{12}s_{23} - c_{12}s_{13}c_{23}e^{i\delta} & -c_{12}s_{23} - s_{12}s_{13}c_{23}e^{i\delta} & c_{13}c_{23} \end{pmatrix} \quad (35)$$

where $s_{ij}, c_{ij} \equiv \sin\theta_{ij}, \cos\theta_{ij}$. This is just same convention adopted to parameterize the quark mixing matrix (CKM). Of course, the values of the parameters are different, as it is evident from Fig. 4 where their values is visualized and compared. Recall incidentally that the leptonic mixing matrix is often called PMNS matrix, to honor the memory of Pontecorvo, Maki, Nakagawa and Sakata.

The values of the parameters of the leptonic mixing matrix, obtained from a global analysis of all oscillation data available in 2016 [65], are presented in Table 2. We present the best fit values and an estimation of the accuracy, obtained from the two sigma ranges as follows: we have estimated the maximum deviation from the best fit value and then divided the result by two, to exhibit a single conservative number. Note that the delta-m-squared are defined as,

$$\delta m^2 = m_2^2 - m_1^2 \ , \quad \Delta m^2 = |m_3^2 - (m_2^2 + m_1^2)/2| \quad (36)$$

The first parameter is probed by solar neutrino experiments (and even more by reactor experiments with a baseline of many 10–100 km) the second one by atmospheric

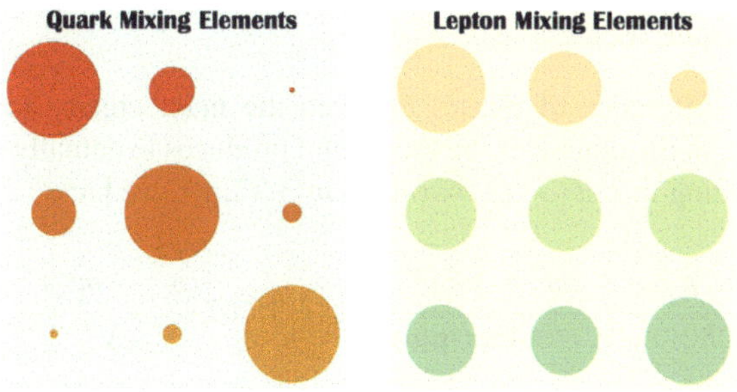

Fig. 4 The surfaces of the circles represent the size of the mixing elements. From top to bottom, from left to right: *Left panel* quark mixing (CKM) elements $|V_{ud}|$, $|V_{us}|$, $|V_{ub}|$, $|V_{cd}|$, ...; *Right panel* lepton mixing (PMNS) elements $|U_{e1}|$, $|U_{e2}|$, $|U_{e3}|$, $|U_{\mu 1}|$, ... in both panels, the mixing matrices are supposed to be unitary. The hierarchical structure of quark mixing elements contrasts with the one of lepton mixing elements

Table 2 Results of the global analysis of oscillation data of the Bari group [65]. The precise meaning of the parameters and the error estimate are discussed in the text

	Normal (inverted)	Error (%)	Units
Δm^2	2.50 (2.46)	1.8	10^{-3} eV^2
δm^2	7.37 (7.37)	2.4	10^{-5} eV^2
$\sin^2\theta_{13}$	2.17 (2.19)	4.8	10^{-2}
$\sin^2\theta_{12}$	2.97 (2.97)	6.2	10^{-1}
$\sin^2\theta_{23}$	4.43 (5.75)	16	10^{-1}
δ	1.39 (1.39)	19	π

neutrino experiments (and even more by long-baseline accelerator experiments). Denoting the lightest neutrino mass (that is not probed by oscillations) with m, these formulae can be inverted to give,

$$\begin{cases} m_1 = m & (= \sqrt{m^2 + \Delta m^2 - \delta m^2/2}) \\ m_2 = \sqrt{m^2 + \delta m^2} & (= \sqrt{m^2 + \Delta m^2 + \delta m^2/2}) \\ m_3 = \sqrt{m^2 + \Delta m^2 + \delta m^2/2} & (= m) \end{cases} \tag{37}$$

where the expression outside (inside) the brackets applies for normal (inverse) mass hierarchy. It is interesting to note that the case of normal mass hierarchy is slightly favored from the present experimental information at $\Delta\chi^2 = 2.8$ (namely, about 1.7σ) from the same analysis [65]. In the next pages, we present several examples of neutrino oscillation formulae, recalling whenever possible how the above set of parameters has been determined.

The phase of vacuum oscillations

In vacuum, the difference of phases between the mass eigenstates $\varphi_{j1} = (E_1 - E_j)t/2$ increases in the course of the time t and produces eventually flavor transformations.[11] Recalling that $E_j - E_1 \approx (m_j^2 - m_1^2)/(2|\mathbf{p}|)$, we have

$$-\varphi_{j1} = \frac{(E_j - E_1)t}{2\hbar} \approx \frac{(m_j^2 - m_1^2)c^4 \times L}{4\hbar c(|\mathbf{p}|c)} = 1.267\frac{(m_j^2 - m_1^2)c^4}{\mathrm{eV}^2} \times \frac{L}{\mathrm{m}} \times \frac{\mathrm{MeV}}{E_\nu} \tag{38}$$

where we included explicit $\hbar$ and c factors, and denoted the distance of propagation with $L \approx ct$ and the kinetic energy of the neutrinos $E_\nu = |\mathbf{p}|c$. Several experiments on atmospheric neutrinos, on artificial neutrino beams, on reactor neutrinos have measured the parameters of neutrino oscillations by using vacuum oscillation formulae.

In order to understand this point, it is sufficient to consider the numerical value of the vacuum oscillation phase φ assuming the two delta-m-squared listed above,

- Consider the largest one. The phase is close to $\pi/2$ when the distance is 500 km and the energy is $E_\nu = 1$ GeV that corresponds to the situation of horizontal atmospheric neutrinos (where the distance is $\sqrt{(R_\oplus + 20\text{ km})^2 - R_\oplus^2} = 500$ km) or to a typical terrestrial experiment with an artificial neutrino beam and a long-baseline. The same value of the phase is for $L = 1$ km and $E_\nu = 2$ MeV, namely, for the conditions of a suitable reactor neutrino experiment.
- Consider the smallest one. If we have a nuclear reactor that produces neutrinos of 3 MeV at $L = 60$ km, we see that also this phase of oscillation becomes large. Therefore, in these conditions, also the associated neutrino oscillation effects become observable.

Comparison of the vacuum and matter terms

The matter term contributes a phase, only for electron neutrinos. The size of the new phase can be compared with the vacuum phase as follows,

$$\varepsilon \equiv \frac{\sqrt{2}G_F n_e}{\Delta m^2/(2E_\nu)} \approx \begin{cases} \left(\frac{7.5\times10^{-5}\mathrm{eV}^2}{\Delta m^2}\right)\left(\frac{E_\nu}{5\text{ MeV}}\right)\left(\frac{\rho_e}{100\text{ mol/cm}^3}\right) \\ \left(\frac{2.4\times10^{-3}\mathrm{eV}^2}{\Delta m^2}\right)\left(\frac{E_\nu}{5\text{ GeV}}\right)\left(\frac{\rho_e}{3\text{ mol/cm}^3}\right) \end{cases} \tag{39}$$

where Δm^2 stands for a difference of squared masses. E.g., let us focus on the solar neutrinos (for which $\Delta m^2 = 7.5 \times 10^{-5}$ eV^2 and that are produced where $\rho_e \sim 100$ mol/cm^3). When the energy $E_\nu \ll 1$ MeV, e.g., for the lowest energies solar neutrinos, the matter effect can be neglected; instead, for highest energy solar neutrinos, with energies above 5 MeV, this is important. The effect exists also for high-energy neutrinos that cross the Earth (when $\rho_e \sim 3$ mol/cm^3 and $\Delta m^2 = 2.4 \times 10^{-3}$ eV^2) but this is still to be clearly observed.

[11] The appearance of the factor half is due to the identity, $e^{i(E_1-E_j)t} - 1 = 2i \sin\phi_{j1} e^{i\varphi_{j1}}$.

Examples of Oscillation Formulae

Below we discuss three specific examples of oscillation formulae, with the purpose of allowing the reader to get familiar with their usage and to understand the phenomenology. More detailed discussions can be found, e.g., in [21], in Appendix 3 and later in this text, wherever needed.

Two flavor vacuum oscillation

The simplest and well-known case is the case when we have only two flavors, say, ν_e and ν_μ. If we have a neutrino with flavor ν_e and momentum **p** that is not a stationary state but rather a superposition of two stationary (mass eigen)states, namely,

$$|\nu_e\rangle = \cos\theta|\nu_1\rangle + \sin\theta|\nu_2\rangle \tag{40}$$

whereas $|\nu_\mu\rangle$ is the orthogonal state. The formula that describes the probability that an ultra relativistic neutrino of type ν_e becomes a neutrino of type ν_μ, called *appearance* probability, is simply,

$$P_{e\mu} = \sin^2 2\theta \ \sin^2\left[\frac{\Delta m^2 L}{4E}\right] \tag{41}$$

where L is the distance between production and detection, E is the neutrino energy, Δm^2 is the difference of the squared masses, and we used the natural units $\hbar = c = 1$. The other probabilites of appearance are $P_{e\mu} = P_{\mu e} = P_{\bar{e}\bar{\mu}} = P_{\bar{\mu}\bar{e}}$, the latter two being those concerning antineutrinos. The probabilities of *disappearance* are fixed by the conservation of probability, e.g., $P_{ee} + P_{e\mu} = 1$; this is a specific feature of the case with two neutrinos. These formulae alone allow one to understand the main facts of neutrino oscillations, such as the main behaviors of atmospheric neutrinos, those of reactor neutrinos on short scale, low-energy solar neutrinos, etc., simply using the appropriate energies, distances and parameters of neutrino oscillations.

Fully averaged oscillations

There is a specific three-flavor case that is very easy to be treated but yet relevant for the interpretation of the high-energy neutrinos from cosmic sources. This is the case of fully averaged oscillations, also called Gribov-Pontecorvo limit [66]. In this case, the neutrino of flavor $\ell = e, \mu$ or τ is produced in (excellent approximation in) vacuum, and the de Broglie's waves of the three mass components, upon propagation on cosmic distances, are supposed to pick very large phases $e^{-iE_i t}$. Therefore, the overlap with a state of different flavor ℓ' gives the amplitude $A_{\ell\ell'} = U^*_{\ell i} U_{\ell i} e^{-iE_i t}$. When we square this amplitude we obtain the probability of conversion. The rapidly oscillatory terms can be thrown away, thus we get the expression,

$$P_{\ell\ell'} = |U_{\ell i}|^2 |U^2_{\ell i}| \tag{42}$$

that is symmetric in the exchange $\ell \leftrightarrow \ell'$ and also identical for neutrinos and antineutrinos, just as in the two flavor case discussed above. Note that any information on the masses disappeared in this limits, and moreover, there are only three independent probabilities, owing to the conservation of the probability and to the symmetry condition. Therefore, these expressions are described in terms of three parameters only, see [67] for a detailed discussion.

Two flavor matter oscillation–costant density

This example is meant to illustrate the impact of matter effect. Let us begin consider oscillations with a single scale, and moreover in constant matter density. This is an approximation that applies to some extent to neutrinos that cross a fraction of the mantle of the Earth, and we use this case for illustrations. We give directly the formula (few more details are given in Appendix 3 for the interested reader).

In these approximations, the probability that a muon converts into an electron is simply,

$$P_{\mu e} = \sin^2\theta_{23}\ \sin^2 2\widetilde{\theta_{13}}\ \sin^2\widetilde{\varphi} \text{ with } \widetilde{\varphi} = \frac{\widetilde{\Delta m^2} L}{4E} \tag{43}$$

(note that, in this case, we have used the notation of the angles from the standard parameterization). In Eq. 43, we introduced the matter-modified mixing angle and squared-mass-difference

$$\begin{cases} \sin 2\widetilde{\theta_{13}} = \sin 2\theta_{13}\ /\Delta \\ \cos 2\widetilde{\theta_{13}} = (\cos 2\theta_{13} - \varepsilon)/\Delta \\ \widetilde{\Delta m^2} = \Delta m^2 \times \Delta \end{cases} \text{where } \Delta = \pm\sqrt{(\cos 2\theta_{13} - \varepsilon)^2 + \sin^2 2\theta_{13}} \tag{44}$$

The sign of Δ is matter of convention. We rewrite the ratio between matter and vacuum term as,

$$\varepsilon \equiv \pm\frac{\sqrt{2}G_F n_e}{\Delta m^2/(2E)} \approx \pm\frac{\rho}{5.5\text{g/cm}^3} \times \frac{Y_e}{1/2} \times \frac{2.4\times 10^{-3}\text{eV}^2}{\Delta m^2} \times \frac{E}{5.5\text{ GeV}} \tag{45}$$

where we identify Δm^2 with Δm^2_{23}. In the expression of ε, instead, the sign is important: it is plus for normal hierarchy and minus for inverted hierarchy. A very important feature of matter effect is here evident: the formulae are not symmetric in the exchange $\theta \to 90° - \theta$, and therefore it is possible to determine the angle more precisely by measuring the effect.

Considering the average matter density of the Earth $\rho = 5.5$ g/cm^3 and $Y_e = 1/2$, we get $n_e = 1.7 \times 10^{24}$ e$^-$/cm^3 for the electronic density. Thus, the characteristic length of MSW theory is,

$$L_* \equiv \frac{1}{\sqrt{2}G_F n_e} \sim 1000 \text{ km} \tag{46}$$

We see that, for normal hierarchy, the maximum of $P_{\mu e}$ obtains when: (1) Δ is as small as possible, in order to maximize $\sin \widetilde{2\theta_{13}}$; moreover, (2) the phase of propagation is $\widetilde{\varphi} \sim \pi/2$. These conditions are met when the neutrino energy and the propagation distance are,

$$E_{\max} = \frac{\Delta m^2 L_*}{2} \cos 2\theta_{13} \sim 5.5 \text{ GeV and } L_{\max} = \frac{\pi L_*}{\tan 2\theta_{13}} \sim 9000 \text{ km} \quad (47)$$

In the case of inverted hierarchy, the matter effect depresses $P_{\mu e}$ that becomes negligible.

Neutrino Telescopes

Neutrinos are the lightest known massive particles, they are neutral and can interact only by weak interactions. Due to these properties they can act as a unique "messenger of information" from remote regions of the Universe. Any of us is reached continuously by a huge flux of neutrinos which energies span over more that 20 order of magnitude (Fig. 5). Neutrinos can pass through the interstellar matter, escape from the dense cores of astrophysical sources and travel without any deflections even through strong magnetic fields. Neutrinos can provide us with information complementary to photons. For this reason, since the second half of the twentieth-century, several neutrino detectors have been proposed and built in order to use these particles as astrophysical probes. Neutrinos allow us to look deeper in space and in time, to observe the cores of the stars, to discover new phenomena. For all purposes, these detectors are often referred to as *neutrino telescopes*. In this section, we introduce the general principles of these detectors, and will discuss them in greater details in the subsequent chapters.

General references:

For general references on neutrinos, see the first paragraph of section "Basic Considerations". A review that introduces nicely to the topic of neutrino telescopes is [25]. However, in the recent times we had important achievements and progresses, especially those due to Borexino and IceCube that we will discuss. Recall finally that one of the main series of conference series of the field of neutrino science, held in Venice and begun by M. Baldo-Ceolin, is called exactly *Neutrino Telescopes* see Appendix 1.

General Considerations

What is a neutrino telescope?

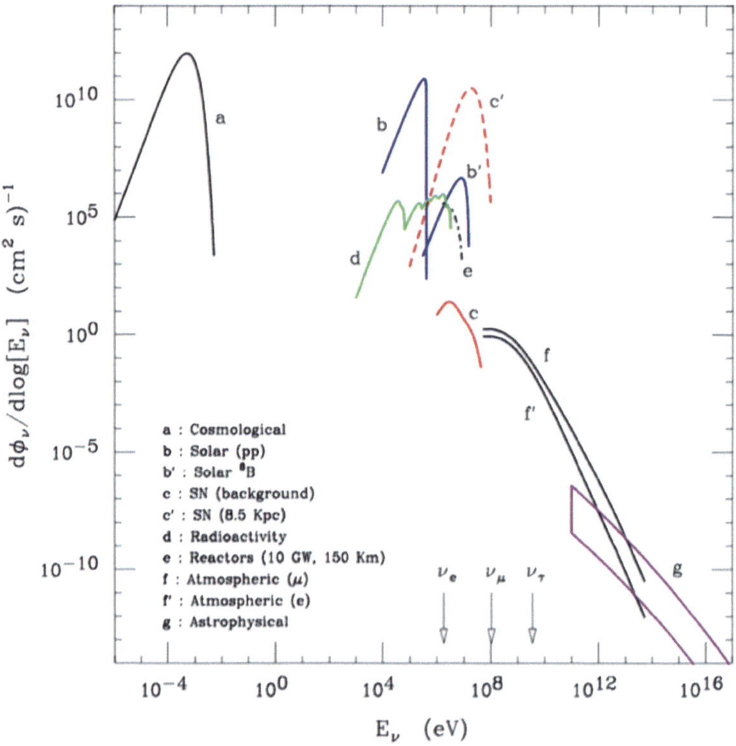

Fig. 5 Flux of neutrinos at the surface of the Earth [339]

In this section, we discuss the detector principles that allow us to observe neutrinos. The main characteristics of an impinging individual neutrino in which we are interested are,

$$\text{energy, arrival time, direction, flavor.} \tag{48}$$

A general feature of neutrino astronomy is that we observe neutrinos as isolated particles. There is nothing analogous to an electromagnetic or gravitational wave for neutrinos, neutrinos are always seen as *quantized* radiation.

Neutrinos are revealed when some secondary particle or radiation is produced in their (weak) interaction with some target. We are not able to follow the trajectory of a neutrino. We can see its manifestations in a single interaction, either in the moment when it happens or subsequently: E.g., one can see the charged lepton associated to the specific type of neutrinos by charged current interactions, or the hadronic (or nuclear) transformation caused by neutral current interactions. In this case we talk of a neutrino event or of a neutrino signal.

Occasionally, certain observable manifestations can be attributed uniquely to neutrinos, or stated more explicitly, the events due to background processes are negligible in comparison to those due to neutrino signals. However, most commonly there other manifestations (besides those due to the signal due to the neutrinos) that have different causes and that need to be accounted for to interpret the observations properly. For example, the cosmic rays radiation that is omnipresent on Earth's surface has to be kept under control and this is the reason why the neutrino telescopes, as a rule, are located underground, or underwater, or under ice.[12] Also the presence of radioactive elements in or around the detector can spoil the chances of a neutrino telescope to observe a signal; we recall that an ordinary glass of water produces as a rule one Bequerel (=one disintegration per second) or more. We refer to the problem of the presence of background events—or, for short, the *background*-on top of the signal events.

Even when we are in an underground location, and we have a telescope built with very pure materials, we have to remember (besides the residual natural radioactivity) of the presence of atmospheric muons and of the particles produced by these muons in "spallation" reactions with the surrounding material. Sometimes, neutrinos from artificial sources might spoil the possibility to observe neutrinos from a natural source. Generally, the problem of the background (or of the residual background) is particularly acute for neutrino telescopes.

Low Energy Neutrino Telescopes

Here, we discuss the principles to detect astrophysical sources of "low-energy neutrinos" that according to the previous discussion, denote the region of energies below 100 MeV.

Owing to the much higher counting rate due to cosmic rays at ground, these detectors have to be located deep under the Earth surface; the existing ones are in mines (sometimes abandoned) or tunnels. Typically they are also endowed by anti-counting devices, to increase confidence that the events occasionally observed in the detector can be attributed the neutrinos. Moreover, it is rather essential that the materials constituting the detector (target, photo-tubes and shield) are as radiopure as possible, to avoid internal contamination and sources of background events.

From neutrinos to observable particles

Generally, neutrinos become visible when they interact with some target and produce one or more observable secondary particles in an elementary reaction.

In ordinary matter, the possible target for neutrinos are first of all the atomic electrons that for several reasons are very important and are discussed below, and/or the nuclei, beginning with the simplest one, those of hydrogen (protons). Note that for protons, the only possible reactions are $\bar{\nu}_e + p \to e^+ + n$ (IBD) and $\nu + p \to \nu + p$. For more complex nuclei, it is possible to produce an excited nucleus that then will emit some type of observable radiation.

[12]The exception is the (planned) telescopes aiming to use the terrestrial atmosphere or its surface to convert the highest energy neutrinos into observable radiation that we discuss later.

Of course, muon or tau charged leptons cannot be produced at these low energies. Thus, the only charged lepton that is detectable is the *electron* or the *positron*—even if, as a rule, it is difficult to measure their charge and distinguish them. There are various techniques to see $e^{\pm}$. For what concerns neutrino astronomy, the two most successful techniques so far are those that use scintillation light or Cherenkov radiation: In fact, as we discuss below, solar neutrinos have been seen mostly with scintillating detectors, geoneutrinos have been only with scintillating detectors, while SN1987A neutrinos have been seen mostly with Cherenkov detectors. However, also radiation from ionization or other methods can be used.

Interestingly neutrino interactions can produce several particles, that are potentially observable; one example is IBD (Eq. 24), another one $\nu_e + {}^{12}\text{C} \to e^- + {}^{12}\text{N}$ with a threshold of 17.3 MeV followed by ${}^{12}\text{N} \to {}^{12}\text{C} + \nu_e + e^+$ (lifetime 15.9 ms). Other particles that permit the detection of neutrinos, are: photons from an excited nucleus previously produced by the neutrino, neutrons extracted from the nucleus or produced by the neutrino, recoiling nuclei. The photon is similar even if not identical to an electron at the relevant energies, whereas a successful detection of the hadrons requires special techniques and arrangements.

We consider below, in their general aspects, the main techniques that have been used to see these secondary particles and have lead to successful observations of neutrinos. Then we will mention briefly the other techniques and the future ones. A more detailed discussion of the specific detectors and of the results obtained is given in later sections.

Large Cherenkov Telescopes/Super-Kamiokande

There is a bluish radiation emitted by a charged particle that travels in water at a speed larger than the phase velocity of the light $\beta > 1/1/n(H_2O) = 0.75$. This was predicted by Heaviside and Vavilov, was first observed by Cherenkov and can be seen in the photos of the core of a nuclear reactor surrounded by water. The light is emitted in a cone of radiation forms that becomes narrower when the velocity increases. Its emission angle θ satisfies the relation $\cos\theta = 1/(\beta n)$; the limiting angle is 42° in water (43° in ice). Cherenkov radiation is not specific of this material, but water or ice are convenient for this type of low-energy neutrino telescopes, being easy to find and cheap. The water Cherenkov are the biggest low-energy neutrino telescopes.

The minimum condition that should be satisfied by a low-energy neutrino telescope, based on these detection principles, is to be able to detect this radiation and to associate it to neutrino reactions. For instance, even a set of individual photo-tubes can work as a low-energy neutrino telescope during the instants when a supernova neutrino burst reaches the Earth. This consideration applies to the gigantic IceCube detector that (in these conditions) can measure the counting rate of a galactic supernova over background (even if cannot measure the energy of the individual event nor their arrival direction). Much more information can be obtained if the neutrino reaction happens in an enclosure, whose walls are instrumented by photo-tubes that detect and measure the Cherenkov radiation. This is the case of Kamiokande, IMB and Super-Kamiokande; for definiteness and in view that it is the only one still

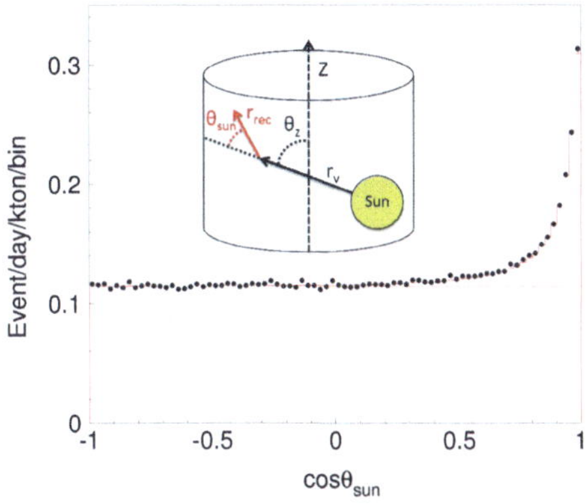

Fig. 6 The principle of solar neutrino detection in Super-Kamiokande and the distribution of the observed events, as a function of the cosine of the angle with the Sun. From [77]

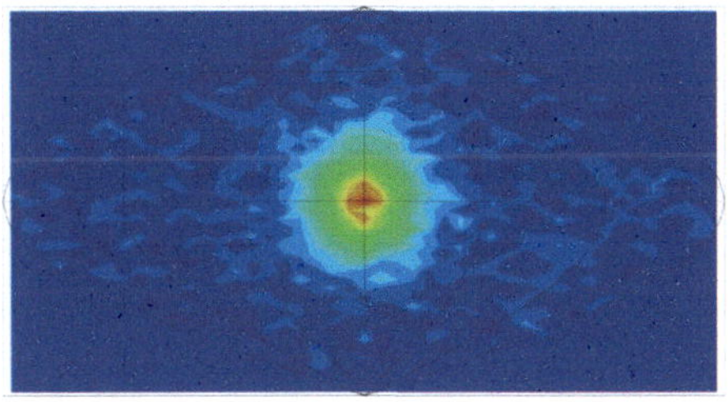

Fig. 7 Image of the Sun obtained by Super-Kamiokande, using the elastic scattering neutrino-electron events

operating, in the following we focus on the latter one, Super-Kamiokande [76]. The electron propagates and stops in the inner part of the detector, and the associated cone of radiation is projected on the walls, forming an observable ring. By measuring Cherenkov photons one can reconstruct not only the time of arrival but also the direction and the energy of the electron. The principle of this method and the results of the observations are shown in Fig. 6.

The Super-Kamiokande is currently the biggest[13] low-energy neutrino telescope of this type (50 kton) and there are plans to enlarge it significantly by building Hyper-Kamiokande [78]. Due to the presence of background processes, the minimum energy for the detection of electrons is not given by $\beta > 0.75$, namely $T_e = 260$ keV, but it is several MeV: the best value, recently achieved, is 3.49 MeV [79]. For this reason, only the higher energy solar neutrinos and supernova neutrinos, for what concerns neutrino astronomy, are currently observable. An impressive and well-known image of the Sun, obtained only by the electrons attributed to solar neutrino reactions, is shown in Fig. 7. Similar images will be obtained in the case of a galactic supernova [80–82]. It is hoped that something similar can be realized also with high-energy neutrinos, leading us to the identification of the sources of the events that have been seen by IceCube. These are the cases when neutrino astronomy is closer to ordinary astronomy even if the angular resolutions that can be attained, as a rule, are much worse.

[13]Only in the restricted sense discussed above, one could consider that IceCube is bigger.

Photons lines from nuclear reactions are also observable, if they have sufficiently high energy. This is the case of the lines from the neutrino-induced reaction when the oxygen nuclei are excited [83] that could allow us to see neutral current reactions of supernova neutrinos. Neutrons can be seen only with difficulties[14]: E.g., in the present configuration of Super-Kamiokande, the IBD reaction can be double tagged seeing $n + p \to D + \gamma(2.2\ \text{MeV})$ only in about 20% of the cases [85]. However, the idea of adding a "wave shifter" to make neutrons more easily visible, such as Gadolinium that absorbs neutrons and releases gammas of higher energy [84], is seriously considered [85].

It should be recalled that this type of detectors has been enormously successful for several achievements, including the study of solar and atmospheric neutrinos, the discovery (and study) of neutrinos oscillations, the long-baseline neutrino experiments, the investigation of matter stability, etc. However, the search for the signal from a (galactic) supernova is still considered as one of the main goals still to be achieved.

(Ultrapure) liquid scintillators/Borexino

Scintillator detectors are the older technique used to detect neutrinos. They enjoy of a rather high light yield, so that they can measure the energy of the events quite precisely, and can reach high levels of radiopurity that allow them to exploit a very low-energy threshold for detection. Unfortunately, it is not possible to reconstruct the direction of a single elastic scattering events.

There are several low-energy telescopes, based on the principle of liquid scintillation that are operating. These include Artyomosk (Ukraina), Baksan (Russia), LVD (Italy), SNO+ (Canada), KamLAND (Japan), and Borexino Italy). The largest telescopes in operation are LVD, SNO+ and KamLAND, and have one kton mass each and among them LVD is the only one dedicated to neutrino astronomy: Its superior stability is aimed to the purpose of observing a galactic supernova event.

Some emphasis should be given also to another dedicated neutrino telescope, Borexino, that despite being among the smallest ones, has achieved the highest levels of radiopurity and has attained very important results in neutrino astronomy. The main reasons are the excellent precision in the energy measurement and an impressively low-energy threshold,

$$\sigma(E = 1\ \text{MeV}) = 5\% \quad E_{\text{thr.}} = 150\ \text{keV} \tag{49}$$

Thanks to these achievements, all solar neutrinos can be studied (see [86] and the subsequent chapters for more discussion); in principle, also the elastic scattering reaction on proton $\nu + p \to \nu + p$ induced by a future supernova neutrino event can be studied. A recent review paper on Borexino, where an overview of the purification

[14]The detection of neutrons is not a problem for Borexino that is based on an ultra-pure scintillator or for SNO that used with special apparata for this purpose.

levels and of the software tools to reject backgroud is [87]. Borexino is considered a model for future developments, e.g., the proposed detectors LENA [88] and/or JUNO [89].

Comparing IBD performances

Let us compare the number of proton targets and the threshold of the previous two types of detectors.

For water Cherenkov telescopes, the fraction of protons per nucleon in water is just $f_{H_2O} = 1/9$, while for scintillators, it depends upon the composition. E.g., for linear alkyl benzene (LAB) with chemical formula $C_6H_5C_nH_{2n+1}$, thus $f_{\mathrm{LAB}}(n = 12) = 5/41$ that is just slightly larger than f_{H_2O}. The mass of the water Cherenkov detector Super-Kamiokande that can be used for the detection of supernova neutrinos is $M_{\mathrm{SK}} = 32$ kton, that has the same number of protons as 29.2 kton of LAB, namely

$$N_p = 2.14 \times 10^{33} \tag{50}$$

We see that, to date, Super-Kamiokande is the low-energy neutrino telescope that has (by far) the largest number of target protons.

Now, we discuss the energy threshold for the detection. For scintillators, we observe the kinetic energies of the final particles and the annihilation of the positron,

$$E_{\mathrm{prompt}} = E_{e^+}^{\mathrm{kin}} + 2m_e + E_n^{\mathrm{kin}} = E_{\bar{\nu}_e} - (m_n - m_p - m_e) = E_{\bar{\nu}_e} - 0.782\ \mathrm{MeV} \tag{51}$$

followed by the 2.22 MeV delayed signal from the formation of deuterium, upon neutron absorption, with a time $\tau = 200–250\ \mu$s that depends upon scintillators. For Cherenkov telescopes, the neutron is not as a rule observable (see also the discussion above) while the positron is considered just as an electron, thereby expressing the visible energy as,

$$E_{\mathrm{vis}} = E_{e^+} = E_{\bar{\nu}_e} - \frac{m_n^2 - m_p^2 - m_e^2}{2m_p} = E_{\bar{\nu}_e} - 1.294\ \mathrm{MeV} \tag{52}$$

The detection threshold in Super-Kamiokande can be kept as low as few MeV but ultra-pure scintillators as Borexino have proved that this can be lowered to 150 keV. The better performance at low-energy (due to higher light yield and lower background counting rate) makes scintillator detector very important or just essential for some experimental applications, even if at present their mass is comparably small.

Other low-energy neutrino telescopes

In the past, several types of low-energy neutrino telescopes have been used and they will be discussed later on to illustrate their importance for neutrino astronomy. This includes for instance radiochemical neutrino detectors (where neutrino reactions with nuclei produced new, radioactive species, used for their detection), that provided crucial information for the discussion of solar neutrinos but whose interest nowadays is considered mostly historical. The SNO detector was a special type of detector, since it was based on heavy water, and deuterons work as targets for neutrinos. It is widely recognized that such a detector would be in good approximation an *ideal* supernova neutrino detector.[15] Many other detectors were considered in the past in connection with solar neutrinos, see e.g., [2, 90]. Among the new detectors, we mention just the very large argon-based detectors that have special potential toward electron neutrinos. Moreover, also the new generation detectors devoted to dark matter search have an interesting application for supernova neutrinos, when their energy threshold allow the detection of the reaction of elastic scattering $\nu + (A, Z) \rightarrow \nu + (A, Z)$ [91].

High-Energy Neutrino Telescopes

Here, we discuss the principles to detect "high-energy neutrinos" from astrophysical sources that according to the previous discussion, denote the region of energies above 100 GeV-1 TeV. However, the same principles can be applied to search other hypothetical sources of neutrinos.

Let us begin with a citation: *We propose setting up apparatus in an underground lake or deep in the ocean in order to separate charged particle directions by Cherenkov radiation. We consider μ mesons produced in the ground layers under the apparatus* [295]. With these words the Russian physicist Moisey Alexandrovich Markov introduced, in 1960, to the scientific community the idea of underwater Cherenkov neutrino telescopes. Neutrinos can be detected only when they interact with matter inside the volume of a detector (or, sometimes, as we will see, in hits proximity).

High-energy neutrino interactions

A neutrino-nucleon interaction can occur via the exchange of a charged boson ($W^{\pm}$) between the leptonic and the hadronic currents:

$$\nu_\ell(\bar{\nu_\ell}) + N \rightarrow \ell^-(\ell^+) + X \tag{53}$$

the charged current (CC) interaction where ℓ indicates the "lepton family" (e, μ, τ) and X the hadronic shower, or via the exchange of the neutral boson Z:

$$\nu_\ell(\bar{\nu_\ell}) + N \rightarrow \nu_\ell(\bar{\nu_\ell}) + X \tag{54}$$

the neutral current (NC) interaction.

[15] However, it is not easy to gather large masses of heavy water for this purpose and there are no plans to operate a similar type detector in the near future.

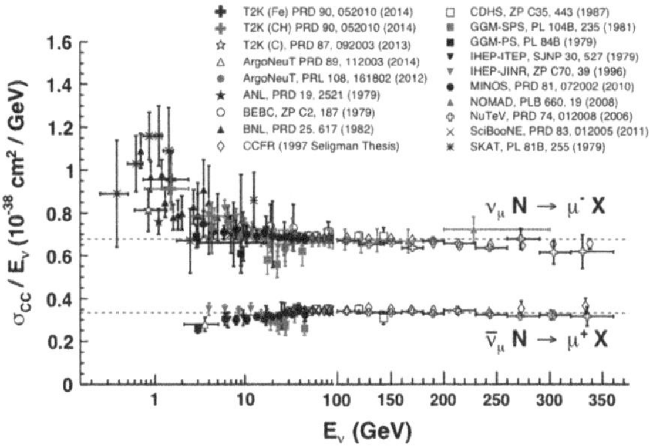

Fig. 8 Measurements of the neutrino-Nucleon (and antineutrino-Nucleon) interaction cross sections divided by the neutrino Energy

Assuming that the target is isoscalar (namely, the nuclei consist of an equal amount of protons and neutrons), the leading order differential cross section for the CC deep-inelastic scattering (DIS) per nucleon, (given previously for parton) can be expressed as function of the Bjorken variables:

$$\frac{d^2\sigma^{\nu}_{CC}}{dxdy} = \frac{2G_F^2 m_N E_\nu}{\pi}\left(\frac{m_W^2}{Q^2+m_W^2}\right)^2\left[x\cdot q(x,Q^2)+x\cdot\bar{q}(x,Q^2)(1-y)^2\right]$$
$$\text{where } x = \frac{Q^2}{2m_N(E_\nu - E_\ell)} \text{ and } y = \frac{E_\nu - E_\ell}{E_\nu} \tag{55}$$

where Q^2 is the invariant squared 4-momentum transferred between the incident neutrino and the outgoing lepton, $G_F = 1.1663210^{-5}\,\text{GeV}^{-2}$ is the Fermi constant and $q(x, Q^2)$, $\bar{q}(x, Q^2)$ are sums of parton density functions for quarks and anti-quarks, respectively. Accordingly to the above expression the ν-nucleon cross section, when $Q^2 \ll m_W^2$ (i.e., for energies $< 10^4$ GeV), has a very small value that rises linearly with the energy. In Fig. 8 are shown the values of $\frac{\sigma^{\nu}_{CC}}{E_\nu}$ and $\frac{\sigma^{\bar{\nu}}_{CC}}{E_{\bar{\nu}}}$ measured with neutrino beams: for $E_\nu \sim 100$ GeV the $\sigma^{\nu}_{CC} \sim 10^{-36}$ cm^2/nucleon so, in order to intercept the faint flux of astrophysical neutrinos a neutrino telescope has to be huge and massive, since a large amount of target nucleons are necessary to produce a neutrino interaction. The order of magnitude of the neutrino flux expected by an astrophysical source and potentially detectable is,

$$\Phi(\nu) = 10^{-11}\cdot E_\nu^{-2} \text{ neutrinos } TeV/(\text{cm}^2\,\text{s}) \tag{56}$$

Rate of interaction

As discussed previously (see, e.g., Fig. 5) the flux of astrophysical neutrinos is expected to decrease with energy much less than the atmospheric neutrino flux that decreases as $\frac{d\Phi^{atm.}_{\nu}}{E_\nu} \sim E_\nu^{-3.7}$ (in astrophysical parlance, one says that the former is harder and the latter is softer). Thus, we expect that the flux of astrophysical neutrinos exceeds the one due to atmospheric neutrinos at sufficiently high energies.

We can estimate quickly (even if not very precisely) the dimension of the detector needed to have a few neutrino events. Let us consider astrophysical neutrinos with $E_\nu = 1$ TeV, and let us assume the flux indicated in Eq. 56. Let us evaluate the number of interactions that we can collect in 1 year ($T = 3.15 \times 10^7$ s) in a volume V of water, with $\rho = 1$ gr/cm^3,

$$\begin{aligned} N_{\text{events}}(1\ \text{TeV}) &= \Phi(\nu) \times \sigma_{CC}^{1TeV} \times N_A \times V \times \rho \times T \\ &= 10^{-11} \times 0.7 \times 10^{-35} \times 6.02 \times 10^{23} \times V \times 1 \times 3.15 \times 10^7 \\ &= 1.3 \times 10^{-15} \cdot (V/\text{cm}^3) = 1.3 \cdot (V/\text{km}^3) \end{aligned} \quad (57)$$

Several considerations make the detection of astrophysical neutrinos easier at energies ≥ 1 TeV:

- The neutrino interaction cross section grows linearly with energy up to ~ 10 TeV.
- The larger energy of the escaping lepton (in particular for ν_μ CC-interactions) makes it easier its identification and the reconstruction of its direction.
- If the flux of astrophysical neutrinos decreases with energy $\frac{d\Phi_\nu^{cosmic}}{E_\nu} \sim E_\nu^{-(2\div 2.2)}$ at high energy the signal/background ratio improves greatly.

Water or ice: a forced choice

Let us follow Markov suggestions, and consider the possibility to detect Cherenkov light induced by relativistic muons produced in neutrino interactions in a transparent medium like *an underground lake or deep in the ocean* (or, also ice, now we known it!). Taking into account that the detector has to be very large, we arrive easily to the decision to employ as a natural target the oceanic water, or the Antarctic ice: no man-made laboratory could ever be large enough to host larger apparata.

A large volume ($\approx$km^3) in deep-sea, or in the ice of the South Pole, is sufficiently shielded from atmospheric radiation to allow the detection of the neutrinos. The Cherenkov light, due to the passage of charged particles created after neutrino interactions, is possible thanks to an arrays of photo-multiplier tubes (PMTs). The PMTs are housed in transparent pressure spheres which are spread over a large volume in oceans, lakes or glacial ice. Each PMT records the arrival time and the amplitude of the signal, sometimes even its full waveform, due to the Cherenkov light released by the muons or by other charged particles. In the typical design of a detectors, the spheres that host the PMTs are attached to vertical strings. The typical PMT spacing along a string is 10–20 m, and the distance between adjacent strings 60–150 m.

A detailed description of the detector principles is given in Appendix 5.

Possible signals

An underwater (or under-ice) neutrino telescope in principle is able to detect all neutrino types (ν_μ, ν_e, ν_τ) by means of CC or NC interactions taking place inside or near the instrumented volume. It is possible to distinguish two main types of

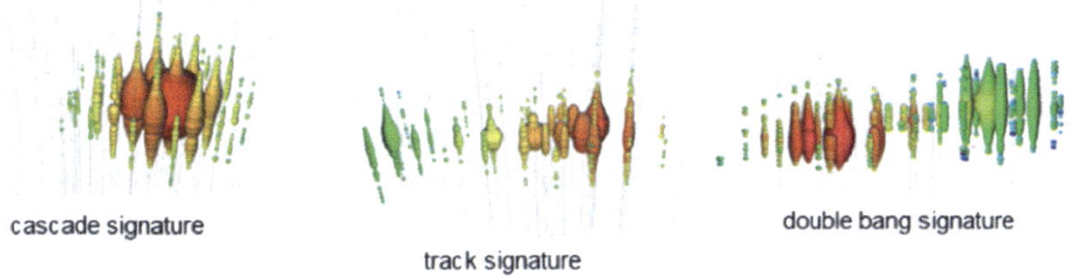

Fig. 9 Three different neutrino event patterns in IceCube according to their simulations. Each symbol indicates a PMT hit by the Cherenkov light. The size of each symbol is proportional to the deposited energy, the color coding (from *red* = early hits) to *green* and to *blue* (later hits) illustrates the hits arrival time. On the *left* is a "cascade event", at the *center* a "track like event", on the *right* "a double bang" like event that could be the signature of a Ultra High-Energy tau neutrino charged current interaction [362]. The first two classes of events have been actually observed that last one not yet

signatures: the track-like events, mainly from ν_μ CC-interaction, and the shower-like ones from $\nu_{\mu,e,\tau}$ NC-interactions and $\nu_{e,\tau}$ CC-interactions. All these interactions can lead to a signal visible in the detector. Figure 9 sketches the experimental signatures of this type of signal in an underwater/ice neutrino telescope—in this case, the IceCube detector.

When a ν_μ has sufficient energy, it gives rise to a high energy muon that moves almost in the same direction as the initial neutrino. The muon does not ionize much and being long-lived, can propagate for long distances, as discussed in Appendix 5; thus, it can be used as a detectable signal, called passing muons, or through-going muons[16] even when the interaction vertex is not contained inside the detector. However, two conditions have to be verified: (1) the interaction vertex should happen close enough to the detector, so that the muon arrives in the detector with sufficient energy to be seen and (2) the muon should arrive from the region under be the detector, namely, the region where the atmospheric muon background is rare or absent; in fact, the flux of atmospheric muons crossing the detector exceeds by several order of magnitude the flux of muons originated by cosmic and atmospheric neutrinos, see Fig. 11.

Still it is not possible to use this signal to search for cosmic neutrinos, unless the background due to atmospheric neutrinos is kept under control. The signal events will be searched in the sample of tracks reconstructed as "up-going" and for instance characterized by their correlation to a specific celestial source, or having a sufficiently high energy that makes it unlikely that they are due to atmospheric neutrinos. The above discussion is summarized in Fig. 10 and quantified in a subsequent chapter; a detailed discussion of the behavior of high-energy muon in water/ice is instead given in Appendix 5 (Fig. 11).

[16]This is the first type of signal that has been considered and studied.

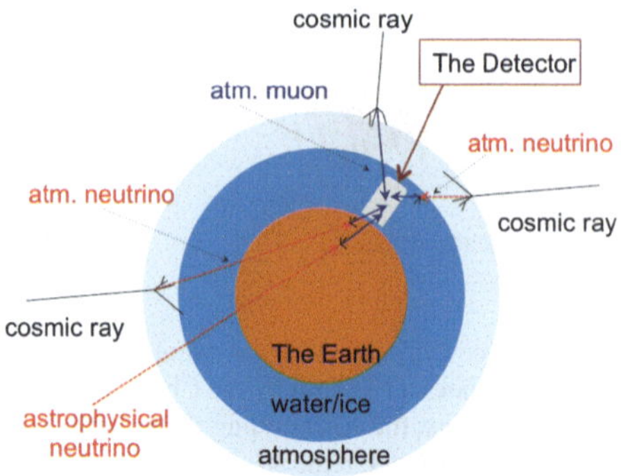

Fig. 10 Schematic view of signal and background events: muons (*blue tracks*) originated in atmospheric showers can reach the detector only from the "above space" while the signal and muons from atmospheric neutrinos can be detected as "up-going" as well "down-going" tracks

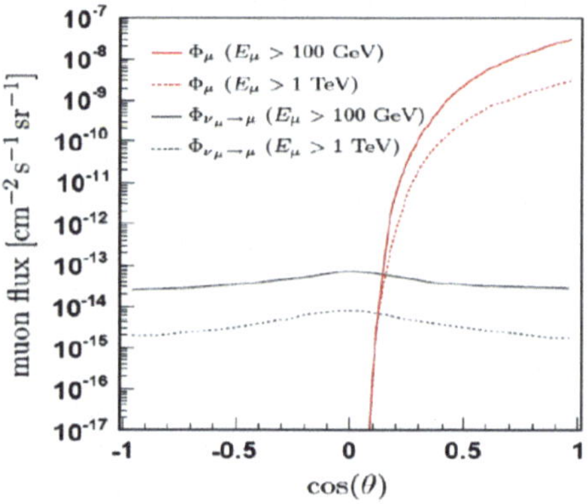

Fig. 11 Flux of atmospheric muons (*red*) and of muons from CC interactions of atmospheric and cosmic neutrinos in a submarine detector at 2.1 km water equivalent depth, as function of the cosine of the muon zenith angle

Effective areas for neutrino detection

In this paragraph, we describe in some details the key quantity to evaluate the signal in a high-energy neutrino detector: the effective area.

The relation between the flux of neutrinos from a point source and the rate of events of a certain class in a detector, $\dot{N}_{\rm c}$, can be written as:

$$\dot{N}_{\rm c} = \int dE_\nu \; \phi_{\nu_\alpha}(E_\nu) \; A_{\rm c}^{\nu_\alpha}(E_\nu, \Omega) \; . \tag{58}$$

In this equation $\phi_{\nu_\alpha}(E_\nu)$ is the differential flux for neutrinos of flavor ν_α in the detector, after accounting for Earth absorption. The quantity $A_{\rm c}^{\nu_\alpha}(E_\nu, \Omega)$ has the dimension of an area, and is called the "effective area"; the subscript "c" is a label that identifies the event class, e.g., "shower events", "track events", "double-bang events" of Fig. 9. The effective area is a function of the neutrino energy (E_ν) and direction (Ω), and it is determined by the class of events considered, the neutrino type and the detector properties.

To compute the events from the source in a finite time interval $[t_i, t_f]$ one has to integrate the event rate in time taking into account the fact that the neutrino flux can be time dependent. For a source with fixed celestial coordinates the direction Ω is a function of time because of the Earth rotation. We have,

$$N_{\rm c}(t_1, t_2) = \int_{t_1}^{t_2} dt \int dE_\nu \, \phi_{\nu_\alpha}(E_\nu, t) \, A_{\rm c}^{\nu_\alpha}(E_\nu, \Omega(t)) \, . \tag{59}$$

The time dependence of the flux can be due to a time variation of the source, but even in the absence of such an intrinsic variation one has to take into account the fact that the flux in Eq. (58) is the flux at the detector position and at high energy it is distorted by absorption effects for neutrinos that traverse the Earth. This is a time dependent effect, except in the case when the detector is located in the Pole, such as IceCube, when the amount of terrestrial matter traversed remains constant during the day.

The rate associated to a diffuse flux can be calculated with an obvious generalization of Eqs. (58) and (59) integrating over the desired angular region.

The calculation of the effective area requires in general a detailed simulation of the detector response, but in several cases it is possible to obtain a reasonably accurate estimate with simple arguments. An example is the observation of "shower events" generated by ν_e, with visible energy above $E_{\rm min}$ and interaction vertex inside the fiducial volume $V_{\rm fid}$. The effective area vanishes for $E_\nu < E_{\rm min}$, and at higher energy can be estimated in reasonable approximation as:

$$A_{\rm showers}(E_\nu; E_{\rm min}) \simeq [\sigma_{\rm cc}(E_\nu) + \sigma_{\rm nc}(E_\nu) \, f_{\rm nc}(E_\nu, E_{\rm min})] \, \frac{M_{\rm fid}}{m_N} \, . \tag{60}$$

In the expression $M_{\rm fid}$ is the mass contained in a (suitably defined) fiducial volume, and m_N is the nucleon mass, so that the ratio $M_{\rm fid}/m_N$ is the number of nucleons in the fiducial volume. The effective area has two contributions from charged current and neutral current interactions. In the first case the entire neutrino energy is "visible", in the form of an electromagnetic shower generate by the e^- in the final state plus an hadronic shower; in the second case the fraction of the ν energy that is transfered to the final state neutrino is not detectable and we see only the hadronic fraction. The factor $f_{\rm nc}(E_\nu, E_{\rm min})$ is the fraction of NC interactions where the energy of the final state neutrino is sufficiently low ($E_{\nu,{\rm fin}} < E_\nu - E_{\rm min}$). In conclusion one obtains the simple result that the effective area is simply the appropriate cross section (per nucleon) times the number of nucleons in the fiducial volume. In this approximation, the effective area is independent from the neutrino direction and from the geometry of the detector and depends only upon the detector fiducial mass.

The calculation of the effective area for ν_μ to generate "track events" with minimum visible energy $E_{\rm min}$ is more difficult. The event selection criteria require the creation of a muon (and therefore a charged current interaction) and also that the muon has a sufficient high energy to form an identifiable "track"; moreover, in most cases the muon will exit the detector, so that only part of the neutrino energy is deposited in the detector and measurable. In this case, it is easy to convince oneself that, differently from the previous case, the effective area depends in an essential manner upon the geometry of the detector and the direction of the neutrino flux.

A very important case is the effective area for neutrino induced muons that is for those events where the neutrino (ν_μ or $\overline{\nu}_\mu$) interacts (with a charged current) interaction outside the detector and the $\mu^\mp$ is observed entering the detector volume. To a good approximation the effective area can be calculate in two basic steps. The first one is the calculation of the "muon yield" that is the average number of muons generated by a ν_μ (or $\overline{\nu}_\mu$) of a certain energy. For the second step, one notes that the muons (of direction Ω) are observed when their trajectories cross the detector in a (transverse) area $A_{\rm det}(\Omega)$.

The effective area for the observation of muons with minumum energy $E_{\rm min}$ can then be estimated as:

$$A_{[\mu^-\uparrow]}(E_\nu, E_{\rm min}, \Omega) \simeq P_{\nu_\mu \to \mu^-}(E_\nu, E_{\rm min})\, A_{\rm det}(\Omega) \tag{61}$$

The muon yield can be calculated as:

$$P_{\nu_\mu \to \mu^-}(E_\nu, E_{\rm min}) = \frac{\rho}{m_N} \int dE_0\, \frac{d\sigma_{\rm cc}^{\nu_\mu}}{dE_0}(E_\nu, E_0)\, R(E_0, E_{\rm min}) \tag{62}$$

In this equation ρ is the density of the material below the detector, $d\sigma_{\rm cc}/dE_0$ is the differential charged current interaction for the production of a muon of energy E_0 and $R(E_0, E_{\rm min})$ is the range[17] for a muon of initial energy E_0 to have final energy $E_{\rm min}$. Equation (62) has a very simple physical meaning. Each muon energy in the differential cross section is weighted by its range. Qualitatively then (and for $E_{\rm min} \ll E_\nu$) :

$$P_{\nu_\mu \to \mu^-}(E_\nu, E_{\rm min}) \approx \frac{\rho}{m_N}\, \sigma_{\rm cc}(E_\nu) \times R(\langle E_\mu \rangle) \tag{63}$$

Using Eq. (61) one finds that the effective area for neutrino-induced muons is the neutrino cross section times the number of nucleons contained the volume formed by the transverse area of the detector (projected in the direction Ω) times the length of the muon range at $E_\mu \approx E_\nu/2$. The energy dependence of the effective area for neutrino-induced muons is therefore the product of the energy dependence of the cross sections and the muon range. For $E_{\nu,\mu} \lesssim 1$ TeV, $\sigma_{\rm cc}(E_\nu)$ and $R(E_\mu)$ are both linear, and the effective area grows quadratically.

Note also that in many cases the number of nucleons contained in the "external volume", where the interactions that generate the muons take place, is larger than the number of nucleons in the detector fiducial volume: this happens when the muon range is longer that the linear size of the detector. In these cases the rate of ν-induced muon events is larger than the rate of events where the neutrino interacts in the detector.

[17]This is measured, e.g., in cm, so that the product $x = \rho \times R$ is measured in g/cm^2.

Local (=Terrestrial) Phenomena

In the next three sections, we will examine several type of neutrinos coming from extraterrestrial sources. However, there are several terrestrial phenomena that give rise to neutrinos. These have their own interest, they are ground for some unsolved problems, and most important for us, in some occasions they are also important backgrounds for the observation of cosmic neutrino. We will refer to them generically as *local phenomena* and we discuss them in this section.

At relatively high energies, neutrinos is mostly due to the collisions of cosmic rays with the Earth's atmosphere—*atmospheric neutrinos*—studied and observed since the sixties (section "Atmospheric Neutrinos"). At low (MeV) energies, there are the antineutrinos from the disintegration (beta decay) of radioactive nuclei—*geoneutrinos*—predicted in the same years but the seen only in the last decade (section "Neutrinos from Terrestrial Radioactivity"). Moreover, there are various *artificial sources* of neutrinos, ranging from those produced in reactors as byproducts (but important also for several scientific investigations) to those produced for the purpose of studying neutrinos in accelerators at larger energies (section "Man-Made Sources").

General references:

A book to understand the basic aspects of atmospheric neutrinos is [9]. Excellent reviews on geoneutrinos are [26–28]. A useful reference on the relevant aspects of reactor neutrinos is [29] while a recent review on artificial neutrino beams and experiments is [30].

Atmospheric Neutrinos

Atmospheric neutrinos are produced in the showers generated by primary cosmic ray protons and nuclei in the Earth atmosphere. Some of the unstable secondary particles produced in the hadronic interactions of the shower particles with the air nuclei (in reactions such as $p + A_{\rm Air} \to p, n, \pi^{\pm}, \pi^{\circ}, K^{\pm}, \ldots$) and decay into neutrinos. The dominant source of neutrinos is the chain decay of $\pi^{\pm}$:

$$\begin{array}{ccl} \pi^{+} \to & \mu^{+} + \nu_{\mu} & \\ & \downarrow & \\ & e^{+} + \nu_e + \overline{\nu}_{\mu} & \end{array} \tag{64}$$

(and the charge conjugate channel). The second important neutrino source is the decay of charged and neutral kaons. At very large energy ($E \gtrsim 10^5$ GeV), when the decay probability of pions and kaons is strongly suppressed, one expects that the Weak decays of particles containing charm (D mesons and the Λ_c baryon) become the dominant source of atmospheric neutrinos.

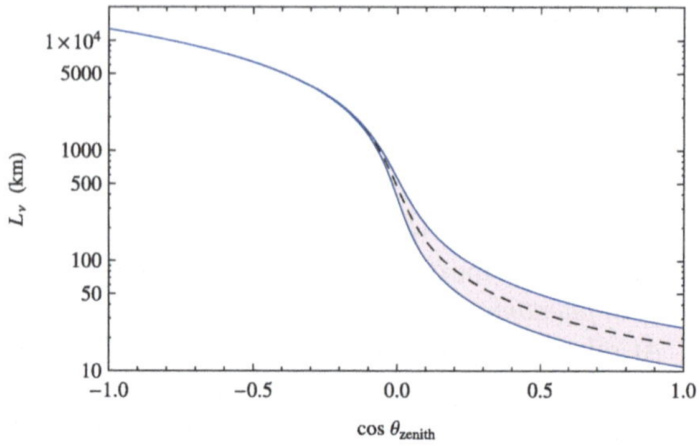

Fig. 12 Pathlength of atmospheric neutrinos as a function of the zenith angle

The interaction cross section for neutrinos with energy below $E_\nu \simeq 10^5$ GeV is sufficiently small, so that they can traverse the Earth with a negligible absorption probability. This has the consequence that a neutrino detector can observe particles coming from all directions that is both down-going and up-going neutrinos. Down-going particles have an average pathlength of order $\simeq$10 km for vertical neutrinos and $\simeq$500 km for horizontal ones, while the pathlength for up-going neutrinos is to a good approximation $L \simeq 2\, R_\oplus\, |\cos\theta_{\rm zenith}|$ (where $R_\oplus \simeq 6371$ km is the Earth radius). This situation, where it is possible to compare measurements of neutrinos that have traveled different path-lengths is ideal for the study of neutrino flavor oscillations, see Fig. 12. In fact the measurements of atmospheric neutrinos performed by Super-Kamiokande [92] and other detectors [93–96] have allowed the discovery of the existence of neutrino oscillations.

In order to extract the neutrino properties from the observations of the neutrino fluxes, one needs to compare the experimental results to theoretical predictions calculated with and without the effect of oscillations. Detailed discussion of such calculations can be found in [97, 98] It is however possible to obtain the main properties of the atmospheric neutrino fluxes: the flavor composition, and the energy and angular distributions, with simple and robust arguments.

Flavor Composition, Energy Spectra and Angular Distributions

About the angular distribution, a crucially important point is that in the absence of oscillations the atmospheric neutrino fluxes, to a good approximation, up–down symmetric:

$$\phi_{\nu_\alpha}(E_\nu, \theta_\nu) = \phi_{\nu_\alpha}(E_\nu, \pi - \theta_\nu) \tag{65}$$

This result can be derived as a theorem (see Fig. 13) using three assumptions: (i) the Earth is a sphere, (ii) the primary cosmic ray fluxes are isotropic, and (iii) neutrino absorption in the Earth is negligible. Neutrinos that enter a sphere with zenith angle $\theta_z^{\rm down}$ will exit the sphere with zenith angle $\theta_z^{\rm up} = \pi - \theta_z^{\rm down}$ (see Fig. 13).

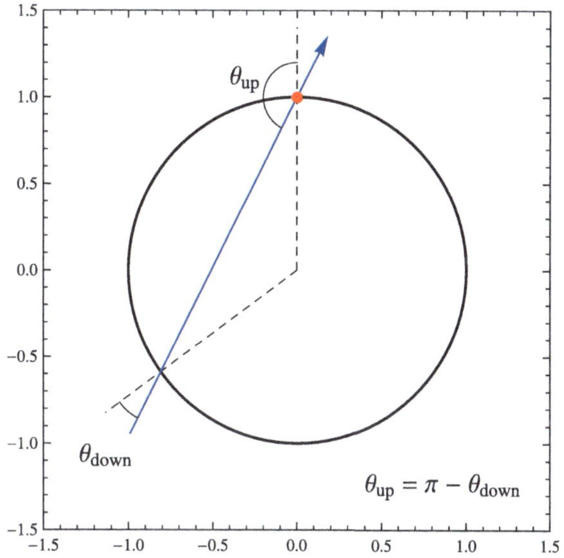

Fig. 13 Geometrical origin of the up–down symmetry of the atmospheric neutrino fluxes (in the absence of flavor oscillations)

Neglecting the absorption of neutrinos inside the Earth (an effect order $\sim 10^{-6}$ for vertical $\nu_{\mu,e}$ at 1 GeV), the rate of neutrinos that enter the Earth with a given energy and zenith angle θ_z is therefore identical to the number of neutrinos of the same energy that exits with zenith angle $\pi - \theta_z$:

$$\dot{N}^{\rm in}_{\nu_\alpha}(E, \theta_z) = \dot{N}^{\rm out}_{\nu_\alpha}(E, \pi - \theta_z) \tag{66}$$

If the flux of primary cosmic rays is isotropic, the flux of neutrinos is equal for all points on the Earth, and can be related to the rate of particles that enter (or exit) the surface of the Earth:

$$\phi_{\nu_\alpha}(E, \theta_z) = \frac{\dot{N}^{\rm in(out)}_{\nu_\alpha}(E, \theta_z)}{4\pi\, R^2_\oplus\, |\cos\theta_z|} \tag{67}$$

where $4\pi\, R^2_\oplus$ is the Earth surface and the factor $|\cos\theta_z|$ takes into account the orientation of the surface with respect to the neutrino direction. The equality of the fluxes follows immediately.

It is clear that this prediction of up–down symmetry for the neutrino fluxes, offers a gold-plated method to study flavor oscillations, since the path-lengths for up-going and down-going neutrinos differ by a large factor ($\approx 10^3$ for the vertical directions), and in the presence of oscillations the up-going and down-going fluxes can be modified in observably different ways, with the result of a violation of the up–down symmetry. Small deviations from exact up–down asymmetry are present also in the absence of oscillations, and are the consequence of the fact that the assumptions outlined above are not rigorously correct. The largest effect is due to the fact at low energy ($E_0 \lesssim$ few GeV) the primary cosmic rays are not isotropic because of the effects of the geomagnetic field that prevents low rigidity particles from reaching the surface of the Earth, generating the so called latitude and east–west effects. introducing anisotropies in the primary cosmic ray flux. The cosmic ray flux is in fact larger at the magnetic poles (where particles or arbitrarily small rigidity can arrive), than at the magnetic equator (where the flux of primary particles is reduced up to maximum rigidity $R \lesssim 60$ GV). For a detector placed near the magnetic equator (like Super-Kamiokande in Japan), this results in a suppression of the down-going hemisphere relative to the up-going one. The exact location of a detector should also be considered, including the effects of the topography of the mountains above it (that introduce a small suppression of low energy the down-going flux). The latitude (and time) dependence of the atmosphere density profile also introduce small but observable effects. At very large energy ($E_\nu \gtrsim 10^5$ GeV), one has to take into account the effects of neutrino absorption in the Earth.

From the fact that charged pion decay is the dominant source of atmospheric neutrinos (see Eq. (64), it is simple to deduce the flavor composition of the fluxes. The crucial point is that the muon decay channel ($\pi^\pm \to \mu^\pm \nu_\mu(\overline{\nu}_\mu)$) accounts to nearly 100% of the charged pion decays, with the mode $e^\pm \nu_e(\overline{\nu}_e)$ having only a branching ratio $\simeq 1.2 \times 10^{-4}$. At low energy, when all muons decay in air, one finds that the atmospheric neutrino fluxes have in reasonably good approxmation the flavor ratios:

$$\nu_\mu : \overline{\nu}_\mu : \nu_e : \overline{\nu}_e \simeq 1 : 1 : 0.55 : 0.45 \tag{68}$$

The ratio $\nu_\mu/\overline{\nu}_\mu$ is approximately unity because each charged pion contributes one ν and one $\overline{\nu}$ of muon flavor. The muon electron ratio $(\nu_\mu + \overline{\nu}_\mu)/(\nu_e + \overline{\nu}_e) \simeq 2$ reflecting the fact that after chain decay each pion contributes two muon neutrinos and one electron neutrino. Finally the ratio $\nu_e/\overline{\nu}_e$ is larger than unity because in the cosmic ray showers there is an excess of π^+ over π^- (this follows from the fact that cosmic rays are mostly protons). These results are less obvious than one can naively think, because they do not refer simply to the total (energy integrated) number of neutrinos, but describe the atmospheric ν flavor composition at any energy (and direction). The ratios in Eq. (68) are the consequence of the fact that in the two body decay of a relativistic pion the neutrino takes approximately one quarter of the energy (more exactly $\langle E_{\nu_\mu}\rangle/E_{\pi^+} = (1 - r_\pi)/2 \simeq 0.213$, with $r_\pi = (m_\mu^2/m_\pi)^2$), while the muon carries the rest of the energy that is that is then shared approximately equally among the three particles of the muon decay final state. (more exactly, and taking into account the effects of muon polarization, one has: $\langle E_{\nu_e}\rangle/E_{\pi^+} = (2 + r_\pi)/10 \simeq 0.257$) and $\langle E_{e^+}\rangle/E_{\pi^+} = \langle E_{\overline{\nu}_\mu}\rangle/E_{\pi^+} = (3 + 4\,r_\pi)/20 \simeq 0.265$) so that the three neutrinos end up having approximately the same energy.

The muon rest frame lifetime is $c\,\tau_\mu \simeq 0.65$ km, and therefore most muons with energy above a few GeV reach the ground before decay, and therefore for $E_\nu \gtrsim 10$ GeV the fluxes of ν_e and $\overline{\nu}_e$ are suppressed, and the ratios of Eq. (68) are not valid. At these energies the dominant source of electron (anti)-neutrinos is the decay of charged and neutral kaons in the three body modes $K^\pm \to \pi^\circ e^\pm \nu_e(\overline{\nu}_e)$ and $K_L \to \pi^\mp e^\pm \nu_e(\overline{\nu}_e)$ (kaons contribute also the muon (anti)-neutrino fluxes with the analogous three body modes, and the two body decay of charged kaons).

Fig. 14 The *top panel* shows the energy spectra of vertical down-going neutrinos calculated by Honda et al. [99]. The *four curves* are for ν_μ, $\overline{\nu}_\mu$ ν_e and $\overline{\nu}_e$. The *botton panel* shows the zenith angle dependence of the atmospheric ν_μ flux in the same calculation

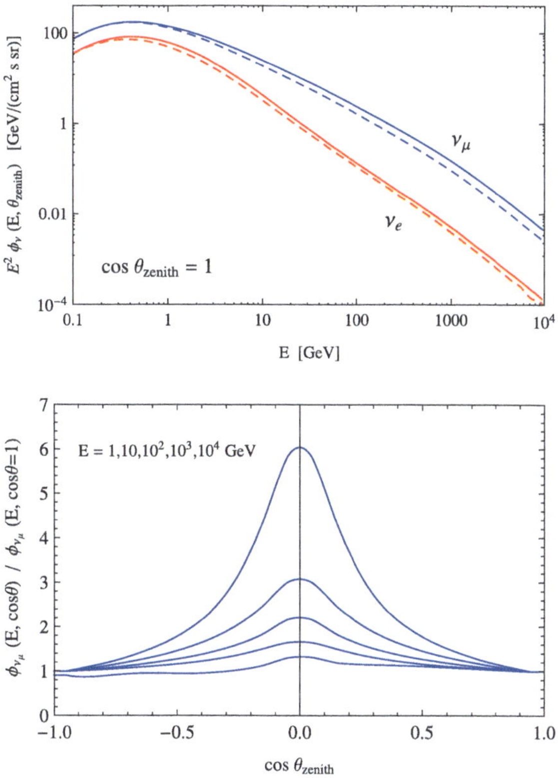

Figure 14 shows the energy spectra and the angular distribution of atmospheric neutrinos in the calculation of Honda et al. [99]. One can see that the spectra of the neutrinos above $E_\nu \gtrsim 1$ TeV falls approximately $\propto E^{-3.7}$ that is they have a spectral index that is one unit larger than the spectral index of the primary cosmic rays. This is a consequence of the fact that high-energy mesons have a small decay probability that (at high energy) decreases approximately $P_{\rm dec} \propto E_\nu^{-1}$. Also the angular distribution that shows a maximum for horizontal directions, is a consequence of a larger decay probability for horizontal mesons. In fact, in the calculation of the atmospheric neutrino fluxes it is essential to consider the competition between decay and interaction for the particles that generate the neutrinos. As an illustration, the interaction length in air of charged pions of energy $E_\pi \simeq 100$ GeV is of order $\lambda_{\rm int} \simeq 120$ g/cm^2, this corresponds at an height of 15 km, representative of the maximum of pion production, to a length $\ell_{\rm int} \simeq 5.60$ km (at sea level where the air density is higher, $\lambda_{\rm int}$ is approximately five times shorter). The pion interaction length at 100 GeV is approximately equal to the decay length $\ell_{\rm dec} \simeq 5.44$ km $E_\pi/(100$ GeV). The interaction length varies only slowly with energy, while the decay length is linear in E_π because of the Lorentz time dilatation, accordingly the pion decay probability $P_{\rm dec} = (1 + \ell_{\rm dec}/\ell_{\rm int})^{-1}$ is approximately unity for $E_\pi \le 100$ GeV, while it decreases

$P_{\rm dec} \propto E_\pi^{-1}$ at large energy. A more detailed calculation has to take into account the distribution in height of the mesons generated in cosmic ray showers and the profile of the air density. The results are reasonably well fitted by the expression:

$$P_{\rm dec}^{(a)} \simeq \left[1 + \frac{E\,\cos\theta(h^*)}{\varepsilon_a}\right]^{-1} . \qquad (69)$$

In this expression

$$\varepsilon_a \simeq \frac{m_a\,h_0}{c\,\tau_a} \qquad (70)$$

where $h_0 \simeq 6.4$ km the scale height of the stratosphere (where $\rho(h)$ is approximately exponential), and m_a and τ_a are the mass and lifetime of the particle, and $\theta(h^*)$ is the zenith angle of the particle trajectory at the height $h^* \simeq 20$ km that is related to the zenith angle of the trajectory at sea level θ_0 by:

$$\cos\theta(h) = \sqrt{1 - \sin^2\theta_0\,\frac{R_\oplus^2}{(R_\oplus + h)^2}} . \qquad (71)$$

The critical energies for $\pi^\pm$, $K^\pm$ and K_L are 115, 840, and 210 GeV.

Atmospheric Neutrino Detection

The event rate for atmospheric neutrinos is of order ~ 0.5 events/(kton day), and to have appreciable rates one needs very massive detectors, moreover, to reduce the background due to secondary cosmic rays that at the surface (at sea level) have a flux ~ 200 particles/(m^2 s), the detectors have to be placed deep underground. The component of γ, $e^\pm$ and hadrons can be absorbed with few meters of rock, but to reduce significantly the flux of the penetrating muons, one needs a coverage of order of 1 km of rock (or equivalently 2.7 km of water). Several type of detectors for atmospheric neutrinos have been constructed such as iron calorimeters, water Cherenkov such as Kamiokande and Super-Kamiokande.

The most abundant class of atmospheric neutrino events are quasi-elastic interactions, with an energy E_ν around one GeV.

$$\begin{aligned} \nu_\ell + n &\to \ell^- + p \\ \overline{\nu}_\ell + p &\to \ell^+ + n \end{aligned} \qquad (72)$$

These interactions can be detected as fully contained events in the detector fiducial volume. With increasing E_ν events with the production of additional particles in the final state become first possible, and then dominant.

A different method to detect fluxes of ν_μ and $\overline{\nu}_\mu$ is the observation of neutrino-induced upward-going muons. In these events the neutrino interaction happens in the rock below the detector, and the $\mu^{\mp}$ produced in charged current interactions propagate to the detector. In through-going events the muons cross the detector, while in stopping events the muons range out inside the detector volume. Because of the up–down symmetry of the atmospheric neutrino fluxes one expects (in the absence of oscillations, and if the detector is placed at a sufficient depth) an equal number of ν-induced up-going and down-going muon events, however down-going events cannot be separated from the flux of atmospheric muon events, where the muon is produced directly in the atmosphere, and therefore only up-going events can be used to study the neutrino fluxes, with the exception of exceptionally deep detectors. For a threshold energy of order 1 GeV, The flux of ν—induced up-going events is of order:

$$\langle \phi_{\mu\uparrow} \rangle \simeq 2 \times 10^{-13}\ (\text{cm}^2\ \text{s sr})^{-1} \simeq 60\ [(1000\ \text{m}^2)\ \text{year sr}]^{-1} \tag{73}$$

For a rock overburden of order $h \sim 1$ km, the atmospheric muon flux is still approximately 10^5 times larger than the ν-induced one. The two fluxes become comparable for a thickness of approximately 5 km of rock.

The range of a muon in rock is $R_\mu \simeq 1.7\ E_{\text{GeV}}$ meters, therefore high-energy neutrinos can produce up-going muons even if they interact far from the detector. For ν_μ and $\overline{\nu}_\mu$ with energy larger than $\gtrsim 10$ GeV (the precise number depends of course on the dimension and geometry of the detector) the event rate in up-going muons is larger than the rate of interactions inside the detector.

Up-going muons have been the first type of atmospheric neutrino events to be observed with simple detectors placed in very deep mines is South Africa [106] and in India [105]. Larger samples of up-going muons have been then obtained by several detectors.

A discussion of the entire data set of Super-Kamiokande on atmospheric and its interpretation can be found in [107–109].

In recent years very interesting measurements of atmospheric neutrinos have been performed by the new high-energy telescopes like IceCube [110–112] and ANTARES [113, 114].

Atmospheric Neutrinos and Flavor Oscillations

The study of atmospheric neutrinos has lead to the discovery of neutrino flavor oscillations. In fact we now know that the neutrino squared mass difference $\Delta m^2_{23} \simeq \Delta m^2_{13}$ has a value of approximately $\pm 2.45 \times 10^{-3}$ eV^2. This corresponds to an oscillation length

$$\ell_{\text{osc}} \simeq \frac{4\pi\ E}{|\Delta m^2_{23}|} \simeq 1012\ E_{\text{GeV}}\ \text{km} \tag{74}$$

For a neutrino energy of order $E_\nu \simeq 1$ GeV, this corresponds to a situation where down-going neutrinos ($L \lesssim 100$ km) do not oscillate, while up-going neutrino ($L \gtrsim 1000$ km) undergo rapid oscillations that can be averaged.

The most precise results on atmospheric neutrinos have been obtained by the Super-Kamiokande (as recognized with the Nobel prize of Takaaki Kajita in 2015), but other detectors, such as Kamiokande, IMB and MACRO have also given important contributions. Detailed reviews of these results can be found for example in [115, 116].

The study of the relevant parameters in the neutrino mass matrix has continued with the use of accelerator and reactor neutrinos, however the potential of atmospheric ν in determining the fundamental properties of the neutrinos is not exhausted. In particular, the very interesting possibility to use atmospheric ν's to determine the neutrino mass hierarchy [117] will be the used by a new generation of atmospheric neutrino detectors [78, 118–121].

Neglecting mass effects the non-diagonal flavor oscillations that involve electron neutrinos and antineutrinos with long wavelengths ($\ell_{\rm osc} \propto 4\pi\ E/|\Delta m^2_{23}| \simeq 4\pi\ E/|\Delta m^2_{13}|$) are small because they are controlled by the small mixing angle θ_{13}. However, for neutrino propagating in matter the effective neutrino mixing matrix is modified. In the case of the normal mass hierarchy, the flavor transition probabilities $\nu_e \leftrightarrow \nu_\mu$ and $\nu_e \leftrightarrow \nu_\tau$ for neutrinos propagating inside the Earth with an energy of few GeV can become large (the energy and zenith angle dependence of the transition probability is shown in Fig. 15), while the transition probabilities $\overline{\nu}_e \leftrightarrow\leftrightarrow \nu_\mu$ and $\overline{\nu}_e \leftrightarrow \overline{\nu}_\tau$ remain small (and in fact are suppressed by the matter effects). The effect is reversed in the case of the inverse mass hierarchy when the electron antineutrinos that can have large transition probabilities (to a good approximation the oscillation probabilities of neutrinos and antineutrinos are simply interchanged passing from the normal to the inverse mass hierarchy).

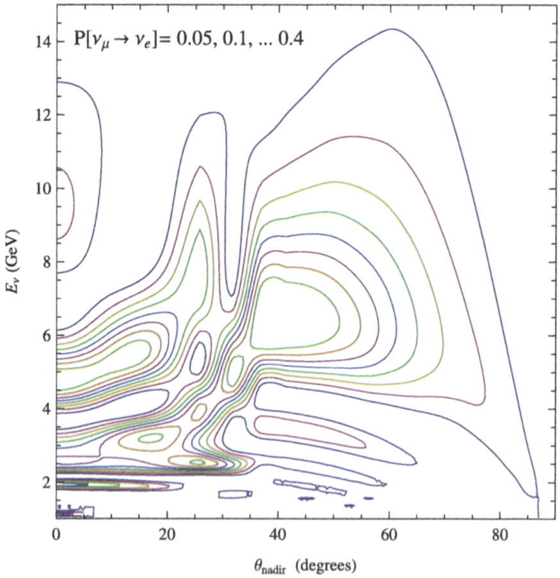

Fig. 15 Contour plot of the oscillation probability $P_{\nu_\mu \to \nu_e}$ in the plane $\{E_\nu, \cos\theta_{\rm nadir}\}$. The different lines are calculated for current the best fit of the oscillation parameters and the Normal Hierarchy for the netrino masses

These effects are more difficult to observe than what one could expect simply inspecting the oscillation probabilities as outlined in Fig. 15 because one has a cancellation between the effects of disappearance and appearance. For example, the flux of ν_e, after considering the effect of oscillation becomes:

$$\phi_{\nu_e}(E, \theta_z) = \phi^\circ_{\nu_e}(E, \theta_z)\, [1 - P_{\nu_e \to \nu_\mu} - P_{\nu_e \to \nu_\tau}] + \phi^\circ_{\nu_\mu}(E, \theta_z)\, P_{\nu_\mu \to \nu_e} \tag{75}$$

To a very good approximation (for $\sin^2 \theta_{23} \simeq 1$) $P_{\nu_e \to \nu_\tau} \simeq P_{\nu_e \to \nu_\mu} \simeq P_{\nu_\mu \to \nu_e}$, so that the disappearance and appearance effects cancel exactly when $\phi_{\nu_\mu} \simeq 2\, \phi_{\nu_e}$. In the energy region where the electron neutrino transitions are enhanced one has that the muon neutrino flux is in fact larger that twice the electron neutrino flux ($\phi_{\nu_\mu} > 2\, \phi_{\nu_e}$) and the net effect is an appearance of electron neutrinos. A similar discussion can be presented for muon neutrinos.

A detector capable of discriminating the sign of the observed leptons and therefore to separate neutrino and anti-neutrino charged current events has a "gold-plated" method to establish the hierarchy, because the oscillation effect is present for neutrinos (antineutrinos) in the case of the normal (inverse) mass hierarchy. This idea is possible for muon event using a magnetized detector, as for example the INO detector in India [121].

The identification of the mass hierarchy is also possible without the separation of neutrino and anti-neutrino events because the sum of the two rates depends on the mass hierarchy. An example of this idea is illustrated in Fig. 16.

Atmospheric Neutrinos as Foreground of Astrophysical Signals

Atmospheric neutrinos are a foreground for the study of neutrinos from astrophysical sources. A straightforward method to determine the astrophysical origin of a set of neutrino events is the observation of a structure in their angular distribution in celestial coordinates (an obvious case is the presence of point-like or quasi point-like sources). Any such anisotropy would be an unambiguous proof that some of the observed neutrinos originate from astrophysical sources outside the solar system. It has however been predicted (see for example [41, 45]) that the first evidence for high-energy astrophysical neutrinos could emerge in the form of an isotropic flux. Since neutrinos can traverse the entire universe with negligible absorption, the sum of all extragalactic sources (most of them very distant and therefore faint and unresolved) form an isotropic flux that can be largest contribution to extraterrestrial neutrino sky. In fact the recent results of IceCube are consistent with the presence of such an isotropic extraterrestrial component (see discussion below).

Atmospheric neutrinos from the conventional source (due to pion and kaon decay) can be separated from the astrophysical contribution because of their characteristic zenith angle distribution, with a strong enhancement for the horizontal plane, and the characteristic flavor composition, very rich in muon neutrinos (with a ratio $(\nu_\mu + \overline{\nu}_\mu)/(\nu_e + \overline{\nu}_e) \simeq 40$) and negligible contribution of τ neutrinos (when, at high

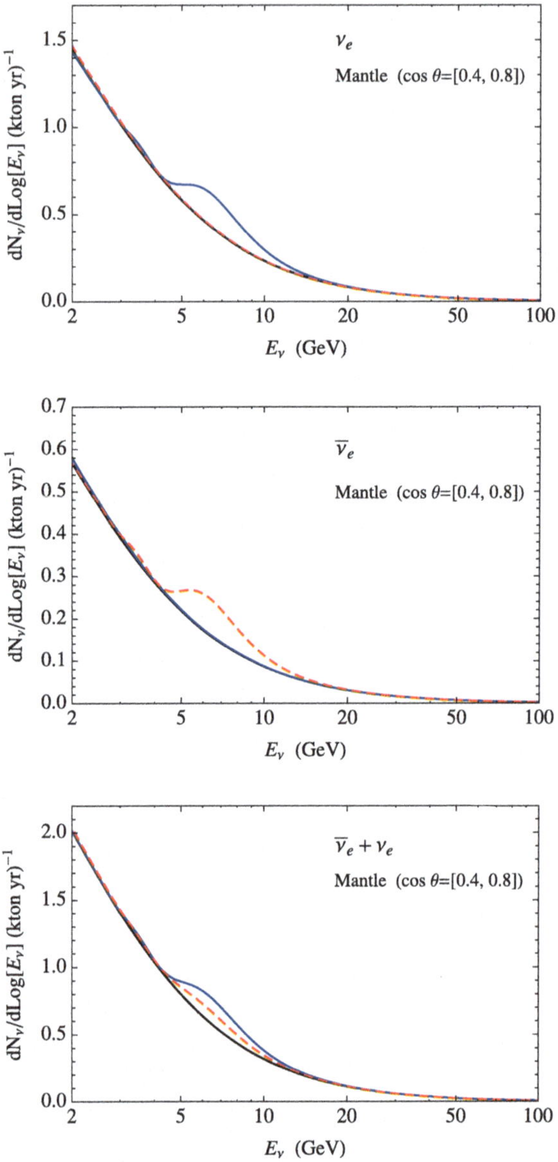

Fig. 16 The figures show the event rate of ν_e (*top panel*), $\overline{\nu}_e$ (*center panel*) and their sum (*bottom panel*) charged current interactions. The *blue* (*balck*) *solid line* is for Normal (inverted) Mass hierarchy. The *dashed line* is for down-going neutrinos

energy, the shortest ν oscillation length $4\pi E_\nu/|\Delta m^2_{23}|$ becomes longer than the Earth diameter).

It is more difficult to disentangle an isotropic astrophysical ν flux from the flux of atmospheric prompt (charm decay) neutrinos. The prompt ν flux is also to a very good approximation isotropic, and has a equal size contributions for the electron and muon flavors. An important difference is that one expects the prompt flux of ν_τ and

$\overline{\nu}_\tau$, to be of order 10% of the electron and muon (anti)-neutrino fluxes, having as main source the chain decay of $D_s^\pm$ mesons (the bound state of the charm quark and a strange anti-quark aor the charge conjugate state) that generate tau-neutrinos via the two body decay mode $D_s^\pm \to \tau^+\nu_\tau(\overline{\nu}_\tau)$, and the following decay of the $\tau^\pm$ (that always creates a tau (anti)-neutrino). This flavor composition is similar closer to the one expected for astrophysical neutrinos after taking into account the expected effect of oscillations however the identification of a large component of τ-neutrinos would be a clear signature for an astrophysical contribution.

The separation atmospheric prompt and astrophysical neutrinos can be performed using the shape of the energy spectra, since one can predict with reasonable accuracy the shape of the atmospheric spectra, from a knowledge of the properties of charmed particles production in hadronic interactions (see [100–102]).

A model-independent way to identify (and eliminate) the atmospheric foreground for down-going neutrinos [103, 104] is the detection, in coincidence with the ν event, other particles associated with the same cosmic ray shower. This veto technique has been used by IceCube in the selection of the "High-Energy Starting Events" (HESE), rejecting those events where the sensors at the boundary of the telescope record a signal consistent with the presence of charged particle that enters the detector in coincidence with the energy deposition in the fiducial volume.

Neutrinos from Terrestrial Radioactivity[18]

The *geoneutrinos* are the electron antineutrinos that are produced by the beta decay of various radioactive elements of the Earth. A few of these neutrinos are observable in ultrapure scintillation detectors; they are those with energies above the threshold for inverse beta decay reactions, $E_{\mathrm{IBD}} = 1806.07$ keV. The experiments KamLAND and Borexino have observed and measured these geoneutrinos beyond any doubt. This offers us new opportunities for studying the energetics of our planet as well as understanding the formation of the Earth.

Motivations

There are several scientific reasons of interest to study geoneutrinos.

- They allow us to infer the amount (abundances) of uranium and thorium together with their ratio in the main terrestrial reservoirs.
- Even more interesting, we can directly estimate the radiogenic contribution to the heat flowing from Earth's surface that is measured conservatively to be in the range from 30 to 50 TW.

---

[18] We thank F. Mantovani for the help in the preparation of this paragraph.

- The measured geoneutrino flux allows us to probe/exclude other exotic radiogenic sources (e.g., georeactors).
- We can directly test the reference geochemical models of our planet, based on the understanding of the early stages of the Earth formation and of the subsequent mechanisms of internal differentiation.

Reactor neutrinos are a well-understood background for geoneutrino search and are mostly visible at higher energies [29]. Conversely, geoneutrinos do not have a particular impact on most of the other neutrino observations, being a stable but weak flux. In any case, geoneutrino detectors will contribute to monitor reactor neutrinos.

The observable antineutrino flux

The U, Th and K decay chains, named after the parent nucleus, include several alpha and beta decays. In these chains, kinetic energy is produced that is eventually degraded to heat; at the end of these processes, stable nuclei are produced. The beta decay processes produce electron antineutrinos, and a few of them have sufficient energy to produce inverse beta decay reactions. This condition includes a few radioisotopes of U and Th decay chains but not the K decay chain that is not observable with the present technologies.

On large scale these chains are plausibly in *secular equilibrium*, therefore the abundances of the species of each chain are expressed in terms of those of the parent nucleus x. E.g., in the simplest case when there is only one decay daughter i decaying in a single channel, we have $N_i/\tau_i = N_x/\tau_x$ (where N are number of nuclei and τ their lifetimes) namely we have equal decay rates, equivalently Hz. The decay rate of the radioisotope i is smaller than the previous expression in two cases,

- when the radioisotope i can decay in more than one way (for instance, bismuth can decay alpha and beta at the same time)
- when the nucleus in the final state can produced in several excited states k.

The two effects can be quantified by the factors $R_i \leq 1$ and $I_{i,k} \leq 1$, respectively, that are normalized by $\sum_i R_i = 1$ and $\sum_k I_{i,k} = 1$ [26]. The total rate of decay of the specific species in the specific decay channel is, $\Gamma_{i,k}^{\text{tot}} = R_i \times I_{i,k} \times N_x/\tau_x$ where again we assume secular equilibrium.

The electron antineutrinos with sufficient energy to produce inverse beta decay reactions are listed in Table 3. Five decay channels belonging to the uranium-238 chain and other three belonging to the thorium-232 chain have indeed a maximum energy larger than E_{IBD}. Therefore, a small fraction of these geoneutrinos (i.e., less than 10%) is visible in those scintillator-based detectors that are able to attain sufficient radiopurity at very low energies. (We recall on passing the lifetimes τ_x of uranium-238 and thorium-232 are, respectively, 4.47×10^9 year and 1.4×10^{10} year.)

Table 3 List of β-decaying nuclei present in Nature that produce a signal visible with the inverse beta decay reaction. We indicate the initial nuclei i (and the chain x to which they belong), the charge of the nucleus in the final state Z, the maximum electron antineutrino energy for the given channel k, the number of signal events assuming the ratio Th/U = 3.9 (see text) with an exposure of 10^{32} protons per year (a unit called Terrestrial Neutrino Unit, TNU). Adapted from [26]

Parent $[x]$	β-decaying species i	Z, i-dependent	$R_i \times I_{i,k}$ (%)	E_{max}, k dependent (in keV)	$S_{\rm tot}$(number)
^{238}U	^{234}Pa$_m$	92	98.2	2268.92	31.21
"	^{214}Bi	84	18.2	3272.00	45.84
"	"	"	1.7	2662.68	1.55
"	"	"	7.4	1894.32	0.14
"	"	"	0.8	1856.51	0.01
^{232}Th	^{212}Bi	84	55.5	2254	20.00
"	^{228}Ac	90	8.0	2069.24	1.21
"	"	"	0.8	1940.18	0.04

All transitions in Table 3 except the two of actinidium are forbidden. For a given beta or beta-gamma transition, the decay spectrum is expected to be reasonably approximated with the universal shape [26],

$$f(E_{\bar{\nu}_e}) = N^{-1}\, E_{\bar{\nu}_e}^2 \;\; p_e E_e \; F(E_e) \text{ where } \begin{cases} p_e = \sqrt{E_e^2 - m_e^2} \\ E_{\bar{\nu}_e} + E_e = E_{max} + m_e \end{cases} \tag{76}$$

where the energies of the electron antineutrino $E_{\bar{\nu}_e}$ and the one of the electron E_e are linked by the condition of energy conservation; the constant N is fixed by normalizing the integral over all possible energies $0 \le E_{\bar{\nu}_e} \le E_{max}$ to 1. One recognizes in this expression the phase space of the antineutrino, $d^3 p_{\bar{\nu}_e} \propto E_{\bar{\nu}_e}^2 dE_{\bar{\nu}_e}$ the one of the electron, $d^3 p_e \propto p_e E_e dE_e$ and finally the effect of the final state interaction between electron and nucleus, described by the Fermi function $F(E_e)$, namely,

$$F(E_e) = \frac{2(\gamma+1)}{\Gamma(2\gamma+1)^2}(2\, r\, p_e)^{2(\gamma-1)} e^{\pi\eta}\, |\Gamma(\gamma + i\eta)|^2 \text{ where } \begin{cases} \gamma = \sqrt{1-(\alpha Z)^2} \\ \eta = \alpha Z\,/\beta_e \end{cases} \tag{77}$$

$\beta_e = p_e/E_e$ is the velocity of the electron (beta ray), Z is the charge of the nucleus in the final state, r its radius, α the fine structure constant and finally η is called the Sommerfeld factor. This function accounts for the fact that the electron is produced near the nucleus and therefore its wavefunction is not a plane wave. Setting $\gamma \to 1$, one recovers the non-relativistic expression, $F(E_e) = 2\pi\eta/(1 - e^{-2\pi\eta})$.

The total antineutrino flux from each isotope i decaying in the channel k and belonging to the chain x is given summing the contributions from the number of parent nuclei $dN_x(\mathbf{r})$ around the given position $\mathbf{r}$ considering an isotropic emission, multiplying by the differential rate per single radioisotope i and in the specific decay channel k namely

$$\frac{d\Gamma_{i,k}}{dE_{\bar{\nu}_e}} = \frac{R_i \times I_{i,k}}{\tau_x} \times f_{i,k}(E_{\bar{\nu}_e}) \tag{78}$$

(in 1/(MeV s)) where τ_x is the lifetime of the parent nucleus[19] and finally including also the survival probability P_{ee}. If $\mathbf{R}$ is the position on Earth where the flux is observed,

$$\Phi_{x,i,k}(\mathbf{R}, E_{\bar{\nu}_e}) = \int \frac{dN_x(\mathbf{r})}{4\pi\,|\mathbf{R}-\mathbf{r}|^2} \times \frac{d\Gamma_{i,k}}{dE_{\bar{\nu}_e}}(E_{\bar{\nu}_e}) \times P_{ee}(E_{\bar{\nu}_e}, |\mathbf{R}-\mathbf{r}|) \tag{79}$$

In order to estimate the various contributions one needs a model for $dN_x(\mathbf{r})$ that is given by the product

$$dN_x(\mathbf{r}) = C_x\, a_x(\mathbf{r})\, \rho(\mathbf{r})/m_x\, d^3\mathbf{r} \tag{80}$$

namely the isotopic composition $C_x = 1$ for thorium-232 and $= 0.993$ for uranium-238; the mass density ρ (in g/cm^3); the abundance in weight $a_x(\mathbf{r})$ (in g/g); the mass of the isotope m_x (in g). Various expressions and useful approximations for the flux are given in [26], which is of the order of few million antineutrinos per cm^2 per second in total; recall that only those above the detection threshold E_{IBD} however, are observable.

Measurements, the BSE model and perspectives

To date, two experiments measured geoneutrinos: they are KamLAND in Japan and Borexino in Italy. They are both ultra-pure scintillator detectors and are located on the continental crust, surrounded by different geological and geochemical formations but leading to similar expectations for the total geoneutrino fluxes. KamLAND is about three times larger but operates in country where the reactor neutrino flux has been very high until 2011 (before the shutdown of the reactors due to Sendai earthquake); Borexino is the cleanest detector ever built and the reactor neutrino flux is quite small instead. Their results are reported in a series of works, namely [70–72] (KamLAND) and [73–75] (Borexino).

The measurements are compatible with the current ideas of the geochemical composition of the Earth, as summarized by different classes of standard models, called Bulk Silicate Earth (BSE) models. These actually describe the composition of the terrestrial primitive mantle on the base of elemental abundances measured in mantle rock samples and in the chondrite meteorites. This is actually a class of models, with a number of variants, named "cosmochemical models", "geochemical models", "geodynamical models", that infer the Earth composition using various assumptions, see [122] for a review. BSE models predict the thorium and uranium abundances $a_x(\mathbf{r})$ and in particular predict that the ratio of thorium and uranium is approximatively

[19]Note that, assuming secular equilibrium, the product $d\Gamma^{\text{tot}}_{i,k}(\mathbf{r}) = dN_x(\mathbf{r}) R_i I_{i,k}/\tau_x$ gives a contribution to the total decay rate of the species i in the specific channel k.

Th/U = 4[20]; this agrees with current findings. The BSE models assume, compatibly with current ideas, that the radioactive elements are located in the crust and in the mantle, but not in the core of the Earth. Radioactive elements are not expected to be uniformly distributed in the crust, but rather, to be more abundant in continental crust than in the oceanic crust. Thus, detectors in different locations on Earth are expected to measure different geoneutrino fluxes.

In the future, it will be interesting to obtain high statistics measurements of geoneutrinos. In this manner, the BSE models will be probed and its parameters measured. Note in particular that, with more precise data, it will be possible to separate geoneutrinos from uranium and thorium decay chains. Moreover it would be important to measure geoneutrinos from various geological locations: It should be remarked that a direct measurement of natural radioactivity of the mantle could be possible only through a detector placed far from continental crust (e.g., in the middle of Pacific Ocean) or if the direction of incoming geoneutrinos could be measured. In conclusion, we note that the study of geoneutrinos links deeply particle and nuclear physics, geochemistry (U and Th abundances) and geophysics (densities and volumes) which makes evident the multidisciplinary character of this new science.

Man-Made Sources

There are several types of man-made neutrino sources of various energies, and in particular

1. nuclear reactors ($\bar{\nu}_e$ of few MeV);
2. "portable" sources, based on electron capture (monochromatic ν_e of few MeV)
3. "accelerator" neutrinos, obtained producing unstable mesons by nucleon collisions (from subGeV region till 100 GeV, mostly ν_μ and $\bar{\nu}_\mu$)

and more one, e.g., the planned neutrino factories', "muon beams" or "beta-beam" (from the radioactive ion beams). The above man-made sources have been important for the investigation of neutrino properties and are currently used to perform further tests of the three-flavor picture and to deepen its understanding.[21] The detectors used in these investigations may play the role of neutrino telescopes, if their depth of installation is not too shallow and if the background rate (linked to the artificial source) is not too high.

The only case when there is an interference between these sources and the activities of neutrino astronomy is the case of reactor neutrinos, whose lower energy tail overlaps with geoneutrinos, and whose higher energy tail overlaps with the region

[20]We note incidentally that, in BSE models, uranium and thorium are expected to be the main radiogenic contributors to the terrestrial heat.

[21]It should be remarked that the evidences of three-flavor neutrino oscillations collected to date have been obtained by studying "natural" neutrino beams and not artificial ones.

of diffuse supernova neutrino background (aka relic supernova neutrinos). For this reasons, some sites on the Earth are less convenient for this type of searches, just as the "seeing" of large cities makes ordinary astronomy impractical.

Low Energy Cosmic Neutrinos

Let us describe the structure of this section devoted to low-energy neutrinos.[22] For various reasons that however include observational and astrophysical aspects, we are especially interested in the low-energy regime of *big-bang neutrinos*, when they decouple from the plasma, and also in the lower energy regime, when they cool down even further, becoming possibly non relativistic; for this reason, we will discuss them in this section (section "Big-Bang Neutrinos") (but we will not discuss the cosmological role of neutrinos in other interesting contexts, such as leptogenesis that have a speculative character or where the observational basis is very limited). Moreover, we will examine also the conventional but very much important types of low-energy neutrinos, namely *solar neutrinos* (section "Sun/Nuclear Fusion"). and *neutrinos from stellar collapses* (section "Core Collapse Supernovae").

General references:

Neutrinos in cosmology have been discussed in the dedicated book [10] and in several useful review papers, e.g., [31–34]. The main reference for solar neutrinos remains the book of J. Bahcall [2]. The model he has originally developed to calculate solar neutrinos is often referred to as the standard solar model (SSM). This is maintained updated by his collaborators, and in particular A. Serenelli [342, 343]; the most recent version of the SSM (2017) is in [172]. See [35, 36] for updated review works and discussion. Chapter 15 of Bahcall's book is titled "Stellar collapses"; this topic is also covered in various recent reviews, concerning the general (theoretical) aspects of supernova neutrinos [38], SN1987A [37], and the electron antineutrinos due to accumulation of cosmic supernova events [39, 40].

Big-Bang Neutrinos

The neutrinos from the big bang are called or *cosmic neutrino background* or sometimes *relic neutrinos*.

[22]We follow the definition of "low energy" given in section "Neutrino Interactions and Properties: Generalities".

Electroweak Plasma and the Number of Neutrinos

In the early instants of the big bang, the neutrinos and antineutrinos of all three species had thermal distributions. They were kept in equilibrium by weak reactions.[23]

Neutrinos participated in weak reactions and affected the amount of light nuclei originally present in the Universe. Big-bang nucleosynthesis is a quantitative theory that has been reviewed [33]. It was used also to provide a determination of the number of relativistic species in the moment of the nucleosynthesis, that is often presented in terms of effective neutrino species at the time of the Big-Bang nucleosynthesis N_ν, BBN. The comparison of the prediction of this theory and the abundances of the primordial helium-4 and deuterium, inferred by the observations, allowed the authors of [33] to deduce the following value,

$$N_\nu,\, \text{BBN} = 3.14^{+0.70}_{-0.65} \tag{81}$$

This agrees very well with the expectations that the only relativistic species are the ordinary three neutrinos.[24] The amount of primordial lithium, instead, is not yet understood.

At a temperature of a couple of MeV, neutrinos exited from equilibrium. In that moment, the other most abundant species in the plasma were photons, electrons and positrons; together with the neutrinos, they were the constituents of the so-called "radiation", namely they were effectively massless fields. Eventually, when the positrons annihilated, this resulted in a sudden heating of the photons but in first (and good) approximation not of the neutrinos; a somewhat better approximation is to assume that the number of neutrinos becomes $N_{\nu,\text{eff}} = 3.046$.

This prediction was tested by the study of the distribution of the anisotropies of the cosmic microwave background (CMB). In fact, the Planck satellite [34] found,

$$N_{\nu,\text{CMB}} = 3.15 \pm 0.23 \tag{82}$$

This observation concerns a much later cosmic time, since the time of the decoupling between matter and radiation was 380,000 year after the big bang ($z_* = 1090$).

Non-relativistic Neutrinos

The distribution of neutrino follows the Fermi-Dirac occupation factor $f(p) = 1/[\exp(p/T_\nu) + 1]$. Thus, each species of neutrino or antineutrino has number density $n_\nu = \int f(p)\, d\mathbf{p}/(2\pi^3)$, namely,

---

[23] It is not expected that they have a measurable chemical potential (i.e., a lepton asymmetry) that distinguishes neutrinos from antineutrinos, and there is no sign of it from the data.

[24] It should be remarked however that the analysis of the primordial helium-4 data is recognized to be difficult and the results are not yet uncontroversial.

$$n_\nu = \frac{3Z(3)}{4\pi^2} T_\nu^3 = \frac{56}{\text{cm}^3} \tag{83}$$

that increases only slightly 56 → 57 if we multiply by 3.046/3 (see discussion just above). Neutrinos and antineutrinos are thought to have a temperature of

$$T_\nu = T_\gamma \left(\frac{4}{11}\right)^{1/3} = 1.95 \text{ K} = 0.168 \text{ meV} \tag{84}$$

that is lower than the one of CMB due to positron annihilation, discussed just above. (See [355] for the conversion factors.)

Probing neutrino masses

It is convenient to think to the massive states from here on. The temperature of neutrinos today is surely smaller than the value of two neutrino masses, as we can see taking the square root of the delta-m-squared values given in Table 2: $\sqrt{\Delta m^2} = 50$ meV and $\sqrt{\delta m^2} = 8.6$ meV. This means that at least two neutrino and antineutrino species are non-relativistic in the present Universe. Moreover, recalling the fact that the lightest neutrino mass is not probed by oscillations, it is interesting to probe neutrino masses by mean of cosmology. Note for comparison that temperature of the neutrinos at the decoupling time is $T_\nu \times (1 + z_*) = 182$ meV.

This can be done using cosmological data and in particular various structures of the Universe that we observe at various scales, from several Gpc to some 10 Mpc: the asymmetries of the CMB, the baryon acoustic oscillations, the Lyman-α forest. In fact, neutrinos have different velocities depending upon their masses that are then associated to different physical scales. The effect of neutrino masses on the observable distributions is not large, only few percent, but the amount of data is impressive and the theory has been refined to match the needs of this comparison. Several recent groups have derived very stringent bounds on neutrino masses, summarized in Table 4. In good approximation this bount applies to the sum of neutrino masses,

Table 4 Tight constraints on the sum of neutrino masses $\Sigma = m_1 + m_2 + m_3$ and obtained in 2015, by analyzing the data on the CMB by the Planck Collaboration [34], polarization included, along with other relevant cosmological data probing smaller scales (Table adapted from [123].)

upper bound	included dataset
on Σ (2σ C. L.)	
153 meV [34]	SNe, BAO, H_0 prior
120 meV [124]	Lyman-α
126 meV [125]	BAO, H_0 & τ priors, Planck SZ clusters
177 meV [126]	BAO
110 meV [127]	BAO, galaxy clustering, lensing

$$\Sigma = m_1 + m_2 + m_3 \tag{85}$$

Note that this parameter should be larger than $\sqrt{\Delta m^2} \approx 50$ meV for normal mass hierarchy and $2\sqrt{\Delta m^2} \approx 100$ meV; therefore the tightest bounds from Table 4 provide already a mild preference for the case of normal mass hierarchy.

Detecting relic neutrinos?

Is it possible to detect in laboratory relic neutrinos? Let us assume to have a target of 100 g of ^{3}H. Electron neutrinos can be seen through the weak reaction [128, 129],

$$\nu_e + {}^3\mathrm{H} \to {}^3\mathrm{He} + \mathrm{e}^-. \tag{86}$$

In the standard assumption of a homogeneous Fermi-Dirac distribution of the cosmic neutrino background, we expect ~8 events per year if neutrinos are Majorana particles and about half if the Dirac hypothesis applies [130]. Indeed, in the former case, the states with positive helicity (by definition, antineutrinos) will act just as neutrinos, since almost at rest. Instead, in the latter case, they will remain antineutrinos and thus they will not react. It can be noticed that the signal rate is not prohibitively small, but the major difficulty consists in attaining a very good energy resolution, sufficient to keep at a manageable level the background from beta decay. The Reader interested in further details can consult Refs. [129, 130].

Sun/Nuclear Fusion[25]

Introduction

A well-known controversy opposed Lord Kelvin to Darwin; the point was that, with the sources of energy known by the time to physicists, the lifetime of the Sun was estimated to be much less than needed for the evolution of the living beings and of our Earth. In fact, the present solar luminosity $L_\odot = 3.84 \times 10^{33}$ erg/s would exhaust the gravitational energy, estimated to be of the order of $G_N M_\odot^2/R_\odot = 3.8 \times 10^{48}$ erg in 30 million years that according to Darwin was at least a factor 10 too short. The stakes were quite high: nothing less than the understanding of the internal engine of the Sun.

The first step in the right direction were remarks, due to Aston and Perrin (1920), that the helium atom weighs less than 4 hydrogen atoms: $f = 4m_\mathrm{H}/m_\mathrm{He} - 1 = 0.7\,\%$. The energy that could be made available thanks to Einstein relation $f M_\odot c^2 = 1.2 \times 10^{52}$, would allow our Sun to radiate energy for 100 billion years with the present power $L_\odot$ which is an enormous amount of time.[26]

[25] We thank G. Bellini for the help in the preparation of this paragraph.

[26] Actually, it is expected that only 10% of these reactions will take place due to absence of convection in the center of our Suns.

The question becomes, what happens actually in the center of the stars. The inner temperature of a star can be roughly estimated from the kinetic energy that one particle gains in the gravitational potential (Eddington). Considering a cold atom of hydrogen in a very diffuse cloud of gas which eventually ends to be trapped, we have $m_p v^2/2 \approx G_N m_p M_\odot / R_\odot = 2$ keV, that is more or less the temperature at the center of our Sun, $T \approx 1.3$ keV. Therefore, the matter is completely ionized there.

Then, two years before the 2nd world war, von Weizsäcker and Bethe-Critchfield studied which are the possible chains of reaction that lead to hydrogen fusion inside a star. They identified the following basic reaction,

$$\mathrm{p} + \mathrm{p} \rightarrow \mathrm{D} + e^+ + \nu_e + 0.42\ \mathrm{MeV} \tag{87}$$

that is the principal reaction of the pp-cycle (that now we know to be the main source of energy generation of our Sun) and also the catalytic CNO cycle that will be discussed later on in some details. Apparently, there is a generic trouble due to the repulsive Coulomb potential $V(r) = Z_1 Z_2 e^2/(4\pi r)$ between two nuclei. Considering, e.g., the above reaction, when two protons touch, namely when their centers are at a few fermi, we have $V(2\ \mathrm{fm}) \approx 0.7$ MeV, that is definitively larger than the available kinetic energy. However, the two hydrogen nuclei may meet (and eventually react) thanks to tunnel effect as pointed out by Gamow.

To obtain a more quantitative understanding, it is sufficient to consider the product of (1) the tail of the statistical (Maxwell-Boltzmann) distribution $dn/dE \propto \exp(-E/T)$, where we consider only the exponential factor and where T is temperature in the center of the Sun, and (2) the penetration factor[27] $P_{\mathrm{Gamow}} = \exp(-\sqrt{E_G/E})$ where the Gamow energy is $E_G = m_p c^2 (\alpha\pi)^2 \approx 490$ keV. It is easy to see that the resulting exponent, $E/T + \sqrt{E_G/E}$, has a maximum that occurs when $E_{\mathrm{peak}} = \sqrt[3]{E_G\, T^2/4} = 6$ keV. This is larger but not *very* far from the typical energies inside the Sun. In fact, now the problem could seem to be reverted, but actually, in the largest majority of the cases the two protons interact elastically due to strong forces and only very rarely undergo the (weak) fusion reaction.

Solar Neutrino Models

Whatever the specific mechanism that converts 4 hydrogen nuclei (i.e., protons) into a nucleus of helium-4, it is necessary to produce 2 electron neutrinos according to the scheme, $2e^- + 4p \rightarrow {}^4\mathrm{He} + 2\nu_e$. For each helium-4 an amount of energy of, $Q \sim 26.7$ MeV should be liberated. If we assume that the all solar energy is due to fusion of hydrogen into helium, the neutrino flux can be estimated in a pretty simple way. Since the Earth receives an energy flux of $K = 8.53 \times 10^{11}\ \mathrm{MeV\,cm^{-2}\,s^{-1}}$,

[27] In semiclassical approximation, the exponent of the tunnel effect is given by $-2\int_0^{r_{\mathrm{cl}}} \sqrt{2\mu(V(r) - E)}dr$, where E is the energy, $\mu = m_p/2$ the reduced mass of the system, and r_{cl} the classical point of inversion defined by $V(r_{\mathrm{cl}}) = E$.

we find $\Phi_\nu = 2K/Q \simeq 6.4 \times 10^{10}$ cm^{-2} s^{-1}: namely, we should receive about 60 billion neutrinos per second per square centimeters on Earth surface.

To calculate their energy distribution it is necessary to specify the chains of reactions responsible for hydrogen fusion. The main mechanisms are those of the PP chain, so called because they are initiated by the weak reaction $p + p \to d + e^+ + \nu_e$, with its possible terminations PPI, PPII and PPIII and the cold CNO cycle, that uses carbon (C), nitrogen (N) and oxygen (O) present in the Sun present in the Sun as catalysts for the formation of helium-4: see Table 5. It is also necessary to know the temperature, the density, and the chemical composition of the inner layers of the Sun, which determine the efficiency of various nuclear reactions.

In order to perform such a calculation a solar model should be built, namely a theoretical description of the Sun, according to known astrophysics and nuclear physics. The name Standard Solar Model (SSM) is usually reserved for the one developed by Bahcall since mid sixties and improved in the course of the years thanks to several collaborators. The physical and chemical input parameters are chosen (within the respective uncertainty intervals) to reproduce the observed properties, as the radius of the Sun, its brightness, its surface chemical composition. The Solar Standard Model was and is an essential tool to investigate the Sun, as it allows to evaluate the goodness of stellar evolution theories to a high level of accuracy.

In half a century, solar evolutionary modeling has made significant progress. A relevant discipline is helioseismology, which studies the seismic properties of the Sun (see [131] for an introduction). Its study allowed to precisely determine the depth of the convective envelope, the abundance of helium in the solar surface and the sound velocity profile in the Sun [132]. Moreover, the determination of some basic physical parameters, such as opacity of the stellar plasma and the astrophysical factors of thermonuclear reactions inside the Sun, has witnessed great progress. LUNA experiment of the National Laboratories of Gran Sasso, for example, determined the speed of the reactions $^3\mathrm{He} + {}^3\mathrm{He} \to {}^4\mathrm{He} + 2p$ [133] and $^3\mathrm{He} + {}^4\mathrm{He} \to {}^7\mathrm{Be} + \gamma$ [134], which determined the relative efficiency of the chain PPI and of the chains PPII and PPIII, and of the process $^{14}\mathrm{N} + p \to {}^{15}\mathrm{O} + \gamma$ [135] that determined the efficiency of the CNO cycle.

The expected spectrum is shown in Fig. 17 and further illustrated in Table 8 given later. About 90 % of solar neutrinos are the *PP* neutrinos, produced by the reaction $p + p \to d + e^+ + \nu_e$, with a continuous spectrum of energy and maximum energy equal to 0.42 MeV. There is also a small component of *pep* neutrinos, about 0.2% of the total, produced from the reaction $p + e^- + p \to d + \nu_e$. The second component of the flux of solar neutrinos, the *beryllium neutrinos*, carry approximately 7 % of the total and are due to electron capture processes on Beryllium $e^- + {}^7\mathrm{Be} \to {}^7\mathrm{Li} + \nu_e$ that produce a quasi-monochromatic spectral energy distribution. In PPIII termination, the boron decay processes $^8\mathrm{B} \to {}^7\mathrm{Be} + e^+ + \nu_e$ produce *boron neutrinos*, correspond to about 0.008 % of the total flux and have a continuous spectrum that extends to relatively high energies. It is should be kept in mind that the instrumental background is less for higher energies, so it is easier to measure the higher energy boron neutrinos than the more abundant and important fluxes of lower energy neutrinos.

Table 5 Nuclear reactions in the Sun. The first 11 reactions form the pp-cycle, grouped in 5 branches; the last 6 is the main branch of the (cold) CNO cycle that contributes (little) to solar luminosity. The 2nd reaction of pp II branch is an electron capture and produces two lines; the 2nd reaction of the pp III branch depends on the energy of the excited Be-8 state E_x that is not known with complete certainty. The energy of the positron is included in Q. Particles or atomic nuclei are indicated; p=^{1}H and D=^{2}H. For the final state, we adopt the notation of Rutherford, $\alpha = {}^4$He and $\beta = e$. Borexino, Super-Kamiokande and the Subdury Neutrino Observatory are indicated by BX, SK, SNO, with question mark when the observation is not yet accomplished. Adapted from [2]

Name	Reaction	Q-value (keV)	E_ν^{max} (keV)	Observed
pp I	$\mathrm{p}+\mathrm{p}\to \mathrm{D}+\beta^+ +\nu_e$	1442	420	BX
	$\mathrm{p}+\mathrm{D}\to {}^3\mathrm{He}+\gamma$	5494	–	
	${}^3\mathrm{He}+{}^3\mathrm{He}\to \alpha+2\mathrm{p}$	12,860	–	
pep	$\mathrm{p}+\mathrm{p}+e\to \mathrm{D}+\nu_e$	1442	1442	BX
pp II	${}^3\mathrm{He}+{}^4\mathrm{He}\to {}^7\mathrm{Be}+\gamma$	1586	–	
	${}^7\mathrm{Be}+e\to {}^7\mathrm{Li}+\nu_e$	862, 384	862, 384	BX, (BX?)
	${}^7\mathrm{Li}+\mathrm{p}\to 2\alpha$	17,347	–	
pp III	${}^7\mathrm{Be}+\mathrm{p}\to {}^8\mathrm{B}+\gamma$	137	–	
	${}^8\mathrm{B}\to {}^8\mathrm{Be}^*+\beta^+ +\nu_e$	$18471 - E_x$	14,600–15,100	SK SNO BX
	${}^8\mathrm{Be}^*\to 2\alpha$	E_x	–	
hep (pp IV)	${}^3\mathrm{He}+\mathrm{p}\to \alpha+\beta^+ +\nu_e$	19,795	18,773	SK? SNO?
CNO-I	${}^{12}\mathrm{C}+\mathrm{p}\to {}^{13}\mathrm{N}+\gamma$	1943	–	
	${}^{13}\mathrm{N}\to {}^{13}\mathrm{C}+\beta^+ +\nu_e$	2221	1199	BX?
	${}^{13}\mathrm{C}+\mathrm{p}\to {}^{14}\mathrm{N}+\gamma$	7551	–	
	${}^{14}\mathrm{N}+\mathrm{p}\to {}^{15}\mathrm{O}+\gamma$	7297	–	
	${}^{15}\mathrm{O}\to {}^{15}\mathrm{N}+\beta^+ +\nu_e$	2754	1732	BX?
	${}^{15}\mathrm{N}+\mathrm{p}\to {}^{12}\mathrm{C}+\alpha$	4966	–	

It should be noted from Fig. 17 that the SSM allows the assessment of the errors in the predictions that sets the ideal level of precision that we would like to match with experiments. A last important remark is that helioseismological measurements allow to perform independent tests of the predictions of the SSM model; more remarks on this aspect later.

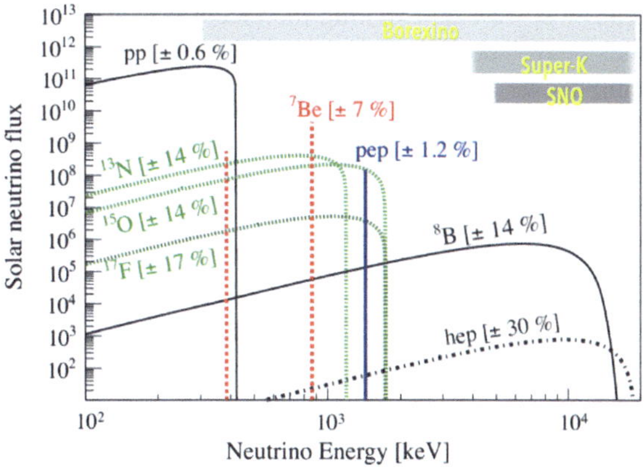

Fig. 17 The neutrino spectrum predicted by the standard solar model of Bahcall and collaborators. The *three gray strips* indicate the energy ranges of operation of experiments able to measure the differential spectrum. See text for discussion. From [36]

Solar Neutrino Oscillations

The discovery of solar neutrino oscillations took a very long time. There are many reasons pertaining to the scientific debate, such as

- The diffidence toward the predictions of astrophysics and nuclear physics (Bahcall) and toward a new experimental technique and result (Homestake)
- A twenty-year long lack of independent experimental tests
- A diffuse credence that neutrinos were massless, subsequently evolved into the prejudice that the mixing angles of neutrinos should be similar to those of the quarks-namely, small
- The fact that the specific oscillation pattern is complicated, since it implies a different behavior at low and at high energy

With hindsight, we see that all these positions are mistaken or at least misleading, except the last one. We will enter a bit into a more detailed discussion of their impact in the scientific discussion, in the section concerning the description of the experiments, section "Solar Neutrino Telescopes/Experiments" for the Reader interested in this type of considerations. Here we will simply describe the properties and features of solar neutrino oscillations, recalling also the role of the terrestrial KamLAND experiment.

The Large Mixing Angle solution

In first approximation, the oscillation pattern of solar neutrinos resembles the adiabatic two-flavor oscillation formula for the electron neutrino survival $P_{ee} = P_{\rm ad}(E_\nu, \delta m^2, \theta_{12})$ described previously. This is called "large mixing angle" solution, making explicit reference to the fact that the angle θ_{12} is large and implicit reference to previous regions of the parameter space that were called "solar neutrino solutions"—as the "small mixing solution", "low mass solution", "vacuum oscillation solution"—that are now discarded. When we consider that there are three neutrinos, we begin writing the electron neutrino as $c_{13}|\nu_L\rangle + s_{13}|\nu_H\rangle$, where only one "heavy" component $|\nu_H\rangle$ that is not affected by matter effect, and the other two components are initially grouped into $|\nu_L\rangle$. It is a good approximation to assume that, just after the production, only the "heavy" component of the electronic neutrino oscillates, producing a state $c_{13}|\nu_L\rangle + s_{13}|\nu_H\rangle e^{-i\infty}$ that describes the fact that the heavy neutrino component oscillates very fast and the other do not. Then, the two components of $|\nu_L\rangle$ undergo the usual two neutrino oscillations, so that the survival probability

$$P_{ee} = \sin^4\theta_{13} + \cos^4\theta_{13} \times P_{\rm ad}(E_\nu, \delta m^2, \theta_{12}) \tag{88}$$

Owing to the properties of the two flavor oscillation formula, we have that this formula approaches the limits

$$P_{ee} \approx \sin^4\theta_{13} + \cos^4\theta_{13} \times \begin{cases} 1 - \frac{1}{2}\sin^2 2\theta_{12} & \text{below 1 MeV [vacuum oscillations]} \\ \sin^2\theta_{12} & \text{above 4 MeV [full MSW effect]} \end{cases} \tag{89}$$

and therefore, the suppression of the expected solar neutrino flux is larger at higher energies. This solution leads us to expect,

1. The possibility of direct investigation in terrestrial laboratories. In fact, the final proof of the correctness of the large angle solution was obtained by the KamLAND experiment [136]. This is a reactor experiment with ∼100 km baseline that was planned to test it. KamLAND contributed to measure precisely the oscillation parameters and in particular the delta-mass-squared [71, 137] and moreover reached other scientific goals, discussed below.
2. A different effect of θ_{13} in high-energy solar neutrinos and vacuum oscillations. This feature allowed global three-flavor data analyses [138] to obtain a hint for this angle, before it was measured in reactor neutrino experiments.
3. A few % day-night effect, implying that the Sun is brighter in neutrino during night time. This due to the fact that the solar neutrinos that reach us are modified by oscillations in matter during their propagation inside the Earth. It was not found (at energies where the effect is not expected) by Borexino [141] while hints of this effect has been measured by Super-Kamiokande [139] and SNO [140].
4. An energy-dependent suppression at intermediate energies, where however we do not have sufficient measurements yet.

It should be recalled that KamLAND has also observed the boron and beryllium branches of the solar neutrino spectrum [142, 143]. For a full discussion of the dedicated solar neutrino experiments and their results, see the next section.

Table 6 Main characteristics of solar neutrino experiments, including the neutrino reactions and the minimum neutrino energy that yields an observable signal. The first four entries refer to experiments that *count the nuclei or neutrons due to neutrino interactions*. The last four entries instead describe experiments *capable to measure the energy* of the final state electrons. See text for a discussion

Experiment name	Experiment type	Main neutrino detection reaction	Energy measurement	Minimum energy (keV)
Homestake	Radiochemical	$\nu_e + {}^{37}\mathrm{Cl} \to e + {}^{37}\mathrm{Ar}$ (CC)	No	814
Gallex/GNO	Radiochemical	$\nu_e + {}^{71}\mathrm{Ga} \to e + {}^{71}\mathrm{Ge}$ (CC)	No	233
SAGE	Radiochemical	$\nu_e + {}^{71}\mathrm{Ga} \to e + {}^{71}\mathrm{Ge}$ (CC)	No	233
SNO	Heavy water	$\nu + D \to \nu + p + n$ (NC)	No	2226
SNO	Heavy water	$\nu_e + D \to p + p + e$ (CC)	Yes	4942
Kamiokande	Water Cherenkov	$\nu + e \to \nu + e$ (CC+NC)	Yes	7747
Super-Kamiokande	Water Cherenkov	$\nu + e \to \nu + e$ (CC+NC)	Yes	3729
Borexino	Ultra-pure scintillator	$\nu + e \to \nu + e$ (CC+NC)	Yes	285

Solar Neutrino Telescopes/Experiments

In this section, we discuss the solar neutrino telescopes.[28] We describe the motivations behind their construction, introducing the principle and the main features of these experiments. Of course, we emphasize which results they obtained and recall the main facts of their scientific life. This discussion will make evident that astrophysics and particle physics have been inextricably intertwined in the scientific discussion of solar neutrinos and this is the reason why we title this section "solar neutrino telescopes/experiments". A synoptic view of these solar neutrino observatories is offered in Table 6, where we stress the neutrino-induced reaction used for the experiment and the *minimum neutrino energy* that triggers it and in Table 7, where other specific properties are recalled.

In the last four rows of Table 6, the electrons in the final state are observed: in other words, these experiments are special since they can probe the differential

[28] The term "neutrino telescopes" emphasizes the astronomical applications; however, they are also named, depending upon the context, detectors, observatories or experiments. Sometimes in the literature the term "neutrino telescopes" is used to refer only to high-energy neutrinos or only to low-energy neutrinos: We do not follow this usage, to stress that we observe the cosmos with neutrinos of different energies/frequencies, just as it happens for ordinary (photon) astronomy.

Table 7 Specific characteristics of solar neutrino experiments. The isotopic abundance of ^{37}Cl is 24.22% and the one of ^{71}Ga is 39.89%. The photo-sensitive masses of Kamiokande and Super-Kamiokande are indicated; the fiducial masses are 680t and 22.5kt, respectively. See also previous table and the text

Detector name	Chemical composition	Target mass	Depth (km w.e.)	Time of operation	Host country
Homestake	C_2Cl_4	149t ^{37}Cl	4.3	67-98	USA
Kamiokande	H_2O	2.1kt	2.7	83-96	Japan
SAGE	Ga (liquid)	20t ^{71}Ga	4.7	89-today	USSR/Russia
Gallex/GNO	$GaCl_3$ (solution)	12t ^{71}Ga	3.4	90-03	Italy
Super-Kamiokande	H_2O	32kt	2.7	96-today	Japan
SNO	D_2O	1kt	6.0	99-06	Canada
Borexino	C_3H_4 (PC)	280t	3.4	07-today	Italy

spectrum of solar neutrinos as emphasized in Fig. 17. Several remarks on these special experiments are in order,

1. Electrons yield an observable signal only if their kinetic energies are larger than

   3.5, 7.5, 3.49, 0.15 MeV in SNO, Kamiokande, Super-Kamiokande, Borexino

   respectively, which increases the minimum neutrino energy from the formal reaction thresholds.[29]
2. The electron energy is measured with finite resolution, and this blurs the minimum neutrino energy that induces a reaction for the same four experiments: Table 6 gives the central value.
3. Thanks the possibility to measure the energy the individual contributions of the various types of solar neutrinos can be distinguished and measured. Kamiokande, Super-Kamiokande and SNO can see boron and hep neutrinos whereas Borexino can explore all types of neutrinos, and did it for those of the pp-cycle.
4. The neutrino-electron elastic scattering reaction (last three cases of the table) produces also an invisible neutrino in the final state carries away energy. The minimum neutrino energies given in the table, that correspond to the electron energies given above, are estimated by the formula $E_\nu^{\min} = (T_e + p_e)/2$.
5. Recall that ν_e cross section on electron is the larger one, whereas those of $\bar{\nu}_e$, $\nu_{\mu,\tau}$ and $\bar{\nu}_{\mu,\tau}$ are about 3/7, 1/6 and 1/7, respectively; the last two are purely due to neutral currents.

See again Fig. 17 for a synoptic view, with emphasis on the considerations here above.

[29] In water Cherenkov detectors, electrons with minimum velocity $\beta > 0.75$ can be in principle detected and in scintillators the formal energy threshold is still less. The true issue that fixes the lowest energy threshold that can be detected, is the presence of background events.

Homestake

The pioneer experiment was conceived in the 1960s mostly thanks to Ray Davis and John Bahcall, who were originally a radiochemist and a (nuclear) physicist, respectively, and due to this enterprise became and are remembered as renowned astrophysicists. The goal was to test the mechanism of energy production in the Sun via the observation of electron neutrinos, although this was not considered particularly appealing by most physicists and astrophysicist by the time [2]. The idea was to track the (small) number of radioactive argon-37 nuclei created by charged current neutrino interactions (see Table 6). Since the preliminary results, obtained in the very late 1960s, the data indicated a very low neutrino counting rate. Eventually, this was measured to be 1/3 of the predictions of Bahcall. These results (both theory and measurement) remained stable in the course of the time, till the publication of the full data set in 1998 [144]. These results were often regarded with diffidence, until Super-Kamiokande provided us in the same years a very strong evidence of similar phenomena (i.e., neutrino oscillations). Homestake gave us the first direct validation, once neutrino oscillations are accounted for, of energy generation in the Sun along with an "inclusive" measurement—namely, due to several types of solar neutrinos at once.

Kamiokande

The original purpose of this experiment was the search for nucleon instability phenomena, very popular among theorists since the 1970s, a period of time sometimes referred as the (beginning of the) "gauge theory era". In fact, Kamioka is the name of the mine whereas NDE is an acronym that means nucleon decay experiment. The idea of the experiment was to use quite inexpensive large mass of water as a supply of nucleons and as a detector. In fact, relativistic charged particles, in particular electrons and muons produced by weak neutrino interactions, produce in water Cherenkov radiation that propagates freely for many tens of meters. The new features of this type of detector, used as a neutrino *solar neutrino telescope*, are the possibilities to measure (1) the direction of the electron, and thus of the neutrino; (2) the energy; (3) the time of the event (=real-time detection). Some 50 events from the direction of the Sun allowed to confirm the deficit of solar neutrinos [145, 146] also when observed by mean of the elastic neutrino scattering reaction and to obtain the first "neutrino image" of the Sun, providing confidence in the successful detection of solar neutrinos.

Gallex/GNO and SAGE

One approach to address the solar neutrino deficit with a minimum bias was to measure the pp- and pep-neutrinos that are reliably known thanks to the luminosity. However, this required to obtain a very low-energy threshold. This is possible also in a radiochemical experiment exploiting the levels of the ^{71}Ga-^{71}Ge transition (see Table 6) as pointed out by Kuzmin in URSS [147] and by Bahcall, Cleveland, Davis and other ones in USA [148]. This approach was implemented in the two experiment

SAGE and Gallex (then, GNO) that obtained consistent results that confimed once more the existence of a deficit. E.g., the expectation from the standard solar model in Gallex was to have some 9 ^{71}Ge atoms each 28 days (=each extraction), mostly due to pp and pep-neutrinos, that yield 55% of the events. The measured counting rate was just slightly larger than pp and pep-neutrinos alone, and it was not possible to claim a deficit of pp- and pep-neutrinos, leaving formally open the possibility that only the high-energy neutrinos were badly misunderstood [149]; as already recalled, the results of SAGE [151, 152] were in very good agreement. However, as emphasized by Bahcall, the disagreement with the standard solar model of Gallex and SAGE results amounted to an impressive figure of 9 sigma. After some tests of the procedure of the extraction of the radioactive Germanium nuclei (produced by mean of an artificial neutrino source) and combining with the data of GNO [150] the measured value of the flux lowered a bit, getting even closer to the one measured in SAGE and making even less plausible the above formal possibility of an alternative interpretation. Moreover, these results and the small errors obtained amounted to a new and important information on solar neutrino fluxes.[30] As mentioned in Table 7 GNO was terminated for non-scientific reasons whereas SAGE is still working.

Super-Kamiokande

The main motivation of this experiment was to clarify as much as possible the issues raised by Kamiokande and other detectors, in particular, those concerning atmospheric neutrino oscillations: Indeed, despite the explicit reference to Kamiokande, the acronym NDE is now supposed to stand for "neutrino detector experiment". However, the experiment was conceived as multipurpose and its target included also solar neutrinos. Large photo-collection efficiency is important for low-energy event detection; photo-coverage is 40%, twice than in Kamiokande. Thanks to the huge mass (see Table 7) the solar neutrino collection rate was for the first time very large, about 15 solar neutrino events/day. In fact the detector collected many 10,000 of events [153–156]. Boron neutrinos have been precisely measured, whereas the highest energy neutrinos, namely, hep neutrinos, have a low flux that is just barely scratched by the present observations of Super-Kamiokande and SNO. The measurement of the rate, of the spectrum and of the temporal distribution obtained by Super-Kamiokande are consistent with those of the other detectors, assuming the correctness of the standard solar model predictions of Bahcall and the existence of neutrino oscillations. Moreover, a precise determination of the oscillation parameters was made possible since the beginning of the new century, identifying the so-called large mixing angle solution.

SNO

There was another very direct and minimum bias approach to address the question of solar neutrino deficit in comparison with the predictions of the SSM. This was the idea

---

[30] However, overall, only some 1000 events were collected and therefore the error in the measurement is still some times larger than the theoretical error on pp- and pep-neutrinos.

of observing neutral current reactions due to all types of neutrinos, and not only the electronic ones seen in radiochemical experiments and having the largest interaction cross section with electrons. As suggested in 1984 [157], the ideal material was heavy water[31]: dissociation of deuterium was the relevant reaction, and therefore the goal became the one to identify isolated neutrons. This required to obtain unprecedented levels of radiopurity. Three different techniques were implemented and neutrons were effectively measured in SNO [140, 158]. In this manner it was understood that the flux measured with neutral current was higher than the one measured by electron neutrinos only, and moreover, in excellent agreement with the prediction of the standard solar model, that after more than 30 year was finally vindicated. Subtracting from the flux measured by SNO the flux extracted using charged current reactions, it was possible to extract the flux of *non-electronic neutrinos* whose measurement was a very strong proof that neutrino oscillations are positively occurring. The charged current measurements also contributed to obtain a rather precise information on the boron neutrinos and their spectrum [159, 160]. Unfortunately, the heavy water had to be eventually returned to the Canadian agency for atomic energy and this is why this telescope was eventually discontinued.

Borexino

Borexino is the first telescope, after Homestake, whose principal aim is the observation of solar neutrinos rather than the investigation of neutrino properties. When we compare the light yield of the same neutrino-electron scattering reaction in a scintillators and in a Cherenkov detector, we realize that we have many more photons in the first case, which makes scintillator detectors potentially sensitive to very low energies. This fact, together with extraordinary success to shield the detector effectively and most of all to eliminate most of the radioactive contaminants, allowed Borexino to accept events with visible energy above 200 keV, the adopted threshold, that could be even lowered till 150 keV. Measurements of the beryllium-7 [161] of the boron-8 (with $E_\nu \geq 3.2$ MeV) [162] of the pep [163] and also of the pp neutrinos [164] were obtained. Summarizing, the observation of all types of pp neutrinos below hep allowed us to reconstruct the entire spectrum and thus to test directly the main features of the large mixing angle solution. A very important point that needs to be stressed is that the measurement of pp neutrinos is in good agreement with the expectations from the measured solar luminosity. Recall that, while neutrinos take just 2.3 s to escape from the center of the Sun, the light takes few 100 thousand years. Therefore, the good agreement between the *present* solar luminosity and the flux of pp neutrinos can be regarded, if a joke is allowed, as a life insurance policy stipulated by mankind with our Sun. The results of the first phase of operation of Borexino are summarized in [165] and [87], see also Table 8. The only significant limitation of Borexino as a neutrino telescope is the relatively small detector mass which reflects

[31] Moreover, heavy water is a transparent liquid and this allowed to adopt the Cherenkov detection technique.

Table 8 Comparison of the solar neutrino fluxes predicted by a couple of SSM (3rd and 4th column), those obtained in a global fit which includes all experimental results obtained till beginning 2016 (5th column) and the results of Borexino (last column). The last two lines show the main components of CNO neutrinos, still to be seen to date. Adapted from [87]

SSM flux component	Units ($cm^{-2} s^{-1} \times$)	High Z/A [GS98]	Low Z/A [AGS09]	Global fit	Borexino result
pp	$\times 10^{10}$	$5.97(1 \pm 0.007)$	$6.04(1 \pm 0.007)$	$5.97^{+0.037}_{-0.033}$	$6.6(1 \pm 0.106)$
^{7}Be	$\times 10^9$	$5.00(1 \pm 0.07)$	$4.56(1 \pm 0.07)$	$4.80^{+0.24}_{-0.22}$	4.43 ± 0.22
pep	$\times 10^8$	$1.44(1 \pm 0.012)$	$1.47(1 \pm 0.012)$	1.448 ± 0.013	1.63 ± 0.35
^{8}B	$\times 10^6$	$5.58(1 \pm 0.14)$	$4.59(1 \pm 0.14)$	$5.16^{+0.13+0.30}_{-0.09-0.26}$	5.2 ± 0.3
^{13}N	$\times 10^8$	$2.96(1 \pm 0.14)$	$2.17(1 \pm 0.14)$	≤ 13.7	≤ 7.7 (sum)
^{15}O	$\times 10^8$	$2.23(1 \pm 0.15)$	$1.56(1 \pm 0.15)$	≤ 2.8	≤ 7.7 (sum)

in the achieved precision of certain measurements. Borexino is still operating as a solar neutrino detector, as will be discussed later on. Thanks to its scientific success, it is becoming a model for a new generation of neutrino detectors and solar neutrino observatories.

Open Problems and Next Steps

Improved solar models

The wide recognition of solar neutrino oscillations culminated in the 2015 Nobel prize in physics awarded to T. Kajita (Super-Kamiokande) and A.B. McDonald (SNO). This translated in a recognition of the correctness of the SSM but occasionally also in a radical shift from the initial suspicious attitude to an acritical acceptance of its predictions. Interestingly, Bahcall himself remained instead rather cautious toward the predictions of SSM. Actually, it was him, along with his collaborators, who clearly formulated one of the major open problems, that concerns the comparison of the SSM and the observations, that we describe briefly. Indeed, new analyses of the solar atmosphere infer lower abundances of the lighter elements (including C, N, O), and imply major modifications in the solar models. The revised SSM disagrees with the solar profiles of sound speed and density as well as the depth of the convective zone and the helium abundance [166, 167]. A lot of useful work has been performed to examine and assess the nature of this discrepancy, see in particular [168–170]. See [171] for a partial summary of the present situation, where it is given less emphasis

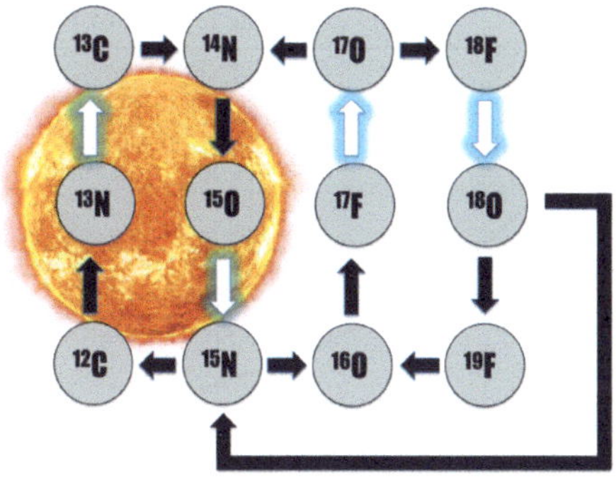

Fig. 18 The complete cold CNO cycle. In very good approximation only the CNO-I branch is active in the Sun; e.g., in units of $10^8/(\text{cm}^2\ \text{s})$, the flux of C13 is 2.96 and the one of N15 is 2.23 whereas F17 adds 0.06 to the latter. Recall the endpoint of C13 is 1.2 MeV and the one of N15 is 1.7 MeV; F17 from CNO-II has a shape very similar to the one of N15

to the issue and more emphasis to the work still to be done. From the point of view of neutrino astronomy, it seems that the observation of a secondary branch of neutrinos, connected to the abundance of the light elements, is the essential step that should be taken: this is discussed in greater details here below. Indeed, the revised model, called AGS09 and dubbed as "Low Z/A", leads to expectations that differ significantly from the old one, called GS98 and dubbed as "High Z/A" as we can see from Table 8.

Detection of CNO neutrinos

Besides the pp-cycle, the Sun produces energy also by subdominant set of reactions, named cold CNO cycle, or sometimes CN-NO bi-cycle, or for short (and in analogy with the pp-cycle) CNO cycle, see Fig. 18. This cycle is very important for stars larger than our Sun and is expected that it is dominant one already when the stellar mass is larger than $1.5 \times M_{\odot}$. Therefore, the investigation of its functioning is rather relevant in astrophysics. Carbon, oxygen and nitrogen isotopes are involved in this catalytic cycle: see Fig. 18. The efficiency of this cycle is (evidently) very sensitive to the actual abundance of these elements. Therefore, the issue of the abundance of light elements has a direct impact on the expectation of the CNO neutrino fluxes, which is a new important motivation for its study by mean of neutrinos.

In order to get a better grasp of the issue, let us consider the spectrum of elastic scattering events due to solar neutrinos in the region of energies around 1 MeV, of interest for CNO neutrinos. We adopt GS98 abundances (see Fig. 18 and Table 8) and calculate the spectrum as a function of the kinetic energy of the electron, considering 3.3×10^{31} target electrons (100 tons of PC), one year of data taking, using the tree level cross section and including vacuum oscillations in the simplest approximation ($P_{ee} = 0.54$). The ideal spectrum is showed by dashed lines in Fig. 19; three of them are from pp cycle, one of them is from the CNO cycle.

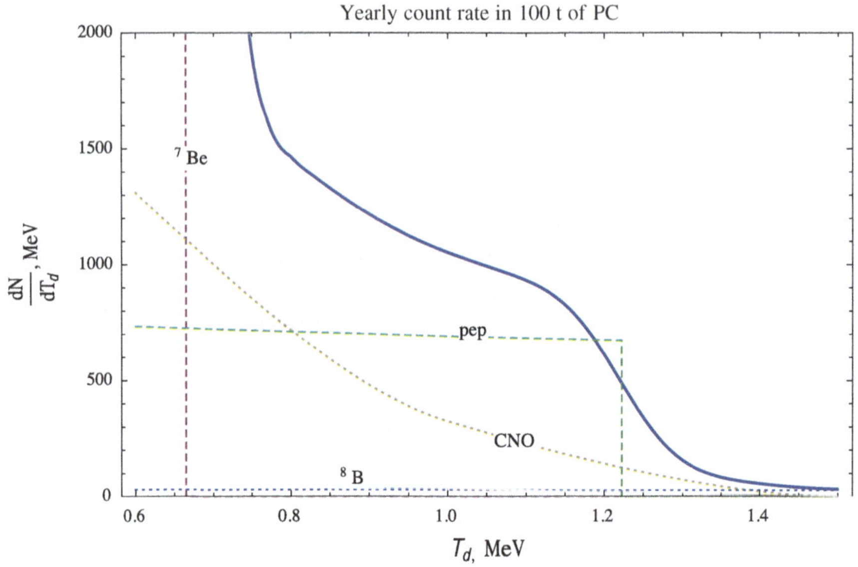

Fig. 19 Theoretical electron recoil spectrum in the region of CNO neutrinos. The individual components of the spectrum are indicated with their label; the total spectrum includes the effect of the energy resolution, as in Borexino detector. See text for a detailed discussion

We show the effect of the energy resolution function of Borexino in the total spectrum $\delta T/T = 5\%/\sqrt{T/\text{MeV}}$. Evidently, the resolution blurs the spectrum, and increases the correlation between the observed signals. However, this effect is understood and its description should only improve (see below). We note that the CNO spectrum is not completely featureless; the plot in linear scale of Fig. 19 shows that a mild change of slope at about 1 MeV due to the endpoint of C13 spectrum.

The only detector at present that has a chance to measure CNO neutrinos is Borexino, as evident from Fig. 17. However, there are three important experimental issues, see [165] and [87],

1. The first is due to ^{11}C present in the scintillators that however Borexino has learned to keep under control.
2. Then, as already noted, the determination of the CNO-flux is correlated with pep and Be neutrinos. Note however that the pep signal is fixed quite reliably by the luminosity constraint, and therefore can be subtracted.
3. Most importantly, there is one isotope of the bismuth (from the radon decay chain) that decays with a lifetime of 138 days as, $^{210}\text{Bi} \to {}^{210}\text{Po}{+}e + \bar{\nu}_e$, and unfortunately, the spectrum of the emitted electrons resembles closely the spectrum of the electrons from CNO neutrino interactions.

The upper bounds obtained by Borexino to date are scratching the higher predictions of the high metallicity, but cannot discriminate them yet.

A new setup to ensure detector stability was implemented in Borexino that together with an extensive calibration by mean of artificial neutrino sources, offers concrete hopes of obtaining soon a first experimental determination. In fact, in order to avoid refurbishing the contaminants, a superior detector stability (especially thermal stability) is needed; but there is also another specific reason [173]. Consider the sequence of β decays that connects lead-210, bismuth-210 and finally polonium-210, *in absence of external sources of contaminants*. The activities of the first two species reach secular equilibrium, $n_{\text{Bi}}(t) \approx n_{\text{Pb}}(0)$ while the activity of of Polonium is,

$$n_{\text{Po}}(t) \approx [n_{\text{Po}}(0) - n_{\text{Bi}}(0)]\, e^{-t/\tau_{\text{Po}}} + n_{\text{Bi}}(0) \tag{90}$$

(recall that the 'activity' is the ratio between the number of nuclei of a certain species and its lifetime $n_i = N_i/\tau_i$). Thus, using the signal from the α decay of Polonium we measure precisely the Bismuth. This provides us with an independent information, precious to extract the signal of CNO neutrinos.

New directions

Two interesting future experiments that evolve from Borexino experiments, are considered at the Gran Sasso National laboratories.

(1) The future experiment SOX [174], aims at investigating the existence of sterile neutrinos using the Borexino experiment along with an intense artificial source of electron neutrinos, produced via electron capture. This experiment will also allow a precise calibration of the Borexino experiment itself, with a potential impact on solar neutrino analyses.

(2) A detector based on liquid argon, depleted of ^{39}Ar isotope, is considered interesting for future steps [175]. In fact the scintillation light yield in liquid argon is 10 times more than in Borexino; thus the energy resolution that can be achieved in this type of detector can reach 1% at 1 MeV. Note that the design of the dark matter detector DarkSide-50, already, was based on Borexino and it is possible to learn from their experience, just as Borexino learned from the CTF. The issue of the background has been studied in simulations, the ^{222}Rn being apparently the most relevant one, but the measurements is anticipated to be feasible with an exposure of some 100 ton $\times$ year.

Finally we recall that besides the astrophysical reasons discussed above, these experiments will allow to investigate various topics, ranging from standard to speculative ones, e.g.: the value of $\sin^2\theta_W$ in a low-energy reaction; the existence of new interactions, e.g., those due to magnetic moment [176]; the presence of new light neutrinos (of various mass) affecting the solar neutrino spectrum [177–179]; etc.

Core Collapse Supernovae

Core collapse supernovae are very important astronomical objects and neutrinos can contribute to their understanding, allowing to monitor the formation of the compact star and the first instants of the explosion. However, while neutrino telescope are ready to collect the signal from the next (galactic) supernova, the astrophysics is difficult and despite the enormous efforts of a wide community of researchers, the theory is still beset with doubts. In this section, we describe the expectations for neutrino astronomy in its general aspects, focussing on those relevant to interpret the only observation we have to date. This is a sample of slightly more than 20 (electron anti)neutrino events, collected from SN1987A. We will argue that, despite the paucity of this dataset, it allows us to extract a lot of useful information.

A few important facts concerning the future galactic supernova should be mentioned—see [21] for more discussion and references. The rate of explosion is not known precisely, but a reasonable expectation is that we have a core collapse supernova each 50 ± 20 years in the Milky Way. The location of the supernova cannot be predicted in advance. However, it is reasonable to model the distributions of the stars, and therefore also the possible supernovae, with an exponential distribution with scale of 3 kpc the distribution of the stars that can produce a supernova. In this manner, we find that 68.2% of the mass is in the inner 7 kpc, and that the typical distance from us is $D = 10 \pm 5$ kpc that are both similar to the distance of the galactic center, 8.5 kpc.

Finally, before proceeding, we recall here some milestones in this field of neutrino astronomy, emphasizing the aspects more relevant for us,

1. Colgate and White 1966 [180] propose the paradigm for the explanation of the core collapse supernovae, where neutrinos are the key agents.
2. Bisnovatyi-Kogan 1970 [181] suggests that magnetic fields and rotations in the core of the collapsing star may have a central role to understand the explosion.
3. Nadyozhin 1978 [182] concludes the first detailed simulation of a supernova that shows that the scale of emission is of several seconds and demonstrates an initial phase of intense neutrino luminosity.
4. Bethe and Wilson 1985 [183] suggest that the energy deposition due to the intense initial neutrino luminosity can re-energize the stalled shock wave.
5. 1987: Kamiokande-II [184], IMB [185], Baksan [186] and LSD [187] observe several events in correlation with SN1987A.
6. Various authors—e.g., Bahcall 1989 [188]—remark that non-LSD data generically meet the expectations.
7. Loredo and Lamb 2002 [189] note that SN1987A data analysis suggests an initial phase of intense luminosity, as the one expected in the delayed scenario.[32]
8. Imshennik and Ryazhskaya 2004 [190] suggest a 2 stage scenario with essential role of rotation and possible explanation of LSD events detected 4.5 h earlier.

In most of the subsequent discussion, we will make direct or indirect reference to the existence of one main phase of neutrino emission lasting several seconds.

[32]With the term "delayed scenario" one refers to the scenario for the explosion put forward by Bethe and Wilson that incorporates the initial phase of intense neutrino luminosity discussed by Nadyozhin (this is also called "standard scenario" or "neutrino assisted explosion").

Expectations

In this section, we recall some generic and rather well-known facts; for a more extensive introduction, see the review paper [38].

Compact star connection and neutrino energetics

The gas expanding from the rest of a supernova (i.e., a supernova remnant) carries a remarkable kinetic energies of the order of 10^{51} erg that is few orders of magnitudes more than the light emitted by supernovae. This is a large amount of energy, but far below what has to be released in order to allow the formation of a compact star. In fact, estimating the potential energy of a star as $\mathscr{E}_{pot} = 3/5 G_N M^2/R$, and using the Chandrasekhar mass $M = 1.44\ M_\odot$ and $R = 12$ km, we have $\mathscr{E}_{pot} = 2.8 \times 10^{53}$ erg, that is about 10% of the rest energy Mc^2.[33] Note however that the value of the total energy not be considered as precisely known, but should be attached a generous error bar, at least in a conservative treatment.

Theoretical astrophysics leads to expect that the massive stars above $\sim 8 M_\odot$ undergo the generic evolutionary sequence: formation of a iron core/ collapse of the iron core/formation of a compact remnant/supernova explosion. Since in the very moment of the collapse the matter is dense and strongly coupled, the neutrinos are the most effective coolant, and they allow the release of the potential energy in a sufficiently short time. In fact, neutrinos are expected to carry 99% or more of the energy (except when a black hole forms and the energy does not emerge).

Models for neutrino emission

The simulations of the process are difficult, however, they pass through some emission phases, that (at least in simulations) can be reasonably well identified. During the collapse, the main radiation is due to electron neutrinos emitted in neutronization processes as $e + p \to n + \nu_e$; a peak of ν_e luminosity is also expected, but the integrated energy is estimated to be less that 1% (thus, the number of events potentially observable is small). Then there is a phase of intense luminosity, lasting some fraction of a second, during the initial rapid accretion onto the nascent neutron star; 10–20% of the energy carried by neutrinos is emitted in this phase. Most of the energy is radiated in quasi-stationary conditions, when the protoneutron star cools by emitting neutrinos and antineutrinos of all species in a time scale from few to 10 s.

Simulations suggest that the six types of neutrinos and antineutrinos, ν_e, ν_μ, ν_τ, $\bar{\nu}_e$, $\bar{\nu}_\mu$, $\bar{\nu}_\tau$ carry the same amount of energy; while this expectation called *equipartition* could be not exact in reality, it is considered reliable within some 30%. Moreover,

[33] For comparison, a uniform sphere of the same mass and radius, rotating with a period of 5 ms, has a rotational energy of 1.3×10^{51} erg; if it moves with a velocity of 1,000 km/s (as for the fastest observed pulsars) its translational energy is 1.4×10^{49} erg; if it is filled by a uniform magnetic field of 10^{15} G (as for the strongest magnetars) it stores 3×10^{47} erg of magnetic energy.

the deviations from the isotropy of the emission should be small, especially in the last phases of emission, while in the initial one could amount to some 10%.

The other crucial information to know is, which is the expected energy distribution. The simplest model is just the one of thermal emission, when neutrinos are emitted from the surface of the protoneutron star, called neutrinosphere. When the protoneutron star reached a stationary situation, it is sensible to assume a (quasi) thermal emission with temperature T, up to small deviations. If each species radiates a luminosity $L \sim R^2 \times T^4$, where $T = 4$ MeV and $R = 15$ km, the neutrinos of all species can carry away most of the stellar energy in some 10 s, since $6 \times L = 2.2 \times 10^{52}$ erg/s.[34]

In the rest of this discussion, we will not consider a detailed description of the flux of neutrinos but only of its integral over the time, namely, the fluence. In other words, we will focus simply on the spectrum of neutrino emission, whose modeling should be plausibly easier. Note that while a value of the temperature around 4 MeV is reasonable and in agreement with recent simulations, the current status of the field does not justify to treat it as a precisely known quantity.

Cautionary remarks

Neutrino oscillations should have a minor role for the observations, owing to the fact that the neutrinos are emitted in similar amount and distribution. The larger effects could obtain for electron neutrinos and/or possibly for the first phases of emission, when there is the maximum dissimilarity between the flavors. It should be noted however (1) that this expectation relies on recent simulation, since in the past it was believed, instead, that the average energies of the three species of neutrinos were very different (2) the theoretical description of neutrino oscillations is difficult and it is considered to be incomplete, since the large amount of neutrinos during the emission affects oscillations, making the problem strongly nonlinear.

Finally, we recall that while the description of neutrino emission is considered to be reliable (even if the errors are not precisely quantified) the nature of the explosion is not. Despite the great improvements, the simulations are not yet able to reproduce the visible manifestation in a clear manner, especially for the stars with the largest masses. Since the energy entailed in the explosion is small, a common attitude is to postpone its understanding to future simulations, focussing for the while on the description of neutrino emission. However, it is evident that such an uncertain theoretical situation makes a cautious attitude quite advisable. For this reason, in the following of the discussion, we will keep theoretical considerations to a minimum, and will instead emphasize the role of the only observation we have to date.

[34]The initial phases of emissions should be modeled in a different manner. For instance, $\bar{\nu}_e$ derive from $n + e^+ \to p + \bar{\nu}_e$, and the luminosity $L \sim N_n\, \sigma_{e^+n}\, T^4_{e^+}$ is one order of magnitude larger than the late time luminosity for reasonable parameters.

Table 9 Relevant features of the four neutrino telescopes, operated at the time of SN1987A, and of the number of events then detected at or above the energy threshold

	Protons (10^{32})	Threshold (MeV)	Observed events	Background events in 30 s	Time of 1st event (UT)
Kamiokande-II	1.4	7.5	11	0.55	7 h 35 min 35 s
IMB	4.6	15	8	0.01	7 h 35 min 41 s
Baksan	0.2	10	5	1	7 h 36 min 12 s
LSD	0.1	5	5	0.3	2 h 52 min 37 s

SN1987A

The data

In the hours before the visible manifestations of SN1987A, bursts of events were observed in four neutrino telescopes, namely Kamiokande-II, IMB, Baksan and LSD. A few significant features are given in Table 9. The observations of the first three telescopes are compatible with a common origin attributed to SN1987A, when we consider the $\sim$1 min inaccuracy in the absolute time measurement of Kamiokande-II and Baksan. Some interpretations of the observations of the last telescope in Table 9 have been proposed, assuming strong deviations from the astrophysics of supernova neutrino emission described above and admitting some tension in the interpretation; however, in the conventional interpretation of the data, the LSD observation is usually left apart.

The data of Kamiokande-II, IMB, Baksan (see [37] for a summary) are compatible with the observation of electron antineutrinos for a core collapse event. This hypothesis is based on the general expectations, including the fact that the IBD reaction of electron antineutrinos has a comparably large cross section in this type of the detector, and it is discussed below. The duration of the burst, as deduced by the data, is of some 10 s; there is a hint of a bright emission phase in the first (fraction of a) second.

Electron antineutrino fluence from SN1987A

We assume the simplest reasonable form of the antineutrino fluence

$$F(E_\nu) = \frac{\mathscr{E}}{4\pi D^2} \frac{E_\nu^2 \, e^{-E_\nu/T}}{6T^4} \tag{91}$$

where $D = 50$ kpc and the two free parameters $\mathscr{E}$ and T describe the total energy radiated and the temperature of the neutrinos, respectively; the average neutrino energy is simply $\langle E_\nu \rangle = 3T$. The best fit values found in a recent accurate analysis of SN1987A observations [37] are

$$\mathscr{E} = 5 \times 10^{52} \text{ erg and } T = 4 \text{ MeV} \tag{92}$$

They rather stable and in good agreement with the values found in previous investigations. If assuming the conservative ranges $\mathscr{E} = (1 - 20) \times 10^{52}$ erg and $T = 1 - 10$ MeV we integrate the likelihood over the other parameter, we find marginalized 1 dimensional likelihoods, and we can estimate the 1 sigma ranges $\mathscr{E} = 4.8^{+2.3}_{-1.0} \times 10^{52}$ erg and $T = 3.9^{+0.5}_{-0.3}$ MeV. If instead we find the various momenta $\langle \mathscr{E}^n \rangle$ and $\langle T^n \rangle$, where $n = 1, 2...$, by using the likelihood as a probability distribution function, and define as usual $\sigma^2_{\mathscr{E}} = \langle \mathscr{E}^2 \rangle - \langle \mathscr{E} \rangle^2$, $\sigma^2_{\mathscr{E}T} = \langle \mathscr{E}T \rangle - \langle \mathscr{E} \rangle \langle T \rangle$ etc., we obtain symmetric ranges $\mathscr{E} = 5.4 \pm 1.7 \times 10^{52}$ erg and $T = 4.0 \pm 0.4$ MeV. Note that the ranges obtained with the two procedures are rather similar. The correlation coefficient $\rho = \sigma^2_{\mathscr{E}T}/(\sigma_{\mathscr{E}} \sigma_T)$ is negative and large, $\rho = -0.7$, since in order to keep the number of events fixed to the observed value, one should vary $\mathscr{E}$ and T in opposite directions. One can consider the value of the fluence for any fixed value of the energy as a random variable and estimate its error by assuming a linear behavior of F and a adopting a Gaussian procedure, i.e., setting $\delta F^2 = \sum_{ij} \partial F/\partial x_i \, \partial F/\partial x_j \, \sigma^2_{ij}$ (in our case $x_1 = \mathscr{E}$ and $x_2 = T$). We find the simple result

$$\frac{\delta F}{F} = 23\% \times \sqrt{1 + 8.3 \left(\frac{E_\nu}{24 \text{ MeV}} - 1 \right)^2} \tag{93}$$

that is quite reasonable; the fluence is best known where more abundant and much less known as we depart from this point; the minimum uncertainty is similar to the statistical uncertainty on the number of observed events, $1/\sqrt{N_{\text{obs}}} \sim 20\%$.

Inference on the next galactic event

The IBD event rate can be estimated easily as [37]

$$\frac{dN_{\text{IBD}}}{dE_e}(E_e) = N_p \, \sigma_{\text{IBD}}(E_\nu) \, F(E_\nu) \, \frac{dE_\nu}{dE_e} \text{ where } E_\nu = \frac{E_e + \Delta}{1 - \frac{E_e}{m_p}} \tag{94}$$

If we rescale the distance of the supernova from $D = 50$ kpc, appropriate to SN1987A, to $D = 10$ kpc, appropriate to a typical galactic event and use the number of protons of Super-Kamiokande $N_p = 2.1 \times 10^{33}$, corresponding to 32 kton of water, we expect $N_{\text{IBD}} = 6,200$ IBD events in the range $E_e = 4$–60 MeV and with the best fit parameters of Eq. 92, while using the range of Eq. 93, i.e., using the fluence $F \pm \delta F$, we have the range,

$$N_{\text{IBD}} = 5{,}500\text{–}7{,}900 \tag{95}$$

The expectations are further illustrated in Fig. 20.

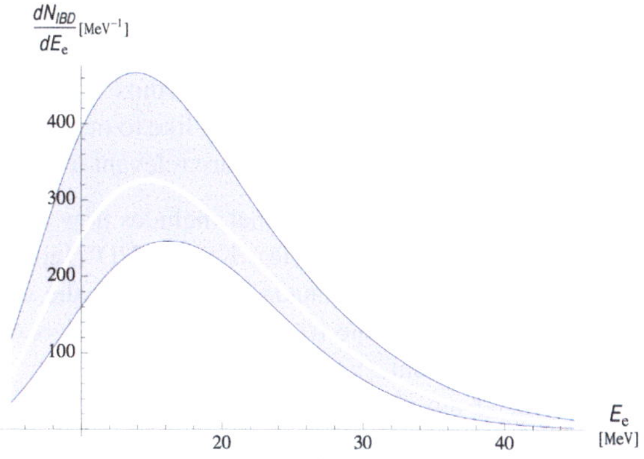

Fig. 20 Expected distribution (spectrum) of the IBD signals in 32 kton of water, for a supernova exploding at 10 kpc. The inner curve has the parameters of Eq. 92. The *purple bands* are uncertainties of the fluence quantified by Eq. 93

Open issues on SN1987A

The fluence discussed above agrees well with the theoretical expectations. More precisely, the value of the radiated energy is in a very good agreement with the typical expectations from a core collapse. Also the temperature agrees with the results of the newest simulations.[35] Two analyses of the data [189, 191] have also found some hint of the initial luminous phase with a significance of about 2 sigma. The remaining issues are three,

1. *directionality:* The events of Kamiokande-II and of IMB, point, on average, in the same direction of arrival of neutrinos. Only the first event of Kamiokande-II, however, has some chance of being due to elastic neutrino scattering; moreover, IMB has some angular bias that favors such an outcome [192]. The significance of this issue is not very strong.
2. *LSD events:* this has been already discussed above. Moreover, we add here that the 5 events observed have very similar energy, not far from the energy threshold, which constrain the interpretation further. For further discussion and references, see [333].
3. *lack of neutron star:* This is perhaps the most serious issue. If the cooling of neutron star is reasonably well modeled, one expects to see, if not a pulsar, at least a X-ray point source. The obtained bounds are several times tighter than the standard expectations [193]. It does not seem impossible to evade this expectation, but the search for a compact remnant remains of prioritary importance.

[35] In the past years, we had several debates on whether the data are in agreement among them and/or with the expectations. Nowadays, both questions are answered affirmatively, and there is a growing awareness of the presence of residual theoretical errors in the expectations.

A Future Galactic Supernova and Neutrino Astronomy

It is evident that the detection of neutrinos from a future galactic supernova will allow us to proceed a lot.[36] We would like to mention here only the main progresses, that we expect to achieve and that are relevant to neutrino astronomy,

- the SNEWS network [361], that includes now Super-Kamiokande (Japan), LVD (italy), IceCube (South Pole), KamLAND (Japan), Borexino (Italy) Daya Bay (China), and HALO (Canada) will provide the astronomical community with a clear and unmistakable trigger and prompt alert of the occurrence of a Galactic core collapse event.
- The Super-Kamiokande detector, thanks to the elastic scattering reaction of neutrinos on electrons, will be able to point to the supernova with a precision of few degrees, several hours before the light will be visible.
- The network of detectors will study the time distribution of events very precisely.
- The electron neutrino flavor will be studied in argon-based detectors (e.g., DUNE) and possibly also exploiting the elastic scattering reaction on electrons.
- The scintillator detectors (LVD, KamLAND, Borexino, SNO+) will study the neutral current events via the 15.11 MeV γ line from neutral current excitation of carbon-14, and in the case of ultra-pure scintillators, also via neutrino elastic scattering on protons. A similar goal will be achieved by the lines of neutral current excitation of oxygen-16 in Super-Kamiokande.

See Fig. 21 for an illustration and [80, 81, 194, 195] for more detailed discussion.

Relic Supernova Neutrinos

The flux due to the core collapse supernovae that happened in the Universe in the course of the time, is called *relic supernova neutrino* flux or also *diffuse supernova neutrino background*: See [39, 40] for reviews and references to the historical literature.

The formalism to account for a cosmological distribution of sources is described in the Appendix 4. Focussing on the antineutrino signal, the formula for the diffuse flux is given by a convolution of the cosmic supernovae with the average, redshifted spectrum $dN_{\bar{\nu}_e}/dE$ of emission, namely

$$\Phi_{\text{relic}\bar{\nu}_e}(E) = D_H \int_0^\infty dz\, n_{\text{sn}}(z)\, \frac{dN_{\bar{\nu}_e}}{dE'}\bigg|_{E'=E(1+z)} \tag{96}$$

[36] Note that the sparse data from SN1987A have been used to extract a lot of useful information, and despite the large amount of work done already, it is not excluded that it will possible to perform further useful work of analysis with these data.

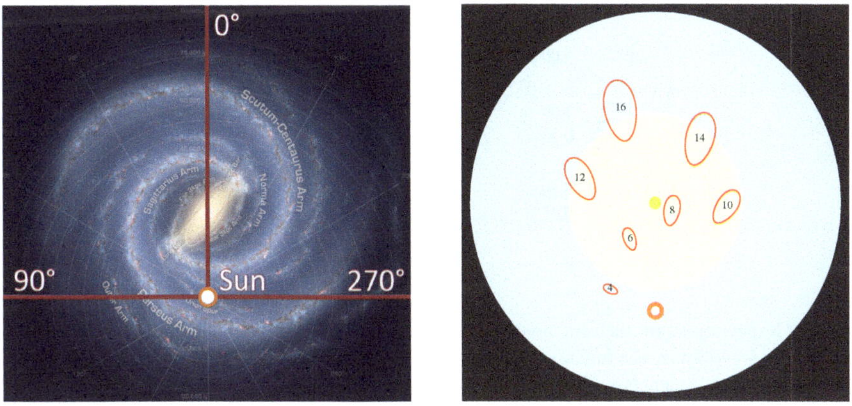

Fig. 21 Face-on views of the Milky Way. *Left* A detailed reconstruction, from Wikipedia. *Right* A sketch: The *orange circle* indicates the Sun, the outer circle is at 15 kpc from the center, the inner one is at 7 kpc. We show a few supernova events, as located by the neutrinos at 1 sigma. We assume to know the total number of events within 20% at 10 kpc and the galactic longitude within 3° thanks to elastic scattering events. The figures give the distance from the Sun in kpc

A reasonable parameterization of the distribution of cosmic supernovae is [196],

$$n_{\rm sn}(z) = n^*_{\rm sn} \times \frac{\left(\sum_{i=0}^{2} k_i (1+z)^{-\eta\alpha_i} \right)^{1/\eta}}{\sqrt{\Omega_\Lambda + \Omega_m \,(1+z)^3}} \tag{97}$$

We have $\eta = 10$, $z_0 = 0$, $z_1 = 1$ and $z_2 = 4$, $k_0 = 1$, the matching condition $k_{i-1}/(1+z)^{\eta\alpha_{i-1}} = k_i/(1+z)^{\eta\alpha_i}$ imposed at $z = z_i$ with the central values,

$$n^*_{\rm sn} = \frac{1.2 \times 10^{-4}}{\rm Mpc^3 year}, \alpha_0 = 3.4, \ \alpha_1 = -0.3, \ \alpha_2 = -2.5 \tag{98}$$

however, conservatively the uncertainties cannot be treated as small.

The expectations on the observable depend upon the properties of the average cosmic supernova and on those on their cosmic distribution. Moreover, there is a strong dependence of the signal upon the detection threshold E_{th} due to the fact that the neutrinos are redshifted. However, the most important quantity is probably just the mass of the detector. The number of events from inverse beta decay per year in 32 kton of water, as the Super-Kamiokande detector,[37] is

$$N_{\rm ev} = 0.3 - 0.7 \text{ for } E_{th} = 20\text{ MeV (resp., } 1.7\text{–}3.5 \text{ for } E_{th} = 10\text{ MeV)} \tag{99}$$

[37]The same number of targets protons is in a detector of 29 kton with "LAB", as used in SNO+. Scintillators have a better tagging of the events, but they have also an high background rate due to the interactions of atmospheric neutrinos on carbon nuclei via neutral currents.

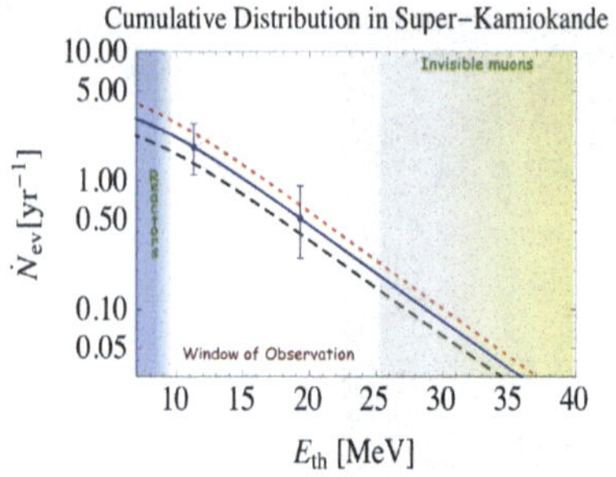

Fig. 22 Expectations for the relic neutrino counting rate in Super-Kamiokande (22.5 kton) based on SN1987A and given as a function of the energy threshold. The *two error bars* are the 2 σ ranges obtained from data analysis. The *three lines* correspond to the extreme models for the cosmic distribution of core collapse supernovae

assuming that the electron antineutrino emission from the average supernova does not differ much from the emission observed from SN1987A [197, 198]. Note that the presence of nuclear reactors on the surface of the Earth makes it impossible to lower the detection threshold below 10 MeV and likewise, those low-energy muons that enter the detector unseen (below the threshold for Cherenkov radiations) will stop and then decay, producing an electron of several ten of MeV, that should be treated as background. See Fig. 22 (from [198]) for a summary.

High-Energy Cosmic Neutrinos: Expectations

The existence of cosmic rays implies the existence of high-energy neutrino sources. The argument is simple: the relativistic protons and nuclei that form the cosmic rays have a non vanishing probability to undergo inelastic hadronic interactions with a target that can be either a gas of ordinary matter or a radiation field. These interactions produce unstable secondary particles that have decay modes that contain neutrinos. The result is that a flux of neutrinos will emerge from all astrophysical sites where cosmic ray protons and nuclei are accelerated or stored [41].

The neutrino emission is accompanied by the radiation of high-energy gamma-rays with a spectrum of similar intensity and spectral shape. This is a consequence of the fact that some of the secondary particles generated in cosmic ray interactions can also decay into photons. In fact the main source of neutrinos is the chain decay of charged pions ($\pi^+ \to \mu^+ \nu_\mu$ followed by $\mu^+ \to e^+ \nu_e \overline{\nu}_\mu$ and charge conjugate modes), while the main source of gamma-rays is the decay of neutral pions ($\pi^\circ \to \gamma\gamma$). The production of the three pion states is related by isospin symmetry and is approximately equal. This implies that the spectra of neutrinos and gamma-rays generated by this "hadronic" mechanism are similar (see more discussion below).

General references:

Two books that examine the cosmic ray connection including neutrinos are [9] and [11]. Besides the review paper mentioned just above, [41], there are several specific review papers that discussed high-energy neutrinos *before knowing the first results* and thereby providing us "predictions". These have been prepared adopting various points of view, e.g., examining the expectations, especially those concerning extragalactic neutrino sources [42–44]; with the goal to collect the formalism needed to discuss extragalactic neutrinos [45]; in order to get prepared to the physics with a new generation of high-energy Cherenkov telescopes (from IceCube onward) [46–49].

The multi-messengers approach

The three fields of cosmic ray studies, gamma astronomy and neutrino astronomy are therefore intimately related, and can in fact be considered as the three faces of a single scientific problem.

High-energy gamma-rays are created not only with the hadronic mechanism discussed above but they can also be generated by relativistic electrons and positrons via bremsstrahlung and inverse Compton scattering. This implies that neutrinos trace the sites where hadronic (protons and nuclei) cosmic rays are present, while gamma-rays trace the sites of both hadronic and leptonic (electrons and positrons) cosmic rays. The understanding of the mechanisms for the generation of the nuclear and electron components of cosmic rays is of fundamental importance in the study of the "High-Energy Universe", and therefore the simultaneous study of the neutrinos and gamma-rays fluxes is crucial for an understanding of the structure and properties of the sources.

High-energy neutrinos propagate with negligible absorption even for very large path-lengths, and when traversing high density environments. This is not the case for gamma-rays, where absorption effects inside the source, or during propagation can significantly reduce the flux (the main mechanism is pair production $\gamma\gamma \to e^+e^-$ in collisions with soft photons). From the two arguments discussed above one can conclude that the ratio ϕ_ν/ϕ_γ can be significantly smaller of unity if the gamma-ray emission is generated by a leptonic mechanism and is unabsorbed, but it can also be much larger than unity if the gamma-ray flux suffers significant absorption.

Components of the Neutrino Flux

The high-energy neutrino flux can be schematically written as the sum of several components:

$$\begin{aligned}\phi_{\nu_\alpha}(E,\Omega) = &\ \phi_{\rm atm}^{\rm standard}(E,\Omega) + \phi_{\rm atm}^{\rm prompt}(E,\Omega) \\ &+ \phi_{\rm Galactic}^{\rm diffuse}(E,\Omega) + \phi_{\rm Extra\ Gal}^{\rm diffuse}(E,\Omega) \\ &+ \sum_{\rm Galactic} \phi_j(E)\,\delta[\Omega - \Omega_j] + \sum_{\rm Extra\ Gal} \phi_k(E)\,\delta[\Omega - \Omega_k]\end{aligned} \tag{100}$$

Atmospheric neutrinos

The first two components describe atmospheric neutrinos generated in cosmic ray showers in the Earth's atmosphere. They are an important foreground to the more interesting observations of the astrophysical components. Atmospheric neutrinos can be split into two components, the first "standard" one is due to the decay of charged pions and kaons, while the second one is due to the weak decays of short lived (hence "prompt") particles containing heavy quarks, with charmed particles accounting for essentially all of the flux. The prompt contribution is expected to be dominant in the atmospheric neutrino fluxes at high energy, when the decay probability of π's and K's is small, because their decay length (proportional to E because of Lorentz time dilatation) become much longer than the interaction length. The atmospheric prompt component has not yet been identified, and its prediction is significantly more difficult than the standard flux, because of our poor knowledge of the dynamics of charmed particles production in hadronic interactions.

Diffuse extraterrestrial flux

The second and third line in Eq. (100) describe astrophysical ν's. "Diffuse" fluxes of neutrinos are associated to ν production in interstellar space in our own Galaxy, and in intergalactic space. The diffuse Galactic emission is due to the interaction of cosmic rays (confined inside the Milky Way by the galactic magnetic fields) with the gas present in interstellar space. The angular distribution of this emission is expected to be concentrated in the galactic plane, very likely with a distribution similar to the one observed for GeV photons by FERMI-LAT [199]. The extragalactic diffuse emission is dominated by the decay of pions created in $p\gamma$ interactions by Ultra High-Energy protons ($E_p \gtrsim 6 \times 10^{19}$ eV) interacting with the 2.7 K cosmic radiation. The significance of these collisions for the shape of the cosmic ray spectrum was first understood by Greisen, Zatsepin and Kuzmin. Because of the very high threshold energy for pion photoproduction, the "GZK" neutrinos are present only at very high energy (with the flux peaking at $E_\nu \sim 10^{18}$ eV).

Point Source extraterrestrial flux

Superimposed to the diffuse ν fluxes one expects the contribution of an ensemble of point-like (or quasi-point like) sources of galactic and extragalactic origin. In Eq. (100) the summation runs over all sources with Ω_j (Ω_k) the celestial coordinates of the galactic (extragalactic) source of index j (k). Neutrinos travel along straight lines and allow the imaging of these sources. It is expected that most of the extragalactic sources will not be resolved, and therefore the ensemble of the extragalactic point

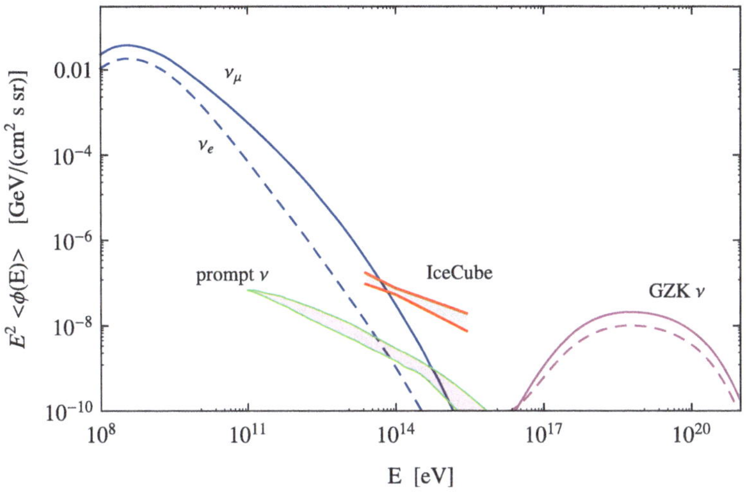

Fig. 23 Neutrino spectra (averaged over all solid angle). The *solid lines* show the sum $\nu_\mu + \overline{\nu}_\mu$, the *dashed lines* the sum $\nu_e + \overline{\nu}_e$). The different curves show the standard atmospheric neutrino fluxes of Honda et al. [288], the atmospheric prompt neutrino flux of Enberg et al. [100] (with systematic uncertainties quoted in the paper. The fluxes of ν_e and ν_μ are approximately equal), the cosmogenic (or GZK) neutrino flux of Engel et al. [254], and the estimated flux of astrophysical neutrinos observed by IceCube [271]

sources (with the exception of the closest and brightest sources) will appear as a diffuse, isotropic flux that can in principle be separated from the atmospheric ν foreground because of a different energy spectrum, and flavor composition.

The different components of the neutrino flux are shown in Fig. 23.

Neutrinos, Photons and Cosmic Rays

The energy distribution and flavor composition of the neutrino emission in the standard hadronic mechanism of production is determined by the spectrum of the interacting primary particles, the nature of the target, and the properties of hadronic interactions and Weak decays.

The main qualitative results are the following:

1. If the spectrum of the interacting primary protons has a power law form: ($N_p(E) \simeq K_p\ E^{-\alpha}$), and the target is a gas of ordinary matter (the so called[38] "pp scenario")

[38] In general neutrinos are created in the interactions of relativistic protons and nuclei, and for collisions with a gas, the target is also formed by protons and (mostly helium) nuclei. The dominant channel is expected to be pp interactions. Adding the (left implicit) contributions of pA, Ap and AA collisions is straightforward.

the energy spectrum of the emitted neutrinos is also a power law with the same spectral index α.

2. In the pp scenario the flavor composition at emission, is in first approximation

$$\nu_e : \overline{\nu}_e : \nu_\mu : \overline{\nu}_\mu : \nu_\tau : \overline{\nu}_\tau \simeq 1 : 1 : 2 : 2 : 0 : 0 \tag{101}$$

It should be stressed that this is only a qualitative result, and the flavor ratios depend from the spectral index α and in general differ by the naive result of equation (101) by 15–20%.

3. In the pp scenario one has an approximately isospin symmetric production of the three pion states (there is a small excess of π^+ over π^- reflecting the isospin of the initial state). After decay of all pions, one obtains approximately equal emissions of neutrino and photons.

$$\dot{N}_\gamma(E) \simeq \dot{N}_\nu(E) . \tag{102}$$

This is again a "rule of thumb" that is only correct to an approximation of 15–20%.

4. If the primary particles interact with a radiation field (the so called "$p\gamma$ scenario) the neutrino emission depends also on the energy distribution $n_\gamma(\varepsilon)$ of the target soft photons. A phenomenologically important case (relevant for models of ν emission in GRB's) is when the target radiation field can be approximated as a power law $n_\gamma(\varepsilon) \propto \varepsilon^{-\beta}$. In this case the neutrino emission is approximately a power law: $\dot{N}_\nu \propto E^{-(\alpha-\beta+1)}$ with a spectral index $\alpha_\nu \simeq \alpha - \beta + 1$ (with α the spectral index of the interacting protons).
5. In the situation discussed above, where in $p\gamma$ interactions the spectrum of the target photons is a power law with spectral index $\beta > 0$, most of the inelastic interactions happen when the c.m. energy of the collision is close to threshold (all protons mostly interact with the more abundant photons at the lowest possible energy, and the c.m. energy of the collisions is approximately independent for E_p). The average pion multiplicity in the final state is therefore small, and the two-body final states $p\gamma \to n\pi^+$ and $p\gamma \to p\pi^\circ$ can be dominant (with the resonant production of a Δ^+ dynamically important). Accordingly one has an excess of π^+ over π^-, that is reflected in an asymmetry $\nu_e/\overline{\nu}_e \gg 1$.
6. In the $p\gamma$ scenario, the importance of the two body final states $n\pi^+$ and $p\pi^\circ$ is reflected in a larger photon/neutrino ratio. If the single pion production is the dominant final state one has $\dot{N}_\gamma(E)/\dot{N}_\nu(E) \approx 2$.

A derivation of the results listed above is outlined in the following. A more detailed discussion can be found in [200] and references therein.

Weak Decays

The chain decay of a π^+ yields three neutrinos of different type (ν_μ, $\overline{\nu}_\mu$ and ν_e). Of course this does not imply that the three energy distributions of the neutrinos

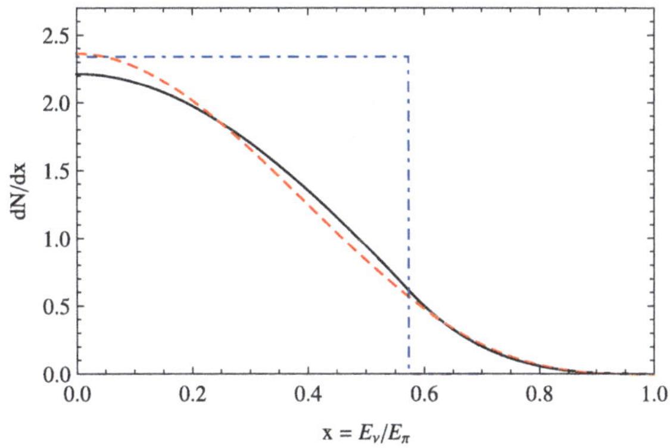

Fig. 24 Neutrino spectra in charged pion (π^+) decay. The three lines are for nu_μ (*dot-dashed*) $\overline{nu}_\mu$ (*solid*) and ν_e (*dashed*)

generated by the decay of an ensemble of π^+ are identical, because for each flavor the neutrino will have a different spectral shape.

Energy distribution of neutrinos from pions

It is easy to show that for the general case of a decay $a \to b + \cdots$, the energy distribution of particle b in a frame where the parent particle is ultra relativistic has the scaling form:

$$\frac{dn_{a\to b}}{dE}(E_b; E_a) = \frac{1}{E_a}\ F_{a\to b}\left(\frac{E_b}{E_a}\right). \tag{103}$$

The functions $F_{a\to\nu_j}(x)$ for all parent particles can be calculated using the measured branching fractions in all the possible final states and the corresponding matrix elements. The neutrino spectra in charged pion decay are shown in Fig. 24.

For the two body decay $\pi^+ \to \mu^+\nu_\mu$, that accounts for approximately 100% of charged pion decay, the scaling distributions are determined by elementary kinematics:

$$F_{\pi^+\to\nu_\mu}(x) = \Theta(1 - r_\pi - x)/(1 - r_\pi) \tag{104}$$

$$F_{\pi^+\to\mu^+}(x) = \Theta(x - r_\pi)/(1 - r_\pi) \tag{105}$$

where $\Theta(x)$ is the step function and $r_\pi = (m_\mu/m_\pi)^2$. The important result is that the muon takes most (approximately 3/4) of the pion energy, and after chain decay the 3 neutrinos take approximately equal fractions of the parent pion energy.

In the study of the muon decay one needs to take into account the matrix element of the Weak decay, and also the polarization of the muon. The muon helicity (the projection of the spin in the direction of the muon momentum) in a frame where the parent pion is ultrarelativistic, is a function of the fractional energy $x = E_\mu/E_\pi$ [201]:

$$h_{\pi^+\to\mu^+}(x) = -h_{\pi^-\to\mu^-}(x) = \frac{2\,r_\pi}{(1 - r_\pi)\ x} - \frac{1 + r_\pi}{1 - r_\pi}. \tag{106}$$

(note that the helicity for the decays $\pi^{\pm} \to \mu^{\pm}$ are related by CP symmetry). The helicity $h_{\pi^+ \to \mu^+}(x)$ takes the value -1 for $x \to 1$ (forward muon emission) and $+1$ for $x \to r_\pi$ (backward muon emission) reflecting the fact that the μ^+ is created as a left-handed particle in the pion rest frame in order to compensate the angular momentum of a left-handed neutrino.

The scaling functions that describe the decay of a muon of helicity h are:

$$F_{\mu^+ \to \overline{\nu}_\mu}(x; h) = \left(\frac{5}{3} - 3\,x^2 + \frac{4\,x^3}{3}\right) + h\,\left(-\frac{1}{3} + 3\,x^2 - \frac{8\,x^3}{3}\right) \tag{107}$$

$$F_{\mu^+ \to \nu_e}(x; h) = \left(2 - 6\,x^2 + 4\,x^3\right) + h\,\left(2 - 12\,x + 18\,x^2 - 8\,x^3\right) \tag{108}$$

The spectra of the neutrino after chain decay are obtained convoluting the muon decay spectrum. Because of CP invariance the spectra for the charged conjugate decays are obtained with the replacement $h \to -h$. Note that since in the decay of $\pi^{\pm}$ the helicity of the muons is $h_{\mu^+}(x) = -h_{\mu^-}(x)$, the neutrino spectra in the chain decays $\pi^{\pm} \to \mu^{\pm} \to \nu_\alpha(\overline{\nu}_\alpha)$ are identical.

Neutrinos from kaons

Significant contributions to the neutrino production are generated by the decay of kaons. Charged kaons contribute with the two body mode $K^+ \to \mu^+\nu_\mu$ (with branching ratio 0.6343), and the three body modes $K^{\pm}_{e3}$ ($K^+ \to \pi^\circ e^+ \nu_e$) and $K^{\pm}_{\mu 3}$ ($K^+ \to \pi^\circ \mu^+ \nu_\mu$) that have branching ratios $B^{\pm}_{e3} \simeq 0.0487$ and $B^{\pm}_{\mu 3} \simeq 0.0327$. The neutral K_L can also produce neutrinos in the K°_{e3} decay mode ($K_L \to \pi^{\mp} e^{\pm} \nu_e(\overline{\nu}_e)$) with combined branching ratio $B^0_{e3} \simeq 0.3881$, and $K^\circ_{\mu 3}$ mode ($K_L \to \pi^{\mp}\mu^{\pm}\nu_\mu(\overline{\nu}_\mu)$) with combined branching ratio $B^0_{\mu 3} \simeq 0.2719$.

The weak decays of particles that contain heavy quarks (c, b and t) are expected to be of negligible importance for astrophysical neutrinos, because the production of these particles is dynamically suppressed.

Gas Target

pp interaction

For a target material made of normal matter at rest the interaction rate of a primary particle of energy E_p is:

$$K_{pp}(E_p) = n_{\rm gas}\,\sigma_{pp}(s). \tag{109}$$

where $n_{\rm gas}$ is the number density of the target material. Equation (109) is written assuming that the primary particles are protons and the target material is also dominated by protons. This is expected to be a good approximation in most circumstances, it is however straightforward to consider the more general case where the relativistic particles and the target material contains also nuclei.

Since the target is at rest, the interaction rate is proportional to the cross section at a well-defined c.m. energy $s = m_p^2 + 2m_p\, E_p$. Hadronic cross sections grow only logarithmically with c.m. energy (or as power law $\propto s^\varepsilon$ with a small exponent ε), the interaction rate of the primary particles changes only slowly with energy.

The multiplicity and energy distributions of the particles produced in hadronic interactions cannot be calculated from first principles, however in reasonably good approximation, the cross section satisfies Feynman scaling, defined by the condition:

$$E\, \frac{d\sigma_j}{dp_\parallel^*}(p_\parallel^*, \sqrt{s}) = F_j(x_F) \tag{110}$$

where $p_\parallel^*$ is the longitudinal momentum in the c.m. frame, $d\sigma_j$ is the inclusive differential cross section for the production of particle type j, and $x_F = 2\, p_\parallel^*/\sqrt{s}$ is the Feynman variable.

For large c.m. energy ($s \gg m_p^2$) the target rest frame energy of secondary particles in the forward hemisphere ($x_F > 0$) is well approximated by the expression $E \simeq E_p\, x_F$ (where E (E_p) is the energy of the final state (projectile) particle in this frame). It follows that the validity of Feynman scaling in the fragmentation region implies also the approximate validity of scaling of the inclusive cross sections in the target rest frame:

$$\frac{dn_{pp\to a}}{dE}(E; E_p) \simeq \frac{1}{E_p}\, F_{pp\to a}\left(\frac{E}{E_p}\right) \qquad (\text{for } E \gg m) \tag{111}$$

The important consequence of Eq. (111) is that if the energy spectrum of the interacting cosmic rays has a power law form, then also the energy spectrum of the secondaries produced in the interactions are also a power law with the same spectral index. For example, the spectrum of charged pions produced per unit time in a source that contains a population of protons $N_p(E_0) = C_p\, E_0^{-\alpha}$ can be calculated as follows:

$$\begin{aligned}
\dot{N}_\pi(E_\nu) &= \int_{E_\nu}^{\infty} dE_0\, N_p(E_0)\, [n_{\rm gas}\, \sigma_{pp}(E_0)\, \beta\, c]\, \frac{dn_{pp\to\pi^+}}{dE_\pi}(E_\pi; E_0) \\
&\simeq [n_{\rm gas}\, \sigma_{pp}\, c] \int_{E_\pi}^{\infty} dE_0\, \left(C_p\, E_0^{-\alpha}\right) \frac{1}{E_0}\, F_{pp\to\pi^+}\left(\frac{E_\pi}{E_0}\right) \\
&\simeq C_p\, [n_{\rm gas}\, \sigma_{pp}\, c] \left[\int_0^1 dx\, x^{\alpha-1}\, F_{pp\to\pi^+}(x)\right] E_\pi^{-\alpha} \\
&= C_p\, [n_{\rm gas}\, \sigma_{pp}\, c]\, Z_{pp\to\pi^+}(\alpha)\, E_\pi^{-\alpha}\, .
\end{aligned} \tag{112}$$

In the second step we have neglected the energy dependence of σ_{pp}, made the approximation $\beta \simeq 1$, and assumed the validity of the scaling given in Eq. (111) for the inclusive cross section for charged pion production. The so called Z factor is given by:

$$Z_{pp\to\pi^+}(\alpha) = \int_0^1 dx\, x^{\alpha-1}\, F_{pp\to\pi^+}(x)\,. \tag{113}$$

If all pions decay, it is straightforward to show that the resulting neutrino spectrum is again a power law with the same spectral index, because the inclusive spectra of the final state particles in the decay of an ultra relativistic parent particle also satisfy a scaling relation. For example, including only the contributions of charged pions (and leaving the dependence of the Z factors on the spectral index implicit) one has:

$$\dot{N}_{\nu_\mu} = C_p\, [n_{\rm gas}\, \sigma_{pp}\, c]\, \left[Z_{pp\to\pi^+}\, Z_{\pi^+\to\nu_\mu} + Z_{pp\to\pi^-}\, Z_{\pi^-\to\nu_\mu}\right]\, E_\nu^{-\alpha} \tag{114}$$

In some practical applications, it is necessary to consider also the case when the population of the protons is not power-law distributed. In these cases, one can resort to the general formulae, or to convenient recipes to relate high energy γ-rays and neutrinos: see the Appendix 6. However, power-law distributions are interesting and useful in several cases, thus we will discuss it in greater details in the following.

Z factors $\pi \to \nu$

The Z factors for weak decays are straightforward to compute. For the $\pi^\pm$ chain decay the three relevant Z-factors are:

$$Z_{\pi^+\to\nu_\mu}(\alpha) = \frac{(1-r_\pi)^{\alpha-1}}{\alpha} \tag{115}$$

$$Z_{\pi^+\to\mu^+\to\overline{\nu}_\mu}(\alpha) = \frac{4\,\left(3-2\,r_\pi - (3+\alpha)\, r_\pi^\alpha + (2+\alpha)\, r_\pi^{1+\alpha}\right)}{\alpha^2\,(2+\alpha)\,(3+\alpha)\,(1-r_\pi)^2} \tag{116}$$

$$Z_{\pi^+\to\mu^+\to\nu_e}(\alpha) = \frac{24\,\left(\alpha\,(1-r_\pi) - r_\pi\,\left(1-r_\pi^\alpha\right)\right)}{\alpha^2\,(1+\alpha)\,(2+\alpha)\,(3+\alpha)\,(1-r_\pi)^2} \tag{117}$$

with $r_\pi = (m_\mu/m_\pi)^2$. The Z-factors for charged conjugate modes are identical because of CP invariance. The Z-factors given above are calculated neglecting the energy loss of muons before decay, and assuming that the muon helicity is exactly conserved. This assumption remains valid, to a very good approximation, also in the presence of a magnetic field, because the bending of the momentum and the spin precession exactly cancel for a particle of electric charge ± 1 and magnetic moment of one Bohr magneton. Neglecting the effects of the muon polarization leads to an overestimate (underestimate) of the Z-factor for the $\pi^+ \to \mu^+ \to \overline{\nu}_\mu$ ($\pi^+ \to \mu^+ \to \nu_e$) channel. For $\alpha = 2$ the Z-factors for pion decay take the values:

$$\{Z_{\pi^+\to\nu_\mu}, Z_{\pi^+\to\mu^+\to\overline{\nu}_\mu}, Z_{\pi^+\to\mu^+\to\nu_e}\}_{\alpha=2} = \left\{\frac{1-r_\pi}{2}, \frac{3+4r_\pi}{20}, \frac{2+r_\pi}{10}\right\}$$
$$\simeq \{0.2135,\ 0.2646,\ 0.2573\}\,, \tag{118}$$

i.e., the three neutrinos carry approximately one quarter of the charged pion energy. This happens because in the first decay the muon carries away a large fraction ($(1 + r_\pi)/2 \simeq 0.787$) of the pion energy.

Because of the different shapes of the energy distributions of the three neutrinos emitted in a pion decay, the flavor ratio

$$R_{\mu e} = \frac{(\nu_\mu + \overline{\nu}_\mu)}{(\nu_e + \overline{\nu}_e)} = \frac{Z_{\pi^+\to\nu_\mu} + Z_{\pi^+\to\mu^+\to\overline{\nu}_\mu}}{Z_{\pi^+\to\mu^+\to\nu_e}} \tag{119}$$

for a power law spectrum of parent pions is a function of its slope.

Photoproduction

In several of the proposed neutrino sources, the target of the primary particles is a radiation field (as for example for models of neutrino emission in Gamma Ray Burst). In this case the interaction rate and the energy distribution of the particles produced in an interaction depend not only on the density, but also on the energy and angular distribution of the target photons. The interaction probability per unit time of a proton of energy E_p traveling in the radiation field described by $n_\gamma(\varepsilon, \Omega_\gamma)$ can be calculated as:

$$K_{p\gamma}\left(E_p\right) = \int d\varepsilon \int_{-1}^{+1} \frac{d\cos\theta_{p\gamma}}{2}\,(1-\cos\theta_{p\gamma})\, n_\gamma(\varepsilon, \cos\theta_{p\gamma})\, \sigma_{p\gamma}(\varepsilon_r) \tag{120}$$

where $\theta_{p\gamma}$ is the angle between the photon and the proton momenta in the interaction, $\sigma_{p\gamma}$ is the photoproduction cross section, and ε_r is the photon energy in the proton rest frame:

$$\varepsilon_r = \frac{E_p\,\varepsilon}{m_p}\,(1-\cos\theta_{p\gamma}) \tag{121}$$

The quantity ε_r is in one to one correspondence with the c.m. energy of the reaction. It is convenient to change the integration variable from $\cos\theta_{p\gamma}$ to ε_r. Restricting ourselves to a situation (and a frame) where the photon distribution is isotropic one can then rewrite (120) as:

$$K_{p\gamma}(E_p) = \frac{1}{2}\,\frac{m_p^2}{E_p^2} \int_{\varepsilon_{\rm th}}^{\infty} d\varepsilon_r\; \varepsilon_r\; \sigma_{p\gamma}(\varepsilon_r) \int_{(m_p\varepsilon_r)/(2E_p)}^{\infty} d\varepsilon\; \frac{n_\gamma(\varepsilon)}{\varepsilon^2} \tag{122}$$

where $\varepsilon_{\rm th}$ is the threshold photon energy for pion production in the proton rest frame:

$$\varepsilon_{\rm th} = m_\pi + \frac{m_\pi^2}{2m_p} \tag{123}$$

Equation (122) can be recast in the form

$$K_{p\gamma}(E_p) = \int_{\varepsilon_{\rm th}}^{\infty} d\varepsilon_r\; f_{p\gamma}\left(\varepsilon_r;\; E_p\right) . \tag{124}$$

This expression shows explicitly the fact that the $p\gamma$ interactions of a proton of energy E_p do not correspond to a single value of the c.m. energy but have a distribution, that in general is a function of E_p, and is determined by the energy (and angular) distribution of the target photons. The probability distribution for ε_r is:

$$p(\varepsilon_r;\; E_p) = \frac{f(\varepsilon_r, E_p)}{K_{p\gamma}(E_p)} . \tag{125}$$

A phenomenologically important case for the target radiation field is the form: $n_\gamma(\varepsilon) = C_\gamma\, \varepsilon^{-\beta}$, that is an (isotropic) power law spectrum of slope β. For this form the last integration in (122) can be performed analytically with the result:

$$K_{p\gamma}(E_p) = C_\gamma\, \frac{2^\beta}{\beta+1} \left(\frac{E_p}{m_p}\right)^{\beta-1} \int_{\varepsilon_{\rm th}}^{\infty} d\varepsilon_r\; \varepsilon_r^{-\beta}\; \sigma_{p\gamma}(\varepsilon_r) = K_0(\beta)\; E_p^{\beta-1} \tag{126}$$

One can see that the interaction rate has the energy dependence $E_p^{\beta-1}$. The growth with energy (for $\beta > 1$) of the interaction rate can be understood observing that a proton of energy E_p can interact inelastically only with photons above a minimum energy $m_p\, \varepsilon_{\rm th}/(2E_p)$. This minimum target energy decreases proportionally to E_p^{-1}, and therefore protons of higher energy can interact with a softer and more abundant photon population.

An important feature of Eq. (126) is that the probability distribution for ε_r takes a form that is independent from E_p:

$$p(\varepsilon_r) \propto \varepsilon_r^{-\beta}\; \sigma_{p\gamma}(\varepsilon_r) \tag{127}$$

This fact has interesting consequences for the spectra of the final state particles. The compute these spectra one has to boost to the laboratory frame the distributions obtained in the c.m. frame:

$$E = \gamma\; (E^* + v\; p_z^*) \tag{128}$$

where v and γ are the velocity and gamma-factor of the c.m. of the reaction in the laboratory frame (where the proton has energy E_p), E^* and p_z^* are the energy and momentum (parallel to v) The Lorentz γ that connects the laboratory and c.m. frames is:

$$\gamma = \frac{E_p + \varepsilon}{\sqrt{s}} \simeq \frac{E_p}{\sqrt{s}} . \tag{129}$$

In the second approximated equality we have neglected the photon energy ε with respect to E_p, this is expected to be an excellent approximation. Similarly one can safely make the approximation $v \simeq 1$. With these approximations one can rewrite the Lorentz boost from the c.m. to the source frame as:

$$E = E_p \, \frac{E^* + p_z^*}{\sqrt{s}} = E_p \, \xi . \tag{130}$$

This equations indicates that to a good approximation all secondary particles of source frame energy E are created in the c.m. frame of the interaction with (to a very good approximation) the same value of the quantity $\xi = (E^* + p_z^*)/\sqrt{s}$. The energy spectrum of (for example) pions created in the interaction of a proton of energy E_p can then be written as:

$$\frac{dn_{p\gamma\to\pi}}{dE_\pi}(E_\pi;\ E_p) \simeq \frac{1}{E_p} \int_{\varepsilon_{\rm th}}^{\infty} d\varepsilon_r \ p(\varepsilon_r;\, E_p) \left[\frac{dn_{p\gamma\to\pi}}{d\xi}(\xi;\ \varepsilon_r)\right]_{\xi=E_\pi/E_p} \tag{131}$$

where $dn_{p\gamma\to\pi}/d\xi(\xi, \varepsilon_r)$ is the ξ distributions of the secondary particles of type π in $p\gamma$ interactions with c.m. energy that correspond to ε_r; this distribution is convoluted for a fixed value $\xi = E_\pi/E_p$ over all possible value of the ε_r with the appropriate distribution. If the energy distribution of the target photon field has a power law form, the function $p(\varepsilon_r;\ E_p)$ is independent from E_p, and Eq. (131) becomes the expression of a scaling law of form:

$$\frac{dn_{p\gamma\to\pi}}{dE_\pi}(E_\pi;\ E_p) \simeq \frac{1}{E_p} \ F_{p\gamma\to\pi}\left(\frac{E_\pi}{E_p};\ \beta\right) . \tag{132}$$

The scaling function $F_{p\gamma\to\pi}$ is not truly universal, but depends on the slope of the target photon spectrum. Integrating over all primary particle energies one can obtain the production rate of pions as:

$$\begin{aligned} Q_\pi(E_\pi) &= \int_{E_\pi}^{\infty} dE_p \ N_p(E_p)\ K_{p\gamma}(E_p)\ \frac{dn_{p\gamma\to\pi}}{dE_\pi}(E_\pi;\, E_p) \\ &= \int_{E_\pi}^{\infty} dE_p \ \left(C_p\ E_p^{-\alpha}\right)\ \left(K_0(\beta)\ E_p^{\beta-1}\right) \frac{1}{E_p}\ F_{p\gamma\to\pi}\left(\frac{E_\pi}{E_p};\ \beta\right) \\ &= C_p\ K_0(\beta) E_\pi^{-(\alpha-\beta+1)} \int_0^1 d\xi \ \xi^{\alpha-\beta}\ F_{p\gamma\to\pi}(\xi;\ \beta) \\ &= C_p\ K_0(\beta)\ Z_{p\gamma\to\pi}[\alpha-\beta+1, \beta]\ E_\pi^{-(\alpha-\beta+1)} . \end{aligned} \tag{133}$$

This equation shows that under the assumptions made, that is: (i) a power law spectrum of protons, and (ii) an isotropic power law spectrum of target photons, the energy distribution of produced secondaries is again a power law with slope $\alpha - \beta + 1$. Note that this result is based on purely kinematical considerations, while the result (112) about the interactions on an ordinary matter target was based on a dynamical assumption about the approximate validity of Feynman scaling in hadronic interactions, and on the weak energy dependence of hadronic cross sections.

This result was first obtained by Waxman and Bahcall in [202], a more detailed discussion of the limits of the derivation given above can be found in [200].

Neutrino Oscillations

The flavor composition of the neutrinos that arrive at the Earth is not equal to the flavor composition at the emission because of the effects of flavor oscillations. In general one can write the equation:

$$\phi_{\nu_\alpha}(E, \text{source}) = \sum_{\nu_\beta} P_{\nu_\beta \to \nu_\alpha}(E, \text{source})\, \phi^{\circ}_{\nu_\beta}(E, \text{source}) \tag{134}$$

where ϕ_{ν_α} is the observable flux of neutrinos of type α at the Earth, $\phi^{\circ}_{\nu_\alpha}$ is the flux from the source calculated assuming standard propagation and constant flavor, and $P_{\nu_\beta \to \nu_\alpha}$ are transition probabilities, that in general can depend on energy, and on the distance (and spatial-structure) of the source.

The transition probabilities have a non-trivial structure because of the existence of standard flavor oscillations, but might depend on additional "new physics" contributions, such as neutrino decay, that become significant only for very long pathlengths. In the standard scenario the neutrino number is conserved and therefore $\sum_\beta P_{\alpha \to \beta} = 1$. More in general, in the presence of a non negligible decay probability or of transitions to additional sterile states, the sum can be less than unity.

Flavor oscillations are a quantum-mechanical phenomenon that is a consequence of the non-identity of the ν flavor $\{\nu_e, \nu_\mu, \nu_\tau\}$ and mass $\{\nu_1, \nu_2, \nu_3\}$ eigenstates. A ν created with energy E and flavor α can be detected after a distance L with a different flavor β with a probability that (in vacuum) depends periodically on the ratio L/E.

The transition probability oscillates according to three frequencies that are proportional to the difference between the ν squared masses Δm^2_{jk}, and different amplitudes related to the mixing matrix $U_{\alpha j}$ that relates the flavor and mass eigenstates. The shortest (longest) oscillation length, corresponding to the largest (smallest) $|\Delta m^2|$) can be written as:

$$\lambda_{12} = (4\pi\, E_\nu)/|\Delta m^2_{12}| \simeq 3.1 \times 10^{12}\; E_{\text{TeV}}\;\; \text{cm},$$
$$\lambda_{23} = (4\pi\, E_\nu)/|\Delta m^2_{23}| \simeq 0.99 \times 10^{11}\; E_{\text{TeV}}\;\; \text{cm}.$$

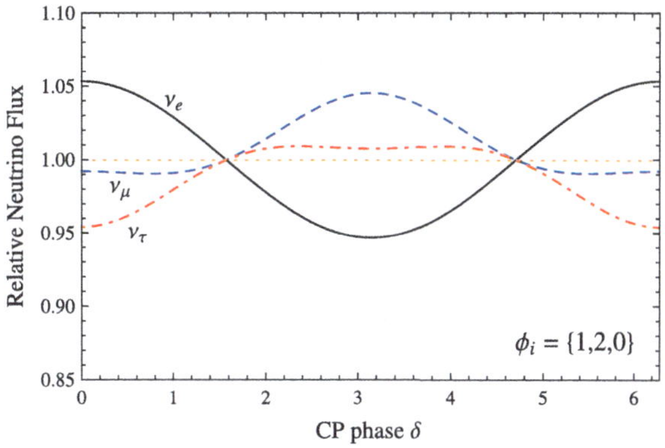

Fig. 25 Relative fluxes of ν_e, ν_μ and ν_τ after averaging over the source distance. The ν flavor composition at the emission is $\{\nu_e, \nu_\mu, \nu_\tau\} = \{1, 2, 0\}$

These lengths are long with respect to the Earth's radius ($R_\oplus \simeq 6.371 \times 10^8$ cm), but are very short with respect to the typical size of astrophysical sources. Therefore oscillations are negligible for atmospheric ν above $E_\nu \simeq 1$ TeV, but can be safely averaged for (essentially all) astrophysical neutrinos. After averaging over space the transition probabilities become independent from energy and take the form [203]:

$$\langle P(\nu_\alpha \to \nu_\beta)\rangle = \langle P(\overline{\nu}_\alpha \to \overline{\nu}_\beta)\rangle = \sum_j |U_{\alpha j}|^2\, |U_{\beta j}|^2 \,. \tag{135}$$

Owing to the symmetry in the exchange $\alpha \leftrightarrow \beta$, these probabilities depend only upon three (natural) parameters, that are combinations of the usual three mixing angle and CP violating phases. For a comprehensive study, see [67].

Flavor composition from pions decay

The mixing matrix $U_{\alpha j}$ depends on three mixing angles (that are now known with good precision) (see for example [204]), and a CP violating phase δ that is not known and can take any value in the interval $[0, 2\pi]$. Figure 25 shows the flavor composition of neutrinos at the Earth, calculated after the space average of standard oscillation, using the best fit values of the mixing angles θ_{12}, θ_{13} and θ_{23}, and plotted as a function of the phase δ. The result is that the fluxes for the three flavors are now approximately equal:

$$\{\nu_e + \overline{\nu}_e, \nu_\mu + \overline{\nu}_\mu, \nu_\tau + \overline{\nu}_\tau\} \approx \{1, 1, 1\} \tag{136}$$

Equation (136) is a robust prediction of the flavor composition of astrophysical neutrinos based on two hypothesis:

- The neutrino flavor composition at the source is approximately what expected for the standard mechanism of ν emission $\{\nu_e, \nu_\mu, \nu_\tau\} \approx \{1, 2, 0\}$.
- The flavor transition probabilities are determined by the extrapolation of the results already established in experiments with accelerator, reactor, atmospheric and solar neutrinos.

The possibility that the one or both the hypothesis listed above are violated has been investigated by many authors.

For the discussion of the flavor composition of the emission, it is interesting to consider a very broad range of possible models for the flavor composition of the neutrino emission, and in this spirit it is interesting to consider "extreme models" of flavor composition where the neutrinos at the emission (summing over ν and $\overline{\nu}$) are pure ν_μ or pure ν_e. These models can be considered as the limits of all realistically possible models of flavor composition models, since a significant production of ν_τ in an astrophysical environment is extraordinary unlikely. Sources emitting pure fluxes of ν_μ or ν_e are in principle possible, and have been in fact advocated in the literature.

Damped muons

The first case (pure ν_μ production) corresponds to the case of a ν source dominated by pion decay, but where the muons lose most of their kinetic energy before decay. In this situation only the ν_μ and $\overline{\nu}_\mu$ created in the direct $\pi^\pm$ decay are emitted at high energy. This situation is in principle possible in a high density environment, or more realistically in the presence of a sufficiently strong magnetic field that results in fast synchrotron energy losses for the muons.

Neutrons decay

A spectrum of $\overline{\nu}_e$ is generated in neutron decay, and it is in principle possible to think of a situation where this is the main neutrino source, and one has a pure $\overline{\nu}_e$ emission. Normally the production of $\overline{\nu}_e$ in neutron decay is of negligible importance. This is because the average fraction of the neutron energy E_n carried away by the $\overline{\nu}_e$ (in a frame where the neutron is ultrarelativistic) is $\langle E_{\overline{\nu}_e}\rangle/E_n \simeq 5.1 \times 10^{-4}$, with end-point $E_{\overline{\nu}_e}^{\max}/E_n \simeq 2(m_n - m_p - m_e)/m_n \simeq 1.66 \times 10^{-3}$. In most cases the flux of the softer n-decay neutrinos is negligible with respect to the contribution of neutrinos from π/K of decay that carry a much larger fraction (of order ~ 0.25) of their parent energy. There are two circumstances where the n contribution can become important. The first one is when the neutrino spectrum has a low-energy cut-off. This happens naturally when the target of the primary particles is a radiation field and there is an interaction energy threshold. In this case the $\overline{\nu}_e$ produced in neutron decay can become the dominant component of the neutrino flux at low energy, because most of the neutrinos from pion and kaon decay have higher energy. A second, more interesting case is when the neutrons are produced in the photo-disintegration of high-energy nuclei [211]. In this case it is in principle possible to have the emission of a pure $\overline{\nu}_e$ flux. In fact the threshold for photo-disintegration of a nucleus of mass number A expressed in terms of energy per nucleon ($E_0 = E_{\rm tot}/A$) is of order:

$$(E_0)_{\rm th}^{\gamma A} \simeq \frac{m_p \varepsilon_{\rm bind}}{2\,\varepsilon_\gamma} \tag{137}$$

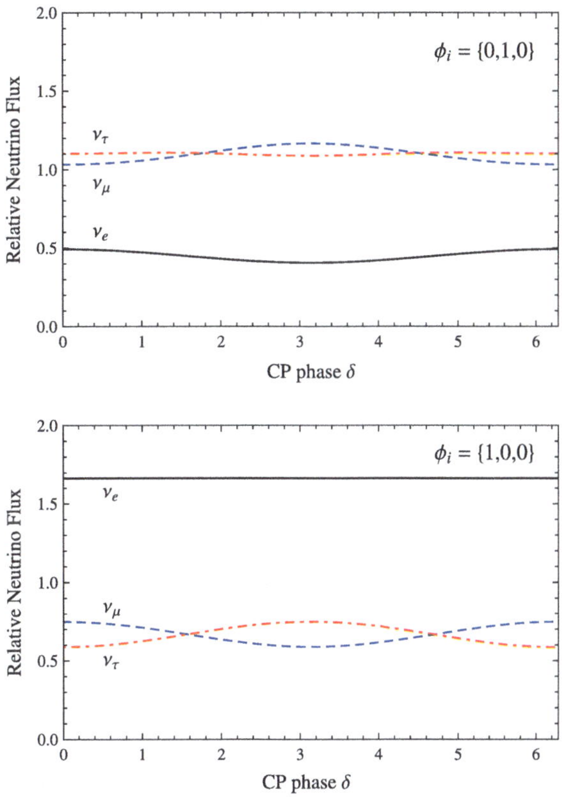

Fig. 26 Relative fluxes of ν_e, ν_μ and ν_τ after averaging over the source distance. In the top panel the ν flavor composition at the emission is $\{\nu_e, \nu_\mu, \nu_\tau\} = \{3, 0, 0\}$; In the bottom panel the initial composition is: $\{\nu_e, \nu_\mu, \nu_\tau\} = \{0, 3, 0\}$

where ε_γ is the energy of the target photons and $\varepsilon_{\rm bind} \simeq 8$ MeV is the binding energy of a nucleon in the nucleus. The threshold for pion photoproduction is:

$$(E_0)^{\pi}_{\rm th} \simeq \frac{m_p\, m_\pi}{2\, \varepsilon_\gamma} \left(1 + \frac{m_\pi^2}{2\, m_p}\right) . \tag{138}$$

Since the binding energy $\varepsilon_{\rm bind}$ is approximately fifteen times smaller than a pion mass, it is in principle possible to have circumstances where the primary particles are below the threshold for pion production, but above the threshold for photo-disintegration. This results in a pure $\overline{\nu}_e$ flux in a certain range of neutrino energy. It should be noted that the photo-disintegration and pion photoproduction thresholds differ by only one order of magnitude, and therefore a pure $\overline{\nu}_e$ flux can only extend for a small interval of energy.

Figure 26 shows (as a function of the CP violating phase δ the flavor composition of neutrinos at the Earth for the two models $\{\nu_e, \nu_\mu, \nu_\tau\} \simeq \{0, 1, 0\}$ and $\{0, 1, 0\}$. One can see that also for these extreme cases, the composition at the Earth is rich in all three flavors, however the range of possibilities becomes broader than what is predicted by the standard production mechanism (illustrated in Fig. 25).

Non-standard Neutrino Properties

A very attractive idea is the potential to use the future observations of astrophysical ν's to obtain information about the fundamental properties of the neutrinos [41, 205–210]. Astrophysical neutrinos travel path-lengths of order 10 Kpc ($\sim 3 \times 10^{22}$ cm) for galactic sources, and as large as several Gigaparsecs ($\sim 10^{28}$ cm) for extragalactic sources. These remarkably long baselines allow the study phenomena such as ν flavor transitions or ν decay in a range of parameters that is unaccessible with other methods. Only observations with SuperNova neutrinos, can provide larger L/E_ν, because the involve much smaller neutrino energies ($E_\nu \sim 10$ MeV).

Several "exotic" processes, beyond standard flavor oscillations, could reveal themselves only in the propagation of neutrinos over very long distances. For example, it has been suggested that some neutrinos could decay into a lighter neutrino and a majoron [212], if the lifetime is sufficiently long this phenomenon could be only detectable for neutrinos propagating over astronomical distances [213, 214]. A second interesting possibility is that neutrinos are pseudo-Dirac states [215] where each generation is actually composed of two maximally mixed Majorana neutrinos separated by a tiny mass difference. If the pseudo-Dirac splittings are sufficiently small, the phenomenology of oscillations on short baselines remains unchanged, however when E_ν/L becomes comparable or smaller than the pseudo-Dirac splittings new transitions become possible, and can in principle be detectable with astrophysical neutrinos [216]. More in general, oscillations into sterile states that are quasi degenerate to the active neutrinos can in principle be investigated down to very small squared mass splittings [217]. Several other mechanisms such as quantum decoherence [218], violations of the equivalence principle [219, 220], neutrinos with varying masses [221, 222] could leave their signature on the propagation of astrophysical neutrinos.

Neutrino decay with a very long lifetime is one of the most interesting possibilities that can be investigated with astrophysical neutrinos. Assuming [209, 214] that only the lowest mass eigenstate is stable, and that the distance of the source is much longer than the decay lengths, the observable flavor ratios depend uniquely on the flavor composition of the lightest eigenstate. Depending on the sign of Δm^2_{23} this is ν_1 (direct mass hierarchy) or ν_3 (inverse hierarchy).

Neutrino Point Sources

The observations of gamma-rays in the GeV range with telescopes on satellites [223–225] and and in the TeV range with Cherenkov telescope at the ground has resulted in the discovery of a large number of point-like and quasi point-like sources.

The neutrino telescopes have dedicated extraordinary efforts to the search of neutrino point sources [226–231, 335]. Limits obtained by the ANTARES and IceCube detector are shown in Fig. 29.

The gamma-ray observations of point sources results are the natural starting point to estimate the possibility to detect astrophysical sources of neutrino emission. The TeV energy range for photons (the highest available today) is very interesting for neutrino astronomy: in fact, the ν signal in present and future telescopes is expected to be due mostly to neutrinos with energies in the TeV-PeV range. Of particular importance has been the scan of the galactic plane performed by the HESS telescope [234, 235], because for the first time a crucially important region of the sky has been observed with an approximately uniform sensitivity with TeV photons.

Sources of γ-rays

The three brightest galactic TeV sources detected by the HESS telescope have integrated fluxes above 1 TeV (in units of 10^{-11} (cm^2 s)$^{-1}$) of approximately 2.1 (CRAB Nebula), 2.0 (RX J1713.7–3946) and 1.9 (Vela Junior).

As already discussed, the fundamental problem in the interpretation of the γ-ray sources is the fact that it is not known if the observed photons have hadronic (π° decay) or leptonic (inverse Compton scattering of relativistic electrons on radiation fields) origin. If the leptonic mechanism is acting, the hadronic component is poorly constrained, and the ν emission can be much smaller than the γ-ray flux. For the hadronic mechanism, the ν flux is at least as large as the photon flux, and higher if the γ-rays in the source are absorbed.

The γ-ray TeV sources, belong to several different classes. The Crab is a Pulsar Wind Nebula, powered by the spin down by the central neutron star. The emission from these objects is commonly attributed to leptonic processes, and in particular the Crab is well described by the Self Synchrotron Comptons model (SSC). The next two brightest sources [236–238] are young SuperNova Remnants (SNR), and it is possible that all or a significant fraction of the photon emission from these objects is of hadronic origin. The γ spectra are power law with a slope $\alpha \simeq 2.0$–2.2 which is consistent with the expectation of the spectra of hadrons accelerated with 1st order Fermi mechanism by the SN blast wave. The extrapolation from the photon to the neutrino flux is then robust, the main uncertainty being the possible presence of a high-energy cut-off in the spectrum.

A source of extraordinary interest is the Galactic Center, with a flux above 1 TeV of order ~ 0.2 (in the same units: 10^{-11} (cm^2 s)$^{-1}$) [239]. Recently the HESS collaboration has also obtained evidence that the emission from the GC extends to high energy [232] and that the source is likely contains a population of relativistic proton with a spectrum that extends to 10^{15} eV. This makes this object a potentially very interesting target for future observations of neutrino telescopes [233].

Another interesting source is the micro-Quasar LS5039 [240], that is not particularly bright in TeV emission with a flux ~ 0.12, but there are reasons to believe that the source could have significant internal absorption for photons, and therefore have a stronger ν emission.

In general, under the hypothesis of (i) hadronic emission and (ii) negligible γ absorption (that correspond to $\phi_\nu \simeq \phi_\gamma$), even we assume that the γ and ν spectra extends as a power law up to 100 TeV or more, the HESS sources are just at (or below) the level of sensitivity of the new ν telescopes as will be discussed in the next section.

For extragalactic sources the constraints from TeV photon observations are less stringent because very high-energy photons are severely absorbed over extragalactic distances due to $\gamma\gamma \rightarrow e^+e^-$ interactions on the infrared photons (from redshifted starlight) that fill intergalactic space. Active Galactic Nuclei (AGN) are strongly variables emitters of high-energy radiation.

Leading candidates as sources of high-energy neutrinos are also Gamma Ray Bursts (GRB) [202, 241–243]. It is possible that individual GRB emits neutrino fluences that are sufficiently high to give detectable rates in the neutrino telescopes.

Point Source Sensitivity

The most promising technique for the detection of neutrino point sources is the detection of ν-induced muons ($\mu \uparrow$). These particles are produced in the charged current interactions of ν_μ and $\overline{\nu}_\mu$ in the matter below the detector. To illustrate this important point we can consider a "reference" ν point source[39] with an unbroken power law spectrum of slope $\alpha = 2.2$, and an absolute normalization (summing over all ν types) $\Phi_\nu(\geq 1 \text{ TeV}) = 10^{-11}$ (cm^2 s)$^{-1}$. The reference source flux corresponds to approximately one half the γ-ray flux of the two brightest SNR detected by HESS.

Using the cross section shown in Fig. 27, (and also taking into account the absorption effects in the Earth illustrated in Fig. 28), the event rates from the reference source of ν interactions with vertex in the detector volume, and for ν-induced muons are shown in Fig. 30, The event rates for e, μ and τ like ν interaction are 10.3, 9.6 and 2.9 events/(km^3 year) (assuming a water filled volume), while the ν-induced muon flux is $\phi_{\mu\uparrow} \simeq 5.6$ (km^2 year)$^{-1}$. The signal depends on the zenith angle of the source, because of ν absorption in the Earth. The rate of e-like events has a small contribution from the "Glashow resonance" (the process $\overline{\nu}_e + e^- \rightarrow W^-$), visible as a peak at $E_\nu \simeq m_W^2/(2m_e) \sim 6 \times 10^6$ GeV. Neglecting the Earth absorption the resonance contributes a small rate $\simeq 0.07$ (km^3 year)$^{-1}$ to the source signal, however this contribution quickly disappears when the source drops much below the horizon, because of absorption in the Earth (Fig. 29).

The $\mu \uparrow$ rate is easier to measure and to disentangle from the atmospheric ν foreground. A crucial advantage is that the detected muon allows a high precision reconstruction of the ν direction, because the angle $\theta_{\mu\nu}$ is small.

---

[39] In this work we have chosen to characterize the normalization of a ν point source as the flux (summed over all ν types) above a minimum energy of $E_{\nu,\text{min}} = 1$ TeV. The reason for this choice is that it allows an immediate comparison with the sources measured by TeV γ-ray telescopes, that are commonly stated as flux above $E_{\gamma,\text{min}} = 1$ TeV. In case of negligible γ absorption one has $\phi_\gamma \simeq \phi_\nu$. Since we consider power law fluxes, it is trivial to restate the normalization in other forms. As discussed later the km^3 telescopes sensitivity peaks at $E_\nu \sim 20$ TeV.

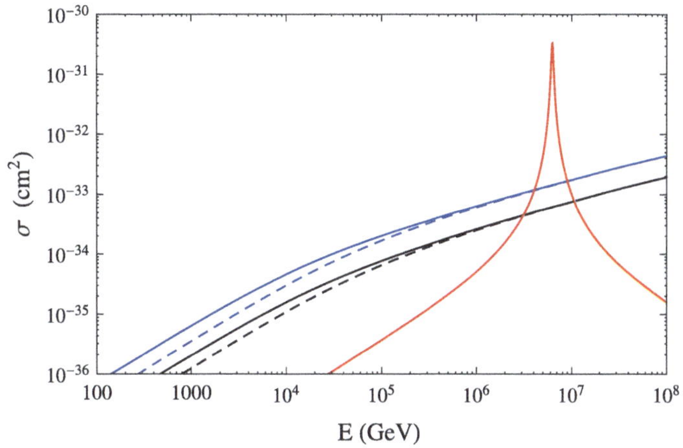

Fig. 27 Neutrino cross section as a function of the energy. The different curves are for charged current and neutral current ν—nucleon (*solid lines*) and $\overline{\nu}$—nucleon interactions. The line with the resonant behavior shows the $\overline{\nu}_e$-e^- cross section

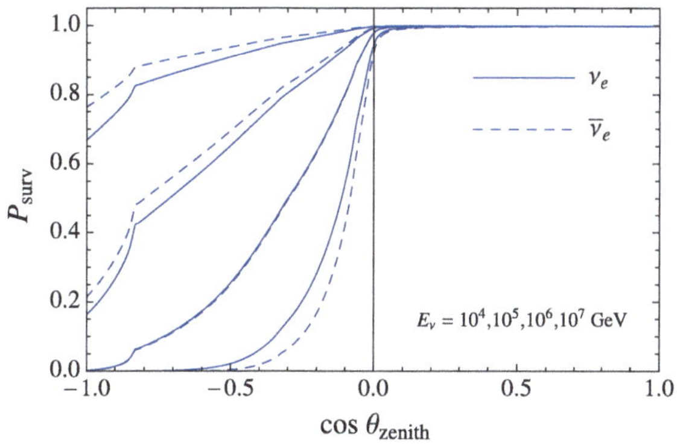

Fig. 28 Absorption probability for ν_e (*solid lines*) and $\overline{\nu}_e$ (*dashed lines*) of different energies ($E_\nu = 10^4$, 10^5, 10^6 and 10^7 GeV) plotted as a function of zenith angle. The calculation was performed using the neutrino cross section of Gandhi et al. [265]

In general the relation between the neutrino and the ν-induced muon fluxes is of order:

$$\Phi_{\mu\uparrow} \simeq (1 \div 5) \left[\frac{\phi_\nu(\geq 1 \text{ TeV})}{10^{-11}\ (\text{cm}^2\ \text{s})^{-1}} \right] (\text{km}^2\ \text{year})^{-1} \tag{139}$$

A more exact calculation is shown in Fig. 31.

Depends on (i) the slope of the ν spectrum, (ii) the presence of a high-energy cut-off, (iii) the threshold energy used for μ detection and (iv) (more weakly) on the zenith angle of the source (because of absorption effects).

The peak of the "response" curve for the ν—induced muons (see Fig. 30) is at $E_\nu \simeq 10$ TeV (20 TeV for a threshold of 1 TeV for the muons), with a total width extending approximately two orders of magnitude. In other words the planned ν telescopes should be understood mostly as telescopes for $\simeq$10–100 TeV neutrino

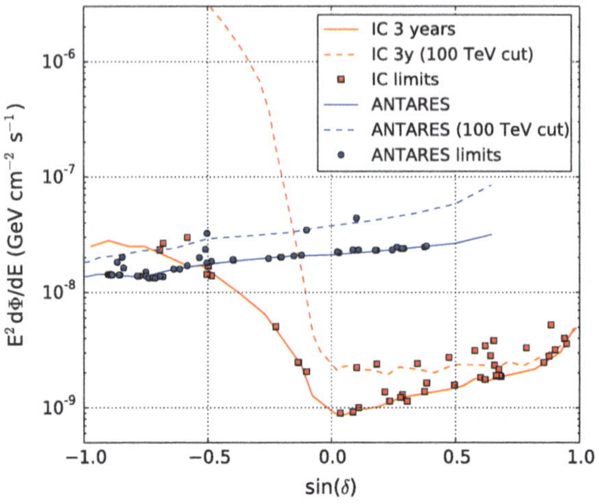

Fig. 29 Figure from [231]. 90% CL limits for selected sources (*squares* and *dots*) and sensitivities a function of the declination (*lines*) reported in the ANTARES 2007-2012 (*blue*) [227] and the IceCube 3 years (*red*) [230] point source analyses. An unbroken E^{-2} power-law source spectrum is assumed for the limits and lower sensitivity curves (*solid lines*). *Dashed lines* indicate the sensitivity for an E^{-2} spectrum with neutrino energies of $E_\nu \leq 100$ TeV

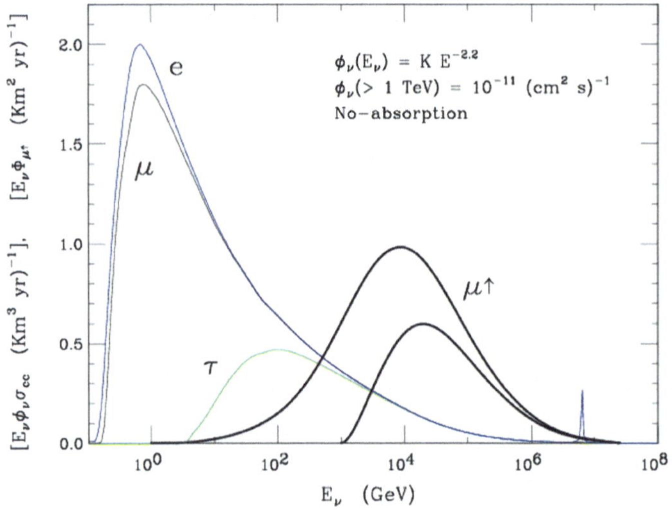

Fig. 30 Event rates from a ν point source. The source has spectrum $\phi_\nu = K\ E_\nu^{-\alpha}$ with slope 2.2 and normalization $\phi_\nu(> 1\ \text{TeV}) = 10^{-11}\ (\text{cm}^2\ \text{s})^{-1}$ equally divided among 6 ν types. The *thin lines* describe the rate of ν-interactions in the detector volume for e, μ and τ like CC-interactions. The *thick lines* are fluxes of ν-induced muons with 1 GeV and 1 TeV threshold

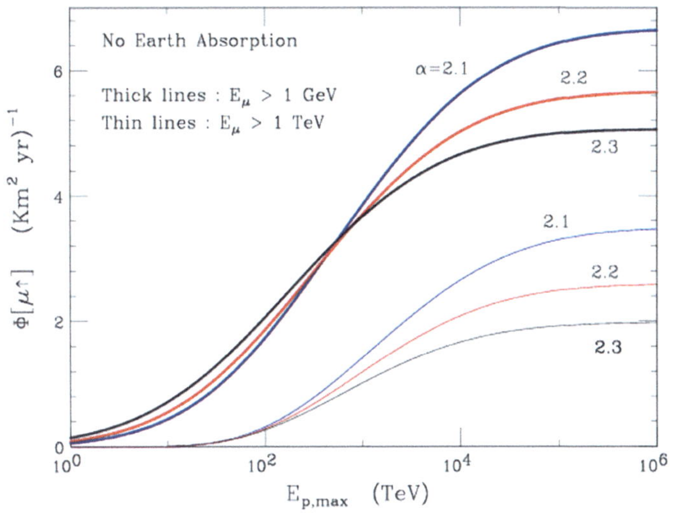

Fig. 31 Flux of ν-induced muons, calculated for a ν point source with a power law spectrum of slope α, plotted as a function of the high-energy cut-off of the parent proton spectrum. The *thick* (*thin*) *lines* are calculated for a threshold $E_{\mu\min} = 1$ GeV (1 TeV). The absolute normalization is chosen so that in the absence of cut-off the total ν flux above 1 TeV has a flux $\Phi_\nu = 10^{-11}$ (cm^2 s)$^{-1}$

sources. This energy range is reasonably well "connected" to the observations of the atmospheric Cherenkov γ-ray observations that cover the 0.1–10 TeV range in E_γ, and the ν sources observable in the planned km^3 telescopes are likely to appear as bright objects for TeV γ-ray instruments.

Because of the background of atmospheric muons, the ν-induced μ's are only detectable when the source is below the horizon. This reduces the sky coverage of a telescope as illustrated in Fig. 32, that gives the time averaged signal obtained by two detectors, placed at the south-pole and in the Mediterranean sea, when the reference source we are discussing is placed at different celestial declinations. For a south pole detector the source remains at a fixed zenith angle $\cos\theta_{\rm zenith} = -\sin\delta$, and the declination dependence of the rate is only caused by difference in ν absorption in the Earth for different zenith angles. The other curves includes the effect of the raising and setting of the source below the horizon.

Some of the most promising galactic sources are only visible for a neutrino telescope in the northern Hemisphere since the Galactic Center is at declination $\delta \simeq -29°$. The interesting source RX J1713.7–3946 is at $\delta = -39°$, Vela Junior at $\delta \simeq -46°$.

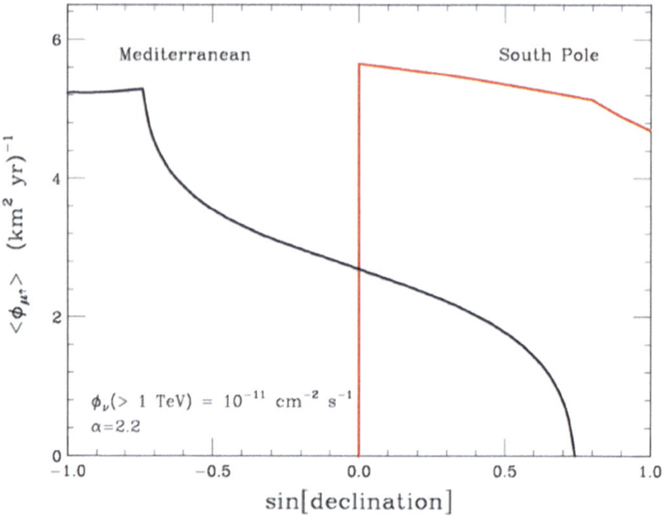

Fig. 32 Average flux of up-going muons from a ν point source of fixed luminosity, plotted as a function of declination for two ν telescopes placed at the south pole (*thin line*) and in the mediterranean sea (*thick line*). The flux is considered detectable only when the source is below the horizon. The effects of the (variable) absorption in the Earth is taken into account

Background Estimates

Since the prediction of the signal size for from the (expected) brightest ν sources is of only few events per year, it is clearly essential to reduce all sources of background to a very small level. This is a possible but remarkably difficult task.

The background problem is illustrated in Fig. 33 that shows the energy spectrum of the muon signal from our "reference" point source, comparing it with the spectrum of the atmospheric ν foreground integrated in a small cone of semi-angle 0.3°. The crucial point is that the energy spectrum from astrophysical sources is harder than the atmospheric ν one, with a median energy of approximately 1 TeV, an order of magnitude higher. It is for this reason that for the detection of astrophysical neutrinos it is planned to use an "offline" threshold of $E_\mu \gtrsim 1$ TeV. The atmospheric background above this threshold is small but still potentially dangerous, it depends on the zenith angle and is maximum (minimum) for the horizontal (vertical) direction at the level of 4 (1) μ/(km^2 year).

The angular window for the integration of the muon signal is determined by three factors: (i) the angular shape of the source, (ii) the intrinsic angle $\theta_{\mu\nu}$, and (iii) the angular resolution of the instrument. The source dimension can be important for the galactic sources, in fact the TeV γ-ray sources have a finite size, in particular the SNR RX J1713.7–3946 has a radius $\sim 1°$, and Vela Junior is twice as large. The detailed morphology of these sources measured by the HESS telescope indicates that most of the emission is coming from only some parts of the shell (presumably

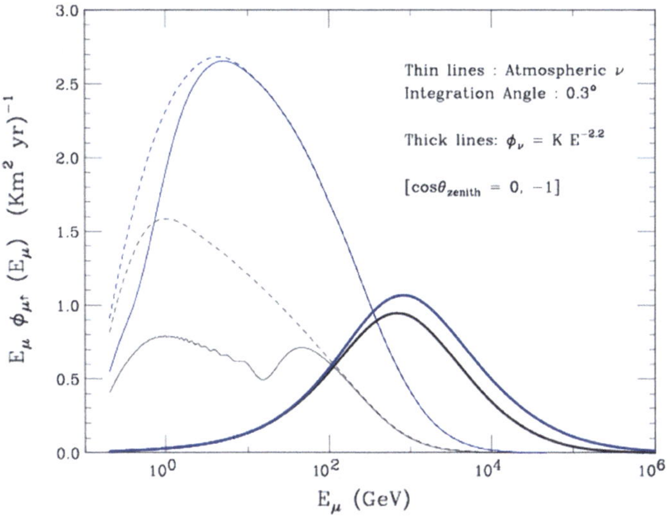

Fig. 33 Energy Spectra of ν-induced muons. The *thin lines* are the atmospheric ν background integrated in a cone of semi-angle 0.3° (*dashed lines* do not consider oscillations). The *thick lines* are for a point source ($\alpha = 2.2$, $\phi_\nu(> 1\ \text{TeV}) = 10^{-11}\ (\text{cm}^2\ \text{s})^{-1}$. The high (low) line is for horizontal (vertical) muons

where the gas density is higher), and the detection of ν emission from these sources could require the careful selection of the angular region of the γ-ray signal.

The distribution of the $\theta_{\mu\nu}$ angle defines the minimum angular dimension of a perfect point source signal. The angle of the muon signal shrinks with increasing energy $\propto (E_\mu)^{-1/2}$. This is easily understood noting that the dominant contribution to $\theta_{\mu\nu}$ is the muon neutrino scattering angle at the ν-interaction point:

$$\cos\theta_{\mu\nu} = 1 - \frac{m_N\, x}{E_{\mu,0}}\left(1 - \frac{E_{\mu,0}}{E_\nu}\right) \tag{140}$$

($E_{\mu,0}$ is the muon energy at the interaction point), and expanding for small angle one finds $\theta_{\mu\nu} \sim \sqrt{m_N\, x/(2\, E_{\mu,0})}$. The 50% containment cone angle shrinks to $\theta_{50\%} \simeq 0.15°$ for $E_\mu = 1$ TeV, that is smaller than the expected experimental resolution in water neutrino telescopes.

Extragalactic Neutrinos

It has been predicted [41] that the most easily detectable signal of astrophysical neutrinos is likely to appear as an isotropic flux that is the unresolved sum of all extragalactic neutrino sources. The identification of this extragalactic flux can rely on three signatures: (i) an energy spectrum harder that the atmospheric flux, (ii) an isotropic angular distribution, and (iii) approximately equal fluxes for all 6 neutrino types. This last point is a consequence of space averaged flavor oscillations.

The flux of prompt (charm decay) atmospheric neutrinos is also approximately isotropic in the energy range considered, and is characterized by equal fluxes of ν_e and ν_μ, with however a significant smaller flux of ν_τ (that are only produced in the decay of $D_s^\pm$). The disentangling of the astrophysical and an prompt atmospheric fluxes is therefore not trivial, and depends on a good determination of the neutrino energy spectrum, together with a convincing model for charmed particle production in hadronic interactions. A model-independent method for the identification of the astrophysical flux requires the separate measurement of the fluxes of all three neutrino flavors, including the ν_τ. The flux of prompt atmospheric ν_μ is also accompanied by an approximately equal flux of $\mu^\mp$ (the differences at the level of 10% are due to the very well understood differences in the spectra produced in weak decays).

Energetics of Extragalactic Neutrinos

The space averaged injection of neutrinos from extragalactic sources can be in general characterized by the injection density $q_\nu(E, t)$, with t the cosmic time. The average density of neutrinos, that fills uniformly the entire universe at the present epoch, (and is related to the ν flux by the simple relation $n_\nu(E) = (4\pi/c)\,\phi_\nu(E)$) can be calculated integrating the ν injection density in time, and taking into account the redshift:

$$\begin{aligned} n_\nu(E) &= \int dE_0 \int dt\; q_\nu(E_0, t)\; \delta\left[E - \frac{E_0}{1+z}\right] \\ &= \int_0^\infty dz \left|\frac{dt}{dz}\right| \;(1+z)\; q_\nu[E(1+z), z] \\ &= \int_0^\infty dz\; \frac{1}{H(z)}\; q_\nu[E(1+z), z] \end{aligned} \tag{141}$$

If the neutrino injection is a power law:

$$q_\nu(E, z) = q_0\; E^{-\alpha}\; F_\nu(z) \tag{142}$$

also the energy spectrum at present epoch is a power law of the same spectral index:

$$\phi_\nu(E) = \frac{c}{4\pi}\, n_\nu(E) = \frac{c}{4\pi}\, \frac{1}{H_0}\, q_0\; \xi\; E^{-\alpha}\,. \tag{143}$$

In this equation the quantity ξ is a constant that depends on the spectral index α and on the cosmological history of the injection $F_\nu(z)$:

$$\xi = \int_0^\infty dz \left[\frac{H_0}{H(z)}\right] \; F_\nu(z)\,(1+z)^{-\alpha} \;. \tag{144}$$

In this equation H_0 is the value of the Hubble constant at the present epoch, and

$$\frac{H(z)}{H_0} = \sqrt{\Omega_\Lambda + \Omega_{\rm m}\,(1+z)^3}\;. \tag{145}$$

For $\alpha = 2$, in an Einstein-De Sitter universe ($\Omega_{\rm m} = 1, \Omega_\Lambda = 0$) with no evolution for the sources one has $\xi = 0.4$, for the "concordance model" cosmology ($\Omega_{\rm m} = 0.3$, $\Omega_\Lambda = 0.7$) this becomes $\xi = 0.53$. For the same concordance model cosmology, if it is assumed that the cosmic time dependence of the neutrino injection is similar to the one fitted to the star-formation history [245, 246], one obtains $\xi(\text{SFR}) \simeq 3.0$, for a time dependence equal to the one fitted to the AGN luminosity evolution [247] one finds $\xi(\text{AGN}) \simeq 2.2$.

The Power density of the neutrino sources per decade of energy is

$$\mathscr{L}_\nu^\circ(E) = \ln 10\; E^2\; q_\nu(E, z=0) \tag{146}$$

is therefore related to the neutrino flux by the relation:

$$\mathscr{L}_\nu^\circ(E) = \ln 10\, \frac{4\pi}{c}\,\frac{\xi}{H_0}\; E^2\; \phi_\nu(E, z=0)\;. \tag{147}$$

An interesting question is the relation between the intensity of the total extragalactic ν contribution, and the potential to identify extragalactic point sources. This clearly depends on the luminosity function and cosmological evolution of the sources. Assuming for simplicity that all sources have energy spectra of the same shape, and in particular power spectra with slope α, then each source can be fully described by its distance and its ν luminosity L (above a fixed energy threshold $E_{\rm min}$). The ensemble of all extragalactic sources is then described by the function $n(L, z)\,dL$ that gives the number of sources with luminosity in the interval between L and $L + dL$ contained in a unit of comoving volume at the epoch corresponding to redshift z. The power density due to the ensemble of all sources is given by:

$$\mathscr{L}(z) = \int dL\; L\; n(L, z) \tag{148}$$

It is is possible (and in fact very likely) that most high-energy neutrino sources are not be isotropic. This case is however contained in our discussion if the luminosity L is understood as an orientation dependent "isotropic luminosity": $L \equiv (4\pi)\,(dL_{\rm true}/d\Omega)$. For a random distribution of the viewing angles it is simple to show that

$$\int dL\; L\; n(L, z) = \int dL_{\rm true}\; L_{\rm true}\; n(L_{\rm true}, z) \tag{149}$$

The ν flux (above the threshold energy $E_{\rm min}$) received from a source described by L and z is:

$$\Phi_{\rm point}(L,z) = \frac{L\;k_\alpha}{4\pi\;E_{\rm min}}\;\frac{(1+z)^{2-\alpha}}{d_L(z)^2} \tag{150}$$

where $d_L(z)$ is the luminosity distance and k_α is an adimensional factor that depends on the spectral index.

If $\Phi_{\rm min}$ is the sensitivity of a neutrino telescope, that is the minimum flux for the detection of a point source, then a source of luminosity L can be detected only if is closer than a maximum distance corresponding to redshift $z_h(L)$. Inspecting Eq. (150) it is simple to see that z_h is a function of the adimensional ratio $x = \sqrt{L/L^*}$ where:

$$L^* = 4\pi\;E_{\rm min}\;\Phi_{\rm min}\;(\alpha-1)/(H_0^2\;k_\alpha) \tag{151}$$

is the order of magnitude of the luminosity of a source that gives the minimum detectable flux when placed at $z \sim 1$. The explicit solution for $z_h(x)$ depends on the cosmological parameters ($\Omega_{\rm m}, \Omega_\Lambda$) and on the spectral slope α. A general closed form analytic solution for $z_h(x)$ does not exist,[40] but it can be easily obtained numerically. It is also easy and useful to write z_h as a power law expansion in x:

$$z_h = x + \frac{1}{4}\,[2(1-\Omega_\Lambda) + \Omega_{\rm m} - 2\alpha]\;x^2 + \cdots \tag{152}$$

The leading term of this expansion is simply x independently from the cosmological parameters and the slope α. This clearly reflects the fact that for small redshift z one probes only the near universe, where and when redshift effects and cosmological evolution are negligible, and the flux simply scales as the inverse square of the distance.

The total flux from all (resolved and unresolved) sources can be obtained integrating over L and z:

$$\Phi_{\rm tot} = \frac{1}{4\pi}\int dL \int_0^\infty dz\;v(z)\,n(L,z)\;\Phi_{\rm point}(L,z) \tag{153}$$

where $v(z)$ is the comoving volume contained between redshift z and $z+dz$:

$$v(z) = 4\pi\;\frac{d_L^2(z)}{(1+z)^2\;H(z)} \simeq \frac{4\pi}{H_0^2}\,z^2\,(1+\cdots) \tag{154}$$

Substituting the definitions (150) and (154), integrating in L and using (148) the total flux corresponds exactly to the result (143). The resolved (unresolved) flux can be obtained changing the limits of integration in redshift to the interval $z \in [0, z_h(L)]$

---

[40] As an example, for $\Omega_{\rm m} = 1, \Omega_\Lambda = 0$ and $\alpha = 2$ the exact solution is $z_h = (x - 1 + \sqrt{1+2x})/2$.

($z \in [z_h(L), \infty]$). The total number of detectable sources is obtained integrating the source density in the comoving volume contained inside the horizon:

$$N_s = \int dL \int_0^{z_h(L)} dz \, v(z) \, n(L, z) \tag{155}$$

A model for $n(L, z)$ to describe the luminosity distribution and cosmic evolution of the sources allows to predict the fraction of the extragalactic associated with the resolved flux, and the corresponding number of sources. Here there is no space for a full discussion, but it may be instructive to consider a simple toy model where all sources have identical luminosity (that is $n(L, z) = n \, \delta[L - \overline{L}(z)]$). In this model, is the (unique) luminosity L of the sources at the present epoch is not too large, (that is for $L \lesssim L^*$ with L^* given in (151)), for a *fixed* total flux for the ensemble of sources, the number of objects that can be resolved is (for $\alpha \simeq 2$):

$$\begin{aligned} N_s &\simeq \frac{\sqrt{4\pi}}{3} \frac{1}{\xi \log(10)} \frac{H_0}{c} \frac{K_\nu \sqrt{L_{\rm decade}}}{(\Phi_{\rm min} \, E_{\rm min})^{3/2}} \\ &\simeq 10 \, (K_{-9}) \left(\Phi_{-11}^{\rm min}\right)^{-\frac{3}{2}} (L_{45})^{\frac{1}{2}} \, \frac{2.5}{\xi} \end{aligned} \tag{156}$$

where K_ν is the coefficient of the diffuse ν flux (K_{-9} is in units 10^{-9} GeV/(cm^2 s sr)), $L_{\rm dec}$ is the power of an individual source per energy decade (L_{45} is in units of 10^{45} erg/s), and $\Phi_{\rm min}$ is the minimum flux above energy $E_{\rm min} = 1$ TeV for source identification (Φ_{-11} is in units 10^{-11} (cm^{2}s)$^{-1}$). It is easy to understand the scaling laws. When the luminosity of the source increases the radius of the source horizon (for L not too large) grows $\propto \sqrt{L/\Phi_{\rm min}}$ and the corresponding volume grows as $V \propto (L/\Phi_{\rm min})^{3/2}$, while (for a fixed total flux) the number density of the sources is $n \propto K_\nu \, L^{-1}$, therefore the number of detectable sources is: $N_s \propto n \, V \propto K_\nu \, L^{1/2} \, \Phi_{\rm min}^{-3/2}$.

Similarly, the ratio of the resolved to the total ν flux can be estimated as:

$$\frac{\phi_{\rm resolved}}{\phi_{\rm total}} \simeq \frac{1}{4\pi} \frac{1}{\xi \log(10)} \frac{H_0^2}{c^2} \frac{L_{\rm dec}}{\Phi_{\rm min} \, E_{\rm min}} \tag{157}$$

$$\simeq 0.005 \, \left(\Phi_{-11}^{\rm min}\right)^{-1} (L_{45}) \, \frac{2.5}{\xi} \tag{158}$$

The scaling $\propto L/\Phi_{\rm min}$ also easily follows from the assumption of an euclidean near universe.

The bottom line of this discussion, is that it is very likely that the ensemble of all extragalactic sources will give its largest contribution as an unresolved , isotropic contribution, with only a small fraction of this total flux resolved in the contribution of few individual point sources. The number of the detectable extragalactic point sources will obviously grow linearly with total extragalactic flux, but also depends crucially on the luminosity function and cosmological evolution of the ν sources. If a reasonable fraction of the individual sources are sufficiently powerful ($L \gtrsim 10^{45}$ erg/s), an interesting number of objects can be detected as point sources. Emission for blazars is a speculative but very exciting possibility.

Cosmic Rays Bounds

Waxman and Bahcall [248] have suggested the existence of an upper bound for the diffuse flux of extragalactic neutrinos, based on the observed flux of ultra high-energy cosmic rays, that is valid if the ν sources are not completely opaque to the cosmic rays that they produce. The logic (and the limits) of the WB bound are simple to grasp. Neutrinos are produced by cosmic ray interactions, and it is very likely (and certainly economic) to assume that the c.r. and ν sources are the same, and that the injection rates for the two type of particles are related. The condition of "transparency" for the source means that a c.r. has a probability $P_{\rm int} \lesssim 0.5$ of interacting in its way out of the source. The transparency condition obviously sets an upper bound on the ν flux, that can be estimated from a knowledge of the cosmic ray extragalactic spectrum, calculating for each observed c.r. the spectrum of neutrinos produced in the shower generated by the interaction of one particle of the same mass and energy, and integrating over the c.r. spectrum. If the c.r. spectrum is a power law of form $\phi_{\rm c.r.}(E) = K\ E^{-\alpha}$, the "upper bound" ν flux obtained saturating the transparency condition (for a c.r. flux dominated by protons) is: $\phi_\nu = (K\ Z_{p\to\nu})\ E^{-\alpha}$ with $Z_{p\to\nu} \sim 0.25$.

The WB upper bound has been the object of several criticisms see for example [249]. There are two problems with it. The first one is conceptual: the condition of transparency is plausible but is not physically necessary. The c.r. sources can be very transparent, for example in SNR's the interaction probability of the hadrons accelerated by the blast wave, is small ($\ll 1\%$) (with a value that depends on the density of the local ISM), but "thick sources" are possible, and have in fact been advocated for a long time, the best example is acceleration in the vicinity of the horizon of a SMBH. The search for "thick" neutrino sources is after all one of the important motivations for ν astronomy.

The second problem is only quantitative. In order to estimate the bound one has to know the spectrum of extragalactic cosmic rays. This flux is "hidden" behind the foreground of galactic cosmic rays, that have a density enhanced by magnetic confinement effects. The separation of the galactic and extragalactic component of the c.r. is a central unsolved problem for cosmic ray science. The extragalactic c.r. is dominant and therefore visible only at the very highest energies, perhaps only above the "ankle" (at $E_{\rm ankle} \simeq 10^{19}$ eV) as assumed in [248]. More recently it has been argued [250] that extragalactic p dominate the c.r. flux for $E_0 \gtrsim 10^{18}$ eV. Waxman and Bahcall have fitted the c.r. flux above the ankle with an a E^{-2} injection spectrum (correcting for energy degradation in the CMBR) and extrapolated the flux with the same form to lower energy. The power needed to generate this c.r. density was $\mathscr{L}_{\rm cr} \simeq 1900\ L_\odot/({\rm Mpc}^3$ decade). The extrapolation of the c.r. flux (that is clearly model dependent) and the corresponding ν upper bound are shown in Fig. 34, where they can be compared to the sensitivity of the ν telescopes.

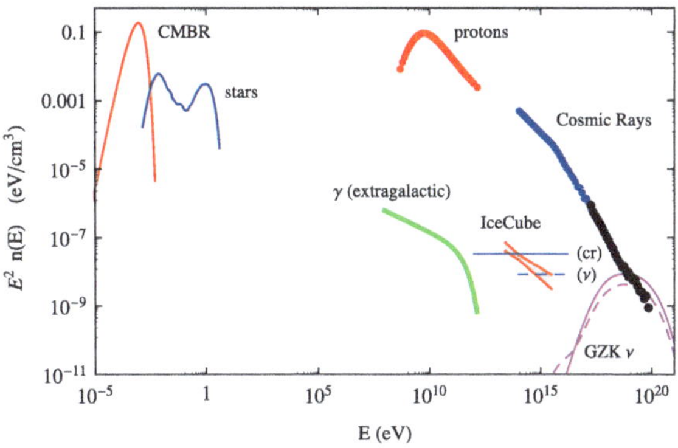

Fig. 34 Different contributions to the average energy density in the universe. The different lines describe the contributions of the 2.7° K CMBR radiation, starlight (with reprocession by dust), as estimated by Franceschini et al. [256]. The extragalactic gamma-ray flux measured by FERMI-LAT [257], and the cosmogenic (or GZK) neutrino flux of Engel et al. [254]. The shaded area is the estimate for astrophysical neutrinos observed by IceCube [271]. The *points* represent a subset of the cosmic ray measurements. CR's are magnetically confined in the Milky Way, and their local energy density is higher that the average density in the universe, with the possible exception of the highest energies when most particles are extragalactic. The *lines* labeled (cr) (ν) are the upper bounds for extragalactic cosmic rays and neutrinos estimated in [248].

Because of the uncertainties in the fitting of the c.r. extragalactic component and its extrapolation to low energy (and the possible loophole of the existence of thick sources) the WB estimate cannot really be considered an true upper bound of the ν flux. However the "WB ν-flux" is important as reasonable (and indeed in many senses optimistic) estimate of the order of magnitude of the true flux. The existence of cosmic rays with energy as large as $E_0 \sim 10^{20}$ eV is the best motivation for neutrino astronomy, since these particles must be accelerated to relativistic energies somewhere in the universe, and *unavoidably* some c.r. will interact with target near (or in) the source producing neutrinos.

Cosmogenic Neutrinos

Soon after the discovery of the cosmic microwave background it was realized that the interactions of ultrahigh-energy protons on the relic photons would suppress the CR flux at energies $\gtrsim 6 \times 10^{19}$ eV, the so-called "GZK cut-off" [251, 252]. It was pointed out subsequently [253] that the GZK interaction also generates a "cosmogenic flux" of neutrinos, through the decay of secondary charged pions. This flux of cosmogenic neutrinos has yet to be detected.

The reaction chain that generates the cosmogenic neutrinos is well known. Above the threshold for pion production most of the $p\gamma_{\rm CMB}$ collisions generate a single pion, with approximately equal number of π^+ and π°. Smaller amount of π^- are created in collisions at higher c.m. energy. Charged pions chain decay into neutrinos.

The calculation of the flux [254, 255] depends critically on the cosmological history of the injections of ultra high-energy protons by extragalactic sources. The observations of the spectrum and composition of UHE cosmic rays are clearly an important constraint, but are not sufficient to determine the flux of cosmogenic neutrinos. This is in part due to our imperfect measurement of the chemical composition of UHE cosmic rays, but more fundamentally, a large fraction (or most) of the protons lose most of their energy before reaching the Milky Way.

Cosmogenic (or occasionally called GZK) neutrinos are a guaranteed source, and their measurement carry important information on the maximum energy and cosmic history of the ultra high-energy cosmic rays.

Cosmogenic neutrinos have energy typically in the range $E_\mu \simeq 10^{18}$–10^{20} eV, and the predicted fluxes are sufficiently small so that their detection is difficult for km^3 detectors such as IceCube. There is therefore a interest in the development of detectors of larger mass, and of new detection methods.

Several interesting ideas are being developed (for a review see [336]), these include acoustic [258], radio [259] and Air Shower [260, 261] detection. Stringent limits on GZK neutrinos have been obtained with different detection methods [262–264].

As photon astronomy is articulated in different fields according to the range of wavelength observed, for neutrino astronomy one can already see the formation of (at least) two different subfields: the "km^3 neutrino Science" that aims at the study of ν in the 10^{12}–10^{16} eV energy range, and "Ultra High-Energy neutrino Science" that studies ν above 10^{18} eV, with the detection of GZK neutrinos as the primary goal.

High-Energy Cosmic Neutrinos: Experiments and First Results

The IceCube Neutrino Telescope

The IceCube Neutrino Observatory [294] is a Cherenkov Neutrino Telescope, located at a depth between 1.5 and 2.5 km under the Antarctic ice at the geographic South Pole (see Fig. 35). The detector consists of 5160 Digital Optical Modules (DOMs), equipped with photomultipliers (PMTs), distributed over a volume of one cubic kilometer, which monitor the ice permanently. The Optical Modules are arranged on 86 detector strings. Each string is buried into the ice using hot water drills. Construction began in 2005, when the first IceCube string was deployed and collected enough data to verify that the optical sensors worked correctly. At the surface IceCube is also equipped with an array of ice filled tanks (named IceTop) that detect extensive atmospheric air showers induced by cosmic rays.

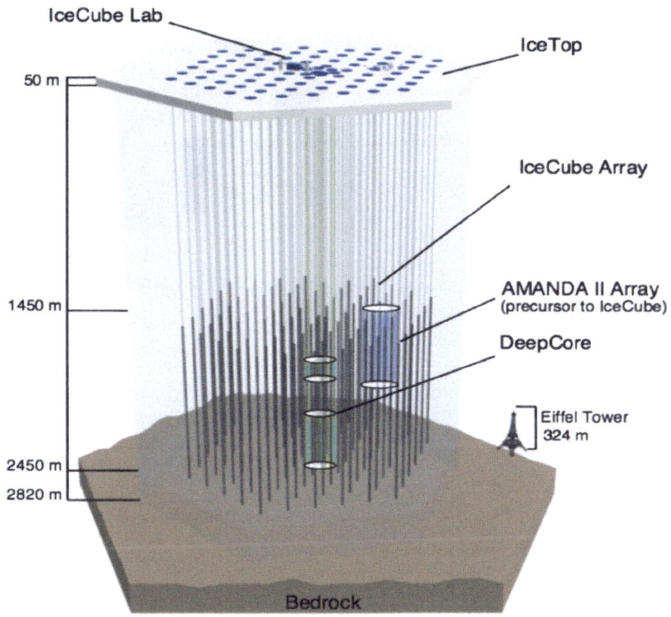

Fig. 35 Schematic view of the IceCube Neutrino Telescope, [340]

The main goal of IceCube is the detection of high-energy cosmic neutrinos. Atmospheric neutrinos are a source of background for this kind of search, as well as a potential signal for other physics items, like the study of neutrino oscillations. In order to study atmospheric neutrinos properties the detector was more densely instrumented towards its bottom using more efficient detector geometry for energies $<100\,\text{GeV}$. The sub-array is indicated in Fig. 35 as 'DeepCore' [297].

Digital optical modules

The main component of an IceCube DOM (Fig. 36) is the 10" Hamamatsu photomultiplier, a mainboard with the necessary electronics to supply power and digitize the signal, an array of light emitting diodes and a 13" glass sphere, resistant to pressure, which encloses all of the above elements [298]. The detection of a photon starts when it arrives at the DOM and crosses the 0.5" thick glass sphere. Leaving the sphere it enters a gel which is used to optically couple the sphere and the PMT. Once in the PMT, the photon can be absorbed by a photo-cathode, emitting an electron which is then multiplied. The resulting signal is readout as a change in the voltage of the PMTs anode, and given to the mainboard electronics.

Strings

The DOMs are arranged in strings. Each string supports 60 DOMs and there are 86 of them. The 80 *standard* IceCube strings are instrumented from a depth of

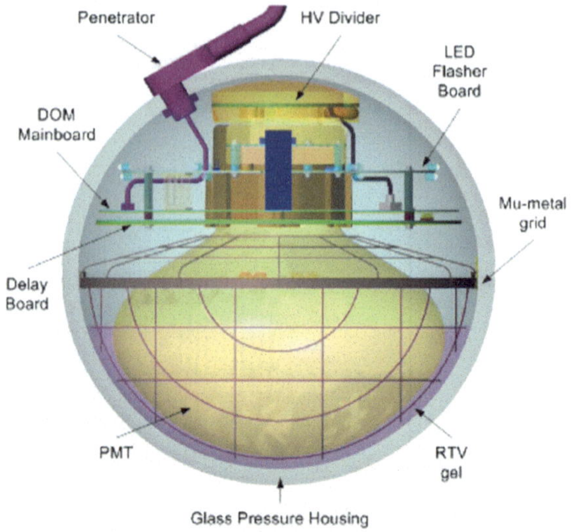

Fig. 36 The IceCube digital optical module

1450–2450 m, with a spacing of 17 m between sensors. The layout of the strings follows a concentric hexagonal pattern, with a typical distance of 125 m to the nearest neighbor. In addition there are 8 dedicated DeepCore strings, which use PMTs with 35% higher quantum efficiency with respect to the standard ones. Six of these surround the central IceCube string. They follow the same hexagonal pattern with a smaller separation, of 70 m. The two remaining strings are deployed inside the DeepCore region. DOMs on DeepCore strings are 7 m apart and only start at a depth of 1.75 km, with a large gap between the positions of the 10th and 11th DOM (see Fig. 37a). Standard IceCube strings have 18 DOMs below this depth; DeepCore strings have 50. Figure 37b is a sketch of the side projection of the detector and the fiducial volume.

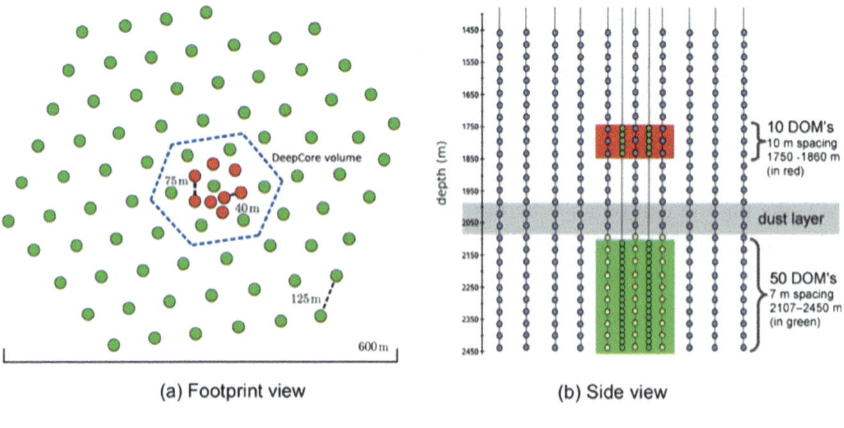

(a) Footprint view

(b) Side view

Fig. 37 Top and lateral projections of the IceCube detector

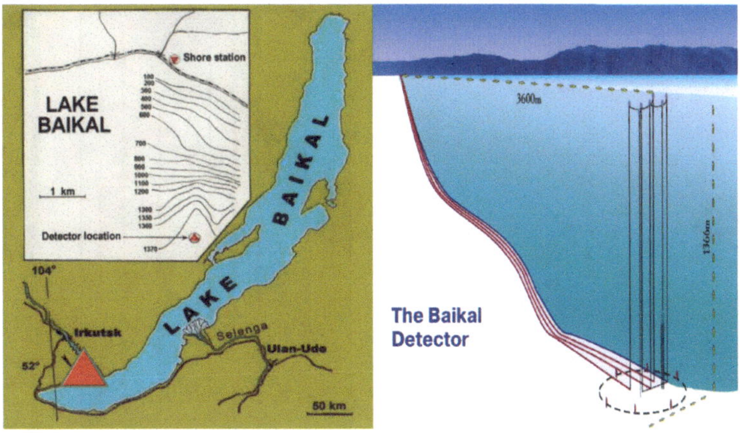

Fig. 38 The location of the BAIKAL neutrino detector and it's schematic representation

The data collection during the construction period allowed publication of physics results for intermediate detector configurations (IC22 with 22 strings, IC40 with 40 strings and, etc.).

Telescopes in the Northern Hemisphere

The BAIKAL neutrino detector

The BAIKAL neutrino detector, deployed at 1100 m depts in the waters of Lake Baikal (the NT200 detector Fig. 38) has been completed in 1998 and consists of 8 strings with 192 PMTs in total.

Three more strings carrying 12 PMTs each where added in 2005, resulting in the NT200+ detector [300]. The next-generation neutrino telescope Baikal-GVD (Fig. 39) is a kilometer-scale deep underwater Cherenkov detector currently under construction in Lake Baikal [301].

Baikal-GVD is designed to detect astrophysical neutrino fluxes at energies from a few TeV up to the highest energies of 10^6 TeV and particularly mapping the high-energy neutrino sky in the Southern Hemisphere including the region of the Galactic center. In April 2015 a first demonstration cluster of Baikal-GVD named Dubna? was deployed and commissioned in Lake Baikal.

It encloses 1.7 Megatons of the waters of Lake Baikal. It comprises a total of 192 optical modules arranged at eight 345 m long strings, as well as an acoustic positioning system and an instrumentation string with equipment for array calibration and monitoring of environmental parameters. By 2020, Baikal-GVD is set to consist of 8 baseline clusters covering a total volume of about 0.4 km^3 (GVD phase-1).

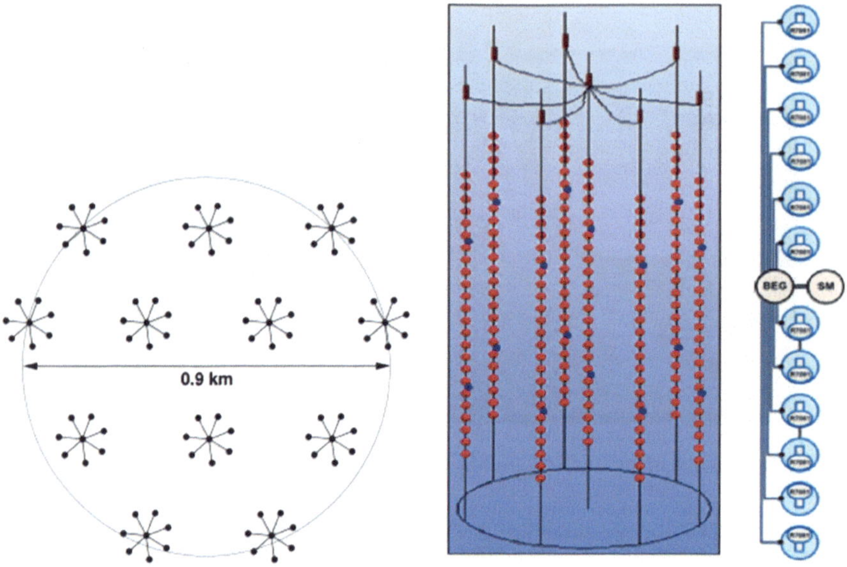

Fig. 39 The Baikal Gigaton Volume Detector (GVD). *Left* arrangement of the 12 clusters; *middle* one cluster; *right* schematic view of a string section

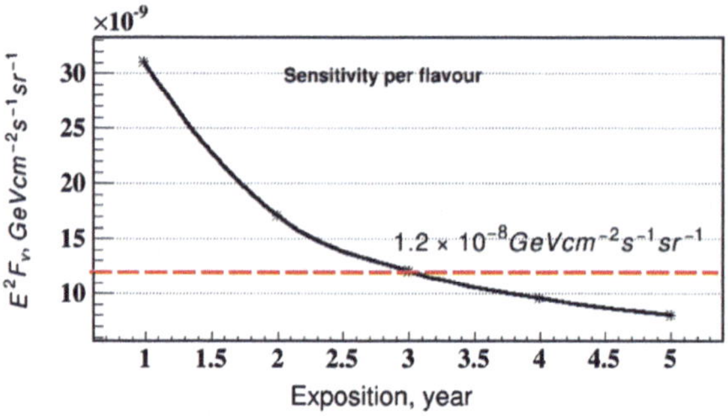

Fig. 40 The Baikal-GVD cluster sensitivity for one flavor neutrino flux with an E^2 spectrum as function of the observation years. The *long-dashed line* indicates the one-avor neutrino ux value obtained by IceCube

Figure 40 shows a preliminary estimation of one cluster sensitivity for one flavor neutrino flux with an E^{-2} spectrum and flavor ration 1 : 1 : 1 for all flavor flux as function of the observation years.

ANTARES

The ANTARES (Astronomy with a Neutrino Telescope and Abyss environmental RESearch) apparatus [299] is located on the bottom of the Mediterranean Sea at a depth of about 2475 m, in French waters, at 40 km South-East off the coast from Toulon.

ANTARES is the first and the only one operating undersea neutrino telescope, and the largest neutrino telescope in the Northern Hemisphere with an instrumented

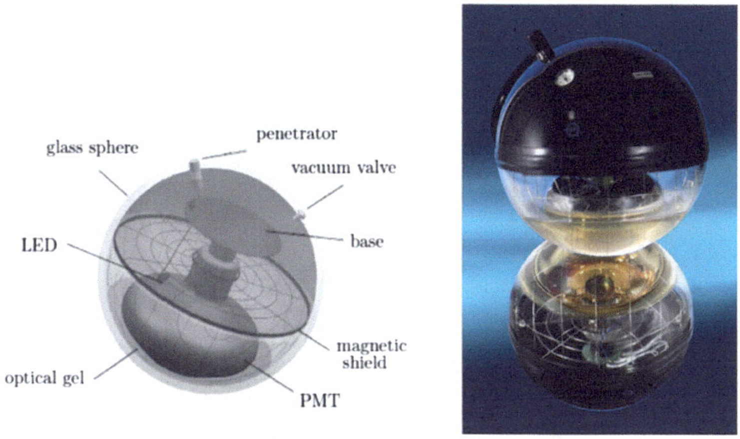

Fig. 41 *Left* schematic view of an ANTARES optical module. *Right* Picture of an ANTARES optical module placed on a mirror

volume of $\geq 0.01\,\text{km}^3$. The basic element of the detector is an Optical Module (OM). A schematic view and a photograph of an OM can be seen in Fig. 41.

Each optical module consists in a pressure resistant glass sphere with a diameter of 43 cm and a wall thickness of 15 mm, which houses a photo-tube (PMT) and the electronics providing the high voltage. The PMT is glued to one of the two hemispheres with transparent silicone rubber gel. The photo-multiplier tube selected for the ANTARES detector is the Hamamatsu R7081-20 which has a diameter of 10 inches, 14 amplification stages and a nominal gain of 5×10^7 at $1760V$. The timing resolution of the PMT, one of the key parameters that determines the angular resolution of the detector, can be characterized by the transit time spread (TTS) that is about 1.3 ns (root mean square). The effective sensitive area of the photo-cathode is $450\,\text{cm}^2$. The PMT is sensitive to single photons with wavelength between 300 and 600 nm. The maximum quantum efficiency ranges from 350 to 450 nm where its value is $\geqslant$25%. The charge resolution for a single photo-electron is about 30%. The Earth's magnetic field ($\sim$46 μT at the ANTARES site) significantly degrades the collection efficiency of photo-tubes by bending the trajectories of electrons, mainly between the photo-cathode and the first dynode. A cage made of 1.1 mm thick high permittivity alloy wire is used to shield the PMT and to minimize the dependence of the OM response with respect to its angle with the magnetic North. Three optical modules are grouped together to form a "storey" or "floor". The schematic view together with a picture of a storey can be seen in Fig. 42.

The optical modules are mounted at the vertexes of an equilateral triangle orthogonal to a titanium structure called Optical Module Frame (OMF). Since the detector is optimized to detect up-going neutrinos, they are oriented downward at 45° with respect to the vertical. The OMs are connected to the Local Control Module (LCM) that is a titanium cylinder placed in the OMF center housing the electronics for the

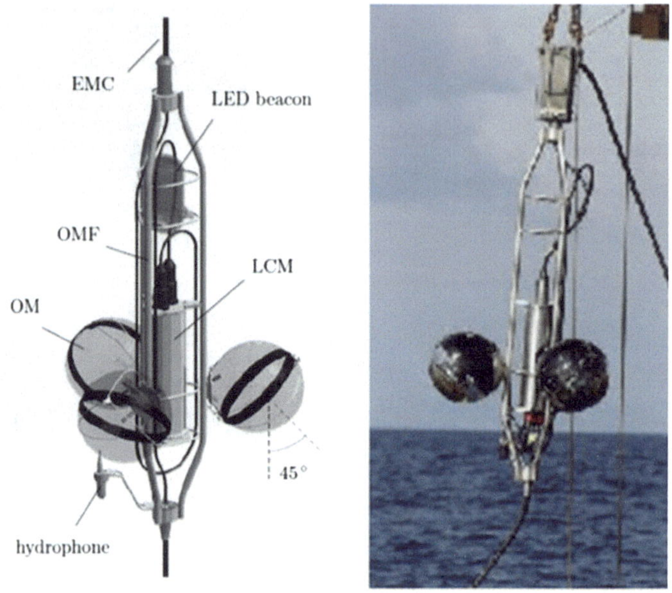

Fig. 42 *Left* schematic view of an ANTARES storey. *Right* picture of an ANTARES storey

data transmission and various instruments for calibration and monitoring. A storey may also contain extra instruments mounted on the OMF, such as a LED beacon or an acoustic hydrophone. The detector consists of approximately 900 photo-multiplier tubes disposed in 13 vertical strings, 12 standard strings plus one containing instruments for environmental properties measurements. These lines are spread over an area of $\sim$0.1 km^2 forming a cylindrical instrumented volume with an active height of $\sim$0.3 km and a diameter of $\sim$0.2 km.

Figure 43 shows a schematic view of the detector array indicating the principal components of the telescope. The basic unit of the detector is the Optical Module (OM). The optical modules are grouped together in storeys of three modules and interconnected via an electro-mechanical cable. Each string is instrumented with several electronics containers. At every storey there is a *Local Control Module* (LCM) and, at the base of each string, there is a *String Control Module* (SCM). Special containers house acoustic and calibration equipment. Each of these containers constitutes a node of the data transmission network, receiving and transmitting data and slow control commands. The functions which they support include reading sensors, adjusting slow-control parameters, the trigger and the distribution of power, clock and reset signals to the front-end electronics. The individual SCM are linked to a common junction box by electro-optical cables. A standard deep-Sea telecommunication cable links the junction box with a shore station where the data are filtered and recorded. The trigger logic in the sea is planned to be as simple and flexible as possible: there are some first levels of trigger that are based on coincidences of hits in

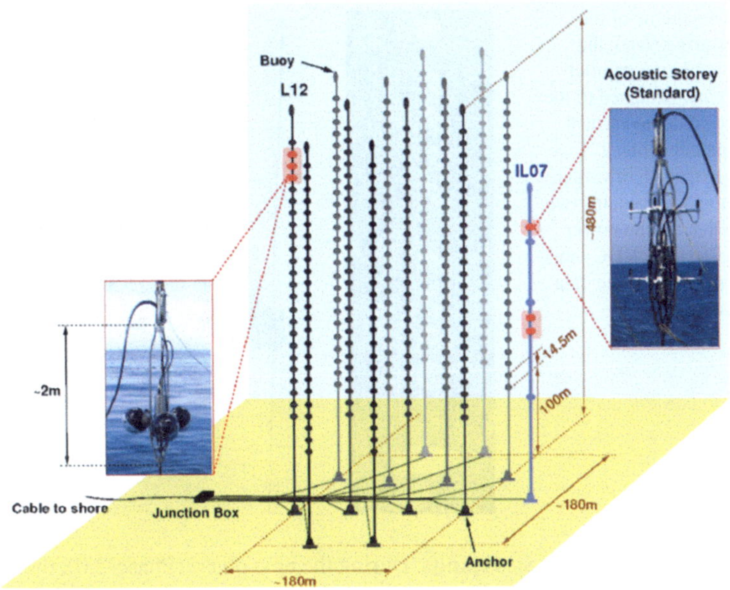

Fig. 43 Scheme of the detector array. The structure of the 12 standard lines and the one containing environmental instruments is shown. In the little photo are showed also the storeys

adjacent OMs and storeys in one or more floors, then a more refined level of trigger, imposing tighter time coincidences over larger numbers of OMs, is elaborated in a farm of processors on shore. The readout rate amounts to several kHz and the corresponding data recording rate less than 100 events per second. Each day ANTARES collects about 3 atmospheric neutrino events.

KM3NeT: a multi-km^3 in Mediterranean Sea

The KM3NeT consortium [302] proposes the construction of a multi-km^3 telescope in the Mediterranean Sea. The main objectives of KM3NeT are the discovery, and subsequent observation, of high-energy neutrino sources in the Universe and the determination of the neutrino mass hierarchy [119]. To meet these objectives, the KM3NeT Collaboration plans to build a new Research Infrastructure consisting of a network of deep-sea neutrino telescopes in the Mediterranean Sea. Three deep-sea sites have been selected for the optical properties of the water, distance to shore and local infrastructure, namely offshore Toulon (KM3NeT-France), Capo Passero in Sicily (KM3NeT-Italy) and Pylos in Peloponnese (KM3NeT-Greece). KM3NeT Phase-1 (started on 2013) already provided several results: the technical design has been validated through in situ prototypes; data analysis tools have been developed; assembly sites for the production of optical modules and strings have been set up; deployment and connection of strings in the deep sea are being optimized for speed and reliability. The Collaboration aims at the deployment of 31 strings before 2018

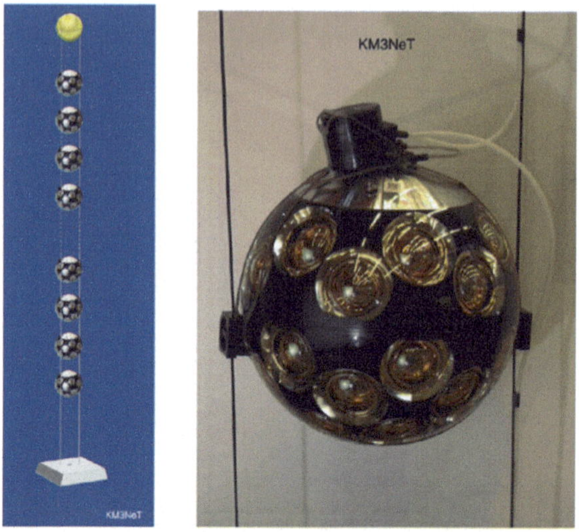

Fig. 44 A scheme of the detection string (*left*), the DOM held by two parallel ropes (*right*). Attached to the ropes is visible the vertical electro-optical cable, a pressure balanced, oil-filled, plastic tube that contains two copper wires for the power transmission (400 VDC) and 18 optical fibers for the data transmission

and to provide the first physics results. During the next phase (Phase-2.0), the Collaboration will complete the construction of ARCA and ORCA by 2020. The ultimate goal is to fully develop the KM3NeT Research Infrastructure to comprise a distributed installation of in total seven blocks in the three foreseen sites (Phase-3) and operate it for ten years or more [119].

The Phase-2.0 goal will consist of three *building blocks* each comprising 115 strings. Each string, anchored to the sea floor and kept close to vertical by a submerged buoy, comprises 18 Digital Optical Modules (DOM). Each DOM (Fig. 44) is made of a 17-inch pressure resistant glass sphere containing 31 3-in. photo-multiplier tubes (PMTs), a number of calibration devices and the read-out electronics. This *multi-PMT* design provides a large photo-cathode area ($\sim$1400 cm^2 per DOM [303]), good separation between single-photon and multiple-photon hits and information on the photon direction.

Each *building block* thus constitutes a three-dimensional array of photo sensors Fig. 45.

Two building blocks (referred as ARCA: Astroparticle Research with Cosmics in the Abyss and locate in the KM3NeT-Italy site) will be sparsely configured to detect high-energy neutrinos of cosmic origin (Fig. 46).

A neutrino telescope in the Mediterranean Sea is ideal for this purpose, since most of the potential Galactic sources are in the Southern sky; in contrast, the IceCube detector at the South Pole is much less sensitive to these individual sources, at least in the energy range of relatively low energies (below few 10 of TeV) where a signal is expected. The KM3NeT/ARCA design has been carefully optimized to maximize the sensitivity to these Galactic sources Fig. 47.

For Phase-2.0, three building blocks are planned: two KM3NeT/ARCA blocks, with a large spacing to target astrophysical neutrinos at TeV energies and above; and

Fig. 45 A schematic representation of one KM3NeT block: 115 strings connected to shore by means of a long electro-optical cable and served by a Sea-floor infrastructure providing power and data transmission

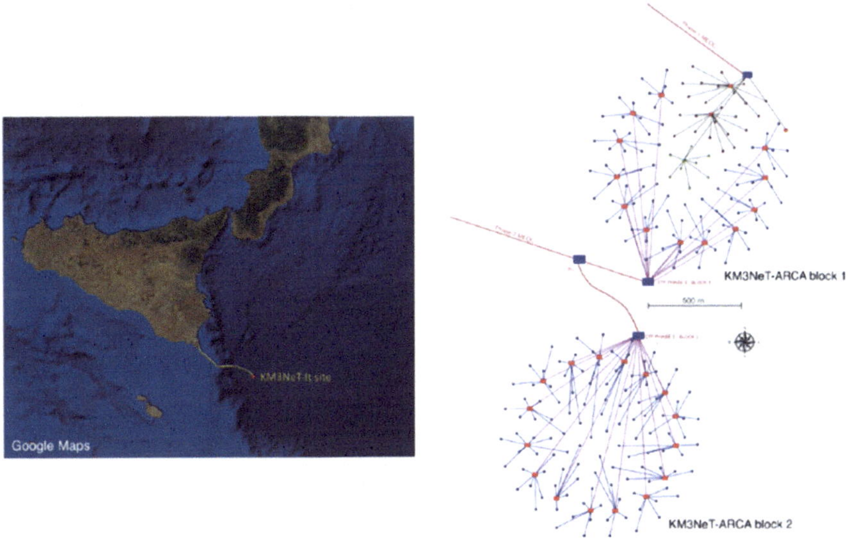

Fig. 46 Map of the Mediterranean Sea close to Sicily, Italy. The cable and the location of the KM3NeT-Italy installation are indicated (*left*). Layout of the two ARCA building blocks (*right*)

one KM3NeT/ORCA block, to target atmospheric neutrinos in the few-GeV range [119] that aims mainly to study neutrino properties.

With respect to IceCube and ANTARES, KM3NeT-ARCA will profit of better optical properties of the Sea-water, of the deeper site, of the exposition to the Galactic Center. Figure 48 to quote two important results that will be possible to reach in

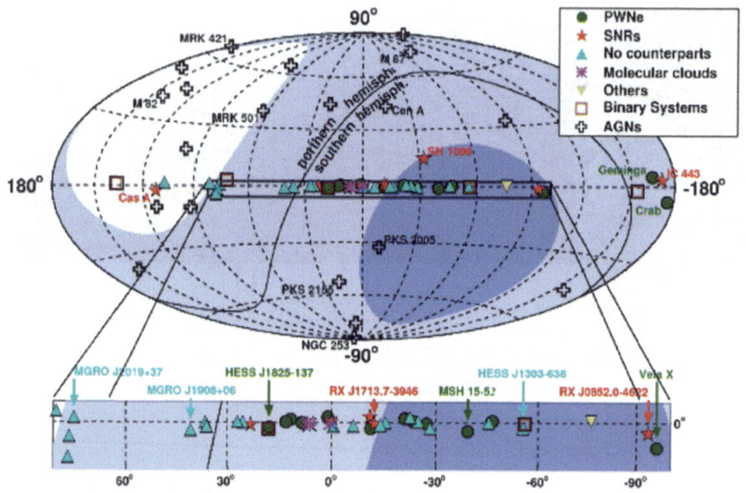

Fig. 47 Sky coverage in Galactic coordinates for detectors located in the Mediterranean Sea and at the South Pole, where only the northern hemisphere can be observed. The *shading* indicates the visibility for a detector in the Mediterranean with 2π downward coverage; *dark* (*light*) *blue areas* are visible at least 75% (25%) of the time.The locations of observed sources of high-energy γ rays are also indicated

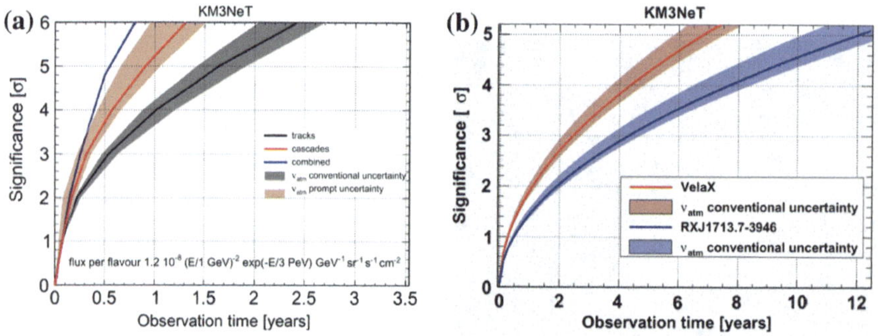

Fig. 48 Significance as a function of KM3NeT/ARCA, 2 building blocks observation time for the detection of: **a** left diffuse flux of neutrinos corresponding to the signal reported by IceCube for the cascade channel (*red line*) and muon channel (*black line*). *Right* **b** right the Galactic sources RX J1713 and Vela-X

short time with this neutrino Telescope: the 5 σ observation, in 1 year, of the *diffuse neutrino flux* observed by IceCube and the 3 σ observation, in few years, of two galactic sources.

One building block (located in KM3NeT-France site Fig. 51) will be densely configured to precisely measure atmospheric neutrino oscillations and probe the *neutrino mass hierarchy*, Fig. 49. This building block is referred to as ORCA: Oscillation Research with Cosmics in the Abyss. We know that neutrino eigenstates of the weak

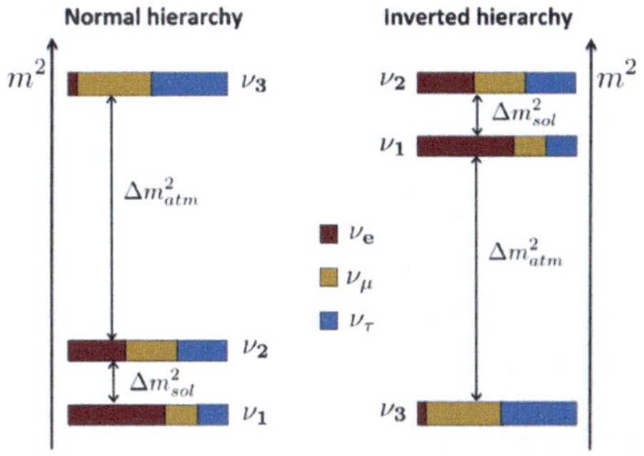

Fig. 49 Scheme of the *normal* and *inverted* neutrino mass hierarchies. The color code indicates the fraction of each flavor (e, μ, τ) present in each of the mass eigenstates (ν_1, ν_2, ν_3)

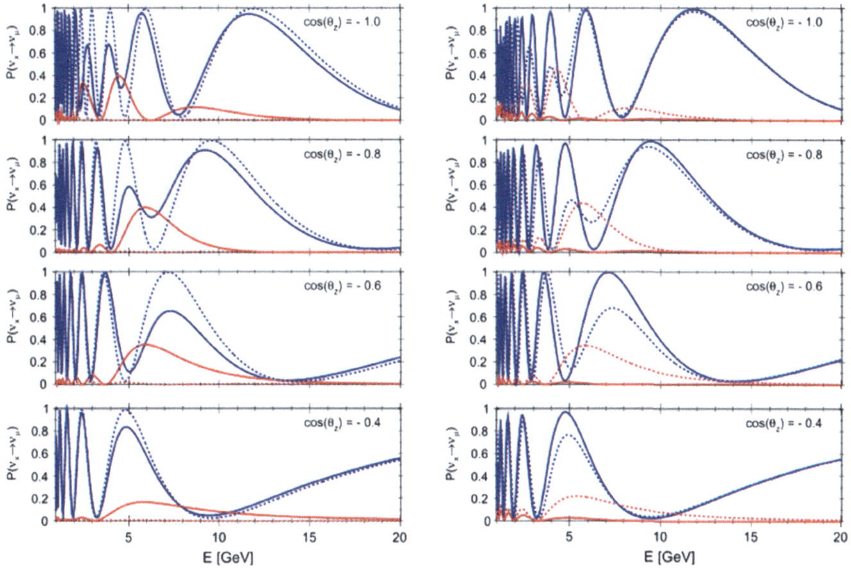

Fig. 50 Oscillation probabilities $\nu_\mu \to \nu_\mu$ (blue lines) and $\nu_e \to \nu_\mu$ (*red lines*) as a function of the neutrino energy for several values of the zenith angle (corresponding to different paths across the Earth). The *solid* (*dashed*) *lines* are for NH (IH). For neutrinos (*left*) and for antineutrinos (*right*). [119]

interactions $(\nu_e, \nu_\mu, \nu_\tau)$ are a linear combination of mass eigenstates (ν_1, ν_2, ν_3). While propagating through the Earth, atmospheric neutrinos, due to matter effects (MSW) are subject to oscillation probabilities that are different if the *mass hierarchy* is *normal* or *inverted*. Different mass orderings lead to different oscillation probabilities, see Fig. 50.

Figure 50 describes the oscillation probabilities $P(\nu_x \to \nu_\mu)$ $(x = e, \mu)$ for several values of the zenith angle θ (i.e., various lengths of the path across the Earth)

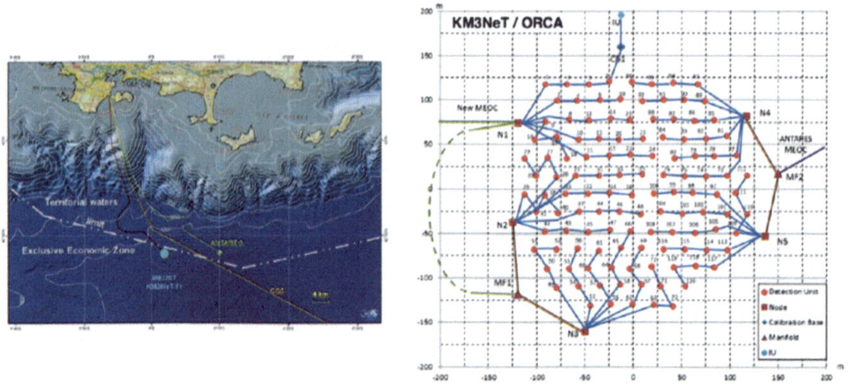

Fig. 51 The location of the KM3NeT-France (close to ANTARES site) installations are indicated (*left*). Layout of the ORCA array (*right*), depicting the 115 Detection Units, cables and connection devices of the full array

as a function of the neutrino energy, both for neutrinos and antineutrinos. In each case, both NMH hypotheses are represented. The strongest impact of the NMH to the oscillation probabilities is in the resonance region $E_\nu \sim (4\!-\!8)$ GeV [119]. For this reason the configuration of KM3NeT-ORCA has to be different with respect to the KM3NeT-ARCA case: the distance between the optical modules has to be substantially reduced: the present choice is for a DOM vertical distance of 9 m along a string and an inter-string distance of 20 m. Due to KM3NeTs flexible design, the technical implementation of ARCA and ORCA is almost identical. The KM3NeT-France infrastructure is located at about 40 km offshore from Toulon, France. The site is outside of the French territorial waters and about 10 km west of the site of the existing ANTARES telescope. Figure 51 illustrates the layout of the full ORCA array; a single KM3NeT building block of 115 strings, corresponding to 5.7 Mton mass. The power/data are transferred to/from the infrastructure via two main electro-optic cables comprising 36/48 optical fibers and a single power conductor (the return is via the sea).

The prelimiary sensitivity of KM3NeT-ORCA is represented in Fig. 52 that shows the significance in the determination of the neutrino mass hierarchy [119]. This significance depends strongly on the true value of θ_{23} and δ_{CP}.

IceCube Evidence for Astrophysical Neutrinos[41]

The first evidence for a flux of extraterrestrial neutrinos has been claimed by the IceCube detector in 2013 [266, 267]. This evidence was obtained by the study of very high-energy events where the neutrino interacts inside the detector fiducial volume.

[41] We thank A. Palladino for the help in the preparation of this paragraph.

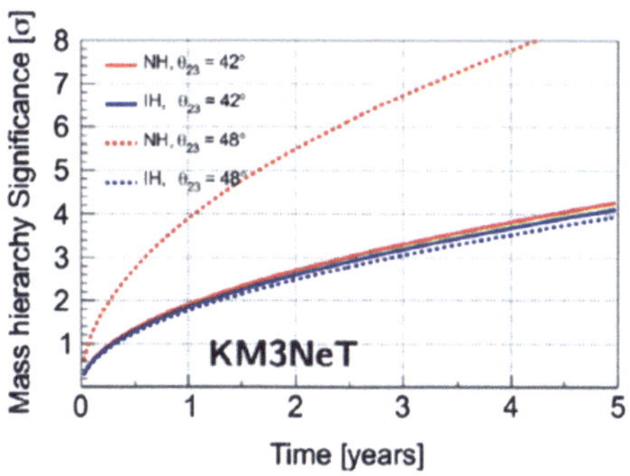

Fig. 52 KM3NeT-ORCA mass hierarchy sensitivity for 9 m vertical spacing between DOMS as a function of the number of years of integrated data and for two selected values of θ_{23} and $\delta_{CP} = 0°$

The IceCube collaboration defines a class of events called "High Energy Starting Events" (HESE), that selects neutrino interactions with the interaction point inside the detector fiducial volume, and where a sufficiently large energy is deposited in the detector volume (the experimental cut requires a minimum number of photoelectrons $N_{p.e.} > 6000$, this corresponds to an energy released in the detector of about $E_{\rm vis} \gtrsim 30$ TeV).

The background to the search of HESE events is largely due to high energy muons created in the decay of pions and kaons (or possibly charm mesons) produced in cosmic ray showers in the atmosphere above the detector. Down-going muons of sufficiently high energy penetrate deep in the ice and can enter the detector, releasing energy in the fiducial volume. An energy cut is required to reduce this muon background that has a rate that falls rapidly with energy. In addition events due to muons can be eliminated using an outer layer of the telescope—i.e., a veto. The rejection algorithm is such that if three or more of the first detected 250 photoelectrons are in the veto region then the event is discarded and considered as associated to a track entering the detector from outside. The efficiency of the veto can be studied experimentally studying the rejection obtained with veto layers of different thickness. Atmospheric neutrino events accompanied by muons, observed in the detector, can be also vetoed. Moreover, it is possible to use the cosmic ray detector just on the vertical direction of IceCube (IceTop) to further veto certain spurious events.

High Energy Starting Events

The very first hint of a signal of astrophysical neutrinos was obtained in the first search of HESE events [266] performed by IceCube on the first 616 days of effective lifetime with a large threshold energy of order 1 PeV (10^{15} eV). This study resulted in the observations of two neutrino events with energy of 1.04 ± 0.16 and 1.14 ± 0.17 PeV, to be compared with an expected background from atmospheric neutrinos of order 0.1 events (where most uncertainty should be attributed to the, still unmeasured, charm component of atmospheric neutrinos).

A more detailed study of the HESE evens on the first two years of data-taking, with the energy threshold lowered to $E_{\rm vis} \gtrsim 30$ TeV, was then published in 2103 [267]. This study yielded 28 events, with an expected background of $10.6^{+5.0}_{-3.6}$, and

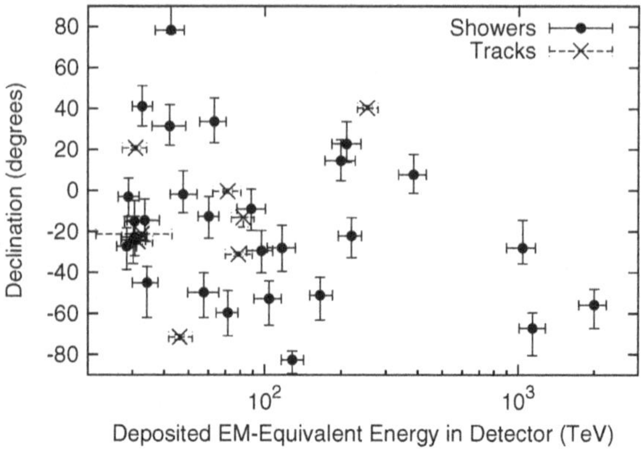

Fig. 53 From [230]. Arrival angles and deposited energies of IceCube "HESE" (High Energy Starting Events) candidates" (3 years exposure). The *disks* are "showers", the *crosses* "tracks". The energy and zenith angle projections of this figure are shown in Figs. 55 and 56

the IceCube collaboration has argued that this represents significant evidence for the existence of an extraterrestrial flux of high-energy neutrinos.

The analysis of the HESE events has been updated including a third year of data taking [268], and then a fourth [337]. The 4 year data sample contains 53 events, the background from atmospheric muons is estimated as 12.6 ± 5.1, while the contribution from atmospheric neutrinos is estimated as $9.0^{+8.0}_{-2.2}$. The significance of an excess of extraterrestrial neutrinos is calculated as $6.5\ \sigma$. The statistical significance for the existence of an extraterrestrial (or astrophysical) component in the high-energy neutrino fluxes is calculated taking into account the flavor composition, energy distribution and zenith angle dependence of the signal.

Information about the flavor of the neutrinos that cause the HESE events is obtained separating the events into two classes using the spatial structure of the energy deposition in the detector (aka, the "topology" of the event). The events are classified as "Tracks" if they contain an identifiable elongated track, and as "Showers" if they do not. In shower events the energy deposition is confined to a volume with linear size of order few meters that is quasi point-like when compared with the distances between optical modules.

The "Track" events are attributed to muons, mostly generated by the charged current interactions of ν_μ and $\overline{\nu}_\mu$; recall that a multi-TeV muon has a range longer than a kilometer of ice, therefore it appears as an elongated energy deposition that can extend up to the entire linear size of the telescope and note that, at very high energies, the transversal size of a muon track become thicker on average. "Shower" events are generated by the charged current interactions of ν_e and $\overline{\nu}_e$, and by the neutral current interactions of all neutrino types. Most of the charged current interactions of ν_τ and $\overline{\nu}_\tau$ also appear as shower events (tau neutrinos-and antineutrinos-give also a very small contribution to track events, when the leptonic decay of the tau lepton, created in charged current interactions, leads to a muon).

The energy and zenith angle distribution in the first three years of IceCube data are shown in Fig. 53.

The correlation between the energy of the interacting neutrino and the visible energy is different in shower and track events. In the "Track" events, the muon exits from the detector volume carrying away an important fraction of the total energy. For shower events one has to consider different cases: in ν_e and $\overline{\nu}_e$ charged current interactions the entire energy E_ν of the neutrino is deposited in the detector, in neutral current interactions the final state contains a neutrino that carries away a large fraction ($\sim$3/4) of the initial energy; also in the charged current interactions of ν_τ and $\overline{\nu}_\tau$ a fraction of the energy is carried away by the neutrino generated in the decay of the $\tau^\mp$ created in the first interaction, even if this is just $\sim$20% on average.

To identify the astrophysical component of the neutrino flux, in the HESE events one needs to subtract the background of non-neutrino events due to down-going muons that penetrate the detector and are not rejected because of inefficiency in the veto algorithm and also to identify the contribution of atmospheric neutrinos. As discussed above, it is possible to estimate the background from muon events studying experimentally the efficiency of the veto as a function of the thickness of the veto layer. A very important effect of the veto system, is the rejection of a large fraction of down-going atmospheric neutrino events. In these events the neutrino can be accompanied by one or more muons, created with sufficiently high energy to reach the detector and to trigger the veto system [269].

Flavor studies

The analysis of the flavor composition of the extra-component of the high-energy neutrino flux is clearly of great interest, since the observed oscillation phenomena lead to neat predictions. The issue was analyzed by several group reaching similar conclusions; the analysis of IceCube collaboration is in [272], An illustration of the study of the flavor composition of the IceCube signal in terms of the track-to-shower ratio of events is given in Fig. 54 following [285]. The top panel of the figure shows the fraction of the neutrinos that have muon flavor, for the three scenarios illustrated in section "Neutrino Oscillations". The bottom panel shows the track/shower ratio for events above a visible energy of 60 TeV (to reduce the background contribution). In the figure the predictions calculated in the three scenarios are compared with the IceCube observations (24 showers and 8 track events). The results are consistent with the prediction of a standard production mechanism, although it is not possible to reject other scenarios, because the statistics is still small. On the other side it is possible to disfavor, with a confidence level of 2σ, the neutron decay mechanism for neutrino production [290]. Also other exotic scenarios, that predict purely ν_e or purely ν_μ at the Earth, are disfavored, respectively, with a confidence level of 3σ and 4σ; this is an important point, because these scenarios would violate the neutrino oscillations.

Spectral and angular studies of the HESE events

The separation between neutrinos of atmospheric and astrophysical origin can also be performed on statistical basis using the fact that atmospheric neutrinos have a well-known flavor composition, energy spectrum, and angular distribution. The method is

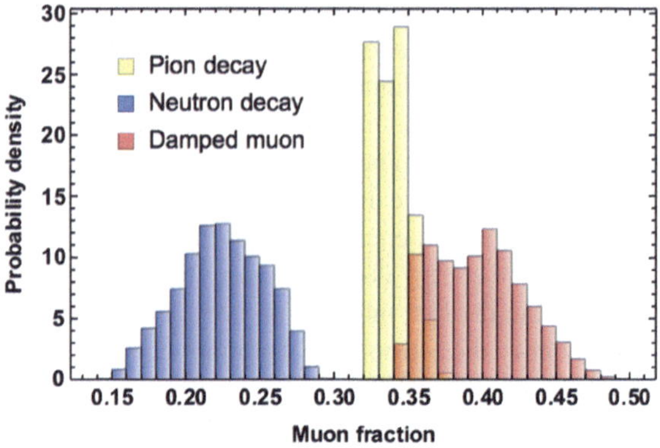

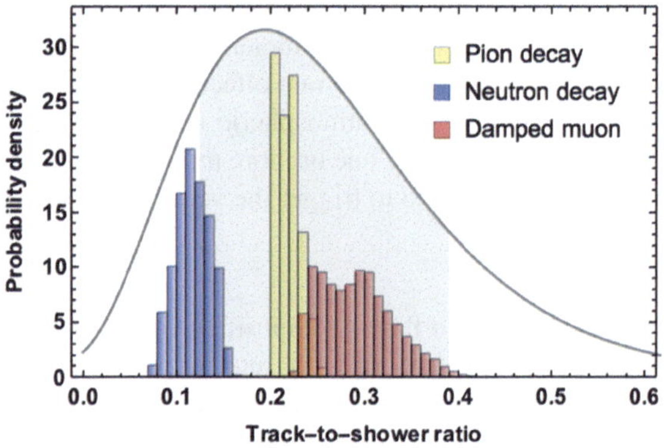

Fig. 54 In the *top panel* the fraction of muon neutrinos expected at the Earth starting from different mechanisms of production. In the *bottom panel* the expected track-to-shower ratios compared with the 4 years IceCube events above 60 TeV (in *light gray* the 1 σ region)

illustrated in Figs. 55 and 56 that show the energy and zenith angle distributions of the IceCube HESE events (for the three years exposure) comparing with the predictions.

The IceCube collaborations has discussed a simple and viable interpretation of the high-energy neutrino data, where the new component is an isotropic extraterrestrial flux of neutrinos, with equal contribution for all six neutrino types (three flavors and $\nu/\overline{\nu}$). A fit to the entire set of data [271] results in a neutrino energy spectrum that is a power law with a slope $\alpha \simeq 2.50 \pm 0.09$. The all flavor neutrino spectrum is

$$\phi_\nu(E) = 6.7^{+1.1}_{-1.2} \times 10^{-18} \left(\frac{E_\nu}{100\,\text{TeV}} \right)^{-2.50\pm0.09} \quad [\text{GeV cm}^2\,\text{s sr}]^{-1}\,. \tag{159}$$

The zenith angle distribution of the HESE Icecube events is shown in Fig. 56, and a 2-dimensional skymap in Fig. 57. The results are statistically consistent with a new isotropic, plausibly extragalactic source of the neutrinos.

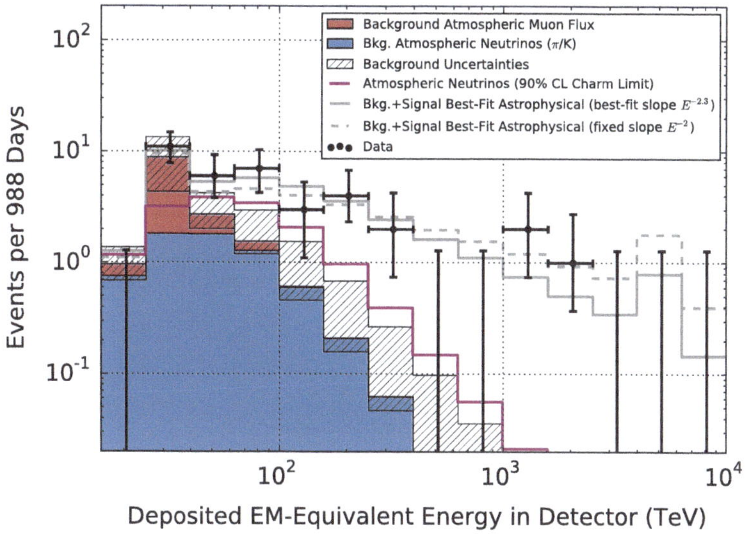

Fig. 55 From [230]. Visible energy distribution of IceCube HESE events compared with predictions

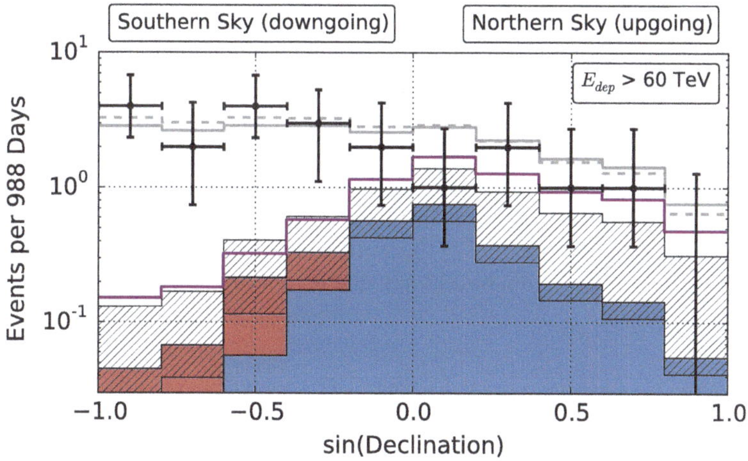

Fig. 56 From [230]. Zenith angle distribution of IceCube HESE events with $E_{dep} > 60$ TeV, The increasing opacity of the Earth to high-energy neutrinos is visible at the right of the plot. Vetoing atmospheric neutrinos by muons from their parent air showers depresses the atmospheric neutrino background on the left. The data are described well by the expected backgrounds and a hard astrophysical isotropic neutrino flux (*gray lines*). Colors as in Fig. 55

Up-going muon events

Further evidence for the existence of a new component of the neutrino flux, besides the atmospheric neutrino component, has been obtained from the flux of up-going muon events [270] and [274]. This type of signal is the one originally considered

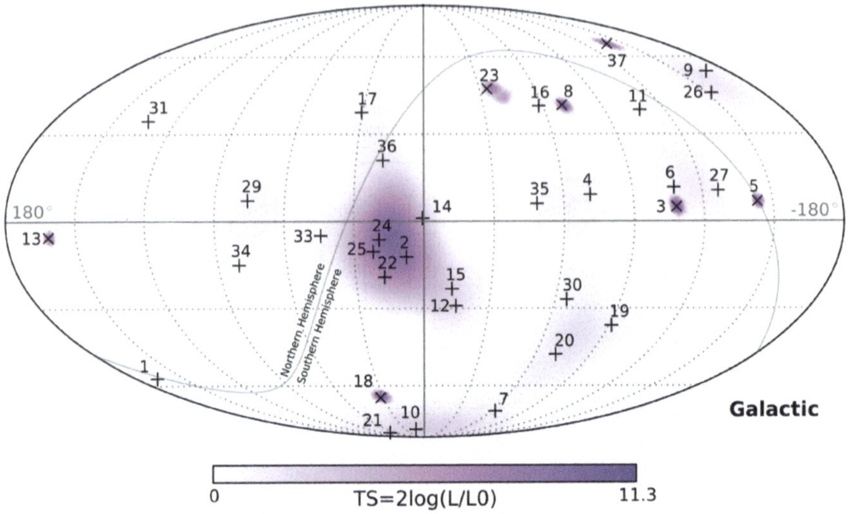

Fig. 57 Arrival directions of the HESE events in galactic coordinates. Shower-like events (median angular resolution $\sim 15°$) are marked with + and those containing muon tracks ($\lesssim 1°$) with ×. Approximately 40% of the events (mostly tracks) are expected to originate from atmospheric backgrounds. The *gray line* denotes the equatorial plane. *Colors* show the test statistic for the point source clustering test at each location. No significant clustering is observed

for the detection of high-energy neutrinos and uses the Earth as a converter and as a shield against atmospheric muons. In IceCube it is detected from the northern sky; the high-energy sample of the most recent analysis [274], includes 29 events with energies above 0.2 PeV. This sample comes mostly from the region with relatively low declinations below the horizon, due to the occurrence of Earth absorption.

The experimental distributions of the corresponding muon neutrinos and antineutrinos are well described with the addition of a hard isotropic neutrino flux, with the following distribution

$$\phi_{\nu_\mu+\bar{\nu}_\mu}(E) = 0.90^{+0.30}_{-0.27} \times 10^{-18} \left(\frac{E_\nu}{100\,\text{TeV}}\right)^{-2.13\pm0.13} (\text{GeV}\,\text{cm}^2\,\text{s}\,\text{sr})^{-1}\,. \quad (160)$$

This is not in good agreement with the slope deduced from HESE events, however we would expect from oscillations that the shape of the spectra are almost independent from the flavor. This has stimulated discussion of the so-called *IceCube spectral anomaly* [275]. In any case, it should be noted the high energy part of HESE events alone (a small sample of events) is *consistent* with the hard slope deduced by the analysis of up-going muon events [274].

An interpretation of this discrepancy in terms of physical signal requires to assume that a fraction of the HESE events collected in the southern sky is of galactic origin; this should correspond to $\sim 1/4$ of the events, namely some 6–7 of them. A similar

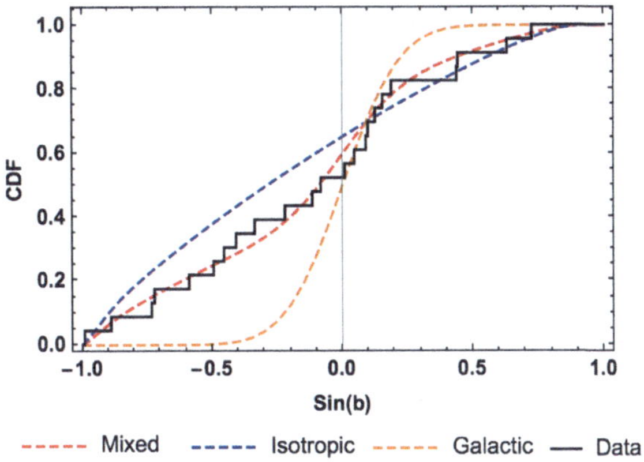

Fig. 58 Distributions of the HESE events in the galactic latitude. The cumulative density function that describes the data collected by IceCube in the southern sky and above 60 TeV is given by the *continuous line*. This is compared with three models for the data: a galactic model, confined at low latitudes (*orange*), an extragalactic model, where the events are isotropically distributed (*blue*), a mixed model determined from the energy spectrum (*red*). The last two models are compatible with the data. From [284]

possibility was in fact considered in many different analyses and astrophysical variants [275, 279–284]. See Fig. 58 for an analysis of the angular distribution.

It should be emphasized that the up-going muon events have a much better angular resolution than the HESE events, about 1 degree in ice. In principle, this could allow the identification of the sources of the events, however to date we do not have any clear evidence of this type; see, e.g., [276–278] and Fig. 59. One possible interpretation is that the sources are far, faint and numerous; another one is that the other type of radiation is heavily reprocessed at the source while neutrinos emerge unscathed. However, it is quite clear that the present data set is too small to derive very reliable conclusions.

The τ neutrino and Glashow resonance

The unambiguous detections of tau neutrino interactions would represent clear evidence of an astrophysical component. The shortest oscillation length for neutrinos of $E_\nu \simeq 30$ TeV, is of order 3.1×10^{12} cm, or more than 5000 Earth radii, therefore ν_τ cannot appear because of flavor oscillations of atmospheric neutrinos.[42] On the other hand flavor oscillations on galactic and extragalactic distances should certainly create a large fraction of tau neutrinos.

The decay length of a relativistic tau lepton is:

$$L_\tau = c\,\tau\,\frac{E}{m_\tau} \simeq 4.89\left[\frac{E}{10\ \text{TeV}}\right]\text{m} \tag{161}$$

[42]Atmospheric neutrinos contain only a very small fraction of ν_τ and $\overline{\nu}_\tau$ created in the two body decay of the $D_s^\pm$ mesons (bound states $[c\overline{s}]$ or $[s\overline{c}]$) $D_s^+ \to \nu_\tau\,\tau^+ \to \nu_\tau\,\overline{\nu}_\tau + \cdots$.

At very high energy the distance between the neutrino interaction point (where the τ lepton is created) and its decay point becomes sufficiently large to be experimentally observable. This particular topology of event is called “double bang”, if the signal manifests itself in different strings of IceCube, or “double pulse”, if the signal manifests itself as a time structure with two bumps in a single optical module. The expected rate is low and it has not been observed yet [273], but it should appear in IceCube as a “double pulse” when a larger data sample becomes available, with a rate of about 0.1 events per year [286].

Electron antineutrinos with an energy of 6.32 PeV can interact with atomic electrons, producing a couple of hadrons, that allow to measure the initial energy of the neutrinos. This was the first method that was suggested to observe the existence of a W boson back in 1960 (!) and it is called the *Glashow resonance* [287]. The study of the Glashow resonance will inform us on the highest energy part of the neutrino spectrum currently probed. The signal is not yet observed and the expected rate depends crucially upon an unexplored region of the neutrino spectrum; see again [286] for a discussion. Recall that electron antineutrinos are inevitably linked by three-flavor oscillations to the other fluxes of neutrinos; in this connection, it is particularly interesting to note that the highest energy passing muon event detected to date has more than 4 PeV [274], which (if correctly interpreted) implies that the primary muon neutrino has even larger energy.

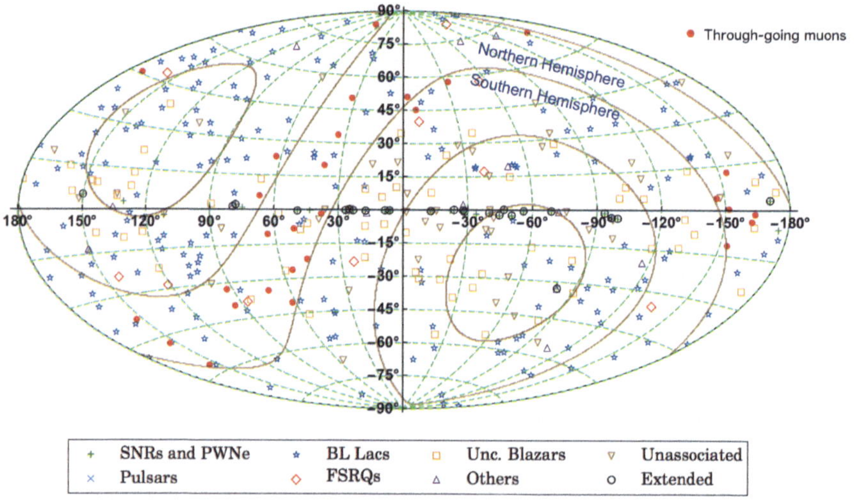

Fig. 59 Map, in Galactic coordinates, of the γ-ray sky observed by Fermi-LAT (2FHL catalog). The *red points* indicate the central position of the 29 up-going muons, with a deposited energy above 200 TeV, observed by IceCube in 6 years [274]. The *brown lines* represent intervals of declination of 30°. From [278]

Neutrinos in Context

Neutrino astronomy/astrophysics has been (and it is) a rather successful discipline-just consider the Nobel prize in physics 2002—that is part of a wide branch of science known as *astroparticle physics*. The latter is in a considerably good state of health, even if (or, perhaps, also because) it is not always clear what it does include and what it does not. Originally, the name 'astroparticle physics' was proposed with the intentions of using particle physics to investigate astrophysical (cosmological) environments and viceversa, and moreover, also to revive cosmic ray physics, that has been to some extent abandoned in favor of accelerator physics. Sometimes the alternative names of "nuclear astrophysics" and "particle astrophysics" have been used in this connection. In the course of the time, the relationships of astroparticle physics with *non-accelerator particle physics* (and more specifically with the physics investigated in underground/underwater/under-ice detectors) have been emphasized.

All these points of view contributed to stress the principal role of neutrinos in multi-messenger astronomy/astrophysics, as discussed and summarized in the first part of the present section (section "Links with Multi-messenger Astronomy"), that expands and elaborates the discussion given in sections "Low Energy Cosmic Neutrinos" and "High Energy Cosmic Neutrinos: Expectations" and in Appendix 6. In the last part (section "Links with Search for New Particles and Phenomena") we will emphasize role of neutrino astronomy and astrophysics for the search of new particle physics phenomena, that is sometimes regarded as one of the principal aims of astroparticle physics.

General references:

A basic book on astroparticle physics is [12]; a more advanced text and rather complete one is [13]. Two classic books where the role of low-energy neutrino astronomy for nuclear and particle astrophysics is stressed are [14] and [15]. The textbook [16] gives strong emphasis to speculative particle physics aspects. Some introductory textbooks to (different aspect of) dark matter are [17, 18]. We point out also a few review works, that emphasize the multi-messenger point of view [50, 51], neutrino interactions [52], and dark matter [53], respectively. A specific model, called νMSM, where the masses of three new (hypothetical) neutrinos explain dark matter and baryogenesis, is described in [54].

Links with Multi-messenger Astronomy

Neutrinos and Gamma Ray Astronomy

The fields of high-energy neutrino astronomy, gamma astronomy and cosmic ray studies are really the three faces of a single scientific field, and the three particle types can be seen as the three "messengers" that carry complementary information about the “high energy universe”. A full understanding of the sites and mechanisms that form the high-energy universe will in fact require simultaneous observations with all three messengers.

The main source for the emission of high-energy particles is the acceleration of charged particles to relativistic energies in astrophysical sites where the dynamics of the system creates electromagnetic fields of sufficient intensity and spatial extension. The relativistic charged particles (the cosmic rays) can then generate neutrinos and gamma-rays interacting with different targets.

Neutrinos are created in the interactions of hadronic cosmic rays (namely, protons and nuclei) with a target that can be formed by an ordinary gas or by a radiation field. Inelastic hadronic interactions create unstable particles (such as charged pions and kaons) that undergo weak decays, yielding neutrinos in the final state. Other secondary particles created in these hadronic interactions (such as neutral pions and η or η' mesons) decay into photons. The results is that any astrophysical site that accelerates or stores cosmic rays, and that it is supplied with sufficient targets to allow collisions, will emit neutrinos and gamma-rays with approximately equal intensities and energy distributions. This can be understood from the dynamics of particle production in hadronic interactions, and an inspection of the branching ratios into photons and neutrinos in the decay of different particles. The simplest way to understand this is that the dominant channel of neutrino emission is the chain decay of $\pi^{\pm}$, while the main channel for the gamma-ray production is the decay of π^0. The production of pions is controlled by isospin symmetry, and at high energy is in good approximation equal for all three charges. The situation is different close to threshold for inelastic interactions, where the ratio π^+/π^- is large, reflecting the fact that the projectile particles are mostly protons.

Gamma-rays can be produced also by high-energy electrons and positrons or by Inverse Compton and bremsstrahlung processes, that are additional channels, besides the hadronic mechanism emphasized above. Depending on the relative sizes (and spectral shapes) of the populations of $e^{\mp}$ and protons (and other nuclei) in a source, and on the properties of the medium that contains the relativistic particles (density in gas and radiation fields) the hadronic or leptonic mechanism can be dominant. Conversely, it should be emphasized that the definitive proof of the occurrence of hadronic processes can be given only by neutrinos.

In addition, one has to consider the fact that the absorption of high-energy photon can be important, with the dominant mechanism the production of e^+e^- pairs in $\gamma\gamma$ collisions. Absorption during propagation in the interstellar or intergalactic space when the photons travels from the source to the Earth can be calculated with reasonable accuracy, on the other hand there is also the possibility of absorption inside a source, or in its immediate vicinity. This requires a theoretical model of the a source, and it is therefore more difficult and subject to uncertainties.

In some cases the absorption of gamma-rays is nearly complete; for example, this is true for those with energies larger than few TeV and coming from extragalactic sources at large distances. In similar cases, neutrinos are the only possible direct probe of the sources. On the other hand gamma-rays can be detected with a much higher event rate, and this allows detailed studies of spectral shapes and time variabilities.

The conclusion is that simultaneous observations of high-energy gamma-rays and neutrinos from astrophysical sources can give informations about the properties of these objects that are very important and also very difficult to infer with different methods, allowing a more accurate reconstruction of the spectra for different types of particles ($e^{\mp}$ versus p), and giving additional information about the structure of the source. It is necessary to study both messengers and to compare the observations.

Neutrinos and Gravity Wave Astronomy

The recent discovery of the first gravitational wave (GW) signal from black-hole mergers certifies the beginning of a new branch of astronomy and establishes the validity of the experimental methods employed. It is expected that GW emission episodes be linked to rapid motions of large masses and/or cataclysmic and massive release of energy, as it happens in compact objects. The typical astrophysical object of this type is a core collapse supernovae, or more extreme objects as a gamma-ray burster and perhaps blazars. In some of these cases, a GW signal could be accompanied by neutrinos. A general point to note is that, while the GW can be thought of as a classical wave far from the zone of radiation, whose amplitude diminishes with the inverse of the distance, the intensity of the neutrino signal diminishes with its square. Therefore, the possibility to reveal a significant neutrino burst decreases more rapidly than the one to reveal a significant GW signal, as the distance of the source increases.

Let us begin considering the possibility of a signal of high-energy neutrinos, correlated with GW. Such a signal could be due to the formation of a jet in the astrophysical environment, or perhaps to some more exotic object such as topological defects, cosmic strongs/cusps, etc., as was pointed out first in [304] (more on this below). It is difficult to obtain clear expectations in this case, but one could suppose that if the neutrinos are strongly beamed, the amount of energy required to produce an observable neutrino signal decreases. Coincidences of high-energy neutrinos and GW, even if without relying on precise expectations, have been in fact searched [305, 306], till now without success.

A more standard case is when the formation of a compact object is accompanied by a large release of quasi-thermal low-energy neutrinos and a burst of GW. However, modeling expectations for the neutrinos on the conventional case, a standard/regular core collapse supernova not too dissimilar from SN1987A, we are bound to observe only the Milky Way and its surrounding, namely, we expect to see a significant amount of neutrinos from few Mpc at most, say, from the local group.[43] So we are discussing events which are rare on human timescale.

[43] In principle, we could detect only one (or few) neutrinos, but tightly correlated in time with other events, such as GW burst.

In the beginning, supernovae were considered a very good target for GW astronomy. The recent astrophysical models of the collapse do not support very optimistic expectations, even if they indicate that there are several possible occasions for a significant emission, with various time scales (from ms to seconds) and intensities. The expectations for the GW signal from supernovae are uncertain by one order of magnitude at least, making unclear whether the detection is likely or not. The mechanism of GW radiation is tightly connected to the multidimensional aspects of the problem, and presumably it is more closely linked to the mechanism of the explosion than the neutrino radiation, and this makes the detection even more important. The observation of a neutrino signal can indicate us the moment of occurrence of the core collapse with a precision of about 10 ms thereby offering the GW observatories an external trigger for the search of a GW burst [81, 307].

Links with Search for New Particles and Phenomena

In this part, we discuss the relationship of new particles (and in particular dark matter) with neutrino astronomy, see section "Dark Matter and New Particles", and also some speculative cases (new oscillations, neutrino decay) when the propagation of neutrino deviates from the standard assumption adopted elsewhere in this chapter, see section "New Oscillations and Propagation Phenomena".

Dark Matter and New Particles

Astrophysics and cosmology have provided us evidences for the dark matter hypothesis, namely, the hypothesis that the Universe and our Galaxy is filled by new and almost invisible particles, that were non-relativistic at the time of the decoupling (cold dark matter hypothesis) and that accounts for 27% of the present energy budget of the Universe [17].

The theoretical literature discussed many specific candidates with various properties [18, 53], including very light axions (with masses from 0.01 to 1 meV), neutral fermions with masses around 1–10 keV, milli-charged particles of mass 1–10 GeV, weakly interacting massive particles (WIMP) with masses around 100 GeV or lighter, and even heavier particles. In the last class, we include topological defects, which are well-motivated candidates in certain gage theories, and could provide us (a fraction of) dark matter.

Very strong anomalies are seen in DAMA experiment, that taken alone would be consistent with a candidate of 10–100 GeV [308][44]; the comparison with the very strong null results from other experiments is not simple, but it seems fair to say that

[44]The hypothesis that these particles populate the Galactic halo with velocities below the escape velocity allows us to estimate that kinematics of nucleon recoil, leading to signals in the few keV region.

the direct search for these particles in laboratory is to date still inconclusive. Signals and bounds from the LHC will be extremely important to assess this hypothesis.

Moreover, further astrophysical investigations of the masses and of the other properties of these hypothetical dark matter particle are also possible. In this respect, there are many connections with neutrino physics and astrophysics that deserve a comment, in particular

- The neutral fermions can be considered a right handed, non-standard neutrino. If they decay in a lighter neutrino and a photon, a monochromatic line at $m_\nu/2 \sim$ few keV is expected [54, 309].
- WIMPs can be trapped in astronomical bodies, such as the Earth, the Sun, the region surround the center of Milky Way. The annihilation (or possibly the decay) of these particles can lead to a variety of annihilation products with energies at or below the mass of this particle. In cases like the Sun or the Earth, only neutrinos will emerge from the source. See, e.g., [310, 311].
- The very high-energy neutrinos seen by IceCube can be attributed to decay or annihilations of very heavy dark matter particles. In this case, a preferential direction of the signal toward the galactic center is expected [291].
- The decay of even heavier particles and/or topological defects could produce neutrinos of interest for radio telescopes or dedicated space missions. For a discussion, see [312, 313].

New Oscillations and Propagation Phenomena

Let us consider some representative source of neutrinos, where L is the distance of propagation and E_ν the energy of the neutrinos; the precise values can differ in specific applications. A few interesting cases are the Sun, SN1987A where $L = 50$ kpc, a generic cosmic source of high-energy neutrinos ($E_\nu = 100$ TeV) at $L = 1$ Gpc. In cosmology, we consider two situations: 1) the neutrinos at the recombination time of 378,000 year, with a photon temperature of $T_\gamma = 4,000$ K, corresponding to neutrino temperature of $T_\gamma \times 1.9/2.7 \times 8.2 \times 10^{-5}$eV/K $= 0.24$ eV 2) the neutrinos from Big-Bang today, at a temperature of 1.9 K = 0.16 meV and associated to a Hubble distance is $L = c/H_0$. Note that in the last case the usual theory of neutrino oscillations does not apply, since at least a pair of neutrino are non-relativistic; thus the figures quoted below should be interpreted accordingly. The various cases are summarized in Table 10.

Suppose that we are interested to consider the existence of new (hypothetical) oscillations phenomena on top of the usual three flavor oscillations. When we consider vacuum oscillations, we need to evaluate, $\varphi = \Delta m^2_{\text{new}}\, L/4\, E$. The *minimum* value of the new mass splitting, for which the new phenomena are potentially relevant, obtaind from the condition $\varphi > \pi/2$, namely from the condition

$$\frac{\Delta m^2_{\text{new}}}{\text{eV}^2} = 1.24\, \frac{E_\nu}{\text{MeV}} \times \frac{\text{m}}{L} \tag{162}$$

Table 10 Representative values of distances and energies of cosmic neutrino sources and corresponding value of a new hypothetical mass splitting that would be relevant for observations

	Sun	SN1987A	Cosmic source	Recombination	Cosmology
L (m)	1.5×10^{11}	1.5×10^{21}	3×10^{25}	3.6×10^{18}	1.4×10^{27}
E_ν (MeV)	5	20	10^8	2.4×10^{-7}	1.6×10^{-10}
Δm^2_{new} (eV^2)	4.1×10^{-11}	1.6×10^{-20}	4.0×10^{-18}	8.4×10^{-20}	1.4×10^{-36}

From the last row of Table 10, we see that solar neutrinos in principle give us access to the study of mass splittings one million of times smaller than those that we know, which is already remarkable; however, neutrinos from cosmic sources are sensitive to masses one billion of times smaller. The other important parameter that should be known is the mixing angle between ordinary neutrinos and the new hypothetical neutrinos. The most interesting cases are when this is large. In fact there are at least two cases, discussed in the scientific literature, when the mixing is maximal: these are the cases of "models with mirror symmetry" and of "quasi-Dirac" neutrinos [314, 315]. In these models, it happens that the usual mass states ν_i are in fact a superposition of mass eigenstates, $\nu_i = (\nu_i^+ + \nu_i^-)/\sqrt{2}$, and therefore the ordinary neutrino fluxes, as a rule, can be halved by the new oscillations. See [177, 210, 217] for a more complete discussion.[45]

In principle, massive neutrinos could be unstable particles. This hypothesis was first considered in connection with solar neutrino observations. Even if we know today that the correct explanation is another one—i.e., neutrino oscillations—it is not impossible that neutrinos do decay after some distance, thereby producing some observable manifestation. Suppose that the heavier neutrinos decay in the lighter ones; one possibility is that this is accompanied by photons (radiative decay); another one is that instead the decay is accompanied by the emission of invisible particle are emitted (non radiative decay). The probability that an unstable neutrino survives is [317, 318],

$$\exp\left(-\frac{T}{E_\nu} \times \frac{m_i c^2}{\tau_i}\right) \tag{163}$$

Here, m_i and τ_i are the masses and lifetime of the i-th neutrino, subject to decay and $T = L/c$ the time since production. Considering the same physical situations, we can compare the sensitivities that it is possible to achieve. Thus, we see the sensitivity is larger when the parameter E_ν/T is smaller. As a reference, we note that in the case of the Sun, we have $E_\nu/T \approx 1{,}000$ eV/s for $E_\nu = 0.5$ MeV. When we note that $E_\nu/T \propto \Delta m^2_{\text{new}}$, we understand that the hierarchy among the different observations is the same one examined previously, see Table 10. Indeed, it was remarked that several observations of neutrino astronomy allow us to probe this hypothesis. This

[45] At cosmological distances, corrections should be applied. The average oscillations phase φ is $\Delta m^2_{\text{new}}(c/H_0)/(4E)$ times $\xi(z_e) = \int_0^{z_e} dz/\left[(1+z)^2\sqrt{\Omega_m(1+z)^3 + \Omega_\Lambda}\right]$ where z_e is the redshift when neutrino have been emitted [316]. The asymptotic value is $\xi(\infty) = 0.53$, the deviations from Hubble law amount to $\xi(z_e)/z_e - 1 = -10\%$ when $z_e = 0.1$.

includes solar neutrinos [319], SN1987A [320, 321] and high energy neutrinos [321]. The value of the neutrino mass, probed by cosmology, could also contribute to assess this hypothesis [341]. None of the existing data analyses to date has produced hints, but only bounds on the hypothesis of neutrino decay.

Above, we have discussed only few representative examples of new (speculative) effects on neutrino propagation. Moreover, we have avoided to discuss what is the expected size of the new effects; as a rule, this is not predicted precisely and therefore we limited the discussion to few phenomenological remarks. However, many more effects are considered in the literature: the possibility that there are observable "wiggles" due to new oscillations; that "matter effects" play a role and amplify a small mixing angle with sterile neutrinos; that there are new sources of matter effects; that, in neutrino oscillations phenomena, deviations from CPT symmetry occur, etc. The interested reader is invited to consult the vast scientific literature that concerns the discussion of these effects, possibly beginning with [314, 322].

Appendix 1: Neutrinos on Web

Besides the web pages of the main Journals, there are several important repositories of papers online, used by various scientific communities that are interested in neutrinos. They include the *arXiv* [344] the searchable *inSPIRE* database [344], and the *The SAO/NASA Astrophysics Data System* [346].

Carlo Giunti and Marco Laveder provide access to an updated selection of references on neutrinos, mostly aimed at particle physicists, called *Neutrino Unbound* [347], while Maury Goodman maintains a dedicated newsletter *Long Baseline News* [348] concerning rumors and references of special interest about neutrinos.

There are several workshops and conference series dedicated to neutrinos, that allow the scientific community to remain informed and updated. The main ones, concerning the themes in which we are interested here, are named,

- *Conference on Neutrino Physics and Astrophysics* (Neutrino)
- *Weak Interaction and Neutrinos* (WIN)
- *Neutrino Oscillation Workshop* (NOW)
- *Neutrino Telescopes in Venice* (NuTel)
- *Neutrino-Nucleus Interactions in Few GeV region* (NuInt)

The above names, already, allow one to have an idea of the main thrusts of the discussion—even if they changes a bit in the course of the years, as it is natural. For each conference, typically, useful web sites are arranged where the files of the presentations are made public before proceedings are issued, allowing a fast, wider and more effective diffusion of the information. More complete and updated lists are in the above web pages of Giunti-Laveder [347] and of Goodman [348].

Finally, we recall that the PDG group [349] releases any two year a very useful report on particle physics, including review works on neutrinos. A database with neutrino cross sections measured below 30 GeV is in [350]. There are several useful sites to obtain parton distribution functions in various formats (Fortran, C++, Mathematica), e.g., [351–353].

Appendix 2: Wavefunction of the Neutrino

In this section we offer a summary of well-known results concerning the description of neutrinos in theoretical particle physics. This summary does not want to be complete and exhaustive, but rather, aims only to give an idea of the description of relativistic fermions to a generic readership, by using the (supposedly known) concepts of quantum/wave mechanics: wavefunctions, hamiltonian, spin, etc.

Plane waves of fermions

The plane waves that describe a relativistic fermion are eigenstates of the energy, momentum and of the helicity (i.e., the projection of the spin over the momentum). In formal terms, using the units $\hbar = c = 1$, so that, e.g., the momentum operator is $\hat{p}^i = -i\partial/\partial x^i$, and denoting the wavefunction with ψ to simplify the notations (or sometimes using the complete notation $\psi(\mathbf{p}\lambda, x)$) we have,

$$\begin{cases} \left[\alpha^i \hat{p}^i + \beta m\right] \psi(x) = E\ \psi(x) & \text{[energy]} \\ \hat{p}^i\ \psi(x) = p^i\ \psi(x) & \text{[momentum]} \\ \Sigma^i \hat{p}^i\ \psi(x) = \lambda\ p\ \psi(x) & \text{[helicity]} \end{cases} \tag{164}$$

where $i = 1, 2, 3$, the contracted indices are summed over. The eigenvalues of the momentum, energy and helicity are, $\mathbf{p}$, $E = \sqrt{p^2 + m^2}$ (with $p = |\mathbf{p}|$), $\lambda = \pm 1$.

With a specific choice of basis (*Dirac representation*) the 4×4 matrices α^i and β of Dirac hamiltonian and the spin matrices Σ^i have the following expressions

$$\alpha^i = \begin{pmatrix} 0 & \sigma^i \\ \sigma^i & 0 \end{pmatrix} \quad \beta = \begin{pmatrix} 1 & 0 \\ 0 & -1 \end{pmatrix} \quad \Sigma^i = \begin{pmatrix} \sigma^i & 0 \\ 0 & \sigma^i \end{pmatrix} \tag{165}$$

where σ^i are the 3 (hermitian, traceless, 2×2 and well-known) Pauli matrices. The normalized wavefunction is given in terms of a spinor u,

$$\psi_\alpha(x) = \frac{e^{-i(Et - \mathbf{px})}}{\sqrt{2EV}}\ u_\alpha(\mathbf{p}\lambda) \text{ where } u_\alpha^*\ u_\alpha = 2E \tag{166}$$

where V is the total volume of the space (that of course plays no role in physical quantities). The explicit expression of the spinor is,

$$u(\mathbf{p}\lambda) = \begin{pmatrix} \sqrt{E+m}\,\chi \\ \lambda\sqrt{E-m}\,\chi \end{pmatrix} \text{ where } \sigma^i p^i\, \chi = \lambda p\, \chi \tag{167}$$

Again in Dirac representation, the charge conjugation matrix and the chirality matrix are

$$C = \begin{pmatrix} 0 & i\sigma^2 \\ i\sigma^2 & 0 \end{pmatrix} \;\; \gamma^5 = \begin{pmatrix} 0 & 1 \\ 1 & 0 \end{pmatrix} \tag{168}$$

The charge conjugated spinor, defined as

$$\psi^c = C\beta^t \psi^* \tag{169}$$

satisfies the same conditions as ψ but with opposite energy, momentum, and spin (thus, with the same helicity as ψ). This wavefunction is used in a very similar manner as one uses progressive and regressive waves in electromagnetism: ψ describes a neutrino, that is ongoing to participate in a reaction, ψ^c represents an antineutrino, outgoing from the site of the reaction.

Helicity and chirality

At this point, we have a very important remark. The neutrino wavefunctions produced in weak interactions have a well-defined helicity in the ultra relativistic limit. In fact applying the chiral projector $P_L = (1-\gamma^5)/2$, that is implied by charged currents, it is easy to prove that

$$\begin{array}{c} P_L\ \psi(\mathbf{p}+, x) \approx 0 \text{ and } P_L\ \psi(\mathbf{p}-, x) \approx \psi(\mathbf{p}-, x) \text{ for neutrinos} \\ P_L\ \psi^c(\mathbf{p}+, x) \approx \psi^c(\mathbf{p}+, x) \text{ and } P_L\ \psi(\mathbf{p}-, x) \approx 0 \text{ for antineutrinos} \end{array} \tag{170}$$

that can be expressed in words saying that the (ultrarelativistic) neutrinos have negative helicity ($\lambda = -1$) while the antineutrinos have positive helicity ($\lambda = +1$). Therefore, whenever the ultra relativistic limit applies[46] we have only two wavefunctions that are produced and that matter: those with wavefunctions $\psi(\mathbf{p}-, x)$ and $\psi^c(\mathbf{p}+, x)$. For this reason, it is often possible to adopt the same definition given in the standard model of electroweak interactions (where neutrinos are massless), namely: neutral leptons with negative helicity are neutrinos, those with positive helicity are antineutrinos.

A last comment on the nature of neutrino masses and quantum field theory. A Dirac massive neutrino has 4 states, but the only 2 of them matter in the above assumptions, those corresponding to the particle oscillator $\hat{a}(\mathbf{p}-)$ and to the antiparticle oscillator $\hat{b}(\mathbf{p}+)$; a Majorana massive neutrino has only two states, which corresponds to the formal replacement $\hat{b}(\mathbf{p}+) \to \hat{a}(\mathbf{p}+)$. See Fig. 1.

[46]This is always true for oscillations in the conventional sense and it is also the typical case in neutrino astronomy. The only practical exception are the Big-Bang neutrinos today.

Appendix 3: Matter Effect, the General Case

Here we collect arguments and technical remarks from [68] that can be useful if one needs to implement oscillations including matter effect numerically, and/or to describe atmospheric neutrino oscillations in the Earth. A web interface, that calls a routine to perform this type of calculations, can be found in [357].

Let us write in full generality the amplitude of three-flavor neutrino oscillations

$$\mathscr{A} = \text{Texp}\left[-i\int dt \mathscr{H}_\nu(t)\right] = R_{23}R_\delta \begin{pmatrix} a_{11} & a_{12} & a_{13} \\ a_{21} & a_{22} & a_{23} \\ a_{31} & a_{32} & a_{33} \end{pmatrix} R_\delta^* R_{23}^t \tag{171}$$

where a_{ij} depend upon Δm^2_{23}, θ_{13}, Δm^2_{12}, θ_{12}, and $H = \pm 1$ (the type of mass hierarchy, normal/inverted): see in particular Eqs. 1, 3, 5, 6 of [69]. By solving numerically the evolution equations, we calculate the complex numbers a_{ij} and therefore the amplitudes and the probabilities. At this level, there is no approximation, except the numerical ones.

When the "solar" Δm^2_{12} is set to zero—i.e., when its effects are negligible—the only non-zero out-of-diagonal elements a_{ij} in Eq. 171 are a_{13} and a_{31}. The CP violating phase δ drops out from the probabilities $P_{\ell\ell'} = |\mathscr{A}_{\ell'\ell}|^2$, that moreover become symmetric, $P_{\ell\ell'} = P_{\ell'\ell}$ for each $\ell, \ell' = e, \mu, \tau$. Therefore, in this approximation we have 3 independent probabilities and all the other ones are fixed. We can chose, e.g.,

$$P_{e\mu} = \sin^2\theta_{23}|a_{13}|^2, \; P_{e\tau} = \cos^2\theta_{23}|a_{13}|^2, \; P_{\mu\tau} = \sin^2\theta_{23}\cos^2\theta_{23}|a_{33} - a_{22}|^2, \tag{172}$$

so that, e.g., $P_{ee} = 1 - P_{\mu e} - P_{\tau e} = |a_{11}|^2$. From these formulae we obtain

$$P_{\mu e} = \sin^2\theta_{23}\,(1 - P_{ee}) \text{ and } P_{\mu\tau} = \frac{1}{4}\sin^2 2\theta_{23}\left|1 - \sqrt{P_{ee}}\,e^{i\hat{\varphi}}\right|^2, \tag{173}$$

where $\hat{\varphi}$ is a (rapidly varying) phase factor. Two important remarks are in order:

1. The last equation shows that $P_{\mu e}$ is large in the region where P_{ee} is small, and that $P_{\mu\tau}$ remains close to zero in the first non-trivial minimum near $\hat{\varphi} = 2\pi$, even when $P_{ee} \approx 0.3$–0.4 due to matter effect.
2. The sign of Δm^2 controls the sign of the vacuum hamiltonian; therefore, switching between the two mass hierarchies or switching between neutrinos and antineutrinos has the same effect; e.g., $P_{e\mu}(\text{IH}) = P_{\bar{e}\bar{\mu}}(\text{NH})$.

The first remark is consistent with the numerical findings, that $P_{\mu e}$ is amplified and $P_{\mu\tau}$ does not deviate strongly from its behavior in vacuum in the conditions that are relevant for our discussion.

Proceeding further with the approximations, and considering at this point the case of constant matter density, we obtain simple and closed expressions. For the case of normal mass hierarchy, they read:

$$\begin{aligned} a_{13} &= a_{31} = -i \sin\widetilde{\varphi} \sin 2\widetilde{\theta_{13}} \\ a_{11} &= \cos\widetilde{\varphi} + i \sin\widetilde{\varphi} \cos 2\widetilde{\theta_{13}} = a_{33}^* \\ a_{22} &= \cos\widetilde{\varphi}' + i \sin\widetilde{\varphi}' \end{aligned} \tag{174}$$

where

$$\widetilde{\varphi}' = \frac{\Delta m^2 L}{4E}(1+\varepsilon) \tag{175}$$

From Eqs. 172 and 174, we recover the expression of Eq. 43, used in the above discussion. In the approximation of constant matter density, the phase $\hat{\varphi}$ entering the expression of the probability $P_{\mu\tau}$ is given by $\sqrt{P_{ee}} \cos\hat{\varphi} \equiv \cos\widetilde{\varphi} \cos\widetilde{\varphi}' - \sin\widetilde{\varphi} \sin\widetilde{\varphi}' \cos 2\widetilde{\theta_{13}}$. This is close to the vacuum phase when ε is large or small in comparison to 1: in fact, we have $\cos 2\widetilde{\theta_{13}} \sim \pm 1$ and $\widetilde{\varphi} \sim \pm \Delta m^2 L/(4E)(1-\varepsilon)$ from Eq. 44, so that $\cos\hat{\varphi} \sim \cos[\Delta m^2 L/(2E)]$.

Appendix 4: Treating Cosmological Neutrino Sources

In this appendix we summarize a few important formulae to treat sources of (low or high energy) neutrinos that are located at cosmological distances. A very useful review of practical cosmology is [323]. A recent determination of cosmological parameters is the one due to the Planck Collaboration, [34].

Covariant distance

Cosmological distances are measured by mean of the redhift z, that is connected to the wavelength λ and the frequency ν as follows,

$$\frac{\lambda_{\text{observed}}}{\lambda_{\text{emitted}}} = \frac{\nu_{\text{emitted}}}{\nu_{\text{observed}}} = 1 + z \tag{176}$$

The Hubble parameter $H_0 = 68 \, \frac{\text{km/s}}{\text{Mpc}}$ corresponds to the Hubble scale,

$$D_H = \frac{c}{H_0} = 1.36 \times 10^{27} \text{ cm} = 4.41 \text{ Gpc} \tag{177}$$

Moreover, we have $\Omega_\Lambda = 0.69$ and $\Omega_m = 0.31$ that fix the covariant distance,

$$D_c(z) = D_H \int_0^z \frac{dz'}{h(z')} \text{ where } h(z) = \sqrt{\Omega_\Lambda + \Omega_m(1+z)^3} \tag{178}$$

For small z, this gives the usual Hubble law case, where the distance is just proportional to z: $D_c(z) \approx D_H \times z$. E.g., we have $D_c(0.1) = 430$ Mpc $\approx D_H \times 0.1$ within 2.5%. The covariant distance can be calculated numerically and tabulated, or even approximated analytically [356]. The comoving volume $V_c(z)$ is a sphere with radius $D_c(z)$, therefore, we have

$$\frac{dV_c}{dz} = 4\pi \frac{D_c(z)^2 D_H}{h(z)} \tag{179}$$

this relation allows us to convert a distribution over the covariant volume in a distribution over the redshift as in the example given below.

Diffuse flux

If we have a source at redshift z that emits neutrinos (or photons) with a spectrum dN_ν/dE (differential in the energy), and if the emission is isotropic, the detected fluence is,

$$\frac{dF_\nu}{dE} = \frac{1}{4\pi D_c(z)^2} \times \frac{dN_\nu}{dE'} \tag{180}$$

where, due to redshift,

$$E' = (1+z)E \tag{181}$$

Note that E' is the emitted (i.e., initial energy), but it is received as E due to the redshift. If there are many of these sources, distributed uniformly and isotropically with a density per comoving volume and per unit time $\dot{\rho}_{\text{source}}(z)$, the total diffuse and stationary flux that we will receive a diffuse flux $\Phi_{\text{diffuse}}(E) = \int dV_c\, \dot{\rho}_{\text{source}}(z)\, dF_\nu/dE$, that is,

$$\Phi_{\text{diffuse}}(E) = D_H \int_0^\infty dz\, n(z)\, \frac{dN_\nu}{dE'} \text{ where } n(z) = \frac{\dot{\rho}_{\text{source}}(z)}{h(z)} \tag{182}$$

Appendix 5: Principles of High Energy Neutrino Telescopes

The Cherenkov Radiation

Quanta of light are produced when a charged particle travels through a dielectric transparent medium at a speed greater than the speed of light in that medium. The medium may be any liquid or solid provided it is transparent. These photons are referred to as Cherenkov radiation in honor of the Russian physicist P.A. Cherenkov for his basic research on this phenomenon. Cherenkov radiation consists of a continuous spectrum of wavelengths extending from the ultraviolet region into the visible part of the spectrum peaking at about 420 nm. Only a negligible amount of photon emissions is found in the infrared or microwave regions. Cherenkov photon emission is the result of local polarization along the path of travel of the charged particle with the emission of electromagnetic radiation when the polarized molecules return to their original states. The Cherenkov effect is depicted clearly in Fig. 60, where a

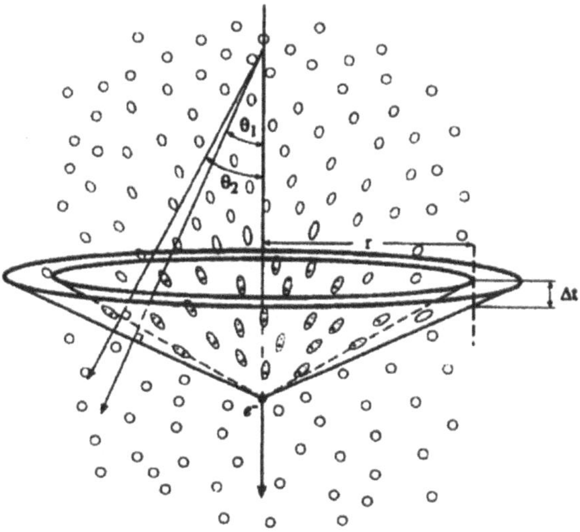

Fig. 60 Depiction of the production of Cherenkov radiation in a dielectric medium and the resulting wave front expansion. The wave front spreading lengthens the excitation pulse on a time scale that is small in comparison to the fluorescence decay. t is the duration of the light pulse along a line parallel to the axis of the particle at a distance r from the axis. Taken from [19]

charged particle traveling along the vertical axis distorts the electron clouds of atoms in close proximity to the high-speed particle traversing a transparent medium.

The Cherenkov radiation is propagated as a conical wave front, that is, the radiation is emitted as a cone having it's axis in the direction of particle path. If charged particles move through an isolator of (initially constant) index of refraction n, the latter is always polarized by the electromagnetic field of the flying particle. Usually these polarization superpose destructively and the perturbation can relax while the particle moves on. But if the speed of the particle β is higher than the phase velocity of light in the medium c/n the perturbation can no longer decay since the reaction time of the medium is not high enough and the perturbation remains behind the particle.

Figure 61 shows the two types of polarization of charged particle traversing it with speed $v < c/n$ (left) or speed $v > c/n$ (right). The threshold condition for the production of Cherenkov radiation in a transparent medium is given by $\beta_{thr} = c/n$ where β is the relative phase velocity of the particle, that is, the velocity of the particle divided by the speed of light in the vacuum c; n is the refractive index of the medium that is, for definition, the ratio of the velocity of light in the vacuum to its velocity in

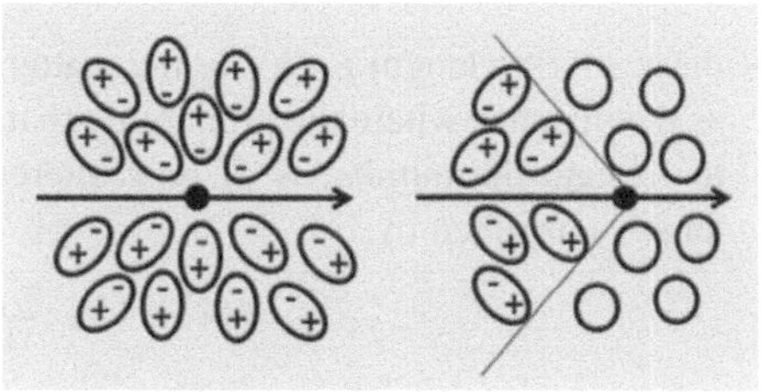

Fig. 61 *Left* polarization of an isolator caused by a charged particle traversing it with speed $v < c/n$. *Right* polarization of an isolator causing by a charged particle traversing it with speed $v > c/n$

the medium. Only charged particles moving with velocity $\beta \geq \beta_{thr} = c/n$ produce Cherenkov photons in transparent media. For particles with velocity $\beta \geq \beta_{thr}$ the number of Cherenkov photons emitted per unit photon energy in a medium of particle path length L

$$\frac{dN}{dE} = L\frac{\alpha}{\hbar c}sin^2\theta$$

where $\alpha = 1/137$ is the fine structure constant, $\hbar = 6.582 \times 10^{-16} eV^{-1}\,\mathrm{cm}^{-1}$ is Planck's constant divided by 2π. The number of photons emitted per path length, in the wavelength interval between λ_1 and λ_2, can be calculated according to the formula:

$$\frac{dN}{d\ell} = 2\pi\alpha z^2\left(\frac{1}{\lambda_2} - \frac{1}{\lambda_1}\right)sin^2\theta$$

where z indicates the particle charge. Over the visible range of wavelengths from $\lambda_1 = 350$ nm to $\lambda_2 = 650$ nm, the number of photons N per path length L is

$$\frac{N}{L} = 490 \cdot sin^2\theta\ \mathrm{cm}^{-1}$$

It is easy to evaluate then that in 1 cm of water, a particle with $\beta = 1$ induces $N \sim 320$ photons in the spectral range of visible light.

Light propagation in the transparent medium

The propagation of light in water is affected by two phenomena: absorption and scattering. Both phenomena depend on the photon wavelength λ. In the first case the photon is lost, in the second one it changes its direction. Moreover, scattering delays the propagation of photons between the points of emission and detection. The parameters generally chosen to describe the two phenomena are:

- the absorption length $L_a(\lambda)$ (or its reciprocal: the absorption coefficient defined as $a(\lambda) = 1/L_a(\lambda)$) that indicates the distance where the probability that a photon has not been absorbed amount to $1/e$, such that the number N of non-absorbed, photons as a function of photon path length x is given by:

$$N(x) = N_0 \cdot exp\left(-\frac{x}{L_a(\lambda)}\right)$$

- the scattering length $L_s(\lambda)$ (or the scattering coefficient $s(\lambda) = 1/L_s(\lambda)$), defined as the distance where the probability that a particle has not been scattered amount to 1/e and the number N of no-scattered photons as a function of photon path length x is given by:

$$N(x) = N_0 \cdot exp\left(-\frac{x}{L_s(\lambda)}\right)$$

- the attenuation length $L_c(\lambda)$ (or the attenuation coefficient $c(\lambda) = a(\lambda) + b(\lambda)$), defined as the distance where the probability that a particle has not been neither absorbed nor scattered amount to 1/e and the number N of surviving photons as a function of photon path length x is given by:

$$N(x) = N_0 \cdot exp\left(-\frac{x}{L_c(\lambda)}\right)$$

- the *volume scattering function* $\chi(\theta, \lambda)$, i.e., the distribution in scattering angle θ;
- when the photon detection area is sufficiently large that *small* scattering angle do not prevent the photon detection is used the *geometrical*, or *effective* scattering length defined as

$$L_s^{eff} = \frac{L_s}{1 - \langle cos(\theta) \rangle}$$

where $\langle cos(\theta) \rangle$ is the mean value of the cosine of the scattering angle.

The knowledge of these parameters is very important for a proper comprehension of the Cherenkov propagation in the detector media. For these reasons all the experimental approaches need a direct measurement of these optical properties (Fig. 62).

Also in the water the propagation of light is quantified, for a given wavelength λ, by the *water inherent optical properties* (IOP): the absorption $a(\lambda)$, scattering $b(\lambda)$ and attenuation $c(\lambda) = a(\lambda) + b(\lambda)$ coefficients. Figure 63 shows the absorption and attenuation lengths measured in the Mediterranean deep Sea site where will be located the KM3NeT-Italy future Cherenkov Neutrino Telescope.

Neutrino interaction detection principle

The detection of the Cherenkov light is made possible by the arrangement of arrays of photo-multiplier tubes (PMTs) housed in transparent pressure spheres which are spread over a large volume in oceans, lakes or glacial ice (Fig. 64).

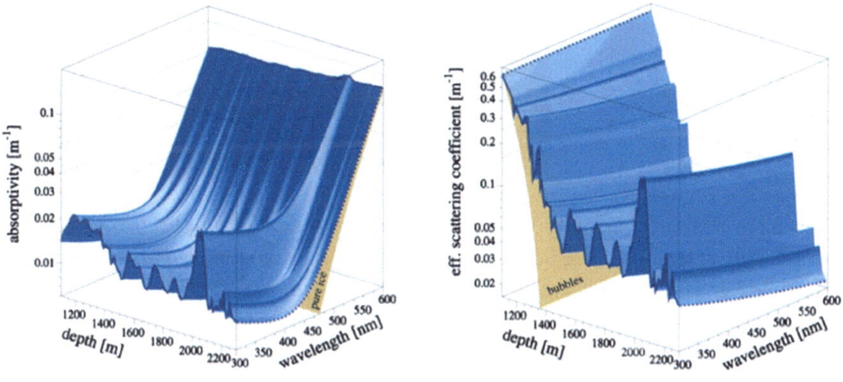

Fig. 62 Absorption coefficient (*left*) and effective scattering coefficient (*right*) in the South Polar ice as functions of depth and wavelength [49]

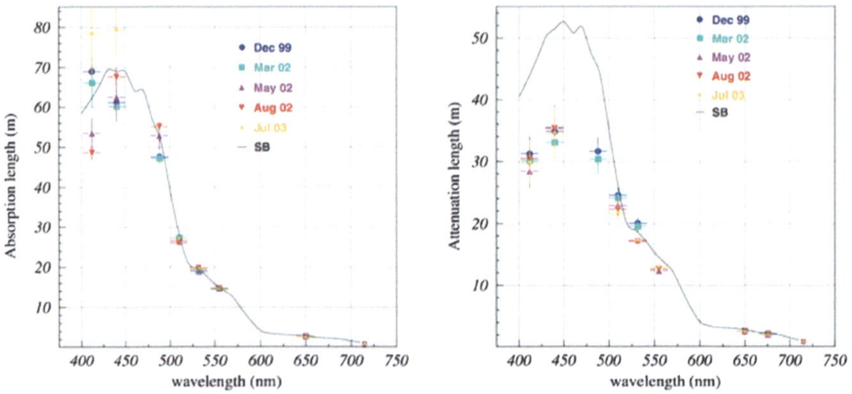

Fig. 63 Average absorption and attenuation lengths [324] measured in December 1999 (*blue circle*), March 2002 (*light blue square*), May 2002 (*purple triangle*), August 2002 (*red upsidedown triangle*) and July 2003 (*dark yellow star*) and in the site where KM3NeT detector will be constructed, at depths 2850 (*left*) and 3250 m (*right*). Statistical errors are plotted. A *solid black line* indicates the values of $L_a(\lambda)$ and $L_c(\lambda)$ for optically pure seawater reported by Smith and Baker [325]

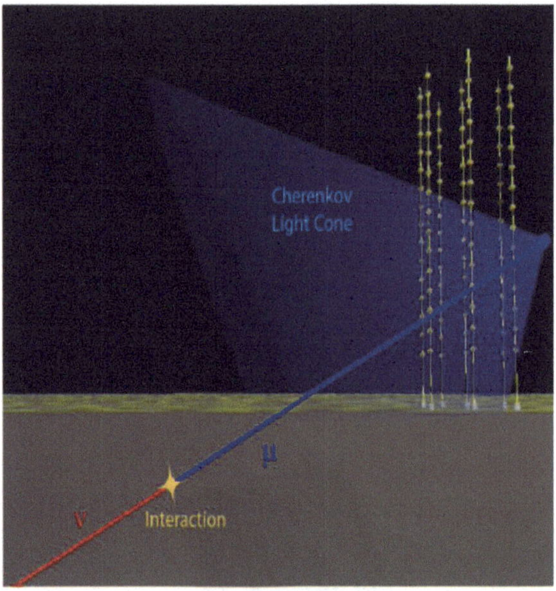

Fig. 64 Detection principle of high-energy muon neutrinos in an underwater/ice Neutrino Telescope. The incoming neutrino interacts with the material around the detector to create a muon. The muon induces the emission of Cherenkov light in the sea water. Photons are then detected by a matrix of light sensors. The original spectrum of light originated by the muon is attenuated in the water such that the dominant wavelength range detected is between 350 and 500 nm

For each PMT it is usually recorded the arrival time and amplitude, sometimes even the full waveform, of Cherenkov light induced by muons or particle showers. In most designs the spheres are attached to vertical strings. The typical PMT spacing along a string is 10–20 m, and the distance between adjacent strings 60–150 m. An underwater (or under-ice) neutrino telescope in principle is able to detect all neutrino types (ν_μ, ν_e, ν_τ) CC or NC interacting inside or near the instrumented volume. It is possible to distinguish two main types of signatures: the track-like events, mainly from ν_μ CC-interaction, and the shower-like ones from $\nu_{\mu,e,\tau}$ NC-interactions and $\nu_{e,\tau}$ CC-interactions.

Muons as high-energy neutrino signals

The CC-interaction of a ν_μ produce a muon track. The passage of these muons through the seawater induces the emission of Cherenkov light, that can be then detected by a three dimensional array of photo-multiplier tubes (PMTs). The most precise measurement of the neutrino direction is achieved reconstructing the trajectory of the leading muon originated in ν_μ CC-interactions; this channel is therefore central to all investigations of astrophysical neutrino sources. Using the time and position information of the detected photons, the muon trajectory can be reconstructed, from which the original neutrino direction can be inferred. Muons propagating in water/ice loose energy mainly by ionization, bremsstrahlung, pair production, photonuclear interactions, knock on electrons. Ionization energy losses are dominant at low energies and they are fairly constant and homogeneous over the track. The dE/dx is about 2 MeV/cm at 1 GeV in water. For muon energies above 100 GeV the main energy losses are due to the radiative processes like e+e− pair production, bremsstrahlung and photonuclear processes. These losses are strongly energy dependent and stochastic. Therefore, only an average total energy loss can be calculated. The value and the contributions of different processes to the energy losses for different muon energies are shown in Fig. 65.

It is convenient to express the average energy loss as:

$$-\langle\frac{dE_\mu}{dx}\rangle = a(E_\mu) + b(E_\mu) \cdot E_\mu \tag{183}$$

It is worth to mention that radiative processes may produce hadronic and electromagnetic cascades along the muon track, other relativistic particles that in water can

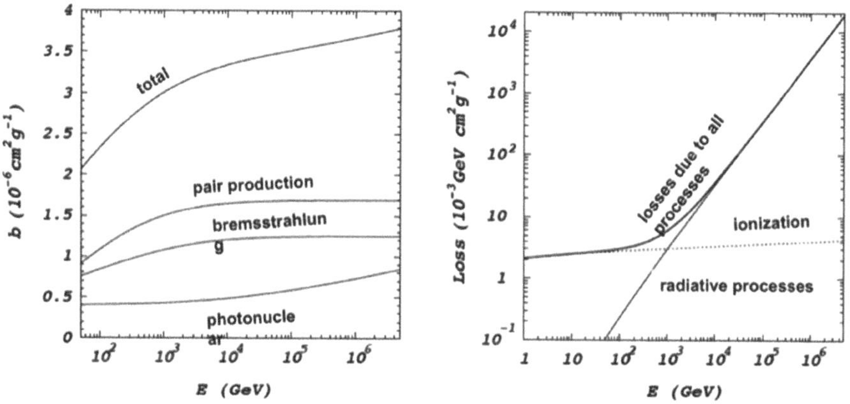

Fig. 65 Muon energy loss in pure water as a function of its energy [296]

induce erenkov light. As already mentioned underwater/ice telescopes are optimized for the detection of muon tracks at energies of the order TeV or above: muons originated by the CC-interactions of such energetic neutrinos may travel for km in water before to be stopped. The properties of the energy losses make possible the muon energy reconstruction: for low muon energies the track length is proportional to the particle energy, instead, for high energies, the amount of the detected light per track length may be used.

Appendix 6: High Energy γ-ray-Neutrino Connection

One can estimate an upper limit on the high-energy neutrino flux from a source of γ-rays, whose γ-rays are not subject to absorption, by assuming that all γ-rays are produced in pp collisions. When the γ-ray obey power-law distributions, we can use the formalism of the Z-factors described in the text [244, 326]. However, as was shown in various works and in particular in [327], more general cases can be treated with relative ease. There are different ways to proceed in the calculations. We mention here two formalisms that are particularly easy-to-use. In both cases, neutrino oscillations are included.

Exponential cut-off

In [328], it was shown that supposing that the primary proton spectrum is given by a power law with index α and an exponential cut-off energy ε_p,

$$\frac{dN_p}{dE_p} = k_p \left(\frac{E_p}{1\ \text{TeV}}\right)^{-\alpha} \exp\left(-\frac{E_p}{\varepsilon_p}\right)$$

the neutrino and gamma-ray spectra are approximatively described by the spectra,

$$\frac{dN_{\gamma/\nu}}{dE_{\gamma/\nu}} = k_{\gamma/\nu} \left(\frac{E_{\gamma/\nu}}{1\ \text{TeV}}\right)^{-\Gamma_{\gamma/\nu}} \exp\left(-\sqrt{\frac{E_{\gamma/\nu}}{\varepsilon_{\gamma/\nu}}}\right)$$

where the parameters are given by,

$$\begin{aligned} k_\nu &\approx (0.71 - 0.16\alpha)k_\gamma \\ \Gamma_\nu &\approx \Gamma_\gamma \approx \alpha - 0.1 \\ \varepsilon_\nu &\approx 0.59\varepsilon_\gamma \approx \varepsilon_p/40 \end{aligned}$$

This method of calculation is based on [329]; it includes pion decays and oscillations in the simplest approximation.

Note the differences among the primary and the secondary spectra: (1) in the region where we have power-law distributions, the secondary spectra are a bit harder, due to scaling violation; (2) the cut-off of the proton spectrum is milder in the spectra of secondary particles, due to the contribution from lower energies particles. The overall coefficient of the secondary particles, however, depends upon the amount of target particles.

General case

Assume that the measured γ-ray spectrum $\phi_\gamma(E)$ is purely hadronic (due to cosmic ray-gas collisions) and that it is not affected by propagation. In [330], based on [327], it was shown that the muon neutrino (resp., antineutrino) spectrum are given by the precise relations

$$\phi_{\nu_\mu}(E) = \alpha_\pi\, \phi_\gamma\left(\tfrac{E}{1-r_\pi}\right) + \alpha_K\, \phi_\gamma\left(\tfrac{E}{1-r_K}\right) + \int_0^1 \frac{dx}{x}\, \mathscr{K}_{\nu_\mu}(x)\, \phi_\gamma\left(\frac{E}{x}\right) \qquad (184)$$

where $\alpha_\pi = 0.380\ (0.278)$ and $\alpha_K = 0.013\ (0.009)$ for ν_μ and $\bar{\nu}_\mu$, respectively, and where $r_x = (m_\mu/m_x)^2$ with $x = \pi, K$. In each expression, the first two contributions describe neutrinos from the two-body decay by pions and kaons, while the third term accounts for neutrinos from muon decay. The kernels for muon neutrinos $\mathscr{K}_{\nu_\mu}(x)$ and for muon antineutrinos $\mathscr{K}_{\bar{\nu}_\mu}(x)$, which account also for oscillations from the source to the Earth, are

$$\mathscr{K}_{\nu_\mu}(x) = \begin{cases} x^2(15.34 - 28.93x) & 0 < x \le r_K \\ 0.0165 + 0.1193x + 3.747x^2 - 3.981x^3 & r_K < x < r_\pi \\ (1-x)^2(-0.6698 + 6.588x) & r_\pi \le x < 1 \end{cases}$$

$$\mathscr{K}_{\bar{\nu}_\mu}(x) = \begin{cases} x^2(18.48 - 25.33x) & 0 < x \le r_K \\ 0.0251 + 0.0826x + 3.697x^2 - 3.548x^3 & r_K < x < r_\pi \\ (1-x)^2(0.0351 + 5.864x) & r_\pi \le x < 1 \end{cases}$$

With this procedure, the expected upper bounds on the neutrino spectra are obtained from the measured γ-ray spectrum in a model-independent way. This method of calculations includes also kaons and it is presumably more accurate than the previous one.

In order to understand the reason why these formulae are valid, we note that the γ-rays from π^0 decay are given by a very simple integral kernel, $F_\gamma(E) = \int_E^\infty dE'\, 2F_{\pi^0}(E')/E'$. This can be easily inverted [327],

$$F_{\pi^0}(E) = -\frac{E}{2}\frac{dF_\gamma}{dE}$$

Then, using the approximate isospin-invariant distribution of pions, $F_{\pi^+} \approx F_{\pi^-} \approx F_{\pi^0} = F_\pi$, for the muon neutrinos from $\pi^+ \to \mu^+\nu_\mu$, we find immediately,

$$F_{\nu_\mu}(E) = \int_{E/(1-r_\pi)}^\infty \frac{dE'}{1-r_\pi}\,\frac{F_\pi(E')}{E'} = \frac{F_\gamma(E/(1-r_\pi))}{2(1-r_\pi)}$$

Some effort is necessary to describe the spectra of the leptons from the muon decay channel in a simple manner (namely to calculate the kernels $\mathscr{K}$ given above). Most other improvements (e.g., oscillations, other two-body decay channels, etc.) can be implemented easily and directly, instead.

References

Books:

1. M.A. Markov, *The Neutrino* (Dubna, 1963)
2. J.N. Bahcall, *Neutrino Astrophysics* (Cambridge University Press, 1989)
3. M. Fukugita, T. Yanagida, Physics *of Neutrinos and Applications to Astrophysics* (Springer, 2003)
4. C. Giunti, C.W. Kim, *Fundamentals of Neutrino Physics and Astrophysics* (Oxford University Press, 2007)
5. Z.-Z. Xing, S. Zhou, *Neutrinos in Particle Physics, Astronomy and Cosmology* (Zhejiang University Press, 2011)
6. F. Halzen, A.D. Martin, *Quarks and Leptons: An Introductory Course in Modern Particle Physics* (Wiley, 1984)
7. W. Greiner, B. Muller, *Gauge Theory of Weak Interactions* (Springer, 1993)
8. L.B. Okun, *Leptons and Quarks* (North Holland, 1982)
9. T.K. Gaisser, *Cosmic Rays and Particle Physics* (Cambridge University Press, 1990)
10. J. Lesgourgues, G. Mangano, G. Miele, S. Pastor, *Neutrino Cosmology* (Cambridge University Press, 2013)
11. V.S. Berezinskii, S.V. Bulanov, V.A. Dogiel, V.L. Ginzburg, V.S. Ptuskin (eds.), *Astrophysics of Cosmic Rays* (North-Holland, Amsterdam, 1990)
12. A. De Angelis, M.J.M. Pimenta, *Introduction to Particle and Astroparticle Physics: Questions to the Universe* (Springer, 2015)
13. M. Spurio, *Particles and Astrophysics: A Multi-messenger Approach* (Springer, 2015)
14. C.E. Rolfs, W.S. Rodney, *Cauldrons in the Cosmos* (University of Chicago Press, 1988)
15. G.G. Raffelt, *Stars as Laboratories for Fundamental Physics: The Astrophysics of Neutrinos, Axions, and Other Weakly Interacting Particles* (University of Chicago Press, 1996)
16. J.W.F. Valle, J.C. Romão, *Neutrinos in High Energy and Astroparticle Physics* (Wiley-VCH, 2015)
17. K. Freeman, G. McNamara, *In Search of Dark Matter* (Springer 2006)
18. G. Bertone, *Particle Dark Matter* (Cambridge University Press, 2013)
19. M. L'Annunziata, *Handbook of Radioactivity Analysis* (Elsevier Science, 2012), https://books.google.it/books?id=mAfH8KEH3j0C

Review Papers:

20. B.M. Pontecorvo, Pages in the development of neutrino physics. Sov. Phys. Usp. **26**, 1087 (1983)
21. A. Strumia, F. Vissani, Neutrino masses and mixings and..., hep-ph/0606054
22. M.C. Gonzalez-Garcia, M. Maltoni, Phenomenology with massive neutrinos. Phys. Rep. **460**, 1 (2008)

23. C.H. Llewellyn, Smith, Neutrino reactions at accelerator energies. Phys. Rep. **3**, 261 (1972)
24. J.A. Formaggio, G.P. Zeller, From eV to EeV: neutrino cross sections across energy scales. Rev. Mod. Phys. **84**, 1307 (2012)
25. A.B. McDonald, C. Spiering, S. Schonert, E.T. Kearns, T. Kajita, Astrophysical neutrino telescopes. Rev. Sci. Instrum. **75**, 293 (2004)
26. G. Fiorentini, M. Lissia, F. Mantovani, Geo-neutrinos and Earth's interior. Phys. Rep. **453**, 117 (2007)
27. S. Dye, Geo-neutrinos and the radioactive power of the Earth. Rev. Geophys. **50**, 3007 (2012)
28. G. Bellini, A. Ianni, L. Ludhova, F. Mantovani, W.F. McDonough, Geo-neutrinos. Prog. Part. Nucl. Phys. **73**, 1 (2013)
29. M. Baldoncini, I. Callegari, G. Fiorentini, F. Mantovani, B. Ricci, V. Strati, G. Xhixha, Reference worldwide model for antineutrinos from reactors. Phys. Rev. D **91**, 065002 (2015)
30. G.J. Feldman, J. Hartnell, T. Kobayashi, Long-baseline neutrino oscillation experiments. Adv. High Energy Phys. **2013**, 475749 (2013)
31. Y.Y.Y. Wong, Neutrino mass in cosmology: status and prospects. Ann. Rev. Nucl. Part. Sci. **61**, 69 (2011)
32. J. Lesgourgues, S. Pastor, Neutrino mass from cosmology. Adv. High Energy Phys. **2012**, 608515 (2012)
33. R.H. Cyburt, B.D. Fields, K.A. Olive, T.H.Yeh, Big bang nucleosynthesis: 2015, arXiv:1505.01076 [astro-ph.CO]
34. P.A.R. Ade [Planck Collaboration], Planck, et al., *results* (XIII, Cosmological parameters, 2015), arXiv:1502.01589 [astro-ph.CO]
35. M. Wurm, Solar neutrino spectroscopy. Phys. Rep. **685**, 1 (2017)
36. F. Vissani, Solar neutrino physics on the beginning of 2017, arXiv:1706.05435 [nucl-th]
37. F. Vissani, Comparative analysis of SN1987A antineutrino fluence. J. Phys. G **42**, 013001 (2015)
38. A. Mirizzi, I. Tamborra, H.T. Janka, N. Saviano, K. Scholberg, R. Bollig, L. Hudepohl, S. Chakraborty, Supernova neutrinos: production, oscillations and detection, arXiv:1508.00785 [astro-ph.HE]
39. J.F. Beacom, The diffuse supernova neutrino background. Ann. Rev. Nucl. Part. Sci. **60**, 439 (2010)
40. C. Lunardini, Diffuse supernova neutrinos at underground laboratories, arXiv:1007.3252 [astro-ph.CO]
41. T.K. Gaisser, F. Halzen, T. Stanev, Particle astrophysics with high-energy neutrinos. Phys. Rep. **258**, 173 (1995) [Erratum-ibid. **271**, 355 (1996)]
42. F. Halzen, D. Hooper, High-energy neutrino astronomy: the cosmic ray connection. Rep. Prog. Phys. **65**, 1025 (2002)
43. F.W. Stecker, Cosmic physics: the high energy frontier. J. Phys. **G29**, R47 (2003)
44. E. Waxman, Neutrino astrophysics: a new tool for exploring the universe. Science **315**, 63 (2007)
45. P. Lipari, Proton and neutrino extragalactic astronomy. Phys. Rev. D **78**, 083011 (2008)
46. T. Chiarusi, M. Spurio, High-energy astrophysics with neutrino telescopes. Eur. Phys. J. C **65**, 649 (2010)
47. T. DeYoung, Neutrino astronomy with IceCube. Mod. Phys. Lett. A **24**, 1543 (2009)
48. L.A. Anchordoqui, T. Montaruli, In search for extraterrestrial high energy neutrinos. Ann. Rev. Nucl. Part. Sci. **60**, 129 (2010)
49. U.F. Katz, C. Spiering, High-energy neutrino astrophysics: status and perspectives. Prog. Part. Nucl. Phys. **67**, 651 (2012)
50. J.K. Becker, High-energy neutrinos in the context of multimessenger physics. Phys. Rep. **458**, 173 (2008)
51. S. Ando et al., Multimessenger astronomy with gravitational waves and high-energy neutrinos. Rev. Mod. Phys. **85**, 1401 (2013)
52. A.M. Stasto, Ultrahigh energy neutrino physics. Int. J. Mod. Phys. A **19**, 317 (2004)
53. C.P. de los Heros, The quest for dark matter with neutrino telescopes, arXiv:1511.03500
54. A. Boyarsky, O. Ruchayskiy, M. Shaposhnikov, The role of sterile neutrinos in cosmology and astrophysics. Ann. Rev. Nucl. Part. Sci. **59**, 191 (2009)

Research Papers:

55. B. Pontecorvo, Mesonium and anti-mesonium. Sov. Phys. JETP **6**, 429 (1957)
56. B. Pontecorvo, Neutrino experiments and the problem of conservation of leptonic charge. Sov. Phys. JETP **26**, 984 (1968)
57. Y. Katayama, K. Matumoto, S. Tanaka, E. Yamada, Prog. Theor. Phys. **28**, 675 (1962)
58. Z. Maki, M. Nakagawa, S. Sakata, Prog. Theor. Phys. **28**, 870 (1962)
59. M. Nakagawa, H. Okonogi, S. Sakata, A. Toyoda, Prog. Theor. Phys. **30**, 727 (1963)
60. L. Wolfenstein, Phys. Rev. D **17**, 2369 (1978)
61. S.P. Mikheyev, A. Yu Smirnov, Nuovo Cim. C **9**, 17 (1986)
62. S.P. Mikheyev, A. Yu Smirnov, Sov. J. Nucl. Phys. **42**, 913 (1986)
63. P. Vogel, J.F. Beacom, Angular distribution of neutron inverse beta decay, $\bar{\nu}_e + p \rightarrow e^+ + n$. Phys. Rev. D **60**, 053003 (1999)
64. A. Strumia, F. Vissani, Precise quasielastic neutrino/nucleon cross-section. Phys. Lett. B **564**, 42 (2003)
65. F. Capozzi, E. Lisi, A. Marrone, D. Montanino, A. Palazzo, Neutrino masses and mixings: Status of known and unknown 3ν parameters, arXiv:1601.07777 [hep-ph] (2016)
66. V.N. Gribov, B. Pontecorvo, Neutrino astronomy and lepton charge. Phys. Lett. B **28**, 493 (1969)
67. A. Palladino, F. Vissani, The natural parameterization of cosmic neutrino oscillations. Eur. Phys. J. C **75**, 433 (2015)
68. C. Lujan-Peschard, G. Pagliaroli, F. Vissani, Counting muons to probe the neutrino mass spectrum. Eur. Phys. J. C **73**, 2439 (2013)
69. G. Battistoni, A. Ferrari, C. Rubbia, P.R. Sala, F. Vissani, Atmospheric neutrinos in a large liquid argon detector, hep-ph/0604182
70. T. Araki et al., [KamLAND Collaboration], Experimental investigation of geologically produced antineutrinos with KamLAND. Nature **436**, 499 (2005)
71. S. Abe et al., [KamLAND Collaboration], Precision measurement of neutrino oscillation parameters with KamLAND. Phys. Rev. Lett. **100**, 221803 (2008)
72. A. Gando et al., [KamLAND Collaboration], Partial radiogenic heat model for Earth revealed by geo-neutrino measurements. Nat. Geosci. **4**, 647 (2011)
73. G. Bellini et al., Borexino Collaboration. Observation of geo-neutrinos. Phys. Lett. B **687**, 299 (2010)
74. G. Bellini et al., [Borexino Collaboration], Measurement of geo-neutrinos from 1353 days of Borexino. Phys. Lett. B **722**, 295 (2013)
75. M. Agostini et al., [Borexino Collaboration], Spectroscopy of geoneutrinos from 2056 days of Borexino data. Phys. Rev. D **92**, 031101 (2015)
76. M. Nakahata [Super-Kamiokande Collaboration], Calibration of Super-Kamiokande using an electron linac. Nucl. Instrum. Meth. A 421, 113 (1999); Y. Fukuda et al. [Super-Kamiokande Collaboration], The Super-Kamiokande detector. Nucl. Instrum. Methods A **501**, 418 (2003)
77. H. Sekiya [Super-Kamiokande Collaboration], Solar neutrino analysis of Super-Kamiokande, arXiv:1307.3686
78. K. Abe et al., Letter of Intent: The Hyper-Kamiokande Experiment—Detector Design and Physics Potential, arXiv:1109.3262 [hep-ex]
79. K. Abe et al. [Super-Kamiokande Collaboration], Solar Neutrino Measurements in Super-Kamiokande-IV, arXiv:1606.07538 [hep-ex]
80. R. Tomas, D. Semikoz, G.G. Raffelt, M. Kachelriess, A.S. Dighe, Supernova pointing with low-energy and high-energy neutrino detectors. Phys. Rev. D **68**, 093013 (2003)
81. G. Pagliaroli, F. Vissani, E. Coccia, W. Fulgione, Neutrinos from supernovae as a trigger for gravitational wave search. Phys. Rev. Lett. **103**, 031102 (2009)
82. K. Abe et al., [Super-Kamiokande Collaboration], Real-time supernova neutrino burst monitor at Super-Kamiokande. Astropart. Phys. **81**, 39 (2016)
83. K. Langanke, P. Vogel, E. Kolbe, Signal for supernova muon-neutrino and tau-neutrino neutrinos in water Cherenkov detectors. Phys. Rev. Lett. **76**, 2629 (1996)

84. J.F. Beacom, M.R. Vagins, GADZOOKS! Anti-neutrino spectroscopy with large water Cherenkov detectors. Phys. Rev. Lett. **93**, 171101 (2004)
85. H. Watanabe et al., [Super-Kamiokande Collaboration], First study of neutron tagging with a water Cherenkov detector. Astropart. Phys. **31**, 320 (2009)
86. OYu. Smirnov, O.A. Zaimidoroga, A.V. Derbin, Search for solar pp neutrinos with an upgrade of CTF detector. Phys. Atom. Nucl. **66**(4), 712 (2003)
87. G. Bellini, The impact of Borexino on the solar and neutrino physics. Nucl. Phys. B **908**, 178 (2016)
88. M. Wurm et al., [LENA Collaboration], The next-generation liquid-scintillator neutrino observatory LENA. Astropart. Phys. **35**, 685 (2012)
89. Z. Djurcic [JUNO Collaboration], JUNO Conceptual Design Report, arXiv:1508.07166 [physics.ins-det]; F. An et al. et al., [JUNO Collaboration], Neutrino physics with JUNO. J. Phys. G **43**(3), 030401 (2016)
90. O.Y. Smirnov et al. [Borexino Collaboration], Measurement of neutrino flux from the primary proton–proton fusion process in the Sun with Borexino detector, arXiv:1507.02432 [hep-ex]
91. M. Biassoni, C. Martinez, Study of supernova ν -nucleus coherent scattering interactions. Astropart. Phys. **36**, 151 (2012)
92. Y. Fukuda et al., [Super-Kamiokande Collaboration], Evidence for oscillation of atmospheric neutrinos. Phys. Rev. Lett. **81**, 1562 (1998)
93. D. Casper et al., Measurement of atmospheric neutrino composition with IMB-3. Phys. Rev. Lett. **66**, 2561 (1991)
94. K.S. Hirata et al., [Kamiokande-II Collaboration], Observation of a small atmospheric muon-neutrino/electron-neutrino ratio in Kamiokande. Phys. Lett. B **280**, 146 (1992)
95. Y. Fukuda et al., [Kamiokande Collaboration], Atmospheric muon-neutrino/electron-neutrino ratio in the multiGeV energy range. Phys. Lett. B **335**, 237 (1994)
96. M. Ambrosio et al., [MACRO Collaboration], Measurement of the atmospheric neutrino induced upgoing muon flux using MACRO. Phys. Lett. B **434**, 451 (1998)
97. M. Honda, T. Kajita, K. Kasahara, S. Midorikawa, Calculation of the flux of atmospheric neutrinos. Phys. Rev. D **52**, 4985 (1995)
98. G.D. Barr, T.K. Gaisser, P. Lipari, S. Robbins, T. Stanev, A Three-dimensional calculation of atmospheric neutrinos. Phys. Rev. D **70**, 023006 (2004)
99. M. Honda, M.S. Athar, T. Kajita, K. Kasahara, S. Midorikawa, Atmospheric neutrino flux calculation using the NRLMSISE-00 atmospheric model. Phys. Rev. D **92**, 023004 (2015)
100. R. Enberg, M.H. Reno, I. Sarcevic, Prompt neutrino fluxes from atmospheric charm. Phys. Rev. D **78**, 043005 (2008)
101. A. Bhattacharya, R. Enberg, M.H. Reno, I. Sarcevic, A. Stasto, Perturbative charm production and the prompt atmospheric neutrino flux in light of RHIC and LHC. JHEP **1506**, 110 (2015)
102. R. Gauld, J. Rojo, L. Rottoli, S. Sarkar, J. Talbert, The prompt atmospheric neutrino flux in the light of LHCb. JHEP **1602**, 130 (2016)
103. S. Schonert, T.K. Gaisser, E. Resconi, O. Schulz, Vetoing atmospheric neutrinos in a high energy neutrino telescope. Phys. Rev. D **79**, 043009 (2009)
104. T.K. Gaisser, K. Jero, A. Karle, J. van Santen, Generalized self-veto probability for atmospheric neutrinos. Phys. Rev. D **90**, 023009 (2014)
105. C.V. Achar et al., Detection of muons produced by cosmic ray neutrinos deep underground. Phys. Lett. **18**, 196 (1965)
106. F. Reines, M.F. Crouch, T.L. Jenkins, W.R. Kropp, H.S. Gurr, G.R. Smith, J.P.F. Sellschop, B. Meyer, Evidence for high-energy cosmic ray neutrino interactions. Phys. Rev. Lett. **15**, 429 (1965)
107. T. Kajita, The measurement of neutrino properties with atmospheric neutrinos. Ann. Rev. Nucl. Part. Sci. **64**, 343 (2014)
108. R. Wendell et al., [Super-Kamiokande Collaboration], Atmospheric neutrino oscillation analysis with sub-leading effects in Super-Kamiokande I, II, and III. Phys. Rev. D **81**, 092004 (2010)

109. E. Richard et al. [Super-Kamiokande Collaboration], Measurements of the atmospheric neutrino flux by Super-Kamiokande: energy spectra, geomagnetic effects, and solar modulation, arXiv:1510.08127 [hep-ex]
110. M.G. Aartsen et al., [IceCube Collaboration], Determining neutrino oscillation parameters from atmospheric muon neutrino disappearance with three years of IceCube DeepCore data. Phys. Rev. D **91**, 072004 (2015)
111. M.G. Aartsen et al., [IceCube Collaboration], Characterization of the Atmospheric Muon Flux in IceCube. Astropart. Phys. **78**, 1 (2016)
112. M.G. Aartsen et al., [IceCube Collaboration], Measurement of the atmospheric ν_e spectrum with IceCube. Phys. Rev. D **91**, 122004 (2015)
113. S. Adrian-Martinez et al., [ANTARES Collaboration], Measurement of atmospheric neutrino oscillations with the ANTARES neutrino telescope. Phys. Lett. B **714**, 224 (2012)
114. S. Adrian-Martinez et al., [ANTARES Collaboration], Measurement of the atmospheric ν_μ energy spectrum from 100 GeV to 200 TeV with the ANTARES telescope. Eur. Phys. J. C **73**, 2606 (2013)
115. T. Kajita, Y. Totsuka, Observation of atmospheric neutrinos. Rev. Mod. Phys. **73**, 85 (2001)
116. T.K. Gaisser, M. Honda, Flux of atmospheric neutrinos. Ann. Rev. Nucl. Part. Sci. **52**, 153 (2002)
117. E.K. Akhmedov, S. Razzaque, A.Y. Smirnov, Mass hierarchy, 2-3 mixing and CP-phase with huge atmospheric neutrino detectors. JHEP **1302**, 082 (2013) [JHEP **1307**, 026 (2013)]
118. M.G. Aartsen et al. [IceCube PINGU Collaboration], Letter of Intent: The Precision IceCube Next Generation Upgrade (PINGU), arXiv:1401.2046 [physics.ins-det]
119. S. Adrian-Martinez et al. [KM3Net Collaboration], Letter of intent for KM3NeT 2.0. J. Phys. G **43**(8), 084001 (2016)
120. U.F. Katz [KM3NeT Collaboration], The ORCA Option for KM3NeT, arXiv:1402.1022 [astro-ph.IM]
121. T. Thakore, A. Ghosh, S. Choubey, A. Dighe, The reach of INO for atmospheric neutrino oscillation parameters. JHEP **1305**, 058 (2013)
122. O. Šrámek et al., Geophysical and geochemical constraints on geoneutrino fluxes from Earths mantle. Earth Planet. Sci. Lett. **361**, 356 (2013)
123. S. Dell'Oro, S. Marcocci, M. Viel, F. Vissani, Neutrinoless double beta decay: 2015 review. Adv. High Energy Phys. **2016**, 2162659 (2016)
124. N. Palanque-Delabrouille et al., Neutrino masses and cosmology with Lyman-alpha forest power spectrum. JCAP **1511**(11), 011 (2015)
125. E. Di Valentino, E. Giusarma, O. Mena, A. Melchiorri and J. Silk, Cosmological limits on neutrino unknowns versus low redshift priors, arXiv:1511.00975 [astro-ph.CO]
126. X. Zhang, Impacts of dark energy on weighing neutrinos after Planck 2015, arXiv:1511.02651 [astro-ph.CO]
127. A.J. Cuesta, V. Niro, L. Verde, Neutrino mass limits: robust information from the power spectrum of galaxy surveys, arXiv:1511.05983 [astro-ph.CO]
128. S. Weinberg, Universal neutrino degeneracy. Phys. Rev. **128**, 1457 (1962)
129. A. Cocco, G. Mangano, M. Messina, Probing low energy neutrino backgrounds with neutrino capture on beta decaying nuclei. J. Cosm. Astropart. Phys. **0706**, 015 (2007)
130. A.J. Long, C. Lunardini, E. Sabancilar, Detecting non-relativistic cosmic neutrinos by capture on tritium: phenomenology and physics potential. J. Cosm. Astropart. Phys. **1408**, 038 (2014)
131. A.G. Kosovichev, Advances in global and local helioseismology: an introductory review. Lect. Notes Phys. **832**, 3 (2011)
132. J.N. Bahcall, M.H. Pinsonneault, S. Basu, Solar models: current epoch and time dependences, neutrinos, and helioseismological properties. Astrophys. J. **555**, 990 (2001)
133. R. Bonetti et al., [LUNA Collaboration], First measurement of the He-3 (He-3, 2p) He-4 cross-section down to the lower edge of the solar Gamow peak. Phys. Rev. Lett. **82**, 5205 (1999)
134. F. Confortola et al., [LUNA Collaboration], Astrophysical S-factor of the He-3(alpha, gamma) Be-7 reaction measured at low energy via prompt and delayed gamma detection. Phys. Rev. C **75**, 065803 (2007)

135. A. Formicola et al., [LUNA Collaboration], Astrophysical S factor of N-14(p, gamma)O-15. Phys. Lett. B **591**, 61 (2004)
136. T. Araki et al., [KamLAND Collaboration], Measurement of neutrino oscillation with KamLAND: evidence of spectral distortion. Phys. Rev. Lett. **94**, 081801 (2005)
137. A. Gando et al., [KamLAND Collaboration], Constraints on θ_{13} from a three-flavor oscillation analysis of reactor antineutrinos at KamLAND. Phys. Rev. D **83**, 052002 (2011)
138. G.L. Fogli, E. Lisi, A. Marrone, A. Palazzo, A.M. Rotunno, SNO, KamLAND and neutrino oscillations: theta(13), arXiv:0905.3549 [hep-ph]
139. A. Renshaw et al., [Super-Kamiokande Collaboration], First indication of terrestrial matter effects on solar neutrino oscillation. Phys. Rev. Lett. **112**(9), 091805 (2014)
140. B. Aharmim et al., [SNO Collaboration], Combined analysis of all three phases of solar neutrino data from the Sudbury Neutrino Observatory. Phys. Rev. C **88**, 025501 (2013)
141. G. Bellini et al., [Borexino Collaboration], Absence of day-night asymmetry of 862 keV ^{7}Be solar neutrino rate in Borexino and MSW oscillation parameters. Phys. Lett. B **707**, 22 (2012)
142. S. Abe et al., [KamLAND Collaboration], Measurement of the 8B solar neutrino flux with the KamLAND liquid scintillator detector. Phys. Rev. C **84**, 035804 (2011)
143. A. Gando et al., [KamLAND Collaboration], ^{7}Be Solar neutrino measurement with KamLAND. Phys. Rev. C **92**(5), 055808 (2015)
144. B.T. Cleveland, T. Daily, R. Davis Jr., J.R. Distel, K. Lande, C.K. Lee, P.S. Wildenhain, J. Ullman, Measurement of the solar electron neutrino flux with the Homestake chlorine detector. Astrophys. J. **496**, 505 (1998)
145. K.S. Hirata et al. [Kamiokande-II Collaboration], Real time, directional measurement of B-8 solar neutrinos in the Kamiokande-II detector. Phys. Rev. D **44**, 2241 (1991) [Phys. Rev. D **45**, 2170 (1992)]
146. Y. Fukuda et al., [Kamiokande Collaboration], Solar neutrino data covering solar cycle 22. Phys. Rev. Lett. **77**, 1683 (1996)
147. V.A. Kuzmin, Detection of solar neutrinos by means of the 71Ga(nu, e-)71Ge reaction. Sov. Phys. JETP **22**, 1051 (1966)
148. J.N. Bahcall et al., Proposed solar neutrino experiment using Ga-71. Phys. Rev. Lett. **40**, 1351 (1978)
149. W. Hampel et al., [GALLEX Collaboration], GALLEX solar neutrino observations: results for GALLEX IV. Phys. Lett. B **447**, 127 (1999)
150. M. Altmann et al., [GNO Collaboration], Complete results for five years of GNO solar neutrino observations. Phys. Lett. B **616**, 174 (2005)
151. J.N. Abdurashitov et al., [SAGE Collaboration], Solar neutrino flux measurements by the Soviet-American Gallium Experiment (SAGE) for half the 22 year solar cycle. J. Exp. Theor. Phys. **95**, 181 (2002)
152. J.N. Abdurashitov et al., [SAGE Collaboration], Measurement of the solar neutrino capture rate with gallium metal. III: Results for the 2002–2007 data-taking period. Phys. Rev. C **80**, 015807 (2009)
153. S. Fukuda et al., [Super-Kamiokande Collaboration], Solar B-8 and hep neutrino measurements from 1258 days of Super-Kamiokande data. Phys. Rev. Lett. **86**, 5651 (2001)
154. J. Hosaka et al., [Super-Kamiokande Collaboration], Solar neutrino measurements in Super-Kamiokande-I. Phys. Rev. D **73**, 112001 (2006)
155. J.P. Cravens et al., [Super-Kamiokande Collaboration], Solar neutrino measurements in Super-Kamiokande-II. Phys. Rev. D **78**, 032002 (2008)
156. K. Abe et al., [Super-Kamiokande Collaboration], Solar neutrino results in Super-Kamiokande-III. Phys. Rev. D **83**, 052010 (2011)
157. H.H. Chen, Direct approach to resolve the solar neutrino problem. Phys. Rev. Lett. **55**, 1534 (1985)
158. Q.R. Ahmad et al., [SNO Collaboration], Direct evidence for neutrino flavor transformation from neutral current interactions in the Sudbury Neutrino Observatory. Phys. Rev. Lett. **89**, 011301 (2002)

159. B. Aharmim et al., [SNO Collaboration], Electron energy spectra, fluxes, and day-night asymmetries of B-8 solar neutrinos from measurements with NaCl dissolved in the heavy-water detector at the Sudbury Neutrino Observatory. Phys. Rev. C **72**, 055502 (2005)
160. B. Aharmim et al., [SNO Collaboration], Low energy threshold analysis of the Phase I and Phase II data sets of the Sudbury Neutrino Observatory. Phys. Rev. C **81**, 055504 (2010)
161. C. Arpesella et al., [Borexino Collaboration], Direct measurement of the Be-7 solar neutrino flux with 192 days of Borexino data. Phys. Rev. Lett. **101**, 091302 (2008)
162. G. Bellini et al., [Borexino Collaboration], Measurement of the solar 8B neutrino rate with a liquid scintillator target and 3 MeV energy threshold in the Borexino detector. Phys. Rev. D **82**, 033006 (2010)
163. G. Bellini et al., [Borexino Collaboration], First evidence of pep solar neutrinos by direct detection in Borexino. Phys. Rev. Lett. **108**, 051302 (2012)
164. G. Bellini et al. [BorexinoCollaboration], Neutrinos from the primary proton-proton fusion process in the Sun. Nature **512**, no. 7515, 383 (2014)
165. G. Bellini et al. [Borexino Collaboration], Final results of Borexino Phase-I on low energy solar neutrino spectroscopy. Phys. Rev. D **89**, no. 11, 112007 (2014)
166. J.N. Bahcall, S. Basu, M. Pinsonneault, A.M. Serenelli, Helioseismological implications of recent solar abundance determinations. Astrophys. J. **618**, 1049 (2005)
167. J.N. Bahcall, A.M. Serenelli, S. Basu, New solar opacities, abundances, helioseismology, and neutrino fluxes. Astrophys. J. **621**, L85 (2005)
168. F.L. Villante, B. Ricci, Linear solar models. Astrophys. J. **714**, 944 (2010)
169. F.L. Villante, Constraints on the opacity profile of the sun from helioseismic observables and solar neutrino flux measurements. Astrophys. J. **724**, 98 (2010)
170. F.L. Villante, A.M. Serenelli, F. Delahaye, M.H. Pinsonneault, The chemical composition of the Sun from helioseismic and solar neutrino data. Astrophys. J. **787**, 13 (2014)
171. A. Serenelli, Alive and well: a short review about standard solar models. Eur. Phys. J. A **52**(4), 78 (2016)
172. N. Vinyoles et al., A new generation of standard solar models. Astrophys. J. **835**, 202 (2017)
173. F.L. Villante, A. Ianni, F. Lombardi, G. Pagliaroli, F. Vissani, A Step toward CNO solar neutrinos detection in liquid scintillators. Phys. Lett. B **701**, 336 (2011)
174. G. Bellini et al., [Borexino Collaboration], SOX: Short distance neutrino Oscillations with BoreXino. JHEP **1308**, 038 (2013)
175. D. Franco et al., Solar neutrino detection in a large volume double-phase liquid argon experiment, arXiv:1510.04196 [physics.ins-det]
176. A. Ianni, D. Montanino, The Cr-51 and Sr-90 sources in BOREXINO as tool for neutrino magnetic moment searches. Astropart. Phys. **10**, 331 (1999)
177. V. Berezinsky, M. Narayan, F. Vissani, Mirror model for sterile neutrinos. Nucl. Phys. B **658**, 254 (2003)
178. P.C. de Holanda, A.Y. Smirnov, Solar neutrino spectrum, sterile neutrinos and additional radiation in the Universe. Phys. Rev. D **83**, 113011 (2011)
179. S. Gariazzo, C. Giunti, M. Laveder, Y.F. Li, E.M. Zavanin, Light sterile neutrinos. J. Phys. G **43**, 033001 (2016)
180. S.A. Colgate, R.H. White, The Hydrodynamic Behavior of Supernovae Explosions. Astrophys. J. **143**, 626 (1966)
181. G.S. Bisnovatyi-Kogan, The explosion of a rotating star as a supernova mechanism. Astron. Zh. **47**, 813 (1970)
182. D.K. Nadyozhin, The neutrino radiation for a hot neutron star formation and the envelope outburst problem. Astrophys. Space Sci. **53**, 131 (1978)
183. H.A. Bethe, J.R. Wilson, Revival of a stalled supernova shock by neutrino heating. Astrophys. J. **295**, 14 (1985)
184. K. Hirata [Kamiokande-II Collaboration], Observation of a Neutrino Burst from the Supernova SN1987A. Phys. Rev. Lett. 58, 1490 (1987); K.S. Hirata, et al., [Kamiokande-II Collaboration], Observation in the Kamiokande-II Detector of the Neutrino Burst from Supernova SN 1987a. Phys. Rev. D **38**, 448 (1988)

185. R.M. Bionta [IMB Collaboration], Observation of a neutrino burst in coincidence with supernova SN1987A in the large magellanic cloud. Phys. Rev. Lett. 58, 1494 (1987); C.B. Bratton, et al., [IMB Collaboration], Angular distribution of events from SN1987A. Phys. Rev. D **37**, 3361 (1988)
186. E.N. Alekseev, L.N. Alekseeva, I.V. Krivosheina, V.I. Volchenko, Detection of the neutrino signal from SN1987A in the LMC using the INR Baksan underground scintillation telescope. Phys. Lett. B **205**, 209 (1988)
187. M. Aglietta et al., On the event observed in the Mont Blanc Underground Neutrino observatory during the occurrence of Supernova 1987a. Europhys. Lett. **3**, 1315 (1987)
188. J.N. Bahcall, chap. 15 of [2]
189. T.J. Loredo, D.Q. Lamb, Bayesian analysis of neutrinos observed from supernova SN 1987A. Phys. Rev. D **65**, 063002 (2002)
190. V.S. Imshennik, O.G. Ryazhskaya, A rotating collapsar and possible interpretation of the LSD neutrino signal from SN 1987A. Astron. Lett. **30**, 14 (2004). [arXiv:astro-ph/0401613]
191. G. Pagliaroli, F. Vissani, M.L. Costantini, A. Ianni, Improved analysis of SN1987A antineutrino events. Astropart. Phys. **31**, 163 (2009)
192. M.L. Costantini, A. Ianni, F. Vissani, SN1987A and the properties of neutrino burst. Phys. Rev. D 70, 043006 (2004); G. Pagliaroli, F. Vissani, Features of Kamiokande-II, IMB and Baksan observations and their interpretation in a two-component model for the signal. Astron. Lett. **35**, 1 (2009)
193. P.S. Shternin, D.G. Yakovlev, A young cooling neutron star in the remnant of Supernova 1987A. Astron. Lett. **34**, 675 (2008)
194. C. Lujan-Peschard, G. Pagliaroli, F. Vissani, Spectrum of supernova neutrinos in ultra-pure scintillators. JCAP **1407**, 051 (2014)
195. R. Laha, J.F. Beacom, S.K. Agarwalla, New power to measure supernova ν_e with large liquid scintillator detectors, arXiv:1412.8425 [hep-ph]
196. H. Yuksel, J.F. Beacom, Neutrino spectrum from SN 1987A and from cosmic supernovae. Phys. Rev. D **76**, 083007 (2007)
197. M. Fukugita, M. Kawasaki, Constraints on the star formation rate from supernova relic neutrino observations. Mon. Not. Roy. Astron. Soc. **340**, L7 (2003)
198. F. Vissani, G. Pagliaroli, The diffuse supernova neutrino background: Expectations and uncertainties derived from SN1987A. Astron. Astrophys. **528**, L1 (2011)
199. M. Ackermann et al., [Fermi-LAT Collaboration], Fermi-LAT observations of the diffuse gamma-ray emission: implications for cosmic rays and the interstellar medium. Astrophys. J. **750**, 3 (2012)
200. P. Lipari, M. Lusignoli, D. Meloni, Flavor composition and energy spectrum of astrophysical neutrinos. Phys. Rev. D **75**, 123005 (2007)
201. P. Lipari, Lepton spectra in the earth's atmosphere. Astropart. Phys. **1**, 195 (1993)
202. E. Waxman, J.N. Bahcall, High energy neutrinos from cosmological gamma-ray burst fireballs. Phys. Rev. Lett. **78**, 2292 (1997)
203. S.M. Bilenky, B. Pontecorvo, Lepton mixing and neutrino oscillations. Phys. Rep. **41**, 225 (1978)
204. F. Capozzi, G.L. Fogli, E. Lisi, A. Marrone, D. Montanino, A. Palazzo, Status of three-neutrino oscillation parameters, circa 2013. Phys. Rev. D **89**, 093018 (2014)
205. J.G. Learned, S. Pakvasa, Detecting tau-neutrino oscillations at PeV energies. Astropart. Phys. **3**, 267 (1995)
206. H. Athar, M. Jezabek, O. Yasuda, Effects of neutrino mixing on high-energy cosmic neutrino flux. Phys. Rev. D **62**, 103007 (2000)
207. D.V. Ahluwalia, Ambiguity in source flux of cosmic/astrophysical neutrinos: effects of bi-maximal mixing and quantum-gravity induced decoherence. Mod. Phys. Lett. A **16**, 917 (2001)
208. S. Pakvasa, Neutrino properties from high energy astrophysical neutrinos. Mod. Phys. Lett. A **19**, 1163 (2004)

209. G. Barenboim, C. Quigg, Neutrino observatories can characterize cosmic sources and neutrino properties. Phys. Rev. D **67**, 073024 (2003)
210. J.F. Beacom, N.F. Bell, D. Hooper, S. Pakvasa, T.J. Weiler, Measuring flavor ratios of high-energy astrophysical neutrinos, Phys. Rev. D **68**, 093005 (2003) [Erratum-ibid. D **72**, 019901 (2005)]
211. L.A. Anchordoqui, H. Goldberg, F. Halzen, T.J. Weiler, Galactic point sources of TeV antineutrinos. Phys. Lett. B **593**, 42 (2004)
212. Y. Chikashige, R.N. Mohapatra, R.D. Peccei, Spontaneously broken lepton number and cosmological constraints on the neutrino mass spectrum. Phys. Rev. Lett. 45, 1926 (1980); G.B. Gelmini, M. Roncadelli, Left-handed neutrino mass scale and spontaneously broken lepton number. Phys. Lett. B **99**, 411 (1981)
213. A. Acker, S. Pakvasa, J.T. Pantaleone, Decaying Dirac neutrinos. Phys. Rev. D **45**, 1 (1992)
214. J.F. Beacom, N.F. Bell, D. Hooper, S. Pakvasa, T.J. Weiler, Decay of high-energy astrophysical neutrinos. Phys. Rev. Lett. **90**, 181301 (2003)
215. L. Wolfenstein, Nucl. Phys. B 186, 147 (1981); S.T. Petcov, On pseudodirac neutrinos, neutrino oscillations and neutrinoless double beta decay. Phys. Lett. B 110, 245 (1982); S.M. Bilenky, B. Pontecorvo, Neutrino oscillations with large oscillation length in spite of large (Majorana) neutrino masses? Sov. J. Nucl. Phys. 38, 248 (1983); C. Giunti, C.W. Kim, U.W. Lee, Oscillations of pseudoDirac neutrinos and the solar neutrino problem. Phys. Rev. D 46, 3034 (1992); M. Kobayashi, C.S. Lim, Pseudo-Dirac scenario for neutrino oscillations. Phys. Rev. D 64, 013003 (2001); K.R. Balaji, A. Kalliomaki, J. Maalampi, Revisiting pseudo-Dirac neutrinos. Phys. Lett. B **524**, 153 (2002)
216. J.F. Beacom, N.F. Bell, D. Hooper, J.G. Learned, S. Pakvasa, T.J. Weiler, Pseudo-Dirac neutrinos, a challenge for neutrino telescopes. Phys. Rev. Lett. **92**, 011101 (2004)
217. P. Keranen, J. Maalampi, M. Myyrylainen, J. Riittinen, Effects of sterile neutrinos on the ultrahigh-energy cosmic neutrino flux. Phys. Lett. B **574**, 162 (2003)
218. D. Hooper, D. Morgan, E. Winstanley, Probing quantum decoherence with high-energy neutrinos. Phys. Lett. B **609**, 206 (2005)
219. M. Gasperini, Experimental constraints on a minimal and nonminimal violation of the equivalence principle in the oscillations of massive neutrinos. Phys. Rev. D **39**, 3606 (1989)
220. H. Minakata, A.Y. Smirnov, High energy cosmic neutrinos and the equivalence principle. Phys. Rev. D **54**, 3698 (1996)
221. R. Fardon, A.E. Nelson, N. Weiner, Dark energy from mass varying neutrinos. JCAP **0410**, 005 (2004)
222. P.Q. Hung, H. Päs, Cosmo MSW effect for mass varying neutrinos. Mod. Phys. Lett. A **20**, 1209 (2005)
223. R.C. Hartman et al., [EGRET Collaboration], The third EGRET catalog of high-energy gamma-ray sources. Astrophys. J. Suppl. **123**, 79 (1999)
224. C. Pittori et al., [AGILE Collaboration], First AGILE catalog of high confidence gamma-ray sources. Astron. Astrophys. **506**, 1563 (2009)
225. F. Acero, Fermi large area telescope third source catalog. Astrophys. J. Suppl. **218**, no. 2, 23 (2015)
226. S. Adrian-Martinez et al., [Antares Collaboration], First search for point sources of high energy cosmic neutrinos with the ANTARES neutrino telescope. Astrophys. J. **743**, L14 (2011)
227. S. Adrian-Martinez et al., [ANTARES Collaboration], Search for cosmic neutrino point sources with four year data of the ANTARES telescope. Astrophys. J. **760**, 53 (2012)
228. S. Adrian-Martinez et al., [ANTARES Collaboration], Searches for point-like and extended neutrino sources close to the Galactic Centre using the ANTARES neutrino telescope. Astrophys. J. **786**, L5 (2014)
229. A. Achterberg et al., [IceCube Collaboration], Five years of searches for point sources of astrophysical neutrinos with the AMANDA-II neutrino telescope. Phys. Rev. D **75**, 102001 (2007)
230. M.G. Aartsen et al., [IceCube Collaboration], Searches for small-scale anisotropies from neutrino point sources with three years of IceCube data. Astropart. Phys. **66**, 39 (2015)

231. S. Adrian-Martinez et al. [ANTARES and IceCube Collaborations], The First combined search for neutrino point-sources in the southern hemisphere with the ANTARES and Icecube neutrino telescopes. Astrophys. J. **823**, no. 1, 65 (2016)
232. HESS Collaboration, Acceleration of petaelectronvolt protons in the Galactic Centre. Nature **531**, 476 (2016)
233. S. Celli, A. Palladino, F. Vissani, Multi-TeV gamma-rays and neutrinos from the Galactic Center region, arXiv:1604.08791 [astro-ph.HE]
234. HESS Collaboration, A new population of very high energy gamma-ray sources in the Milky Way. Science **307**, 1938 (2005)
235. HESS Collaboration, The HESS survey of the inner galaxy in very high-energy gamma-rays. Astrophys. J. **636**, 777 (2006). [astro-ph/0510397]
236. HESS Collaboration, **437**, L7 (2005). [astro-ph/0505380]
237. HESS Collaboration, High-energy particle acceleration in the shell of a supernova remnant. Nature **432**, 75 (2004)
238. HESS Collaboration, A detailed spectral and morphological study of the gamma-ray supernova remnant RX J1713.7-3946 with HESS, [astro-ph/0511678]
239. HESS Collaboration, Very high energy gamma rays from the direction of Sagittarius A*. Astron. Astrophys. **425**, L13 (2004)
240. HESS Collaboration, Science **309**, 746 (2005)
241. T. Piran, The physics of gamma-ray bursts. Rev. Mod. Phys. **76**, 1143 (2004)
242. B. Zhang, P. Meszaros, Gamma-ray bursts: progress, problems & prospects. Int. J. Mod. Phys. A **19**, 2385 (2004)
243. A. Dar, A. De Rujula, Towards a complete theory of Gamma Ray Bursts. Phys. Rept. **405**, 203 (2004)
244. M.L. Costantini, F. Vissani, Expected neutrino signal from supernova remnant RX J1713.7-3946 and flavor oscillations. Astropart. Phys. **23**, 477 (2005)
245. A. Heavens, B. Panter, R. Jimenez, J. Dunlop, The complete star formation history of the Universe. Nature **428**, 625 (2004)
246. M. Fukugita, P.J.E. Peebles, The cosmic energy inventory. Astrophys. J. **616**, 643 (2004)
247. Y. Ueda et al., Cosmological evolution of the hard X-ray AGN luminosity function: formation history of supermassive black holes. Prog.Th. Phys. Supp **155**, 209 (2004)
248. E. Waxman, J.N. Bahcall, High energy neutrinos from astrophysical sources: an upper bound. Phys. Rev. D **59**, 023002 (1999)
249. K. Mannheim, R.J. Protheroe, J.P. Rachen, On the cosmic ray bound for models of extragalactic neutrino production. Phys. Rev. D **63**, 023003 (2001)
250. V. Berezinsky, A.Z. Gazizov, S.I. Grigorieva, Dip in UHECR spectrum as signature of proton interaction with CMB. Phys. Lett. B **612**, 147 (2005)
251. K. Greisen, End to the cosmic ray spectrum? Phys. Rev. Lett. **16**, 748 (1966)
252. G.T. Zatsepin, V.A. Kuzmin, Upper limit of the spectrum of cosmic rays. JETP Lett. **4**, 78 (1966)
253. V.S. Berezinsky, G.T. Zatsepin, Cosmic rays at ultrahigh-energies (neutrino?). Phys. Lett. B **28**, 423 (1969)
254. R. Engel, D. Seckel, T. Stanev, Neutrinos from propagation of ultrahigh-energy protons. Phys. Rev. D **64**, 093010 (2001)
255. M. Ahlers, L.A. Anchordoqui, M.C. Gonzalez-Garcia, F. Halzen, S. Sarkar, GZK Neutrinos after the Fermi-LAT diffuse photon flux measurement. Astropart. Phys. **34**, 106 (2010)
256. A. Franceschini, G. Rodighiero, M. Vaccari, The extragalactic optical-infrared background radiations, their time evolution and the cosmic photon-photon opacity. Astron. Astrophys. **487**, 837 (2008)
257. M. Ackermann et al., [Fermi-LAT Collaboration], The spectrum of isotropic diffuse gamma-ray emission between 100 MeV and 820 GeV. Astrophys. J. **799**, 86 (2015)
258. D. Saltzberg, Status of radio and acoustic detection of ultra-high energy cosmic astro-ph/0501364

259. H. Falcke, P. Gorham, R.J. Protheroe, Prospects for radio detection of ultra-high energy cosmic rays and New Astron. Rev. **48**, 1487 (2004)
260. A. Letessier-Selvon, Auger: a large air shower array and neutrino telescope. Nucl. Phys. Proc. Suppl. **118**, 399 (2003)
261. Euso Collaboration, Extreme Universe Space Observatory–EUSO: an innovative project for the detection of extreme energy cosmic rays and neutrinos. Nuovo Cim. **24C**, 445 (2001)
262. A. Aab et al. [Pierre Auger Collaboration], Improved limit to the diffuse flux of ultrahigh energy neutrinos from the Pierre Auger Observatory. Phys. Rev. D **91**, no. 9, 092008 (2015)
263. S.W. Barwick et al., [ARIANNA Collaboration], A first search for cosmogenic neutrinos with the ARIANNA hexagonal radio array. Astropart. Phys. **70**, 12 (2015)
264. P. Allison et al., [ARA Collaboration], First constraints on the ultra-high energy neutrino flux from a prototype station of the Askaryan Radio Array. Astropart. Phys. **70**, 62 (2015)
265. R. Gandhi, C. Quigg, M.H. Reno, I. Sarcevic, Neutrino interactions at ultrahigh-energies. Phys. Rev. D **58**, 093009 (1998)
266. M.G. Aartsen et al., [IceCube Collaboration], First observation of PeV-energy neutrinos with IceCube. Phys. Rev. Lett. **111**, 021103 (2013)
267. M.G. Aartsen et al., [IceCube Collaboration], Evidence for high-energy extraterrestrial neutrinos at the IceCube detector. Science **342**, 1242856 (2013)
268. M.G. Aartsen et al., [IceCube Collaboration], Observation of high-energy astrophysical neutrinos in three years of IceCube data. Phys. Rev. Lett. **113**, 101101 (2014)
269. M.G. Aartsen et al. [IceCube Collaboration], Atmospheric and astrophysical neutrinos above 1 TeV interacting in IceCube. Phys. Rev. D **91**, no. 2, 022001 (2015)
270. M.G. Aartsen et al., [IceCube Collaboration], Evidence for astrophysical muon neutrinos from the Northern Sky with IceCube. Phys. Rev. Lett. **115**(8), 081102 (2015)
271. M.G. Aartsen et al. [IceCube Collaboration], A combined maximum-likelihood analysis of the high-energy astrophysical neutrino flux measured with IceCube. Astrophys. J. **809**, no. 1, 98 (2015)
272. M.G. Aartsen et al., [IceCube Collaboration], Flavor ratio of astrophysical neutrinos above 35 TeV in IceCube. Phys. Rev. Lett. **114**(17), 171102 (2015)
273. M.G. Aartsen et al., [IceCube Collaboration], Search for astrophysical tau neutrinos in three years of IceCube data. Phys. Rev. D **93**(2), 022001 (2016)
274. M.G. Aartsen et al. [IceCube Collaboration], Observation and characterization of a cosmic muon neutrino flux from the Northern Hemisphere using six years of IceCube data. Astrophys. J. **833**, no.1, 3 (2016)
275. A. Palladino, M. Spurio, F. Vissani, On the IceCube spectral anomaly. JCAP **1612**(12), 045 (2016)
276. P. Padovani, E. Resconi, P. Giommi, B. Arsioli, Y.L. Chang, Extreme blazars as counterparts of IceCube astrophysical neutrinos. Mon. Not. Roy. Astron. Soc. **457**(4), 3582 (2016)
277. M.G. Aartsen et al., [IceCube Collaboration], The contribution of Fermi-2LAC blazars to the diffuse TeV-PeV neutrino flux. Astrophys. J. **835**(1), 45 (2017)
278. A. Palladino, F. Vissani, Can BL Lac emission explain the neutrinos above 0.2 PeV?, arXiv:1702.08779 [astro-ph.HE]
279. M. Ahlers, K. Murase, Probing the Galactic Origin of the IceCube excess with gamma-rays. Phys. Rev. D **90**, no. 2, 023010 (2014)
280. L.A. Anchordoqui, H. Goldberg, T.C. Paul, L.H.M. da Silva, B.J. Vlcek, Estimating the contribution of Galactic sources to the diffuse neutrino flux. Phys. Rev. D **90**(12), 123010 (2014)
281. A.M. Taylor, S. Gabici, F. Aharonian, Galactic halo origin of the neutrinos detected by IceCube. Phys. Rev. D **89**, no. 10, 103003 (2014)
282. A. Neronov, D.V. Semikoz, Evidence the Galactic contribution to the IceCube astrophysical neutrino flux. Astropart. Phys. **75**, 60 (2016)
283. S. Troitsky, Search for Galactic disk and halo components in the arrival directions of high-energy astrophysical neutrinos. JETP Lett. **102**(12), 785 (2015)

284. A. Palladino, F. Vissani, Extragalactic plus Galactic model for IceCube neutrino events,' arXiv:1601.06678 [astro-ph.HE], accepted for publication in ApJ
285. A. Palladino, G. Pagliaroli, F.L. Villante, F. Vissani, What is the flavor of the cosmic neutrinos seen by IceCube? Phys. Rev. Lett. **114**, 171101 (2015)
286. A. Palladino, G. Pagliaroli, F.L. Villante, F. Vissani, Double pulses and cascades above 2 PeV in IceCube. Eur. Phys. J. C **76**, 52 (2016)
287. S.L. Glashow, Resonant scattering of antineutrinos. Phys. Rev. **118**, 316 (1960)
288. M. Honda, T. Kajita, K. Kasahara, S. Midorikawa, T. Sanuki, Calculation of atmospheric neutrino flux using the interaction model calibrated with atmospheric muon data. Phys. Rev. D **75**, 043006 (2007)
289. S. Adrian-Martinez et al. [ANTARES Collaboration], Constraints on the neutrino emission from the Galactic Ridge with the ANTARES telescope, arXiv:1602.03036 [astro-ph.HE]
290. G. Pagliaroli, A. Palladino, F.L. Villante, F. Vissani, Testing nonradiative neutrino decay scenarios with IceCube data. Phys. Rev. D **92**, 113008 (2015)
291. A. Esmaili, P.D. Serpico, Are IceCube neutrinos unveiling PeV-scale decaying dark matter? JCAP **1311**, 054 (2013)
292. C. Lunardini, S. Razzaque, K.T. Theodoseau, L. Yang, Neutrino events at IceCube and the Fermi Bubbles. Phys. Rev. D **90**, 023016 (2014)
293. P. Padovani, E. Resconi, Are both BL Lacs and pulsar wind nebulae the astrophysical counterparts of IceCube neutrino events? Mon. Not. Roy. Astron. Soc. **443**, no. 1, 474 (2014)
294. A. Achterberg et al., [IceCube Collaboration], First year performance of The IceCube neutrino telescope. Astropart. Phys. **26**, 155 (2006)
295. M.A. Markov, On high energy neutrino physics, in *Proceedings, 10th International Conference on High-Energy Physics (ICHEP 60),* pp. 578–581 (1960), http://inspirehep.net/record/1341439/files/C60-08-25-p578.pdf
296. S.I. Klimushin, E.V. Bugaev, I.A. Sokalski, Precise parametrizations of muon energy losses in water, hep-ph/0106010
297. R. Abbasi et al., [IceCube Collabboration], The design and performance of IceCube DeepCore. Astropart. Phys. **35**, 615 (2012)
298. K. Hanson, O. Tarasova, [IceCube Collaboration], Design and production of the IceCube digital optical module. Nucl. Instrum. Methods **A567**, 214 (2006)
299. M. Ageron et al., ANTARES: the first undersea neutrino telescope. Nucl. Instrum. Methods **A656**, 11 (2011)
300. V. Aynutdinov et al., [BAIKAL Collaboration], The BAIKAL neutrino experiment: physics results and perspectives. Nucl. Instrum. Meth. A **602**, 14 (2009)
301. A. Avrorin et al., The prototyping/early construction phase of the BAIKAL-GVD project. Nucl. Instrum. Meth. A **742**, 82 (2014)
302. http://www.km3net.org/home.php
303. S. Adrián-Martinez et al., (KM3NeT Collaboration), Expansion cone for the 3-inch PMTs of the KM3NeT Optical Modules. JINST **8**, T03006 (2013)
304. T. Damour, A. Vilenkin, Gravitational wave bursts from cosmic strings, Phys. Rev. Lett. 85, 3761 (2000); V.S. Berezinsky, A. Vilenkin, Ultrahigh-energy neutrinos from hidden sector topological defects. Phys. Rev. D **62**, 083512 (2000)
305. S. Adrian-Martinez et al., [LIGO Scientific and VIRGO Collaborations], A first search for coincident gravitational waves and high energy neutrinos using LIGO, Virgo and ANTARES data from 2007. JCAP **1306**, 008 (2013)
306. S. Adrian-Martinez et al. [ANTARES and IceCube and LIGO Scientific and Virgo Collaborations], High-energy Neutrino follow-up search of gravitational wave event GW150914 with ANTARES and IceCube, arXiv:1602.05411 [astro-ph.HE]
307. F. Halzen, G.G. Raffelt, Reconstructing the supernova bounce time with neutrinos in IceCube. Phys. Rev. D **80**, 087301 (2009)
308. R. Bernabei et al., [DAMA and LIBRA Collaborations], New results from DAMA/LIBRA. Eur. Phys. J. C **67**, 39 (2010)

309. S.H. Hansen, J. Lesgourgues, S. Pastor, J. Silk, Constraining the window on sterile neutrinos as warm dark matter. Mon. Not. Roy. Astron. Soc. **333**, 544 (2002)
310. M. Cirelli, N. Fornengo, T. Montaruli, I.A. Sokalski, A. Strumia, F. Vissani, Spectra of neutrinos from dark matter annihilations. Nucl. Phys. B **727**, 99 (2005) [Err.: Nucl. Phys. B **790**, 338 (2008)]
311. M. Blennow, J. Edsjo, T. Ohlsson, Neutrinos from WIMP annihilations using a full three-flavor Monte Carlo. JCAP **0801**, 021 (2008)
312. V. Berezinsky, E. Sabancilar, A. Vilenkin, Extremely high energy neutrinos from cosmic strings. Phys. Rev. D **84**, 085006 (2011)
313. C. Lunardini, E. Sabancilar, Cosmic strings as emitters of extremely high energy neutrinos. Phys. Rev. D **86**, 085008 (2012)
314. R. Foot, R.R. Volkas, Neutrino physics and the mirror world: how exact parity symmetry explains the solar neutrino deficit, the atmospheric neutrino anomaly and the LSND experiment. Phys. Rev. D **52**, 6595 (1995)
315. M. Kobayashi, C.S. Lim, Pseudo Dirac scenario for neutrino oscillations. Phys. Rev. D **64**, 013003 (2001)
316. J.F. Beacom, N.F. Bell, D. Hooper, J.G. Learned, S. Pakvasa, T.J. Weiler, PseudoDirac neutrinos: a challenge for neutrino telescopes. Phys. Rev. Lett. **92**, 011101 (2004)
317. J.N. Bahcall, N. Cabibbo, A. Yahil, Are neutrinos stable particles? Phys. Rev. Lett. **28**, 316 (1972)
318. S. Pakvasa, K. Tennakone, Neutrinos of non-zero rest mass. Phys. Rev. Lett. **28**, 1415 (1972)
319. K. Eguchi et al., [KamLAND Collaboration], A high sensitivity search for anti-nu(e)'s from the sun and other sources at KamLAND. Phys. Rev. Lett. **92**, 071301 (2004)
320. P. Baerwald, M. Bustamante, W. Winter, Neutrino decays over cosmological distances and the implications for neutrino telescopes. JCAP **1210**, 020 (2012)
321. G. Pagliaroli, A. Palladino, F.L. Villante, F. Vissani, Testing nonradiative neutrino decay scenarios with IceCube data. Phys. Rev. D **92**, no. 11, 113008 (2015)
322. J.F. Beacom, N.F. Bell, D. Hooper, S. Pakvasa, T.J. Weiler, Measuring flavor ratios of high-energy astrophysical neutrinos. Phys. Rev. D **68** (2003) 093005 [Err.: Phys. Rev. D **72** (2005) 019901]
323. D.W. Hogg, Distance measures in cosmology, arXiv:astro-ph/9905116 (1999)
324. G. Riccobene et al., Deep seawater inherent optical properties in the Southern Ionian Sea [NEMO Collaboration]. Astropart. Phys. **27**, 1 (2007)
325. R.C. Smith, K.S. Baker, Optical properties of the clearest natural waters (200800 nm). Appl. Opt. **20**, 177 (1981)
326. V. Berezinsky, High energy neutrino astronomy, Proceedings of Neutrino 1990. Nucl. Phys. B Proc. Suppl 19 (1991) 375
327. F. Vissani, Neutrinos from galactic sources of cosmic rays with known γ-ray spectra. Astropart. Phys. **26**, 310 (2006)
328. A. Kappes, J. Hinton, C. Stegmann, F.A. Aharonian, Potential neutrino signals from galactic gamma-ray sources. Astrophys. J. **656**, 870 (2007) Erratum: [Astrophys. J. **661**, 1348 (2007)]
329. S.R. Kelner, F.A. Aharonian, V.V. Bugayov, Energy spectra of gamma-rays, electrons and neutrinos produced at proton-proton interactions in the very high energy regime. Phys. Rev. D **74**, 034018 (2006) Erratum: [Phys. Rev. D **79**, 039901 (2009)]
330. F.L. Villante, F. Vissani, How precisely neutrino emission from supernova remnants can be constrained by gamma ray observations? Phys. Rev. D **78**, 103007 (2008)
331. F. Vissani, G. Pagliaroli, F.L. Villante, On the goals of neutrino astronomy. Nuovo Cim. **32**, 353 (2009); G. Pagliaroli, F.L. Villante, F. Vissani, Neutrini dallo spazio (discutendo gli obiettivi di una nuova astronomia). Nuovo Saggiatore **25**(3–4), 5–18 (2009)
332. F. Vissani, Neutrino Sources and Properties, Gran Sasso Summer Institute 2014, Hands-On Experimental Underground Physics at LNGS (2014)
333. F. Vissani, M.L. Costantini, W. Fulgione, A. Ianni, G. Pagliaroli, What is the issue with SN1987A neutrinos?, arXiv:1008.4726 [hep-ph]

334. F. Vissani, A. Boeltzig, Supernova neutrinos: risks and opportunities. PoS NEUTEL **2015**, 008 (2015)
335. M.G. Aartsen et al. [IceCube Collaboration], The IceCube Neutrino Observatory—Contributions to ICRC 2015 Part I: Point Source Searches, arXiv:1510.05222 [astro-ph.HE]

Proceedings:

336. J. Learned, Alternative techniques for high energy neutrino astronomy. Nucl. Phys. Proc. Suppl. **118**, 405 (2003)
337. M.G. Aartsen et al. [IceCube Collaboration], The IceCube Neutrino Observatory—Contributions to ICRC 2015 Part II: Atmospheric and Astrophysical Diffuse Neutrino Searches of All Flavors, arXiv:1510.05223 [astro-ph.HE]
338. A.D. Avrorin et al. [BAIKAL-GVD Collaboration], Status of the Baikal-GVD Project, doi:https://doi.org/10.1142/9789814663618-0019
339. P. Lipari, Introduction to Neutrino Physics, in *Proceedings of CERN-CLAF School of Physics* (2001), Edited by E. Nick, B. Marechal. Geneva, CERN, 2003 available as http://cds.cern.ch/record/677618/files/p115.pdf
340. A. Karle, Neutrino astronomy–a review of future experiments. Nucl. Phys. Proc. Suppl. **235–236**, 364 (2013)
341. C. Pena-Garay, Neutrino Physics faces Cosmology, talk given at GSSI (L'Aquila), on January 19 (2016)
342. J. Bahcall webpage: http://www.sns.ias.edu/~jnb/
343. A. Serenelli webpage: http://www.ice.csic.es/personal/aldos/Home.html
344. The *arXiv*, http://arxiv.org
345. The *inSPIRE* database, http://inspirehep.net
346. The *SAO/NASA Astrophysics Data System*, http://esoads.eso.org

Online Material:

347. *Neutrino Unbound*, http://www.nu.to.infn.it/
348. *Long Baseline News*, http://www.hep.anl.gov/ndk/longbnews/
349. The *Particle Data Group* report, http://pdg.lbl.gov
350. The *Durham database* on neutrino cross sections, http://durpdg.dur.ac.uk/review/neutrino/
351. The *Martin-Stirling-Thorne-Watt* parton distribution functions, https://mstwpdf.hepforge.org
352. The *Durham HepData Project*, http://hepdata.cedar.ac.uk/pdfs/
353. *LHAPDF6*, http://lhapdf.hepforge.org/pages.html
354. The *National Institute of Standards and Technology* (NIST), http://www.nist.gov/pml/
355. Fundamental Physical Constants http://physics.nist.gov/cuu/Constants/Table/allascii.txt
356. Numerical approximation for the comoving distance, https://www.ssl.berkeley.edu/~mlampton/ComovingDistance.pdf
357. Calculation of the probability of oscillations in the Earth, http://pcbat1.mi.infn.it/~battist/cgi-bin/oscil/index.r
358. Enomoto Sanshiro's *Geoneutrino Spectrum and Luminosity* webpage, http://www.awa.tohoku.ac.jp/~sanshiro/research/geoneutrino/spectrum/
359. For geochemical and geodynamical models, see wiki page on *Carbonaceous chondrite*, https://en.wikipedia.org/wiki/Carbonaceous_chondrite#CI_group
360. For cosmogeophysical models, see wiki page on *Enstatite chondrite*, https://en.wikipedia.org/wiki/Enstatite_chondrite
361. SNEWS: SuperNova Early Warning System, http://snews.bnl.gov
362. The *IceCube Web Page*, https://icecube.wisc.edu/masterclass/neutrinos/

Gravitational Physics: From Quantum to Waves

Salvatore Capozziello, Eugenio Coccia, Mariafelicia De Laurentis and Viviana Fafone

Abstract Gravitational interaction is widely considered one of the most challenging topics of physics. Despite of the undoubtful successes of General Relativity, the lack of a self-consistent theory, capable of encompassing gravitational aspects of quantum phenomena, local scales, galactic and extragalactic astrophysics up to cosmology, it is one of the most frustrating aspects of this long ranging interaction. Here, with no claim to completeness, we review some of these aspects considering the issues of foundation of General Relativity, quantum gravity, strong and weak field regimes, alternative theories of gravity, gravitational waves and multimessanger astronomy.

S. Capozziello (✉)
Dipartimento di Fisica, Universitá di Napoli "Federico II", Napoli, Italy
e-mail: capozzie@na.infn.it

S. Capozziello
INFN Sez. di Napoli, Compl. Univ. di Monte S. Angelo, Edificio G,
Via Cinthia, 80126 Napoli, Italy

S. Capozziello · E. Coccia · V. Fafone
Gran Sasso Science Institute, Viale F. Crispi 7, 67100 L'aquila, Italy

E. Coccia · V. Fafone
Dipartimento di Fisica, Universitá di Tor Vergata, Via della Ricerca Scientifica,
00133 Roma, Italy
e-mail: eugenio.coccia@lngs.infn.it

V. Fafone
e-mail: viviana.fafone@roma2.infn.it

E. Coccia · V. Fafone
INFN Sez. Roma Tor Vergata, Via della Ricerca Scientifica, 00133 Roma, Italy

M. De Laurentis
Institute for Theoretical Physics, Goethe University, Max-von-Laue-Str. 1,
60438 Frankfurt, Germany
e-mail: laurentis@th.physik.uni-frankfurt.de

M. De Laurentis
Tomsk State Pedagogical University, ul. Kievskaya, 60, 634061 Tomsk, Russia

M. De Laurentis
Lab. Theor. Cosmology, Tomsk State University of Control Systems
and Radioelectronics (TUSUR), 634050 Tomsk, Russia

R. Aloisio et al. (eds.), *Multiple Messengers and Challenges in Astroparticle Physics*, https://doi.org/10.1007/978-3-319-65425-6_5

The goal is to provide a quick summary and guideline for these open issues. Particular emphasis is given to the theory and the discovery gravitational waves that have been the *experimentum crucis* to confirm General Relativity and definitely opened the era of the so called *gravitational astronomy*.

Introduction

Several questions and shortcomings emerged in the last thirty years leading to the conclusion that General Relativity, in the former Einstein's formulation, is not the final theory of gravitational interaction also if a wide range of phenomena can be described in its framework. The main issue of any theory of gravity is to encompass, under the same standard, the positive results achieved by the General Relativity and the behavior of gravitational interaction at infrared and ultraviolet scales. In other words, this ambitious program would like to join under the same standard phenomena ranging from quantum scales to cosmology. Such a requirement comes out from the fact that gravity is a long range interaction acting at any scale, from elementary particles up to the whole Universe. The shortcoming emerges because General Relativity is essentially a "classical theory" formulated to recover the Newtonian results in the weak field limit. The needs to describe the Universe as a dynamical system and the issue of quantization of space-time have been following problems. However, despite these puzzles, General Relativity is an excellent theory that can be considered the basic paradigm to any approach aimed to describe the gravitational interaction. In this first section, we will outline the main ingredients that reliable theory of gravity has to match. Furthermore, we will give a quick overview of problems related to General Relativity.

A Short History of Theories of Gravity

It is remarkable that gravity is probably the fundamental interaction which still remains the most enigmatic, even though it is related to phenomena experienced in everyday life and is the one most easily conceived without any sophisticated knowledge. As a matter of fact, the gravitational interaction was the first one to have been put under the microscope of experimental investigation, obviously due to the simplicity of constructing a suitable experimental apparatus.

Galileo Galilei was the first to introduce pendula and inclined planes to the study of terrestrial gravity at the end of the 16th century [1–5]. Gravity played an important role in the development of Galileo's ideas about the necessity of experiment in the study of Science, which had a great impact on modern scientific thinking. However, it was not until 1665, when Isaac Newton introduced the now renowned "inverse-square gravitational force law", that terrestrial gravity was actually related to celestial gravity in a single theory [6, 7]. Newton's theory made correct predictions for a variety of phenomena at different scales, including both terrestrial experiments and planetary motion.

Obviously, Newton's contribution to gravity, quite apart from his enormous contribution to physics overall, is not restricted to the expression of the inverse square law. Much attention should be paid to the conceptual basis of his gravitational theory, which incorporates two key ideas:

1. The idea of absolute space, i.e. the view of space as a fixed, unaffected structure; a rigid arena where physical phenomena take place.
2. The idea of what was later called the Weak Equivalence Principle which, expressed in the language of Newtonian theory, states that the inertial and the gravitational mass coincide.

Asking whether Newton's theory, or any other physical theory, is right or wrong, would be an ill-posed question to begin with, since any consistent theory is apparently "right". A more appropriate way to pose the question would be to ask how suitable this theory is to describe the physical world or, even better, how large a portion of the physical world is sufficiently described by such a theory. Also, one could ask how unique the specific theory is for the description of the relevant phenomena. It was obvious, in the first 20 years after the introduction of Newtonian gravity, that it did manage to explain all of the aspects of gravity known at that time. However, all of the questions above were posed sooner or later.

In 1855, Urbain Le Verrier observed a 35 arc-second excess precession of Mercury's orbit and later on, in 1882, Simon Newcomb measured this precession more accurately to be 43 arc-seconds [8–12]. This experimental fact was not predicted by Newton's theory. It should be noted that Le Verrier initially tried to explain the precession within the context of Newtonian gravity, attributing it to the existence of another, yet unobserved, planet whose orbit lies within that of Mercury. He was apparently influenced by the fact that examining the distortion of the planetary orbit of Uranus in 1846 had led him, and, independently, John Couch Adams, to the discovery of Neptune and the accurate prediction of its position and momenta. However, this innermost planet was never found.

On the other hand, in 1893, Ernst Mach stated what was later called by Albert Einstein "Mach's principle". This is the first constructive attack to Newton's idea of absolute space after the eighteenth century debate between Gottfried Wilhelm von Leibniz and Samuel Clarke (Clarke was acting as Newton's spokesman) on the same subject, known as the Leibniz–Clarke Correspondence [13]. Mach's idea can be considered as rather vague in its initial formulation and it was essentially brought into the mainstream of physics later on by Einstein along the following lines:

> ...inertia originates in a kind of interaction between bodies ....

This is obviously in contradiction with Newton's ideas, according to which inertia was always relative to the absolute frame of space. There exists also a later, probably clearer interpretation of Mach's Principle, which, however, also differs in substance. This was given by Dicke [14]:

> The gravitational constant should be a function of the mass distribution in the Universe.

This is different from Newton's idea of the gravitational constant as being universal and unchanging. Now Newton's basic axioms have to be reconsidered.

But it was not until 1905, when Albert Einstein completed Special Relativity, that Newtonian gravity would have to face a serious challenge. Einstein's new theory, which managed to explain a series of phenomena related to non-gravitational physics, appeared to be incompatible with Newtonian gravity. Relative motion and all the linked concepts had gone well beyond Galileo and Newton ideas and it seemed that Special Relativity should somehow be generalised to include non-inertial frames. In 1907, Einstein introduced the equivalence between gravitation and inertia and successfully used it to predict the gravitational redshift. Finally, in 1915, he completed the theory of General Relativity (GR), a generalisation of Special Relativity which included gravity and any accelerated frame. Remarkably, the theory matched perfectly the experimental result for the precession of Mercury's orbit, as well as other experimental findings like the Lense-Thirring [15–18] gravitomagnetic precession (1918) and the gravitational deflection of light by the Sun, as measured in 1919 during a Solar eclipse by Arthur Eddington [19].

GR overthrew Newtonian gravity and continues to be up to now an extremely successful and well-accepted theory for gravitational phenomena. As mentioned before, and as often happens with physical theories, Newtonian gravity did not lose its appeal to scientists. It was realised, of course, that it is of limited validity compared to GR, but it is still sufficient for most applications related to gravity. What is more, in weak field limit of gravitational field strength and velocities, GR inevitably reduces to Newtonian gravity. Newton's equations for gravity might have been generalised and some of the axioms of his theory may have been abandoned, like the notion of an absolute frame, but some of the cornerstones of his theory still exist in the foundations of GR, the most prominent example being the Equivalence Principle, in a more suitable formulation of course.

This brief chronological review, besides its historical interest, is outlined here also for a practical reason. GR is bound to face the same questions as were faced by Newtonian gravity and many people would agree that it is actually facing them now. In the forthcoming sections, experimental facts and theoretical problems will be presented which justify that this is indeed the case. Remarkably, there exists a striking similarity to the problems which Newtonian gravity faced, i.e. difficulty in explaining several observations, incompatibility with other well-established theories and lack of uniqueness.

What We Mean for a "Good Theory of Gravity"

From a phenomenological point of view, there are some minimal requirements that any relativistic theory of gravity has to match. First of all, it has to explain the astrophysical observations (e.g. the orbits of planets, self-gravitating structures).

This means that it has to reproduce the Newtonian dynamics in the weak-energy limit. Besides, it has to pass the classical Solar System tests which are all experimentally well founded [20]. As second step, it should reproduce Galactic dynamics considering the observed baryonic constituents (e.g. luminous components as stars, sub-luminous components as planets, dust and gas), radiation and Newtonian potential which is, by assumption, extrapolated to Galactic scales.

Thirdly, it should address the problem of large scale structure (e.g. clustering of galaxies) and finally cosmological dynamics, which means to reproduce, in a self-consistent way, the cosmological parameters as the expansion rate, the Hubble constant, the density parameter and so on. Observations and experiments, essentially, probe the standard baryonic matter, the radiation and an attractive overall interaction, acting at all scales and depending on distance: the gravity.

The simplest theory which try to satisfies the above requirements is the GR [21]. It is firstly based, on the assumption that space and time have to be entangled into a single space-time structure, which, in the limit of no gravitational forces, has to reproduce the Minkowski space-time structure. Einstein profited also from ideas earlier put forward by Riemann, who stated that the Universe should be a curved manifold and that its curvature should be established on the basis of astronomical observations [22].

In other words, the distribution of matter has to influence point by point the local curvature of the space-time structure. The theory, eventually formulated by Einstein in 1915, was strongly based on three assumptions that the Physics of Gravitation has to satisfy.

The "*Principle of Relativity*", that requires all frames to be good frames for Physics, so that no preferred inertial frame should be chosen a priori (if any exist).

The "*Principle of Equivalence*", that amounts to require inertial effects to be locally indistinguishable from gravitational effects (in a sense, the equivalence between the inertial and the gravitational mass).

The "*Principle of General Covariance*", that requires field equations to be "generally covariant" (today, we would better say to be invariant under the action of the group of all space-time diffeomorphisms) [23].

And - on the top of these three principles - the requirement that causality has to be preserved (the "*Principle of Causality*", i.e. that each point of space-time should admit a universally valid notion of past, present and future).

Let us also recall that the older Newtonian theory of space-time and gravitation, that Einstein wanted to reproduce at least in the limit of weak gravitational forces (what is called today the "post-Newtonian approximation"), required space and time to be absolute entities, particles moving in a preferred inertial frame following curved trajectories, the curvature of which (e.g. the acceleration) had to be determined as a function of the sources (i.e. the "forces").

On these bases, Einstein was led to postulate that the gravitational forces have to be expressed by the curvature of a metric tensor field $ds^2 = g_{\mu\nu}dx^\mu dx^\nu$ on a four-dimensional space-time manifold, having the same signature of Minkowski metric, e.g., the so-called "Lorentzian signature", herewith assumed to be $(-, +, +, +)$. He also postulated that space-time is curved in itself and that its curvature is locally determined by the distribution of the sources, e.g., being space-time a continuum, by the four-dimensional generalization of what in Continuum Mechanics is called the "matter stress-energy tensor", e.g. a rank-two (symmetric) tensor $T^{(m)}_{\mu\nu}$.

Hilbert and Einstein [23] proved that the field equations for a metric tensor $g_{\mu\nu}$, related to a given distribution of matter-energy, can be achieved by starting from the Ricci curvature scalar R which is an invariant. We will give details below.

The choice of Hilbert and Einstein was completely arbitrary (as it became clear a few years later), but it was certainly the simplest one both from the mathematical and the physical point of view. As it was later clarified by Levi–Civita in 1919, curvature is not a "purely metric notion" but, rather, a notion related to the "linear connection" to which "parallel transport" and "covariant derivation" refer [24].

In a sense, this is the precursor idea of what in the sequel would be called a "gauge theoretical framework" [25], after the pioneering work by Cartan in 1925 [26]. But at the time of Einstein, only metric concepts were at hands and his solution was the only viable.

It was later clarified that the three principles of relativity, equivalence and covariance, together with causality, just require that the space-time structure has to be determined by either one or both of two fields, a Lorentzian metric g and a linear connection Γ, assumed at the beginning to be torsionless for the sake of simplicity.

The metric g fixes the causal structure of space-time (the light cones) as well as its metric relations (clocks and rods); the connection Γ fixes the free-fall, i.e. the locally inertial observers. They have, of course, to satisfy a number of compatibility relations which amount to require that photons follow null geodesics of Γ, so that Γ and g can be independent, a priori, but constrained, *a posteriori*, by some physical restrictions. These, however, do not impose that Γ has necessarily to be the Levi-Civita connection of g [27].

This justifies—at least on a purely theoretical basis—the fact that one can envisage the so-called "alternative theories of gravitation", that we prefer to call "*Extended Theories of Gravitation*" (ETGs) since their starting points are exactly those considered by Einstein and Hilbert: theories in which gravitation is described by either a metric (the so-called "purely metric theories"), or by a linear connection (the so-called "purely affine theories") or by both fields (the so-called "metric-affine theories", also known as "first order formalism theories"). In these theories, the Lagrangian is a scalar density of the curvature invariants constructed out of both g and Γ.

The choice of Hilbert-Einstein Lagrangian is by no means unique and it turns out that the Hilbert-Einstein Lagrangian is in fact the only choice that produces an invariant that is linear in second derivatives of the metric (or first derivatives of the connection). A Lagrangian that, unfortunately, is rather singular from the Hamiltonian point of view, in much than same way as Lagrangians, linear in canonical momenta, are rather singular in Classical Mechanics (see e.g. [28]).

A number of attempts to generalize GR (and unify it to Electromagnetism) along these lines were followed by Einstein himself and many others (Eddington, Weyl, Schrödinger, just to quote the main contributors; see, e.g., [29]) but they were eventually given up in the fifties of XX Century, mainly because of a number of difficulties related to the definitely more complicated structure of a non-linear theory (where by "non-linear" we mean here a theory that is based on non-linear invariants of the curvature tensor), and also because of the new understanding of physics that is currently based on four fundamental forces and requires the more general "gauge framework" to be adopted (see [30]).

Still a number of sporadic investigations about "alternative theories" continued even after 1960 (see [20] and Refs. quoted therein for a short history). The search for a coherent quantum theory of gravitation or the belief that gravity has to be considered as a sort of low-energy limit of string theories [31], something that we are not willing to enter here in detail, has more or less recently revitalized the idea that there is no reason to follow the simple prescription of Einstein and Hilbert and to assume that gravity should be classically governed by a Lagrangian linear in the curvature.

Further curvature invariants or non-linear functions of them should be also considered, especially in view of the fact that they have to be included in both the semi-classical expansion of a quantum Lagrangian or in the low-energy limit of a string Lagrangian.

Moreover, it is clear from the recent astrophysical observations and from the current cosmological hypotheses that Einstein equations are no longer a good test for gravitation at Solar System, Galactic, extra-galactic and cosmic scale, unless one does not admit that the matter side of field equations contains some kind of exotic matter-energy which is the "dark matter" and "dark energy" side of the Universe.

The idea which we propose here is much simpler. Instead of changing the matter side of Einstein field equations in order to fit the "missing matter-energy" content of the currently observed Universe (up to the 95% of the total amount!), by adding any sort of inexplicable and strangely behaving matter and energy, we claim that it is simpler and more convenient to change the gravitational side of the equations, admitting corrections coming from non-linearities in the effective Lagrangian. However, this is nothing else but a matter of taste and, since it is possible, such an approach should be explored. Of course, provided that the Lagrangian can be conveniently tuned up (i.e., chosen in a huge family of allowed Lagrangians) on the basis of its best fit with all possible observational tests, at all scales (Solar, Galactic, extragalactic and cosmic).

Something that, in spite of some commonly accepted but disguised opinion, can and should be done before rejecting a priori a non-linear theory of gravitation (based on a non-singular Lagrangian) and insisting that the Universe has to be necessarily described by a rather singular gravitational Lagrangian (one that does not allow a coherent perturbation theory from a good Hamiltonian point of view) accompanied by matter that does not follow the behaviour that standard baryonic matter, probed in our laboratories, usually satisfies.

General Relativity and Its Shortcomings

Considering the above discussion it is worth noticing that in the last 30 years several shortcomings came out in the Einstein theory and people began to investigate whether GR is the only fundamental theory capable of explaining the gravitational interaction. Such issues come, essentially, from cosmology and quantum field theory. The shortcomings are related both to many theoretical aspects and to observational results. In this section we will try to summarize these problems. An important issue has to be underlined: even if there are many problems, the reaction of scientific community is not uniform. In a very simple scheme we can summarize the guide lines.

Many people will agree that modern physics is based on two main pillars: GR and Quantum Field Theory. Each of these two theories has been very successful in its own arena of physical phenomena: GR in describing gravitating systems and non-inertial frames from a classical point of view on large enough scales, and Quantum Field Theory at high energy or small scale regimes where a classical description breaks down. However, Quantum Field Theory assumes that space-time is flat and even its extensions, such as Quantum Field Theory in curved space time, consider space-time as a rigid arena inhabited by quantum fields. GR, on the other hand, does not take into account the quantum nature of matter. Therefore, it comes naturally to ask what happens if a strong gravitational field is present at quantum scales. How do quantum fields behave in the presence of gravity? To what extent are these amazing theories compatible?

Let us try to pose the problem more rigorously. Firstly, what needs to be clarified is that there is no final proof that gravity should have some quantum representation at high energies or small scales, or even that it will retain its nature as an interaction. The gravitational interaction is so weak compared with other interactions that the characteristic scale under which one would expect to experience non-classical effects relevant to gravity, the Planck scale, is 10^{-33} cm. Such a scale is not of course accessible by any current experiment and it is doubtful whether it will ever be accessible to future experiments either.[1] However, there are a number of reasons for which one would prefer to fit together GR and Quantum Field Theory [32, 33]. Let us list some of the most prominent ones here and leave the discussion about how to address them for the next subsection. Curiosity is probably the motivation leading scientific research. From this perspective it would be at least unusual if the gravity research community was so easily willing to abandon any attempt to describe the regime where both quantum and gravitational effects are important. The fact that the Planck scale seems currently experimentally inaccessible does not, in any way, imply that it is physically irrelevant. On the contrary, one can easily name some very important open issues of contemporary physics that are related to the Planck scale.

A particular example is the Big Bang scenario in which the Universe inevitably goes through an era in which its dimensions are smaller than the Planck scale (Planck

[1]This fact does not imply, of course, that imprints of Quantum Gravity phenomenology cannot be found in lower energy experiments.

era). On the other hand, space-time in GR is a continuum and so in principle all scales are relevant. From this perspective, in order to derive conclusions about the nature of space-time one has to answer the question of what happens on very small and very large scales.

The Quantum Gravity Problem

One of the main challenges of modern physics is to construct a theory able to describe the fundamental interactions of nature as different aspects of the same theoretical construct. This goal has led, in the past decades, to the formulation of several unification schemes which, *inter alia*, attempt to describe gravity by putting it on the same footing as the other interactions. All these schemes try to describe the fundamental fields in terms of the conceptual apparatus of Quantum Mechanics. This is based on the fact that the states of a physical system are described by vectors in a Hilbert space $\mathcal{H}$ and the physical fields are represented by linear operators defined on domains of $\mathcal{H}$. Until now, any attempt to incorporate gravity in this scheme has either failed or been unsatisfactory. The main conceptual problem is that the gravitational field describes simultaneously the gravitational degrees of freedom and the background space-time in which these degrees of freedom live.

Owing to the difficulties of building a complete theory unifying interactions and particles, during the last decades the two fundamental theories of modern physics, GR and Quantum Mechanics, have been critically re-analyzed. On the one hand, one assumes that the matter fields (bosons and fermions) come out from superstructures (e.g. Higgs bosons or superstrings) that, undergoing certain phase transitions, have generated the known particles. On the other hand, it is assumed that the geometry (e.g. the Ricci tensor or the Ricci scalar) interacts directly with quantum matter fields which back-react on it. This interaction necessarily modifies the standard gravitational theory, that is, the Lagrangian of gravity plus the effective fields is modified with respect to the Hilbert-Einstein one, and this fact can directly lead to the ETGs.

From the point of view of cosmology, the modifications of standard gravity provide inflationary scenarios of interest. In any case, a condition that must be satisfied in order for such theories to be physically acceptable is that GR is recovered in the low-energy limit.

Although remarkable conceptual progress has been made following the introduction of generalized gravitational theories, at the same time the mathematical difficulties have increased. The corrections introduced into the Lagrangian augment the (intrinsic) non-linearity of the Einstein equations, making them more difficult to study because differential equations of higher order than second are often obtained and because it is impossible to separate the geometric from the matter degrees of freedom. In order to overcome these difficulties and simplify the equations of motion, one often looks for symmetries of the Lagrangian and identifies conserved quantities which allow exact solutions of dynamics to be discovered. The key step in the implementation of this program consists of passing from the Lagrangian of the relevant fields to a point-like Lagrangian or, in other words, in going from a system with an infinite number of degrees of freedom to one with a finite number of degrees of freedom. Fortunately, this is feasible in cosmology because most models of physical interest are spatially homogeneous Bianchi models and the observed Universe is spatially homogeneous and isotropic to a high degree (Friedmann-Lemaitre-Robertson-Walker (FLRW) models).

The need for a quantum theory of gravity was recognized at the end of the 1950s, when physicist tried for the first time to treat all interactions at a fundamental level and to describe them in terms of Quantum Field Theory. Naturally, the first attempts to quantize gravity used the canonical approach and the covariant approach, which had been applied with remarkable success to Electromagnetism. In the first approach applied to Electromagnetism, one considers the electric and magnetic fields satisfying the Heisenberg Uncertainty Principle and the quantum states are gauge-invariant functionals generated by the vector potential defined on three-surfaces of constant time. In the second approach, one quantizes the two degrees of freedom of the Maxwell field without any $3 + 1$ decomposition of the metric, while the quantum states are elements of a Fock space [34]. These procedures are equivalent in Electromagnetism. The former allows for a well-defined measure, whereas the latter is more convenient for perturbative calculations such as, for example, the computation of the S-matrix in Quantum Electro Dynamics (QED).

These methods have been applied also to GR, but many difficulties arise in this case due to the fact that Einstein's theory cannot be formulated in terms of a quantum field theory on a fixed Minkowski background. To be more specific, in GR the geometry of the background space-time cannot be given a priori: space-time is the dynamical variable itself. In order to introduce the notions of causality, time, and evolution, one must first solve the equations of motion and then "build" the space-time. For example, in order to know if particular initial conditions will give rise to a black hole, it is necessary to fully evolve them by solving the Einstein equations. Then, taking into account the causal structure of the obtained solution , one has to study the asymptotic metric at future null infinity in order to assess whether it is related, in the causal past, with those initial conditions. This problem becomes intractable at the quantum level. Due to the Uncertainty Principle, in non-relativistic Quantum Mechanics, particles do not move along well-defined trajectories and one can only calculate the probability amplitude $\psi(t, x)$ that a measurement at time t detects a particle around the spatial point x. Similarly, in Quantum Gravity, the evolution of an initial state does not provide a specific space-time. In the absence of a space-time, how is it possible to introduce basic concepts such as causality, time, elements of the scattering matrix, or black holes?

The canonical and covariant approaches provide different answers to these questions. The canonical approach is based on the Hamiltonian formulation of GR and aims at using the canonical quantization procedure. The canonical commutation relations are the same that lead to the Uncertainty Principle; in fact, the commutation of certain operators on a spatial three-manifold of constant time is imposed, and this three-manifold is fixed in order to preserve the notion of causality. In the limit of asymptotically flat space-time, the motion generated by the Hamiltonian must be interpreted as temporal evolution (in other words, when the background becomes the Minkowski space-time, the Hamiltonian operator assumes again its role as the generator of translations). The canonical approach preserves the geometric features of GR without the need to introduce perturbative methods [35–39].

The covariant approach, instead, employs Quantum Field Theory concepts and methods. The basic idea is that the problems mentioned above can be easily circumvented by splitting the metric $g_{\mu\nu}$ into a kinematical part $\eta_{\mu\nu}$ (usually flat) and a dynamical part $h_{\mu\nu}$, as in

$$g_{\mu\nu} = \eta_{\mu\nu} + h_{\mu\nu}\,. \tag{1}$$

The geometry of the background is given by the flat metric tensor and is the same as in Special Relativity and ordinary Quantum Field Theory, which allows one to define the concepts of causality, time, and scattering. The quantization procedure is then applied to the dynamical field, considered as a (small) deviation of the metric from the Minkowski background metric. Quanta are discovered to be particles with spin two, called *gravitons*, which propagate in flat space-time and are defined by $h_{\mu\nu}$. Substituting the metric $g_{\mu\nu}$ into the Hilbert-Einstein action, it follows that the Lagrangian of the gravitational sector contains a sum whose terms represent, at different orders of approximation, the interaction of gravitons and, eventually, terms describing matter-graviton interaction (if matter is present). Such terms are analyzed by using the standard techniques of perturbative Quantum Field Theory.

These quantization programs were both pursued during the 1960s and 1970s. In the canonical approach, Arnowitt, Deser, and Misner [36] provided a Hamiltonian formulation of GR using methods proposed earlier by Dirac and Bergmann. In this Hamiltonian formalism, the canonical variables are the three-metric on the spatial submanifolds obtained by foliating the four-dimensional manifold (note that this foliation is arbitrary). The Einstein equations give constraints between the three-metrics and their conjugate momenta and the evolution equation for these fields, known as the *Wheeler-DeWitt* (WDW) *equation*. In this way, GR is interpreted as the dynamical theory of the three-geometries (*geometrodynamics*). The main difficulties arising from this approach are that the quantum equations involve products of operators defined at the same space-time point and, in addition, they entail the construction of distributions whose physical meaning is unclear. In any case, the main problem is the absence of a Hilbert space of states and, as consequence, a probabilistic interpretation of the quantities calculated is missing.

The covariant quantization approach is closer to the known physics of particles and fields in the sense that it has been possible to extend the perturbative methods of QED to gravitation. This has allowed the analysis of the mutual interaction between gravitons and of the matter-graviton interactions. The formulation of Feynman rules for gravitons and the demonstration that the theory might be unitary at every order of the expansion was achieved by DeWitt [40].

Further progress was achieved with Yang-Mills theories, which describe the strong, weak, and electromagnetic interactions of quarks and leptons by means of symmetries. Such theories are renormalizable because it is possible to give the fermions a mass through the mechanism of Spontaneous Symmetry Breaking. Then, it is natural to attempt to consider gravitation as a Yang-Mills theory in the covariant perturbation approach and check whether it is renormalizable. However, gravity does not fit into this scheme; it turns out to be non-renormalizable when one considers the graviton-graviton interactions (two-loops diagrams) and graviton-matter interactions (one-loop diagrams).[2] The covariant method allows one to construct a theory of gravity which is renormalizable at one-loop in the perturbative series [41]. Due to the non-renormalizability of gravity at different orders, its validity is restricted only to the low-energy domain, i.e., to large scales, while it fails at high energy and small scales. This implies that the full unknown theory of gravity has to be invoked near or at the Planck era and that, sufficiently far from the Planck scale, GR and its first loop corrections describe the gravitational interactions. In this context, it makes sense to add higher order terms to the Hilbert-Einstein action as we will do in the second part of this Report. Besides, if the free parameters are chosen appropriately, the theory has a better ultraviolet behavior and is asymptotically free. Nevertheless, the Hamiltonian of these theories is not bounded from below and they are unstable. In particular, unitarity is violated and probability is not conserved.

An alternative approach to the search for a theory of Quantum Gravity comes from the study of the Electroweak interaction. In this approach, gravity is treated neglecting the other fundamental interactions. The unification of the Electromagnetic and the weak interactions suggests that it might be possible to obtain a consistent theory when gravitation is coupled to some kind of matter. This is the basic idea of *Supergravity* [42]. In this class of theories, the divergences due to the bosons (gravitons in this case) are cancelled exactly by those due to the fermions, leading to a renormalized theory of gravity. Unfortunately, this scheme works only at the two-loop level and for matter-gravity couplings. The Hamiltonian is positive-definite and the theory turns out to be unitary. But, including higher order loops, the infinities re-appear, destroying the renormalizability of the theory.

Perturbative methods are also used in String Theories. In this case, the approach is completely different from the previous one because the concept of particle is replaced by that of an extended object, the fundamental string. The usual physical particles, including the spin two graviton, correspond to excitations of the string. The theory has only one free parameter (the string tension) and the interaction couplings are determined uniquely. It follows that string theory contains all fundamental physics and it is therefore considered as a candidate for the *Theory of Everything*. String Theory seems to be unitary and the perturbative series converges implying finite terms. This property follows from the fact that strings are intrinsically extended objects, so that ultraviolet divergencies coming from small scales or from large transfer impulses, are naturally cured. In other words, the natural cutoff is given by the string length, which is of Planck size l_{Pl}. At scales larger than l_{Pl}, the effective string action can be rewritten in terms of non-massive vibrational modes, i.e., in terms of scalar and tensor fields (*tree-level effective action*). This eventually leads to

[2]Higher order terms in the perturbative series imply an infinite number of free parameters. At the one-loop level it is sufficient to renormalize only the effective constants G_{eff} and Λ_{eff} which, at low energy, reduce to Newton's constant G_N and the cosmological constant Λ.

an effective theory of gravity non-minimally coupled with scalar fields, the so-called *dilaton fields*.

To conclude, let us summarize the previous considerations:

- a consistent (i.e., unitary and renormalizable) theory of gravity does not exist yet.
- In the quantization program for gravity, two approaches have been used: the *covariant approach* and the *perturbative approach*. They do not lead to a definitive theory of Quantum Gravity.
- In the low-energy regime (with respect to the Planck energy) at large scales, GR can be generalized by introducing into the Hilbert-Einstein action terms of higher order in the curvature invariants and non-minimal couplings between matter and gravity. These lead, at the one-loop level, to a consistent and renormalizable theory.

A part the lack of final theory, the Quantum Gravity Problem already contains some issues and shortcomings which could be already addressed by the today physics. We will summarize them in the forthcoming section.

Considered the status of art, are some predictions of Quantum Gravity already available? Can remnants of Planck scale be detected at lower energy couplings and masses? As it is well known, only a fine-tuned combination of the low-energy constants leads to an observable Universe like ours. It would thus appear strange if a fundamental theory possessed just the right constants to achieve this. Hogan [43] has argued that Grand Unified Theories constrain relations among parameters, but leave enough freedom for a selection. In particular, he suggests that one coupling constant and two light fermion masses are not fixed by the symmetries of the fundamental theory.[3] One could then determine this remaining free constants only by the (weak form of the) Anthropic Principle: they have values such that a Universe like ours is possible. The cosmological constant, for example, must not be much bigger than the presently observed value, because otherwise the Universe would expand too fast to allow the formation of galaxies. The Universe is, however, too special to be explainable on purely anthropic grounds. We know that the maximal entropy would be reached if all the matter in the observable Universe were collected into a single giant black hole. This entropy would be about 10^{123}, which is exceedingly more than the observed entropy of about 10^{88}. The "probability" for our Universe would then be about $\frac{exp(10^{88})}{exp(10^{123})}$. From the Anthropic Principle alone one would not need such a special Universe. As for the cosmological constant, for example, one could imagine its calculation from a fundamental theory. Taking the presently observed value for Λ, one can construct a mass according to

$$\left(\frac{\hbar\Lambda^{\frac{1}{2}}}{G}\right) \simeq 15\,\mathrm{MeV}\,,$$

[3] String theory contains only one fundamental dimensionfull parameter, the string length. The connection to low energies may nonetheless be non-unique due to the existence of many different possible "vacua".

which in elementary particle physics is not an unusually big or small value. The observed value of Λ could thus emerge together with medium-size particle mass scales. Since fundamental theories are expected to contain only one dimensionfull parameter, low-energy constants emerge from fundamental quantum fields. An important example in string theory, is the dilaton field from which one can calculate the gravitational constant. In order that these fields mimic physical constants, two conditions have to be satisfied. First, decoherence must be effective in order to guarantee a classical behaviour of the field. Second, this "classical" field must then be approximately constant in large-enough space-time regions, within the limits given by experimental data. The field may still vary over large times or large spatial regions and thus mimic a "time- or space-varying constant". The last word on any physical theory has to be spoken by experiments (observations). Apart from the possible determination of low-energy constants and their dependence on space and time, what could be the main tests for Quantum Gravity?

- Black-hole evaporation: A key test would be the final evaporation phase of a black hole. To this end, it would be useful to observe signatures of primordial black holes. These objects are forming not at the end of stellar collapse, but they can result from strong density perturbations in the early Universe. In the context of inflation, their initial mass can be as small as 1 g. Primordial black holes with initial masses of about 5×10^{14} g would evaporate at the present age of the Universe. Unfortunately, no such object has yet been observed. Especially promising may be models of inflationary cosmology acting at different scales [44].
- Cosmology: Quantum aspects of gravitational field may be observed in the anisotropy spectrum of the cosmic microwave background. First, future experiments may be able to observe the contribution of the gravitons generated in the early Universe. This important effect was already emphasized in [45]. The production of gravitons by the cosmological evolution would be an effect of Quantum Gravity. Second, quantum-gravitational correction terms from the Wheeler-DeWitt equation or its generalization in loop quantum cosmology may leave their impact on the anisotropy spectrum. Third, a discreteness in the inflationary perturbations could manifest itself in the spectrum [43].
- Discreteness of space and time: Both in String Theory and Quantum Gravity there are hints of a discrete structure of space-time. This quantum foam could be seen through the observation of effects violating local Lorentz invariance [46], for example, in the dispersion relation of the electromagnetic waves coming from gamma-ray bursts. It has even been suggested that space-time fluctuations could be seen in atomic interferometry [47]. However, there exist severe observational constraints [48].
- Signatures of higher dimensions: An important feature of String Theory is the existence of additional space-time dimensions. They could manifest themselves in scattering experiments at the Large Hadron Collider (LHC) at CERN. It is also imaginable that they cause observable deviations from the standard cosmological scenario [49].

Some of these features are discussed in detail in [50]. Of course, there may be other possibilities which are not yet known and which could offer great surprises. It is, for example, imaginable that a fundamental theory of Quantum Gravity is intrinsically non-linear [51, 52]. This is in contrast to most currently studied theories of Quantum Gravity, which are linear. Quantum Gravity has been studied since the end of the 1920s. No doubt, much progress has been made since then. The final goal has not yet been reached. The belief expressed here is that a consistent and experimentally successful theory of Quantum Gravity will be available in the future. However, it may still take a while before this time is reached. In any case, ETGs could constitute a serious approach in this direction.

Dark Energy and Dark Matter

The revision of standard early cosmological scenarios leading to inflation implies that a new approach is necessary also at later epochs: ETGs could play a fundamental role also in this context. In fact, the increasing bulk of data accumulated over the past few years has paved the way for a new cosmological model usually referred to as the *Concordance Model* or Λ Cold Dark Matter (ΛCDM) model.

The Hubble diagram of type Ia supernovae (hereafter SNeIa) measured by both the Supernova Cosmology Project [53, 54] and the High-z Team [55, 56] up to redshift $z \sim 1$, was the first piece of evidence that the Universe is currently undergoing a phase of accelerated expansion. Besides, balloon-born experiments such as *BOOMERANG* [57] and *MAXIMA* [58] determined the location of the first two Doppler peaks in the spectrum of Cosmic Microwave Background (CMB) anisotropies, strongly suggesting a Universe with flat spatial sections. When combined with the constraints on the matter density parameter Ω_M, these data indicate that the Universe is dominated by an unclustered fluid with negative pressure commonly referred to as *dark energy*, which drives the accelerated expansion. This picture has been further strengthened by the recent precise measurements of the CMB spectrum by the *WMAP* satellite experiment [59–61], and by the extension of the SNeIa Hubble diagram to redshifts larger than one [62].

An overwhelming number of papers appeared following these observational evidences, which present a large variety of models attempting to explain the cosmic acceleration. The simplest explanation would be the well known cosmological constant Λ [63]. Although the latter provides the best-fit to most of the available astrophysical data [59], the ΛCDM model fails egregiously in explaining why the inferred value of Λ is so tiny (120 orders of magnitude lower) in comparison with the typical value of the vacuum energy density predicted by particle physics, and why its present value is comparable to the matter density, this is the so-called *coincidence problem*.

As a tentative solution, many authors have replaced the cosmological constant with a scalar field ϕ rolling slowly down a flat section of a potential $V(\phi)$ and giving rise to the models known as *quintessence* [64, 65]. Albeit successful in fitting the data with many models, the quintessence approach to dark energy is still plagued by the coincidence problem since the dark energy and dark matter densities evolve differently and reach comparable values only during a very short time of the history of the Universe, coinciding in order of magnitude right at the present era. In other words, the quintessence dark energy is tracking matter and evolves in the same way for a long time; at late times, somehow it changes its behaviour and no longer tracks the dark matter but begins to dominate in the fashion of a (dynamical) cosmological constant. This is the coincidence problem of quintessence.

Furthermore, the origin of this quintessence scalar field is mysterious, although various (usually rather exotic) models have been proposed, leaving a great deal of uncertainty on the choice of the scalar field potential $V(\phi)$ necessary to achieve the late-time acceleration of the Universe. The subtle and elusive nature of dark energy has led many authors to look for a completely different explanation of the acceleration behaviour of the cosmos without introducing exotic components. To this end, it is worth stressing that the present-day cosmic acceleration only requires a negative pressure component that comes to dominate the dynamics late in the matter era, but does not tell us anything about the nature and the number of the cosmic fluids advocated to fill the Universe. This consideration suggests that it could be possible to explain the accelerated expansion with a single cosmic fluid characterized by an equation of state causing it to act like dark matter at high densities, while behaving as dark energy at low densities. An attractive feature of these models, usually referred to as *Unified Dark Energy* (UDE) or *Unified Dark Matter* (UDM) models, is that the coincidence problem is solved naturally, at least at the phenomenological level, with such an approach [66, 67]. Examples are the generalized Chaplygin gas [68], the tachyon field [69], and condensate cosmology [70]. A different class of UDE models with a single fluid has been proposed [71, 72]: its energy density scales with the redshift z in such a way that a radiation-dominated era, followed by a matter era and then by an accelerating phase can be naturally achieved. These models are extremely versatile since they can be interpreted both in the framework of UDE or as two-fluid scenarios containing dark matter and scalar field dark energy. A characteristic feature of this approach is that a generalized equation of state can always be obtained and the fit to the observational data can be attempted. However, such models explain the phenomenology but cannot be addressed to some fundamental physics.

There is another, different, way to approach the problem of the cosmic acceleration. As stressed in [73], it is possible that the observed acceleration is not the manifestation of yet another ingredient of the cosmic pie, but rather the first signal of a breakdown, in the infrared limit, of the laws of gravitation as we understand them. From this point of view, it is tempting to modify the Einstein-Friedmann equations to see whether it is still possible to fit the astrophysical data with models containing only standard matter without exotic fluids. Examples are the Cardassian expansion [74] and Dvali-Gabadadze-Porrati (DGP) gravity [75]. Within the same conceptual framework, it is possible to find alternative schemes in which a quintessential behaviour is obtained by incorporating effective models coming from fundamental physics and giving rise to generalized or higher order gravity actions [76] (see Refs. [77–80]. For instance, a cosmological constant may be recovered as a consequence of a non-vanishing torsion field, leading to a model consistent with both the SNeIa Hubble diagram and observations of the Sunyaev–Zel'dovich effect in galaxy clusters [81]. SNeIa data could also be efficiently fitted by including in the gravitational sector higher order curvature invariants [77, 82–84]. These alternative models provide naturally a cosmological component with negative pressure originating in the geometry of the Universe, thus overcoming the problematic nature of quintessence scalar fields. Cosmological models coming from ETGs are on the track of this philosophy.

The variety of cosmological models which are phenomenologically viable candidates to explain the observed accelerated expansion is clear from this short review. This overabundance signals that only a limited number of cosmological tests are available to discriminate between competing theories, and it is clear that there is a high degeneracy of models. Let us remark that both the SNeIa Hubble diagram and the angular size-redshift relation of compact radio sources [85] are distance-based probes of the cosmological model and, therefore, systematic errors and biases could be iterated. With this point in mind, it is interesting to search for tests based on time-dependent observables. For example, one can take into account the *lookback time* to distant objects since this quantity can discriminate between different cosmological models. The lookback time is observationally estimated as the difference between the present-day age of the Universe and the age of a given object at redshift z. This estimate is possible if the object is a galaxy observed in more than one photometric band since its colour is determined by its age as a consequence of stellar evolution. Hence, it is possible to obtain an estimate of the galaxy's age by measuring its magnitude in different bands and then using stellar evolutionary codes to choose the model that best reproduces the observed colours [86, 87].

Coming to the weak-field limit, which essentially means considering Solar System scales, any alternative relativistic theory of gravity is expected to reproduce GR which, in any case, is firmly tested only in this limit and at these scales [20]. Even this limit is a matter of debate since several relativistic theories do not reproduce exactly the Einsteinian results in their Newtonian limit but, in some sense, generalize them. As was first noticed by Stelle [88], R^2-gravity gives rise to Yukawa-like corrections to the Newtonian potential with potentially interesting physical consequences. For example, it is claimed by some authors that the flat rotation curves of galaxies can be explained by such terms [89]. Others [90] have shown that a conformal theory of gravity is nothing else but a fourth order theory containing such terms in the Newtonian limit. Reports of an apparent anomalous long-range acceleration in the data of the Pioneer 10/11, Galileo, and Ulysses spacecrafts could be accommodated in a general theoretical scheme incorporating Yukawa corrections to the Newtonian potential [91, 92].

In general, any relativistic theory of gravitation yields corrections to the weak-field gravitational potentials (e.g., [93]) which, at the post-Newtonian level and in the Parametrized Post-Newtonian (PPN) formalism, could constitute a test of these theories [20]. Furthermore, the newborn *gravitational lensing astronomy* [94] is providing additional tests of gravity over small, large, and very large scales which will soon provide direct measurements of the variation of the Newton coupling [95], the potential of galaxies, clusters of galaxies, and several other features of self-gravitating systems. Very likely, such data will be capable of confirming or ruling out as GR or ETGs.

This short overview shows that several shortcomings point out that GR cannot to be the final theory of gravity notwithstanding its successes in addressing a large amount of theoretical and experimental issues. ETGs could be a viable approach to solve some of these problems at IR and UV scales without pretending to be the comprehensive and self-consistent fundamental theory of gravity but in the track outlined by GR and then in the range of gauge theories. This review paper is mainly devoted to the theoretical foundation of ETGs trying to insert them in the framework of gauge theories and showing that they are nothing else but a straightforward extension of GR. The cosmological phenomenology and the genuinely astrophysical aspects of ETGs are not faced here. We refer the readers to the excellent reviews and books quoted in the bibliography [65, 77–79, 96–103].

The Physical Meaning of General Relativity

GR, as we discussed above, is a theory of gravitation that was developed by Albert Einstein between 1907 and 1915. According to GR, the observed gravitational attraction between masses results from their warping of space and time. Up to the beginning of the 20th century, Newton's law of universal gravitation had been accepted for more than two hundred years as a valid description of the gravitational force between masses. In Newton's model, gravity is the result of an attractive force between massive objects. Although even Newton was bothered by the unknown nature of that force, the basic framework was extremely successful for describing motions. Experiments and observations show that Einstein's description of gravitation accounts for several effects that are unexplained by Newton's law, such as anomalies in the orbits of Mercury and other planets. GR also predicts novel effects of gravity, such as gravitational waves, gravitational lensing and an effect of gravity on time known as gravitational time dilation. Many of these predictions have been confirmed by experiments, while others are the subject of ongoing research. For example, the direct evidence of gravitational waves has been an amazing confirmation of GR, as we will going to discuss in detail below. Furthermore GR developed as an essential tool in modern astrophysics. It provides the foundation for the current understanding of black holes, regions of space where gravitational attraction is so strong that light can not escape. Their strong gravity is thought to be responsible for the intense radiation emitted by certain types of astronomical objects (such as active galactic nuclei or quasars). GR is also part of the framework of the standard Big Bang model of cosmology. Although GR is not the only relativistic theory of gravity, it is the simplest theory that is consistent with the experimental data. Nevertheless, a number of open questions remain, the most fundamental of which is how GR can be reconciled with the laws of quantum physics to produce a complete and self-consistent theory of quantum gravity. In this section we discuss the physical foundation of GR outlining some of the issues that must be recovered in any relativistic theory of gravity, in particular the ETGs.

The Equivalence Principle

Equivalence principle is the physical foundation of any metric theory of gravity [20, 108] and the starting point of our considerations.

The first formulation of Equivalence principle comes out from the theory of gravitation formulates by Galileo and Newton; it is called the Weak Equivalence Principle and it states that the "inertial mass" m and the "gravitational mass" M of any object are equivalent. In Newtonian physics, the "inertial mass" m is a coefficient which appears in the second Newton law: $\vec{F} = m\,\vec{a}$ where $\vec{F}$ is the force exerted on a mass m with acceleration $\vec{a}$; in Special Relativity (without gravitation) the "inertial mass" of a body appears to be proportional to the rest energy of the body: $E = m\,c^2$. Considering the Newtonian gravitational attraction, one introduces the "gravitational mass" M: the gravitational attraction force between two bodies of "gravitational mass" M and M' is $F = G_N MM'/r^2$ where G_N is the Newtonian gravitational constant and r the distance between the two bodies. Various experiments [109] demonstrate that $m \equiv M$. The present accuracy of this relation in laboratory is of the order of 10^{-13}; spatial projects are currently designed to achieve precision of 10^{-15} [110] and 10^{-18} [111].

The Weak Equivalence Principle statement implies that it is impossible to distinguish between the effects of a gravitational field from those experienced in uniformly accelerated frames, using the simple observation of the free-falling particles behavior. The Weak Equivalence Principle can be formulated again in the following statement [20]:

> If an uncharged test body is placed at an initial event in space-time and given an initial velocity there, then its subsequent trajectory will be independent of its internal structure and composition.

A generalization of Weak Equivalence Principle claims that the Special Relativity is only locally valid. It has been achieved by Einstein after the formulation of Special Relativity theory where the concept of mass looses some of its uniqueness: the mass is reduced to a manifestation of energy and momentum. According to Einstein, it is impossible to distinguish between uniform acceleration and an external gravitational field, not only for free-falling particles but whatever is the experiment. This equivalence principle has been called Einstein Equivalence Principle ; its main statements are the following [20]:

- Weak Equivalence Principle is valid;
- the outcome of any local non-gravitational test experiment is independent of velocity of free-falling apparatus;

- the outcome of any local non-gravitational test experiment is independent of where and when in the Universe it is performed.

One defines as "local non-gravitational experiment" an experiment performed in a small-size[4] freely falling laboratory. From the Einstein Equivalence Principle, one gets that the gravitational interaction depends on the curvature of space-time, i.e. the postulates of any metric theory of gravity have to be satisfied [20]:

- space-time is endowed with a metric $g_{\mu\nu}$;
- the world lines of test bodies are geodesics of the metric;
- in local freely falling frames, called local Lorentz frames, the non-gravitational laws of physics are those of Special Relativity.

One of the predictions of this principle is the gravitational red-shift, experimentally verified by Pound and Rebka in 1960 [112]. It is worth stressing that gravitational interactions are specifically excluded from Weak Equivalence Principle and Einstein Equivalence Principle. In order to classify alternative theories of gravity, the Gravitational Weak Equivalence Principle and the Strong Equivalence Principle has to be introduced. The Strong Equivalence Principle extends the Einstein Equivalence Principle by including all the laws of physics in its terms [20]:

- Weak Equivalence Principle is valid for self-gravitating bodies as well as for test bodies (Gravitational Weak Equivalence Principle);
- the outcome of any local test experiment is independent of the velocity of the free-falling apparatus;
- the outcome of any local test experiment is independent of where and when in the Universe it is performed.

Therefore, the Strong Equivalence Principle contains the Einstein Equivalence Principle, when gravitational forces are ignored. Many authors claim that the only theory coherent with the Strong Equivalence Principle is GR. An extremely important issue related to the consistency of Equivalence Principle with respect to the Quantum Mechanics. Some phenomena, like neutrino oscillations could violate it if induced by the gravitational field. In Part III of this Report, we discuss possible neutrino oscillation effects induced by corrections to GR.

The Geodesic Structure

In GR the space-time metric is related to geodesic motion because the Equivalence Principle requires that the motion of a point-like body in free fall be described by the geodesic equation. The latter can be derived from the variational principle

$$\delta S = \delta \int_A^B ds = 0\,, \tag{2}$$

[4] In order to avoid the inhomogeneities.

where ds is the line element and A and B are the initial and final points along the space-time trajectory, respectively. It is instructive to deduce the geodesic equation in order to show how it is related to the metric structure (i.e. the casual structure) of GR.

The line element is written as

$$ds = \left|g_{\alpha\beta}dx^\alpha dx^\beta\right|^{1/2} = \left|g_{\alpha\beta}\frac{dx^\alpha}{ds}\frac{dx^\beta}{ds}\right|^{1/2} ds\,, \tag{3}$$

with s playing the role of an affine parameter, and from which it follows that

$$g_{\alpha\beta}u^\alpha u^\beta = -1\,, \tag{4}$$

where $u^\alpha = \dfrac{dx^\alpha}{ds}$ is the four-velocity of the particle. Substitution into Eq. (2) yields

$$\delta S = \delta\int_A^B \left|g_{\alpha\beta}\frac{dx^\alpha}{ds}\frac{dx^\beta}{ds}\right|^{1/2} ds = 0\,. \tag{5}$$

By performing this variation, one obtains

$$\delta S = \int_A^B \frac{1}{2\sqrt{\left|g_{\alpha\beta}u^\alpha u^\beta\right|}}\left[g_{\alpha\beta,\lambda}\delta x^\lambda\frac{dx^\alpha}{ds}\frac{dx^\beta}{ds} + 2g_{\alpha\beta}\frac{d}{ds}\left(\delta x^\alpha\right)\frac{dx^\beta}{ds}\right] ds = 0\,. \tag{6}$$

The second term in square brackets is $g_{\alpha\beta}\delta\left(\frac{dx^\alpha}{ds}\frac{dx^\beta}{ds}\right)$ as a consequence of the fact that $\delta\left(ds\right) = d\left(\delta s\right)$, hence

$$g_{\alpha\beta}\delta\left(\frac{dx^\alpha}{ds}\frac{dx^\beta}{ds}\right) = g_{\alpha\beta}\frac{dx^\alpha}{ds}\delta\left(\frac{dx^\beta}{ds}\right) + g_{\alpha\beta}\frac{dx^\beta}{ds}\delta\left(\frac{dx^\alpha}{ds}\right) = 2g_{\alpha\beta}\frac{dx^\beta}{ds}\frac{d}{ds}\left(\delta x^\alpha\right)\,. \tag{7}$$

Using $g_{\alpha\beta}u^\alpha u^\beta = -1$, it is

$$\delta S = \int_A^B \frac{1}{2}\left[g_{\alpha\beta,\lambda}\delta x^\lambda\frac{dx^\alpha}{ds}\frac{dx^\beta}{ds} + 2g_{\alpha\beta}\frac{dx^\beta}{ds}\frac{d}{ds}\left(\delta x^\alpha\right)\right] ds = 0 \tag{8}$$

and integration by parts of the second term yields

$$\delta S = \int_A^B \frac{1}{2}\left(g_{\alpha\beta,\lambda}\delta x^\lambda\frac{dx^\alpha}{ds}\frac{dx^\beta}{ds}\right) ds + \left[g_{\alpha\beta}\frac{dx^\beta}{ds}\delta x^\alpha\right]_A^B - \int_A^B \frac{d}{ds}\left(g_{\alpha\beta}\frac{dx^\beta}{ds}\right)\delta x^\alpha ds = 0\,. \tag{9}$$

By imposing that, at the endpoints, it is $\delta x^\alpha\left(A\right) = \delta x^\alpha\left(B\right) = 0$, the second term vanishes and

$$\delta S = \int_A^B \frac{1}{2}\left(g_{\alpha\beta,\lambda}\frac{dx^\alpha}{ds}\frac{dx^\beta}{ds}\delta x^\lambda\right)ds - \int_A^B \left(g_{\alpha\beta}\frac{d^2x^\beta}{ds^2} + g_{\alpha\beta,\lambda}\frac{dx^\lambda}{ds}\frac{dx^\beta}{ds}\right)\delta x^\alpha ds = 0\,. \quad (10)$$

This equation can be written as

$$\delta S = \int_A^B \left[\left(\frac{1}{2}g_{\alpha\beta,\lambda} - g_{\lambda\beta,\alpha}\right)\frac{dx^\alpha}{ds}\frac{dx^\beta}{ds} - g_{\lambda\beta}\frac{d^2x^\beta}{ds^2}\right]\delta x^\lambda ds = 0\,. \quad (11)$$

This integral vanishes for all variations δx^λ with fixed endpoints if

$$g_{\lambda\beta}\frac{d^2x^\beta}{ds^2} = \left(\frac{1}{2}g_{\alpha\beta,\lambda} - g_{\lambda\beta,\alpha}\right)u^\alpha u^\beta\,. \quad (12)$$

Since

$$g_{\lambda\beta,\alpha}u^\alpha u^\beta = g_{\lambda\alpha,\beta}u^\beta u^\alpha = \frac{1}{2}\left(g_{\lambda\beta,\alpha} + g_{\lambda\alpha,\beta}\right)u^\alpha u^\beta\,, \quad (13)$$

whereas

$$g_{\lambda\beta}\,\frac{d^2x^\beta}{ds^2} = \frac{1}{2}\left(g_{\alpha\beta,\lambda} - g_{\lambda\beta,\alpha} - g_{\lambda\alpha,\beta}\right)u^\alpha u^\beta \quad (14)$$

we have

$$\{\lambda, \alpha\beta\} = \frac{1}{2}\left(g_{\lambda\alpha,\beta} + g_{\lambda\beta,\alpha} - g_{\alpha\beta,\lambda}\right) \quad (15)$$

and

$$g_{\lambda\beta}\frac{d^2x^\beta}{ds^2} + \{\lambda, \alpha\beta\}\,u^\alpha u^\beta = 0\,. \quad (16)$$

Multiplying by $g^{\lambda\tau}$ and remembering that

$$g^{\lambda\tau}g_{\lambda\beta} = \delta^\tau_\beta\,, \qquad g^{\lambda\tau}\,\{\lambda, \alpha\beta\} = \Gamma^\tau_{\alpha\beta}\,, \quad (17)$$

one has

$$\frac{d^2x^\tau}{ds^2} + \Gamma^\tau_{\alpha\beta}u^\alpha u^\beta = 0\,, \quad (18)$$

which is the geodesic equation describing the free fall motion of a point-like body in the gravitational field represented by the Christoffel symbols $\Gamma^\tau_{\alpha\beta}$.

This result means that the connection is Levi-Civita and then g and Γ are intimately related in GR. This feature, as stressed by Palatini [27] and Einstein [113], shows that geodesic structure (Γ) and casual structure (g) are not independent. This is not true, in general, in ETGs as we will discuss in the next part of this Report.

The Einstein Field Equations in Metric and Palatini Formalisms

The Einstein field equations are a particular case of those deduced in previous section for the Einstein-Cartan theory. Here we report their derivation in details in view of the variational principles that we will adopt for ETGs. In other words, the field equations in the cases of non-minimal couplings and non-linear functions of curvature invariants. We illustrate the derivation of the Einstein field equations in vacuo as the starting point.

Let us consider

$$\delta \int d\Omega \sqrt{-g}\, \mathcal{L} = 0, \tag{19}$$

where $\sqrt{-g}\, d\Omega$ is the invariant volume element and $\mathcal{L}$ is the Lagrangian density. In fact, under the coordinate transformation $\overline{x}^{\alpha} \to x^{\alpha} = x^{\alpha}\left(\overline{x}^{\mu}\right)$, where $\overline{x}^{\mu}$ are the "initial" local coordinates, we have

$$d\Omega = J\, d\overline{\Omega}\,, \qquad J = \det\left(\frac{\partial x^{\alpha}}{\partial \overline{x}^{\mu}}\right)\,, \tag{20}$$

with J the Jacobian determinant of the transformation. Moreover, we have

$$\overline{g}_{\alpha\beta} = \text{diag}\ (-1, 1, 1, 1)\ , \tag{21}$$

$$\overline{g}_{\alpha\beta} = \frac{\partial x^{\mu}}{\partial \overline{x}^{\alpha}} \frac{\partial x^{\nu}}{\partial \overline{x}^{\beta}}\, g_{\mu\nu}\,, \tag{22}$$

$\overline{g} = -1 = J^2 g$ and, therefore,

$$d\overline{\Omega} = \frac{d\Omega}{J} = \sqrt{-g}\, d\Omega\,. \tag{23}$$

Since we want the Euler-Lagrange equations deriving from the variational principle to be of second order, the Lagrangian must be quadratic in the first order derivatives of $g_{\mu\nu}$. These first order derivatives contain the Christoffel symbols, which are not coordinate-invariant. Then we have to choose for the Lagrangian density $\mathcal{L}$ expressions containing higher order derivatives and, a priori, this brings the danger that the field equations could become of order higher than second (we will discuss in detail this point for ETGs). The obvious choice of Hilbert and Einstein for the Lagrangian density $\mathcal{L}$ was the Ricci scalar curvature R. The variational principle is then

$$\delta \int d\Omega\, \sqrt{-g}\, R = 0\,. \tag{24}$$

The relations

$$\delta g = g\, g^{\mu\nu}\delta g_{\mu\nu} = -g\, g_{\mu\nu}\delta g^{\mu\nu} \tag{25}$$

yield

$$\delta\left(\sqrt{-g}\right) = -\frac{\delta g}{2\sqrt{-g}} = -\frac{1}{2}\sqrt{-g}\, g_{\alpha\beta}\delta g^{\alpha\beta}\,, \tag{26}$$

from which it follows that

$$\begin{aligned}
&\int \left[(\delta\sqrt{-g})\, R + \sqrt{-g}\, R_{\mu\nu}\delta g^{\mu\nu} + \sqrt{-g}\, g^{\mu\nu}\delta R_{\mu\nu}\right] d\Omega = \\
&= \int \sqrt{-g}\,\delta g^{\mu\nu}\left[R_{\mu\nu} - \tfrac{1}{2}Rg_{\mu\nu}\right]d\Omega + \int \sqrt{-g}\, g^{\mu\nu}\delta R_{\mu\nu} d\Omega = 0\,.
\end{aligned} \tag{27}$$

The second integral can be evaluated in the local inertial frame, obtaining

$$R_{\mu\nu}(0) = G_{\mu\nu^{\alpha},\alpha} - G_{\mu\alpha^{\alpha},\nu}\,, \tag{28}$$

$$\delta R_{\mu\nu}(0) = \frac{\partial}{\partial x^{\alpha}}\left(\delta G^{\alpha}_{\mu\nu}\right) - \frac{\partial}{\partial x^{\nu}}\left(\delta G^{\alpha}_{\mu\alpha}\right)\,, \tag{29}$$

$$\begin{aligned}
g^{\mu\nu}(0)\delta R_{\mu\nu}(0) &= g^{\mu\nu}(0)\frac{\partial}{\partial x^{\alpha}}\left(\delta G^{\alpha}_{\mu\nu}\right) - g^{\mu\nu}(0)\frac{\partial}{\partial x^{\nu}}\left(\delta G^{\alpha}_{\mu\alpha}\right) \\
&= g^{\mu\nu}(0)\frac{\partial}{\partial x^{\rho}}\left(\delta G^{\rho}_{\mu\nu}\right) - g^{\mu\rho}(0)\frac{\partial}{\partial x^{\rho}}\left(\delta G^{\alpha}_{\mu\alpha}\right) \\
&= \frac{\partial}{\partial x^{\rho}}\left[g^{\mu\nu}(0)\delta G^{\rho}_{\mu\nu} - g^{\mu\rho}(0)\delta G^{\alpha}_{\mu\alpha}\right]\,.
\end{aligned} \tag{30}$$

Then, we can write

$$g^{\mu\nu}(0)\delta R_{\mu\nu}(0) = \frac{\partial W^{\rho}}{\partial x^{\rho}}\,, \qquad W^{\rho} = g^{\mu\nu}(0)\delta G^{\rho}_{\mu\nu} - g^{\mu\rho}(0)\delta G^{\alpha}_{\mu\alpha}\,. \tag{31}$$

The second integral in Eq. (27) can be discarded since its argument is a pure divergence; in fact, in general coordinates it is

$$\int \sqrt{-g}\, g^{\mu\nu}\delta R_{\mu\nu} d\Omega = \int \sqrt{-g}\,\frac{\partial W^{\rho}}{\partial x^{\rho}}\, d\Omega = \int \sqrt{-g}\; W^{\rho}{}_{;\rho} d\Omega = \int \frac{\partial}{\partial x^{\rho}}\left(\sqrt{-g}\, W^{\rho}\right) d\Omega = 0\,, \tag{32}$$

and then

$$\int \sqrt{-g}\,\delta g^{\mu\nu}\left(R_{\mu\nu} - \frac{1}{2}Rg_{\mu\nu}\right) d\Omega = 0\,, \tag{33}$$

from which we obtain the vacuum field equations of GR

$$G_{\mu\nu} \equiv R_{\mu\nu} - \frac{1}{2} g_{\mu\nu} R = 0\,, \tag{34}$$

as Euler-Lagrange equations of the Hilbert-Einstein action. Vice-versa, starting from Eq. (34) and retracing the previous steps in inverse order (i.e., integrating the Einstein equations), one can re-obtain the Hilbert-Einstein action (24), thus demonstrating the equivalence between this action and the field equations (34). Introducing matter fields as sources is straightforward, once we introduce suitable forms of stress-energy tensor. A suitable choice is the stress-energy tensor of a perfect fluid, that is

$$T^{(m)}_{\mu\nu} = (P + \rho)\, u_\mu u_\nu + P\, g_{\mu\nu}\,, \tag{35}$$

where u^μ is the four-velocity of the fluid particles and P and ρ are the pressure and energy density of the fluid, respectively, the continuity equation requires $T^{(m)}_{\mu\nu}$ to be covariantly constant, i.e., to satisfy the conservation law

$$\nabla^\mu T^{(m)}_{\mu\nu} = 0\,, \tag{36}$$

where ∇_α denotes the covariant derivative operator of the metric $g_{\mu\nu}$. In fact, $\nabla^\mu R_{\mu\nu}$ does not vanish, except in the special case $R \equiv 0$. Einstein and Hilbert independently concluded that the field equations had to be

$$G_{\mu\nu} = \kappa\, T^{(m)}_{\mu\nu}\,, \tag{37}$$

where $G_{\mu\nu}$ is now called the *Einstein tensor* of $g_{\mu\nu}$. These equations satisfy the conservation law (36) since the relation

$$\nabla^\mu G_{\mu\nu} = 0\,, \tag{38}$$

holds as a contraction of the Bianchi identities that the curvature tensor of $g_{\mu\nu}$ has to satisfy [114].

The Lagrangian that, when varied, produces the field equations (37) is the sum of a "matter" Lagrangian density $\mathcal{L}^{(m)}$, the variational derivative of which is

$$T^{(m)}_{\mu\nu} = -\frac{2}{\sqrt{-g}} \frac{\delta\left(\sqrt{-g}\,\mathcal{L}_{(m)}\right)}{\delta g^{\mu\nu}}\,, \tag{39}$$

and of the above gravitational (*Hilbert-Einstein*) *Lagrangian*

$$L_{HE} \equiv \sqrt{-g}\,\mathcal{L}_{HE} = \sqrt{-g}\,R\,, \tag{40}$$

where g is the determinant of the metric $g_{\mu\nu}$. Solving the Einstein field equations means to determine the form of the metric $g_{\mu\nu}$ starting from the distribution of matter-energy in a given region of space. In other words, this achievement fixes the (causal) metric structure of the space-time. This is the derivation of the Einstein field equations in the so called metric formalism.

However it is possible to obtain the field equations without resorting to a local inertial frame [115]. Following the Weyl attempt to unify Gravity with Electromagnetism [116], as discussed in section "The Physical Meaning of General Relativity", an affine connection defined over the space-time, can be assumed as a dynamical field non-trivially depending on a metric. As we have seen, Weyl's idea failed because of a wrong choice of the Lagrangian and few more issues, but it generated however a keypoint: connections may have a physically interesting dynamics.

Einstein soon showed a great interest in Weyl's idea. He too began to play with connections, in order to seek for a "geometrically" Unified Theory. But he never arrived to "dethronize" g in the description of the gravitational field. He was not so happy with the fact that the gravitational field is not the fundamental object, but just a by-product of the metric; however, he never really changed his mind about the physical and mathematical role of g.

In 1925 Einstein constructed a theory depending on a metric g and a symmetric affine connection Γ to be varied independently (the so-called "Palatini method", because of a misunderstanding with W. Pauli, see [117, 118]); he defined a Lagrangian theory in which the Palatini-Einstein gravitational Lagrangian is

$$\sqrt{-g}\mathcal{L}_{\mathrm{PE}}(g, \Gamma, \partial\Gamma) :\equiv \overset{\Gamma}{R}\,\sqrt{-g}\,. \tag{41}$$

The Lagrangian density $\mathcal{L}_{\mathrm{PE}}$ contains (first order) derivatives of Γ but no derivatives of g. In other words, it is of order zero in the metric while it is first order in the connection. By a well-known feature of the Lagrangian formalism, this implies that a priori g has no independent dynamics, being a sort of "Lagrange multiplier"; if we would assume that Γ is frozen from the beginning, g would be frozen too. On the other hand since g has dynamics, this happens because g is related to Γ, which is the true dynamical variable of this Lagrangian theory. This time is the metric g to gain a dynamical meaning from Γ, that plays the role of a fundamental field.

In 4 dimension, there are $10+40$ independent variables and the field equations, in vacuum, are:

$$\begin{cases} \overset{\Gamma}{R}_{(\mu\nu)} - \frac{1}{2}\overset{\Gamma}{R}\, g_{\mu\nu} = 0 \\ \\ \overset{\Gamma}{\nabla}_{\alpha}\,(\sqrt{g}\, g^{\mu\nu}) = 0 \end{cases} \tag{42}$$

where $\overset{\Gamma}{R}_{(\mu\nu)}$ is the symmetric part of $\overset{\Gamma}{R}_{\mu\nu}$ and $\nabla\Gamma$ denotes the covariant derivative with respect to Γ. The second field equation (42) constrains the connection Γ, which is a priori arbitrary, to coincide a posteriori with the Levi-Civita connection of the metric g (Levi-Civita Theorem). By substituting this information into the first field equation

(42), the vacuum Einstein equation for g is obtained. In the "Palatini formalism", the metric g determines rods, clocks and protractors, while the connection Γ determines the free-fall, but since *a posteriori* the same result of GR is found, Einstein soon ceased to show a real interest in this formalism.

The situation does not change if matter is present through a matter Lagrangian $\mathcal{L}_{(m)}$ (independent of Γ but just depending on g and other external matter fields), that generates the above energy-momentum tensor $T^{(m)}_{\mu\nu}$. If the total Lagrangian is then assumed to be $\mathcal{L}_{\text{tot}} :\equiv \mathcal{L}_{\text{PE}} + \mathcal{L}_{(m)}$, field equations (42) are replaced by

$$\overset{\Gamma}{R}_{(\mu\nu)} - \frac{1}{2}\overset{\Gamma}{R}\, g_{\mu\nu} = \kappa\, T_{\mu\nu} \tag{43}$$

and again (42) implies, a posteriori, that (43) reduces to the Einstein equations.

Let us remark that the dynamical coincidence between Γ and the Levi-Civita connection of g is entirely due to the particular Lagrangian considered by Einstein, which is the *simplest*, but not the only possible one. Furthermore, the Palatini method privileges the affine structure with respect to the metric structure. Notice that, in this case (i.e. in "Palatini formalism"), the relations

$$\Gamma^{\lambda}_{\mu\nu} = \left\{ {}^{\lambda}_{\mu\nu} \right\}_g \tag{44}$$

are field equations: the fact that Γ is the Levi-Civita connection of g is no longer an assumption a priori but it is the outcome of the field equations [115]. As soon as the Lagrangian is not the Palatini-Einstein but a more general one, the affine connection does not coincide, in general, with the Levi-Civita one. This fact gives rise to extremely interesting results. Below we will discuss this point considering the curvature quantities derived in the Palatini formalism as different from those derived in the metric approach.

The Newtonian Limit

GR is not the only theory of gravitation and, several alternative theories of gravity have been investigated from the 60's, considering the space-time to be "special relativistic" at a background level and treating gravitation as a Lorentz-invariant field on the background.

Two different classes of experiments have been studied: the first ones testing the foundations of gravitation theory—among them the Equivalence Principle—the second one testing the metric theories of gravity where space-time is endowed with a metric tensor and where the Einstein Equivalence Principle is valid. However as discussed in the Introduction, for several fundamental reasons extra fields might be necessary to describe the gravitation, e.g. scalar fields or higher-order corrections in curvature invariants.

Two sets of equations can be distinguished [109]. The first ones couple the gravitational fields to the non-gravitational contents of the Universe, i.e. the matter distribution, the electromagnetic fields, etc. The second set of equations gives the evolution of non-gravitational fields. Within the framework of metric theories, these laws depend only on the metric: this is a consequence of the EEP and the so-called "minimal coupling". In most theories, including GR, the second set of equations is derived from the first one, which is the only fundamental one; however, in many situations, the two sets are decoupled. The gravitational field studied in these approaches (without cosmological considerations) is mainly due to the Sun and the Eddington-Robertson expansion gives the corresponding metric. Assuming spherical symmetry and a static gravitational field, one can prove that there exists a coordinate system such as

$$ds^2 = -A(r)\,dt^2 + B(r)\,dr^2 + r^2(\sin^2\theta\, d\phi^2 + d\theta^2)\,, \tag{45}$$

dt being the proper time between two neighbouring events. The Newtonian gravitational field does not exceed $G_N M_\odot / R_\odot c^2 \sim 2\times 10^{-6}$, where c is the speed of light, $M_\odot$ is the mass of the Sun and $R_\odot$ its radius. The metric is quasi-Minkowskian, $A(r)$ and $B(r)$ are dimensionless functions which depend only on G_N, M, c and r. Indeed, the only pure number that can be built with these four quantities is $G_N M / rc^2$. The Eddington-Robertson metric is a Taylor expansion of A and B which gives

$$\begin{aligned} ds^2 \simeq & -\left(1 - 2\alpha\frac{G_N M}{r c^2} + 2\beta\left(\frac{G_N M}{rc^2}\right)^2 + \cdots\right)dt^2 \\ & + \left(1 + 2\gamma\frac{G_N M}{rc^2} + \ldots\right)dr^2 + r^2(\sin^2\theta\, d\phi^2 + d\theta^2)\,. \end{aligned} \tag{46}$$

which in isotropic coordinates is

$$\begin{aligned} ds^2 \simeq & -\left(1 - 2\alpha\frac{G_N M}{r c^2} + 2\beta\left(\frac{G_N M}{rc^2}\right)^2 + \cdots\right)dt^2 \\ & + \left(1 + 2\gamma\frac{G_N M}{rc^2} + \cdots\right)\left(dx^2 + dy^2 + dz^2\right)\,. \end{aligned} \tag{47}$$

where $r,\ \theta,\ \phi$ are related to x, y, z in the usual manner:

$$r = \sqrt{x^2+y^2+z^2}\,, \qquad \theta = tan^{-1}\left(\frac{z}{\sqrt{x^2+y^2+z^2}}\right), \qquad \phi = tan^{-1}\left(\frac{y}{x}\right) \tag{48}$$

where r is the new "isotropic" radial coordinate, not to be confused with the Schwarzschild r (see [115] for a detailed discussion of this important issue). The coefficients $\alpha,\ \beta,\ \gamma$ are called the post- Newtonian parameters and their values depend on the considered theory of gravity: in GR, one has $\alpha = \beta = \gamma = 1$. The post Newtonian parameters can be measured not only in the Solar System but also in relativistic binary neutron stars such as *PSR* $1913 + 16$.

A generalization of the previous formalism is known as *Parametrized Post-Newtonian* (PPN) formalism. Comparing metric theories of gravity among them and with experimental results becomes particularly easy if the PPN approximation is used. The following requirements are needed:

- particles are moving slowly with respect to the speed of light;
- gravitational field is weak and considered as a perturbation of the flat space-time;
- gravitational field is also static, i.e. it does not change with time.

The PPN limit of metric theories of gravity is characterized by a set of 10 real-valued parameters; each metric theory of gravity corresponds to particular values of PPN parameters. The PPN framework has been used first for the analysis of Solar System gravitational experiments, then for the definition and the analysis of new tests of gravitation and finally for the analysis and the classification of alternative metric theories of gravity [20].

By the middle 1970s, the Solar System was no more considered as the unique testing ground of gravitation theories. Many alternative theories of gravity agree with GR in the Post-Newtonian limit and thus with Solar System experiments; nevertheless, they do not agree with other predictions (such as cosmology, neutron stars, black holes and gravitational radiation) for which the post-Newtonian limit is not adequate. In addition, the possibility that experimental probes, such as gravitational radiation detectors, would be available in the future to perform extra-Solar System tests led to the abandon of the Solar System as the only arena to test gravitation theories.

The study of the binary pulsar *PSR* $1913 + 16$, discovered by Hulse and Taylor [119], showed that this system combines large post-Newtonian gravitational effects, highly relativistic gravitational fields (associated with the pulsar) with the evidence of an emission of gravitational radiation (by the binary system itself). Relativistic gravitational effects allowed one to do accurate measurements of astrophysical parameters of the system, such as the mass of the two neutron stars. The measurement of the orbital period rate of change agreed with the prediction of the gravitational waves (GW) emission in the GR framework, in contradiction with the predictions of most alternative theories, even those with PPN limits identical to GR. However, the evidence was not conclusive to rule out other theories since several shortcomings remain, up to now, unexplained and, as we see below, other forms of gravitational radiation (e.g. polarizations and massive states) with respect to those predicted by GR.

The Minkowskian Limit

Replacing the Newtonian limit by a less restrictive hypothesis leads to the weak field approximation: practically, the field is still weak, but it is allowed to change in time and there is no more restriction on the test particles motion. New phenomena are associated with this hypothesis like the emission of gravitational radiation and the deflection of light. This framework allows one to split the metric $g_{\mu\nu}$ into two parts: the flat Minkowski metric $\eta_{\mu\nu} = \text{diag}(-1, 1, 1, 1)$ plus a perturbative term $h_{\mu\nu}$, assumed to be small. This linearized version of GR describes the propagation of a symmetric tensor $h_{\mu\nu}$ on a flat background space-time. So, the metric reads

$$g_{\mu\nu} = \eta_{\mu\nu} + h_{\mu\nu} \qquad \text{with} \qquad \left|h_{\mu\nu}\right| \ll 1 \,. \tag{49}$$

As $h_{\mu\nu}$ is small, one can neglect terms higher than the first order in the perturbation $h_{\mu\nu}$; in particular, one can raise/lower indexes with $\eta_{\mu\nu}$ and $\eta^{\mu\nu}$ as the corrections are of higher order in the perturbation

$$g^{\mu\nu} = \eta^{\mu\nu} - h_{\mu\nu} \qquad \text{with} \qquad h^{\mu\nu} = \eta^{\mu\rho}\eta^{\nu\sigma}h_{\rho\sigma} \,. \tag{50}$$

The aim is to find the equations of motion to which the perturbations $h_{\mu\nu}$ obey by investigating the Einstein equations to the first order. Inserting the new metric (49) in the Einstein tensor, we obtain

$$G_{\mu\nu} = \frac{1}{2}\left(\partial_\sigma\partial_\nu h^\sigma_\mu + \partial_\sigma\partial_\mu h^\sigma_\nu - \partial_\mu\partial_\nu h - \ h_{\mu\nu} - \eta_{\mu\nu}\,\partial_\rho\partial_\sigma h^{\rho\sigma} + \eta_{\mu\nu} h\right) \,, \tag{51}$$

where $h = \eta^{\mu\nu}h_{\mu\nu} = h^\mu_\mu$ is the trace of the perturbation and $= -\frac{1}{2}\partial_{tt} + \partial_{xx} + \partial_{yy} + \partial_{zz}$ is the d'Alembertian of the flat space-time, using from now on (unless otherwise stated) geometrical units for which $c = 1$.
The stress-energy tensor is computed at the 0-order in $h_{\mu\nu}$: the energy and the momentum have to be small too, according to the weak field approximation and the lowest non-vanishing order in $T_{\mu\nu}$ is of the same order of magnitude as the perturbation. Therefore, the conservation law becomes $\partial^\mu T_{\mu\nu} = 0$.

Gravitational Waves

GW are weak ripples in the curvature of space-time, produced by the motions of matter. They propagate at the speed of light. The linearized Einstein equations allow wave solutions, in a way similar to Electromagnetism. These GW are transverse to the propagation direction and show two independent polarization states.

The new metric (49) does not fix the space-time frame completely; two possible gauges can be applied in addition to simplify the Einstein equations. Using the

Lorentz gauge $\partial_\mu h^\mu_\lambda - \frac{1}{2}\partial_\lambda h = 0$, the Einstein equations are linearized and can be written as

$$\Box\, h_{\mu\nu} - \frac{1}{2}\,\eta_{\mu\nu}\,\Box\, h \;=\; \frac{16\pi G_N}{c^4}\,T^{(m)}_{\mu\nu}\,. \tag{52}$$

where standard units have been restored. The trace-reversed perturbation is defined as

$$\bar{h}_{\mu\nu} \;=\; h_{\mu\nu} - \frac{1}{2}\,\eta_{\mu\nu}\,h\,. \tag{53}$$

One can choose a frame in which the harmonic gauge condition, $\partial_\mu \bar{h}^\mu_\nu = 0$, is verified. Then, the Einstein field equations become

$$\Box\bar{h}_{\mu\nu} \;=\; \frac{16\,\pi\,G_N}{c^4}\,T^{(m)}_{\mu\nu}\,. \tag{54}$$

and, in vacuum, one has simply

$$\Box\bar{h}_{\mu\nu} \;=\; 0\,. \tag{55}$$

These equations are similar to of the Electromagnetism field equations and one can use the same method to solve them. Indeed, looking at the Einstein equations in vacuum, Eq. (55), one can note that they are in the form of a wave equation for $\bar{h}_{\mu\nu}$, the d'Alembertian reduces to the form $\Box = -\frac{1}{2}\partial_{tt} + \nabla^2$. Therefore, in the absence of matter, one looks for plane waves solutions

$$\bar{h}_{\mu\nu} \;=\; C_{\mu\nu}e^{ik_\sigma x^\sigma}\,, \tag{56}$$

where $C_{\mu\nu}$ is a constant and symmetric tensor of rank 2 and k is a time-like vector, the wave vector. The plane wave in Eq. (56) is a solution of the linearized equations in vacuum, Eq. (55), if the wave vector k is null, i.e. satisfies the condition $k^\sigma k_\sigma = 0$ and shows that GW propagate to the speed of light.

The four conditions of the harmonic gauge $k_\mu C^{\mu\nu} = 0$ lead to six independent components for the symmetric tensor $C^{\mu\nu}$. As there are still some unused degrees of freedom, one can make another gauge choice on the tensor $C_{\mu\nu}$:

$$C^\mu_\mu = 0 \quad \text{Traceless}\,; \tag{57}$$

$$C^{0\mu} = 0 \quad \text{Transverse}\,. \tag{58}$$

One has, in this way, the so called Transverse-Traceless (TT) gauge. These five relations give four new constraints on $C^{\mu\nu}$ in addition to the harmonic gauge condition; therefore, only two independent components remain in $C^{\mu\nu}$. As the wave is traceless, one can check from Eq. (53) that $\bar{h}^{TT}_{\mu\nu} = h^{TT}_{\mu\nu}$. Therefore, the general form of the symmetric tensor $C_{\mu\nu}$ is finally

$$C_{\mu\nu} = \begin{pmatrix} 0 & 0 & 0 & 0 \\ 0 & C_{11} & C_{12} & 0 \\ 0 & C_{12} & -C_{11} & 0 \\ 0 & 0 & 0 & 0 \end{pmatrix} \tag{59}$$

Let us define $C_{11} = h_+$ and $C_{12} = h_\times$; GW appears to have two polarized states, "+" and "×", which modify the space-time curvature in different ways. In tensorial form, one can write

$$h = \left[h_+ (\vec{e}_1 \otimes \vec{e}_1 - \vec{e}_2 \otimes \vec{e}_2) + 2\, h_\times (\vec{e}_1 \otimes \vec{e}_2)\right] e^{i\omega(t - x/c)} . \tag{60}$$

Being $\vec{\xi} = (\xi^1, \xi^2, \xi^3)$ the separation between two free particles and taking into account the geodesic deviation [120] which describes the evolution of two free-falling particles, if the GW propagates in the direction x^3, only ξ^1 and ξ^2 are involved in its passage. Assuming a polarized GW, the integration of the geodesic deviation equation gives:

- Polarization "+" ($h_\times = 0$) :

$$\begin{pmatrix} \xi^1(t) \\ \xi^2(t) \end{pmatrix} = \begin{pmatrix} 1 + \frac{1}{2}\, h_+ e^{ik^\sigma x_\sigma} & 0 \\ 0 & 1 - \frac{1}{2}\, h_+ e^{ik^\sigma x_\sigma} \end{pmatrix} \begin{pmatrix} \xi^1(0) \\ \xi^2(0) \end{pmatrix} \tag{61}$$

- Polarization "×" ($h_+ = 0$) :

$$\begin{pmatrix} \xi^1(t) \\ \xi^2(t) \end{pmatrix} = \begin{pmatrix} 1 & \frac{1}{2}\, h_\times e^{ik^\sigma x_\sigma} \\ \frac{1}{2}\, h_\times e^{ik^\sigma x_\sigma} & 1 \end{pmatrix} \begin{pmatrix} \xi^1(0) \\ \xi^2(0) \end{pmatrix} \tag{62}$$

Let us consider now a test-mass ring (a massless and free-falling set of particles) interacting with GW, lying in a plane orthogonal to the direction of the wave propagation. Its oscillations depend on the GW polarization.

After having found a solution to Einstein equations in vacuum, let us solve now Eq. (54) with a non-zero source term. The solution is computed using the retarded Green function

$$\bar{h}_{\mu\nu}(t, \vec{x}) = \frac{4\, G_N}{c^4} \int_{\vec{y} \in \text{Source}} \frac{1}{|\vec{x} - \vec{y}\,|}\, T^{(m)}_{\mu\nu}(t_r, \vec{y})\, d^3\vec{y}\,, \tag{63}$$

with $|\vec{x} - \vec{y}\,| = \sqrt{\delta_{ij}(x^i - y^i)(x^j - y^j)}$ (Euclidean distance) and $t_r = t - |\vec{x} - \vec{y}\,|/c$ (retarded time). Let us consider an isolated source with a density ρ and a characteristic dimension δR, located at a distance R from the observation point $\vec{x}$. One assumes $\delta R \ll R$ so, in particular, one gets $|\vec{x} - \vec{y}\,| \approx R$ and one can move this constant term outside the integral in Eq. (63). As the stress-energy tensor verifies the conservation of energy $\partial^\mu T^{(m)}_{\mu\nu} = 0$, the harmonic gauge condition $\partial_\mu \bar{h}^\mu_\nu = 0$ is also verified. Moreover, the radiation is mostly emitted at frequencies $\omega/2\pi$, so that $\frac{\delta R}{c} \ll \frac{1}{\omega}$. Then, it is possible to demonstrate that only the spatial coordinates of tensor $\bar{h}_{\mu\nu}$ are different from zero. The quadrupole momentum tensor q_{ij} of the source energy density is defined as

$$q_{ij}(t) = \int_{\vec{y}\,\epsilon\,\text{source}} y_i\, y_j\, T^{(m)}_{00}(t,\vec{y})\, d^3\vec{y} \qquad \text{with} \qquad T^{(m)}_{00} \approx \rho c^2 \, . \tag{64}$$

The metric perturbation is given by

$$\bar{h}_{ij}(t,\mathbf{x}) = \frac{2\,G_N}{R\,c^4}\,\ddot{q}_{ij}(t_r)\, . \tag{65}$$

So, GW, generated by an isolated non-relativistic object, is proportional to the second derivative of the quadrupolar momenta of the energy density. Equation (65) shows that the metric perturbation amplitude h varies as the inverse of distance to the source R; a faster decreasing with the distance, e.g. $1/R^2$, would make vain the hope of any GW detection. Fortunately, GW detectors are sensitive to the amplitude of the signal.

The energy emitted by gravitational radiation is difficult to define. A way to overcome this difficulty is to define the stress-energy tensor by developing the metric $g_{\mu\nu}$ and the Einstein tensor $G_{\mu\nu}$ at the second order:

$$g_{\mu\nu} = \eta_{\mu\nu} + h_{\mu\nu} + h^{(2)}_{\mu\nu}\, , \tag{66}$$

$$G_{\mu\nu} = G^{(1)}_{\mu\nu}\,[\eta + h^{(2)}] + G^{(2)}_{\mu\nu}\,[\eta + h]\, . \tag{67}$$

Einstein equations in the vacuum $G_{\mu\nu} = 0$ can be written in the form

$$G^{(1)}_{\mu\nu}\,[\eta + h^{(2)}] = \frac{8\,\pi\,G_N}{c^4}\, t_{\mu\nu}\, , \tag{68}$$

with the definition

$$t_{\mu\nu} = \frac{c^4}{8\,\pi\,G_N}\, G^{(2)}_{\mu\nu}\,[\eta + h]\, . \tag{69}$$

The Bianchi identity says that $\partial_\mu t^{\mu\nu} = 0$, therefore $t_{\mu\nu}$ can be considered as the stress-energy tensor of a gravitational field, yet, it is only a pseudo-tensor [121]. One can compute the energy density t_{00} by averaging over many cycles (because the energy is not localized) the GW energy:

$$t_{00} = \frac{c^2}{16\,\pi\,G}\,\langle\, \dot{h}_+^{\,2} + \dot{h}_\times^{\,2}\,\rangle\, . \tag{70}$$

Then, the source luminosity is $L = r^2 \int_S c\, t_{00}\, d\Omega$, where the integration is performed on sphere S of radius r. Introducing the reduced quadrupolar momenta

$$Q_{ij} = q_{ij} - \frac{1}{3}\delta_{ij}\delta^{kl}q_{kl}\,, \tag{71}$$

one obtains the Einstein quadrupole formula

$$L = \frac{G}{c^5}\langle \frac{d^3Q_{ij}}{dt^3}\frac{d^3Q_{ij}}{dt^3}\rangle\,. \tag{72}$$

It is worth noticing that a GW emission requires a variation of the quadrupolar momentum, as shown already in Eq. (65). This typical feature of gravitational radiation, added to the weakness of the coupling constant between gravitation and matter, here $G_N/c^5 \approx 10^{-53}\,\mathrm{W}^{-1}$, explains why GW amplitudes are so small compared to those produced by electromagnetic radiation. In conclusion, the gravitational radiation is quadrupolar and a symmetric spherical body does not emit GWs because its reduced quadrupolar momenta are zero.

The corresponding quantum field is the graviton with zero mass and spin 2. Within the framework of more general theories, the gravitation can be described as a combination of states with a spin 2 and spin zero or as a particle with spin maximum 2. One can also imagine that the mass of graviton is not zero or that the state mass spectrum is complex. Presently, there is no observational reason to doubt that the present observational bounds on the mass of the graviton are severe.

The discovery of GWs, following a prediction by Thorne, opened "*a new window onto the Universe*" [122]. It is clear that being sensitive to an additional radiation it will lead to major discoveries like when the Universe became observed through radio, X or gamma waves. Then, it will allow physicists to test GR in the strong field regime, to check the gravity velocity (assumed to be the speed of light in the Einstein theory) or to verify that GWs only change distances perpendicular to their direction of propagation.

Alternative relativistic theories of gravity are also consistent with the existence of GWs. However, many essential features of the radiation are different: the number of polarization states, the propagation speed, the efficiency of wave generation, the possible existence of massive states [122, 123].

In this section, we have outlined, with no claim to completeness, some physical issues which constitute the test-bed of GR and any alternative relativistic theory of gravity. As we shall see below, ETGs enlarge the possibilities of viable theories in the track of GR with the aim to address, partially, some of the open questions at infrared and ultra-violet scales.

A Summary of Gauge Symmetries

GR can be considered as a gauge theory. Here we want to outline the importance of gauge symmetries with respect to both gravitational interaction and other fundamental forces. Modern gauge theory has emerged as one of the most significant developments of physics of XX century. It has allowed us to realize partially the issue of unifying the fundamental interactions of nature. We now believe that the electromagnetism, which has been long studied, has been successfully unified with the nuclear weak interaction, the force to which radioactive decay is due. What is the most remarkable about this unification is that these two forces differ in strength by a factor of nearly 10^5. This important accomplishment by the Weinberg-Salam gauge theory [114], and insight gained from it, have encouraged the hope that also the other fundamental forces could be unified within a gauge theory framework. At the same time, it has been realized that the potential areas of application for gauge theory extended far beyond elementary particle physics. Although much of the impetus for a gauge theory came from new discoveries in particle physics, the basic ideas behind gauge symmetry have also appeared in other areas as seemingly unrelated, such as condensed matter physics, non-linear wave phenomena and even pure mathematics. This great interest in gauge theory indicates that it is in fact a very general area of study and not only limited to elementary particles.

Gauge invariance was recognised only recently as the physical principle governing the fundamental forces between the elementary particles. Yet the idea of gauge invariance was first proposed by Hermann Weyl in 1919 when only the electron and proton [116] were known as fundamental particles. It required nearly 50 years for gauge invariance to be "rediscovered" and its significance to be understood. The reason for this long hiatus was that Weyl's physical interpretation of gauge invariance was shown to be incorrect soon after he had proposed the theory. Gauge invariance only managed to survive because it was known to be a symmetry of Maxwell's equations and thus became a useful mathematical help in order to simplify equations and thus became a useful mathematical device for simplifying many calculations in the electrodynamics. In view of present success of gauge theory, we can say that gauge invariance was the classical case of a good idea which was discovered long before its time.

In this section, we present a brief summary of gauge theory in view of the fact that any theory of gravity can be considered under the same standard. The early history of gauge theory can be divided into old and new periods where the dividing can be set in the 1950s. The most important question is what motivated Weyl to propose the idea of gauge invariance as a physical symmetry? How did he manage to express it in a mathematical form that has remained almost the same today although the physical interpretation has altered radically? And, how did the development of Quantum Mechanics lead Weyl himself to a rebirth of a gauge theory? The new period of gauge theory begins with the pioneering attempt of Yang and Mills to extend gauge symmetry beyond the narrow limits of electromagnetism [124]. Here we will review the radically new interpretation of gauge invariance required by Yang-Mills theory and the reasons for the failure of original theory. By comparing the new theory with that of Weyl, we can see that many original ideas of Weyl have been rediscovered and incorporated into the modern theory [125].

In these next subsection, our purpose is to present an elementary introduction to a gauge theory in order to show that any relativistic theory of gravity is a gauge theory.

What Is a Gauge Symmetry?

In physics, gauge invariance (also called gauge symmetry) is the property of a field theory where different configurations of the underlying fundamental but unobservable fields result in identical observable quantities. A theory with such a property is called a gauge theory. A transformation from a field configuration to another is called a gauge transformation. Modern physical theories describe nature in terms of fields, e.g., the electromagnetic field, the gravitational field, and fields for the electron and all other elementary particles. A general feature of these theories is that none of these fundamental fields, which are the fields that change under a gauge transformation, can be directly measured. On the other hand, the observable quantities, namely the ones that can be measured experimentally as charges, energies, velocities, etc. do not change under a gauge transformation, even though they are derived from the fields that do change. This (and any) kind of invariance under a transformation is called a symmetry.

For example, in classical electromagnetism the Electric field, E, and the magnetic field, B, are observable, while the underlying and more fundamental electromagnetic potentials V and A are not. Under a gauge transformation which jointly alters the two potentials, no change occurs either in E or B or in the motion of charged particles. In this example, the gauge transformation was just a mathematical feature without any physical relevance, except that gauge invariance is intrinsically connected to the fundamental law of charge conservation. As shown above, with the advent of Quantum Mechanics in the 1920s, and with successive Quantum Field Theory, the importance of gauge transformations has steadily grown. Gauge theories constrain the laws of physics, because of the fact that all the changes induced by a gauge transformation have to cancel each other out when written in terms of observable quantities. Over the course of the 20th century, physicists gradually realized that all forces (fundamental interactions) arise from the constraints imposed by local gauge symmetries, in which case the transformations vary from point to point in space and time. Perturbative quantum field theory (usually employed for scattering theory) describes forces in terms of force mediating particles called gauge bosons. The nature of these particles is determined by the nature of the gauge transformations. The culmination of these efforts is the Standard Model, a quantum field theory explaining all of the fundamental interactions except gravity.

The Einstein Connection

In 1919, people thought that only two fundamental forces of nature existed, Electromagnetism and Gravitation. In that same year, a group of scientists also made the first experimental observation of starlight bending in the gravitational field of the sun during a total eclipse [126]. The brilliant confirmation of Einstein's General Theory of Relativity inspired Weyl to propose his own revolutionary idea of gauge invariance in 1919. To see how this came about, let us first briefly recall some basic ideas on which Relativity was built. The fundamental concept underlying both Special and GR is that are no absolute frames of reference in the Universe. The physical motion of any system must be described relatively to some arbitrary coordinate frame specified by an observer, and the laws of physics must be independent of the choice of frame. In Special Relativity, one usually, defines convenient reference frames, which are called "inertial", in motion with uniform velocity. For example, consider a particle which is moving with constant velocity v with respect to an observer. Let S be the rest frame of the observer and S' be an inertial frame which is moving at the same velocity as the particle. The observer can either state that the particle is moving with velocity v in S or that it is at rest S'. The important point to be noted from this trivial example is that the inertial frame S' can always be related by a simple Lorentz transformation to the observer frame S. The transformation depends only on the relative velocity between the and observers, not on their positions in space-time. The particle and observer can be infinitesimally close together or at opposite ends of the Universe; the Lorentz transformation is still the same. Thus the Lorentz transformation, or rather the Lorentz symmetry group of Special Relativity, is an example of "global" symmetry. In GR, the description of relative motion is much more complicated because now one is dealing with the motion of a system embedded in a gravitational field. As an illustration, let us consider the following "gedanken" experiment for measuring the motion of a test particle which is moving through a gravitational field. The measurement is to be performed by a physicist in an elevator. The elevator cable as broken so that the elevator and the physicist are falling freely [129]. As the particle moves through the field, the physicist determines its motion with respect to the elevator. Since both particle and elevator are falling in the same field, the physicist can describe the particle motion as if there were no gravitational field. The acceleration of the elevator cancels out the acceleration of particle due to gravity. This example of the Principle of Equivalence, follows from the well-known fact that all bodies accelerate at the same rate given the gravitational field (e.g. $9.8\,\mathrm{m/s^2}$ on the surface of the Earth). Let us now compare the physicist in the falling elevator with the observer in the inertial frame in Special Relativity. It could appear that the elevator corresponds to an accelerating or "non-inertial" frame that is analogous to the frame S' in which the particle appeared to be at rest. However, it is not true that a real gravitational field does not produce the same acceleration at every point in space. As one moves infinitely far away from the source, the gravitational field will eventually vanish. Thus, the falling elevator can only be used to define a reference frame within an infinitesimally small region where the gravitational field can be considered uniform. Over a finite region, the variation of the field may be sufficiently large for the acceleration of the particle not to be completely cancelled. We see that an important difference between Special Relativity and GR is that a reference frame can only be defined "locally" or at a single point in a gravitational field. This creates a fundamental problem. To illustrate this difficult point, let us now suppose that there are many more physicists in nearby falling elevators. Each physicist makes an independent measurement so that the path of particle in the gravitational field can be determined. The measurements are made in separate elevators at different locations in the field. Clearly, one cannot perform an ordinary Lorentz transformation between the elevators. If the different elevators were related only by Lorentz transformation, the acceleration would have to be independent of position and the gravitational field could not decrease with distance from the source. Einstein solved the problem of relating nearby falling frames by defining a new mathematical relation known as "connection". To understand the meaning of a connection, let us consider a 4-vector A_μ which represents some physically measured quantity. Now suppose that the physicist in the elevator located at x observes that A_μ changes by an amount dA_μ and a second physicist in a different elevator at x' observes a change in dA'_μ. In Special Relativity, the differential dA_μ is also a vector like A_μ itself. Thus, the differential A_ν in the elevator at x' is given by the relation [5]

$$dA'_\nu = \frac{\partial x^\mu}{\partial x'^\nu} dA_\mu \,, \tag{73}$$

where $\mu, \nu = 0, 1, 2, 3$. The relation (73) follows from the fact that the Lorentz transformation between x and x' is a linear transformation. We can no longer assume that the transformation from x to x' is linear in GR. Thus, we must write for dA'_ν the general expression

$$dA'_\nu = \frac{\partial x^\mu}{\partial x'^\nu} dA_\mu + A_\mu d\left(\frac{\partial x^\mu}{\partial x'^\nu}\right) = \frac{\partial x^\mu}{\partial x'^\nu} dA_\mu + A_\mu \frac{\partial^2 x^\mu}{\partial x'^\nu \partial x'^\lambda} dx'^\lambda \,. \tag{74}$$

Clearly, the second derivatives $\dfrac{\partial^2 x^\mu}{\partial x'^\nu \partial x'^\lambda}$ will vanish if the x^μ are linear functions of the x'^ν. Such terms are actually quite familiar in physics. They occur in "curvilinear" coordinate systems. These curvilinear coefficients are denoted by the symbol

$$\Gamma^\mu_{\nu\lambda} = \frac{\partial^2 x^\mu}{\partial x'^\nu \partial x'^\lambda} \,, \tag{75}$$

and are called the components of a "connection". They are also called affine connections or Christoffel symbols [115]. It is important to note that the gravitational

[5] The components of the 4-vector $A_\mu = (A^0, \mathbf{A})$ and $A^\mu = (A_0, \mathbf{A})$ with $A^0 = -A_0$. Vector components with upper and lower indices are related by $x_\mu = g_{\mu\nu} x^\nu$, where $g_{\mu\nu}$ is the metric tensor which appears in the definition of the invariant space-time interval $ds^2 = g_{\mu\nu} dx^\mu dx^\nu$. The components of $g_{\mu\nu}$ are $g_{11} = g_{22} = g_{33} = 1$, $g_{00} = -1$ and all other components are zero.

connection is not simply the result of using a curvilinear coordinate system. The value of the connection at each point in space-time is dependent on the properties of the gravitational field. The field is important in the determination of the relative orientation of the different falling elevators in the same way that the "upward" direction on the surface of the earth varies from one position to another. The analogy with curvilinear coordinate systems merely indicates that the mathematical descriptions of free-falling frames and curvilinear coordinates are similar. Einstein generalized this similarity and arrived at the revolutionary idea of replacing gravity by the curvature of space-time [127]. Let us briefly summarize the essential characteristics of GR that Weyl would have utilized for his new gauge theory. First of all, GR involves a specific force, gravitation, which is not studied in Special Relativity. However, by studying the properties of coordinates frames just as in Special Relativity, one learns that only local coordinates can be defined in a gravitational field. This local property is required by the physical behavior of the field and leads naturally to the idea of a connection between local coordinate frames. Thus the essential difference between Special Relativity and GR is that the former is a global theory while the latter is a local theory. This local property was the key to Weyl's gauge theory [125]. In section "A Summary of Gauge Symmetries" we will develope extensively this idea.

The Weyl Gauge

Weyl went a step beyond GR and asked the following question: if the effects of a gravitational field can be described by a connection which gives the relative orientation between local frame in space-time, can other forces of nature such as Electromagnetism also be associated with similar connections? Generalizing the idea that all physical measurements are relative, Weyl proposed that absolute magnitude or norm of a physical vector also should not be an absolute quantity but should depend on its location in space-time. A new connection would then be necessary in order to relate the lengths of vectors at different positions. This is the scale or "gauge" invariance. It is important to note here that the true significance of Weyl's proposal lies on the local property of gauge symmetry and not in a special choice of the norm or "gauge" as a physical variable. The assumption of locality is a powerful condition that determines not only the general structure but many of the detailed features of gauge theory. Weyl's gauge invariance can be easily expressed in mathematical form [128]. Let us consider a vector at position x with norm given by $f(x)$. If we shift the vector or transform the coordinates so that the vector is now at $x + dx$, the norm becomes $f(x + dx)$. Expanding to first order in dx, we can write the norm as

$$f(x + dx) = f(x) + \partial_\mu f dx^\mu \, . \tag{76}$$

We now introduce a gauge change through a multiplicative scaling factor $S(x)$. The factor $S(x)$ is defined for convenience to equal unity at the position x. The scale factor at $x + dx$ is then given by

$$S(x+dx) = 1 + \partial_\mu S dx^\mu \, . \tag{77}$$

The norm of the vector at $x + dx$ is then equal to the product of Eqs. (76) and (77). Keeping only first order terms in dx, we obtain

$$Sf = f + \left(\partial_\mu S\right) f dx^\mu + \partial_\mu f dx^\mu \, . \tag{78}$$

For a constant vector, we see that the norm has changed by an amount

$$\left(\partial_\mu + \partial_\mu S\right) f dx^\mu \, . \tag{79}$$

Te derivative $\partial_\mu S$ is the new mathematical "connection" associated with the gauge change. Weyl identified the gauge connection $\partial_\mu S$ with the electromagnetic potential A_μ. It is straightforward to show that a second gauge change with a scale factor Λ will transform the connection as follows,

$$\partial_\mu S \to \partial_\mu S + \partial_\mu \Lambda \, . \tag{80}$$

From classical Electromagnetism, we know that the potential behaves under gauge transformation like

$$A_\mu \to A_\mu + \partial_\mu \Lambda \, , \tag{81}$$

which leaves the electric and magnetic fields unchanged. Since the forms of (80) and (81) are identical, it appeared that Weyl's new interpretation of the potential as a gauge connection was perfectly compatible with Electromagnetism. Unfortunately, it was soon pointed out that the basic idea of scale invariance itself would lead to conflict with known physical facts [129]. Some years later, Bergmann noted, that Weyl's original interpretation of gauge invariance would also be in conflict with Quantum Theory. The wave description of matter defines a natural scale for a particle through its Compton wavelength $\lambda = \frac{h}{Mc}$. Since the wavelength is determined by the particle mass M, it cannot depend on position and thus contradicts Weyl's original assumption about scale invariance. Despite the initial failure of Weyl's gauge theory, the idea of a local gauge symmetry survived. It was well known that Maxwell's equations were invariant under a gauge change. However, without an acceptable interpretation, gauge invariance was regarded as only an "accidental" symmetry of Electromagnetism. The gauge transformation property in Eq. (81) was interpreted as just a statement of the well known arbitrariness of the potential in classical physics. Only the electric and magnetic fields were considered to be real and observable. Gauge symmetry was retained largely because it was useful for calculations in both classical and quantum electrodynamics. In fact a lot of problems in electrodynamics can often be most easily solved by first choosing a suitable gauge, such as the Coulomb gauge or Lorentz gauge, in order to make the equations more tractable [125].

Electromagnetism as a Gauge Theory

It is clear that the electromagnetic interaction of charged particles could be interpreted as a local gauge theory within the context of Quantum Mechanics. In analogy with Weyl's first theory, the phase of a particle wavefunction can be identified as a new physical degree of freedom which is dependent on the space-time position. The value can be changed arbitrarily by performing purely mathematical phase transitions on the wavefunction at each point. Therefore, there must be some connections between phase value nearby points. The role of this connection is payed by the electromagnetic potential. This strict relation between potential and the change in phase is clearly demonstrated by the Aharonov–Bohm effect [130]. Thus by using the phase of wavefunction as the local variable instead of the norm of a vector, Electromagnetism can be interpreted as a local gauge theory very much as Weyl envisioned. Gauge transformations can be viewed as merely phase changes so that they look more like a property of Quantum Mechanics than Electromagnetism. In addition, the symmetry defined by the gauge transformations does not appear to be "natural". The set of all gauge transformations forms a one-dimensional unitary group known as the $U(1)$ group. This group does not arise from any form of coordinate transformation like the more familiar spin-rotation group $SU(2)$ or Lorentz group. Thus, one has lost the original interpretation proposed by Weyl of a new space-time symmetry. The status of gauge theory was also influenced by the historical fact that Maxwell had formulated Electromagnetism long before Weyl proposed the idea of gauge invariance. Therefore, unlike the GR, the gauge symmetry group did not play any essential role in defining the dynamical content of Electromagnetism. This sequence of events was to be completely reversed in the development of modern gauge theory [125, 131, 132].

The Yang-Mills Gauge

In 1954, C.N. Yang and R. Mills proposed that the strong nuclear interaction can be described by a field theory like Electromagnetism. They postulated that the local gauge group was the $SU(2)$ isotopic-spin group. This idea was revolutionary because it changed the very concept of the "identity" of an elementary particle. If the nuclear interaction is a local gauge theory like Electromagnetism, then there is a potential conflict with the notion of how a particle state. For example, let us assume that we can "turn off" the electromagnetic interaction so that we cannot distinguish the proton and neutron by electric charge. We also ignore the small mass difference. We must then label the proton as the "up" state of isotopic spin $\frac{1}{2}$ and the neutron as the "down" state. But if isotopic spin invariance is an independent symmetry at each point in space-time, we cannot assume that the "up" state is the same at any other point. The local isotopic spin symmetry allows to choose arbitrarily which direction is "up" at each point without reference to any other point. Given that the labelling of a proton or a neutron is arbitrary at each point, once the choice has been made at one location, it is clear that some rule is then needed in order to make a comparison with the choice at any other position. The required rule, as Weyl proposed originally, is supplied by a connection. A new isotopic spin potential field was therefore postulated by Yang and Mills in analogy with the electromagnetic potential. However, the greater complexity of the $SU(2)$ isotopic-spin group as compared to the $U(1)$ phase group means that Yang-Mills potential will be quite different from the electromagnetic field. In Electromagnetism, the potential provides a connection between the phase values of the wavefunction at different positions. In the Yang-Mills theory, the phase is replaced by a more complicated local variable that specifies the direction of the isotopic spin. In order to understand qualitatively how this leads to a connection, we need only to recall that the $SU(2)$ isotopic-spin group is also the group of rotations in a 3-dimensional space.[6] As an example, let us visualize the "up" component of isotopic spin $\frac{1}{2}$ as a vector in an abstract "isotopic spin space". An obvious way to relate the "up" states at different locations x and y is to ask how much the "up" state at x needs to be rotated so that it is oriented in the same direction as the "up" state at y. This suggests that the connection between isotopic spin states at different points must act like isotopic spin rotation itself. In other words, if a test particle in the "up" state at x is moved through the potential field to position y, its isotopic spin direction must be rotated by the field so that it is pointing in the "up" direction corresponding to the position y. We can immediately generalize this idea to states of arbitrary isotopic spin. Since the components of an isotopic spin state can be transformed into another one by elements of the $SU(2)$ group, we can conclude that the connection must be capable of performing the same isotopic spin transformations as the $SU(2)$ group itself. This idea that the isotopic spin connection, and therefore the potential, acts like the $SU(2)$ symmetry group is the most important result of the Yang–Mills theory. This concept lies at the heart of the local gauge theory. It shows explicitly how the gauge symmetry group is built into the dynamics of the interaction between particles and fields. How is ti possible for a potential to generate a rotation in an internal symmetry space? To answer this question, we must define the Yang–Mills potential more carefully in the language of the rotation group. A 3-dimensional rotation $R(\theta)$ of a wavefunction is written as

$$R(\theta)\psi = e^{i\theta\mathbf{L}}\psi\,, \tag{82}$$

where θ is the angle of rotation and $\mathbf{L}$ is the angular momentum operator. Let us compare this rotation with the phase change of wavefunction after a gauge transformation. The rotation has the same mathematical form as the phase factor of the wavefunction. However, this does not mean that the potential itself is a rotation operator like $R(\theta)$. We noted earlier that the amount of phase change must also be proportional to the potential in order to ensure that Schrödinger equation remains gauge invariant. To

---

[6]Technically, the $SU(2)$ group is different from the group of 3-dimensional rotations, $O(3)$; the $SU(2)$ group is the "covering group" of $O(3)$.

satisfy this condition, the potential must be proportional to the angular momentum operator **L** in (82). Thus, the most general form of the Yang–Mills potential is a linear combination of the angular momentum operators

$$A_\mu = \sum_i A^i_\mu(x) L_i \,, \tag{83}$$

where the coefficients $A^i_\mu(x)$ depend on the space-time position and we explicitly write the sum over the components. This relation indicates that the Yang–Mills is not a rotation, but rather is a "generator" of a rotation. For the case of Electromagnetism, the angular momentum operator is replaced by a unit matrix and $A^i_\mu(x)$ is just proportional to the phase change $\partial_\mu \lambda$. The relation (83) explicitly displays the dual role of the Yang–Mills potential as both a field in space-time and an operator in the isotopic-spin space.

We can immediately deduce some interesting properties of the Yang–Mills potential. For example, the potential must have three charge components corresponding to the three independent angular momentum operators L_+, L_- and L_3. The potential component which acts like a raising operator L_+ can transform a "down" state into a "up" state. We can associate this formal operation with a real process where a neutron absorbs a unit of isotopic spin from the gauge field and turns into a proton. This example indicates that the Yang–Mills gauge field must itself carry electric charge unlike electromagnetic potential. The Yang–Mills field also differs in other respects from the electromagnetic field but they both have one property in common, namely, they have zero mass. The zero mass of the photon is well known from Maxwell's equations, but local gauge invariance requires that the mass of the gauge potential field be identically zero for any gauge theory. The reason is that the mass of the potential must be introduced into a Lagrangian through a term of the form

$$m^2 A_\mu A^\mu \,. \tag{84}$$

This guarantees that the correct equation of motion for a vector field will be obtained from the Euler-Lagrange equations. Unfortunately, the term given by (84) is not invariant under a gauge transformation. The special transformation property of the potential will introduce extra terms in (84) proportional to A_μ, which are not cancelled by the transformation of the wavefunction. Thus, the standard mass term is not allowed in the Yang–Mills gauge field must have exactly zero mass like the photon. The Yang–Mills field will therefore exhibit long-range behaviour like Coulomb field and cannot reproduce the observed short range of the nuclear force. Since this conclusion appeared to be an inescapable consequence of a local gauge invariance, the Yang–Mills theory was not considered to be an improvement on the already existing theories of the strong nuclear interaction.

Although the Yang–Mills theory field in its original purpose, it established the foundation for modern gauge theory. The $SU(2)$ isotopic-spin gauge transformation could not be regarded as a mere phase change; it required an entirely new interpretation of a gauge invariance. Yang and Mills showed for the first time that local gauge symmetry was a powerful fundamental principle that could provide new insight into the newly discovered "internal" quantum numbers like isotopic spin. In the Yang–Mills theory, isotopic spin was not just a label for the charge states of particles, but it was crucially involved in determining the fundamental form of the interaction [125].

Geometry and Gauge

The Yang–Mills theory revived the old ideas that elementary particles have degrees of freedom in some "internal" space. By showing how these internal degrees of freedom could be unified in a non-trivial way with the dynamical motion in space-time, Yang and Mills discovered a new type of geometry in physics.

The geometrical structure of a gauge theory can be seen by comparing the Yang–Mills theory with of GR. The essential role of the connection is evident in both gauge theory and relativity. There is an analogy between non-inertial coordinate frame and gauge theory but the local frame has to be is located in an abstract space associated with the gauge symmetry group. To see how the gauge group defines an internal space, let us examine the examples of the $U(1)$ phase group and the $SU(2)$ isotopic spin group. For the $U(1)$ group, the internal space consists of all possible values of the phase of the wavefunction. These phase values can be interpreted as angular coordinates in a 2-dimensional space. The internal symmetry space of $U(1)$ thus looks like a ring, and the coordinate of each point in this space is just the phase value itself. The internal space defined by $SU(2)$ group is more complicated because it describes rotation in a 3-dimensional space.

We recall that the orientations of an isotopic spin state can be generated by starting from a fixed initial isotopic direction, which can be chosen as the isotopic spin "up" direction, and then rotating to the desired final direction. The values of the three angles which specify the rotation can be considered as the coordinates for a point inside an abstract 3-dimensional space. Each point corresponds to a distinct rotation so that the isotopic spin states themselves can be identified with the points in this angular space. Thus, the internal symmetry space of the $SU(2)$ group looks like the interior of a 3-dimensional sphere.

The symmetry space of a gauge group provides the local non-inertial coordinate frame for the internal degrees of freedom. To an imaginary observer inside this internal space, the interaction between a particle and an external gauge field looks like a rotation of the local coordinates. The amount of the rotation is determined by the strength of the external potential, and the relative change in the internal coordinates between two space-time points is just given by the connection as stated before. Thus, we see that there is a similarity between the geometrical description of relativity and the internal space picture.

The internal space is called a "fiber" by mathematicians [133, 134]. The idea of using a gauge potential to "link" together space-time with internal symmetries space is a new concept in physics. The new space formed by the union of 4-dimensional space time with an internal space is called a "fiber bundle" space. The reason for this name is that the internal spaces or "fibers" at each space-time point can all be viewed as the same space because they can be transformed into each other by a gauge transformation. Hence, the total space is a collection or "bundle" of fibers.

Given that Yang and Mills developed their theory using the same terminology as electrodynamics, it is relevant to ask if there are good reasons to describe gauge theory in geometrical terms, other than to establish a historical link with relativity. The best reason for doing so is that the geometrical picture provides a valuable aid to the standard language of field theory. Most of the pedagogical aids in field theory are based on a long familiarity with electrodynamics. Modern gauge theory, on the other hand, requires a new approach in order to deal with all of the fundamental forces between elementary particles. The geometrical picture can provide a common arena for discussing electromagnetism, the strong and weak nuclear forces, and even gravity, because it depends on only very general properties of gauge theory.

Local Gauge Transformations

We have described qualitatively how the gauge group is associated with a connection. Any particle or system which is localized in a small volume and carries an internal quantum number like isotopic spin has a direction in the internal symmetry space. This internal direction can be arbitrarily chosen at each point in space-time. In order to compare these internal space directions at two different space-time points x and $x + dx$, we need to define an appropriate connection which can tell us how much the internal direction at x differs from the direction $x + dx$. This connection must be capable of relating all possible directions in the internal space to each other. The most obvious way to relate two directions is to find out how much one direction has to rotate so that it agrees with the other direction. The set of all such rotations forms a symmetry group; thus, the connection between inertial space directions at different points act like a symmetry group as well. Our problem now is to see how a symmetry group transformation can lead to a connection which we identify with a gauge potential field. Let us begin by writing the general form of a local symmetry transformation for an arbitrary (non Abelian) group,

$$U\psi = exp\left(-iq\sum_k \theta^k(x)F_k\right)\psi\,. \tag{85}$$

The "local" nature of the transformation is indicated by the parameters $\theta^k(x)$ which are continuous of x. The constant q is the electric charge for the $U(1)$ gauge group or a general "coupling constant" for an arbitrary gauge group. This is the only way in which the charge enters directly into the calculation. The general transformation (85) is identical to the usual form of an ordinary spatial rotation if we identify the position-dependent parameter $\theta^k(x)$ with rotation angles. The F_k are the generators of the internal symmetry group and satisfy the usual commutation relations

$$\left[F_i, F_j\right] = ic_{ijk}F_k\,, \tag{86}$$

where the structure constant c_{ijk} depends on the particular group. For the isotopic-spin rotation group $SU(2)$, the generators, F_k are the angular momentum operators. To see how the transformation (85) defines a connection, let us consider the following simple operation. We will take a test particle described by a wavefunction $\psi(x)$ and move it between two points x and $x + dx$ in space-time, and analyze how its direction changes in the internal symmetry space. The internal direction at x is initially chosen to have the angles $\theta^k(x)$. As the test particle moves away form x, the internal direction changes in some continuous way until it reaches $x + dx$ where it has new internal direction given angles $\theta^k(x + dx)$. For an infinitesimal distance dx, this change can be described by the transformation (85) acting on $\psi(x)$ and producing a rotation of the internal direction equal to the difference $d\theta^k = \theta^k(x + dx) - \theta^k(x)$. This rotation gives us what we need, namely, a connection between internal space directions at different points in space-time. We also see that this connection involves the derivative of a quantity just like the connection defined by Weyl. In this case, the quantities are the internal rotation angles $\theta^k(x)$. This is a straightforward generalization of the phase of a wavefunction to a set of angles which specify the internal direction.

Connections and Potentials

Let us see how to calculate the connection from the symmetry transformation (85) by moving the charged test particle through an external potential field. We will explicitly separate the particle wavefunction $\psi(x)$ into external and internal parts. Let us write

$$\psi(x) = \sum_\alpha \psi_\alpha(x)u_\alpha\,, \tag{87}$$

where u_α form a set o a "basis vectors" in the internal space. The index α is an internal label such as the components of isotopic spin. The basis u_α is analogous to the local non-inertial frame in relativity. The external part $\psi_\alpha(x)$ is then a "component of $\psi(x)$ in the basis u_α. Under an inertial symmetry transformation, they transform in the usual way

$$\psi_\beta = U_{\alpha\beta}\psi_\alpha\,, \tag{88}$$

where $U_{\alpha\beta}$ is the matrix representation of the symmetry group. We assume, that the representations is irreducible so that the particle has a unique charge or isotopic spin. The decomposition in Eq. (87) is particularly useful because it will allow us to interpret the effect of the external potential field on the particle as a precession of the internal basis.

Now, when the test particle moves from x to $x + dx$ through the external potential field, $\psi(x)$ changes by an amount $d\psi$ given

$$d\psi = \psi(x + dx) - \psi(x) . \tag{89}$$

In general, $d\psi$ must contain both the change in the external x-dependent part of ψ_α and change in the internal space basis u_α. From Eq. (87), we can expand $d\psi$ to first order in dx as

$$d\psi = \sum_\alpha \left[\left(\partial_\mu \psi_\alpha\right) dx^\mu u_\alpha + \psi_\alpha du_\alpha \right] . \tag{90}$$

The second terms contains the change du_α in the internal space basis. This term is given by the connection which we discussed above; it describes the effect of the external potential field on the internal space direction of the particle.

We now need to calculate the change du_α in the internal space basis. The connection between the internal space direction at different space-time points is given by an internal rotation. In this case, the internal directions are specified by a set of basis vectors, so we must calculate the change du_α from an infinitesimal internal rotation $U(dx)$ which is associated with the external displacement dx.

From Eq. (85), we calculate the infinitesimal internal rotation $U(dx)$,

$$U(dx) = exp\left[-iq \sum_k d\theta^k F_k \right] , \tag{91}$$

$$d\theta^k = (\partial_\mu \theta^k) dx_\mu \tag{92}$$

which rotates the internal basis u by an amount du,

$$U(dx)u = u + du . \tag{93}$$

The generators F_k act like matrix operators on the column basis vector u_α so we can write

$$U(dx)u_\alpha = exp\left[-iq \sum_k \left(\partial_\mu \theta^k\right) dx^\mu \left(F_k\right)_{\alpha\beta} \right] u_\beta . \tag{94}$$

Expanding $U(x)$ to first order in dx, we obtain

$$u_\alpha + du_\alpha = \left[\delta_{\alpha\beta} - iq \sum_k \left(\partial_\mu \theta^k\right) dx^\mu \left(F_k\right)_{\alpha\beta} \right] u_\beta , \tag{95}$$

which then gives for the change in the basis,

$$du_\alpha = -iq \sum_k \left(\partial_\mu \theta^k\right) dx^\mu \, (F_k)_{\alpha\beta} \, u_\beta \,. \tag{96}$$

The net change du_α will give us the connection that we have been seeking. Let us therefore introduce the new "connection operator"

$$\left(A_\mu\right)_{\alpha\beta} = \sum_k \left(\partial_\mu \theta^k\right) (F_k)_{\alpha\beta} \,. \tag{97}$$

We thus finally obtain for the total change $d\psi$,

$$d\psi = \sum_{\alpha\beta} \left[\left(\partial_\mu \psi_\alpha\right) \delta_{\alpha\beta} - iq \left(A_\mu\right)_{\alpha\beta} \psi_\alpha \right] dx^\mu u_\beta \,, \tag{98}$$

where we have put in $\delta_{\alpha\beta}$ in order to factor out the basis vector u_β. Now, we can factor the change $d\psi$ into its own external and internal parts

$$d\psi = \sum_\beta (d\psi)_\beta \, u_\beta \equiv \sum_\beta \left(D_\mu \psi_\beta\right) dx^\mu u_\beta \,. \tag{99}$$

The operator D_μ the gauge covariant derivative which describes the changes in both the external and internal parts of $\psi(x)$. Thus we get from Eq. (98)

$$D_\mu \psi_\beta = \sum_\alpha \left[\delta_{\alpha\beta} \partial_\mu - iq \left(A_\mu\right)_{\alpha\beta} \right] \psi_\alpha \,. \tag{100}$$

For the case of the electromagnetic gauge group $U(1)$, the internal space is one-dimensional so that Eq. (100) reduces to

$$D_\mu \psi = \left(\partial_\mu - iqA_\mu\right) \psi \,. \tag{101}$$

This is the "canonical momentum" which is familiar from Electromagnetism. We can also deduce from the example of the $U(1)$ group, that the connection operator defined in Eq. (97), should be identified as the generalized version of the vector potential field A_μ. Thus, we conclude that the external potential field is indeed a connection in the internal symmetry space.

Choosing a Gauge

We know that the usual reason for selecting a particular gauge is to simplify a calculation or to explicitly display an interesting feature of a problem. This freedom to choose a gauge is another example of the arbitrariness in the vector potential.

The choice of a gauge actually involves both gauge invariance and Lorentz invariance simultaneously. A particular gauge usually imposes constraints on the vector potential, such as $\nabla \cdot \mathbf{A} = 0$ for the Coulomb gauge. In general, such equations are obviously not Lorentz invariant. Yet we know that two observers, each in a different inertial frame, can choose the Coulomb gauge for the same electromagnetism problem. The electric and magnetic fields observed in the two different frames can then be related by the usual Lorentz transformations between two frames. This example points out the fact that the space-time location x at which the internal coordinate $\theta(x)$ is evaluated is also not a fixed position. A Lorentz transformation which changes the spatial coordinate in the inertial frame also affects the value of the internal angles ad can be interpreted as an internal rotation. Thus, regardless of whether the coordinate change is associated with Lorentz transformation of observers or an actual movement of the particle in the external field, the effect on the internal space is the same: it is rotated by a gauge transformation.

A Lorentz transformation between two inertial frames is therefore always associated with a gauge transformation. Thus, the vector potential observed in the two frames are related by

$$A'_\mu = L_\mu{}^\nu A_\nu - \partial_\mu \lambda \,, \tag{102}$$

where $L_\mu{}^\nu$ is the Lorentz transformation. This shows that the vector potential actually does not transform like an ordinary four-vector under a Lorentz transformation. It picks up an extra term $\partial_\mu \lambda$ due to the rotation in the internal space. This interesting fact is well known in quantum field theory but it is rarely mentioned in ordinary electromagnetism [135].

We can now see exactly what is involved for a particular choice of gauge. In the Lorentz gauge, $\partial^\mu A_\mu$ is required to be invariant,

$$\partial'^\mu A'_\mu = \partial^\mu A_\mu \,, \tag{103}$$

even though A_μ is not a true four-vector. From Eq. (102), we see that this is possible only if

$$\partial^\mu \partial_\mu \lambda = 0 \,, \tag{104}$$

or, equivalently,

$$\nabla^2 \lambda - \frac{\partial^2 \lambda}{\partial t^2} = 0\,, \tag{105}$$

which is the familiar equation for λ in the Lorentz gauge. Thus we see that the choice of Lorentz gauge is a requirement that the effect of the internal space precession be eliminated so that $\partial^\mu A_\mu$ can be treated as if it were a relativistic invariant. By the same reasoning, other gauges like the Coulomb gauge are not invariant because the gauge condition does not completely eliminate the extra internal precession term. Thus in the Coulomb gauge

$$\nabla' \cdot \mathbf{A}' = \nabla \cdot \mathbf{A} = 0\,, \tag{106}$$

but $\mathbf{A}$ and $\mathbf{A}'$ are not related by a simple Lorentz transformation. An additional gauge rotation is required [136].

Gravity from Gauge Invariants

The above considerations on gauge symmetry can be developed for gravity. In this section, before discussing the physical meaning of GR, we will derive in a very general way the field equations considering manifolds equipped with curvature and torsion. Specifically, we want to show the role of global and local Poincaré invariance and the relevance of conservation laws in any theory of gravity. The approach is completely general and suitable for spinor, vector, bivector and tetrad fields independently of their specific physical meaning. Our first issue is what can "generate" gravity.

What Can "Generate" Gravity?

Since the perturbative scheme is unsatisfactory because it fails over one loop level and cannot be renormalized [137], as we have seen above, we can ask what can to produce gravity or in other words if there exists invariance principles leading to the gravitation [138]. Following the prescriptions of GR, the physical space-time is assumed to be a four-dimensional differential manifold. In Special Relativity, this manifold is the Minkwoski flat-space-time M_4 while, in GR, the underlying space-time is assumed to be curved in order to describe the effects of gravitation.

Utiyama [139] was the first to propose that GR can be seen as a gauge theory based on the local Lorentz group $SO(3,1)$ in much the same way that the Yang-Mills gauge theory [124] was developed on the basis of the internal isospin gauge group $SU(2)$. In this formulation, the Riemannian connection is the gravitational counterpart of the Yang-Mills gauge fields. While $SU(2)$, in the Yang-Mills theory, is an internal symmetry group, the Lorentz symmetry represents the local nature of space-time rather than internal degrees of freedom. The Einstein Equivalence Principle, asserted for GR, requires that the local space-time structure can be identified with the Minkowski space-time possessing Lorentz symmetry. In order to relate local Lorentz symmetry to the external space-time, we need to solder the local space to the external space. The soldering tools are the tetrad fields. Utiyama regarded the tetrads as objects given a priori. Soon after, Sciama [140] recognized that space-time should necessarily be endowed with torsion in order to accommodate spinor fields. In other words, the gravitational interaction of spinning particles requires the modification of the Riemann space-time of GR to be a (non-Riemannian) curved space-time with torsion. Although Sciama used the tetrad formalism for his gauge-like handling of gravitation, his theory fell shortcomings in treating tetrad fields as gauge fields. Kibble [141] made a comprehensive extension of the Utiyama gauge theory of gravitation by showing that the local Poincaré symmetry $SO(3, 1) \rtimes T(3, 1)$ ($\rtimes$ represents the semi-direct product) can generate a space-time with torsion as well as curvature. The gauge fields introduced by Kibble include the tetrads as well as the local affine connection. There have been a variety of gauge theories of gravitation based on different local symmetries which gave rise to several interesting applications in theoretical physics [142–151].

Following the Kibble approach, it can be demonstrated how gravitation can be formulated starting from a pure gauge point of view. In particular, the aim of this section is to show, in details, how a theory of gravitation is a gauge theory which can be obtained starting from the local Poincaré symmetry and this feature works not only for GR but also for ETGs.

A gauge theory of gravity based on a nonlinear realization of the local conformal-affine group of symmetry transformations can be formulated in any case [152]. This means that the coframe fields and gauge connections of the theory can be always obtained. The tetrads and Lorentz group metric have been used to induce a space-time metric. The inhomogeneously transforming (under the Lorentz group) connection coefficients give rise to gravitational gauge potentials used to define covariant derivatives accommodating minimal coupling of matter and gauge fields. On the other hand, the tensor valued connection forms can be used as auxiliary dynamical fields associated with the dilation, special conformal and deformation (shear) degrees of freedom inherent to the bundle manifold. This allows to define the bundle curvature of the theory. Then boundary topological invariants have been constructed. They served as a prototype (source free) gravitational Lagrangian. Finally the Bianchi identities, covariant field equations and gauge currents are obtained.

Here, starting from the Invariance Principle, we consider first the Global Poincaré Invariance and then the Local Poincaré Invariance. This approach lead to construct a given theory of gravity as a gauge theory. This point of view, if considered in detail, can avoid many shortcomings and could be useful to formulate self-consistent schemes for any theory of gravity.

Invariance Principles and the Noether Theorem

As it is well-known, field equations and conservation laws can be obtained from a least action principle. The same principle is the basis of any gauge theory so we start from it to develop our considerations. Let us start from a least action principle and the Noether theorem.

Let $\chi(x)$ be a multiplet field defined at a space-time point x and $\mathcal{L}\{\chi(x), \partial_j\chi(x); x\}$ be the Lagrangian density of the system. In order to make a distinction between the global transformations and the local transformations, for the moment we use the Latin indices ($i, k = 0, 1, 2, 3$) for the former and the greek indices ($\mu, \nu = 0, 1, 2, 3$) for the latter. The action integral of the system over a given space-time volume Ω is defined by

$$I(\Omega) = \int_\Omega \mathcal{L}\{\chi(x), \partial_j\chi(x); x\}\, d^4x. \tag{107}$$

Now let us consider the infinitesimal variations of the coordinates

$$x^i \to x'^i = x^i + \delta x^i, \tag{108}$$

and the field variables

$$\chi(x) \to \chi'(x') = \chi(x) + \delta\chi(x). \tag{109}$$

Correspondingly, the variation of the action is given by

$$\delta I = \int_{\Omega'} \mathcal{L}'(x')\, d^4x' - \int_\Omega \mathcal{L}(x)\, d^4x = \int_\Omega \left[\mathcal{L}'(x')||\partial_j x'^j|| - \mathcal{L}(x)\right] d^4x. \tag{110}$$

Since the Jacobian for the infinitesimal variation of coordinates becomes

$$||\partial_j x'^j|| = 1 + \partial_j(\delta x^j), \tag{111}$$

the variation of the action takes the form,

$$\delta I = \int_\Omega \left[\delta\mathcal{L}(x) + \mathcal{L}(x)\, \partial_j(\delta x^j)\right] d^4x \tag{112}$$

where

$$\delta\mathcal{L}(x) = \mathcal{L}'(x') - \mathcal{L}(x)\,. \tag{113}$$

For any function $\Phi(x)$ of x, it is convenient to define the fixed point variation δ_0 by,

$$\delta_0\Phi(x) := \Phi'(x) - \Phi(x) = \Phi'(x') - \Phi(x'). \tag{114}$$

Expanding the function to first order in δx^j as

$$\Phi(x') = \Phi(x) + \delta x^j\, \partial_j \Phi(x), \tag{115}$$

we obtain

$$\delta\Phi(x) = \Phi'(x') - \Phi(x) = \Phi'(x') - \Phi(x') + \Phi(x') - \Phi(x) = \delta_0\Phi(x) + \delta x^j\, \partial_j \Phi(x), \tag{116}$$

or

$$\delta_0\Phi(x) = \delta\Phi(x) - \delta x^j \partial_j \Phi(x). \tag{117}$$

The advantage of having the fixed point variation is that δ_0 commutes with ∂_j:

$$\delta_0 \partial_j \Phi(x) = \partial_j \delta_0 \Phi(x). \tag{118}$$

For $\Phi(x) = \chi(x)$, we have

$$\delta\chi = \delta_0\chi + \delta x^i \partial_i \chi, \tag{119}$$

and

$$\delta\partial_i\chi = \partial_i(\delta_0\chi) - \partial(\delta x^j)\partial_i\chi. \tag{120}$$

Using the fixed point variation in the integrand of (112) gives

$$\delta I = \int_\Omega \left[\delta_0 \mathcal{L}(x) + \partial_j(\delta x^j\, \mathcal{L}(x))\right] d^4x. \tag{121}$$

If we require the action integral defined over any arbitrary region Ω be invariant, that is, $\delta I = 0$, then we must have

$$\delta\mathcal{L} + \mathcal{L}\partial_j(\delta x^j) = \delta_0\mathcal{L} + \partial_j(\mathcal{L}\delta x^j) = 0. \tag{122}$$

If $\partial_j(\delta x^j) = 0$, then $\delta\mathcal{L} = 0$, that is, the Lagrangian density $\mathcal{L}$ is invariant. In general, however, $\partial_j(\delta x^j) \neq 0$, and $\mathcal{L}$ transforms like a scalar density. In other words, $\mathcal{L}$ is a Lagrangian density unless $\partial_j(\delta x^j) = 0$.

For convenience, let us introduce a function $h(x)$ that behaves like a scalar density, namely

$$\delta h + h\partial_j(\delta x^j) = 0. \tag{123}$$

We further assume $\mathcal{L}(\chi, \partial_j\chi; x) = h(x)L(\chi, \partial_j\chi; x)$. Then we see that

$$\delta\mathcal{L} + \mathcal{L}\partial_j(\delta x^j) = h\delta L. \tag{124}$$

Hence the action integral remains invariant if

$$\delta L = 0. \tag{125}$$

The newly introduced function $L(\chi, \partial_j\chi; x)$ is the scalar Lagrangian of the system.

Let us calculate the integrand of (121) explicitly. The fixed point variation of $\mathcal{L}(x)$ is a consequence of a fixed point variation of the field $\chi(x)$,

$$\delta_0\mathcal{L} = \frac{\partial\mathcal{L}}{\partial\chi}\delta_0\chi + \frac{\partial\mathcal{L}}{\partial(\partial_j\chi)}\delta_0(\partial_j\chi) \tag{126}$$

which can be cast into the form,

$$\delta_0\mathcal{L} = [\mathcal{L}]_\chi\delta_0\chi + \partial_j\left(\frac{\partial\mathcal{L}}{\partial(\partial_j\chi)}\delta_0\chi\right) \tag{127}$$

where

$$[\mathcal{L}]_\chi \equiv \frac{\partial\mathcal{L}}{\partial\chi} - \partial_j\left(\frac{\partial\mathcal{L}}{\partial(\partial_j\chi)}\right). \tag{128}$$

Consequently, we have the action integral in the form

$$\delta I = \int_\Omega \left\{[\mathcal{L}]_\chi\delta_0\chi + \partial_j\left(\frac{\partial\mathcal{L}}{\partial(\partial_j\chi)}\delta\chi - T^j_k\,\delta x^k\right)\right\} d^4x, \tag{129}$$

where

$$T^j{}_k := \frac{\partial\mathcal{L}}{\partial(\partial_j\chi)}\partial_k\chi - \delta^j_k\,\mathcal{L} \tag{130}$$

is the canonical energy-momentum tensor density. If the variations are chosen in such a way that $\delta x^j = 0$ over Ω and $\delta_0\chi$ vanishes on the boundary of Ω, then $\delta I = 0$ gives us the Euler-Lagrange equation,

$$[\mathcal{L}]_\chi = \frac{\partial\mathcal{L}}{\partial\chi} - \partial_j\left(\frac{\partial\mathcal{L}}{\partial(\partial_j\chi)}\right) = 0. \tag{131}$$

On the other hand, if the field variables obey the Euler-Lagrange equation, $[\mathcal{L}]_\chi = 0$, then we have

$$\partial_j\left(\frac{\partial\mathcal{L}}{\partial(\partial_j\chi)}\delta\chi - T^j{}_k\,\delta x^k\right) = 0, \tag{132}$$

which gives rise, considering also the Noether theorem, to conservation laws. These very straightforward considerations are at the basis of our following discussion.

The Global Poincaré Invariance

As standard, we assert that our space-time in the absence of gravitation is a Minkowski space M_4. The isometry group of M_4 is the group of Poincaré transformations (PT) which consists of the Lorentz group $SO(3,1)$ and the translation group $T(3,1)$. The Poincaré transformations of coordinates are

$$x^i \overset{PT}{\rightarrow} x'^i = a^i{}_j x^j + b^i \,, \tag{133}$$

where a^i_j and b^i are real constants, and a^i_j satisfy the orthogonality conditions $a^i_k a^k_j = \delta^i_j$. For infinitesimal variations,

$$\delta x'^i = \chi'(x') - \chi(x) = \varepsilon^i{}_j x^j + \varepsilon^i \tag{134}$$

where $\varepsilon_{ij} + \varepsilon_{ji} = 0$. While the Lorentz transformation forms a six parameter group, the Poincaré group has ten parameters. The Lie algebra for the ten generators of the Poincaré group is

$$\begin{aligned} [\Xi_{ij}, \Xi_{kl}] &= \eta_{ik}\, \Xi_{jl} + \eta_{jl}\, \Xi_{ik} - \eta_{jk}\, \Xi_{il} - \eta_{il}\, \Xi_{jk}, \\ [\Xi_{ij}, T_k] &= \eta_{jk} T_i - \eta_{ik} T_j,\ [T_i, T_j] = 0, \end{aligned} \tag{135}$$

where Ξ_{ij} are the generators of Lorentz transformations, and T_i are the generators of four-dimensional translations. Obviously, $\partial_i(\delta x^i) = 0$ for the Poincaré transformations (133). Therefore, our Lagrangian density $\mathcal{L}$, which is the same as L with $h(x) = 1$ in this case, is invariant; namely, $\delta\mathcal{L} = \delta L = 0$ for $\delta I = 0$.

Suppose that the field $\chi(x)$ transforms under the infinitesimal Poincar é transformations as

$$\delta\chi = \frac{1}{2}\varepsilon^{ij} S_{ij}\chi, \tag{136}$$

where the tensors S_{ij} are the generators of the Lorentz group, satisfying

$$S_{ij} = -S_{ji},\ [S_{ij}, S_{kl}] = \eta_{ik}\, S_{jl} + \eta_{jl}\, S_{ik} - \eta_{jk}\, S_{il} - \eta_{il}\, S_{jk}. \tag{137}$$

Correspondingly, the derivative of χ transforms as

$$\delta(\partial_k\chi) = \frac{1}{2}\varepsilon^{ij} S_{ij}\partial_k\chi - \varepsilon^i{}_k \partial_i\chi. \tag{138}$$

Since the choice of infinitesimal parameters ε^i and ε^{ij} is arbitrary, the vanishing variation of the Lagrangian density $\delta\mathcal{L} = 0$ leads to the identities,

$$\frac{\partial\mathcal{L}}{\partial\chi} S_{ij}\chi + \frac{\partial\mathcal{L}}{\partial(\partial_k\chi)}(S_{ij}\partial_k\chi + \eta_{ki}\partial_j\chi - \eta_{kj}\partial_i\chi) = 0. \tag{139}$$

We also obtain the following conservation laws

$$\partial_j T^j_k = 0,\ \partial_k\left(S^k{}_{ij} - x_i T^k{}_j + x_j T^k{}_i\right) = 0, \tag{140}$$

where

$$S^k{}_{ij} := -\frac{\partial \mathcal{L}}{\partial(\partial_k \chi)} S_{ij}\chi. \tag{141}$$

These conservation laws imply that the energy-momentum and angular momentum, respectively

$$P_l = \int T_l^0 \, d^3x, \; J_{ij} = \int \left[S^0{}_{ij} - \left(x_i T^0{}_j - x_j T^0{}_i \right) \right] d^3x, \tag{142}$$

are conserved. This means that the system invariant under the ten parameter symmetry group has ten conserved quantities. This is an example of Noether symmetry. The first term of the angular momentum integral corresponds to the spin angular momentum while the second term gives the orbital angular momentum. The global Poincaré invariance of a system means that, for the system, the space-time is homogeneous (all space-time points are equivalent) as dictated by the translational invariance and is isotropic (all directions about a spacc-time point are equivalent) as indicated by the Lorentz invariance. It is interesting to observe that the fixed point variation of the field variables takes the form

$$\delta_0 \chi = \frac{1}{2} \varepsilon^j{}_k \Xi_j{}^k \chi + \varepsilon^j T_j \chi, \tag{143}$$

where

$$\Xi_j{}^k = S_j{}^k + \left(x^j \partial_k - x^k \partial_j \right), \; T_j = -\partial_j. \tag{144}$$

We remark that $\Xi_j{}^k$ are the generators of the Lorentz transformation and T_j are those of the translations.

The Local Poincaré Invariance

As next step, let us consider a modification of the infinitesimal Poincaré transformations (134) by assuming that the parameters ε^j_k and ε^j are functions of the coordinates and by writing them altogether as

$$\delta x^\mu = \varepsilon^\mu{}_\nu(x) \, x^\nu + \varepsilon^\mu(x) = \xi^\mu, \tag{145}$$

which we call the local Poincaré transformations (or the general coordinate transformations). As before, in order to distinguish between global transformations and local transformation, we use the Latin indices (j, $k = 0, 1, 2, 3$) for the former and the Greek indices ($\mu, \nu = 0, 1, 2, 3$) for the latter. The variation of the field variables $\chi(x)$ defined at a point x is still the same as that of the global Poincaré transformations,

$$\delta\chi = \frac{1}{2}\varepsilon_{ij}S^{ij}\,\chi. \tag{146}$$

The corresponding fixed point variation of χ takes the form,

$$\delta_0\chi = \frac{1}{2}\varepsilon_{ij}S^{ij}\chi - \xi^\nu\partial_\nu\chi. \tag{147}$$

Differentiating both sides of (147) with respect to x^μ, we have

$$\delta_0\partial_\mu\chi = \frac{1}{2}\varepsilon^{ij}S_{ij}\partial_\mu\chi + \frac{1}{2}(\partial_\mu\varepsilon^{ij})\,S^{ij}\chi - \partial_\mu(\xi^\nu\partial_\nu\chi). \tag{148}$$

By using these variations, we obtain the variation of the Lagrangian L,

$$\delta\mathcal{L} + \partial_\mu(\delta x^\mu)\mathcal{L} = h\delta L = \delta_0\mathcal{L} + \partial_\nu(\mathcal{L}\delta x^\nu) = -\frac{1}{2}(\partial_\mu\varepsilon^{ij})\,S^\mu{}_{ij} - \partial_\mu\xi^\nu\,T^\mu_\nu, \tag{149}$$

which is no longer zero unless the parameters ε^{ij} and ξ^ν become constants. Accordingly, the action integral for the given Lagrangian density $\mathcal{L}$ is not invariant under the local Poincaré transformations. We notice that while $\partial_j(\delta x^j) = 0$ for the local Poincaré transformations, $\partial_\mu\xi^\mu$ does not vanish under local Poincaré transformations. Hence, as expected $\mathcal{L}$ is not a Lagrangian scalar but a Lagrangian density. As mentioned earlier, in order to define the Lagrangian L, we have to select an appropriate non-trivial scalar function $h(x)$ satisfying

$$\delta h + h\partial_\mu\xi^\mu = 0. \tag{150}$$

Now we consider a minimal modification of the Lagrangian so as to make the action integral invariant under the local Poincaré transformations. It is rather obvious that if there is a covariant derivative $\nabla_k\chi$ which transforms as

$$\delta(\nabla_k\chi) = \frac{1}{2}\varepsilon^{ij}S_{ij}\nabla_k\chi - \varepsilon^i{}_k\nabla_i\chi, \tag{151}$$

then a modified Lagrangian $L'(\chi, \partial_k\chi, x) = L(\chi, \nabla_k\chi, x)$, obtained by replacing $\partial_k\chi$ of $L(\chi, \partial_k\chi, x)$ by $\nabla_k\chi$, remains invariant under the local Poincaré transformations, that is

$$\delta L' = \frac{\partial L'}{\partial\chi}\delta\chi + \frac{\partial L'}{\partial(\nabla_k\chi)}\delta(\nabla_k\chi) = 0. \tag{152}$$

To find such a k-covariant derivative, we introduce the gauge fields $A^{ij}{}_\mu = -A^{ji}{}_\mu$ and define the μ-covariant derivative

$$\nabla_\mu\chi := \partial_\mu\chi + \frac{1}{2}A^{ij}{}_\mu S_{ij}\chi, \tag{153}$$

in such a way that the covariant derivative transforms as

$$\delta_0 \nabla_\mu \chi = \frac{1}{2} S_{ij} \nabla_\mu \chi - \partial_\mu (\xi^\nu \nabla_\nu \chi). \tag{154}$$

The transformation properties of $A^{ab}{}_\mu$ are determined by $\nabla_\mu \chi$ and $\delta \nabla_\mu \chi$. Making use of

$$\delta \nabla_\mu \chi = \frac{1}{2} \varepsilon^{ij}{}_{,\mu} S_{ij} \chi + \frac{1}{2} \varepsilon^{ij} S_{ij} \partial_\mu \chi - (\partial_\mu \xi^\nu) \partial_\nu \psi + \frac{1}{2} \delta A^{ij}{}_\mu S_{ij} \chi + \frac{1}{4} A^{ij}{}_\mu S_{ij} \varepsilon^{kl} S_{kl} \chi \tag{155}$$

and comparing with (153) we obtain,

$$\delta A^{ij}{}_\mu S_{ij} \chi + \varepsilon^{ij}{}_{,\mu} S_{ij} \chi + \frac{1}{2} \left(A^{ij}{}_\mu \varepsilon^{kl} - \varepsilon^{ij} A^{kl}{}_\mu \right) S_{ij} S_{kl} \chi + \left(\partial_\mu \xi^\nu \right) A^{ij}{}_\nu S_{ij} \chi = 0. \tag{156}$$

Using the antisymmetry in ij and kl to rewrite the term in parentheses on the *rhs* of (156) as $\left[S_{ij}, S_{kl} \right] A^{ij}{}_\mu \varepsilon^{kl} \chi$, we see the explicit appearance of the commutator $\left[S_{ij}, S_{kl} \right]$. Using the expression for the commutator of Lie algebra generators

$$\left[S_{ij}, S_{kl} \right] = \frac{1}{2} c^{[ef]}{}_{[ij][kl]} S_{ef}, \tag{157}$$

where $c^{[ef]}{}_{[ij][kl]}$ (the square brackets denote anti-symmetrization) is the structure constants of the Lorentz group (deduced below), we have

$$\left[S_{ij}, S_{kl} \right] A^{ij}{}_\mu \varepsilon^{kl} = \frac{1}{2} \left(A^{ic}_\mu \varepsilon^j_c - A^{cj}_\mu \varepsilon^i_c \right) S_{ij}. \tag{158}$$

The substitution of this equation and the consideration of the antisymmetry of $\varepsilon_c{}^b = -\varepsilon^b{}_c$ enable us to write

$$\delta A^{ij}{}_\mu = \varepsilon^i{}_k A^{kj}{}_\mu + \varepsilon^j{}_k A^{ik}{}_\mu - (\partial_\mu \xi^\nu) A^{ij}{}_\nu - \partial_\mu \varepsilon^{ij}. \tag{159}$$

We require the k-derivative and μ-derivative of χ to be linearly related as

$$\nabla_k \chi = V_k{}^\mu(x) \nabla_\mu \chi, \tag{160}$$

where the coefficients $V_k{}^\mu(x)$ are position-dependent and behave like a new set of field variables. From (160) it is evident that $\nabla_k \chi$ varies as

$$\delta \nabla_k \chi = \delta V_k^\mu \nabla_\mu \chi + V_k^\mu \delta \nabla_\mu \chi. \tag{161}$$

Comparing with $\delta \nabla_k \chi = \frac{1}{2} \varepsilon^{ab} S_{ab} \nabla_k \chi - \varepsilon^j{}_k \nabla_j \chi$ we obtain,

$$V^k_\alpha \delta V^\mu_k \nabla_\mu \chi - \xi^\nu{}_{,\alpha} \nabla_\nu \chi + V^k_\alpha \varepsilon^j{}_k \nabla_j \chi = 0. \tag{162}$$

Exploiting $\delta \left(V^k_\alpha V^\mu_k \right) = 0$ we find the quantity $V_k{}^\mu$ transforms according to

$$\delta V_k{}^\mu = V_k{}^\nu \partial_\nu \xi^\mu - V_i{}^\mu \varepsilon^i{}_k. \tag{163}$$

It is also important to recognize that the inverse of $\det(V_k{}^\mu)$ transforms like a scalar density as $h(x)$ does. For our minimal modification of the Lagrangian density, we utilize this available quantity for the scalar density h; namely, we let

$$h(x) = [\det(V_k{}^\mu)]^{-1}. \tag{164}$$

In the limiting case, when we consider Poincaré transformations, that are not space-time dependent, $V_k{}^\mu \to \delta^\mu_k$ so that $h(x) \to 1$. This is a desirable property. Then we replace the Lagrangian density $\mathcal{L}(\chi, \partial_k \chi, x)$, invariant under the global Poincaré transformations, by a Lagrangian density

$$\mathcal{L}(\chi, \partial_\mu \chi; x) \to h(x) L(\chi, \nabla_k \chi). \tag{165}$$

The action integral with this modified Lagrangian density remains invariant under the local Poincaré transformations. Since the local Poincaré transformations $\delta x^\mu = \xi^\mu(x)$ are nothing else but generalized coordinate transformations, the newly introduced gauge fields V^λ_i and $A^{ij}{}_\mu$ can be interpreted, respectively, as the tetrad (*vierbein*) fields which set the local coordinate frame and as a local affine connection with respect to the tetrad frame (see also [153]).

Spinors, Vectors, Bivectors and Tetrads

Let us consider first the case where the multiplet field χ is the Dirac field $\psi(x)$ which behaves like a four-component spinor under the Lorentz transformations and transforms as

$$\psi(x) \to \psi'(x') = S(\Lambda)\psi(x), \tag{166}$$

where $S(\Lambda)$ is an irreducible unitary representation of the Lorentz group. Since the bilinear form $v^k = i\overline{\psi}\gamma^k\psi$ is a vector, it transforms according to

$$v^j = \Lambda^j{}_k v^k, \tag{167}$$

where $\Lambda^j{}_i$ is a Lorentz transformation matrix satisfying

$$\Lambda_{ij} + \Lambda_{ji} = 0. \tag{168}$$

The invariance of v^i (or the covariance of the Dirac equation) under the transformation $\psi(x) \to \psi'(x')$ leads to

$$S^{-1}(\Lambda)\gamma^{\mu}S(\Lambda) = \Lambda^{\mu}_{\nu}\gamma^{\nu}, \tag{169}$$

where the $\gamma's$ are the Dirac γ-matrices satisfying the anticommutator,

$$\gamma_i\gamma_j + \gamma_j\gamma_i = \eta_{ij}\mathbf{1}. \tag{170}$$

Furthermore, we notice that the γ-matrices have the following properties:

$$\begin{cases} (\gamma_0)^{\dagger} = -\gamma_0,\ \left(\gamma^0\right)^2 = (\gamma_0)^2 = -1,\ \gamma_0 = -\gamma^0 \text{ and } \gamma_0\gamma^0 = 1 \\ (\gamma_k)^{\dagger} = \gamma_k\ ,\left(\gamma^k\right)^2 = (\gamma_k)^2 = 1;\ (k = 1, 2, 3) \text{ and } \gamma_k = \gamma^k \\ (\gamma_5)^{\dagger} = -\gamma_5,\ (\gamma_5)^2 = -1 \text{ and } \gamma^5 = \gamma_5. \end{cases} \tag{171}$$

We assume the transformation $S(\Lambda)$ can be put into the form $S(\Lambda) = e^{\Lambda_{\mu\nu}\gamma^{\mu\nu}}$. Expanding $S(\Lambda)$ about the identity and only retaining terms up to the first order in the infinitesimals and expanding $\Lambda_{\mu\nu}$ to the first order in $\varepsilon_{\mu\nu}$

$$\Lambda_{\mu\nu} = \delta_{\mu\nu} + \varepsilon_{\mu\nu},\ \varepsilon_{ij} + \varepsilon_{ji} = 0, \tag{172}$$

we get

$$S(\Lambda) = 1 + \frac{1}{2}\varepsilon^{ij}\gamma_{ij}. \tag{173}$$

In order to determine the form of γ_{ij}, we substitute (172) and (173) into (169) to obtain

$$\frac{1}{2}\varepsilon_{ij}\left[\gamma^{ij}, \gamma^k\right] = \eta^{ki}\varepsilon_{ji}\gamma^j. \tag{174}$$

Rewriting the *rhs* of (174) using the antisymmetry of ε_{ij} as

$$\eta^{ki}\varepsilon_{ji}\gamma^j = \frac{1}{2}\varepsilon_{ij}\left(\eta^{ki}\gamma^j - \eta^{kj}\gamma^i\right), \tag{175}$$

yields

$$\left[\gamma^k, \gamma^{ij}\right] = \eta^{ki}\gamma^j - \eta^{kj}\gamma^i. \tag{176}$$

Assuming the solution to have the form of an antisymmetric product of two matrices, we obtain the solution

$$\gamma^{ij} := \frac{1}{2}\left[\gamma^i, \gamma^j\right]. \tag{177}$$

If $\chi = \psi$, the group generator S_{ij} appearing in (137) is identified with

$$S_{ij} \equiv \gamma_{ij} = \frac{1}{2}(\gamma_i\gamma_j - \gamma_j\gamma_i). \tag{178}$$

To be explicit, the Dirac field transforms under Lorentz transformations (LT) as

$$\delta\psi(x) = \frac{1}{2}\varepsilon^{ij}\gamma_{ij}\psi(x). \tag{179}$$

The Pauli conjugate of the Dirac field is denoted $\overline{\psi}$ and defined by

$$\overline{\psi}(x) := i\psi^{\dagger}(x)\,\gamma_0,\, i \in \mathbb{C}. \tag{180}$$

The conjugate field $\overline{\psi}$ transforms under LTs as,

$$\delta\overline{\psi} = -\overline{\psi}\frac{1}{2}\varepsilon^{ij}\overline{\psi}\gamma_{ij}. \tag{181}$$

Under local LTs, $\varepsilon_{ab}(x)$ becomes a function of space-time. Now, unlike $\partial_\mu\psi(x)$, the derivative of $\psi'(x')$ is no longer homogenous due to the occurrence of the term $\gamma^{ab}\left[\partial_\mu\varepsilon_{ab}(x)\right]\psi(x)$ in $\partial_\mu\psi'(x')$, which is non-vanishing unless ε_{ab} is constant. When going from locally flat to curved space-time, we must generalize ∂_μ to the covariant derivative ∇_μ to compensate for this extra term, allowing to gauge the group of LTs. Thus, by using ∇_μ, we can preserve the invariance of the Lagrangian for arbitrary local LTs at each space-time point

$$\nabla_\mu\psi'(x') = S(\Lambda(x))\nabla_\mu\psi(x). \tag{182}$$

To determine the explicit form of the connection belonging to ∇_μ, we study the derivative of $S(\Lambda(x))$. The transformation $S(\Lambda(x))$ is given by

$$S(\Lambda(x)) = 1 + \frac{1}{2}\varepsilon_{ab}(x)\gamma^{ab}. \tag{183}$$

Since $\varepsilon_{ab}(x)$ is only a function of space-time for local Lorentz coordinates, we express this infinitesimal LT in terms of general coordinates only by shifting all space-time dependence of the local coordinates into tetrad fields as

$$\varepsilon_{ab}(x) = V_a^{\ \lambda}(x)V_b^{\nu}(x)\varepsilon_{\lambda\nu}. \tag{184}$$

Substituting this expression for $\varepsilon_{ab}(x)$, we obtain

$$\partial_\mu\varepsilon_{ab}(x) = \partial_\mu\left[V_a^{\ \lambda}(x)V_b^{\nu}(x)\varepsilon_{\lambda\nu}\right]. \tag{185}$$

However, since $\varepsilon_{\lambda\nu}$ has no space-time dependence, this reduces to

$$\partial_\mu\varepsilon_{ab}(x) = V_a^{\ \lambda}(x)\partial_\mu V_{b\lambda}(x) - V_b^{\ \nu}(x)\partial_\mu V_{a\nu}(x). \tag{186}$$

Letting

$$\omega_{\mu ba} := V_b{}^{\nu}(x)\partial_\mu V_{a\nu}(x), \tag{187}$$

the first and second terms in Eq. (186) become $V_a^{\lambda}(x)\partial_\mu V_{b\lambda}(x) = \frac{1}{2}\omega_{\mu ab}$ and $V_b^{\nu}(x)$ $\partial_\mu V_{a\nu}(x) = \frac{1}{2}\omega_{\mu ba}$ respectively. Using the identification

$$\partial_\mu \varepsilon_{ab}(x) = \omega_{\mu ab}, \tag{188}$$

we write

$$\partial_\mu S(\Lambda(x)) = -\frac{1}{2}\gamma^{ab}\omega_{\mu ab}\,. \tag{189}$$

According to (153), the covariant derivative of the Dirac spinor is

$$\nabla_\mu \psi = \partial_\mu \psi + \frac{1}{2}A^{ij}{}_{\mu}\gamma_{ij}\psi. \tag{190}$$

Correspondingly, the covariant derivative of $\bar{\psi}$ is given by

$$\nabla_\mu \overline{\psi} = \partial_\mu \overline{\psi} - \frac{1}{2}A^{ij}{}_{\mu}\bar{\psi}\gamma_{ij}. \tag{191}$$

Using the covariant derivatives of ψ and $\bar{\psi}$, we can show that

$$\nabla_\mu v_j = \partial_\mu v_j - A^i{}_{j\mu}v_i. \tag{192}$$

The same covariant derivative should be used for any covariant vector v_k under the Lorentz transformation. Since $\nabla_\mu(v_i v^i) = \partial_\mu(v_i v^i)$, the covariant derivative for a contravariant vector v^i must be

$$\nabla_\mu v^i = \partial_\mu v^i + A^i{}_{j\mu}v^j. \tag{193}$$

Since the tetrad $V_i{}^{\mu}$ is a covariant vector under Lorentz transformations, its covariant derivative must transform according to the same rule. Using $\nabla_a = V_a^{\mu}(x)\nabla_\mu$, the covariant derivatives of a tetrad in local Lorentz coordinates read

$$\nabla_\nu V_i{}^{\mu} = \partial_\nu V_i{}^{\mu} - A^k{}_{i\nu}V_k{}^{\mu},\ \nabla_\nu V^i{}_{\mu} = \partial_\nu V^i{}_{\mu} + A^i{}_{k\nu}V^k{}_{\mu}. \tag{194}$$

The inverse of $V_i{}^{\mu}$ is denoted by $V^i{}_{\mu}$ and satisfies

$$V^i{}_{\mu}V_i{}^{\nu} = \delta_\mu{}^{\nu}, \quad V^i{}_{\mu}V_j{}^{\mu} = \delta^i{}_j. \tag{195}$$

To allow the transition to curved space-time, we take account of the general coordinates of objects that are covariant under local Poincaré transformations. Here we define the covariant derivative of a quantity v^λ which behaves like a controvariant vector under the local Poincaré transformation. Namely

$$D_\nu v^\lambda \equiv V_i{}^\lambda \nabla_\nu v^i = \partial_\nu v^\lambda + \Gamma^\lambda{}_{\mu\nu} v^\mu,\ D_\nu v_\mu \equiv V^i{}_\mu \nabla_\nu v_i = \partial_\nu v_\mu - \Gamma^\lambda{}_{\mu\nu} v_\lambda, \quad (196)$$

where

$$\Gamma^\lambda{}_{\mu\nu} := V_i{}^\lambda \nabla_\nu V^i{}_\mu \equiv -V^i{}_\mu \nabla_\nu V_i{}^\lambda. \quad (197)$$

The definition of $\Gamma^\lambda{}_{\mu\nu}$ implies

$$D_\nu V_i{}^\lambda = \nabla_\nu V_i{}^\lambda + \Gamma^\lambda{}_{\mu\nu} V_i{}^\mu = \partial_\nu V_i{}^\lambda - A^k{}_{i\nu} V_k{}^\lambda + \Gamma^\lambda{}_{\mu\nu} V_i{}^\mu = 0, \quad (198)$$

$$D_\nu V^i{}_\mu = \nabla_\nu V^i{}_\mu - \Gamma^\lambda{}_{\mu\nu} V^i{}_\lambda = \partial_\nu V^i{}_\mu + A^i{}_{k\nu} V^k{}_\mu - \Gamma^\lambda{}_{\mu\nu} V^i{}_\lambda = 0.$$

From (198) we find,

$$A^i{}_{k\nu} = V^i{}_\lambda \partial_\nu V_k{}^\lambda + \Gamma^\lambda{}_{\mu\nu} V^i{}_\lambda V_k{}^\mu = -V_k{}^\lambda \partial_\nu V^i{}_\lambda + \Gamma^\lambda{}_{\mu\nu} V^i{}_\lambda V_k{}^\mu. \quad (199)$$

or, equivalently, in terms of ω defined in (187),

$$A^i{}_{k\nu} = \omega^i{}_{\nu k} + \Gamma^\lambda{}_{\mu\nu} V^i{}_\lambda V_k{}^\mu = -\omega_{k\nu}{}^i + \Gamma^\lambda{}_{\mu\nu} V^i{}_\lambda V_k{}^\mu. \quad (200)$$

Using this in (190), we may write

$$\nabla_\mu \psi = (\partial_\mu - \Gamma_\mu)\psi, \quad (201)$$

where

$$\Gamma_\mu = \frac{1}{4}\left(\omega^i{}_{j\mu} - \Gamma^\lambda{}_{\mu\nu} V^i{}_\lambda V_j{}^\nu\right)\gamma_i{}^j, \quad (202)$$

which is known as the Fock-Ivanenko connection.

We now study the transformation properties of $A_{\mu ab}$. Recall $\omega_{\mu ab} = V_a{}^\lambda(x)\partial_\mu V_{\beta\lambda}(x)$ and since $\partial_\mu \eta_{ab} = 0$, we write

$$\Lambda^a{}_{\bar{a}} \eta_{ab} \partial_\mu \Lambda^b_{\bar{b}} = \Lambda^a{}_{\bar{a}} \partial_\mu \Lambda_{a\bar{b}}. \quad (203)$$

Note that barred indices are equivalent to the primed indices used above. Hence, the spin connection transforms as

$$A_{\bar{a}\bar{b}\bar{c}} = \Lambda_{\bar{a}}^{\ a} \Lambda_{\bar{b}}^{\ b} \Lambda_{\bar{c}}^{\ c} A_{abc} + \Lambda_{\bar{a}}^{\ a} \Lambda_{\bar{c}}^{\ c} V^\mu_a(x) \partial_\mu \Lambda_{\bar{b}c}. \quad (204)$$

To determine the transformation properties of

$$\Gamma_{abc} = A_{abc} - \left[V^\mu_a(x)\partial_\mu V^\nu_b(x)\right] V_{\nu c}(x), \quad (205)$$

we consider the local LT of $\left[V_a{}^\mu(x)\partial_\mu V^\nu_b(x)\right] V_{\nu c}(x)$ which is,

$$\left[V^{\mu}_{\bar{a}}(x)\partial_{\mu}V^{\nu}_{\bar{b}}\right]V_{\nu\bar{c}}(x) = \Lambda_{\bar{a}}{}^{a}\Lambda_{\bar{b}}{}^{b}\Lambda_{\bar{c}}{}^{c}\left[A^{\nu}_{ab}V_{\nu c}(x)\right] + \Lambda_{\bar{a}}{}^{a}\Lambda_{\bar{c}}{}^{c}V^{\mu}_{a}(x)\partial_{\mu}\Lambda_{c\bar{b}}. \tag{206}$$

From this result, we obtain the following transformation law,

$$\Gamma_{\bar{a}\bar{b}\bar{c}} = \Lambda_{\bar{a}}{}^{a}\Lambda_{\bar{b}}{}^{b}\Lambda_{\bar{c}}{}^{c}\Gamma_{abc}. \tag{207}$$

We now explore the consequence of the antisymmetry of ω_{abc} in bc . Recalling the equation for Γ_{abc}, exchanging b and c and adding the two equations, we obtain

$$\Gamma_{abc} + \Gamma_{acb} = -V^{\mu}_{a}(x)\left[\left(\partial_{\mu}V^{\nu}_{b}(x)\right)V_{\nu c}(x) + \left(\partial_{\mu}V^{\nu}_{c}(x)\right)V_{\nu b}(x)\right]. \tag{208}$$

We know however, that

$$\partial_{\mu}\left[V^{\nu}_{b}(x)V_{\nu c}(x)\right] = V_{\nu c}(x)\partial_{\mu}V^{\nu}_{b}(x) + V_{\lambda b}(x)\partial_{\mu}V^{\lambda}_{c}(x) + V_{b}{}^{\nu}(x)V^{\lambda}_{c}(x)\partial_{\mu}g_{\lambda\nu}. \tag{209}$$

Letting $\lambda \to \nu$ and exchanging b and c, we obtain

$$\partial_{\mu}\left[V^{\nu}_{b}(x)V_{\nu c}(x)\right] = -V_{b}{}^{\lambda}(x)V^{\nu}_{c}(x)\partial_{\mu}g_{\nu\lambda}, \tag{210}$$

so that, finally,

$$\Gamma_{abc} + \Gamma_{acb} = V_{a}{}^{\mu}(x)V_{b}{}^{\lambda}(x)V_{c}{}^{\nu}(x)\partial_{\mu}g_{\nu\lambda}. \tag{211}$$

This, however, is equivalent to

$$\Gamma_{\bar{a}\bar{b}\bar{c}} + \Gamma_{\bar{a}\bar{c}\bar{b}} = V^{\bar{\mu}}_{\bar{a}}(x)V^{\bar{\lambda}}_{\bar{b}}(x)V^{\bar{\nu}}_{\bar{c}}(x)\partial_{\bar{\mu}}g_{\bar{\nu}\bar{\lambda}}, \tag{212}$$

and then

$$\Gamma_{\mu\lambda\nu} + \Gamma_{\mu\nu\lambda} = \partial_{\mu}g_{\nu\lambda}, \tag{213}$$

which we recognize as the general coordinate connection. It is known that the covariant derivative for general coordinates is

$$\nabla_{\mu}A_{\nu}^{\lambda} = \partial_{\mu}A_{\nu}^{\lambda} + \Gamma^{\lambda}_{\mu\sigma}A_{\nu}^{\sigma} - \Gamma^{\sigma}_{\mu\nu}A_{\sigma}^{\lambda}. \tag{214}$$

In a Riemannian manifold, the connection is symmetric under the exchange of $\mu\nu$, that is, $\Gamma^{\lambda}_{\mu\nu} = \Gamma^{\lambda}_{\nu\mu}$. Using the fact that the metric is a symmetric tensor we can now determine the form of the Christoffel connection by cyclically permuting the indices of the general coordinate connection equation (213) yielding

$$\Gamma_{\mu\nu\lambda} = \frac{1}{2}\left(\partial_{\mu}g_{\nu\lambda} + \partial_{\nu}g_{\lambda\mu} - \partial_{\lambda}g_{\mu\nu}\right). \tag{215}$$

Since $\Gamma_{\mu\nu\lambda} = \Gamma_{\nu\mu\lambda}$ is valid for general coordinate systems, it follows that a similar constraint must hold for local Lorentz transforming coordinates as well, so we expect $\Gamma_{abc} = \Gamma_{bac}$. Recalling the equation for Γ_{abc} and exchanging a and b, we obtain

$$\omega_{abc} - \omega_{bac} = V_{\nu c}(x)\left[V^{\mu}_{a}(x)\partial_{\mu}V^{\nu}_{b}(x) - V^{\mu}_{b}(x)\partial_{\mu}V^{\nu}_{a}(x)\right]. \tag{216}$$

We now define the *objects of anholonomicity* as

$$\Omega_{cab} := V_{\nu c}(x)\left[V^{\mu}_{a}(x)\partial_{\mu}V^{\nu}_{b}(x) - V^{\mu}_{b}(x)\partial_{\mu}V^{\nu}_{a}(x)\right]. \tag{217}$$

Using $\Omega_{cab} = -\Omega_{cba}$, we permute indices in a similar manner as was done for the derivation of the Christoffel connection above yielding,

$$\omega_{ab\mu} = \frac{1}{2}\left[\Omega_{cab} + \Omega_{bca} - \Omega_{abc}\right]V^{c}_{\ \mu} \equiv \Delta_{ab\mu}. \tag{218}$$

For completeness, we determine the transformation law of the Christoffel connection. Making use of $\Gamma^{\lambda}_{\mu\nu}e_{\lambda} = \partial_{\mu}e_{\nu}$ where

$$\partial_{\overline{\mu}}e_{\overline{\nu}} = X^{\mu}_{\overline{\mu}}X^{\nu}_{\overline{\nu}}\partial_{\mu}e_{\nu} + X^{\mu}_{\overline{\mu}}\left(\partial_{\mu}X^{\nu}_{\overline{\nu}}\right)e_{\nu}, \tag{219}$$

we can show

$$\Gamma^{\overline{\lambda}}_{\overline{\mu}\overline{\nu}} = X^{\mu}_{\overline{\mu}}X^{\nu}_{\overline{\nu}}X_{\lambda}^{\ \overline{\lambda}}\Gamma^{\lambda}_{\ \mu\nu} + X^{\mu}_{\overline{\mu}}X_{\nu}^{\ \overline{\lambda}}X^{\nu}_{\ \mu\overline{\nu}}, \tag{220}$$

where

$$X^{\nu}_{\ \mu\overline{\nu}} \equiv \partial_{\mu}\partial_{\overline{\nu}}x^{\nu}. \tag{221}$$

In the light of the above considerations, we may regard infinitesimal local gauge transformations as local rotations of basis vectors belonging to the tangent space [151, 154] of the manifold. For this reason, given a local frame on a tangent plane to the point x on the base manifold, we can obtain all other frames on the same tangent plane by means of local rotations of the original basis vectors. Reversing this argument, we observe that by knowing all frames residing in the horizontal tangent space to a point x on the base manifold enables us to deduce the corresponding gauge group of symmetry transformations. These arguments are completely general and can be adopted for any theory of gravity coming from gauge symmetry.

Curvature, Torsion and Metric

From the definition of the Fock-Ivanenko covariant derivative, we can find the second order covariant derivative

$$D_\nu D_\mu \psi = \partial_\nu \partial_\mu \psi + \frac{1}{2} S_{cd} \left(\psi \partial_\nu A_\mu^{\ cd} + A_\mu^{\ cd} \partial_\nu \psi \right) + \Gamma^\rho_{\ \mu\nu} D_\rho \psi + \frac{1}{2} S_{ef} A_\nu^{\ ef} \partial_\mu \psi$$
$$+ \frac{1}{4} S_{ef} S_{cd} A_\nu^{\ ef} A_\mu^{\ cd} \psi. \tag{222}$$

Recalling $D_\nu V^{c\mu} = 0$, we can solve for the spin connection in terms of the Christoffel connection

$$A_\mu^{\ cd} = -V^d_{\ \lambda} \partial_\mu V^{c\lambda} - \Gamma_\mu^{\ cd}. \tag{223}$$

The derivative of the spin connection is then

$$\partial_\mu A^{cd}_{\ \ \nu} = -V^d_{\ \lambda} \partial_\mu \partial_\nu V^{c\lambda} - \left(\partial_\nu V^{c\lambda} \right) \partial_\mu V_\lambda^{\ d} - \partial_\mu \Gamma^{cd}_{\ \ \nu}. \tag{224}$$

Noting that the Christoffel connection is symmetric and partial derivatives commute, we find

$$\left[D_\mu, D_\nu \right] \psi = \frac{1}{2} S_{cd} \left[\left(\partial_\nu A^{cd}_{\ \ \mu} - \partial_\mu A^{cd}_{\ \ \nu} \right) \psi \right] + \frac{1}{4} S_{ef} S_{cd} \left[\left(A^{ef}_{\ \ \nu} A^{cd}_{\ \ \mu} - A^{ef}_{\ \ \mu} A^{cd}_{\ \ \nu} \right) \psi \right], \tag{225}$$

where

$$\partial_\nu A^{cd}_{\ \ \mu} - \partial_\mu A^{cd}_{\ \ \nu} = \partial_\mu \Gamma^{cd}_{\ \ \nu} - \partial_\nu \Gamma^{cd}_{\ \ \mu}. \tag{226}$$

Relabeling running indices, we can write

$$\frac{1}{4} S_{ef} S_{cd} \left(A^{ef}_{\ \ \nu} A^{cd}_{\ \ \mu} - A^{ef}_{\ \ \mu} A^{cd}_{\ \ \nu} \right) \psi = \frac{1}{4} \left[S_{cd}, S_{ef} \right] A^{ef}_{\ \ \mu} A^{cd}_{\ \ \nu} \psi. \tag{227}$$

Using $\{\gamma_a, \gamma_b\} = 2\eta_{ab}$ to deduce

$$\{\gamma_a, \gamma_b\} \gamma_c \gamma_d = 2\eta_{ab} \gamma_c \gamma_d, \tag{228}$$

we find that the commutator of bi-spinors is given by

$$\left[S_{cd}, S_{ef} \right] = \frac{1}{2} \left[\eta_{ce} \delta^a_d \delta^b_f - \eta_{de} \delta^a_c \delta^b_f + \eta_{cf} \delta^a_e \delta^b_d - \eta_{df} \delta^a_e \delta^b_c \right] S_{ab}. \tag{229}$$

Clearly the terms in brackets on the *rhs* of (229) are antisymmetric in *cd* and *ef* and also antisymmetric under the exchange of pairs of indices *cd* and *ef*. Since the alternating spinor is antisymmetric in *ab*, so it must be the terms in brackets: this means that the commutator does not vanish. Hence, the term in brackets is totally antisymmetric under interchange of indices *ab*, *cd* and *ef* and exchange of these pairs of indices. We identify this as the structure constant of the Lorentz group [40]

$$\left[\eta_{ce} \delta^a_d \delta^b_f - \eta_{de} \delta^a_c \delta^b_f + \eta_{cf} \delta^a_e \delta^b_d - \eta_{df} \delta^a_e \delta^b_c \right] = c_{[cd][ef]}^{\ \ \ \ [ab]} = c^{[ab]}_{\ \ \ [cd][ef]}, \tag{230}$$

with the aid of which we can write

$$\frac{1}{4}\left[S_{cd}, S_{ef}\right] A^{ef}{}_{\mu} A^{cd}{}_{\nu}\psi = \frac{1}{2} S_{ab}\left[A^{a}{}_{e\nu}A^{eb}{}_{\mu} - A^{b}{}_{e\nu}A^{ae}{}_{\mu}\right]\psi, \tag{231}$$

where

$$A^{a}{}_{e\nu}A^{eb}{}_{\mu} - A^{b}{}_{e\nu}A^{ae}{}_{\mu} = \Gamma^{a}_{\nu e}\Gamma^{eb}{}_{\mu} - \Gamma^{b}_{\nu e}\Gamma^{ea}{}_{\mu}. \tag{232}$$

Combining these results, the commutator of two μ-covariant differentiations gives

$$[\nabla_{\mu}, \nabla_{\nu}]\chi = -\frac{1}{2} R^{ij}{}_{\mu\nu} S_{ij}\chi, \tag{233}$$

where

$$R^{i}{}_{j\mu\nu} = \partial_{\nu}A^{i}{}_{j\mu} - \partial_{\mu}A^{i}{}_{j\nu} + A^{i}{}_{k\nu}A^{k}{}_{j\mu} - A^{i}{}_{k\mu}A^{k}{}_{j\nu}. \tag{234}$$

Using the Jacobi identities for the commutator of covariant derivatives, it follows that the field strength $R^{i}{}_{j\mu\nu}$ satisfies the Bianchi identity

$$\nabla_{\lambda}R^{i}{}_{j\mu\nu} + \nabla_{\mu}R^{i}{}_{j\nu\lambda} + \nabla_{\nu}R^{i}{}_{j\lambda\mu} = 0. \tag{235}$$

Permuting indices, this can be put into the cyclic form

$$\varepsilon^{\alpha\beta\rho\sigma}\nabla_{\beta}R^{ij}{}_{\rho\sigma} = 0, \tag{236}$$

where $\varepsilon^{\alpha\beta\rho\sigma}$ is the Levi-Civita alternating symbol. Furthermore, $R^{ij}{}_{\mu\nu} = \eta^{jk}R^{i}{}_{k\mu\nu}$ is antisymmetric with respect to both pairs of indices,

$$R^{ij}{}_{\mu\nu} = -R^{ji}{}_{\mu\nu} = R^{ji}{}_{\nu\mu} = -R^{ij}{}_{\nu\mu}. \tag{237}$$

This condition is known as the first curvature tensor identity.

To determine the analogue of $[\nabla_{\mu}, \nabla_{\nu}]\chi$ in local coordinates, we start from $\nabla_k\psi = V^{\mu}_{k}\nabla_{\mu}\psi$. From $\nabla_k\psi$ we obtain,

$$\nabla_l\nabla_k\psi = V^{\nu}_{l}\left(\nabla_{\nu}V^{\mu}_{k}\right)\nabla_{\mu}\psi + V^{\nu}_{l}V^{\mu}_{k}\nabla_{\nu}\nabla_{\mu}\psi. \tag{238}$$

Permuting indices and recognizing

$$V_{\mu}{}^{a}\nabla_{\nu}V^{\mu}_{k} = -V_{k}{}^{\mu}\nabla_{\nu}V^{a}_{\mu}, \tag{239}$$

(which follows from $\nabla_{\nu}\left(V^{a}_{\mu}V^{\mu}_{k}\right) = 0$), we arrive at

$$V^{\nu}_{l}\left(\nabla_{\nu}V^{\mu}_{k}\right)\nabla_{\mu}\psi - V^{\mu}_{k}\left(\nabla_{\mu}V^{\nu}_{l}\right)\nabla_{\nu}\psi = \left(V^{\mu}_{l}V^{\nu}_{k} - V^{\mu}_{k}V^{\nu}_{l}\right)\left(\nabla_{\nu}V_{\mu}{}^{a}\right)\nabla_{a}\psi. \tag{240}$$

Defining

$$C^{a}{}_{kl} := \left(V^{\mu}_{k}V^{\nu}_{l} - V^{\mu}_{l}V^{\nu}_{k}\right)\nabla_{\nu}V_{\mu}{}^{a}, \tag{241}$$

the commutator of the k-covariant differentiations takes the final form [141]

$$[\nabla_k, \nabla_l]\chi = -\frac{1}{2}R^{ij}{}_{kl}S_{ij}\chi + C^i{}_{kl}\nabla_i\chi, \tag{242}$$

where

$$R^{ij}{}_{kl} = V_k{}^{\mu}V_l{}^{\nu}R^{ij}{}_{\mu\nu}. \tag{243}$$

As done for $R^i{}_{j\mu\nu}$ using the Jacobi identities for the commutator of covariant derivatives, we find the Bianchi identity in Einstein-Cartan space-time [158, 160]

$$\varepsilon^{\alpha\beta\rho\sigma}\nabla_\beta R^{ij}_{\rho\sigma} = \varepsilon^{\alpha\beta\rho\sigma}C_{\beta\rho}{}^{\lambda}R^{ij}{}_{\sigma\lambda}. \tag{244}$$

The second curvature identity

$$R^k{}_{[\rho\sigma\lambda]} = 2\nabla_{[\rho}C_{\sigma\lambda]}{}^k - 4C_{[\rho\sigma}{}^bC_{\lambda]b}{}^k \tag{245}$$

leads to,

$$\varepsilon^{\alpha\beta\rho\sigma}\nabla_\beta C_{\rho\sigma}{}^k = \varepsilon^{\alpha\beta\rho\sigma}R^k{}_{j\rho\sigma}V^j_\beta. \tag{246}$$

Notice that if

$$\Gamma^\lambda{}_{\mu\nu} = V_i{}^\lambda\nabla_\nu V^i{}_\mu = -V_\mu{}^i\nabla_\nu V^\lambda_i, \tag{247}$$

then

$$\Gamma^\lambda{}_{\mu\nu} - \Gamma^\lambda{}_{\nu\mu} = V_i^\lambda\left(\nabla_\nu V^i_\mu - \nabla_\mu V^i_\nu\right). \tag{248}$$

Contracting by $V_k^\mu V_l^\nu$, we obtain [141],

$$C^a{}_{kl} = V_k{}^\mu V_l{}^\nu V_\lambda{}^a\left(\Gamma^\lambda{}_{\mu\nu} - \Gamma^\lambda{}_{\nu\mu}\right). \tag{249}$$

We therefore conclude that C^a_{kl} is related to the antisymmetric part of the affine connection

$$\Gamma^\lambda{}_{[\mu\nu]} = V_\mu{}^kV_\nu{}^lV_a{}^\lambda C^a{}_{kl} \equiv T^\lambda{}_{\mu\nu}, \tag{250}$$

which is usually interpreted as space-time torsion $T^\lambda{}_{\mu\nu}$. Considering $\Delta_{ab\mu}$ defined in (218), we see that the most general connection in the Poincaré gauge approach to gravitation is

$$A_{ab\mu} = \Delta_{ab\mu} - K_{ab\mu} + \Gamma^\lambda{}_{\nu\mu}V_{a\lambda}V_b{}^\nu, \tag{251}$$

where

$$K_{abc} = -\left(T^\lambda{}_{\nu\mu} - T_{\nu\mu}{}^\lambda + T_\mu{}^\lambda{}_\nu\right)V_{a\lambda}V_b{}^\nu V_c{}^\mu, \tag{252}$$

is the contorsion tensor. Now, the quantity $R^\rho_{\sigma\mu\nu} = V_i{}^\rho R^i{}_{\sigma\mu\nu}$ may be expressed as

$$R^{\rho}{}_{\sigma\mu\nu} = \partial_\nu \Gamma^{\rho}{}_{\sigma\mu} - \partial_\mu \Gamma^{\rho}{}_{\sigma\nu} + \Gamma^{\rho}{}_{\lambda\nu}\Gamma^{\lambda}{}_{\sigma\mu} - \Gamma^{\rho}{}_{\lambda\mu}\Gamma^{\lambda}{}_{\sigma\nu}. \tag{253}$$

Therefore, we can regard $R^{\rho}{}_{\sigma\mu\nu}$ as the curvature tensor with respect the affine connection $\Gamma^{\lambda}{}_{\mu\nu}$. By using the inverse of the tetrad, we define the metric of the space-time manifold by

$$g_{\mu\nu} = V^{i}{}_{\mu} V^{j}{}_{\nu} \eta_{ij}. \tag{254}$$

From (198) and the fact that the Minkowski metric is constant, it is obvious that the metric so defined is covariantly constant, that is,

$$D_\lambda g_{\mu\nu} = 0. \tag{255}$$

The space-time thus specified by the local Poincaré transformation is said to be metric. It is not difficult to show that

$$\sqrt{-g} = [\det V^{i}{}_{\mu}] = [\det V_{i}{}^{\mu}]^{-1}, \tag{256}$$

where $g = \det g_{\mu\nu}$. Hence we may take $\sqrt{-g}$ for the density function $h(x)$.

The Field Equations of Gravity

Finally, we are able to deduce the field equations for the gravitational field. From the curvature tensor $R^{\rho}{}_{\sigma\mu\nu}$, given in (253), the Ricci tensor follows

$$R_{\sigma\nu} = R^{\mu}{}_{\sigma\mu\nu}. \tag{257}$$

and the scalar curvature

$$R = R^{\nu}{}_{\nu} = \overset{\mathrm{L}}{R} + \partial_i K_{a}{}^{ia} - T_{a}{}^{bc} K_{bc}{}^{a} \tag{258}$$

where $\overset{\mathrm{L}}{R}$ denotes the usual Ricci scalar of GR. Using this scalar curvature R, we choose the Lagrangian density for free Einstein-Cartan gravity

$$\mathcal{L}_G = \frac{1}{2\kappa}\sqrt{-g}\left(\overset{\mathrm{L}}{R} + \partial_i K_{a}{}^{ia} - T_{a}{}^{bc} K_{bc}{}^{a} - 2\Lambda\right), \tag{259}$$

where κ is a gravitational coupling constant, and Λ is the cosmological constant. These considerations can be easily extended to any function of $\overset{\mathrm{L}}{R}$. Observe that the second term is a divergence and may be ignored. The field equation can be obtained from the total action,

$$S = \int \left\{ \mathcal{L}_{\text{field}}(\chi, \partial_\mu \chi, V_i{}^\mu, A^{ij}{}_\mu) + \mathcal{L}_G \right\} d^4x, \tag{260}$$

where the matter Lagrangian density is taken to be

$$\mathcal{L}_{\text{field}} = \frac{1}{2} \left[\overline{\psi} \gamma^a D_a \psi - \left(D_a \overline{\psi} \right) \gamma^a \psi \right]. \tag{261}$$

Modifying the connection to include Christoffel, spin connection and contorsion contributions so as to operate on general, spinorial arguments, we have

$$\Gamma_\mu = \frac{1}{4} g_{\lambda\sigma} \left(\Delta^\sigma{}_{\mu\rho} - \overset{\text{L}}{\Gamma}{}^\sigma{}_{\rho\mu} - K^\sigma{}_{\rho\mu} \right) \gamma^{\lambda\rho}. \tag{262}$$

It is important to keep in mind that $\Delta^\sigma{}_{\mu\rho}$ act only on multi-component spinor fields, while $\overset{\text{L}}{\Gamma}{}^\sigma{}_{\rho\mu}$ act on vectors and arbitrary tensors. The gauge covariant derivative for a spinor and adjoint spinor is then given by

$$D_\mu \psi = \left(\partial_\mu - \Gamma_\mu \right) \psi, \; D_\mu \overline{\psi} = \partial_\mu \overline{\psi} - \overline{\psi} \Gamma_\mu. \tag{263}$$

The variation of the field Lagrangian is

$$\delta \mathcal{L}_{\text{field}} = \overline{\psi} \left(\delta \gamma^\mu D_\mu + \gamma^\mu \delta \Gamma_\mu \right) \psi. \tag{264}$$

We know that the Dirac gamma matrices are covariantly vanishing, so

$$D_\kappa \gamma_\iota = \partial_\kappa \gamma_\iota - \Gamma^\mu_{\iota\kappa} \gamma_\mu + \left[\gamma_\iota, \widehat{\Gamma}_\kappa \right] = 0. \tag{265}$$

The 4×4 matrices $\widehat{\Gamma}_\kappa$ are real matrices used to induce similarity transformations on quantities with spinor transformation [155] properties according to

$$\gamma'_i = \widehat{\Gamma}^{-1} \gamma_i \widehat{\Gamma}. \tag{266}$$

Solving for $\widehat{\Gamma}_\kappa$ leads to,

$$\widehat{\Gamma}_\kappa = \frac{1}{8} \left[\left(\partial_\kappa \gamma_\iota \right) \gamma^\iota - \Gamma^\mu_{\iota\kappa} \gamma_\mu \gamma^\iota \right]. \tag{267}$$

Taking the variation of $\widehat{\Gamma}_\kappa$,

$$\begin{aligned} \delta \widehat{\Gamma}_\kappa &= \frac{1}{8} \left[\begin{matrix} \left(\partial_\kappa \delta \gamma_\iota \right) \gamma^\iota + \left(\partial_\kappa \gamma_\iota \right) \delta \gamma^\iota - \left(\delta \Gamma^\mu_{\iota\kappa} \right) \gamma_\mu \gamma^\iota \\ - \Gamma^\mu_{\iota\kappa} \left(\left(\delta \gamma_\mu \right) \gamma^\iota + \gamma_\mu \delta \gamma^\iota \right) \end{matrix} \right] \\ &= \frac{1}{8} \left[\left(\partial_\kappa \delta \gamma_\iota \right) \gamma^\iota - \left(\delta \Gamma^\mu_{\iota\kappa} \right) \gamma_\mu \gamma^\iota \right]. \end{aligned} \tag{268}$$

Since we require the anticommutator condition on the gamma matrices $\{\gamma^{\mu}, \gamma^{\nu}\} = 2g^{\mu\nu}$ to hold, the variation of the metric gives

$$2\delta g^{\mu\nu} = \{\delta\gamma^{\mu}, \gamma^{\nu}\} + \{\gamma^{\mu}\delta\gamma^{\nu}\}. \tag{269}$$

One solution to this equation is,

$$\delta\gamma^{\nu} = \frac{1}{2}\gamma_{\sigma}\delta\gamma^{\sigma\nu}. \tag{270}$$

With the aid of this result, we can write

$$\left(\partial_{\kappa}\delta\gamma_{\iota}\right)\gamma^{\iota} = \frac{1}{2}\partial_{\kappa}\left(\gamma^{\nu}\delta g_{\nu\iota}\right)\gamma^{\iota}. \tag{271}$$

Finally, exploiting the anti-symmetry in $\gamma_{\mu\nu}$ we obtain

$$\delta\widehat{\Gamma}_{\kappa} = \frac{1}{8}\left[g_{\nu\sigma}\delta\Gamma_{\mu\kappa}^{\ \ \sigma} - g_{\mu\sigma}\delta\Gamma_{\nu\kappa}^{\ \ \sigma}\right]\gamma^{\mu\nu}. \tag{272}$$

The field Lagrangian defined in the Einstein-Cartan space-time can be written [156–160] explicitly in terms of its Lorentzian and contorsion components as

$$\mathcal{L}_{\text{field}} = \frac{1}{2}\left[\left(\overset{\text{L}}{D}_{\mu}\overline{\psi}\right)\gamma^{\mu}\psi - \overline{\psi}\gamma^{\mu}\overset{\text{L}}{D}_{\mu}\psi\right] - \frac{\hbar c}{8}K_{\mu\alpha\beta}\overline{\psi}\left\{\gamma^{\mu}, \gamma^{\alpha\beta}\right\}\psi. \tag{273}$$

Using the following relations

$$\begin{cases} -\frac{1}{4}K_{\mu\alpha\beta}\overline{\psi}\left\{\gamma^{\mu}, \gamma^{\alpha\beta}\right\}\psi = \frac{1}{4}K_{\mu\alpha\beta}\overline{\psi}\gamma^{\beta\alpha}\gamma^{\mu}\psi - \frac{1}{4}K_{\mu\alpha\beta}\overline{\psi}\gamma^{\mu}\gamma^{\alpha\beta}\psi, \\ \gamma^{\mu}\gamma^{\nu}\gamma^{\lambda}\varepsilon_{\mu\nu\lambda\sigma} = \left\{\gamma^{\mu}, \gamma^{\nu\lambda}\right\}\varepsilon_{\mu\nu\lambda\sigma} = 3!\gamma_{\sigma}\gamma_{5}, \\ \left\{\gamma^{\mu}, \gamma^{\nu\lambda}\right\} = \gamma^{[\mu}\gamma^{\nu}\gamma^{\lambda]}, \end{cases} \tag{274}$$

we obtain

$$K_{\mu\alpha\beta}\overline{\psi}\left\{\gamma^{\mu}, \gamma^{\alpha\beta}\right\}\psi = \frac{1}{2i}K_{\mu\alpha\beta}\varepsilon^{\alpha\beta\mu\nu}\left(\overline{\psi}\gamma_{5}\gamma_{\nu}\psi\right). \tag{275}$$

Here we define the contorsion axial vector

$$K_{\nu} := \frac{1}{3!}\varepsilon^{\alpha\beta\mu\nu}K_{\alpha\beta\mu}. \tag{276}$$

Multiplying through by the axial current $j_{\nu}^{5} = \overline{\psi}\gamma_{5}\gamma_{\nu}\psi$, we obtain

$$\left(\overline{\psi}\gamma_{5}\gamma_{\nu}\psi\right)\varepsilon^{\alpha\beta\mu\nu}K_{\mu\alpha\beta} = -6ij_{\nu}^{5}K^{\nu}. \tag{277}$$

Thus, the field Lagrangian density becomes

$$\mathcal{L}_{\text{field}} = \frac{1}{2}\left[\left(\overset{\text{L}}{D}_\mu\overline{\psi}\right)\gamma^\mu\psi - \overline{\psi}\gamma^\mu\overset{\text{L}}{D}_\mu\psi\right] + \frac{3i\hbar c}{8}K_\mu j_5^\mu. \tag{278}$$

The total action reads

$$\begin{aligned}\delta I &= \delta\int\mathcal{L}_G\sqrt{-g}d^4x + \delta\int\mathcal{L}_{\text{field}}\sqrt{-g}d^4x \\ &= \int\left(\delta\mathcal{L}_G + \delta\mathcal{L}_{\text{field}}\right)\sqrt{-g}d^4x.\end{aligned} \tag{279}$$

Writing the metric in terms of the tetrads $g^{\mu\nu} = V_i^\mu V^{\nu i}$, we observe

$$\delta\sqrt{-g} = -\frac{1}{2}\sqrt{-g}\left(\delta V_i^\mu V_\mu{}^i + V_{\nu i}\delta V^{\nu i}\right). \tag{280}$$

By using

$$\delta V^{\nu i} = \delta\left(\eta^{ij}V_j^\nu\right) = \eta^{ij}\delta V_j^\nu, \tag{281}$$

we are able to deduce

$$\delta\sqrt{-g} = -\sqrt{-g}V_\mu{}^i\delta V_i{}^\mu. \tag{282}$$

For the variation of the Ricci tensor $R_{i\nu} = V_i{}^\mu R_{\mu\nu}$ we have

$$\delta\overset{\text{L}}{R}_{i\nu} = \delta V_i{}^\mu\overset{\text{L}}{R}_{\mu\nu} + V_i{}^\mu\delta\overset{\text{L}}{R}_{\mu\nu}. \tag{283}$$

In an inertial frame, the Ricci tensor reduces to

$$\overset{\text{L}}{R}_{\mu\nu} = \partial_\nu\overset{\text{L}}{\Gamma}{}^\beta_{\ \beta\mu} - \partial_\beta\overset{\text{L}}{\Gamma}{}^\beta_{\ \nu\mu}, \tag{284}$$

so that

$$\delta\overset{\text{L}}{R}_{i\nu} = \delta V_i{}^\mu\overset{\text{L}}{R}_{\mu\nu} + V_i{}^\mu\left(\partial_\nu\delta\overset{\text{L}}{\Gamma}{}^\beta_{\ \beta\mu} - \partial_\beta\delta\overset{\text{L}}{\Gamma}{}^\beta_{\ \nu\mu}\right). \tag{285}$$

The second term can be converted into a surface term, so it may be ignored. Collecting our results, we have

$$\begin{cases}\delta g^{\mu\nu} = -g^{\mu\rho}g^{\nu\sigma}\delta g_{\rho\sigma}, \\ \delta\sqrt{-g} = -\frac{1}{2}\sqrt{-g}g_{\mu\nu}\delta g^{\mu\nu} = -\sqrt{-g}V_\mu{}^i\delta V_i{}^\mu, \\ \delta R_{\mu\nu} = g_{\rho\mu}\left(\nabla_\lambda\delta\Gamma^{\lambda\rho}_{\ \ \nu} - \nabla_\nu\delta\Gamma^{\lambda\rho}_{\ \ \lambda}\right) + T_{\lambda\mu}{}^\rho\delta\Gamma^\lambda{}_{\rho\nu},\ \delta\overset{\text{L}}{R}_{i\nu} = \delta V_i{}^\mu\overset{\text{L}}{R}_{\mu\nu} \\ \delta R = \overset{\text{L}}{R}{}^{\mu\nu}\delta g_{\mu\nu} + g^{\mu\nu}\left(\nabla_\lambda\delta\overset{\text{L}}{\Gamma}{}^\lambda{}_{\mu\nu} - \nabla_\nu\delta\overset{\text{L}}{\Gamma}{}^\lambda{}_{\mu\lambda}\right) - T_a{}^{bc}\delta K_{bc}{}^a.\end{cases} \tag{286}$$

From the above results, we obtain

$$\delta I_G = \frac{1}{16\pi}\int \left[\begin{array}{c} \left(R_i^{\ \mu} - \frac{1}{2}V_i^{\ \mu}R - V_i^{\ \mu}\Lambda\right)\delta V^i_{\ \mu} + 2g^{\rho\lambda}T_{\mu\lambda}^{\ \ \sigma}\delta\Gamma^{\mu}_{\ \rho\sigma} \\ +g^{\mu\nu}\left(\nabla_\lambda \delta \overset{\mathrm{L}}{\Gamma}{}^{\lambda}_{\ \ \mu\nu} - \nabla_\nu \delta \overset{\mathrm{L}}{\Gamma}{}^{\lambda}_{\ \ \mu\lambda}\right) \end{array} \right] \sqrt{-g}d^4x. \tag{287}$$

The last term in the action can be converted into a surface term, so it may be ignored. Using the four-current υ^μ introduced earlier, the action for the matter fields read [155]

$$\begin{aligned} \delta I_{\text{field}} &= \int \left[\overline{\psi}\delta\gamma^\mu \nabla_\mu \psi + \overline{\psi}\gamma^\mu \delta\widehat{\Gamma}_\mu \psi\right]\sqrt{-g}d^4x \qquad (288) \\ &= \int \left\{ \begin{array}{c} \left[\frac{1}{2}g^{\mu\nu}\overline{\psi}\gamma_i\left(\nabla_\nu\psi\right) + T^{\mu}_{\ \rho\sigma}T_i^{\ \rho\sigma} - \delta^\mu_i T_{\lambda\rho\sigma}T^{\lambda\rho\sigma}\right]\delta V^i_{\ \mu} \\ +\frac{1}{8}\left(g^{\rho\nu}\upsilon^\mu - g^{\rho\mu}\upsilon^\nu\right)\left(g_{\mu\sigma}\delta\overset{\mathrm{L}}{\Gamma}{}^{\sigma}_{\ \ \nu\rho} - g_{\nu\sigma}\delta\overset{\mathrm{L}}{\Gamma}{}^{\sigma}_{\ \ \mu\rho}\right) \end{array} \right\}\sqrt{-g}d^4x. \end{aligned}$$

Removing the derivatives of variations of the metric appearing in $\delta\Gamma^{\sigma}_{\ \nu\rho}$ via partial integration, and equating to zero the coefficients of $\delta g^{\mu\nu}$ and $\delta T^{\sigma}_{\ \nu\rho}$ in the variation of the action integral, we obtain

$$\begin{aligned} 0 = &\frac{1}{16\pi}\left(R_{\mu\nu} - \frac{1}{2}g_{\mu\nu}R - g_{\mu\nu}\Lambda\right) + \left(\frac{1}{2}\overline{\psi}\gamma_\nu\nabla_\mu\psi - \frac{1}{4}\nabla_\mu\upsilon_\nu\right) \\ &+\nabla_\sigma T_{\mu\nu}^{\ \ \sigma} + T_{\mu\rho\sigma}T_\nu^{\ \rho\sigma} - g_{\mu\nu}T_{\lambda\rho\sigma}T^{\lambda\rho\sigma} \end{aligned} \tag{289}$$

and

$$T_{\rho\sigma\lambda} = \kappa\tau_{\rho\sigma\lambda}. \tag{290}$$

Equation (289) have the form of Einstein equations

$$G_{\mu\nu} - g_{\mu\nu}\Lambda = \kappa\Sigma_{\mu\nu}, \tag{291}$$

where $\kappa = \dfrac{8\pi G_N}{c^4}$ and the Einstein tensor and non-symmetric energy-momentum tensors are

$$G_{\mu\nu} = R_{\mu\nu} - \frac{1}{2}g_{\mu\nu}R, \tag{292}$$

$$\Sigma_{\mu\nu} = \Theta_{\mu\nu} + \mathfrak{T}_{\mu\nu}, \tag{293}$$

respectively.[7] Here we identify $\Theta_{\mu\nu}$ as the canonical energy-momentum

$$\Theta^\mu_{\ \nu} = \frac{\partial\mathcal{L}_{\text{field}}}{\partial(\nabla_\mu\chi)}\nabla_\nu\chi - \delta^\mu_{\ \nu}\mathcal{L}_{\text{field}}, \tag{294}$$

[7] We will deduce below these some equations in the framework of GR.

while $\mathfrak{T}_{\mu\nu}$ is the stress-tensor form of the non-Riemannian manifold. For the case of spinor fields being considered here the explicit form of the energy-momentum components [161] are (after symmetrization of corresponding canonical source terms in the Einstein equation),

$$\Theta_{\mu\nu} = -\left[\overline{\psi}\gamma_\mu \nabla_\nu \psi - \left(\nabla_\nu \overline{\psi}\right)\gamma_\mu \psi + \overline{\psi}\gamma_\nu \nabla_\mu \psi - \left(\nabla_\mu \overline{\psi}\right)\gamma_\nu \psi\right] \tag{295}$$

and by using the second field equation (290), we determine

$$\mathfrak{T}_{\mu\nu} = \nabla_\sigma T_{\mu\nu}{}^{\sigma} + T_{\mu\rho\sigma}\tau_\nu{}^{\rho\sigma} - g_{\mu\nu} T_{\lambda\rho\sigma}\tau^{\lambda\rho\sigma}, \tag{296}$$

where $\tau_{\mu\nu}{}^{\sigma}$ is the so-called spin-energy potential [157, 158]

$$\tau_{\mu\nu}{}^{\sigma} := \frac{\partial \mathcal{L}_{\text{field}}}{\partial(\nabla_\sigma \chi)}\gamma_{\mu\nu}\chi. \tag{297}$$

Explicitly, the spin energy potential reads $\tau^{\mu\nu\sigma} = \overline{\psi}\gamma^{[\mu}\gamma^{\nu}\gamma^{\sigma]}\psi$. The equation of motion obtained from the variation of the action with respect to $\overline{\psi}$ reads [157, 158]

$$\gamma^\mu \nabla_\mu \psi + \frac{3}{8} T_{\mu\nu\sigma}\gamma^{[\mu}\gamma^{\nu}\gamma^{\sigma]}\psi = 0. \tag{298}$$

It is interesting to observe that this generalized curved space-time Dirac equation can be recast into the nonlinear equation of the Heisenberg-Pauli type

$$\gamma^\mu \nabla_\mu \psi + \frac{3}{8}\varepsilon\left(\overline{\psi}\gamma^\mu \gamma_5 \psi\right)\gamma_\mu \gamma_5 \psi = 0. \tag{299}$$

Although the gravitational field equation is similar in form to the Einstein field equation, it differs from the original Einstein equations because the curvature tensor, containing space-time torsion, is non-Riemannian. Assuming that the Euler-Lagrange equations for the matter fields are satisfied, we obtain the following conservation laws for the angular-momentum and energy-momentum

$$\begin{aligned} V_i^\mu V_j^\nu \Sigma_{[\mu\nu]} &= \nabla_\nu \tau_{ij}{}^{\nu}, \\ V_\mu{}^k \nabla_\nu \Sigma^\nu{}_\kappa &= \Sigma^\nu{}_\kappa T^k{}_{\mu\nu} + \tau^\nu_{ij} R^{ij}{}_{\mu\nu}. \end{aligned} \tag{300}$$

We have shown how all the necessary ingredients for a theory of gravitation can be obtained from a gauge theory of local Poincaré symmetry. In fact gauge fields can be obtained by requiring the invariance of the Lagrangian density under local Poincaré invariance. The resulting gravity theory describes a space endowed with non-vanishing curvature and torsion. The simplest prototype of this approach is the Einstein-Cartan Theory which is the lowest order gravitational action linear in the curvature scalar and quadratic in torsion. This scheme, as will see in the second part of this Report, can be extended to more general gravitational theories as the ETGs. The Dirac spinors can be introduced as matter sources which couple with gravity via the torsion stress-tensor $\mathfrak{T}_{\mu\nu}$ which is a component of the total energy- momentum $\Sigma_{\mu\nu}$.

In view of the structure of the generalized energy-momentum-tensor, we remark that the gravitational field equations here obtained are of Yang-Mills type. It is worth noticing that the torsion tensor plays the role of the Yang-Mills field strength. Beside this gauge approach, we want to show now that gravitational field can be recovered also by space-time deformations.

Quantum Field Theory in Curved Space-Time

At small scales and high energies, an hydrodynamic description of matter as a perfect fluid is inadequate: a more accurate description requires quantum field theory formulated on a curved space, in the framework of either GR or another relativistic theory of gravity. Since, at scales comparable to the Compton wavelength of the relevant particles, matter must be quantized, one can employ a semiclassical description of gravitation in which the Einstein equations assume the form

$$G_{\mu\nu} \equiv R_{\mu\nu} - \frac{1}{2} g_{\mu\nu} R = < T_{\mu\nu} > , \tag{301}$$

where the usual Einstein tensor $G_{\mu\nu}$ appears on the left-hand side whereas the right-hand side contains the expectation value of a quantum stress-energy tensor sourcing the gravitational field. Here the coupling constant has been incorporated in the average process so we have not to distinguish between effective and bare couplings. More precisely, if $|\psi >$ is a quantum state describing the early Universe, then

$$< T_{\mu\nu} > \equiv < \psi | \hat{T}_{\mu\nu} | \psi > , \tag{302}$$

where $\hat{T}_{\mu\nu}$ is the quantum operator associated with the classical energy-momentum tensor of the matter field and the right-hand side is an appropriately regularized expectation value. In general, a quantized matter field $\hat{\phi}$ is subject to self-interactions and it interacts also with other fields and with the gravitational background. Such interaction terms may be included in the definition of an effective potential[8]

$$V_{eff}(\phi) = < a | \hat{\mathcal{H}} | a > \tag{303}$$

[8]Hereafter, scalar fields and potentials are understood as their effective values, obtained averaging over quantum states. In this sense, classical fields and potentials are the expectation values of quantum fields and potentials.

with

$$\phi = < a|\hat{\phi}|a > , \tag{304}$$

where $|a>$ represents a normalized state of the theory under consideration (i.e., $< a|a > = 1$) and $\hat{\mathcal{H}}$ is the Hamiltonian operator satisfying $\delta < a|\hat{\mathcal{H}}|a > = 0$, where δ is the variation on the average of $\mathcal{H}$-eigenstates. This condition corresponds to energy conservation.

In a curved space-time, even in the absence of classical matter and radiation, quantum fluctuations of matter fields give non-vanishing contributions to $< T_{\mu\nu} >$, an effect similar to the vacuum of QED [41, 162]. When matter fields are free, massless and conformally invariant, these corrections assume the form

$$< T_{\mu\nu} > = k_1\, {}^{(1)}H_{\mu\nu} + k_3\, {}^{(3)}H_{\mu\nu} . \tag{305}$$

Here k_1 and k_3 are numerical coefficients, while

$${}^{(1)}H_{\mu\nu} = 2R_{;\mu\nu} - 2g_{\mu\nu}\Box R + 2RR\_\mu\nu - \frac{1}{2}\, g_{\mu\nu}R^2 , \tag{306}$$

$${}^{(3)}H_{\mu\nu} = R^{\sigma}{}_{\mu}R_{\nu\sigma} - \frac{2}{3}\, RR_{\mu\nu} - \frac{1}{2}\, g_{\mu\nu}R^{\sigma\tau}R_{\sigma\tau} + \frac{1}{4}\, g_{\mu\nu}R^2 . \tag{307}$$

The divergence of the tensor ${}^{(1)}H_{\mu\nu}$ vanishes,

$${}^{(1)}H^{\nu}_{\mu;\nu} = 0 . \tag{308}$$

This tensor can be obtained by varying a quadratic contribution to the local action,

$${}^{(1)}H_{\mu\nu} = \frac{2}{\sqrt{-g}}\frac{\delta}{\delta g^{\mu\nu}}\left(\sqrt{-g}\, R^2\right) . \tag{309}$$

In order to remove the infinities coming from $< T_{\mu\nu} >$ and obtain a renormalizable theory, one has to introduce infinitely many counterterms in the Lagrangian density of gravity. One of these terms is $CR^2\sqrt{-g}$, where C is a parameter that diverges logarithmically. Equation (301) cannot be generated by a finite action because then the gravitational field would be completely renormalizable, i.e., it would suffice to eliminate a finite number of divergences to make gravity similar to QED. Instead, one can only construct a truncated quantum theory of gravity. The expansion in loops is done in terms of $\hbar$, so the truncated theory at the one-loop level contains all terms of order $\hbar$. In this sense, this is the first quantum correction to GR. It assumes that matter fields are *free* and, due to the Equivalence Principle, all forms of matter couple in the same way to gravity. It also implies an *intrinsic* non-linearity of gravity, so that a number of loops are needed in order to take into account self-interactions or mutual interactions between matter and gravitational fields. At the one-loop level, divergences can be removed by renormalizing the cosmological constant Λ_{eff} and the gravitational constant G_{eff}. The one-loop contributions to $< T_{\mu\nu} >$ are the quantities ${}^{(1)}H_{\mu\nu}$ and ${}^{(3)}H_{\mu\nu}$ above. In addition, one has to consider

$$ {}^{(2)}H_{\mu\nu} = 2R^{\sigma}{}_{\mu;\nu\sigma} - \Box R_{\mu\nu} - \frac{1}{2}g_{\mu\nu}\Box R + R^{\sigma}{}_{\mu}R_{\sigma\nu} - \frac{1}{2}R^{\sigma\tau}R_{\sigma\tau}g_{\mu\nu} . \tag{310} $$

It is shown in Refs. [41, 162] that the relation

$$ {}^{(2)}H_{\mu\nu} = \frac{1}{3}{}^{(1)}H_{\mu\nu} \tag{311} $$

holds in conformally flat space-times. In this case, only the first and third terms of $H_{\mu\nu}$ are present in Eq. (305). Since one can add to the parameter C in the Lagrangian term $C\sqrt{-q}R^2$ an arbitrary constant, the coefficient k_1 can assume any value—the latter should be determined experimentally [41, 162].

The tensor ${}^{(3)}H_{\mu\nu}$ is conserved only in conformally flat space-times (for example, FLRW spaces) and it cannot be obtained by varying a local action. Finally, one has

$$ k_3 = \frac{1}{1440\pi^2}\left(N_0 + \frac{11}{2}N_{1/2} + 31N_1\right) , \tag{312} $$

where the N_i's ($i = 0, 1/2, 1$) are determined by the number of quantum fields with spin 0, 1/2, and 1. Vector fields contribute more to k_3 due to the larger coefficient 31 of N_1. These massless fields, as well as the spinorial ones, are described by conformally invariant equations and appear in $< T_{\mu\nu} >$ in the form (305). The trace of the energy-momentum tensor vanishes for conformally invariant classical fields while owing to the term weighted by k_3, one finds that the expectation value of the tensor (305) has non-vanishing trace. This fact is at the origin of the so-called *trace anomaly*.

Let us discuss briefly how the conformal anomalies are generated when the origin of the tensor $T_{\mu\nu}$ is not classical, i.e., when quantum field theories are formulated in curved space-time. As we will see in more detail later, if a theory is conformally invariant, under the conformal transformation

$$ g_{\mu\nu}(x) \to \tilde{g}_{\mu\nu}(x) \equiv \Omega^2(x)g_{\mu\nu}(x) . \tag{313} $$

the action in $(n+1)$ space-time dimensions satisfies the functional equation

$$ S[\tilde{g}_{\mu\nu}] = S[g_{\mu\nu}] + \int d^{n+1}x \, \frac{\delta S[\tilde{g}_{\mu\nu}]}{\delta \tilde{g}^{\rho\sigma}}\delta\tilde{g}^{\rho\sigma} , \tag{314} $$

where the use of

$$ \delta\tilde{g}^{\mu\nu}(x) = -2\Omega^{-1}(x)\,\tilde{g}^{\mu\nu}(x)\delta\Omega(x) , \tag{315} $$

and of the classical variational principle

$$T^{(m)}_{\mu\nu} = -\frac{2}{\sqrt{-g}}\frac{\delta S^{(m)}}{\delta g}, \tag{316}$$

yields

$$S[\tilde{g}^{\mu\nu}] = S[g_{\mu\nu}] - \int d^{n+1}x\sqrt{-\tilde{g}}\; T^{\rho}{}_{\rho}(\tilde{g}_{\mu\nu})\Omega^{-1}\delta\Omega\,. \tag{317}$$

From this, it follows that

$$T^{\rho}{}_{\rho}[g_{\mu\nu}(x)] = -\frac{\Omega(x)}{\sqrt{-g}}\frac{\delta S[\tilde{g}^{\mu\nu}]}{\delta\Omega(x)}\bigg|_{\Omega=1}\,. \tag{318}$$

Hence, if the classical action is invariant under conformal transformations, the trace of the energy-momentum tensor vanishes. At the quantum level this situation could not occur for the following reason. A conformal transformation is, essentially, a rescaling of lengths with a different rescaling factor at each space-time point x; the presence of a mass, and hence of a length scale, in the theory breaks conformal invariance and generates the trace anomaly. To preserve conformal invariance one has to consider massless fields, as done in (305). In this case one obtains the condition

$$< T^{\rho}_{\rho} > = 0\,, \tag{319}$$

which allows one to consider a conformally invariant theory. Note that gravity is not renormalizable in the usual way; because of this, divergences appear as soon as quantum effects are considered. A loop expansion yields

$$< T^{\rho}_{\rho} > = < T^{\rho}_{\rho} >_{div} + < T^{\rho}_{\rho} >_{ren} = 0\,, \tag{320}$$

confirming the validity of Eq. (319). In this case conformal invariance is preserved only if the divergent part is equal (up to the sign) to the renormalized tensor. An anomalous trace term will appear on the right-hand side of the field equations (301) which, at one-loop and in the zero mass limit of the fields, is given by

$$< T^{\rho}_{\rho} >_{div} = \left[\tilde{k}_1\left(M - \frac{2}{3}\Box R\right) + \tilde{k}_3\mathcal{G}\right] = - < T^{\rho}_{\rho} >_{ren}\,, \tag{321}$$

for a four-dimensional theory. Here $\tilde{k}_1$ and $\tilde{k}_3$ are proportional to k_1 and k_3, while M and $\mathcal{G}$ are obtained from ${}^{(1)}H_{\mu\nu}$ and ${}^{(3)}H_{\mu\nu}$ as

$$M = R^{\alpha\beta\gamma\delta} R_{\alpha\beta\gamma\delta} - 2R^{\alpha\beta} R_{\alpha\beta} + \frac{1}{3} R^2 , \tag{322}$$

$$\mathcal{G} = R^2 - 4R^{\alpha\beta} R_{\alpha\beta} + R^{\alpha\beta\gamma\delta} R_{\alpha\beta\gamma\delta} . \tag{323}$$

$\mathcal{G}$ is the *Gauss-Bonnet* term. In four dimensions, the integral

$$\int d^4x \sqrt{-g}\, \mathcal{G} \tag{324}$$

is an invariant (*Euler characteristic*) which provides information about the topology of the space-time manifold on which the theory is formulated (*Gauss-Bonnet theorem*). In a FLRW background M vanishes identically but $\mathcal{G}$ gives non-vanishing contributions to (305) even if the variation of (324) is zero (in four dimensions).

In general, by summing all the geometric terms deduced from the Riemann tensor and of the same order in $< T^{\rho}_{\rho} >_{ren}$, one derives the right-hand side of (305). If the background metric is conformally flat, this can be expressed by means of Eqs. (306) and (307). Then, one can conclude that the trace anomaly due to the geometric terms arises because the one-loop approach is an attempt to formulate quantum field theories on curved space-time.[9] Cosmological models arising from (305) are studied in [163]. The masses of the matter fields and their mutual interactions can be neglected in the high curvature limit because $R \gg m^2$. The matter-graviton interactions generate non-minimal coupling terms in the effective Lagrangian. The one-loop contributions of such terms are comparable to the ones due to the trace anomaly and generate, from the conformal point of view, the same effects on gravity. The simplest effective Lagrangian that takes into account these corrections is

$$\mathcal{L}_{NMC} = -\frac{1}{2} \nabla^{\alpha}\phi \nabla_{\alpha}\phi - V(\phi) - \frac{\xi}{2} R\phi^2 , \tag{325}$$

where ξ is a dimensionless coupling constant between the scalar and the gravitational fields. The scalar field stress-energy tensor will be modified accordingly but a conformal transformation can be found such that the modifications due to curvature terms can, at least formally, be cast in the form of a matter-curvature interaction. The same argument holds for the trace anomaly. Certain Grand-Unified theories lead to a polynomial coupling of the form $1 + \xi\phi^2 + \zeta\phi^4$ generalizing the one of (325), while an exponential coupling $e^{-\alpha\varphi} R$ between a scalar field (dilaton) φ and the Ricci scalar appears instead in the effective Lagrangian of string theories. The field equations obtained by varying the Lagrangian density $\sqrt{-g}\, \mathcal{L}_{NMC}$ are

[9] Equations (306) and (307) can include terms containing derivatives of the metric of order higher than fourth (fourth order corresponding to the R^2 term) if all possible Feynman diagrams are included. For example, corrections such as $R\Box R$ or $R^2\Box R$ can be present in ${}^{(3)}H_{\mu\nu}$ implying equations of motion that contain sixth order derivatives of the metric. Also these terms can be treated by making use of conformal transformations [223].

$$\left(1-\kappa\xi\phi^2\right)G_{\mu\nu}=\kappa\left\{\nabla_\mu\phi\nabla_\nu\phi-\frac{1}{2}g_{\mu\nu}\nabla^\alpha\phi\,\nabla_\alpha\phi-V\,g_{\mu\nu}\xi\left[g_{\mu\nu}\Box\left(\phi^2\right)-\nabla_\mu\nabla_\nu\left(\phi^2\right)\right]\right\}, \tag{326}$$

$$\Box\phi-\frac{dV}{d\phi}-\xi R\phi=0\,. \tag{327}$$

The non-minimal coupling of the scalar field is reminiscent of that exhibited by the four-vector potential of curved space Maxwell theory. below. Motivation for the non-minimal coupling in the Lagrangian $\mathcal{L}_{NMC}$ comes from many directions. A nonzero ξ is generated by first loop corrections even if it is absent in the classical action [41, 162, 164–168]. Renormalization of a classical theory with $\xi=0$ shifts this coupling constant to a value which is typically small [169, 170] but can, however, affect drastically an inflationary cosmological scenario and determine its success or failure [96, 171–175]. A non-minimal coupling term is expected at high curvatures [166, 168], and it has been argued that classicalization of the Universe in quantum cosmology indeed requires $\xi\neq 0$ [177]. Moreover, non-minimal coupling can solve potential problems of primordial nucleosynthesis [178, 179] and the absence of pathologies in the propagation of ϕ-waves seems to require conformal coupling for all non-gravitational scalar fields ([180–183], see also [184, 185]).[10] The conformal value $\xi=1/6$ is also an infrared fixed point of the renormalization group in finite GUTs [137, 186–196]. Non-minimally coupled scalar fields have been widely used in relation to specific inflationary scenarios [197–208]. The approach adopted was largely one in which ξ is regarded as a free parameter to be used at will in order to fix possible problems of specific inflationary scenarios; see [96, 176] for more general treatments. Geometric reheating of the Universe with strong coupling $\xi\gg 1$ has also been studied [209, 210] and non-minimally coupled scalar fields have been considered in relation with wormholes [211–213], black holes [214, 215], and boson stars [216–218]. The value of the coupling ξ is not, in general, a free parameter but it depends on the physical nature of the particular scalar field ϕ [166, 168, 219–222] (see [96, 174, 176] for reviews of the available theoretical prescriptions for the value of ξ).

To conclude, any attempt to formulate quantum field theory on a curved space-time necessarily leads to modifying the Hilbert-Einstein action. This means adding terms containing non-linear invariants of the curvature tensor or non-minimal couplings between matter and the curvature originating in the perturbative expansion. In cosmology, all these modifications may affect deeply inflationary scenarios originally proposed using minimally coupled scalars [174, 176]. Although rare and very speculative alternatives have been proposed to the inflationary paradigm, the latter is currently accepted by most authors as the "canonical" cure to the shortcomings of the Standard Big Bang Model, with the added bonus of providing a viable mechanism for the generation of density perturbations to seed the structures observed today in the Universe. However, the effects of non-minimal coupling on the inflationary paradigm need to be assessed carefully.

[10]Note, however, that the distinction between gravitational and non-gravitational fields becomes representation-dependent in ETGs, together with the various formulations of the EP.

On the other hand, the vacuum energy of free quantized fields of very low masses can significantly alter also the recent expansion of the Universe as shown in [224, 225]. In fact, the effective action can be obtained from non-perturbative sums of scalar curvature terms in the propagator. As a result of non-perturbative quantum effects, the scalar curvature of the matter-dominated Universe stops decreasing and approaches a constant value. The Universe evolves from an open matter-dominated epoch to a mildly inflating de Sitter expansion. The Hubble constant, during the present de Sitter epoch, as well as the time at which the transition occurs from matter-dominated to de Sitter expansion, are determined by the mass of the field and by the present matter density. These models provide a theoretical explanation of the observed recent acceleration of the Universe, and gives a good fit to data from high-redshift Type Ia Supernovae, with masses of about 10^{-33} eV, and a current ratio of matter density to critical density, $\Omega_0 < 0.4$. The age of the Universe then follows with no further free parameters in the theory, and turns out to be greater than 13 Gyr. The Universe is spatially open and consistent with the possibility of inflation in the very early Universe. Furthermore, such models arise from standard renormalizable theories of free quantum fields in curved space-time, and do not require a cosmological constant or the associated fine-tuning. In this perspective, ETGs represent also a valid alternative to Dark Energy models.

Extended Theories of Gravity

General Remarks

As widely discussed above, due to the problems of Standard Cosmological Model, and, first of all, to the lack of a definitive Quantum Gravity Theory, alternative theories have been considered in order to attempt, at least, a semi-classical scheme where GR and its positive results could be recovered. One of the most fruitful approaches is that of *Extended Theories of Gravity* (ETGs) which have become a sort of paradigm in the study of gravitational interaction. They are based on corrections and enlargements of the Einstein theory. The paradigm consists, essentially, in adding higher-order curvature invariants and minimally or non-minimally coupled scalar fields into dynamics which come out from the effective action of Quantum Gravity [77, 97]. This approach is coherent to the fact that these generalized theories emerge, like Einstein's gravity, from the Gauge Theory, as we have seen above, and can be framed in a bundle structure as we will show below.

Other motivations to modify GR, as discussed, come from the issue of a full recovering of the Mach Principle which leads to assume a varying gravitational coupling. This principle states that the local inertial frame is determined by some average of the motion of distant astronomical objects [226]. This fact implies that the gravitational coupling can be scale-dependent and related to some scalar field. As a consequence, the concept of "inertia" and the Equivalence Principle have to be revised.

Besides, every unification scheme as Superstrings, Supergravity or Grand Unified Theories, takes into account effective actions where non-minimal couplings to the geometry or higher-order terms in the curvature invariants are present. Such contributions are due to one-loop or higher-loop corrections in the high-curvature regimes near the full (not yet available) Quantum Gravity regime [77]. Specifically, this scheme was adopted in order to deal with the quantization on curved space-times and the result was that the interactions among quantum scalar fields and background geometry or the gravitational self-interactions yield corrective terms in the Hilbert-Einstein Lagrangian [41]. Moreover, it has been realized that such corrective terms are inescapable in order to obtain the effective action of Quantum Gravity at scales closed to the Planck one [227]. All these approaches are not the "*full quantum gravity*" but are needed as working schemes toward it. In the next subsection, we will discuss the Quantum Field Theory approach in curved space-time and the emergence of curvature corrections.

In summary, higher-order terms in curvature invariants (such as R^2, $R_{\mu\nu}R^{\mu\nu}$, $R^{\mu\nu\alpha\beta}R_{\mu\nu\alpha\beta}$, $R\,\Box R$, or $R\,\Box^k R$) or non-minimally coupled terms between scalar fields and geometry (such as $\phi^2 R$) have to be added to the effective Lagrangian of gravitational field when quantum corrections are considered. For instance, one can notice that such terms occur in the effective Lagrangian of strings or in Kaluza-Klein theories, when the mechanism of dimensional reduction is used [228].

On the other hand, from a conceptual point of view, there is no a priori reason to restrict the gravitational Lagrangian to a linear function of the Ricci scalar R, minimally coupled with matter [97]. Furthermore, the idea that there are no "exact" laws of physics could be taken into serious account: in such a case, the effective Lagrangians of physical interactions are "stochastic" functions. This feature means that the local gauge invariances (i.e. conservation laws) are well approximated only in the low energy limit and the fundamental physical constants can vary [229].

Beside fundamental physics motivations, all these theories have acquired a huge interest in cosmology due to the fact that they "naturally" exhibit inflationary behaviours able to overcome the shortcomings of Cosmological Standard Model (based on GR). The related cosmological models seem realistic and capable of matching with the Cosmic Microwave Background Radiation observations [230–232]. Furthermore, it is possible to show that, via conformal transformations, the higher-order and non-minimally coupled terms always correspond to the Einstein gravity plus one or more than one minimally coupled scalar fields [233–238].

More precisely, higher-order terms appear always as contributions of order two in the field equations in metric formalism. For example, a term like R^2 gives fourth order equations [239], $R\,\Box R$ gives sixth order equations [223, 237, 238, 240, 241], $R\,\Box^2 R$ gives eighth order equations [242] and so on. By a conformal transformation, any 2nd-order derivative term corresponds to a scalar field[11]: for example, fourth-order gravity gives Einstein plus one scalar field, sixth-order gravity gives Einstein plus two scalar fields and so on [237, 243].

[11] The dynamics of such scalar fields is usually given by the corresponding Klein-Gordon Equation, which is second order.

Considering a mathematical point of view, the problem of reducing more general theories to Einstein standard form has been extensively treated; one can see that, through a "Legendre" transformation, higher-order theories, under suitable regularity conditions on the Lagrangian, take the form of the Einstein one in which a scalar field (or more than one) is the source of the gravitational field (see for example [97, 244, 246, 247]); on the other side, as discussed above, it has been studied the mathematical equivalence between models with variable gravitational coupling with the Einstein gravity through suitable conformal transformations (see [14, 248, 249, 251]).

In any case, the debate on the physical meaning of conformal transformations is far to be solved [see [100, 250] and references therein for a comprehensive review]. Several authors claim for a true physical difference between Jordan frame (higher-order theories and/or variable gravitational coupling) since there are experimental and observational evidences which point out that the Jordan frame could be suitable to better match solutions with data. Others state that the true physical frame is the Einstein one according to the energy theorems [247]. However, the discussion is open and no definitive statement has been formulated up to now.

The problem should be faced from a more general point of view and the Palatini approach to gravity, introduced in section "The Einstein Field Equations in Metric and Palatini Formalisms", could be useful to this goal [113, 117, 252]. In [253], this approach is widely discussed for ETGs and several important applications are reported.

The fundamental idea of the Palatini formalism, as we have seen in section "Quantum Field Theory in Curved Space-Time", is to consider the connection Γ, entering the definition of the Ricci tensor, to be independent of the metric g defined on the space-time $\mathcal{M}$. The Palatini formulation for the standard Hilbert-Einstein theory results to be equivalent to the purely metric theory: this follows from the fact that the field equations for the connection Γ, firstly considered to be independent of the metric, give the Levi-Civita connection of the metric g. As a consequence, there is no reason to impose the Palatini variational principle in the standard Hilbert-Einstein theory instead of the metric variational principle.

However, the situation completely changes if we consider the ETGs, depending on functions of curvature invariants, as $f(R)$, or non-minimally coupled to some scalar field. In these cases, the Palatini and the metric variational principle provide different field equations and the theories thus derived differ [245, 247]. The relevance of Palatini approach, in this framework, has been recently proven in relation to cosmological applications [76, 77, 253–258].

It has also been studied the crucial problem of the Newtonian potential in alternative theories of gravity and its relations with the conformal factor [259]. In [253] interesting revision of the literature on the Newtonian limit of Palatini theories is provided. From a physical point of view, considering the metric g and the connection Γ as independent fields means to decouple the metric structure of space-time and its geodesic structure (being, in general, the connection Γ not the Levi-Civita connection of g). The chronological structure of space-time is governed by g while the trajectories of particles, moving in the space-time, are governed by Γ.

This decoupling enriches the geometric structure of space-time and generalizes the purely metric formalism. This metric-affine structure of space-time is naturally translated, by means of the same (Palatini) field equations, into a bi-metric structure of space-time. Beside the *physical* metric g, another metric h is involved. This new metric is related, in the case of $f(R)$-gravity, to the connection. As a matter of fact, the connection Γ results to be the Levi-Civita connection of h and thus provides the geodesic structure of space-time [260].

If we consider the case of non-minimally coupled interaction in the gravitational Lagrangian (scalar-tensor theories), the new metric h is related to the non-minimal coupling. The new metric h can be thus related to a different geometric and physical aspect of the gravitational theory. Thanks to the Palatini formalism, the non-minimal coupling and the scalar field, entering the evolution of the gravitational fields, are separated from the metric structure of space-time. The situation mixes when we consider the case of higher-order-scalar-tensor theories. Due to these features, the Palatini approach could greatly contribute to clarify the physical meaning of conformal transformation [260]. A part the issue of the physical frame, as we have said before, higher-order corrections in curvature invariants and non-minimal couplings emerge from the formulation of Quantum Field Theory in a curved space-time.

Ghosts, Massless and Massive Gravitational Modes

Detecting new gravitational modes could be a crucial experiment able to discriminate among theories since these modes would constitute evidence that GR must be enlarged or modified [261, 262]. In general, field equations containing higher order terms describe, in addition to the massless spin two field (the standard graviton of GR), also spin zero and spin two massive modes, the latter possibly being ghosts. This result is general and can be obtained by means of a straightforward generalization of the above discussion for $f(R)$-gravity.

Let us generalize the Hilbert-Einstein action by adding curvature invariants different from the Ricci scalar,

$$S = \int d^4x \sqrt{-g} f(R, P, Q) \,, \tag{328}$$

where

$$P \equiv R_{\mu\nu} R^{\mu\nu} \,, \tag{329}$$

$$Q \equiv R_{\mu\nu\rho\sigma} R^{\mu\nu\rho\sigma} \,. \tag{330}$$

By varying the action (328) with respect to $g^{\mu\nu}$, one obtains the field equations [263]

$$
\begin{aligned}
F\, G_{\mu\nu} &= \frac{1}{2}\, g_{\mu\nu}\, (f - R\, F) - \left(g_{\mu\nu}\Box - \nabla_\mu \nabla_\nu\right) F \\
&- 2\left(f_P R^\alpha_\mu R_{\alpha\nu} + f_Q\, R_{\alpha\beta\gamma\mu} R^{\alpha\beta\gamma}{}_\nu\right) \\
&- g_{\mu\nu} \nabla_\alpha \nabla_\beta \left(f_P R^{\alpha\beta}\right) - \Box \left(f_P R_{\mu\nu}\right) \\
&+ 2\nabla_\alpha \nabla_\beta \left(f_P\, R^\alpha{}_{(\mu} \delta^\beta{}_{\nu)} + 2 f_Q\, R^\alpha{}_{(\mu\nu)}{}^\beta\right),
\end{aligned} \tag{331}
$$

where

$$
F \equiv \frac{\partial f}{\partial R}, \qquad f_P \equiv \frac{\partial f}{\partial P}, \qquad f_Q \equiv \frac{\partial f}{\partial Q}. \tag{332}
$$

The trace of Eq. (331) yields

$$
\begin{aligned}
&\Box \left(F + \frac{f_P}{3} R\right) \\
&= \frac{1}{3} \left\{2f - RF - 2\nabla_\alpha \nabla_\beta \left[(f_P + 2 f_Q) R^{\alpha\beta}\right] - 2\left(f_P P + f_Q Q\right)\right\}.
\end{aligned} \tag{333}
$$

Expanding the third term on the right-hand side of (333) and using the contracted Bianchi identities, one obtains

$$
\begin{aligned}
&\Box \left(F + \frac{2}{3}(f_P + f_Q) R\right) \\
&= \frac{1}{3}\Bigg[2f - RF - 2R^{\mu\nu} \nabla_\mu \nabla_\nu \left(f_P + 2 f_Q\right) - R\Box(f_P + 2 f_Q) \\
&- 2\left(f_P P + f_Q Q\right)\Bigg].
\end{aligned} \tag{334}
$$

By defining

$$
\Phi \equiv F + \frac{2}{3}\left(f_P + f_Q\right) R \tag{335}
$$

and

$$
\frac{dV}{d\Phi} \equiv \frac{1}{3}\left[2f - RF - 2R^{\mu\nu} \nabla_\mu \nabla_\nu \left(f_P + 2 f_Q\right) - R\Box(f_P + 2 f_Q) - 2\left(f_P P + f_Q Q\right)\right], \tag{336}
$$

the Klein-Gordon equation

$$
\Box \Phi - \frac{dV}{d\Phi} = 0 \tag{337}
$$

is obtained. In order to find the modes of the gravity waves of this theory, we linearize around the Minkowski background,

$$g_{\mu\nu} = \eta_{\mu\nu} + h_{\mu\nu}\,, \tag{338}$$

$$\Phi = \Phi_0 + \delta\Phi\,; \tag{339}$$

then Eq. (335) yields

$$\delta\Phi = \delta F + \frac{2}{3}\left(\delta f_P + \delta f_Q\right) R_0 + \frac{2}{3}\left(f_{P0} + f_{Q0}\right)\delta R\,, \tag{340}$$

where $R_0 \equiv R(\eta_{\mu\nu}) = 0$ and, similarly, $f_{P0} = \frac{\partial f}{\partial P}\mid_{\eta_{\mu\nu}}$, which is either constant or zero (a zero subscript denoting quantities evaluated with the Minkowski metric). δR denotes the first order perturbation of the Ricci scalar which, together with the perturbed parts of the Riemann and Ricci tensors, is given by

$$\delta R_{\mu\nu\rho\sigma} = \frac{1}{2}\left(\partial_\rho\partial_\nu h_{\mu\sigma} + \partial_\sigma\partial_\mu h_{\nu\rho} - \partial_\sigma\partial_\nu h_{\mu\rho} - \partial_\rho\partial_\mu h_{\nu\sigma}\right)\,, \tag{341}$$

$$\delta R_{\mu\nu} = \frac{1}{2}\left(\partial_\sigma\partial_\nu h^\sigma{}_\mu + \partial_\sigma\partial_\mu h^\sigma{}_\nu - \partial_\mu\partial_\nu h - \Box h_{\mu\nu}\right)\,, \tag{342}$$

$$\delta R = \partial_\mu\partial_\nu h^{\mu\nu} - \Box h\,, \tag{343}$$

where $h \equiv \eta^{\mu\nu}h_{\mu\nu}$. The first term of Eq. (340) is

$$\delta F = \frac{\partial F}{\partial R}\mid_0 \delta R + \frac{\partial F}{\partial P}\mid_0 \delta P + \frac{\partial F}{\partial Q}\mid_0 \delta Q\,, \tag{344}$$

however since δP and δQ are second order, it is $\delta F \simeq F_{,R0}\,\delta R$ and

$$\delta\Phi = \left[F_{,R0} + \frac{2}{3}\left(f_{P0} + f_{Q0}\right)\right]\delta R\,. \tag{345}$$

Eq. (334) then yields the Klein-Gordon equation for the scalar perturbation $\delta\Phi$

$$\begin{aligned}\Box\delta\Phi &= \frac{1}{3}\,\frac{F_0}{F_{,R0} + \frac{2}{3}\left(f_{P0} + f_{Q0}\right)}\,\delta\Phi \\ &\quad - \frac{2}{3}\,\delta R^{\alpha\beta}\partial_\alpha\partial_\beta\left(f_{P0} + 2f_{Q0}\right) - \frac{1}{3}\,\delta R\,\Box(f_{P0} + 2f_{Q0}) \\ &= m_s^2\,\delta\Phi\,. \end{aligned} \tag{346}$$

The second line of Eq. (346) vanishes because f_{P0} and f_{Q0} are constant and the scalar mass is defined as

$$m_s^2 \equiv \frac{F_0}{3F_{,R0} + 2\left(f_{P0} + f_{Q0}\right)} \,. \tag{347}$$

Perturbing the field equations (331), one obtains

$$\begin{aligned}
&F_0\left(\delta R_{\mu\nu} - \frac{1}{2}\eta_{\mu\nu}\delta R\right) \\
&= -\left(\eta_{\mu\nu}\Box - \partial_\mu\partial_\nu\right)\left[\delta\Phi - \frac{2}{3}\left(f_{P0} + f_{Q0}\right)\delta R\right] \\
&- \eta_{\mu\nu}\partial_\alpha\partial_\beta\left(f_{P0}\delta R^{\alpha\beta}\right) - \Box\left(f_{P0}\delta R_{\mu\nu}\right) \\
&+ 2\,\partial_\alpha\partial_\beta\left(f_{P0}\,\delta R^{\alpha}{}_{(\mu}\delta^{\beta}{}_{\nu)} + 2f_{Q0}\,\delta R^{\alpha}{}_{(\mu\nu)}{}^{\beta}\right) .
\end{aligned} \tag{348}$$

It is convenient to work in Fourier space so that, for example, $\partial_\gamma h_{\mu\nu} \to ik_\gamma h_{\mu\nu}$ and $\Box h_{\mu\nu} \to -k^2 h_{\mu\nu}$, where now $k^2 \equiv k^\mu k_\mu$. Then, Eq. (348) becomes

$$\begin{aligned}
&F_0\left(\delta R_{\mu\nu} - \frac{1}{2}\eta_{\mu\nu}\delta R\right) \\
&= \left(\eta_{\mu\nu}k^2 - k_\mu k_\nu\right)\left[\delta\Phi - \frac{2}{3}\left(f_{P0} + f_{Q0}\right)\delta R\right] \\
&+ \eta_{\mu\nu}k_a k_b(f_{P0}\delta R^{ab}) + k^2(f_{P0}\delta R_{\mu\nu}) \\
&- 2k_\alpha k_\beta\left(f_{P0}\,\delta R^{\alpha}{}_{(\mu}\delta^{\beta}{}_{\nu)}\right) - 4k_\alpha k_\beta\left(f_{Q0}\,\delta R^{\alpha}{}_{(\mu\nu)}{}^{\beta}\right) .
\end{aligned} \tag{349}$$

We rewrite the metric perturbation as

$$h_{\mu\nu} = \bar{h}_{\mu\nu} - \frac{\bar{h}}{2}\,\eta_{\mu\nu} + \eta_{\mu\nu}h_f \tag{350}$$

and use the gauge freedom to demand that the usual conditions $\partial_\mu \bar{h}^{\mu\nu} = 0$ and $\bar{h} = 0$ hold. The first condition implies that $k_\mu \bar{h}^{\mu\nu} = 0$, while the second one gives

$$h_{\mu\nu} = \bar{h}_{\mu\nu} + \eta_{\mu\nu}h_f \,, \tag{351}$$

$$h = 4h_f \,. \tag{352}$$

With these conditions in mind, we have

$$\delta R_{\mu\nu} = \frac{1}{2}\left(2k_\mu k_\nu h_f + k^2\eta_{\mu\nu}h_f + k^2\bar{h}_{\mu\nu}\right) , \tag{353}$$

$$\delta R = 3k^2 h_f , \tag{354}$$

$$k_\alpha k_\beta\, \delta R^{\alpha}{}_{(\mu\nu)}{}^{\beta} = -\frac{1}{2}\left[\left(k^4\eta_{\mu\nu} - k^2 k_\mu k_\nu\right) h_f + k^4\bar{h}_{\mu\nu}\right] , \tag{355}$$

$$k_\alpha k_\beta\, \delta R^{\alpha}{}_{(\mu}\delta^{\beta}{}_{\nu)} = \frac{3}{2}\, k^2 k_\mu k_\nu h_f . \tag{356}$$

Using Eqs. (350)–(356) in Eq. (349), a little algebra yields

$$\frac{1}{2}\left(k^2 - k^4\frac{f_{P0} + 4f_{Q0}}{F_0}\right)\bar{h}_{\mu\nu}$$
$$= (\eta_{\mu\nu}k^2 - k_\mu k_\nu)\frac{\delta\Phi}{F_0} + (\eta_{\mu\nu}k^2 - k_\mu k_\nu)h_f . \tag{357}$$

Defining now $h_f \equiv -\delta\Phi/F_0$, we find the perturbation equation

$$k^2\left(1 + \frac{k^2}{m^2_{spin\ 2}}\right)\bar{h}_{\mu\nu} = 0 , \tag{358}$$

where

$$m^2_{spin\ 2} \equiv -\frac{F_0}{f_{P0} + 4f_{Q0}} , \tag{359}$$

while Eq. (346) gives

$$\Box h_f = m_s^2\, h_f . \tag{360}$$

It is easy to see from Eq. (358) that we have a modified dispersion relation corresponding to a massless spin two field ($k^2 = 0$) and a massive spin two ghost mode with

$$k^2 = \frac{2F_0}{f_{P0} + 4f_{Q0}} \equiv -m^2_{spin\ 2} \tag{361}$$

with mass $m^2_{spin\ 2}$. In fact, the propagator of $\bar{h}_{\mu\nu}$ can be rewritten as

$$G(k) \propto \frac{1}{k^2} - \frac{1}{k^2 + m^2_{spin2}} . \tag{362}$$

The negative sign of the second term indicates its ghost nature, which agrees with the results found in the literature for this class of theories [264–269]. As a check, we can see that for the Gauss-Bonnet Lagrangian density $\mathcal{G} = Q - 4P + R^2$, we have $f_{P0} = -4$ and $f_{Q0} = 1$, then Eq. (358) simplifies to $k^2 \bar{h}_{\mu\nu} = 0$ and in this case we have no ghosts, as expected.

The solution of Eqs. (358) and (360) can be expanded in plane waves as

$$\bar{h}_{\mu\nu} = A_{\mu\nu}(\overrightarrow{p}) \exp(ik^\alpha x_\alpha) + \text{c.c.}\,, \tag{363}$$

$$h_f = a(\overrightarrow{p}) \exp(iq^\alpha x_\alpha) + \text{c.c.}\,, \tag{364}$$

where

$$k^\alpha \equiv \left(\omega_{m_{spin\ 2}}, \overrightarrow{p}\right)\,, \qquad \omega_{m_{spin\ 2}} = \sqrt{m_{spin\ 2}^2 + p^2}\,, \tag{365}$$

$$q^\alpha \equiv \left(\omega_{m_s}, \overrightarrow{p}\right)\,, \qquad \omega_{m_s} = \sqrt{m_s^2 + p^2}\,, \tag{366}$$

and where $m_{spin\ 2}$ is zero (respectively, non-zero) in the case of massless (respectively, massive) spin two modes and the polarization tensor $A_{\mu\nu}(\overrightarrow{p})$ is given by Eqs. (21)–(23) of Ref. [270]. In Eqs. (358) and (363), the equation and the solution for the standard waves of GR [122] have been obtained while Eqs. (360) and (364) are the equation and the solution for the massive mode, respectively (see also Ref. [271]).

The fact that the dispersion law for the modes of the massive field h_f is not linear has to be emphasized. The velocity of every "ordinary" (i.e., arising from GR) mode $\bar{h}_{\mu\nu}$ is the light speed c, but the dispersion law (366) for the modes of h_f is that of a massive field which can be discussed like a wave packet [271]. The group velocity of a wave packet of h_f centered in $\overrightarrow{p}$ is

$$\overrightarrow{v_g} = \frac{\overrightarrow{p}}{\omega}\,, \tag{367}$$

which is exactly the velocity of a massive particle with mass m and momentum $\overrightarrow{p}$. From Eqs. (366) and (367), it is easy to obtain

$$v_g = \frac{\sqrt{\omega^2 - m^2}}{\omega}\,. \tag{368}$$

In order for the wave packet to have constant speed, it must be [271]

$$m = \sqrt{(1 - v_g^2)}\,\omega\,. \tag{369}$$

Before proceeding, we discuss the phenomenological constraints on the mass of the gravitational wave field. For frequencies in the range relevant for space-based and terrestrial gravitational antennas, i.e., 10^{-4} Hz $\leq f \leq 10$ kHz [105, 272–277], a strong constraint is available. For a massive gravitational wave it is

$$\omega = \sqrt{m^2 + p^2}\,, \tag{370}$$

and then

$$0\ \text{eV} \leq m \leq 10^{-11}\ \text{eV}\,. \tag{371}$$

A stronger bound comes from cosmology and Solar System tests, which provide

$$0\ \text{eV} \leq m \leq 10^{-33}\ \text{eV}\,. \tag{372}$$

The effects of these light scalars can be discussed as those of a coherent gravitational wave.

New Polarization States of Gravitational Radiation

Looking at Eq. (346) we see that we can have a $k^2 = 0$ mode corresponding to a massless spin two field with two independent polarizations plus a scalar mode while, if $k^2 \neq 0$, we have a massive spin two ghost mode ("poltergeist") and there are five independent polarization tensors plus a scalar mode. First, let us consider the case in which the spin two field is massless.

Taking $\overrightarrow{p}$ in the z-direction, a gauge in which only A_{11}, A_{22}, and $A_{12} = A_{21}$ are different from zero can be chosen. The condition $\bar{h} = 0$ gives $A_{11} = -A_{22}$. In this frame, we can take the polarization bases[12]

$$e^{(+)}_{\mu\nu} = \frac{1}{\sqrt{2}}\begin{pmatrix} 1 & 0 & 0 \\ 0 & -1 & 0 \\ 0 & 0 & 0 \end{pmatrix}, \qquad e^{(\times)}_{\mu\nu} = \frac{1}{\sqrt{2}}\begin{pmatrix} 0 & 1 & 0 \\ 1 & 0 & 0 \\ 0 & 0 & 0 \end{pmatrix}, \tag{373}$$

$$e^{(s)}_{\mu\nu} = \frac{1}{\sqrt{2}}\begin{pmatrix} 0 & 0 & 0 \\ 0 & 0 & 0 \\ 0 & 0 & 1 \end{pmatrix}. \tag{374}$$

Substituting these expressions into Eq. (350), it follows that

[12] These polarizations are defined in the physical three-space. The polarization vectors are orthogonal to each another and are normalized according to $e_{\mu\sigma}e^{\sigma\nu} = 2\delta^{\nu}_{\mu}$. The other modes are not traceless, in contrast to the ordinary "plus" and "cross" polarization modes of GR.

$$h_{\mu\nu}(t,z) = A^{+}(t-z)\,e^{(+)}_{\mu\nu} + A^{\times}(t-z)\,e^{(\times)}_{\mu\nu}$$

$$+\, h_s(t - v_g z)\,e^{s}_{\mu\nu}\,. \tag{375}$$

The terms $A^{+}(t-z)\,e^{(+)}_{\mu\nu}$ and $A^{\times}(t-z)\,e^{(\times)}_{\mu\nu}$ describe the two standard polarizations of gravitational waves which arise in GR, while the term $h_s(t - v_g z)\,\eta_{\mu\nu}$ is the massive field arising from the generic $f(R)$ theory.

When the spin two field is massive, the bases of the six polarizations are defined by

$$e^{(+)}_{\mu\nu} = \frac{1}{\sqrt{2}}\begin{pmatrix}1 & 0 & 0\\ 0 & -1 & 0\\ 0 & 0 & 0\end{pmatrix}, \qquad e^{(\times)}_{\mu\nu} = \frac{1}{\sqrt{2}}\begin{pmatrix}0 & 1 & 0\\ 1 & 0 & 0\\ 0 & 0 & 0\end{pmatrix}, \tag{376}$$

$$e^{(B)}_{\mu\nu} = \frac{1}{\sqrt{2}}\begin{pmatrix}0 & 0 & 1\\ 0 & 0 & 0\\ 1 & 0 & 0\end{pmatrix}, \qquad e^{(C)}_{\mu\nu} = \frac{1}{\sqrt{2}}\begin{pmatrix}0 & 0 & 0\\ 0 & 0 & 1\\ 0 & 1 & 0\end{pmatrix}, \tag{377}$$

$$e^{(D)}_{\mu\nu} = \frac{\sqrt{2}}{3}\begin{pmatrix}\frac{1}{2} & 0 & 0\\ 0 & \frac{1}{2} & 0\\ 0 & 0 & -1\end{pmatrix}, \qquad e^{(s)}_{\mu\nu} = \frac{1}{\sqrt{2}}\begin{pmatrix}0 & 0 & 0\\ 0 & 0 & 0\\ 0 & 0 & 1\end{pmatrix}, \tag{378}$$

and the amplitude can be written in terms of the six polarization states as

$$h_{\mu\nu}(t,z) = A^{+}(t - v_{g_{s2}}z)\,e^{(+)}_{\mu\nu} + A^{\times}(t - v_{g_{s2}}z)\,e^{(\times)}_{\mu\nu}$$

$$+\, B^{B}(t - v_{g_{s2}}z)\,e^{(B)}_{\mu\nu} + C^{C}(t - v_{g_{s2}}z)\,e^{(C)}_{\mu\nu}$$

$$+\, D^{D}(t - v_{g_{s2}}z)\,e^{(D)}_{\mu\nu} + h_s(t - v_g z)\,e^{s}_{\mu\nu}\,, \tag{379}$$

where

$$v_{g_{s2}} = \frac{\sqrt{\omega^2 - m_{s2}^2}}{\omega} \tag{380}$$

is the group velocity of the massive spin two field. The first two polarizations are the same as in the massless case, inducing tidal deformations of the (x, y) plane. Figure 1 illustrates how each gravitational wave polarization affects test masses arranged in a circle before the wave impinges on them.

From a purely quantum-mechanical point of view, the presence of the ghost mode may seem as a pathology of the theory. There are several reasons why this mode is problematic in the particle interpretation of the metric perturbations. The ghost mode can be viewed as either a particle state with positive energy and negative probability density, or as a positive probability density state with negative energy. In the first

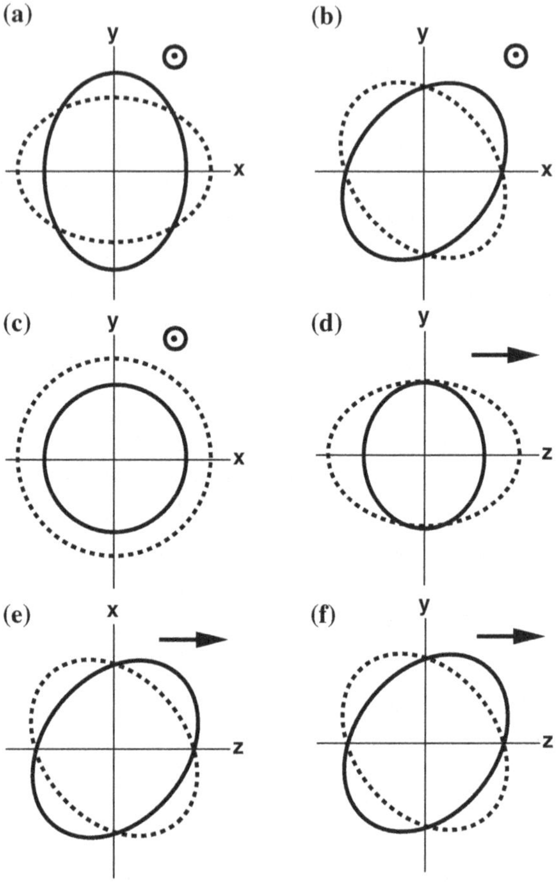

Fig. 1 The six polarization modes of gravitational waves. We illustrate the displacement induced at phases spaced by π radians by each mode on a circle of test particles at rest before the wave impinges upon them. The wave propagates out of the plane of the page in (**a**), (**b**), and (**c**) and into this plane in (**d**), (**e**), and (**f**). While **a** and **b** describe the "plus" and "cross" modes, respectively, **c** corresponds to the scalar mode, and **d**, **e**, and **f** to the D, B, and C modes

case, allowing the presence of such a particle will induce violations of unitarity, while the negative energy scenario leads to a theory without ground state and the system becomes unstable. Vacuum can decay into pairs of ordinary and ghost gravitons leading to a catastrophic instability.

A way out of these problems consists of imposing a very weak coupling of the ghost with the other particles in the theory, such that the decay rate of the vacuum becomes comparable to the inverse of the Hubble time. The present vacuum state will then appear to be sufficiently stable. This is not a viable option in our theory because the ghost state appears in the gravitational sector, which is bound to couple to all forms of matter present and it seems physically and mathematically unlikely for the ghost graviton to couple differently than the ordinary massless graviton does.

Another possibility consists of assuming that this picture does not hold up to arbitrarily high energies and that at some cutoff scale M_{cutoff} the theory gets modified appropriately to ensure a ghost-free behavior and a stable ground state. This can happen, for example, if we assume that Lorentz-invariance is violated at M_{cutoff}, thereby restricting any potentially harmful decay [278]. However, there is no guarantee that modified gravities like the one investigated here are valid to arbitrarily high energies. Such models are plagued at the quantum level by the same problems of ordinary GR, i.e., they are not renormalizable. It is, therefore, not necessary for them to be considered as genuine candidates for a quantum gravity theory and the corresponding ghost particle interpretation becomes ambiguous. At the classical level, the perturbation $h_{\mu\nu}$ should be viewed as nothing more than a tensor representing the stretching of space-time away from flatness. A ghost mode then makes sense as just another way of propagating this perturbation of the space-time geometry, one which, in the propagator, carries a sign opposite to that of an ordinary massive graviton. Viewed in this way, the presence of the massive ghost graviton will induce on an interferometer the same effects as an ordinary massive graviton transmitting the perturbation, but with the opposite sign of the displacement. Tidal stretching of the polarization plane by a polarized wave will turn into shrinking and vice-versa. Eventually, the signal will be a superposition of the displacements coming from the ordinary massless spin two graviton and the massive ghost. Since these two modes induce competing effects, their superposition will lead to a less pronounced signal than the one expected were the ghost mode absent, setting less stringent constraints on the theory. However, the presence of the new modes will also affect the total energy density carried by the gravitational waves and this may also appear as a candidate signal in stochastic gravitational wave backgrounds.

The Detector Response

Let us consider now the possible response of a detector in the presence of gravitational waves coming from a definite direction. The detector output depends on the gravitational wave amplitude, which is determined by specific theoretical models. However, one can study the detector response to each gravitational wave polarization without specifying a priori the theoretical model. Following Refs. [279–281, 286–289], the angular pattern function of a detector of gravitational waves is given by

$$F_A(\hat{\Omega}) = \mathbf{D} : \mathbf{e}_A(\hat{\Omega}) \,, \tag{381}$$

$$\mathbf{D} = \frac{1}{2}\left(\hat{\mathbf{u}} \otimes \hat{\mathbf{u}} - \hat{\mathbf{v}} \otimes \hat{\mathbf{v}}\right) \,, \tag{382}$$

where $A = +, \times, B, C, D, s$ and : denotes a contraction between tensors. $\mathbf{D}$ is the *detector tensor* representing the response of a laser-interferometric detector. It maps the metric perturbation in a signal on the detector. The vectors $\hat{\mathbf{u}}$ and $\hat{\mathbf{v}}$ are unitary and orthogonal to each other, they are directed to each detector arm, and they form an orthonormal coordinate basis together with the unit vector $\hat{\mathbf{w}}$ (see Fig. 2). $\hat{\Omega}$ is the unit vector directed along the direction of propagation of the gravitational wave.

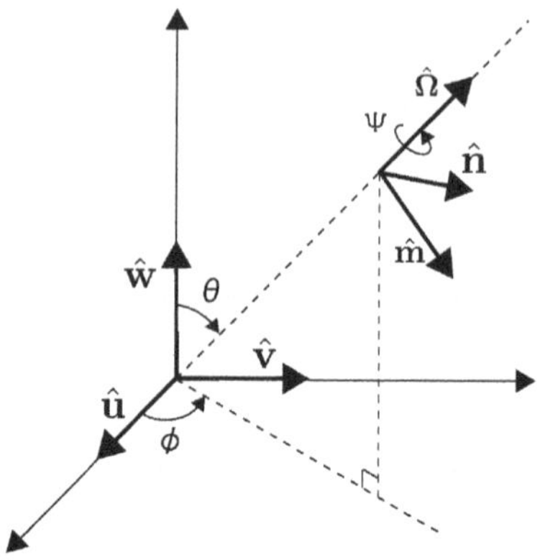

Fig. 2 The coordinate systems used to calculate the polarization tensors and a view of the coordinate transformation

Equation (381) holds only when the arm length of the detector is much smaller than the gravitational wave wavelength, a condition satisfied by ground-based laser interferometers but not by space interferometers such as *LISA*. A standard orthonormal coordinate system for the detector is

$$\hat{\mathbf{u}} = (1, 0, 0) \ , \tag{383}$$

$$\hat{\mathbf{v}} = (0, 1, 0) \ , \tag{384}$$

$$\hat{\mathbf{w}} = (0, 0, 1) \ , \tag{385}$$

and the coordinate system for the gravitational wave, rotated by (θ, ϕ), is given by

$$\hat{\mathbf{u}}' = (\cos\theta\cos\phi, \cos\theta\sin\phi, -\sin\theta) \ , \tag{386}$$

$$\hat{\mathbf{v}}' = (-\sin\phi, \cos\phi, 0) \ , \tag{387}$$

$$\hat{\mathbf{w}}' = (\sin\theta\cos\phi, \sin\theta\sin\phi, \cos\theta) \ . \tag{388}$$

A rotation by the angle ψ around the direction of propagation of the gravitational wave gives the most general choice of coordinates, that is

$$\hat{\mathbf{m}} = \hat{\mathbf{u}}'\cos\psi + \hat{\mathbf{v}}'\sin\psi \, , \tag{389}$$

$$\hat{\mathbf{n}} = -\hat{\mathbf{v}}'\sin\psi + \hat{\mathbf{u}}'\cos\psi \, , \tag{390}$$

$$\hat{\Omega} = \hat{\mathbf{w}}' \, . \tag{391}$$

The coordinates $\left(\hat{\mathbf{u}}, \hat{\mathbf{v}}, \hat{\mathbf{w}}\right)$ are related to $\left(\hat{\mathbf{m}}, \hat{\mathbf{n}}, \hat{\Omega}\right)$ by the rotation angles $(\phi,\ \theta,\ \psi)$, as shown in Fig. 2. Using the vectors $\hat{\mathbf{m}}$, $\hat{\mathbf{n}}$, and $\hat{\Omega}$, the polarization tensors are

$$\mathbf{e}_{+} = \frac{1}{\sqrt{2}}\left(\hat{\mathbf{m}} \otimes \hat{\mathbf{m}} - \hat{\mathbf{n}} \otimes \hat{\mathbf{n}}\right) , \tag{392}$$

$$\mathbf{e}_{\times} = \frac{1}{\sqrt{2}}\left(\hat{\mathbf{m}} \otimes \hat{\mathbf{n}} + \hat{\mathbf{n}} \otimes \hat{\mathbf{m}}\right) , \tag{393}$$

$$\mathbf{e}_{B} = \frac{1}{\sqrt{2}}\left(\hat{\mathbf{m}} \otimes \hat{\Omega} + \hat{\Omega} \otimes \hat{\mathbf{m}}\right) , \tag{394}$$

$$\mathbf{e}_{C} = \frac{1}{\sqrt{2}}\left(\hat{\mathbf{n}} \otimes \hat{\Omega} + \hat{\Omega} \otimes \hat{\mathbf{n}}\right) . \tag{395}$$

$$\mathbf{e}_{D} = \frac{\sqrt{3}}{2}\left(\frac{\hat{\mathbf{m}}}{\mathbf{2}} \otimes \frac{\hat{\mathbf{m}}}{\mathbf{2}} + \frac{\hat{\mathbf{n}}}{\mathbf{2}} \otimes \frac{\hat{\mathbf{n}}}{\mathbf{2}} + \hat{\Omega} \otimes \hat{\Omega}\right) , \tag{396}$$

$$\mathbf{e}_{s} = \frac{1}{\sqrt{2}}\left(\hat{\Omega} \otimes \hat{\Omega}\right) . \tag{397}$$

Taking into account Eqs. (381) and (382), the angular patterns for each polarization are

$$F_{+}(\theta, \phi, \psi) = \frac{1}{\sqrt{2}}(1 + \cos^2\theta)\cos 2\phi \cos 2\psi \\ - \cos\theta \sin 2\phi \sin 2\psi , \tag{398}$$

$$F_{\times}(\theta, \phi, \psi) = -\frac{1}{\sqrt{2}}(1 + \cos^2\theta)\cos 2\phi \sin 2\psi \\ - \cos\theta \sin 2\phi \cos 2\psi , \tag{399}$$

$$F_{B}(\theta, \phi, \psi) = \sin\theta\,(\cos\theta \cos 2\phi \cos\psi - \sin 2\phi \sin\psi) , \tag{400}$$

$$F_{C}(\theta, \phi, \psi) = \sin\theta\,(\cos\theta \cos 2\phi \sin\psi + \sin 2\phi \cos\psi) , \tag{401}$$

$$F_{D}(\theta, \phi) = \frac{\sqrt{3}}{32}\cos 2\phi \left[6\sin^2\theta + (\cos 2\theta + 3)\cos 2\psi\right] , \tag{402}$$

$$F_{s}(\theta, \phi) = \frac{1}{\sqrt{2}}\sin^2\theta \cos 2\phi . \tag{403}$$

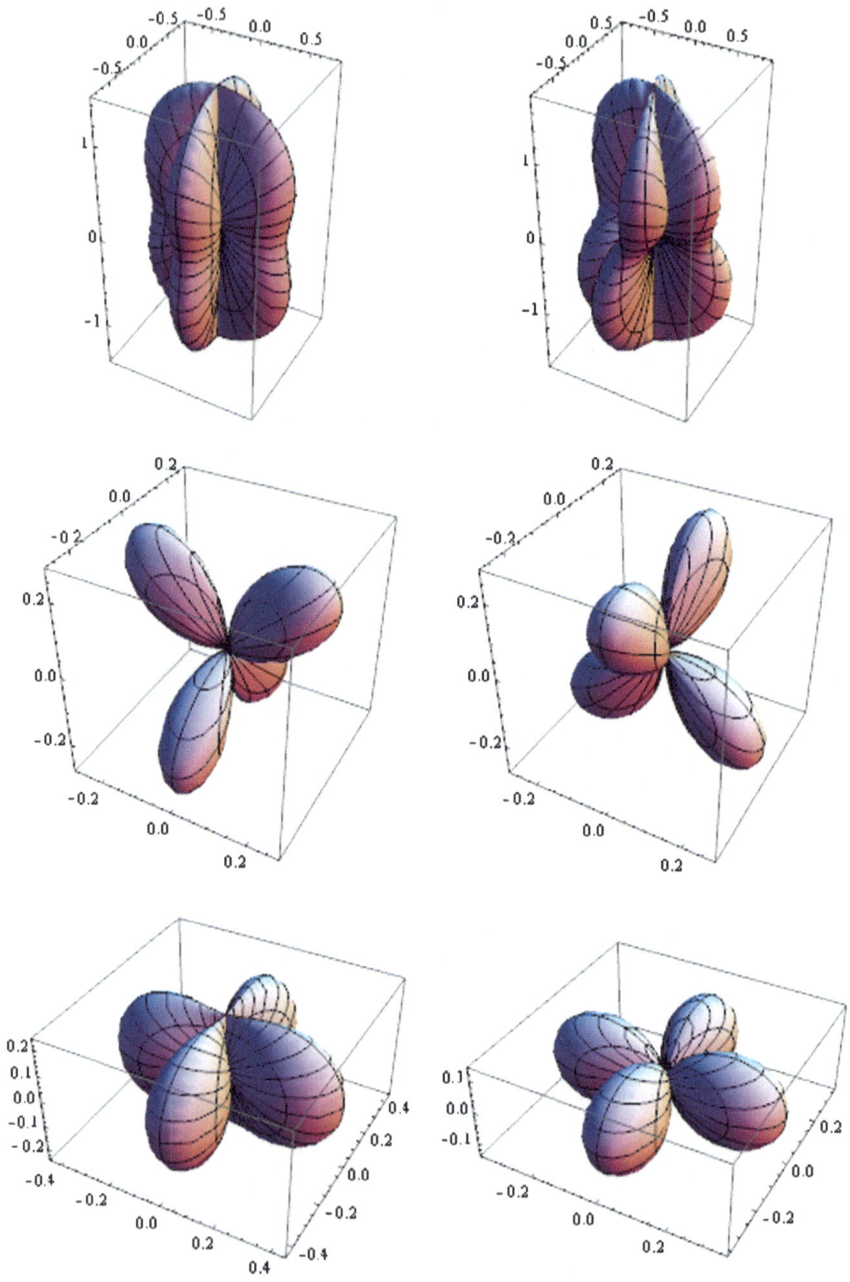

Fig. 3 Angular pattern functions of an interferometric detector for the various polarizations. From *left* to *right* and from *top* to *bottom*, one sees constant level surfaces corresponding to the "plus", "cross", B, C, D, and scalar modes

The angular pattern functions for each polarization are plotted in Fig. 3. Even if we have considered a different model, these results are consistent, for example, with those of Refs. [279, 280, 282–285].

Another area of research which we do not discuss here consists of the study of the stochastic background of gravitational waves which may contain the possible signature of extra gravitational wave modes and be relevant for the detectability of these contributions to gravitational radiation.

The above analysis covers extended gravity models with a generic class of higher order Lagrangian densities and Lagrangian terms of the form $f(R, P, Q)$. We have linearized the field equations of these theories around a Minkowski background and found that, in addition to a massless spin two field, the theory contains also spin zero and two massive modes with the latter being, in general, ghosts. If the interferometer is directionally sensitive and we also know the orientation of the source (and, of course, if the source is coherent) the discussion is straightforward. In this case, the massive mode coming from the simplest extension of GR, i.e., $f(R)$ gravity, would induce longitudinal displacements along the direction of propagation of the wave, which should be detectable, and only the scalar mode would be the detectable truly new signal [271]. But, even in this case, there could be a second scalar mode inducing a similar effect and representing a massive ghost, although with a negative sign.

For the situation considered here, massive modes are certainly of interest for the *LISA* space interferometer. It is in principle possible that massive gravitational wave modes could be produced in more significant quantities in cosmological or early astrophysical processes in alternative theories of gravity, a possibility which is still largely unexplored. This situation should be kept in mind when looking for a signature capable of distinguishing these theories from GR, and it seems to deserve further investigation.

The weak-field limit of ETGs shows new aspects of gravitation which are not present in GR. The Newtonian and post-Newtonian limits give weak-field potentials which are not of the standard Newtonian form. The corrections, in general, are Yukawa-like terms which could explain in a very natural fashion several astrophysical and cosmological observations [290–292].

The post-Minkowskian limit of ETGs exhibits new gravitational field modes which can easily be interpreted as massive gravitons.

The study of generation, propagation, and detection of GWs in the weak-field limit of a given relativistic theory of gravity is an important part of astrophysics. Primordial gravitational waves generated during the early epochs of the Universe (especially during inflation) would allow, when detected, to rule out or constrain certain theories and investigate others. The detection of GWs of astrophysical or cosmological origin opens a new branch of astronomy providing information which is not accessible with visible, infrared, optical, X-ray, or γ-ray astronomy. In fact, GWs can be generated in regions deep inside supernovae, near black hole horizons, or very early in the history of the Universe when the latter is completely opaque to photons. The study of relativistic astrophysics related to GWs in ETGs is a broad and complex subject that open new interesting scenarios. In the forthcoming sections, we will discuss these points and, in particular, the multimessenger approach related to the discovery of GWs.

The Discovery of Gravitational Waves: The Turning Point

The last decades have seen breakthroughs in many fields of astronomy, fostered by technological improvements enabling the development of very large optical telescopes, radio antennas and arrays, infrared, X-ray, and gamma-ray satellites.

The first direct detection of gravitational waves by the LIGO and Virgo Collaborations on September 14th, 2015, opened a new window to explore the Universe, complementary to the electromagnetic observations [314]. Gravitational waves are in fact emitted by the cumulative mass and momentum of entire systems, so they have long wavelengths and convey direct information about large-scale regions. Furthermore, gravitational waves couple extremely weakly to matter, making them very hard to detect but also allowing them to travel to us substantially unaffected by intervening matter, even from the earliest moments of the Big Bang. Efforts to detect gravitational waves made a decisive step with the construction of second generation interferometric detectors, the advanced versions of the km baseline ground-based instruments. At the same time, progress in relativistic astrophysics and numerical relativity have improved our understanding of expected amplitudes and rates of a number of classes of gravitational wave signals. The purpose of this section is to provide an overview of the present status of gravitational wave search and give an outlook of the physics of gravitational waves, their impact in fundamental physics, astrophysics and cosmology.

The first indirect evidence for gravitational wave emission was provided by the famous Hulse-Taylor binary system, PSR1913+16, consisting of an observed pulsar with 17 Hz radio emission in an 8 h orbit with an unseen neutron star companion, which showed a quadratic decrease in the orbital period (40 s over 30 years), in remarkably good agreement with expectation from GWs energy loss [294].

It is worth stressing again that GWs are produced by accelerated masses. In analogy with the electromagnetic case, the GWs produced by a mass distribution can be expressed as an expansion in multipole mass moments. Due to momentum and angular momentum conservation laws, the lowest non-zero term in the expansion is the quadrupole term. The amplitude of a wave at a distance r from a source is proportional to the second time derivative of the quadrupole moment of the source and inversely proportional to r, while the luminosity of the source is proportional to the square of the third time derivative of the quadrupole moment.

GWs are expected to be emitted over a wide range of frequencies, spanning from 10^{-17} Hz in the case of ripples in the cosmological background to 10^3 Hz for those processes where compact objects such as neutron stars are involved (the generation of a neutron star in a supernova explosion or the merger of two neutron stars in a binary system).

It is useful to classify GW sources (and the methods for detecting their waves) by the frequency band in which they radiate.

As far as compact sources are concerned, the band in which GWs are generated is typically related to the source's size R and mass M. R is meant to set the scale over which the source's dynamics vary; for example, it could be the actual size of a particular body, or the separation of members of a binary. The GW frequency of such a source can be written as $f_{GW} \simeq \frac{1}{2\pi}\sqrt{\frac{GM}{R^3}}$. Since the radius of the source must be larger than its Schwartschild radius $R_S = 2GM/c^2$, we can estimate an upper bound for the frequency of a compact source:

$$f_{GW}(M) < \frac{1}{4\sqrt{2}\pi}\frac{c^3}{GM} \simeq 10^4\,\mathrm{Hz}\left(\frac{M_\odot}{M}\right). \tag{404}$$

This is a stringent upper limit, since many interesting sources are quite a bit larger than R_S, or else evolve through a range of sizes before terminating their emission at $R \sim R_S$. Nonetheless, this frequency gives some sense of the types of compact sources that are likely to be important in each band. High-frequency compact sources are of stellar mass (up to several solar masses); low frequency compact sources are of thousands to millions of solar masses, or else contain widely separated stellar mass bodies.

The challenge for the experimental physicist is that the predicted GW amplitude (equal in magnitude to twice the induced strain) in the vicinity of the Earth is extremely small, typically not higher than 10^{-21} [295]. Different methods have been proposed to detect gravitational waves, depending on the signal frequency: from the polarization of the CMB in the very low frequency range to Pulsar Timing, atom interferometers, space-based and ground based interferometers and resonant-mass detectors for the highest frequencies [296–304].

In the following section, we will focus on optical interferometric GW detectors.

Interferometric GW Detection: An Overview

Interferometric detectors measure the passage of a GW by observing its action upon widely-separated test masses. The space-based detector eLISA has test masses in a large equilateral triangle that orbits the sun. On the ground, each mass is suspended with a sophisticated isolation system made by a cascade of pendula, to eliminate the effect of local ground noise. Above the resonant frequency of the pendulum (typically about 1 Hz), the mass moves freely. (In space, the masses are actually free floating.) In the absence of a GW, the interferometer arms L1 and L2 have about the same length L. As an example we can consider the interferometer arms along the x and y directions and a + polarised wave impinging on the detector down the z axis. For the sake of simplicity, we can further consider the axes of the + polarization to be aligned with the detector. The tidal force of this wave will stretch one arm while squeezing the other; each arm oscillates between stretch and squeeze as the wave itself oscillates. The wave is thus detectable by measuring the separation between the test masses in each arm and watching for this oscillation. In particular, since one arm is always stretched while the other is squeezed, we can monitor the difference in length of the two arms: $\delta L(t) \equiv L_1(t) - L_2(t)$. By applying the equation of geodesic deviation to the separation of the test masses and using the weak field approximation, the change in length for the case considered here turns out to be [305] $\delta L(t) = h_+(t)L$. For a generic GW:

$$\frac{\delta L(t)}{L} = F^+ h_+(t) + F^\times h_\times(t) \tag{405}$$

where F^+ and $F^\times$ are named antenna pattern functions and weight the two polarisations as a function of the source's position and orientation relative to the detector.

Following preliminary studies by Pirani in 1956 [301] and Gertsenshtein and Pustovoit in 1962 [302], the detailed analysis of the possibility to measure GWs with laser interferometry was performed and published by Rainer Weiss in 1972 [303]. Since then, impressive technological improvements have been achieved, which found their expression in the current large-scale interferometers LIGO and Virgo and paved the way to future projects, both ground and space-based.

Ground-Based Interferometers

From the first to the second generation. At the beginning of the 21st century, a set of initial detectors was put in operation, including TAMA300 in Japan [306], GEO600 in Germany [307], LIGO in the United States [308], and Virgo in Italy [309]. Combinations of these detectors made joint science runs from 2002 through 2011, setting upper limits on GWs emitted by a variety of sources.

Then, the construction of the second generation instruments, Advanced LIGO [310] and Advanced Virgo [311], started. The goal of these upgraded interferometers is to improve the sensitivity by a factor of 10 in the Hz-kHz frequency band with respect to initial detectors, thus increasing the volume of observable universe and the rate of detectable sources by a factor of a thousand. More recently, other international projects have started: the Japanese KAGRA [312] and LIGO India [313] are expected to join Advance Virgo and Advanced LIGO in the 2020s, thus nourishing the world network of second generation detectors.

Third generation interferometers. The detection of the first GW signals announced by the Virgo and LIGO Collaborations [314, 315] has provided a new tool to observe the universe. The expected detections of the second generation instruments at their design sensitivity are of the order of a few tens per year. However, the expected signal-to-noise ratios may not be sufficient for precise astrophysics and cosmology. For this reason, the possibility to realise new detectors, with a drastically improved sensitivity with respect to second generation interferometers, is being considered.

The first studies in this direction were funded by the EU within the Framework Program 7 and resulted in the proposal of the Einstein Telescope (ET) [316, 317]. ET is a large underground and cryogenic interferometer, with the goal to improve the sensitivity of the advanced detectors by one order of magnitude and extend the low-frequency bandwidth. The effort toward the definition of a third generation interferometer design has more recently started within the US GW community leading to the Cosmic Explorer project [318]. The sensitivity target of these instruments is the observation of compact binary sources with high signal-to-noise ratio throughout the universe.

Space-Based Interferometers

It is not possible to measure GWs with frequencies below 1 Hz using a ground-based instrument: even with a "perfect" isolation system, gravitational coupling to fluctuations in the local mass distribution would ultimately limit the sensitivity to these frequencies. The only way to measure these signals is to build a GW observatory in the quiet environment of space, far away from low-frequency noise sources.

The reference project for such an instrument is the LISA mission [319], consisting of three spacecrafts orbiting the Sun in a triangular configuration, connected by the arms of a laser interferometer. The interaction with the GW is measured by monitoring the changes in distance between free falling test masses inside the spacecrafts.

In order to test some key concept of GW detection in space, a technology demonstrator was conceived: the LISA Pathfinder (LPF) mission [320]. LPF satellite comprises two test masses in a gravitational free-fall, similar to those projected for use in eLISA, at about 30 cm distance, interferometrically read. The LPF goal was set to test LISA technologies to within less than one order of magnitude of the LISA requirements. LPF was launched successfully on December 3, 2015. The results show that the two test masses can be put in free fall with a residual relative acceleration noise of $(0.54 \pm 0.01) \times 10^{-15}\,\mathrm{g}/\sqrt{\mathrm{Hz}}$ (g is the standard gravity acceleration), for frequencies between 0.7 and 20 mHz. This value is within a factor 1.25 of the requirement for the LISA mission.

Science with GWs

Because of their extremely weak interaction with matter, GWs can bring us information from the innermost parts of compact objects and, after travelling cosmological distances without relevant interactions, can reveal features of their sources unaccessible with e.m. radiation or neutrinos.

The science reach achievable with GW observations promises to be very rewarding, shedding light on many crucial open questions in fundamental physics, astronomy and cosmology. There is a wide literature on these topics (see for example [321]). In this section, a summary will be given, starting from the various source types.

Binary Systems

Binary systems can be made of neutron stars, black holes, super-massive black holes, white dwarfs, variously combined by types and masses [321]. The dynamics of these systems is governed by energy loss through GW emission and the signal is typically characterized by three different phases:

- inspiral: the amplitude and frequency increase as the two object get closer and closer. This is the status in which the system spends most of its lifetime. For instance the pulsars of PSR1913+16 will still have to orbit 300 millions years around each other.
- merger: this is the phase when the two objects coalesce into a single one
- ring down: the final object relaxes by damped oscillations, characterized by a superposition of quasi-normal modes.

The possibility to detect the emitted radiation depends on the compactness of the components, the mass of the system and on its evolutionary stage. Neutron stars and stellar mass black holes systems can reach high frequencies before merging, so their inspiral, coalescence and ring down can be observed by ground-based interferometers. On the other hand, space-based detectors can measure the GWs emitted during all phases by binary systems involving high mass and supermassive black holes and white dwarfs, and during the inspiral by neutron stars and stellar mass black holes binaries.

Binary systems are extremely valuable GW sources for the amount of scientific outcomes they can provide in fundamental physics, astrophysics and cosmology. The emitted waveform is known very accurately, thanks to the progress achieved in the Post-Newtonian formalism and in numerical relativity. This feature makes binary systems "standard candles" [322]: the intrinsic luminosity, and thus the luminosity distance, of an inspiral GW event can be inferred directly from the amplitude of the observed waves and from the component masses, which govern the structure of the signal. As a consequence, from the observation of the GWs emitted by binary systems in galaxies or clusters of galaxies at high known redshift, it will be possible to measure the Hubble constant or the dark energy equation of state. An overview on some of the other possible science outcomes from binary systems is given in the following. For a more detailed review, see for instance [316, 317, 319].

Tests of General Relativity. The accurate knowledge of the emitted waveform makes binary systems the ideal sources to test the validity of General Relativity.

In Einstein's theory, GWs are supposed to travel at the speed of light, being the graviton a massless particle. On the other hand, in a massive graviton theory GWs would obey a dispersion relation and the propagation speed of the waves would depend on its wavelength. As a consequence, the phasing of the signal seen by a detector far from the source (hundreds of Mpc) would be distorted because of the shifted times-of-arrival of the waves emitted with different wavelengths during the inspiral. The effect would increase by increasing the distance from the source. Thus, observations of inspiralling binaries can be used to place bounds on the mass of the graviton, or equivalently the Compton wavelength of the graviton [323]. Results in this perspective and for gravitational Lorentz violation are reported in [324].

From the time evolution of the phase in the inspiralling signal it is also possible to put constraints on theories alternative to GR, such as the Brans-Dicke theory, or to test for violations of GR without assuming a particular alternative model according to the method proposed in [326–328], based on the post-Newtonian (PN) expansion of the phase of an inspiral signal in the frequency domain.

Black hole physics. GR black holes are expected to fulfill the no-hair theorem, i.e. their behaviour is completely characterised by mass and spin. All other information (for which "hair" is a metaphor) about the matter which formed the black hole or is falling into it, "disappears" behind the event horizon and is therefore permanently inaccessible to external observers. Different methods have been proposed to test the validity of this theorem [329–332], for example by measuring the BH oscillations. In fact, a perturbed Kerr BH (as that resulting after the coalescence) emits gravitational radiation which consists of a superposition of damped sinusoids named quasi-normal modes. As a consequence of the no-hair theorem, the frequencies and time-constants of these modes depend only on the BH mass and spin. Therefore, if the observed gravitational signal from a ringing BH can be decomposed into a superposition of different modes, then the frequencies and time-constants of each of the modes could be used to infer the mass and spin of the BH. If the object is a GR BH, then the masses and spins obtained from the different modes should all be consistent within the measurement errors. Otherwise, the measurement could suggest the failure of GR.

Another interesting measurement concerns the test of the uniqueness theorem which states that the Kerr metric is the only possible description for a BH. In binary black hole systems where one component is much heavier than the other (as the $(10 - 10^6)\,M_\odot$ for a LISA-like detector or the (1–100) $M_\odot$ for ground-based interferometers), many cycles of the GW radiation are emitted while the smaller object is in the strong field region close to the larger object. These GWs encode a map of the spacetime structure in the vicinity of the large BH, which can be used to measure properties of the central object [329, 333, 334].

So far, the existence of intermediate mass black holes IMBHs with mass 10^2–$10^4 M_\odot$, has not been confirmed experimentally, being rather far in the future of e.m. observations. The GW detection and characterization of a binary system containing an IMBH would corroborate the existence of such objects, improving our understanding of the formation and evolution of supermassive BH, as well as galaxy evolution modelling and cosmography. Signals from IMBH binaries would start in the LISA band and sweep through to the ground based detectors band, allowing the observation of different phases of the coalescence.

As a last example, we can refer to the issue of initial seeds of massive black holes (MBHs) located in the center of many galaxies. The open questions concern their masses, formation history, localization. Future GW interferometers will be able to discriminate between the two current scenarios: the light seed scenario, in which $100\,M_\odot$ BH seeds form at redshift $z \sim 20$ from the collapse of Population III stars [335, 336], and the heavy seed scenario, in which BH of mass $10^5 M_\odot$ form from direct collapse of dust clouds [337]. The possibility to observe binary systems containing BHs with these masses could definitely rule out one of the two scenarios.

Spinning Neutron Stars

Besides emitting GWs as components of binary systems, neutron stars are interesting GW sources by themselves. They can emit a continuous signal due to non-axisymmetric distortions or free precession. A spinning non-axisymmetric neutron star at distance d, rotating with frequency ν emits monochromatic GWs at a frequency $f = 2\nu$ of amplitude

$$h_0 \simeq 10^{-27}\left(\frac{I_3}{10^{38}\,\mathrm{kg\,m}^2}\right)\left(\frac{10\,\mathrm{kpc}}{d}\right)\left(\frac{f}{100\,\mathrm{Hz}}\right)^2\left(\frac{\epsilon}{10^{-6}}\right) \tag{406}$$

where I_3 is the momentum of inertia with respect to the rotational axis and ϵ is the star ellipticity. This last parameter has large uncertainties, because it strongly depends on the internal composition and magnetic field of the star. For a canonical 1.4 $M_\odot$ neutron star, the ellipticity is estimated to be (1–8) $\times 10^{-6}$, depending on the model used for the crust and the EOS, and can go up to a few $\times 10^{-3}$ for stars containing solid quark matter. The maximum expected signal frequency is below 2 kHz.

The other emission mechanism for continuous GWs from spinning neutron stars is the free precession, i.e. the "wobble" of a neutron star with a misaligned rotation axis with respect to its symmetry axis (defining the wobble angle θ_w). A large-angle wobble would typically produce GWs of amplitude [325]

$$h_0 \sim 10^{-28}\left(\frac{\theta_W}{0.1\,\mathrm{rad}}\right)\left(\frac{10\,\mathrm{kpc}}{d}\right)\left(\frac{\nu}{500\,\mathrm{Hz}}\right)^2. \tag{407}$$

The radiation is emitted at $f = \nu + \nu_{prec}$ (where ν_{prec} is the precession frequency) and $f = 2\nu$. The smallness of the emitted amplitude is compensated by its long lived duration, which makes it possible integrating over long period of times, thus improving the SNR. The measurement of the signal emitted by neutron stars can provide information about their internal structure and their demography. In fact, current models of stellar evolution suggest that our Galaxy contains 10^9 neutron stars, while about 10^5 are expected to be active pulsars. Up to now about 2000 pulsars have been observed [325]; there are numerous reasons for this, including selection effects and because many produce only faint emission.

Apart from the continuous emission, neutron stars can produce burst-like signals as those generated by instabilities and glitches. Many radio pulsars exhibit glitches, that are events in which the star is seen to suddenly spin-up and then relax towards stable secular spin-down. So far, over a hundred pulsars are known to have glitched at least once [316, 317]. Glitches have also been observed in magnetars. Despite the large observational data, the glitch remains an enigma from a theoretical point of view [338], possibly related to the existence of superfluids in the interior of mature neutron stars. The signal emitted is expected to be weak, but its strength is largely unknown, being dependent on the glitch mechanism. For some models, the amplitude should be in the reach of third generation ground-based interferometers. An observed signal would immediately constrain the cold-matter nuclear EOS [339, 340]. A lack of detection would set relevant limits on the asymmetry and fluid dynamics associated with pulsar glitches.

Finally, neutron stars may undergo a number of instabilities, which can be directly associated with unstable modes of oscillation [341]. Detection of these signals would allow us to put constraints on the interior structure of the star and the extreme physics associated with the high-density region. The most promising instability scenarios concerns rotating stars. Of particular interest are the GW driven instabilities of the f-modes and the r-modes.

Supernovae

Core-Collapse Supernovae (CCSNe) are among the most energetic events in the Universe, releasing 10^{53} erg of energy, most of which (about 99%) emitted through neutrinos. The precise mechanism of explosion operating in CCSNe is uncertain [342–344]. The observation of SN1987a in the Large Magellanic Cloud, performed both in the e.m. spectrum and with neutrinos, made it possible to identify some basic features of the explosion, such as the amount of delivered energy (3×10^{52} erg) which is consistent with the prediction for a progenitor mass around 22 $M_{\odot}$ [345]. There are strong indications that the CCSNe explosion involves a multitude of multi-dimensional processes, including rotation, convection, turbulence, and various hydrodynamic instabilities. The general framework includes a burst emission at bounce and a post-bounce emission, during which the actual explosion takes place [346]. The emitted waveform encodes the physics behind the collapse: the expected GW amplitude is in the range $10^{-24} - 10^{-22}(d/1\text{Mpc})$ and most of the emission takes place at frequencies between 200 and 1000 Hz, but the various explosion scenarios exhibit unique spectral distributions and vary in total emitted energies [342, 347].

With advanced interferometric GW detectors, CCSNe will be observable in the Milky Way ($d \sim$ 10–15 kpc) and in the Magellanic Clouds [348] ($d \sim$ 50–70 kpc), but the expected rates are no more than 1–2 events per century, even in the most optimistic scenarios. At distances between 3 and 10 Mpc (in the reach of third generation ground based interferometers) the expected rates range between (0.5–1) year^{-1}.

GW observations will provide a unique tool to shed light on the actual SN mechanism. Constraining the CCSNe mechanism will represent a breakthrough in our understanding of the large range of phenomena associated with stellar collapse, BH and NS formation, and gamma-ray bursts (GRBs). In particular, for this last family of objects, the investigation through GWs will contribute to better discriminate between long and short GRBs being the first class related essentially to SN or hypernovae events and the second to coalescing systems. The simultaneous observation of GWs, neutrinos and e.m. radiation emitted in these events, will provide insight in the dynamics of the collapse, timing of the GW emission and additional astrophysical information, such as the progenitor type and mass and the explosion features. The contribution of GW observations in this multimessenger framework is fundamental: GWs carry information on the high-density nuclear EoS, explosion asymmetries and pulsar kicks, the formation of a BH in a failing CCSN, and can help uncover rare events such as the accretion-induced collapse of a white dwarf to a NS, or weak or failing CCSNe that have very weak or absent e.m. signatures.

Stochastic Background

The strength of this signal is expressed through the quantity Ω_{GW} defined as the ratio of its energy density per logarithmic frequency interval and the critical density. It can be either of astrophysical and cosmological origin [289].

Astrophysical background. A GW stochastic background of astrophysical origin may generate from the superposition of a large number of unresolved sources since the beginning of stellar activity, such as magnetars, binary systems, r-modes and dynamical bar modes in proto neutron stars, Pop II core collapse to NS and to BH. The detection of such a background would put very strong constraints on the physical properties of compact objects, the initial mass function and/or the star formation history. On the other hand, it could be a "noise" that would mask the stochastic background of cosmological origin.

Cosmological background. The GW background of cosmological origin, once disentangled from the astrophysical background, can carry extraordinary information on the state of the very early universe and on physics at correspondingly high energies. In fact, a background of relic particles gives a snapshot of the state of the universe at the time when these particles decoupled from the primordial plasma. For this reason, the small cross section of the gravitons made them decouple decoupled below the Planck scale $M_{Pl} \sim 10^{19}$ GeV, that is already 10^{-44} s after the big-bang. This means that a background of GWs produced in the very early universe encodes still today, in its frequency spectrum, all the information about the conditions in which it was created. For comparison, the photons that we observe in the CMB decoupled when the temperature was of order $T \simeq 0.2$ eV, or 3×10^5 year after the big bang. This difference in scales simply reflects the difference in the strength of the gravitational and electromagnetic interactions.

Primordial GW backgrounds can be wide-band, where $\Omega_{GW}(f)$ is approximately constant over a large range of frequencies, or peaked, where $\Omega_{GW}(f)$ varies strongly in frequency. Wide-band sources are processes that extend over a large range of the cosmological scale factor $a(t)$, such as inflation and cosmic string evolution. Both these sources depend on unknown fundamental physics, and also have an approximate scaling symmetry [349–353]. Possible alternatives to exponential inflation as a source of primordial perturbations are the 'pre-big-bang' scenario in string cosmology and the ekpyrotic and cyclic models involving a contracting phase and subsequent brane collision [354]. Peaked sources of stochastic background result from an event localised in cosmic time, typically a phase transition or reheating after inflation [355–364].

The signal deriving from a cosmological background is expected to spread across a huge range of frequencies (from 10^{-18} to 10^{12} Hz, with frequency bands more or less populated depending on the specific process considered [365]). For this reason, besides interferometric detectors, also other instruments working at lower frequencies may give a valuable contribution to the detection of this signal. The pulsar timing array projects will be sensitive at nHz frequencies, while in the range 10^{-18}–10^{-16} Hz the imprint of the GW cosmological background can be looked for in the B-modes polarisation of CMB [366].

The First Direct GW Detection

A century after the fundamental predictions of Einstein [293] and Schwarzschild [367], on February 11, 2016, the LIGO Scientific Collaboration and the VIRGO Collaboration reported the first direct detection of gravitational waves [314] and the first direct observation of the merger of a binary black hole system into a single black hole. This measurement has given the possibility to study the properties of space-time in the strong-field regime and confirm predictions of general relativity for the nonlinear dynamics of highly disturbed black holes [368].

The Advanced LIGO detectors [310] (in Hanford, WA, and Livingston, LA), on September 14th 2015 at 09:50:45 UTC, reported the coincident observation of a signal, initially detected by a low-latency search for generic gravitational-wave transients [369]. The signal reached first Livingston and, after about 6.9 ms, arrived at Hanford. The signal has been then analysed with a matched-filter, constructed from relativistic models of compact binary objects [370] and found to be the most significant event in each detector in the first part of the observing run, with a combined signal-to-noise ratio (SNR) of 24 [371]. The time evolution of GW150914, shown in Fig. 4, suggests that this signal has been produced by the coalescence of a binary black hole system: the inspiral and merger, and subsequent final black hole ringdown. In about eigth cycles, lasting 0.2 s, the frequency increases from 35 to 150 Hz, where also the amplitude is maximum. The evolution of two inspiralling masses, m_1 and m_2, is characterized by the chirp mass [372]:

$$M_c = \frac{(m_1 m_2)^{3/5}}{(m_1+m_2)^{1/5}} = \frac{c^3}{G}\left[\frac{5}{96}\pi^{-8/3} f^{-11/3} \dot{f}\right]^{3/5} \tag{408}$$

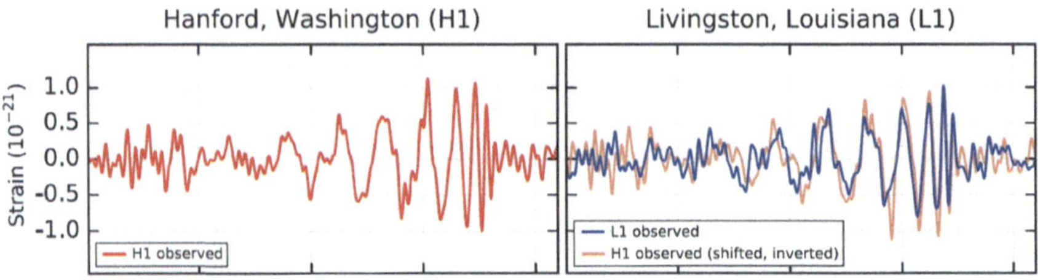

Fig. 4 Times relative to September 14, 2015 at 09:50:45 UTC. *Left* H1 strain. *Right* L1 strain and, for a visual comparison, the H1 data are also shown, shifted in time by the arrival time difference and inverted (to account for the detectors relative orientations). Figure adapted from [314]

Table 1 Source parameters for GW150914, given in the source frame; to convert to the detector frame multiply by $(1+z)$ [373]. The evaluation of the source redshift assumes standard cosmology [374]

Primary black hole mass	36^{+5}_{-4} $M_\odot$
Secondary black hole mass	29^{+4}_{-4} $M_\odot$
Final black hole mass	62^{+4}_{-4} $M_\odot$
Final black hole spin	$0.67^{+0.05}_{-0.07}$
Luminosity distance	410^{+160}_{-180} Mpc
Source redshift z	$0.09^{+0.03}_{-0.04}$

where G and c are the gravitational constant and the speed of light; f and $\dot{f}$ are the observed frequency and its time derivative and can be both estimated from the data. This yields the result $M_c \simeq 30$ $M_\odot$, which implies that, in the detector frame, the total mass $M = m_1 + m_2$ is larger than 70 $M_\odot$. While this rules out the possibility that the signal may have been generated by a binary neutron star system, the idea that the source could have been a neutron star orbiting around a black hole cannot be excluded. However, in this case, to reach a chirp mass of 30 $M_\odot$, the mass of the black hole must have been of the order of 3000 $M_\odot$ and the coalescence would have occurred at much lower frequencies, unobservable by ground based detectors (Fig. 1).

To evaluate the source parameters, general relativity-based models [375–378] have been used, in some cases including also spin precession, and, for each model, a coherent Bayesian analysis has been performed to derive the distributions of the source parameters [379], discussed in detail in [380] in the source frame. The uncertainties include statistical and systematic errors deriving from the average of the results of different waveform models. Using the fits to numerical simulations of binary black hole mergers provided in [381, 382], the mass and spin of the final black hole, the total energy radiated in gravitational waves, and the peak gravitational-wave luminosity [380] have been computed. The total energy estimated to be radiated in gravitational waves is $3.0^{+0.5}_{-0.5}$ $M_\odot c^2$, and the peak gravitational-wave luminosity has been $3.6^{+0.5}_{-0.4} \times 10^{56}$ erg/s, equivalent to 200^{+30}_{-20} $M_\odot c^2$/s.

Around the time of the event, both detectors were in steady state operation since several hours. Instrumental and environmental disturbances have been investigated to rule out the possibility that GW150914 could be an instrumental artefact [383]. The detectors' sensitivity to environmental disturbances was measured by evaluating their

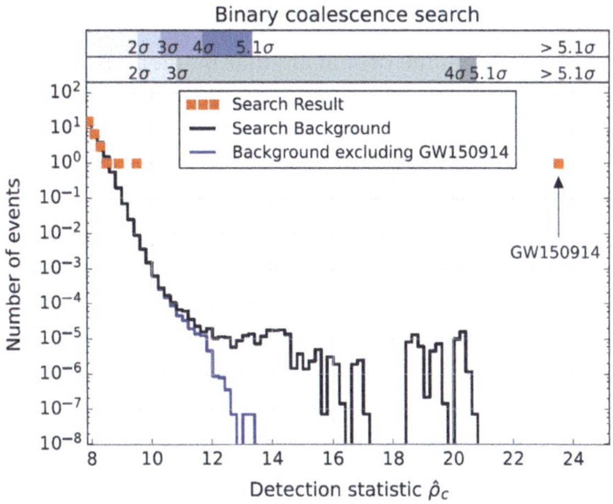

Fig. 5 Search results from the binary coalescence search. The histogram shows the number of candidate events (*orange markers*) and the mean number of background events (*black lines*) in the search class where GW150914 was found. The scales on the *top* give the significance of an event in Gaussian standard deviations based on the corresponding noise background. The tail in the *black-line* background is due to random coincidences of GW150914 in one detector with noise in the other detector. The *purple curve* is the background excluding those coincidences. Figure from [314]

response to magnetic, radio-frequency, acoustic, and vibration excitations. Finally, there is no evidence for instrumental transients that are temporally correlated between the two detectors.

Sixteen days of coincident observations between the two LIGO detectors, from September 12th to October 20th, 2015, have been analysed to assess the statistical significance of GW150914. This is a portion of the first science run of Advanced LIGO, that ended on January 12th, 2016. GW150914 has been independently detected by two different types of searches. One is targeted to the search of signals from coalescing compact objects, using optimal matched filtering with waveforms predicted by general relativity. The other search is optimized for generic transient signals, with minimal assumptions about waveforms. The methods used by the two searches are independent, thus their response to detector noise results different, uncorrelated, events. However, strong signals are expected to be detected by both searches (see also Fig. 5).

The results of the unmodeled search are deeply presenter in [369] and will not be further discussed here. We shall focus on the results from the binary coalescence method [370].

This search has targeted gravitational-wave signal from binary systems with individual masses ranging between 1 and 99 $M_\odot$, total mass smaller than 100 $M_\odot$, and dimensionless spins up to 0.99 [370]. The effective-one-body formalism [384], combining the post-Newtonian approach [372, 385] with black hole perturbation theory and numerical relativity, has been used for modelling systems with total mass larger than 4 $M_{\odot}$. In the waveform model [375, 376], the spins of the coalescing objects are assumed to be aligned with the orbital angular momentum, nonetheless, systems with misaligned spins in the parameter space of GW150914 [370] can be effectively recovered. About 250,000 template waveforms have been used to cover the whole parameter space. For each template in each detector, the matched-filter signal-to-noise ratio $\rho(t)$ is evaluated and the maxima of $\rho(t)$ with respect to the time of arrival of the signal is identified [386–388]. The lists of events in each detector are then compared looking for coincidences within a time window of 15 ms: 10 ms due to the intersite travel time plus 5 ms for uncertainty in arrival time of low SNR signals. The coincident events have been ranked on the basis of the quadrature sum $\hat{\rho}_c$ of the ρ from both detectors [371]. To estimate the background of this search the list of events of one detector are time shifted with respect to the list of the other detector and a new set of coincident events is computed. This procedure has been repeated $\sim 10^7$ times, equivalent to an observation time of about 608,000 years. Both candidate and background events are divided into three search classes according to the template length. The background for the search class of GW150914 is shown in Fig. 5.

Since the GW150914 detection statistic, $\hat{\rho}_c = 23.6$, is larger than any background event, only an upper limit on its false alarm rate can be placed. Considering all the three search classes, this bound is 1 in 203,000 years, which corresponds to a false alarm probability $<2 \times 10^{-7}$, or a detection probability larger than 5.1 σ.

Tests of General Relativity

The detection of GW150914 provides an unprecedented opportunity to study the motion of a compact binary system in the large velocity, highly nonlinear regime, and to observe the final merger of the binary and the excitation of relativistic modes of the gravitational field. Several investigations have been carried on to determine whether GW150914 is consistent with the merger of a binary black hole system in general relativity [368].

The first consistency check performed concerns the mass and spin of the final black hole. In the general relativity framework, the product of a black hole binary coalescence is a Kerr black hole, completely described by its mass and spin. For this family of inspirals, these quantities can be evaluated with Einstein's equations and are a function of the masses and spins of the progenitor black holes. Using the relations between the initial black holes and final black hole masses and spins evaluated through numerical relativity simulations [381], the estimates of the final mass and spin obtained from the low-frequency part of the waveform have been compared to those retrieved from the high-frequency component of the waveform. The test of the inspiral-merger-ringdown consistency shows no evidence of discrepancies with the predictions of general relativity.

Within the post-Newtonian formalism, the phase of the gravitational signal during the inspiral can be expressed as a power series in $f^{1/3}$. The values of these coefficients can also be computed in general relativity. Thus, a test of consistency with general relativity [389, 390] can be performed. The coefficients have been made to deviate from the nominal values and it has been checked whether the resulting waveform was consistent with the data. Thus, empirical bounds on several high-order post-Newtonian coefficients have been determined in the dynamical regime [368].

Finally, the data from GW150914 have been used to constrain the Compton wavelength of the graviton, λ_g. General relativity assumes massless gravitons that travel at the speed of light $v_g = c$. On the contrary, if the graviton has a (small) mass, the dispersion relation becomes $E^2 = p^2c^2 + m_g^2c^4$, where E is the energy, p the momentum, and m_g is the graviton rest mass related to its Compton wavelength as $\lambda_g = h/(m_g c)$, where h is the Planck constant. Thus, the ratio v_g^2/c^2 becomes equal to $c^2p^2/E^2 = 1 - h^2c^2/(\lambda_g^2 E^2)$, and a massive graviton would propagate at a speed dependent on energy (or frequency). Or, in other words, the lower frequencies propagate slower compared to higher frequencies. This dispersion can be included in the phase of the gravitational wave signal from a coalescing binary as [394] $\Phi_{MG} = -(\pi D c)/[\lambda_g^2(1+z)f]$, where z is the redshift and D a cosmological distance defined in [394]. The signal from GW150914 shows no evidence for dispersion, thus it has been possible to constrain the Compton wavelength of the graviton to be $\lambda_g > 10^{13}$ km, which is equivalent to a bound on the graviton mass $m_g < 1.2 \times 10^{-22}$ eV/c^2 at 90% confidence level. This observation improves the Solar System bound [395] by a small factor and that from binary pulsar observations [396] by a factor of a thousand.

To summarize, all these tests are consistent with general relativity in the strong-field regime.

Astrophysics with GW150914

This observation provides the first robust confirmation of several theoretical predictions: "heavy" black holes do exist, binary black hole systems form in nature and merge within the age of the universe at a detectable rate [397]. Two main types of formation models, involving isolated binaries in galactic fields [398] or dynamical interactions in young and old dense stellar environments [399], predict such mergers. The progenitor black holes of the GW150914 coalescence are more massive than the those in known XRBs with reliably measured masses: this discovery provides the most robust evidence for the existence of "heavy" ($\geq$25 $M_\odot$) stellar-mass black holes. This discovery implies relatively weak massive-star winds and thus the formation of GW150914 in a low-metallicity environment [400]: below $\simeq$ 1/2 $Z_\odot$ and possibly below $\simeq$ 1/4 $Z_\odot$. The rate of binary black hole mergers inferred from this observation is consistent with the higher end of rate predictions ($\geq$1 Gpc^{-3} year^{-1}) from both types of formation models. The low measured redshift ($z \simeq 0.1$) and the low inferred metallicity of the stellar progenitors imply two different scenarios for the formation of binary black hole systems:

- in a low-mass galaxy in the local universe and a prompt merger;
- at high redshift with a time delay between formation and merger of the order of several Gyr.

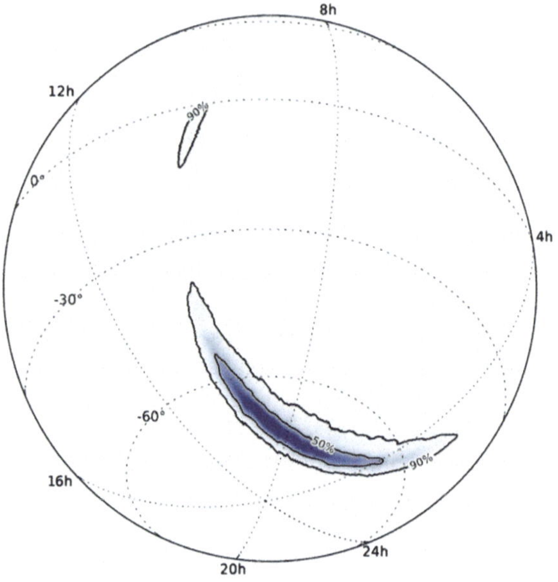

Fig. 6 An orthographic projection of the posterior probability density function (PDF) for the sky location of GW150914 showing contours of the 50 and 90% credible regions plotted over a colour-coded PDF. The sky localization forms part of an annulus, set by the time delay of $6.9^{+0.5}_{-0.4}$ ms between the two advanced LIGO detectors. Figure from [380]

Ground-based gravitational waves detectors are all-sky monitors with no intrinsic capability to determine the direction of incoming transient signals. Thus, a network of instruments is required to reconstruct the location of a source in the sky, through the time-of-arrival, and the relative amplitude and phase at different detectors [380]. The measured time-delay of GW150914 between the Livingston and Hanford sites was $6.9^{+0.5}_{-0.4}$ ms. Since at the time of the event, only the two LIGO instruments were in observational mode, the source location could only be reconstructed to approximately an annulus, due to this time-delay [401, 402]. Figure 6 shows the sky map for GW150914: it corresponds to a projected 2-dimensional credible region 590 deg^2 wide at 90% confidence level [380]. The associated 3-dimensional comoving volume probability region is $10^{-}2$ Gpc^3, which includes about 10^5 Milky Way-equivalent galaxies. This area of the sky was targeted by follow-up observations covering radio, optical, near infra-red, X-ray, and gamma-ray wavelengths [403] and searches for coincident high energy neutrinos [404].

Neutrino candidates coincident with GW150914 were searched within the data recorded by the IceCube [405] and Antares [406] detectors. No neutrino candidate was found to be in both temporal and spatial coincidence with the gravitational wave event. Within ± 500 s of the gravitational wave event, three neutrino candidates were detected by IceCube and zero by Antares. This result is consistent with the expected atmospheric background. Furthermore, none of the neutrino candidates were spatially coincident with GW150914.

Since this event was due to a binary black hole merger, there was little expectation of detectable electromagnetic or neutrino signatures.

As a concluding remark, we can see that the discovery of GWs opens a new window in observational astrophysics and cosmology where information carried by gravitational interaction can be matched with that carried by electromagnetic (EM) interaction and neutrinos. These researches are under the standard of the so called "multimessenger astronomy" that we are going to discuss below.

Multimessenger Astronomy with Gravitational Waves

One of the most promising field of contemporary astrophysics is the investigation of the most powerful and violent events in the Universe, taking advantage of the simultaneous observations of all possible cosmic messengers: photons at all wavelengths, cosmic rays, neutrinos and gravitational waves. The goals of multimessenger observations are to gain a more complete understanding of cosmic processes through a combination of information from different probes, and to increase search sensitivity with respect to single messenger analysis. The recent detection of the first gravitational wave signals from binary black hole (BBH) mergers by the two scientific collaborations LIGO and Virgo has added a new unique probe and has created exciting new opportunities. The LIGO Hanford and Livingston sites are just the first two advanced detectors nodes of a growing global network of highly sensitive GW facilities, soon to include Advanced Virgo, later KAGRA, and in the future LIGO-India and even a space-based interferometer. Some of the most promising astrophysical sources of GWs are expected to produce broadband EM emission and also neutrinos. The presence or absence of any EM or neutrino signature will provide constraints on emission mechanisms, progenitors and energetics of the GW source, as well as its environment. The benefit of joint GW, EM and neutrino observations will be tremendous, and will bring the study of the GW signals fully into the realm of astrophysics and cosmology.

Multimessenger Search for Gravitational Waves Signal and Electromagnetic Counterparts

The simultaneous and coincident observations of photons and GWs from astrophysical sources will open new windows in unexplored domains of the physics of supranuclear density matter and very strong, time varying gravitational fields. EM observations are key to localize and characterize the astrophysical source, to probe the physics of its environment and the distribution of magnetic fields, while GWs provide insight into its mass distribution and gravitational fields in the strong regime. Several detectable GW sources, like core-collapse supernovae, binary NS (BNS) or NS–BH (NSBH) mergers, and the early evolution of new born highly magnetized NSs, are expected to be accompanied by EM emission across the spectrum and over time scales ranging from seconds to years. For transient GW sources, multiwavelength observations are crucial to find an EM counterpart and improve the source localization down to the arcsecond level, leading to the identification of the host galaxy and measurement of the redshift. This will allow us not only to determine the EM intrinsic luminosity of the GW source, but also improve the estimates of the the parameters of the GW source, such as the GW luminosity distance. In the long term, redshift and GW luminosity distance will allow us to obtain for the first time an independent measure of the Hubble constant using general relativity as only calibrator [408, 409].

In compact binary coalescence (CBC) events, a binary formed by two tight neutron stars, two black holes, or a NS and a BH experience a runaway orbital decay due to gravitational radiation, followed by the merger of the compact objects. When the mergers involve at least one NS, EM signatures at different timescales and wavelengths due to energetic outflows are expected.

These mergers are thought to be the progenitor of short Gamma Ray Bursts (GRBs). If a relativistic jet forms, there is a prompt short GRB lasting a few seconds or less, followed by X-ray, optical, and radio afterglows of hours to days duration (e.g., [410, 411]). Furthermore, in the sub-relativistic ejecta, rapid neutron capture (e.g., [412]) is hypothesized to produce a macronova, an optical and near-infrared signal lasting hours to weeks (e.g., [391–393, 413]). Eventually, we may observe a radio blast wave from this sub-relativistic outflow, detectable for months to years (e.g., [414, 415]). Furthermore, several seconds prior to or tens of minutes after merger, we may see a coherent radio burst lasting milliseconds (e.g., [416]). On the other hand, in the case of a stellar-mass binary black hole (BBH) the current consensus is that no significant EM counterpart emission is expected, except for those in highly improbable environments pervaded by large ambient magnetic fields or baryon densities. The coincident detection of a short gamma-ray burst (GRB) and GW signal would provide the first direct evidence that short GRBs are associated to the merging of two compact objects and will discriminate on their nature (NSBH or binary NS mergers). A joint GRB/GW detection will yield a wealth of information on the mechanism powering the GRB. This motivated the search in low-latency of gravitational-wave candidates to obtain prompt electromagnetic observations [417–419], and vice versa the use of GRBs detected through gamma-ray satellites are also being used for searches for coincident GW signals [420, 421].

The main challenges to detect the electromagnetic counterparts of gravitational wave signals are: (1) the expected electromagnetic signals are transient, possibly rapid, and faint [422] and (2) the gravitational-wave sky localisation is poor covering hundreds of square degrees [423, 424]. Overcoming these challenges require to develop low-latency data analysis to identify gravitational wave candidate signals and to provide alerts within minutes to the astronomical community [425–428]. One then needs to coordinate observations with multi-wavelength wide field of view instruments able to cover large region of the sky, followed by rapid image analysis, which allow us to identify counterpart candidates and the use of larger telescopes for deep photometry and spectroscopy. This last step is crucial to characterise the nature of the counterparts. The optimisation of observational strategies, data analysis and the interpretation of the observations require theoretical work consisting in precise source modelling for the GW and EM emissions [429].

GW150914 and GW151226 were identified by the low-latency pipelines, and their event times, significances, and sky locations were sent to many worldwide groups of astronomers [430, 431]. Ground- and space-based facilities were able to cover many hundreds of square degrees with radio, optical, near-infrared, X-ray, and gamma-ray instruments [432–437]. No convincing electromagnetic counterparts were detected, but these first broadband campaigns to search for a counterpart of an Advanced LIGO GW source represent a milestone, showing the capabilities and challenges of different steps of the search described above.

The detection of an electromagnetic counterpart and the identification of the galaxy hosting the GW source will have an enormous scientific return on different astrophysical fields, going from formation and evolution of stars and compact objects, e.g. [438], to different messenger emission processes, e.g. [439], and nuclear astrophysics, e.g. [440, 441]. Organizing joint observations and data analysis will have high impact in terms of development and coordination of observatory networks, strenghten collaborations among different scientific communities, and also for computational science.

EM observations will also help in supporting and guiding searches for continuous GW sources, like asymmetric spinning NSs (e.g. pulsars), both isolated and in binary systems, ranging from the radio to the gamma-ray band, and will be crucial to probe the physics of matter at supra-nuclear densities that can not be tested on Earth laboratories and the environment of strongest gravitational fields in the Universe, leading to a breakthrough of paramount importance to both physics, astrophysics and cosmology.

Multimessenger Search for Sources of Gravitational Waves and High-Energy Neutrinos

The new field of multimessenger astronomy includes also neutrino observations. GWs and neutrinos represent important novel opportunities in studying a variety of cosmic processes, providing complementary information to electromagnetic searches [457]. They carry information about the origin of high-energy emission as well as the dynamics and structure of the process driving the emission [458–464], which may be difficult to decipher via electromagnetic observations ([465, 466] and references therein). Both GWs and neutrinos carry information from the source's central region. While GWs are produced by the bulk motion of the progenitor, high-energy neutrinos require hadron acceleration in relativistic outflows from the central engine. Astrophysical processes that produce GWs are expected to drive relativistic outflows, which can emit GeV-PeV neutrinos.

The search for common sources of GWs and high-energy-neutrino has recently become possible with the construction and upgrade of large-scale observatories.

High-energy neutrino observatories currently in operation include IceCube [467–469] a cubic-kilometer detector at the geographic South Pole and Antares [470] in the Mediterranean sea. Antares is planned to be followed by a multi-cubic-kilometer detector in the Mediterranean sea called KM3NeT in the following years [471]. The Baikal Neutrino Telescope, operating at Lake Baikal, is also planned to be upgraded to a km3 volume [472]. IceCube is also sensitive to low energy (MeV) thermal neutrinos from nearby supernovae [473], and contributes to the Supernova Early Warning System (SNEWS) network along with several other neutrino detectors, including Super-Kamiokande [474], Borexino [475], the Large Volume Detector [476] and the Sudbury Neutrino Observatory [477]. High-energy neutrinos of cosmic origin have recently been observed, for the first time, by IceCube [478, 479]. Their detection represents a major step towards multimessenger astronomy. In an all-sky search for the most energetic (>30 TeV) neutrinos, a total of 28 astrophysical neutrino candidates were identified over a period of 2 years, significantly above the number of expected atmospheric background (10 events). The neutrinos have no clear directional or temporal clustering. While many sources of astrophysical origin have been suggested (see, e.g., [480–489], so far there is no compelling evidence for a specific source population. Several classes of astrophysical transients that are detected electromagnetically are also associated with significant GW and high-energy neutrino emission. These processes include the gamma-ray bursts [465, 490–499], core-collapse supernovae with rapidly rotating cores [500–502], flares from soft gamma repeaters [503–505], or even cosmic string cusps [506–509]. Gamma-ray bursts (GRBs) are one of the most promising sources for joint GW+neutrino observations. The intense flashes of MeV photons are thought to be produced in relativistic outflows likely driven by accretion onto a compact object [511, 512], or by a rapidly rotating neutron star [513]. Hadrons accelerated in the relativistic outflows will result in the production of high-energy neutrinos through interaction with photons or other hadrons. Another promising source type, soft-gamma repeaters (SGR), are sources of short bursts of gamma rays emitted at irregular intervals. They occasionally emit giant gamma-ray/X-ray flares that can be detected from the Milky Way or nearby galaxies within a few megaparsecs, at a rate lower than the GRB rate (e.g., [524]). SGRs are thought to be highly magnetized neutron stars. Their gamma outbursts may be a consequence of the tectonic activity of the neutron star crust (starquake). The seismic vibration of a neutron star following a starquake gives rise to GW emission (e.g., [505, 525, 526]). Tectonic activity in the neutron star is followed by the reconfiguration of the neutron star's magnetic fields, resulting in X-ray and gamma-ray radiation [527, 528]. Sudden magnetic reconfiguration may also accelerate protons and other nuclei, leading to the production of high-energy neutrinos [503, 529]. For joint GW+neutrino detection, a particularly interesting subset of these sources are those that have faint electromagnetic emission, such as choked GRBs [530–532] for which the relativistic outflow stalls before it can break out of the stelar envelope, and low-luminosity GRBs with mildly relativistic ejecta [493, 533]. The coincident detectability of GWs and high-energy neutrinos, and the scientific potential of such searches, was first recognized in [534], and later independently in [535]. The population of cosmic events producing both GWs and high-energy neutrino emission was first constrained observationally [536] with data from the initial LIGO-Virgo detectors and the partially completed Ice-Cube detector (see section "Sec16due.2The Einstein connection"). While the population of cosmic events expected to produce detectable joint emission falls below these early observational limits, the results were used to estimate the projected upper limits achievable with advanced GW detectors and the completed IceCube, which can provide meaningful constraints or detection. A search for coincident GWs and high-energy neutrinos was carried out using the initial LIGO-Virgo observatories and the partially completed Antares detector [537]. The search found no significant coincident events. Antares is mainly sensitive to TeV neutrino sources from the southern hemisphere, making this a complementary search to that performed with Ice-Cube, which is mainly sensitive to the northern hemisphere at TeV energies. Recognizing the scientific potential of multimessenger observations, various initiatives aim to use GWs and neutrinos in combination with electromagnetic observations. For instance, multiple astrophysical event candidates from both the initial LIGO-Virgo detectors [538, 539] and IceCube [540, 541] have been followed up by EM telescopes: QUEST, TAROT, ZADKO, Pi of the sky, ROTSE, SkyMapper, the Palomar Transient Factory, and the Swift satellite [542–545]. Sub-threshold GRB candidates detected through gamma-rays are also being used for searches for coincident GW signals [546–548].

Joint Analysis of GWs and Low Energy Neutrinos from a Supernova

A supernova is an explosion of a massive supergiant star. No events in nature surpasses its raw power: about 10^{53}erg/s (equivalent to 10^{46}J/s) is released as neutrinos from a "core-collapse" supernova, which is as much instantaneous power as all the rest of the luminous, visible Universe combined. They give birth to the most exotic states of matter known: neutron stars and black holes. Supernovae have been at the forefront of astronomical research for the better part of a century, and yet no one is sure how they work. There are important scientific motivations for a joint analysis of GWs and low-energy neutrino data to probe the processes powering a supernova explosion. Several mechanisms in a core-collapse supernova can give rise to GW bursts. Signals may last from milliseconds to seconds. While the astrophysical motivation for expecting GWs to accompany core-collapse supernovae is strong, the expected rate, GW strength and waveform morphology are uncertain. The expected energy going into GWs would be $10^{-10} \div 10^{-4}$ of a solar mass, or $10^{44} \div 10^{50}$ ergs. This is only a small fraction of the energy liberated in neutrinos: the core collapse of a massive star is expected to produce a huge flux of neutrinos. Actually nearly all (about 99%) of the binding energy $O(10^{53}$ ergs) of the resulting neutron star (or black hole) ends up in neutrinos escaping on a short timescale (a few tens of seconds). The neutrinos (and antineutrinos) produced are of all flavours, and have energies in the few to tens of MeV range. The burst of neutrinos from SN1987A in the Large Magellanic

Cloud, detected by various neutrino detectors, confirmed the baseline model of stellar collapse. On the contrary, the estimates of GW bursts associated with supernova rely on models. Most such models are not yet three-dimensional, do not incorporate the entire set of possibly relevant physics, and do not predict robust supernova explosions as observed in the electromagnetic universe. So, despite the availability of multiple potential explosion mechanisms and their associated multi-dimensional dynamics and GW signatures, the current picture is unlikely to be complete. It is well worth to study the supernova signals with GWs and neutrino detectors with a multi-messenger approach. The first important point is that GWs and neutrinos from core-collapse supernovae are emitted in the inner-most, high-density region of the supernova core which cannot be probed electromagnetically. GWs and neutrinos thus are the only messengers that can carry "live" dynamical information from deep inside a dying massive star and constrain the detailed, yet unknown, mechanism driving the core-collapse supernova explosion. Another point is that both GWs and neutrinos leave the core-collapse event on approximately the same time scale (seconds) in contrast to electromagnetic radiation, which can take hours or days to become visible to astronomers. Moreover, a visible supernova near the edge of sensitivity may be a rather poor tag of core collapse for both GW and neutrino detectors separately, in the presence of detector background, given what will likely be a large uncertainty on the occurrence itself of a collapse and on its timing. Let us develop these considerations. Several of the world's present neutrino detectors are sensitive to a neutrino burst from a galactic supernova. Super-K would observe some $\sim$8000 events for a core collapse at the center of the Milky Way, $\sim$8.5 kpc away. The LVD and Borexino scintillation detectors at Gran Sasso, and KamLAND in Japan, would observe hundreds of interactions. The IceCube detector, although nominally a multi-GeV neutrino detector, would observe a coincident increase in count rate in its phototubes due to a diffuse burst of Cherenkov photons in the ice, and has sensitivity to a galactic supernova. The duty factor of neutrino detectors in observing mode has typically been 90% or larger. As reported in the previous paragraph, Super-K, LVD, IceCube and Borexino detectors are operating as part of the SNEWS (SuperNova Early Warning System) network, which has the aim of providing a prompt alert to astronomers in the case of a coincident supernova neutrino burst. The SNEWS network utilizes the fact that neutrinos arrive up to several hours earlier than the detectable electromagnetic signal from supernovae, as was the case for the first confirmed astronomical MeV neutrino source, SN 1987A [550, 551]. A current initiative for multimessenger searches in the weak-signal limit is the proposed Astrophysical Multimessenger Observatory Network (AMON; [552]). AMON is planned to perform a real-time correlation analysis of sub-threshold signals from available astronomical messengers; photons, low and high-energy neutrinos, cosmic rays, and GWs. These sub-threshold triggers will be used to increase the detectable number of sources and to trigger follow-up searches that can further enhance the significance of a potential coincident detection.

Final Remarks and Perspectives

Far from being detailed and self-consistent, this review paper is just aimed to present a route on the ongoing researches in gravitational physics, starting from the foundation of General Relativity, the problem of Quantum Gravity, the formulation and the extension of alternative theories up to the recent discovery of gravitational waves and their possible applications to astrophysics and cosmology. The status of art clearly shows that even though gravitation is the most ancient interaction framed into a scientific scheme, it is the most elusive: this situation depends on the fact that gravity behavior is essentially unknown at infrared and ultraviolet scales. In the first case, issues related to dark matter and dark energy need to be addressed in a self-consistent scheme, from a theoretical point of view, and, mainly, these ingredients have to be supported by experimental evidences. On the other hand, at ultraviolet scales, gravitation misses to be dealt under the same standard of other interactions. In fact, the same concepts of space and time loose their meaning at quantum level, so that usual approaches for quantization and renormalization need to be revised for gravity.

However, despite of this situation, the overwhelming amount of excellent experimental results related to fundamental gravitational physics (e.g. tests of equivalence principle, atomic physics), precision cosmology (detailed surveys of supernovae, large-scale structure and cosmic microwave background), relativistic astrophysics (neutron stars, black holes, GRBs, high-energy astrophysics), and gravitational waves are contributing towards a comprehensive picture of gravitational interaction.

We can conclude that understanding gravity in view of a self-consistent theory at all scales will be one of the main challenge of physics for the forthcoming years.

References

1. M. Artigas, W.R. Shea, *Galileo in Rome: The Rise and Fall of a Troublesome Genius* (Oxford University Press, Oxford, 2004)
2. F. Daxecker, *The Physicist and Astronomer, Christoph Scheiner: Biography, Letters, Works*. Veroffentlichungen der Universitat Innsbruck 246 (2004)
3. S. Drake, *Discoveries and Opinions of Galileo*. (1610 Letter to the Grand Duchess Christina) (Anchor Books, 1957)
4. S. Drake, *Galileo at Work: His Scientific Biography* (The University of Chicago Press, Chicago, 1978)
5. S. Drake, Galileo's discovery of the law of free fall. Sci. Am. **228**(5), 84 (1973)
6. J. Gleick, *Isaac Newton* (Pantheon Publisher, New York, 2003)
7. R.S. Westfall, *Never at Rest* (Cambridge University Press, Cambridge, UK, 1980)
8. J.C. Adams, *On the Perturbations of Uranus* (Cambridge University Press, Cambridge, 1846)
9. M. Grosser, *The Discovery of Neptune* (Harvard University Press, 1962)
10. S. Newcomb, *The Elements of the Four Inner Planets and the Fundamental Constants of Astronomy* (American Ephemeris, 1897)
11. E. Mach, *Die Mechanik in ihrer Entwicklung historisch-kritisch dargestellt* (Princeton University Press, F.A. Brockhaus, Princeton, 1912)

12. E.C. Banks, *Ernst Mach's World Elements* (Kluwer Academic Publishers, Dordrecht, The Netherlands, 2003)
13. H.G. Alexander, *The Leibniz-Clarke Correspondence* (Manchester University Press, Manchester, UK, 1998)
14. R.H. Dicke, The many faces of Mach, in *Gravitation and Relativity*, ed. by H.Y. Chiu, W.F. Hofmann (Benjamin, New York, 1964)
15. H. Thirring, *Physikalische Zeitschrift* **19**, 204 (1918)
16. H. Pfister, Gen. Relativ. Gravit. **39**(11), 1735 (2007)
17. H. Thirring, *Physikalische Zeitschrift* **22**, 29 (1921)
18. J. Lense, H. Thirring, Physikalische Zeitschrift **19**, 156 (1918)
19. J. Eisberg, Eddington's stellar models and early twentieth century astrophysics. Ph.D. thesis (Harvard University, 1991)
20. C.M. Will, *Theory and Experiment in Gravitational Physics*, 2nd edn. (Cambridge University Press, Cambridge, UK, 1993)
21. A. Einstein, Ann. der Physik **49**, 769 (1916)
22. L.P. Eisenhart, *Riemannian Geometry* (Princeton University Press, Princeton, 1955)
23. E. Schrödinger, *Space-Time Structure* (Cambridge University Press, Cambridge, 1960)
24. T. Levi-Civita, *The Absolute Differential Calculus* (Blackie and Son, London, 1929)
25. O. Klein, *New Theories in Physics, 77* (Intern. Inst. of Intellectual Co-operation, League of Nations, 1938)
26. E. Cartan, Ann. Ec. Norm. **42**, 17 (1925)
27. A. Palatini, Rend. Circ. Mat. Palermo **43**, 203 (1919)
28. V.I. Arnold, *Mathematical Methods of Classical Mechanics* (Springer, Berlin, 1978)
29. G. Basini, S. Capozziello, Gen. Relativ. Gravit. **35**, 2217 (2003)
30. M. Kaku, *Quantum Field Theory* (Oxford University Press, Oxford, 1993)
31. M.B. Green, J.H Schwarz, E. Witten, *Superstring Theory* (Cambridge University Press, Cambridge, 1987)
32. D.R. Brill, R.H. Gowdy, Rep. Prog. Phys. **33**, 413 (1970)
33. C.J. Isham, in *Quantum Gravity 2: A Second Oxford Symposium*, ed. by C.J. Isham, R. Penrose, D.W. Sciama (Clarendon Press, Oxford, 1981)
34. C. Itzykson, J.C. Zuber, *Quantum Field Theory* (McGraw-Hill, New York, 1980)
35. J.A. Wheeler, Rev. Mod. Phys. **29**, 463 (1957)
36. R. Arnowitt, S. Deser, C.W. Misner, in *Gravitation: An Introduction to Current Research*, ed. by L. Witten (Wiley, New York, 1962)
37. B.S. De Witt, in *Battelle Rencontres*, ed. by C. DeWitt, J.A. Wheeler (Benjamin, New York, 1968)
38. C.W. Misner, in *Relativity*, ed. by M. Carmeli, L. Fickler, L. Witten (Plenum, San Francisco, 1970)
39. C.W. Misner, in *Magic Without Magic*, ed. by J. Klauder (Freeman, San Francisco, 1972)
40. B.S. De Witt, Dynamical theory of groups and fields (Les Houches Lectures 1963), in *Relativity, Groups and Topology* (Gordon and Breach Science Publishers, New York, 1965)
41. N.D. Birrell, P.C.W. Davies, *Quantum Fields in Curved Space* (Cambridge University Press, Cambridge, UK, 1982)
42. P. van Nieuwenhuizen, Phys. Rep. **68**, 189 (1981)
43. C.J. Hogan, Rev. Mod. Phys. **72**, 1149 (2000)
44. T. Bringmann, C. Kiefer, D. Polarski, Phys. Rev. D **65**, 024008 (2002)
45. A.A. Starobinsky, JETP Lett. **30**, 682 (1979)
46. G. Amelino-Camelia, Mod. Phys. Lett. A **17**, 899 (2002)
47. I. Percival, W. Strunz, Proc. R. Soc. Lond. A **453**, 431 (1997)
48. A. Peters, K.Y. Chung, S. Chu, Metrologia **38**, 25 (2001)
49. http://lhc.web.cern.ch/lhc/
50. D. Kimberly, J. Magueijo, Cosmology and gravitation: XIth Brazilian School of Cosmology and Gravitation. AIP Conf. Proc. **782**, 241 (2005)
51. R. Penrose, Gen. Rel. Grav. **28**, 581 (1996)

52. T.P. Singh, Bulg. J. Phys. **33**, 217 (2006)
53. R.A. Knop et al., Astrophys. J. **598**, 102 (2003)
54. S. Perlmutter et al., Astrophys. J. **517**, 565 (1999)
55. A.G. Riess et al., Astron. J. **116**, 1009 (1998)
56. J.L. Tonry et al., Astrophys. J. **594**, 1 (2003)
57. P. de Bernardis et al., Nature **404**, 955 (2000)
58. R. Stompor et al., Astrophys. J. (Lett.) **561**, L7 (2001)
59. D.N. Spergel et al., Astrophys. J. (Suppl.) **148**, 175 (2003)
60. G. Hinshaw et al., Astrophys. J. (Suppl.) **148**, 135 (2003)
61. D.N. Spergel et al., [WMAP Collaboration], Astrophys. J. (Suppl.) **170**, 377 (2007)
62. A.G. Riess et al., Astrophys. J. **607**, 665 (2004)
63. V. Sahni, A.A. Starobinsky, Int. J. Mod. Phys. D **9**, 373 (2000)
64. T. Padmanabhan, Phys. Rep. **380**, 235 (2003)
65. E.J. Copeland, M. Sami, S. Tsujikawa, Int. J. Mod. Phys. D **15**, 1753 (2006)
66. S. Nojiri, S.D. Odintsov, Gen. Relativ. Gravit. **38**, 1285 (2006)
67. S. Nojiri, S.D. Odintsov, Phys. Rev. D **74**, 086005 (2006)
68. A. Kamenshchik, U. Moschella, V. Pasquier, Phys. Lett. B **511**, 265 (2001)
69. T. Padmanabhan, Phys. Rev. D **66**, 021301 (2002)
70. B.A. Bassett, M. Kunz, D. Parkinson, C. Ungarelli, Phys. Rev. D **68**, 043504 (2003)
71. V.F. Cardone, A. Troisi, S. Capozziello, Phys. Rev. D **69**, 083517 (2004)
72. S. Capozziello, V.F. Cardone, E. Elizalde, S. Nojiri, S.D. Odintsov, Phys. Rev. D **73**, 043512 (2006)
73. A. Lue, R. Scoccimarro, G. Starkman, Phys. Rev. D **69**, 044005 (2004)
74. K. Freese, M. Lewis, Phys. Lett. B **540**, 1 (2002)
75. G.R. Dvali, G. Gabadadze, M. Porrati, Phys. Lett. B **485**, 208 (2000)
76. S. Capozziello, Int. J. Mod. Phys. D **11**, 483 (2002)
77. S. Nojiri, S.D. Odintsov, Int. J. Geom. Meth. Mod. Phys. **4**, 115 (2007)
78. S. Capozziello, V. Faraoni, *Beyond Einstein Gravity: A Survey of Gravitational Theories for Cosmology and Astrophysics, Fundamental Theories of Physics 170* (Springer, New York, 2011)
79. A. De Felice, S. Tsujikawa, Living Rev. Relativ. **13**, 3 (2010)
80. S. Capozziello, M. De Laurentis, *Invariance Principles and Extended Gravity: Theory and Probes* (Nova Science Publishers, New York, 2011)
81. S. Capozziello, V.F. Cardone, E. Piedipalumbo, M. Sereno, A. Troisi, Int. J. Mod. Phys. D **12**, 381 (2003)
82. S. Capozziello, V.F. Cardone, S. Carloni, A. Troisi, Int. J. Mod. Phys. D **12**, 1969 (2003)
83. B. Li, J.D. Barrow, Phys. Rev. D **75**, 084010 (2007)
84. B. Li, J.D. Barrow, D.F. Mota, Phys. Rev. D **76**, 104047 (2007)
85. S. Podariu, R.A. Daly, M.P. Mory, B. Ratra, Astrophys. J. **584**, 577 (2003)
86. S. Capozziello, F.V. Cardone, M. Funaro, S. Andreon, Phys. Rev. D **70**, 123501 (2004)
87. S. Capozziello, P.K.S. Dunsby, E. Piedipalumbo, C. Rubano, Astr. Astrop. **472**(1), 51 (2007)
88. K.S. Stelle, Phys. Rev. D **16**, 953 (1977)
89. R.H. Sanders, Ann. Rev. Astron. Astrophys. **2**, 1 (1990)
90. P.D. Mannheim, D. Kazanas, Astrophys. J. **342**, 635 (1989)
91. J.D. Anderson et al., Phys. Rev. D **65**, 082004 (2002)
92. O. Bertolami, C.G. Böhmer, F.S.N. Lobo, Phys. Rev. D **75**, 104016 (2007)
93. I. Quandt, H.J. Schmidt, Astron. Nachr. **312**, 97 (1991)
94. P. Schneider, J. Ehlers, E.E. Falco, *Gravitational Lenses* (Springer, Berlin, 1992)
95. L.M. Krauss, M. White, Astrophys. J. **397**, 357 (1992)
96. V. Faraoni, *Cosmology in Scalar-Tensor Gravity* (Kluwer Academic, Dordrecht, 2004)
97. S. Capozziello, M. Francaviglia, Gen. Rel. Grav.: Special Issue on Dark Energy **40**, 357 (2007)
98. L. Amendola, S. Tsujikawa, *Dark Energy Theory and Observations* (Cambridge University Press, Cambridge, UK, 2010)

99. S. Capozziello, V. Salzano, Adv. Astron. Special Issue "Dark Sides of Our Universe" **217420**, 34 (2009)
100. S. Capozziello, M. De Laurentis, V. Faraoni, The Special Issue in Cosmol. Open Astron. J. **2**, 1874 (2009)
101. F.S.N. Lobo, arXiv:0807.1640v1 [gr-qc] (2008)
102. S. Nojiri, S.D. Odintsov, Phys. Rep. **505**, 59 (2011)
103. T. Clifton, P.G. Ferreira, A. Padilla, C. Skordis, e-Print: arXiv:1106.2476 [astro-ph.CO]
104. S. Sengupta Anand for the LIGO Scientific Collaboration and the Virgo Collaboration, LIGO-Virgo searches for gravitational waves from coalescing binaries: a status update, in *Proceedings of the 8th E. Amaldi Conference on Gravitational Waves (Amaldi8)*, New York (2009); J. Phys.: Conf. Ser
105. F. Acernese et al., Class. Quantum Gravity **24**, S381 (2007)
106. R.N. Lang, S.A. Hughes, Class. Quantum Gravity **26**, 094035 (2009)
107. B. Abbott et al., LIGO Scientific Collaboration, Class. Quantum Gravity **27**, 173001 (2010)
108. S. Capozziello, in *Quantum Gravity Research Trends*, ed. by A. Reimer, vol. 227 (Nova Science Publishers, Inc., 2005)
109. Ph. Tourrenc, General relativity and gravitational waves, in *Proceedings of the International Summer School on Experimental Physics of Gravitational Waves*, 6–18 Sept 1999, Urbino (Italy), ed. by M. Barone et al. (World Scientific, Singapore, 2000)
110. http://www.onera.fr/prix-en/brun2002/edmond-brun2002.html
111. http://astro.estec.esa.nl/astrogen/COSPAR/step/
112. R.V. Pound, G.A. Rebka, Phys. Rev. Lett. **4**, 337 (1960)
113. A. Einstein, Sitzung-ber. Preuss. Akad. Wiss. 414 (1925)
114. S. Weinberg, *Gravitation and Cosmology* (Wiley, New York, 1972)
115. C.W. Misner, K.S. Thorne, J.A. Wheeler, *Gravitation*, ed. W.H. Freeman (San Francisco, 1973)
116. H. Weyl, Ann. der Physik **364**, 101 (1919)
117. M. Ferraris, M. Francaviglia, C. Reina, Gen. Relativ. Gravit. **14**, 243 (1982)
118. G. Magnano, M. Ferraris, M. Francaviglia, Gen. Relativ. Gravit. **19**, 465 (1987)
119. R.A. Hulse, J.H. Taylor, ApJ **195**, L51 (1975)
120. R.M. Wald, *General Relativity* (Chicago University Press, Chicago, 1984)
121. L.D. Landau, E.M. Lifschitz, *The Classical Theory of Fields* (Pergamon Press, Oxford, 1989)
122. K.S. Thorne, The theory of gravitational radion: an introductory review, in *Gravitational Radiation*, ed. by N. Deruelle, T. Piran (North-Holland, Amsterdam, 1983), pp. 1–57
123. M. Maggiore, *Gravitational Wawes: Theory and Experiments* (Oxford University Press, Oxford, 2007)
124. C.N. Yang, R.L. Mills, Phys. Rev. **96**, 191 (1954)
125. K. Moriyasu, *Gauge Theory* (World Scientific Publishing, Singapore, 1983)
126. H. Von Kluber, Vistas Astron. **3**, 47 (1960)
127. A. Einstein, The Meaning of Relativity (Princeton University, New Jersey, 1955)
128. C.N. Yang, Phys. Today **42** (1980)
129. P.G. Bergmann, Phys. Today **44** (1979)
130. Y. Aharonov, D. Bohm, Phys. Rev. **123**, 1511 (1961)
131. J.D. Jackson, *Classical Electrodynamics* (Wiley, New York, 1962)
132. H. Goldstein, *Classical Mechanics* (Addison-Wesley, Reading, Mass, 1959)
133. Y. Choque-Bruhat, C. de Witt-Morette, M. Dillard-Bleick, *Analysis, Manifolds and Physics*, revised edn. (North-Holland, New York, 1982)
134. W. Dreschler, M.E. Mayer, *Fiber Bundle Thechniques in Gauge Theories* (Springer, Heidelberg, 1977)
135. J.D. Bjorken, S. Drell, *Relativistic Quantum Fields* (McGraw-Hill, New York, 1965)
136. P.W. France, Am. J. Phys. **44**, 798 (1976)
137. I.L. Buchbinder, S.D. Odintsov, I. Shapiro, *Effective Action in Quantum Gravity* (IOP Publishing, Bristol, 1992)
138. S. Capozziello, M. De Laurentis, Int. J. Geom. Meth. Mod. Phys. **6**, 1 (2009)

139. R. Utiyama, Phys. Rev. **101**, 1597 (1956)
140. D.W. Sciama, *On the analog between charge and spin in General Relativity, in Recent Developments in General Relativity* (Pergamon Press, Festschrift for Leopold Infeld, New York, 1962)
141. T.W. Kibble, J. Math. Phys. **2**, 212 (1961)
142. G. Grignani et al., Phys. Rev. D **45**, 2719 (1992)
143. F.W. Hehl, J.D. McCrea, E.W. Mielke, Y. Ne'eman, Phys. Rep. **258**, 1 (1995)
144. A. Inomata et al., Phys. Rev. D **19**, 1665 (1979)
145. E.A. Ivanov, J. Niederle, Phys. Rev. D **25**, 976 (1982)
146. E.A. Ivanov, J. Niederle, Phys. Rev. D **25**, 988 (1982)
147. F. Mansouri et al., Phys. Rev. D **13**, 3192 (1976)
148. F. Mansouri, Phys. Rev. Lett. **42**, 1021 (1979)
149. D. Ivanenko, G.A. Sardanashvily, Phys. Rep. **94**, 1 (1983)
150. G. Sardanashvily, arXiv: gr-qc/0201074 (2002)
151. L.N. Chang et al., Phys. Rev. D **13**, 235 (1976)
152. S.A. Ali, S. Capozziello, Int. J. Geom. Meth. Mod. Phys. **4**, 1041 (2007)
153. G. Basini, S. Capozziello, Mod. Phys. Lett. A **20**, 251 (2005)
154. F. Mansouri, Phys. Rev. D **16**, 2456 (1977)
155. D.R. Brill et al., Rev. Mod. Phys. **29**, 465 (1957)
156. S. Carroll et al., Phys. Rev. D **50**, 3867 (1994)
157. F.W. Hehl et al., J. Math. Phys. **12**, 1334 (1971)
158. F.W. Hehl, P. von der Heyde, G.D. Kerlick, J.M. Nester, Rev. Mod. Phys. **48**, 393 (1976)
159. I.L. Shapiro, Phys. Rep. **357**, 113 (2002)
160. M. Blagojevic, Three lectures on Poincaré Gauge theory. SFIN **A1**, 147 (2003)
161. J. Schwinger, Phys. Rev. **130**, 800 (1963)
162. L. Parker, D.J. Toms, *Quantum Field Theory in Curved Space-Time: Quantized Fields and Gravity* (Cambridge University Press, Cambridge, UK, 2009)
163. N. Bilić, B. Guberina, R. Horvat, H. Nikolić, H. Stefancić, Phys. Lett. B **657**, 232 (2007)
164. N.D. Birrell, P.C.W. Davies, Phys. Rev. D **22**, 322 (1980)
165. B. Nelson, P. Panangaden, Phys. Rev. D **25**, 1019 (1982)
166. L.H. Ford, D.J. Toms, Phys. Rev. D **25**, 1510 (1982)
167. L. Parker, D.J. Toms, Phys. Rev. D **29**, 1584 (1985)
168. L.H. Ford, Phys. Rev. D **32**, 1409 (1987)
169. B. Allen, Nucl. Phys. B **226**, 232 (1983)
170. J. Ishikawa, Phys. Rev. D **28**, 2445 (1983)
171. L.F. Abbott, Nucl. Phys. B **185**, 233 (1981)
172. T. Futamase, K. Maeda, Phys. Rev. D **39**, 399 (1989)
173. T. Futamase, T. Rothman, R. Matzner, Phys. Rev. D **39**, 405 (1989)
174. V. Faraoni, Phys. Rev. D **53**, 6813 (1996)
175. V. Faraoni, *Proceedings of the 7th Canadian Conference on General Relativity and Relativistic Astrophysics* (Calgary, Canada, 1997)
176. V. Faraoni, Int. J. Theor. Phys. **40**, 2259 (2001)
177. T. Okamura, in *Proceedings of the Asia-Pacific Conference on Gravitation and Cosmology*, Seoul, Korea 1996, ed. by Y.M. Cho, C.H. Lee, S.-W. Kim (World Scientific, Singapore, 1998)
178. X. Chen, in *Proceedings COSMO 2000, 4th International Workshop on Particle Physics and the Early Universe*, Cheju Island, Korea, 2000, ed. by J.E. Kim, P. Ko, K. Lee (World Scientific, Singapore, 2001)
179. X. Chen, R.J. Scherrer, G. Steigman, Phys. Rev. D **63**, 123504 (2001)
180. S. Sonego, V. Faraoni, Class. Quantum Gravity **10**, 1185 (1993)
181. V. Faraoni, S. Sonego, in *Proceedings of the 5th Canadian Conference on General Relativity and Relativistic Astrophysics*, vol. 386, Waterloo, Canada 1993, ed. by R.B. Mann, R.G. McLenaghan (World Scientific, Singapore, 1994)
182. A.A. Grib, E.A. Poberii, Helv. Phys. Acta **68**, 380 (1995)
183. A.A. Grib, W.A. Rodrigues, Grav. Cosmol. **1**, 273 (1995)

184. S. Deser, R.I. Nepomechie, Ann. Phys. (NY) **154**, 396 (1984)
185. V. Faraoni, E. Gunzig, Int. J. Mod. Phys. D **8**, 177 (1998)
186. I.L. Buchbinder, Fortschr. Phys. **34**, 605 (1986)
187. I.L. Buchbinder, S.D. Odintsov, Sov. J. Nucl. Phys. **40**, 848 (1984)
188. I.L. Buchbinder, S.D. Odintsov, Lett. Nuovo Cimento **42**, 379 (1985)
189. I.L. Buchbinder, S.D. Odintsov, I. Shapiro, in *Group-Theoretical Methods in Physics*, ed. by M. Markov (Nauka, Moscow, 1986)
190. S.D. Odintsov, Fortschr. Phys. **39**, 621 (1991)
191. T.S. Muta, S.D. Odintsov, Mod. Phys. Lett. A **6**, 3641 (1991)
192. E. Elizalde, S.D. Odintsov, Phys. Lett. B **333**, 331 (1994)
193. I.L. Buchbinder, S.D. Odintsov, I. Lichzier, Class. Quantum Gravity **6**, 605 (1989)
194. M. Reuter, Phys. Rev. D **49**, 6379 (1994)
195. A. Bonanno, Phys. Rev. D **52**, 969 (1995)
196. A. Bonanno, D. Zappalá, Phys. Rev. D **55**, 6135 (1997)
197. A. Barroso, J. Casasayas, P. Crawford, P. Moniz, A. Nunes, Phys. Lett. B **275**, 264 (1992)
198. R. Fakir, S. Habib, Mod. Phys. Lett. A **8**, 2827 (1993)
199. J. Garcia-Bellido, A. Linde, Phys. Rev. D **52**, 6730 (1995)
200. E. Komatsu, T. Futamase, Phys. Rev. D **58**, 023004 (1998)
201. E. Komatsu, T. Futamase, Phys. Rev. D **59**, 064029 (1999)
202. J. Lee et al., Phys. Rev. D **61**, 027301 (1999)
203. D.S. Salopek, J.R. Bond, J.M. Bardeen, Phys. Rev. D **40**, 1753 (1989)
204. R. Fakir, W.G. Unruh, Phys. Rev. D **41**, 3012 (1990)
205. R. Fakir, W.G. Unruh, Phys. Rev. D **41**, 1792 (1990)
206. R. Fakir, S. Habib, W.G. Unruh, Astrophys. J. **394**, 396 (1992)
207. J.C. Hwang, H. Noh, Phys. Rev. Lett. **81**, 5274 (1998)
208. T. Futamase, M. Tanaka, Phys. Rev. D **60**, 063511 (1999)
209. S. Tsujikawa, K. Maeda, T. Torii, Phys. Rev. D **61**, 103501 (2000)
210. S. Tsujikawa, B. Bassett, Phys. Lett. B **536**, 9 (2002)
211. J.J. Halliwell, R. Laflamme, Class. Quantum Gravity **6**, 1839 (1989)
212. D.H. Coule, K. Maeda, Class. Quantum Gravity **7**, 955 (1990)
213. D.H. Coule, Class. Quantum Gravity **9**, 2353 (1992)
214. W.A. Hiscock, Class. Quantum Gravity **7**, L35 (1990)
215. J.J. Van der Bij, E. Radu, Nucl. Phys. B **585**, 637 (2000)
216. P. Jetzer, Phys. Rep. **220**, 163 (1993)
217. J.J. Van der Bij, M. Gleiser, Phys. Lett. B **194**, 482 (1987)
218. A.R. Liddle, M.S. Madsen, Int. J. Mod. Phys. D **1**, 101 (1992)
219. M.B. Voloshin, A.D. Dolgov, Sov. J. Nucl. Phys. **35**, 120 (1982)
220. C.T. Hill, D.S. Salopek, Ann. Phys. (NY) **213**, 21 (1992)
221. Y. Hosotani, Phys. Rev. D **32**, 1949 (1985)
222. L. Parker, D.J. Toms, Phys. Rev. D **32**, 1409 (1985)
223. L. Amendola, A. Battaglia-Mayer, S. Capozziello, S. Gottlöber, V. Müller, F. Occhionero, H.J. Schmidt, Class. Quantum Gravity **10**, L43 (1993)
224. L. Parker, A. Raval, Phys. Rev. D **60**, 063512 (1999)
225. L. Parker, W. Komp, D.A.T. Vanzella, ApJ **588**, 663 (2003)
226. H. Bondi, *Cosmology* (Cambridge University Press, Cambridge, UK, 1952)
227. G.A. Vilkovisky, Class. Quantum Gravity **9**, 895 (1992)
228. M. Gasperini, G. Veneziano, Phys. Lett. B **277**, 256 (1992)
229. J.D. Barrow, A.C. Ottewill, J. Phys. A **16**, 2757 (1983)
230. A.A. Starobinsky, Phys. Lett. B **91**, 99 (1980)
231. J.P. Duruisseau, R. Kerner, Gen. Relativ. Gravit. **15**, 797 (1983)
232. D. La, P.J. Steinhardt, Phys. Rev. Lett. **62**, 376 (1989)
233. P. Teyssandier, P. Tourrenc, J. Math. Phys. **24**, 2793 (1983)
234. K. Maeda, Phys. Rev. D **39**, 3159 (1989)
235. D. Wands, Class. Quantum Gravity **11**, 269 (1994)

236. S. Capozziello, R. de Ritis, A.A. Marino, Gen. Relativ. Gravit. **30**, 1247 (1998)
237. S. Gottlöber, H.-J. Schmidt, A.A. Starobinsky, Class. Quantum Gravity **7**, 893 (1990)
238. F. Adams, K. Freese, A. Guth, Phys. Rev. D **43**, 965 (1991)
239. T.V. Ruzmaikina, A.A. Ruzmaikin, JETP **30**, 372 (1970)
240. H.A. Buchdahl, Acta Math. **85**, 63 (1951)
241. A. Berkin, K. Maeda, Phys. Lett. B **245**, 348 (1990)
242. A. Battaglia-Mayer, H.J. Schmidt, Class. Quantum Gravity **10**, 2441 (1993)
243. H.J. Schmidt, Class. Quantum Gravity **7**, 1023 (1990)
244. L.M. Sokolowski, Class. Quantum Gravity **6**, 2045 (1989)
245. M. Ferraris, M. Francaviglia, I. Volovich, Class. Quantum Gravity **11**, 1505 (1994)
246. M. Ferraris, M. Francaviglia, G. Magnano, Class. Quantum Gravity **5**, L95 (1988)
247. G. Magnano, L.M. Sokolowski, Phys. Rev. D **50**, 5039 (1994)
248. R.H. Dicke, Phys. Rev. D **125**, 2163 (1962)
249. T. Damour, G. Esposito-Farèse, Class. Quantum Gravity **9**, 2093 (1992)
250. N. Lanahan-Trembley, V. Faraoni, Class. Quantum Gravity **24**, 5667 (2007)
251. T. Damour, G.W. Gibbons, C. Gundlach, Phys. Rev. Lett. **64**, 123 (1990)
252. H.A. Buchdahl, J. Phys. A **12**(8), 1229 (1979)
253. G.J. Olmo, Int. J. Mod. Phys. D **20**, 413 (2011)
254. S. Nojiri, S.D. Odintsov, Phys. Rev. D **68**, 123512 (2003)
255. D.N. Vollik, Phys. Rev. D **68**, 063510 (2003)
256. B. Li, M.C. Chu, Phys. Rev. D **74**, 104010 (2006)
257. B. Li, K.C. Chan, M.C. Chu, Phys. Rev. D **76**, 024002 (2007)
258. S. Capozziello, M. De Laurentis, M. Francaviglia, S. Mercadante, Found. Phys. **39**, 1161 (2009)
259. X. Meng, P. Wang, Gen. Rel. Grav. **36**(8), 1947 (2004); X. Meng, P. Wang, Gen. Relativ. Gravit. **36**, 2673 (2004)
260. G. Allemandi, M. Capone, S. Capozziello, M. Francaviglia, Gen. Relativ. Gravit. **38**, 33 (2006)
261. S. Bellucci, S. Capozziello, M. De Laurentis, V. Faraoni, Phys. Rev. D **79**, 104004 (2009)
262. S. Capozziello, E. Elizalde, S. Nojiri, S.D. Odintsov, Phys. Lett. B **671**, 193 (2009)
263. S.M. Carroll, A. De Felice, V. Duvvuri, D.A. Easson, M. Trodden, M.S. Turner, Phys. Rev. D **71**, 063513 (2005)
264. A. Nunez, S. Solganik, Phys. Lett. B **608**, 189 (2005)
265. T. Chiba, J. Cosmol. Astropart. Phys. **0503**, 008 (2005)
266. K.S. Stelle, Gen. Relativ. Gravit. **9**, 343 (1978)
267. S. Nojiri, S.D. Odintsov, M. Sasaki, Phys. Rev. D **71**, 123509 (2005)
268. G. Cognola, E. Elizalde, S. Nojiri, S.D. Odintsov, S. Zerbini, Phys. Rev. D **73**, 084007 (2006)
269. D. Bazeia, R. Menezes, AYu. Petrov, Eur. Phys. J. C **58**, 171 (2008)
270. H. Van Dam, M.J.G. Veltman, Nucl. Phys. B **22**, 397 (1970)
271. S. Capozziello, C. Corda, M. De Laurentis, Phys. Lett. B **699**, 255 (2008)
272. B. Willke et al., Class. Quantum Gravity **23**, S207 (2006)
273. D. Sigg et al., www.ligo.org/pdfpublic/P050036.pdf
274. B. Abbott et al., Phys. Rev. D **72**, 042002 (2005)
275. M. Ando et al., Class. Quantum Gravity **19**, 1615 (2002)
276. D. Tatsumi, Y. Tsunesada, Class. Quantum Gravity **21**, S451 (2004)
277. http://www.lisa.esa.int
278. R. Emparan, J. Garriga, J. High Energy Phys. **0603**, 028 (2006)
279. L.S. Finn, P.J. Sutton, Phys. Rev. D **65**, 044022 (2002)
280. B. Abbott et al., Phys. Rev. D **76**, 082003 (2007)
281. N. Bonasia, M. Gasperini, Phys. Rev. D **71**, 104020 (2005)
282. A. Nishizawa, A. Taruya, K. Hayama, S. Kawamura, M. Sakagami, Phys. Rev. D **79**, 082002 (2009)
283. M.E. Tobar, T. Suzuki, K. Kuroda, Phys. Rev. D **79**, 082002 (2009)
284. D. di Serafino, S. Gomez, L. Milano, F. Riccio, G. Toraldo, J. Glob. Optim. **48**, 41 (2010)

285. D. di Serafino, G. Liuzzi, V. Piccialli, F. Riccio, G. Toraldo, J. Glob. Optim. https://doi.org/10.1007/s10898-010-9525-9
286. D. Babusci, L. Baciotti, F. Fucito, A. Nagar, Phys. Rev. D **64**, 062001 (2001)
287. A. Viceré, Phys. Rev. D **66**, 062002 (2002)
288. P. Leaci et al., Class. Quantum Gravity **25**, 195018 (2008)
289. M. Maggiore, Phys. Rep. **331**, 283 (2000)
290. S. Capozziello, V.F. Cardone, A. Troisi, J. Cosmol. Astropart. Phys. **0608**, 001 (2006)
291. S. Capozziello, E. De Filippis, V. Salzano, Mon. Not. R. Ast. Soc. **394**, 947 (2009)
292. S. Capozziello, V.F. Cardone, S. Carloni, A. Troisi, Phys. Lett. A **326**, 292 (2004)
293. A. Einstein, Sitzungsber. K. Preuss. Akad. Wiss. **1**, 688 (1916)
294. R.A. Hulse, J.H. Taylor, Astroph. J. **195**, L51 (1975)
295. L. Grishchuk, *Astrophysics Update* (Springer, Berlin, 2004), p. 281
296. U. Seljak, M. Zaldarriaga, Phys. Rev. Lett. **78**, 2054 (1997)
297. S. Detweiler, Astrophys. J. **234**, 1100 (1979)
298. M.V. Sazhin, Sov. Astron. **22**, 35 (1978)
299. B. Linet, P. Tourrenc, Can. J. Phys. **54**, 1129 (1976)
300. P. Bender et al., MPQ Report No. 233 (1998)
301. F.A.E. Pirani, Acta Physica Polonica **25**, 389 (1956)
302. M.E. Gertsenshtein, V.I. Pustovoit, Sov. Phys.—JETP **16**, 433 (1962)
303. R. Weiss, Quarterly Progress Report of RLE, MIT **105**, 54 (1972)
304. J. Weber, Phys. Rev. **117**, 306 (1960)
305. K.S. Thorne, in *300 years of Gravitation*, ed. by Hawking, Israel (Cambridge University Press, Cambridge, 1987), p. 300
306. R. Takahashi et al. (TAMA Collaboration), Class. Quantum Gravity **25**, 114036 (2008)
307. H. Grote (LIGO Scientific Collaboration), Class. Quantum Gravity **27**, 084003 (2010)
308. B. Abbott et al., Rep. Prog. Phys. **72**, 076901 (2009)
309. T. Accadia et al. (The Virgo Collaboration), JINST **7**, 3012 (2012)
310. J. Aasi et al., Class. Quantum Gravity **32**, 074001 (2015)
311. F. Acernese et al., Class. Quantum Gravity **32**, 024001 (2015)
312. Y. Aso, Y. Michimura, K. Somiya, M. Ando, O. Miyakawa, T. Sekiguchi, D. Tatsumi, H. Yamamoto (The KAGRA Collaboration), Phys. Rev. D **88**, 043007 (2013)
313. C.S. Unnikrishnan, Int. J. Mod. Phys. D **22**, 1341010 (2013)
314. B.P. Abbott et al. (LIGO Scientific Collaboration and Virgo Collaboration), Phys. Rev. Lett. **116**, 061102 (2016)
315. B.P. Abbott et al. (LIGO Scientific Collaboration and Virgo Collaboration), Phys. Rev. X **6**, 041015 (2016)
316. M. Abernathy et al. (The ET Science Team), Einstein gravitational wave telescope conceptual design study. https://tds.ego-gw.it/itf/tds/file.php?callFile=ET-0106C-10.pdf (2010)
317. M. Punturo et al., Class. Quantum Gravity **27**, 084007 (2010)
318. B.P. Abbott et al., Class. Quantum Gravity **34**, 044001 (2017)
319. LISA Gravitational Wave Observatory, https://www.elisascience.org/
320. LISA Pathfinder, https://sci.esa.int/lisa-pathfinder/
321. K. Riles, Prog. Part. Nucl. Phys. **68**, 1–54 (2013)
322. B.F. Schutz, Determining the Hubble constant from gravitational wave observations. Nature **323**, 310 (1986)
323. G. Steigman, Primordial nucleosynthesis in the precision cosmology era. Annu. Rev. Nucl. Part. Sci. **57**, 463–491 (2007)
324. N. Yunes, K. Yagi, F. Pretorius, Phys. Rev. D **94**, 084002 (2016)
325. G. Ashton, D.I. Jones, R. Prix, Phys. Rev. D **91**, 062009 (2015)
326. A. Królak, K. Kokkotas, G. Schäfer, Estimation of the post-Newtonian parameters in the gravitational-wave emission of a coalescing binary. Phys. Rev. D **52**, 2089–2111 (1995)
327. C.M. Will, *Theory and Experiments in Gravitational Physics* (Cambridge University Press, New York, USA, 1981)

328. K. Arun, M. Iyer, Qusailah, B. Sathyaprakash, Testing post-Newtonian theory with gravitational waves. Class. Quantum Gravity **23**, L37–L43 (2006)
329. E. Komatsu et al., Seven-year Wilkinson Microwave Anisotropy Probe (WMAP) observations: cosmological interpretation. Astrophys. J. Suppl. Ser. **192**, 18 (2011)
330. D.A. Brown et al., Prospects for detection of gravitational waves from intermediate-mass-ratio inspirals. Phys. Rev. Lett. **99**, 201102 (2007)
331. E.A. Huerta, J.R. Gair, Intermediate-mass-ratio inspirals in the Einstein Telescope. I. Signal-to-noise ratio calculations. Phys. Rev. D **83**, 044020 (2011)
332. O. Dreyer et al., Black hole spectroscopy: testing general relativity through gravitational wave observations. Class. Quantum Gravity **21**, 787 (2004)
333. J.M. Fregeau, S.L. Larson, M.C. Miller, R. O'Shaughnessy, F.A. Rasio, Observing IMBH-IMBH binary coalescences via gravitational radiation. Astrophys. J. Lett. **646**, L135–L138 (2006)
334. B. Carter, Axisymmetric black hole has only two degrees of freedom. Phys. Rev. Lett. **26**, 331–333 (1971)
335. S. Wilkins, N. Trentham, A. Hopkins, MNRAS **385**, 687 (2008)
336. A. Sesana, M. Volonteri, F. Haardt, The imprint of massive black hole formation models on the LISA data stream. Mon. Not. R. Astron. Soc. **377**, 1711–1716 (2007)
337. P. Madau, M.J. Rees, Massive black holes as population III remnants. Astrophys. J. Lett. **551**, L27–L30 (2001)
338. C.M. Espinoza, A.G. Lyne, B.W. Stappers, M. Kramer, A study of 315 glitches in the rotation of 102 pulsars. MNRAS **414**, 1679–1704 (2011)
339. A.L. Piro, E. Pfahl, Fragmentation of collapsar disks and the production of gravitational waves. Astrophys. J. **658**, 1173 (2007)
340. N. Andersson, Gravitational waves from instabilities in relativistic stars. Class. Quantum Gravity **20**, 105 (2003)
341. M.H.P.M. van Putten et al., Gravitational radiation from gamma-ray burst-supernovae as observational opportunities for LIGO and VIRGO. Phys. Rev. D **69**, 044007 (2004)
342. E. Chassande-Mottin, M. Hendry, P. Sutton, S. Máarka, Multimessenger astronomy with the Einstein Telescope. Gen. Relativ. Gravit. **43**, 437–464 (2011)
343. K. Ioka, Mon. Not. R. Astron. Soc. **327**, 639 (2001)
344. H.A. Bethe, Supernova mechanisms. Rev. Mod. Phys. **62**, 801–866 (1990)
345. C.L. Fryer, V. Kalogera, Astrophys. J. **554**, 548 (2011)
346. C.D. Ott, Class. Quantum Gravity **26**, 204015 (2009)
347. H.-T. Janka, K. Langanke, A. Marek, G. Martínez-Pinedo, B. Müller, Phys. Rep. **442**, 38 (2007)
348. R.E. Frey (LIGO Scientific Collaboration), AIP Conf. Proc. **928**, 11 (2007)
349. Y.-I. Sekiguchi, M. Shibata, Axisymmetric collapse simulations of rotating massive stellar cores in full general relativity: numerical study for prompt black hole formation. Phys. Rev. D **71**, 084013 (2005)
350. D.H. Lyth, What would we learn by detecting a gravitational wave signal in the cosmic microwave background anisotropy? Phys. Rev. Lett. **78**, 1861–1863 (1997)
351. M.S. Turner, Detectability of inflation-produced gravitational waves. Phys. Rev. D **55**, 435–439 (1997)
352. S. Kuroyanagi, T. Chiba, N. Sugiyama, Imprint of in ation dynamics on the spectrum of the primordial gravitational wave background. Phys. Rev. D **79**, 103501 (2009)
353. L. Grishchuk, The implications of the microwave background anisotropies for laser-interferometer-tested gravitational waves. Class. Quantum Gravity **14**, 1445–1454 (1997)
354. L.A. Boyle, A. Buonanno, Relating gravitational wave constraints from primordial nucleosynthesis, pulsar timing, laser interferometers, and the CMB: implications for the early universe. Phys. Rev. D **78**, 043531 (2008)
355. S. Sarangi, S.H. Tye, Cosmic string production towards the end of brane in ation. Phys. Lett. B **536**, 185–192 (2002)

356. C. Caprini, R. Durrer, T. Konstandin, G. Servant, General properties of the gravitational wave spectrum from phase transitions. Phys. Rev. D **79**, 083519 (2009)
357. C. Grojean, G. Servant, Gravitational waves from phase transitions at the electroweak scale and beyond. Phys. Rev. D **75**, 043507 (2007)
358. A. Kosowsky, A. Mack, T. Kahniashvili, Gravitational radiation from cosmological turbulence. Phys. Rev. D **66**, 024030 (2002)
359. T. Kahniashvili, A. Kosowsky, G. Gogoberidze, Y. Maravin, Detectability of gravitational waves from phase transitions. Phys. Rev. D **78**, 043003 (2008)
360. L. Randall, G. Servant, Gravitational waves from warped spacetime. J. High Energy Phys. **2007**, 054 (2007)
361. N.J. Craig, *Gravitational Waves from Supersymmetry Breaking* (2009)
362. R. Easther, E.A. Lim, Stochastic gravitational wave production after inflation. JCAP **0604**, 010 (2006)
363. J.-F. Dufaux, A. Bergman, G.N. Felder, L. Kofman, J.-P. Uzan, Theory and numerics of gravitational waves from preheating after inflation. Phys. Rev. D **76**, 123517 (2007)
364. J. Garcia-Bellido, D.G. Figueroa, A. Sastre, A gravitational wave background from reheating after hybrid inflation. Phys. Rev. D **77**, 043517 (2008)
365. B.P. Abbott et al. (LIGO Scientific Collaboration and Virgo Collaboration), An upper limit on the stochastic gravitational-wave background of cosmological origin. Nature **460** (2009)
366. C. Cheng, Q.-G. Huang, S. Wang, JCAP **1412**, 044 (2014)
367. K. Schwarzschild, Sitzungsber. K. Preuss. Akad. Wiss. **1**, 189 (1916)
368. B.P. Abbott et al., Phys. Rev. Lett. **116**, 221101 (2016)
369. B.P. Abbott et al., Phys. Rev. D **93**, 122004 (2016)
370. B.P. Abbott et al., Phys. Rev. D **93**, 122003 (2016)
371. S.A. Usman et al., Class. Quantum Gravity **33**, 215004 (2016)
372. L. Blanchet, T. Damour, B.R. Iyer, C.M. Will, A.G. Wiseman, Phys. Rev. Lett. **74**, 3515 (1995)
373. A. Krolak, B.F. Schutz, Gen. Relativ. Gravit. **19**, 1163 (1987)
374. P.A.R. Ade et al., Astron. Astrophys. **594**, A13 (2016)
375. A. Taracchini et al., Phys. Rev. D **89**, 061502 (2014)
376. M. Prrer, Class. Quantum Gravity **31**, 195010 (2014)
377. M. Hannam, P. Schmidt, A. Bohé, L. Haegel, S. Husa, F. Ohme, G. Pratten, M. Prrer, Phys. Rev. Lett. **113**, 151101 (2014)
378. S. Khan, S. Husa, M. Hannam, F. Ohme, M. Prrer, X. Jimnez Forteza, A. Bohé, Phys. Rev. D **93**, 044007 (2016)
379. J. Veitch et al., Phys. Rev. D **91**, 042003 (2015)
380. B.P. Abbott et al., Phys. Rev. Lett. **116**, 241102 (2016)
381. J. Healy, C.O. Lousto, Y. Zlochower, Phys. Rev. D **90**, 104004 (2014)
382. S. Husa, S. Khan, M. Hannam, M. Prrer, F. Ohme, X. Jimnez Forteza, A. Boh, Phys. Rev. D **93**, 044006 (2016)
383. B.P. Abbott et al., Class. Quantum Gravity **33**, 134001 (2016)
384. A. Buonanno, T. Damour, Phys. Rev. D **62**, 064015 (2000)
385. L. Blanchet, T. Damour, G. Esposito-Farèse, B.R. Iyer, Phys. Rev. Lett. **93**, 091101 (2004)
386. B. Allen, W.G. Anderson, P.R. Brady, D.A. Brown, J.D.E. Creighton, Phys. Rev. D **85**, 122006 (2012)
387. B.S. Sathyaprakash, S.V. Dhurandhar, Phys. Rev. D **44**, 3819 (1991)
388. B.J. Owen, B.S. Sathyaprakash, Phys. Rev. D **60**, 022002 (1999)
389. C.K. Mishra, K.G. Arun, B.R. Iyer, B.S. Sathyaprakash, Phys. Rev. D **82**, 064010 (2010)
390. T.G.F. Li, W. Del Pozzo, S. Vitale, C. Van Den Broeck, M. Agathos, J. Veitch, K. Grover, T. Sidery, R. Sturani, A. Vecchio, Phys. Rev. D **85**, 082003 (2012)
391. B.D. Metzger, G. Martinez-Pinedo, S. Darbha et al., Mon. Not. R. Astron. Soc. **406**, 2650 (2010)
392. J. Barnes, D. Kasen, Astrophys. J. **775**, Article ID 18 (2013)
393. D. Grossman, O. Korobkin, S. Rosswog, T. Piran, Mon. Not. R. Astron. Soc. **439**, 757 (2014)
394. C.M. Will, Phys. Rev. D **57**, 2061 (1998)

395. C. Talmadge, J.P. Berthias, R.W. Hellings, E.M. Standish, Phys. Rev. Lett. **61**, 1159 (1988)
396. L.S. Finn, P.J. Sutton, Phys. Rev. D **65**, 044022 (2002)
397. B.P. Abbott et al., Astrophys. J. Lett. **818**, L22 (2016)
398. V. Kalogera, K. Belczynski, C. Kim, R. O'Shaughnessy, B. Willems, Phys. Rep. **442**, 75–108 (2007)
399. S.F. Portegies Zwart, S.L.W. McMillan, Astrophys. J. Lett. **528**, L17 (2000)
400. K. Belczynski et al., Astrophys. J. **714**, 1217 (2010)
401. L.P. Singer et al., Astrophys. J. **795**, 105 (2014)
402. C.P.L. Berry et al., Astrophys. J. **804**, 1114 (2015)
403. B.P. Abbott et al., Astrophys. J. Lett. **826**, L13 (2016)
404. S. Adrián-Martínez et al., Phys. Rev. D **93**, 122010 (2016)
405. A. Achterberg et al., Astropart. Phys. **26**, 155 (2006)
406. M. Ageron et al., Nucl. Instrum. Meth. A **656**, 11 (2011)
407. B.P. Abbott et al., Living Rev. Relativ. **19**, 1 (2016)
408. B.T. Schutz, Nature **323**, 310 (1986)
409. S. Nissanke et al., ApJ **725**, 496 (2010)
410. D. Eichler, M. Livio, T. Piran, D.N. Schramm, Nature **340**, 126 (1989)
411. E. Berger, Annu. Rev. Astron. Astrophys. **52**, 43 (1994)
412. J.M. Lattimer, D.N. Schramm, Astrophys. J. **210**, 549 (1976)
413. L. Li, B. Paczynski, Astrophys. J. **507**, L59 (1998)
414. E. Nakar, T. Piran, Nature **478**, 82 (2011)
415. K. Hotokezaka, S. Nissanke, G. Hallinan et al., Astrophys. J. **831**, 190 (2016)
416. B.M.S. Hansen, M. Lyutikov, MNRAS **322**, 695 (2001)
417. J. Abadie et al. (LIGO Scientific Collaboration, Virgo Collaboration), Astron. Astrophys. **539**, A124 (2012)
418. J. Aasi, J. Abadie, B.P. Abbott et al., Astrophys. J. Suppl. **211**, 7 (2014)
419. P.A. Evans, J.K. Fridriksson, N. Gehrels et al., Astrophys. J. Suppl. **203, 28** (2012)
420. B.P. Abbott, R. Abbott, T.D. Abbott et al., Astrophys. J. **841**, 89 (2017)
421. J. Aasi, B.P. Abbott, R. Abbott et al., Phys. Rev. D **89**, 122004 (2014)
422. B.D. Metzger, E. Berger, Astrophys. J. **746**, 48 (2012)
423. B.P. Abbott, R. Abbott, T.D. Abbott et al., Living Rev. Relativ. **19**, 1 (2016)
424. L.P. Singer, L.R. Price, B. Farr et al., Astrophys. J. **795**, 105 (2014)
425. S. Klimenko, G. Vedovato, M. Drago et al., Phys. Rev. D **93**, 042004 (2016)
426. K. Cannon et al., Astrophys. J. **748**, 136 (2012)
427. T. Adams et al., Class. Quantum Gravity **33**, 175012 (2015)
428. S.A. Usman et al., Class. Quantum Gravity **33**, 215004 (2016)
429. R. Fernandez, B.D. Metzger, Annu. Rev. Nucl. Part. Sci. **66**, 23 (2016)
430. B.P. Abbott, R. Abbott, T.D. Abbott, Astrophys. J. Lett. **826**, L13 (2016)
431. B.P. Abbott, R. Abbott, T.D. Abbott et al., Astrophys. J. **225**, 8 (2016)
432. V. Connaughton et al., Astrophys. J. Lett. **826**, 6 (2016)
433. V. Savchenko et al., Astrophys. J. Lett. **820**, 36 (2016)
434. M. Tavani et al., Astrophys. J. Lett. **825**, 4 (2016)
435. M.M. Kasliwal et al., Astrophys. J. Lett. **824**, 24 (2016)
436. M. Soares-Santos et al., Astrophys. J. Lett. **823**, 33 (2016)
437. N.T. Palliyaguru et al., Astrophys. J. Lett. **829**, 28 (2016)
438. B.P. Abbott, R. Abbott, T.D. Abbott et al., Astrophys. J. Lett. **818**, L22 (2016)
439. F. Pannarale, F. Ohme, Astrophys. J. Lett. **791**, L7 (2014)
440. K. Hotokezaka, T. Piran, M. Paul, Nat. Phys. **11**, 1042 (2015)
441. S. Rosswog, U. Feindt, O. Korobkin et al., Class. Quantum Gravity **34**, 104001 (2017)
442. Adams et al., arXiv:1512.02864 (2015)
443. Aasi et al., ApJ **785**, 119 (2014)
444. Aasi et al., PRD **91**, 062008 (2015)
445. Aasi et al., PRD **87**, 042001 (2013)
446. Aasi et al., arXiv:1510.03621 (2015)

447. P. Leaci, R. Prix, PRD **91**, 102003 (2015)
448. Howell et al., PASA **32**, 46 (2015)
449. Abadie et al., A&A **539**, 124 (2012)
450. Abadie et al., A&A **541**, 155 (2012)
451. Aasi et al., ApJS **211**, 7 (2014)
452. Evans et al., ApJS **203**, 28 (2012)
453. L. Blanchet, Living Rev. Relativ. **17** (2014)
454. A.H. Mroue et al., PRL **111**, 241104 (2013)
455. V. Ferrari, L. Gualtieri, GRG **40**, 945 (2008)
456. J.A. Faber, F.A. Rasio, Living Rev. Relativ. **15** (2012)
457. The content of this paragraph have been inspired to the recent paper M.G. Aartsen et al., Phys. Rev. D **90**(10), 102002 (2014)
458. B.P. Abbott et al., Rep. Prog. Phys. **72**, 076901 (2009)
459. S. Razzaque, P. Meszaros, E. Waxman, Phys. Rev. D **68**, 083001 (2003). arXiv:astro-ph/0303505
460. I. Bartos, B. Dasgupta, S. Marka, Phys. Rev. D **86**, 083007 (2012)
461. R. Abbasi et al., Nature (London) **484**, 351 (2012). arXiv:1204.4219
462. I. Bartos, A.M. Beloborodov, K. Hurley, S. Marka, Phys. Rev. Lett. **110**, 241101 (2013). arXiv:1301.4232
463. S. Hummer, P. Baerwald, W. Winter, Phys. Rev. Lett. **108**, 231101 (2012)
464. H.-N. He et al., Astrophys. J. **752**, 29 (2012). arXiv:1204.0857
465. S. Ando et al., Rev. Modern Phys. **85**, 1401 (2013). arXiv:1203.5192
466. I. Bartos, P. Brady, S. Marka, Class. Quantum Gravity **30**, 123001 (2013). arXiv:1212.2289
467. IceCube Collaboration et al., Astropart. Phys. **26**, 155 (2006). astro-ph/0604450
468. R. Abbasi et al., Nucl. Instrum. Methods Phys. Res. A **601**, 294 (2009)
469. R. Abbasi et al., Nucl. Instrum. Methods Phys. Res. A **618**, 139 (2010). arXiv:1002.2442
470. M. Ageron et al., Nucl. Instrum. Methods Phys. Res. A **656**, 11 (2011). arXiv:1104.1607
471. M. de Jong, Nucl. Instrum. Methods Phys. Res. A **623**, 445 (2010)
472. A. Avrorin et al., Nucl. Instrum. Methods Phys. Res., Sect. A **626–627**, S13 (2011)
473. R. Abbasi et al., Astron. Astrophys. **535**, A109 (2011)
474. S. Fukuda et al., Nucl. Instrum. Methods Phys. Res. A **501**, 418 (2003)
475. L. Cadonati, F.P. Calaprice, M.C. Chen, Astropart. Phys. **16**, 361 (2002). hep-ph/0012082
476. N.Y. Agafonova et al., Astropart. Phys. **28**, 516 (2008). arXiv:0710.0259
477. J. Boger et al., Nucl. Instrum. Methods Phys. Res. A **449**, 172 (2000). nucl-ex/9910016
478. M.G. Aartsen et al., Phys. Rev. Lett. **111**, 021103 (2013)
479. IceCube Collaboration, Science **342** (2013)
480. I. Cholis, D. Hooper, JCAP6 **30** (2013). arXiv:1211.1974
481. W. Winter, Phys. Rev. D **88**, 083007 (2013). arXiv:1307.2793
482. O.E. Kalashev, A. Kusenko, W. Essey, Phys. Rev. Lett. 111, 041103 (2013). arXiv:1303.0300
483. E. Roulet, G. Sigl, A. van Vliet, S. Mollerach, JCAP1 **28** (2013). arXiv:1209.4033
484. F.W. Stecker, ArXiv e-prints **1305**, 7404 (2013)
485. H.-N. He, T. Wang, Y.-Z. Fan, S.-M. Liu, D.-M. Wei, Phys. Rev. D **87**, 063011 (2013). arXiv:1303.1253
486. K. Murase, M. Ahlers, B.C. Lacki, Phys. Rev. D **88**, 121301 (2013). arXiv:1306.3417
487. L.A. Anchordoqui et al., Phys. Rev. D **89**, 083003 (2014). arXiv:1306.5021
488. M.C. Gonzalez-Garcia, F. Halzen, V. Niro, ArXiv e-Prints **1310**, 7194 (2013)
489. S. Razzaque, Phys. Rev. D **88**, 081302 (2013). arXiv:1309.2756
490. S. Hummer, P. Baerwald, W. Winter, Phys. Rev. Lett. **108**, 231101 (2012). arXiv:1112.1076
491. E. Waxman, J. Bahcall, Phys. Rev. Lett. **78**, 2292 (1997). astro-ph/9701231
492. S. Ando, J.F. Beacom, Phys. Rev. Lett. **95**, 061103 (2005). astro-ph/0502521
493. K. Murase, K. Ioka, S. Nagataki, T. Nakamura, Astrophys. J. Lett. **651**, L5 (2006). astro-ph/0607104
494. S. Kobayashi, P. Meszaros, Astrophys. J. **589**, 861 (2003). astro-ph/0210211

495. M.B. Davies, A. King, S. Rosswog, G. Wynn, Astrophys. J. Lett. **579**, L63 (2002). astro-ph/0204358
496. M.H. van Putten et al., Phys. Rev. D **69**, 044007 (2004). gr-qc/0308016
497. A.L. Piro, E. Pfahl, Astrophys. J. **658**, 1173 (2007). astro-ph/0610696
498. K. Kiuchi, M. Shibata, P.J. Montero, J.A. Font, Phys. Rev. Lett. **106**, 251102 (2011). arXiv:1105.5035
499. A. Corsi, P. Meszaros, Astrophys. J. **702**, 1171 (2009). arXiv:0907.2290
500. C.L. Fryer, K.C.B. New, Living Rev. Relativ. **14**, 1 (2011)
501. K. Kotake, K. Sato, K. Takahashi, Rep. Prog. Phys. **69**, 971 (2006). astro-ph/0509456
502. M. Shibata, Y.-I. Sekiguchi, Phys. Rev. D **71**, 024014 (2005). astro-ph/0412243
503. K. Ioka, S. Razzaque, S. Kobayashi, P. Meszaros, Astrophys. J. **633**, 1013 (2005). astro-ph/0503279
504. B. Zink, P.D. Lasky, K.D. Kokkotas, Phys. Rev. D **85**, 024030 (2012). arXiv:1107.1689
505. D. Murphy et al., Phys. Rev. D **87**, 103008 (2013). arXiv:1302.3915
506. T. Damour, A. Vilenkin, Phys. Rev. Lett. **85**, 3761 (2000). gr-qc/0004075
507. V. Berezinsky, E. Sabancilar, A. Vilenkin, Phys. Rev. D **84**, 085006 (2011). arXiv:1108.2509
508. X. Siemens et al., Phys. Rev. D **73**, 105001 (2006). grqc/0603115
509. J. Aasi et al., Phys. Rev. Lett. **112**, 131101 (2014). arXiv:1310.2384
510. P. Meszaros, Astropart. Phys. **43**, 134 (2013). arXiv:1204.1897
511. S. E. Woosley, Astrophys. J. **405**, 273 (1993)
512. A.I. MacFadyen, S.E. Woosley, Astrophys. J. **524**, 262 (1999). astro-ph/9810274
513. V.V. Usov, Nature (London) **357**, 472 (1992)
514. S.E. Woosley, J.S. Bloom, Ann. Rev. Astron. Astrophys. **44**, 507 (2006). astro-ph/0609142
515. D. Eichler, M. Livio, T. Piran, D.N. Schramm, Nature (London) **340**, 126 (1989)
516. E. Mueller, H.-T. Janka, Astron. Astrophys. **317**, 140 (1997)
517. C.D. Ott, Class. Quantum Gravity **26**, 063001 (2009). arXiv:0809.0695
518. K. Kotake, Comptes Rendus Physique **14**, 318 (2013). arXiv:1110.5107
519. K. Hotokezaka et al., Astrophys. J. Lett. **778**, L16 (2013). arXiv:1310.1623
520. L. Rezzolla et al., Astrophys. J. Lett. **732**, L6 (2011). arXiv:1101.4298
521. B.S. Sathyaprakash, B.F. Schutz, Living Rev. Relativ. **12**, 2 (2009). arXiv:0903.0338
522. F.A. Rasio, S.L. Shapiro, Class. Quantum Gravity **16**, 1 (1999). gr-qc/9902019
523. M. Shibata, K. Taniguchi, K. Uryu, Phys. Rev. D **71**, 084021 (2005). gr-qc/0503119
524. K. Hurley et al., Mon. Not. Roy. Astron. Soc. **403**, 342 (2010). arXiv:0907.2462
525. B.J. Owen, Phys. Rev. Lett. **95**, 211101 (2005). astro-ph/0503399
526. A. Watts, Neutron star oscillations and QPOs during magnetar flares, in *36th COSPAR Scientific Assembly, COSPAR Meeting* 36 (2006), p. 168
527. R.C. Duncan, C. Thompson, Astrophys. J. Lett. **392**, L9 (1992)
528. C. Thompson, R.C. Duncan, Mon. Not. R. Astron. Soc. **275**, 255 (1995)
529. J.D. Gelfand et al., Astrophys. J. Lett. **634**, L89 (2005). astro-ph/0503269
530. P. Meszaros, E. Waxman, Phys. Rev. Lett. **87**, 171102 (2001). arXiv:astro-ph/0103275
531. H.B.J. Koers, R.A.M.J. Wijers, ArXiv e-prints *711* (2007). arXiv:0711.4791
532. S. Horiuchi, S. Ando, Phys. Rev. D **77**, 063007 (2008). arXiv:0711.2580
533. X.-Y. Wang, S. Razzaque, P. Meszaros, Z.-G. Dai, Phys. Rev. D **76**, 083009 (2007). arXiv:0705.0027
534. Y. Aso et al., Class. Quantum Gravity **25**, 114039 (2008). arXiv:0711.0107
535. T. Pradier, Nucl. Instrum. Methods Phys. Res. A **602**, 268 (2009). arXiv:0807.2562
536. I. Bartos, C. Finley, A. Corsi, S. Marka, Phys. Rev. Lett. **107**, 251101 (2011)
537. S. Adrian-Martnez et al., JCAP6 **8** (2013). arXiv:1205.3018
538. J. Piscionere et al., LOOC UP: locating and observing optical counterparts to unmodeled pulses in gravitational waves, in *American Astronomical Society Meeting Abstracts*, vol. 38 (Bulletin of the American Astronomical Society, 2007), pp. 910+
539. J. Kanner et al., Class. Quantum Gravity **25**, 184034 (2008). arXiv:0803.0312
540. R. Abbasi et al., Astron. Astrophys. **539**, A60 (2012). arXiv:1111.7030
541. M. Kowalski, A. Mohr, Astropart. Phys. **27**, 533 (2007). astro-ph/0701618

542. LIGO Scientific Collaboration et al., Astron. Astrophys. **539**, A124 (2012). arXiv:1109.3498
543. J. Abadie et al., Astron. Astrophys. **541**, A155 (2012). arXiv:1112.6005
544. P.A. Evans et al., Astrophys. J. Supp. Ser. **203**, 28 (2012). arXiv:1205.1124
545. The LIGO Scientific Collaboration et al., ArXiv e-prints (2013). arXiv:1310.2314
546. R. Frey, Detecting GW bursts associated with subthreshold Swift GRBs, 2009, (poster) GWDAW13
547. E.D. Harstad, Ph.D. thesis (2013)
548. L. Blackburn et al., ArXiv e-prints **1303**, 2174 (2013)
549. P. Antonioli et al., New J. Phys. **6**, 114 (2004). astro-ph/0406214
550. K. Hirata et al., Phys. Rev. Lett. **58**, 1490 (1987)
551. R.M. Bionta et al., Phys. Rev. Lett. **58**, 1494 (1987)
552. M.W.E. Smith et al., Astropart. Phys. **45**, 56 (2013). arXiv:1211.5602

GW170817: The Dawn of Multi-messenger Astronomy Including Gravitational Waves

Marica Branchesi

Abstract On 2017 August 17 the Advanced LIGO and Advanced Virgo detectors detect for the first time the signal, GW170817, from the coalescence of a binary system of neutron stars Abbott et al. (Phys Rev Lett, 2017, [1]). Exactly 1.7 s after the merger time (12:41:04 UTC) the Fermi Gamma-ray Burst Monitor independently detected a gamma-ray burst, GRB170817A Goldstein et al. (ApJL, 848, L14, 2017, [2]), Abbott et al. (ApJL, 848, 2017, [3]). An extensive observing campaign involving more than 70 world-wide ground and space observatories was performed leading to the discovery of the counterpart signals across all the electromagnetic spectrum Abbott et al. (ApJL, 2017, [4]). This observational campaign marks the birth of multi-messenger astronomy, which uses gravitational waves and electromagnetic emission. The collected multi-messenger data confirmed ten-year-old theoretical models. And at the same time, the richness of details of the taken data will require to develop new theory and to make other observations in the coming years to be interpreted.

The First Gravitational Wave Observation of the Coalescence of a Binary System of Neutron Stars

This gravitational-wave signal GW170817 (about 100 s starting from 24 Hz) was the loudest yet observed, with a combined signal-to-noise ratio (SNR) of 32.4 and a false-alarm-rate estimate of less than one per 8.0×10^4 years [1]. The signal carried information about the astrophysical source: masses, tidal deformability, distance, inclination, and sky localization. The component masses were inferred to be between 0.86 and 2.26 M_{odot}, and when restricting the component spins to a range consistent with the BNS observed population, the component masses result in the range 1.17–1.60 M_{odot}. These masses are in agreement with the dynamically measured masses of known neutron stars in binaries [5], and their inconsistency with the masses of

M. Branchesi (✉)
Gran Sasso Science Institute, Viale Francesco Crispi 7, L'aquila, Italy
e-mail: marica.branchesi@gssi.infn.it

M. Branchesi
Laboratori Nazionali Gran Sasso INFN, Assergi (L'Aquila), Italy

R. Aloisio et al. (eds.), *Multiple Messengers and Challenges in Astroparticle Physics*, https://doi.org/10.1007/978-3-319-65425-6_6

known black holes in galactic binary systems [6], suggesting that the GW170817 source was composed of two neutron stars.

However, the masses and lower limit on the compactness estimated from the gravitational-wave observations does not allow us to exclude objects more compact than neutron stars such as quark stars, black holes, or more exotic objects. The detection of the electromagnetic counterpart demonstrates the presence of matter confirming at least the presence of a neutron star in the system.

Details of the objects' internal structure become important as the orbital separation approaches the size of the bodies. As the gravitational-wave frequency increases, tidal effects in binary neutron stars increasingly affect the phase and become significant above $f_{GW} \simeq 600$ Hz. Unfortunately, the noise of the LIGO and Virgo detectors at these frequencies made it difficult to extract signal information and only a lower bound on the energy emitted $E_{rad} > 0.025\ M_{odot}c^2$ was set before the onset of strong tidal effects. Estimate of the tidal deformability (or polarizability) disfavor equations of state that predict less compact stars.

The luminosity distance estimate of 40^{+8}_{-14} Mpc set this GW signal as the closest ever observed and, by association, the closest short gamma-ray burst with a distance measurement. The viewing angle is constrained to be less than 56°. The source was localized by the Advanced LIGO and Virgo network within a sky region of 28 deg^2 (90% probability) near the southern end of the constellation Hydra. The advanced Virgo interferometer was essential for the sky localization. The low signal amplitude observed ($SNR = 2$) in Virgo with respect to the signal detected by the LIGO detectors indicated that the source was close to a blind spot of the interferometer significantly constraining the sky position. Distance and sky position located the source to within a volume of 380 Mpc^3. A region of the local Universe where properties and distribution of the galaxies are known. This played a key role to drive the successful optical search, which identified the counterpart and the host galaxy.

Binary neutron star mergers may result in a black hole, or short- or long-lived neutron star. In all these cases the remnant is expected to emit gravitational waves. However, the ringdown of a black hole produces gravitational waves at frequencies around 6 kHz, where the interferometer response is strongly reduced making its observation unfeasible. Searches have been made for short (tens of ms) and intermediate duration ($\lesssim$500 s) gravitational-wave signals from a neutron star remnant at frequencies up to 4 kHz without evidence of a post-merger signal of astrophysical origin. The upper limits placed on the strength of the signal cannot definitively rule out the existence of a short- or long-lived postmerger neutron star.

The detection of GW170817 enable to estimate the local coalescence rate density of binary systems of neutron stars to $1540^{3200}_{-1220}\ Gpc^{-3}yr^{-1}$ [1].

Discovery of the High-Energy Counterpart

A hard X-ray electromagnetic (EM) counterpart, short duration, sub-luminous Gamma-ray Burst, GRB170817A, was detected by the Fermi-GBM at 12:41:06 UTC, and announced automatically just 14 s after. It was also detected by the International Gamma-ray Astrophysics Laboratory (INTEGRAL) through an offline search [7] initiated by the LIGO-Virgo and Fermi-GBM notice.

The probability of the near-simultaneous temporal and spatial observation of GRB170817A and GW170817 occurring by chance is 5.0×10^{-8}. The association of GW170817 and GRB170817A provides insight into the origin of short gamma-ray bursts and fundamental physics. These observations are the first firm evidence that at least a fraction of short duration GRBs are associated with compact star mergers [3].

The arrival delay of only $(+1.74 \pm 0.05)$ s between high-energy and GW signal and travelled distance places stringent constraints on the difference between the speed of gravity and the speed of light to be between 3×10^{-15} and $+7 \times 10^{-16}$ times the speed of light. The measurement is consistent with GWs propagating at the speed of light as predicted by general relativity. The strong constraint on the speed of GWs ruled out many classes of modified gravity models (see, e.g. [8, 9]).

GRB170817A is the closest short GRB with known distance. Its fluence of 2.2×10^{-7} ergcm^{-2} in the 10–1000 keV energy range corresponds to a gamma-ray isotropic equivalent energy Eiso 4.3×10^{46} erg, which makes it three to four orders of magnitude smaller than the average energy of short GRBs with known redshift. The low luminosity has been interpreted as:

1. Uniform jet observed at large angles, provided that the jet bulk Lorentz factor is significantly smaller than usually assumed (e.g. [10]).
2. Emission from a structured jet, with a fast and energetic inner part surrounded by a slower component [11]. The slower component can be a cocoon, formed by the deposit of the energy of the jet when it excavates its way through the merger (dynamical) and post-merger (disk viscosity driven and neutrino driven) ejecta. This shocked jet material may produce a quasi-isotropic prompt, high energy EM signal (e.g. [12, 13]).
3. Jet-less scenario where an isotropic fireball expands ahead of the kilonova ejecta [14].

The Multi-wavelength Electromagnetic Follow-Up Campaign

The LIGO-Virgo sky localization of GW170817 (placing the source in a region of about 30 deg^2 and at a distance of 40 Mpc) were sent to the astronomers at 17:54:51 UTC and became observable to telescopes in Chile about 10 h after the merger. Immediately, the One-Meter, Two- Hemisphere (1M2H) team using the 1 m Swope Telescope started to observe the galaxies in volume corresponding to the GW170817 signal. The galaxies were ranked based on stellar mass and star formation rate. In one of them, NGC4993, an early-type galaxy at the same distance of the gravitational-wave signal, they discovered a new and bright blue optical transient (SSS17a, now with the International Astronomical Union identification of AT 2017gfo) at 23:33 UTC. The optical transient was independently detected by multiple teams during the same night. Then, all the telescopes and satellites around the world started the most extensive observing campaign ever across the electromagnetic spectrum pointing at AT 2017gfo for weeks [4]. The observations taken by a network of ground-based telescopes, from 40 cm to 10 m, and space-based observatories in the ultraviolet (UV), optical, and near-infrared showed a peculiar spectral energy evolution: a rapid dimming of initial UV-blue emission (see e.g [15]), brightening of the red and near-infrared emission, which began to fade after a few days to a week (see e.g [16]). The telescopes involved in the photometric observation include: CTIO1.3 m, DECam, IRSF, the Gemini-South FLAMINGO2, Gemini-South GMOS, GROND, HST, iTelescope.Net telescopes, the Korea Microlensing Telescope Network, Las Cumbres Observatory, the Lee Sang Gak Telescope, the Magellan-Baade and Magellan-Clay 6.5 m telescopes, the Nordic Optical Telescope, Pan-STARRS1, REM/ROS2 and REM/REMIR, SkyMapper, Subaru Hyper Suprime-Cam, ESO-VISTA, ESO-VST/OmegaCAM, and ESO-VLT/FORS2.

The first spectrum of AT 2017gfo was taken by the Magellan-Clay telescope. It showed a blue and featureless continuum between 4000 and 10000 Å. It appeared as an unusual spectrum, but common to cataclysmic variable stars and young core-collapse supernovae. The following spectra taken 24–48 h later (by the SALT-RSS, ePESSTO with the EFOSC2 instrument in spectroscopic mode at the ESO New Technology Telescope, the X-shooter spectrograph on the ESO Very Large Telescope, and the Goodman Spectrograph on the 4 m SOAR telescope) were crucial to exclude any transient contaminant and firmly associate AT 2017gfo with GW170817. The spectra showed an exceptionally fast spectral evolution in the blue part and absence of any features identifiable with absorption lines common in supernova-like transients. The rapid fading in the blue, the spectral energy evolution and appearance of broad spectral features in the red near-IR reproduced the expectations of disk wind outflow and dynamical ejecta kilonova models (e.g. [17–20]). During the coalescence of two neutron stars, the mass ejected at extremely high velocity in the interstellar medium is the ideal site for the formation of heavy elements for rapid neutron capture. While the radioactive decay of such elements determines the emission called kilonova, the universe is enriched with elements heavier than iron, the r-process nucleosynthesis elements. The ESO-VLT/X-shooter spectra, which simultaneously cover the wavelength range 320024800 Å and were taken over 2 weeks with a close to daily sampling revealed signatures of the radioactive decay of these elements [10, 21]. Coalescences of two neutron stars are very rare events; in Milky Way equivalent galaxies, they can happen with a rate of 32–474 Myr^{-1} [22], but the quantity of r-process materials that can be formed are enough to explain the abundance of heavy elements in the Universe (see e.g [23–26]).

The UV/optical/infrared observations of GW170817 compared with kilonova models and numerical relativity results were used to estimate a lower bound on the tidal deformability parameter. Together with the upper bound from the GW signal, both, extremely stiff and soft NS equations of state, can be ruled out [27].

After significant upper limit on the emission associated to no detections by Swift-XRT and the Chandra X-ray Observatory, on 26 Aug 2017 Chandra detected X-ray emission at the position of AT 2017gfo.

The observed X-ray flux implies an isotropic luminosity of $9 \times 10^{38}\,\mathrm{ergs}^{-1}$ if located in NGC 4993 [28]. Further Chandra observations, performed between 1 and 2 Sep 2017, confirmed the presence of continued X-ray activity, and suggested a slight increase in luminosity.

On September 2 and September 3 (16 days after the merger) a radio emission appeared in the Karl G. Jansky Very Large Array (VLA) data at a frequency of 3 GHz and at a frequency of 6 GHz. The Australia Telescope Compact Array (ATCA) also detected the source on September 5 in the 5.5–9 GHz band. In observations at 3 GHz with the VLA, the source shows evidence of an increase in flux density over a timescale of 2 weeks [29]. The observed X-ray radio emission are consistent with the emission from a relativistic collimated jet viewed off-axis. The observer not being aligned with the opening angle of the jetted outflow is expected to observe the afterglow when the jet has spread and decelerated enough so that the beaming cone of the emission includes the observers line of sight; the onset of the afterglow is delayed by several days or weeks and the emission weaker with respect to an on-axis observer. The X-ray and radio detection were interpreted as the first observations of an off-axis GRB (see e.g. [28, 30]), scenario theorized for about two decades but never observed before. Both the X-ray and radio emission were also consistent with a cocoon that breaks out when the jet transfers a large fraction of its energy into the surrounding ejecta. The cocoon expands over a wide angle at mildly relativistic velocities [31].

No ultra-high-energy gamma-rays and no neutrino candidates consistent with the source were found in follow-up searches.

The proximity to the Sun stopped the optical and X-ray observations of the GW source for about 3 months. Observations of GW170817 continued in the radio band (0.618 GHz frequency range) with the VLA, the ATCA and the upgraded Giant Metrewave Radio Telescope, showing a steady rise of the light curve and a spectrum consistent with optically thin synchrotron emission [32]. These observations completely exclude the collimated ultra-relativistic uniform jet viewed off-axis and an isotropic thin and uniform fireball, and requires a mildly relativistic wide angle outflow moving towards the observer. This outflow could be the high velocity tail of the neutron-rich material dynamically ejected during the merger (isotropic outflow with velocity profile) but the most likely scenario remains the cocoon able to explain also the prompt gamma-rays. Chandra observations during the first days of December 2017 showed brightening X-ray emission [28, 30, 33]. The similar, slow rising of both the X-ray and radio emissions indicate a common (synchrotron) source as origin. On December 6, AT2017gfo was also detected in the optical band by the HST [34].

The multi-messenger campaign of GW170817 marked also the beginning of cosmology with gravitational waves [35]. Combining the distance to the source inferred from the gravitational-wave signal with the recession velocity of the galaxy inferred from measurements of the redshift using electromagnetic data, the present-day expansion rate of the Universe has been determined as given by the Hubble constant of $70.0^{+12.0}_{-8.0}$ kms^{-1}Mpc^{-1} (maximum a posteriori and 68% credible interval). The Hubble constant estimate is consistent with existing measurements, including the cosmic microwave background measurements from Planck [36], the Type Ia supernova measurements from SHoES [37], the baryon acoustic oscillations measurements from SDSS [38], the strong lensing measurements from H0LiCOW [39], while being completely independent of them. The large uncertainty in the H0 measurement with GWs due to the degeneracy between distance and inclination in the GW measurements could improve with precise polarization measurements, and additional standard-siren measurements from future gravitational-wave sources will provide more precise constraints of this cosmological parameter.

As described above, the first multi-messenger campaign had and will have strong impact in many astrophysics fields starting from emission process physics and nuclear astrophysics, to fundamental physics and cosmology. This campaign demonstrated the importance of the development, coordination and collaboration of a network of gravitational-wave, electromagnetic, and neutrino observatories, the importance of joint analysis and interpretation of the multi-messenger data. We are in an exciting new era of multi-messenger, time-domain astronomy.

References

1. B.P. Abbott, R. Abbott, T.D. Abbott, F. Acernese, K. Ackley, C. Adams, T. Adams, P. Addesso, R.X. Adhikari, V.B. Adya, et al., Phys. Rev. Lett. (2017)
2. A. Goldstein, P. Veres, E. Burns, M.S. Briggs, R. Hamburg, D. Kocevski, C.A. Wilson-Hodge, R.D. Preece, S. Poolakkil, O.J. Roberts, C.M. Hui, V. Connaughton, J. Racusin, A. von Kienlin, T. Dal Canton, N. Christensen, T. Littenberg, K. Siellez, L. Blackburn, J. Broida, E. Bissaldi, W.H. Cleveland, M.H. Gibby, M.M. Giles, R.M. Kippen, S. Mc Breen, J. Mc Enery, C.A. Meegan, W.S. Paciesas, M. Stanbro, ApJL **848**, L14 (2017). https://doi.org/10.3847/2041-8213/aa8f41
3. B.P. Abbott, R. Abbott, T.D. Abbott, F. Acernese, K. Ackley, C. Adams, T. Adams, P. Addesso, R.X. Adhikari, V.B. Adya, et al., ApJL **848** (2017)
4. B.P. Abbott, R. Abbott, T.D. Abbott, F. Acernese, K. Ackley, C. Adams, T. Adams, P. Addesso, R.X. Adhikari, V.B. Adya, et al., ApJL (2017)
5. J. Antoniadis, P.C.C. Freire, N. Wex, T.M. Tauris, R.S. Lynch, M.H. van Kerkwijk, M. Kramer, C. Bassa, V.S. Dhillon, T. Driebe, J.W.T. Hessels, V.M. Kaspi, V.I. Kondratiev, N. Langer, T.R. Marsh, M.A. McLaughlin, T.T. Pennucci, S.M. Ransom, I.H. Stairs, J. van Leeuwen, J.P.W. Verbiest, D.G. Whelan, Science **340**, 448 (2013). https://doi.org/10.1126/science.1233232
6. L. Kreidberg, C.D. Bailyn, W.M. Farr, V. Kalogera, ApJ **757**, 36 (2012). https://doi.org/10.1088/0004-637X/757/1/36
7. V. Savchenko, C. Ferrigno, E. Kuulkers, A. Bazzano, E. Bozzo, S. Brandt, J. Chenevez, T.J.L. Courvoisier, R. Diehl, A. Domingo, L. Hanlon, E. Jourdain, A. von Kienlin, P. Laurent, F. Lebrun, A. Lutovinov, A. Martin-Carrillo, S. Mereghetti, L. Natalucci, J. Rodi, J.P. Roques, R. Sunyaev, P. Ubertini, ApJL **848**, L15 (2017). https://doi.org/10.3847/2041-8213/aa8f94
8. T. Baker, E. Bellini, P.G. Ferreira, M. Lagos, J. Noller, I. Sawicki, ArXiv e-prints (2017)
9. P. Creminelli, F. Vernizzi, ArXiv e-prints (2017)
10. E. Pian, P. D'Avanzo, S. Benetti, M. Branchesi, E. Brocato, S. Campana, E. Cappellaro, S. Covino, V. D'Elia, J.P.U. Fynbo, F. Getman, G. Ghirlanda, G. Ghisellini, A. Grado, G. Greco,

J. Hjorth, C. Kouveliotou, A. Levan, L. Limatola, D. Malesani, P.A. Mazzali, A. Melandri, P. Møller, L. Nicastro, E. Palazzi, S. Piranomonte, A. Rossi, O.S. Salafia, J. Selsing, G. Stratta, M. Tanaka, N.R. Tanvir, L. Tomasella, D. Watson, S. Yang, L. Amati, L.A. Antonelli, S. Ascenzi, M.G. Bernardini, M. Boër, F. Bufano, A. Bulgarelli, M. Capaccioli, P. Casella, A.J. Castro-Tirado, E. Chassande-Mottin, R. Ciolfi, C.M. Copperwheat, M. Dadina, G. De Cesare, A. di Paola, Y.Z. Fan, B. Gendre, G. Giuffrida, A. Giunta, L.K. Hunt, G.L. Israel, Z.P. Jin, M.M. Kasliwal, S. Klose, M. Lisi, F. Longo, E. Maiorano, M. Mapelli, N. Masetti, L. Nava, B. Patricelli, D. Perley, A. Pescalli, T. Piran, A. Possenti, L. Pulone, M. Razzano, R. Salvaterra, P. Schipani, M. Spera, A. Stamerra, L. Stella, G. Tagliaferri, V. Testa, E. Troja, M. Turatto, S.D. Vergani, D. Vergani, Nature **551**, 67 (2017). https://doi.org/10.1038/nature24298
11. A. Kathirgamaraju, R. Barniol Duran, D. Giannios, MNRAS **473**, L121 (2018). https://doi.org/10.1093/mnrasl/slx175
12. D. Lazzati, A. Deich, B.J. Morsony, J.C. Workman, MNRAS (2017)
13. D. Lazzati, D. López-Cámara, M. Cantiello, B.J. Morsony, R. Perna, J.C. Workman, ApJL (2017)
14. O.S. Salafia, G. Ghisellini, G. Ghirlanda, MNRAS **474**, L7 (2018). https://doi.org/10.1093/mnrasl/slx189
15. P.A. Evans, S.B. Cenko, J.A. Kennea, S.W.K. Emery, N.P.M. Kuin, O. Korobkin, R.T. Wollaeger, C.L. Fryer, K.K. Madsen, F.A. Harrison, Y. Xu, E. Nakar, K. Hotokezaka, A. Lien, S. Campana, S.R. Oates, E. Troja, A.A. Breeveld, F.E. Marshall, S.D. Barthelmy, A.P. Beardmore, D.N. Burrows, G. Cusumano, A. D'Ai, P. D'Avanzo, V. D'Elia, M. de Pasquale, W.P. Even, C.J. Fontes, K. Forster, J. Garcia, P. Giommi, B. Grefenstette, C. Gronwall, D.H. Hartmann, M. Heida, A.L. Hungerford, M.M. Kasliwal, H.A. Krimm, A.J. Levan, D. Malesani, A. Melandri, H. Miyasaka, J.A. Nousek, P.T. O'Brien, J.P. Osborne, C. Pagani, K.L. Page, D.M. Palmer, M. Perri, S. Pike, J.L. Racusin, S. Rosswog, M.H. Siegel, T. Sakamoto, B. Sbarufatti, G. Tagliaferri, N.R. Tanvir, A. Tohuvavohu, Science **358**, 1565 (2017). https://doi.org/10.1126/science.aap9580
16. V.A. Villar, J. Guillochon, E. Berger, B.D. Metzger, P.S. Cowperthwaite, M. Nicholl, K.D. Alexander, P.K. Blanchard, R. Chornock, T. Eftekhari, W. Fong, R. Margutti, P.K.G. Williams, ApJLl **851**, L21 (2017). https://doi.org/10.3847/2041-8213/aa9c84
17. D. Kasen, N.R. Badnell, J. Barnes, ApJ **774**, 25 (2013). https://doi.org/10.1088/0004-637X/774/1/25
18. B.D. Metzger, Living Rev. Relativ. **20**, 3 (2017). https://doi.org/10.1007/s41114-017-0006-z
19. D. Kasen, B. Metzger, J. Barnes, E. Quataert, E. Ramirez-Ruiz, Naturet **551**, 80 (2017). https://doi.org/10.1038/nature24453
20. M. Tanaka, Y. Utsumi, P.A. Mazzali, N. Tominaga, M. Yoshida, Y. Sekiguchi, T. Morokuma, K. Motohara, K. Ohta, K.S. Kawabata, F. Abe, K. Aoki, Y. Asakura, S. Baar, S. Barway, I.A. Bond, M. Doi, T. Fujiyoshi, H. Furusawa, S. Honda, Y. Itoh, M. Kawabata, N. Kawai, J.H. Kim, C.H. Lee, S. Miyazaki, K. Morihana, H. Nagashima, T. Nagayama, T. Nakaoka, F. Nakata, R. Ohsawa, T. Ohshima, H. Okita, T. Saito, T. Sumi, A. Tajitsu, J. Takahashi, M. Takayama, Y. Tamura, I. Tanaka, T. Terai, P.J. Tristram, N. Yasuda, T. Zenko, PASJ **69**, 102 (2017). https://doi.org/10.1093/pasj/psx121
21. S.J. Smartt, T.W. Chen, A. Jerkstrand, M. Coughlin, E. Kankare, S.A. Sim, M. Fraser, C. Inserra, K. Maguire, K.C. Chambers, M.E. Huber, T. Krühler, G. Leloudas, M. Magee, L.J. Shingles, K.W. Smith, D.R. Young, J. Tonry, R. Kotak, A. Gal-Yam, J.D. Lyman, D.S. Homan, C. Agliozzo, J.P. Anderson, C.R. Angus, C. Ashall, C. Barbarino, F.E. Bauer, M. Berton, M.T. Botticella, M. Bulla, J. Bulger, G. Cannizzaro, Z. Cano, R. Cartier, A. Cikota, P. Clark, A. De Cia, M. Della Valle, L. Denneau, M. Dennefeld, L. Dessart, G. Dimitriadis, N. Elias-Rosa, R.E. Firth, H. Flewelling, A. Flörs, A. Franckowiak, C. Frohmaier, L. Galbany, S. González-Gaitán, J. Greiner, M. Gromadzki, A.N. Guelbenzu, C.P. Gutiérrez, A. Hamanowicz, L. Hanlon, J. Harmanen, K.E. Heintz, A. Heinze, M.S. Hernandez, S.T. Hodgkin, I.M. Hook, L. Izzo, P.A. James, P.G. Jonker, W.E. Kerzendorf, S. Klose, Z. Kostrzewa-Rutkowska, M. Kowalski, M. Kromer, H. Kuncarayakti, A. Lawrence, T.B. Lowe, E.A. Magnier, I. Manulis, A. Martin-Carrillo, S. Mattila, O. McBrien, A. Müller, J. Nordin, D. O'Neill, F. Onori, J.T. Palmerio, A.

Pastorello, F. Patat, G. Pignata, P. Podsiadlowski, M.L. Pumo, S.J. Prentice, A. Rau, A. Razza, A. Rest, T. Reynolds, R. Roy, A.J. Ruiter, K.A. Rybicki, L. Salmon, P. Schady, A.S.B. Schultz, T. Schweyer, I.R. Seitenzahl, M. Smith, J. Sollerman, B. Stalder, C.W. Stubbs, M. Sullivan, H. Szegedi, F. Taddia, S. Taubenberger, G. Terreran, B. van Soelen, J. Vos, R.J. Wainscoat, N.A. Walton, C. Waters, H. Weiland, M. Willman, P. Wiseman, D.E. Wright, Ł. Wyrzykowski, O. Yaron, Nature **551**, 75 (2017). https://doi.org/10.1038/nature24303

22. B.P. Abbott, R. Abbott, T.D. Abbott, F. Acernese, K. Ackley, C. Adams, T. Adams, P. Addesso, R.X. Adhikari, V.B. Adya, et al., ApJL (2017)
23. M.R. Drout, A.L. Piro, B.J. Shappee, C.D. Kilpatrick, J.D. Simon, C. Contreras, D.A. Coulter, R.J. Foley, M.R. Siebert, N. Morrell, K. Boutsia, F. Di Mille, T.W.S. Holoien, D. Kasen, J.A. Kollmeier, B.F. Madore, A.J. Monson, A. Murguia-Berthier, Y.C. Pan, J.X. Prochaska, E. Ramirez-Ruiz, A. Rest, C. Adams, K. Alatalo, E. Bañados, J. Baughman, T.C. Beers, R.A. Bernstein, T. Bitsakis, A. Campillay, T.T. Hansen, C.R. Higgs, A.P. Ji, G. Maravelias, J.L. Marshall, C. Moni Bidin, J.L. Prieto, K.C. Rasmussen, C. Rojas-Bravo, A.L. Strom, N. Ulloa, J. Vargas-González, Z. Wan, D.D. Whitten, Science **358**, 1560 (2017). https://doi.org/10.1126/science.aaq0049
24. M.M. Kasliwal, E. Nakar, L.P. Singer, D.L. Kaplan, D.O. Cook, A. Van Sistine, R.M. Lau, C. Fremling, O. Gottlieb, J.E. Jencson, S.M. Adams, U. Feindt, K. Hotokezaka, S. Ghosh, D.A. Perley, P.C. Yu, T. Piran, J.R. Allison, G.C. Anupama, A. Balasubramanian, K.W. Bannister, J. Bally, J. Barnes, S. Barway, F. Bellm, V. Bhalerao, D. Bhattacharya, N. Blagorodnova, J.S. Bloom, P.R. Brady, C. Cannella, D. Chatterjee, S.B. Cenko, B.E. Cobb, C. Copperwheat, A. Corsi, K. De, D. Dobie, S.W.K. Emery, P.A. Evans, O.D. Fox, D.A. Frail, C. Frohmaier, A. Goobar, G. Hallinan, F. Harrison, G. Helou, T. Hinderer, A.Y.Q. Ho, A. Horesh, W.H. Ip, R. Itoh, D. Kasen, H. Kim, N.P.M. Kuin, T. Kupfer, C. Lynch, K. Madsen, P.A. Mazzali, A.A. Miller, K. Mooley, T. Murphy, C.C. Ngeow, D. Nichols, S. Nissanke, P. Nugent, E.O. Ofek, H. Qi, R.M. Quimby, S. Rosswog, F. Rusu, E.M. Sadler, P. Schmidt, J. Sollerman, I. Steele, A.R. Williamson, Y. Xu, L. Yan, Y. Yatsu, C. Zhang, W. Zhao, ArXiv e-prints (2017)
25. B.P. Abbott, R. Abbott, T.D. Abbott, F. Acernese, K. Ackley, C. Adams, T. Adams, P. Addesso, R.X. Adhikari, V.B. Adya, et al., ApJL (2017)
26. B. Côté, C.L. Fryer, K. Belczynski, O. Korobkin, M. Chruślińska, N. Vassh, M.R. Mumpower, J. Lippuner, T.M. Sprouse, R. Surman, R. Wollaeger, ArXiv e-prints (2017)
27. D. Radice, A. Perego, F. Zappa, S. Bernuzzi, ArXiv e-prints (2017)
28. E. Troja, L. Piro, H. van Eerten, R.T. Wollaeger, M. Im, O.D. Fox, N.R. Butler, S.B. Cenko, T. Sakamoto, C.L. Fryer, R. Ricci, A. Lien, R.E. Ryan, O. Korobkin, S.K. Lee, J.M. Burgess, W.H. Lee, A.M. Watson, C. Choi, S. Covino, P. D'Avanzo, C.J. Fontes, J.B. González, H.G. Khandrika, J. Kim, S.L. Kim, C.U. Lee, H.M. Lee, A. Kutyrev, G. Lim, R. Sánchez-Ramírez, S. Veilleux, M.H. Wieringa, Y. Yoon, Nature **551**, 71 (2017). https://doi.org/10.1038/nature24290
29. G. Hallinan, A. Corsi, K.P. Mooley, K. Hotokezaka, E. Nakar, M. M. Kasliwal, D. Kaplan, D. A. Frail, S. T. Myers, T. Murphy, K. De, D. Dobie, J. Allison, K. W. Bannister, V. Bhalerao, P. Chandra, T. E. Clarke, S. Giacintucci, A.Y.Q. Ho, L. P. Singer, **358** (2017)
30. R. Margutti, E. Berger, W. Fong, C. Guidorzi, K.D. Alexander, B.D. Metzger, P.K. Blanchard, P.S. Cowperthwaite, R. Chornock, T. Eftekhari, M. Nicholl, V.A. Villar, P.K.G. Williams, J. Annis, D.A. Brown, H. Chen, Z. Doctor, J.A. Frieman, D.E. Holz, M. Sako, M. Soares-Santos, ApJL **848**, L20 (2017). https://doi.org/10.3847/2041-8213/aa9057
31. O. Gottlieb, E. Nakar, T. Piran, K. Hotokezaka, ArXiv e-prints (2017)
32. K.P. Mooley, E. Nakar, K. Hotokezaka, G. Hallinan, A. Corsi, D.A. Frail, A. Horesh, T. Murphy, E. Lenc, D.L. Kaplan, K. De, D. Dobie, P. Chandra, A. Deller, O. Gottlieb, M.M. Kasliwal, S.R. Kulkarni, S.T. Myers, S. Nissanke, T. Piran, C. Lynch, V. Bhalerao, S. Bourke, K.W. Bannister, L.P. Singer, Provided by the SAO/NASA Astrophysics Data System, Nature (2017). https://doi.org/10.1038/nature25452, http://cdsads.u-strasbg.fr/abs/2017, arXiv:1711.11573
33. D. Haggard, J.J. Ruan, M. Nynka, V. Kalogera, P. Evans, GCN **22206** (2017)
34. A.J. Levan, J.D. Lyman, N.R. Tanvir, I. Mandel, J. Hjorth, A.S. Fruchter, T. Kangas, B. Gompertz, K.D. Wiersema, D. Steeghs, S. Rosswog, LIGO/Virgo GW170817: Further hubble space telescope observations

35. B.P. Abbott, R. Abbott, T.D. Abbott, F. Acernese, K. Ackley, C. Adams, T. Adams, P. Addesso, R.X. Adhikari, V.B. Adya, et al., Nature (2017)
36. Planck Collaboration, P.A.R. Ade, N. Aghanim, M. Arnaud, M. Ashdown, J. Aumont, C. Baccigalupi, A.J. Banday, R.B. Barreiro, J.G. Bartlett, et al., A & A **594**, A13 (2016). https://doi.org/10.1051/0004-6361/201525830
37. A.G. Riess, L.M. Macri, S.L. Hoffmann, D. Scolnic, S. Casertano, A.V. Filippenko, B.E. Tucker, M.J. Reid, D.O. Jones, J.M. Silverman, R. Chornock, P. Challis, W. Yuan, P.J. Brown, R.J. Foley, ApJ **826**, 56 (2016). https://doi.org/10.3847/0004-637X/826/1/56
38. É. Aubourg, S. Bailey, J.E. Bautista, F. Beutler, V. Bhardwaj, D. Bizyaev, M. Blanton, M. Blomqvist, A.S. Bolton, J. Bovy, H. Brewington, J. Brinkmann, J.R. Brownstein, A. Burden, N.G. Busca, W. Carithers, C.H. Chuang, J. Comparat, R.A.C. Croft, A.J. Cuesta, K.S. Dawson, T. Delubac, D.J. Eisenstein, A. Font-Ribera, J. Ge, J.M. Le Goff, S.G.A. Gontcho, J.R. Gott, J.E. Gunn, H. Guo, J. Guy, J.C. Hamilton, S. Ho, K. Honscheid, C. Howlett, D. Kirkby, F.S. Kitaura, J.P. Kneib, K.G. Lee, D. Long, R.H. Lupton, M.V. Magaña, V. Malanushenko, E. Malanushenko, M. Manera, C. Maraston, D. Margala, C.K. McBride, J. Miralda-Escudé, A.D. Myers, R.C. Nichol, P. Noterdaeme, S.E. Nuza, M.D. Olmstead, D. Oravetz, I. Pâris, N. Padmanabhan, N. Palanque-Delabrouille, K. Pan, M. Pellejero-Ibanez, W.J. Percival, P. Petitjean, M.M. Pieri, F. Prada, B. Reid, J. Rich, N.A. Roe, A.J. Ross, N.P. Ross, G. Rossi, J.A. Rubiño-Martín, A.G. Sánchez, L. Samushia, R.T. Génova-Santos, C.G. Scóccola, D.J. Schlegel, D.P. Schneider, H.J. Seo, E. Sheldon, A. Simmons, R.A. Skibba, A. Slosar, M.A. Strauss, D. Thomas, J.L. Tinker, R. Tojeiro, J.A. Vazquez, M. Viel, D.A. Wake, B.A. Weaver, D.H. Weinberg, W.M. Wood-Vasey, C. Yèche, I. Zehavi, G.B. Zhao, B.O.S.S. Collaboration, Phys. Rev. D **92**(12), 123516 (2015). https://doi.org/10.1103/PhysRevD.92.123516
39. V. Bonvin, F. Courbin, S.H. Suyu, P.J. Marshall, C.E. Rusu, D. Sluse, M. Tewes, K.C. Wong, T. Collett, C.D. Fassnacht, T. Treu, M.W. Auger, S. Hilbert, L.V.E. Koopmans, G. Meylan, N. Rumbaugh, A. Sonnenfeld, C. Spiniello, MNRAS **465**, 4914 (2017). https://doi.org/10.1093/mnras/stw3006

Cosmology: Searching for Deviations from the Standard Cosmological Model

Nicola Bartolo, Sabino Matarrese and Matteo Viel

Abstract Cosmology is experiencing an exciting phase, characterized by the enormous amount of data which are already available or will soon become available. This epoch, which has been dubbed the era of precision cosmology, clearly requires an adequate theoretical effort to provide predictions at high level of accuracy. At the same time, the quality of the data available allows one to face with totally new and challenging questions. In this chapter we will focus on three complementary issues: (i) observable predictions of inflation in the early universe; (ii) a short overview of some cosmological effects of modified gravity scenarios; (iii) neutrinos in cosmology and large-scale structures.

N. Bartolo (✉) · S. Matarrese
Dipartimento di Fisica e Astronomia "G. Galilei", Università degli Studi di Padova, via Marzolo 8, 35131 Padova, Italy
e-mail: nicola.bartolo@pd.infn.it

S. Matarrese
e-mail: sabino.matarrese@pd.infn.it

N. Bartolo · S. Matarrese
INFN, Sezione di Padova, via Marzolo 8, 35131 Padova, Italy

N. Bartolo · S. Matarrese
INAF - Osservatorio Astronomico di Padova, Vicolo dell'Osservatorio 5, 35122 Padova, Italy

S. Matarrese
INFN, Gran Sasso Science Institute, Viale F. Crispi 7, 67100 Laquila, Italy

M. Viel
INAF, Osservatorio Astronomico di Trieste, Via G. B. Tiepolo 11, 34131 Trieste, Italy
e-mail: viel@oats.inaf.it

M. Viel
INFN, Sezione di Trieste, Via Valerio 2, 34127 Trieste, Italy

R. Aloisio et al. (eds.), *Multiple Messengers and Challenges in Astroparticle Physics*, https://doi.org/10.1007/978-3-319-65425-6_7

Introduction

Cosmology has undergone tremendous progress in the last decade, which has been primarily driven by the success of Cosmic Microwave Background (CMB) experiments like WMAP and *Planck* satellites. These efforts have allowed to measure the parameters characterizing structure formation to unprecedented accuracy and also to face completely new challenges in the removal and treatment of systematic effects in the data themselves. A consistent picture from the analysis of large temperature (and polarization) anisotropies in the CMB has emerged: there are no hints of departure from the so-called standard 'vanilla' cosmological model, the so-called ΛCDM model + initial conditions from Inflation. This model is based on a flat space filled with cold dark matter, cosmological constant, baryons and negligible contribution from neutrinos and photons. Basically, a relatively simple, six parameter model with $\Omega_b h^2, \Omega_c h^2, A_s, \tau, n_s, \theta_{MC}$ (baryon and cold dark matter energy densities, amplitude of the power spectrum at CMB scales, optical depth, scalar spectral index and sound horizon at last scattering) provides a very accurate fit to the data including the lensing and polarization information. All the structures we observe in the universe developed via gravitational instability from tiny primordial density (curvature) perturbations. Inflation is part of the standard model of cosmology as the generator of such initial seeds. They are supposed to originate during the inflationary accelerated expansion from quantum fluctuations of one (or more) primordial (scalar) fields at high energy scales in the very early universe. The amplitude of the power spectrum A_s and the power spectrum spectral index n_s can be predicted as primordial initial conditions within the simplest models of inflation. Inflation predicts scalar (density) perturbations that are adiabatic, coherent on very large scales (superhorizon), almost (but not exactly) scale invariant and nearly Gaussian distributed. Crucially, inflation also predicts a stochastic background of primordial gravitational waves. Underlying the standard cosmological model there is the cosmological principle, which leads to a statement of statistical homogeneity and isotropy. Searching for possible extensions/departures from the standard cosmological model is of primary importance and could have profound implications for fundamental physics and structure formation.

Prior to the latest developments, the main observational probes that have allowed to establish this paradigm were the Hubble law, the CMB and Big Bang Nucleosynthesis. These observations, complemented by the study of the large scale structure, have provided evidence for isotropy and homogeneity of our Universe. Nowadays, Cosmology is however facing new important challenges. These challenges touch all the different building blocks on which the cosmological standard model is based on. On one side there are open issues at the largest accessible cosmological scales. Some of these issues are related to the initial conditions and hence to inflation. For example, looking for signatures of non-Gaussianity in the primordial perturbations, which will allow to establish the link with inflationary models. Quantifying whether some 'anomalies' could be present in the data and what fundamental meaning it could have might lead to question one of the pillars of the standard cosmological models such as (statistical) isotropy and homogeneity. Some of these issues deal with the late time evolution of the universe, like, e.g. understanding on firm grounds alternatives to the cosmological constant in terms of modified gravity/dark energy models and how to observationally distinguish them, or addressing whether or not new particles, relativistic or not, can impact on the dynamical state of the Universe. On the other side there are small/medium scale interesting questions: describing the impact of massive neutrinos in the structure formation era and find evidence in the data of their impact; challenge the nature of dark matter at the smallest scales and use astrophysical objects to constrain its properties (thermal velocity and masses) in different particle physics scenarios.

In general, in order to find evidence of new physics beyond the standard cosmological model, it is evident that a comprehensive analysis must proceed in a broad way by performing searches which span different redshifts and scales. Fundamental discoveries are also likely to be hidden in a regime in which either statistical errors (at the largest scales) or systematic errors (at small/medium scales) need to be addressed with new ideas and tools which will also require a good knowledge of astrophysics and galaxy formation.

This review summarizes the recent progress made in some of these topics.

Inflation in the Early Universe

Inflation is recognized to be by far the most promising scenario to understand the origin, in the very early universe, of the initial conditions from which all the structures we observe today originated (from the anisotropies of the Cosmic Microwave Background to galaxies and clusters of galaxies, down to the smallest scales probed by stars and planets). The data revolution that Cosmology underwent in the last 15 years has allowed to sharpen our understanding of the universe and to better address some fundamental questions about the physical mechanism that took place in the very early stages of its expansion: we have been able to rule out various alternatives to inflation and to confirm to a high degree of confidence and accuracy various basic predictions of inflation. Such a revolution will continue over the next decade, offering the opportunity to further scrutinize the nature of primordial cosmological perturbations. We will be able to strongly improve our knowledge of the fundamental mechanisms underlying inflation and of the finest details of the inflationary dynamics. In this chapter, we will review the basic predictions of the inflationary models and the present observational status. At the end the main focus will be on the 'future of Inflation': what are the open issues about the nature of the primordial perturbations still to be addressed? What are the forthcoming/future observables that can allow a breakthrough into our comprehension of the physics of inflation?

Basic Predictions of Inflation

Inflation [1–7] is an initial phase in the evolution of the Universe characterized by accelerated expansion. Such a phase must be long enough as to solve a number of cosmological 'shortcomings' of the standard Hot Big Bang model (the so called monopole, horizon and flatness 'problems'). In fact the idea of inflation was originally introduced as an elegant solution to these issues, by advocating an equation of state with negative pressure ($w \simeq p/\rho < -1/3$) before the radiation epoch of the standard Hot Big Bang model, characterized by an equation of state $w = 1/3$. For sure inflation must occur before primordial nucleosynthesis. A system which realizes such a scenario is the simplest system one could think of: a scalar field which slowly rolls down a flat potential. In this way the potential dominates over the kinetic term ($\frac{1}{2}(\nabla\phi)^2$) thereby allowing the background equation of state

$$w = \frac{\frac{1}{2}\dot{\phi}^2 - V(\phi)}{\frac{1}{2}\dot{\phi}^2 + V(\phi)} \tag{1}$$

to be negative (here a dot represents a derivative w.r.t. to cosmic time t) In such a scenario the accelerated expansion during inflation is driven by the vacuum energy of a scalar field, the 'inflaton'. In the case where $\frac{1}{2}\dot{\phi}^2 \ll V(\phi)$, $w \simeq -1$ and the universe expands almost exponentially.

A useful way to parametrize (in a sort of model independent way) the slow-roll phase of inflation and to guarantee a long duration of the accelerated expansion is to introduce the so-called slow-roll parameters

$$\epsilon_V = \frac{1}{16\pi G}\left(\frac{V'}{V}\right)^2 \simeq 4\pi G\frac{\dot{\phi}^2}{H^2} = -\frac{\dot{H}}{H^2}, \tag{2}$$

$$\eta_V = \frac{1}{8\pi G}\left(\frac{V''}{V}\right) \simeq -\frac{\ddot{\phi}}{H\dot{\phi}} + \epsilon_V, \tag{3}$$

where $\simeq$ means equalities hold at the lowest order of the slow-roll parameters. In order to have inflation $\epsilon_V \ll 1$, and we also require $\eta_V \ll 1$ in such a a way to sustain this condition for a long time.

Since inflation was proposed, the most appealing aspect of such a scenario was soon realized [8–14]: it provides a compelling and simple mechanism to generate the the primordial density perturbations which are the initial seeds for structure formation in the Universe. The basic idea is that on microscopic scales the scalar field has quantum fluctuations whose vacuum state is well understood in terms of the usual flat-space quantum field theory. The underlying accelerated expansion gravitationally amplifies such quantum fluctuations and stretches them from microscopic to large cosmological scales, where they are turned into matter density (curvature) fluctuations at the end of inflation. Here 'microscopic' and 'large cosmological scales' refers to the ratio ($k/(aH)$), i.e. to the ratio between the wavelength of a given perturbation mode in Fourier space and the size of the comoving Hubble radius $(aH)^{-1} = (1/\dot{a})$. The comoving Hubble radius represents the scale beyond which causal processes become ineffective. During inflation the scale factor accelerates, hence the comoving Hubble radius decreases. Therefore a given inflaton fluctuation mode will start inside the Hubble radius and later on will cross the Hubble radius and will be stretched on superhorizon scales, where it becomes 'frozen'.

A crucial aspect of the inflationary scenario is that, by the same mechanism, fluctuations in the gravitational field itself are generated, so that a a spectrum of gravitational waves (tensor perturbations) is produced. This happen independently of the specific inflationary model for the generation of the density (curvature) perturbations, being a generic prediction of an accelerated expansion of the Universe. This is one of the reasons why the observational signature of inflationary gravitational waves is considered as a 'smoking gun' of inflation.

After inflation ends the standard radiation dominated phase starts with the comoving Hubble radius growing in time. Therefore, a given fluctuation mode will eventually come back inside the comoving Hubble radius: the inflaton fluctuations are left imprinted on a given length scale during the inflationary accelerated phase, when a given mode crosses out the comoving Hubble radius. Then they are stretched on super-horizon scales and will be present when the fluctuation modes re-enter the Hubble radius after inflation ends. The structures we observe in the Universe started to form via gravitational instability around these initial seeds. The perturbations at the last scattering surface, that we observe as CMB anisotropies, and the distribution of galaxies all formed from the primordial inflationary seeds.

The models of single-field slow-roll inflation make some very precise predictions

- A Universe which is very close to being spatially flat
- generation of primordial perturbations on superhorizon scales
- scalar curvature (density) perturbations which are adiabatic
- production of a stochastic background of gravitational waves (tensor fluctuations)
- primordial perturbations (curvature and tensor) which have a nearly scale-invariant power spectrum
- primordial perturbations which are almost Gaussian distributed

At present there is strong observational evidence in favour of the inflationary scenario. The best constraints come from the latest observations of the *Planck satellite* on the temperature CMB anisotopries (and a first preliminary analysis of data in polarization). All the data are fully consistent with a number of the above predictions: the spatial flatness of the Universe, the adiabatic nature of the primordial perturbations, their coherence on superhorizon scales, the near scale-invariance of the spectrum of primordial density (curvature) perturbations and their (near) Gaussianity. At the moment we only have upper limits on the amplitude of the inflationary gravitational waves. Deviations from spatial flatness are bounded to be $|\Omega_K| < 0.005$ at 95% CL, for more details, see [15]. Only very small contributions, of a *few percentage*, from non-adiabatic perturbations are allowed by the latest data of the *Planck* satellite on CMB anisotropies [15, 16]. An interesting point, which sometimes is not properly considered, is the large-scale coherence of the perturbations generated from inflation. Such a feature has been observed in particular via the measured cross-correlation between the CMB temperature fluctuations and the (E-mode) polarization $\langle TE \rangle$. Such a cross correlation is a well definite prediction of the standard model and it is a distinctive signature of adiabatic superhorizon fluctuations. It shows that fluctuations had a very strong coherence on scales which were superhorizon at the time of the last-scattering epoch (at a redshift of $z \simeq 1090$, i.e. at the time when CMB started to form). This is exactly what inflation predicts, and therefore this shows that a mechanism of the type inherent in the inflationary mechanism probably took place in the early universe to bring such a pattern. Such a pattern (with a negative cross-correlation at multipoles between 50 and 200) is very different from that predicted by those models that were the initial competitors of inflation as mechanisms for structure formation, such as cosmic strings or textures (or more general causal seed models) which predict a positive cross-correlation. The measurements of the $\langle TE \rangle$ cross-correlation were therefore able to definitely rule out definitely these competitors [17]. Figure 1 shows the measurement of the $\langle TE \rangle$ power spectra obtained with the first-year WMAP satellite data. Notice that multipoles between 50 and 150

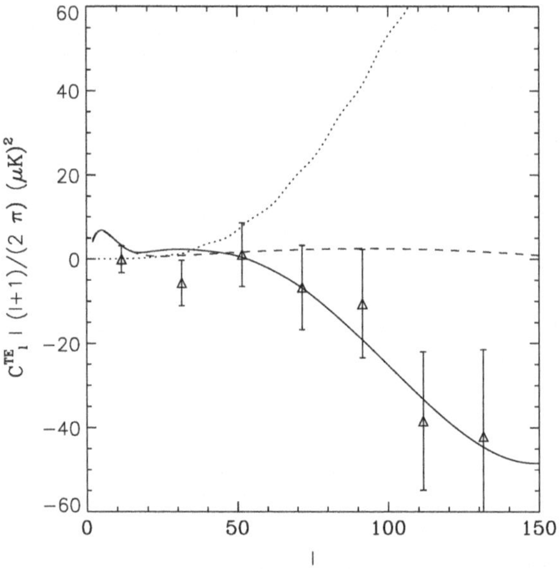

Fig. 1 Temperature-polarization angular power spectrum as measured by the WMAP satellite. The WMAP TE data [18] are shown in bins of $\Delta\ell = 10$. The WMAP detection of a TE *anti*-correlation at $\ell = 50$–150 implies the existence of superhorizon (adiabatic) fluctuations at decoupling, as predicted by inflation, since those scales correspond to superhorizon scales at the epoch of decoupling. Such a detection rules out various classes of causal seed models. The *solid line* shows the large-angle TE power spectrum predicted in typical primordial adiabatic models, while the *dashed lines* correspond to typical primordial isocurvature models, and the *dotted line* shows the typical prediction of causal scaling seed models. Figure taken from [17]

correspond to scales that were superhorizon at the last scattering epoch (i.e. at the time when CMB photons started to travel freely).

In the following we recall in more details some of the current constraints on models of single-field slow-roll inflation.

Constraints on Single-Field Models of Slow-Roll Inflation

Single-field models of slow-roll inflation predict a a power-spectrum of adiabatic superhorizon curvature fluctuations given by

$$\mathcal{P}_{\mathcal{R}}(k) = A_s \left(\frac{k}{k_*}\right)^{n_s - 1 + \frac{1}{2}\, \mathrm{d}n_s/\mathrm{d}\ln k\, \ln(k/k_*) + \frac{1}{6}\, \mathrm{d}^2 n_s/\mathrm{d}\ln k^2 (\ln(k/k_*))^2 + \ldots}, \tag{4}$$

where the comoving curvature perturbation $\mathcal{R}$ is directly related to the (gauge-invariant) inflaton field fluctuation as $\mathcal{R} = -H\delta\phi/\dot{\phi}$.[1] The quantity $\mathcal{R}$ can be loosley thought of as a (gauge-invariant) measure of the primordial density fluctuation at Hubble radius crossing. Here A_s is the scalar amplitude and n_s, $\mathrm{d}n_s/\mathrm{d}\ln k$ and $\mathrm{d}^2 n_s/\mathrm{d}\ln k^2$ are the scalar spectral index, the running of the scalar spectral index, and the running of the running of the scalar spectral index, respectively.

Similarly, the tensor (gravitational waves) power spectrum is predicted to be of the form

$$\mathcal{P}_t(k) = A_t \left(\frac{k}{k_*}\right)^{n_t + \frac{1}{2}\, \mathrm{d}n_t/\mathrm{d}\ln k\, \ln(k/k_*) + \ldots}, \tag{5}$$

with a meaning of the tensor parameters analogous to the ones of the scalar counterparts. At leading order in the slow-roll parameters the various parameters of Eq. (4) are predicted to be

$$A_s = \frac{V}{24\pi^2 M_{\mathrm{pl}}^4 \epsilon_V}, \tag{6}$$

$$n_s - 1 = 2\eta_V - 6\epsilon_V, \tag{7}$$

$$\mathrm{d}n_s/\mathrm{d}\ln k = +16\epsilon_V\eta_V - 24\epsilon_V^2 - 2\xi_V^2, \tag{8}$$

$$\mathrm{d}^2 n_s/\mathrm{d}\ln k^2 = -192\epsilon_V^3 + 192\epsilon_V^2\eta_V - 32\epsilon_V\eta_V^2 \tag{9}$$

$$- 24\epsilon_V\xi_V^2 + 2\eta_V\xi_V^2 + 2\varpi_V^3, \tag{10}$$

where the *slow-roll parameters* ϵ_V and η_V are defined in Eqs. (2) and (3), and the higher order parameters are given by

---

[1]For simplicity we have written the comoving curvature perturbation $\mathcal{R}$ in the so called spatially flat gauge, where the inflaton field fluctuations are gauge-invariant, corresponding to the so called Sasaki-Mukhanov variable [19, 20].

$$\xi_V^2 = \frac{M_{\rm pl}^4 V_\phi V_{\phi\phi\phi}}{V^2}, \tag{11}$$

and

$$\varpi_V^3 = \frac{M_{\rm pl}^6 V_\phi^2 V_{\phi\phi\phi\phi}}{V^3}. \tag{12}$$

Here $M_{\rm pl} = (8\pi G)^{-1/2}$ is the reduced Planck mass. The tensor parameters in Eq. (5) are predicted to be

$$A_{\rm t} = \frac{2V}{3\pi^2 M_{\rm pl}^4}, \tag{13}$$

$$n_{\rm t} = -2\epsilon_V, \tag{14}$$

$$\mathrm{d}n_{\rm t}/\mathrm{d}\ln k = +4\epsilon_V\eta_V - 8\epsilon_V^2. \tag{15}$$

In the previous expressions the slow-roll parameters must be evaluated at the time the mode k_* crosses out the Hubble radius for the first time, i.e. when $k_* = a_* H_*$. Notice in particular that, contrary to the curvature (scalar) amplitude A_s, the tensor amplitude A_t depends only on the Hubble parameter during inflation. Therefore it provides a direct measurement of the energy scale at which inflation takes place. In particular it turns out to be useful to quantify the gravitational waves amplitude relative to the curvature (density) perturbation amplitude, introducing the so-called *tensor-to-scalar perturbation ratio* (at some pivot scale k_*)

$$r = \frac{\mathcal{P}_t(k_*)}{\mathcal{P}_{\mathcal{R}}(k_*)}. \tag{16}$$

At lowest order in the slow-roll parameters it turns out that

$$r = 16\epsilon_V \approx -8n_{\rm t}, \tag{17}$$

which is called *the consistency relation*, since it is a relation that must hold true for *all* single-field models of slow-roll inflation. Such a consistency relation is sometimes referred to as the 'holy grail' of inflation, in the sense that an observational confirmation of this relation would strongly indicate that inflation did took place in the early universe, since it is very difficult that such a relation might be realized by other mechanisms of structure formation.

The tensor-to-scalar ratio is directly related to the energy scale of inflation via the relation

$$V_*^{1/4} = \left(\frac{3\pi^2 A_{\rm s}}{2}\right)^{1/4} r^{1/4} M_{\rm pl} = (1.94 \times 10^{16}\ \mathrm{GeV}) \left(\frac{r_*}{0.12}\right)^{1/4}. \tag{18}$$

Moreover, through the tensor-to-scalar ratio there is a simple way to understand the connection between the inflaton field excursion during inflation and the (relative) amplitude of the gravitational waves produced during inflation, given by the following relation [21]

$$\frac{\Delta\phi}{M_{\rm pl}} = \int_{\phi_{end}}^{\phi_*} d\phi \approx \frac{1}{\sqrt{8}} \int_0^{N_*} \mathrm{d}N \sqrt{r}. \tag{19}$$

Here $N_* = \int_{t_*}^{t_{end}} Hdt = \ln(a_{end}/a_*)$ is a 'natural' measure of time during inflation, and specifically it corresponds the number of e-folds between the time of horizon-crossing of the mode k_* and the end of inflation. From this relation we see that, if the the inflaton field moves along its potential of the order the Planck mass (or larger) then $r \geq 2 \times 10^{-3}$–0.01, according to the precise evolution of r during inflation (which is however quite constrained in the slow-roll paradigm, rising at second order in the slow-roll parameters). Such a theoretical threshold sets a benchmark between the so-called 'large-field' models of inflation and 'small-field' models of inflation.

As we have already mentioned, the inflationary stochastic background of gravitational waves can be considered as a smoking gun of inflation, both because it is directly related to the accelerated expansion of the universe (and to the energy scale at which it happens), and because most of the models of the early universe that have been proposed as alternative to inflation do in general predict a very low (unmeasurable) level of gravitational waves (see, e.g. the discussion in [22]). For these reasons a detection of the primordial gravitational waves from inflation would be of outmost importance. The privileged signature would be a specific pattern (curl-like) in the polarization of the CMB, the so called B-mode polarization [23, 24]. The reason is that such a signal could be produced *only* by gravitational waves, and not by scalar curvature (density) perturbations (which can generate only E-mode curl free polarization), making the B-mode on its own a smoking gun of primordial tensor modes.[2] Finally, a detection of inflationary gravitational waves would have a revolutionary impact, not only for Cosmology, but also for Fundamental Physics. It would clearly point towards physics beyond the Standard model of particle physics and moreover, it would be the first observational evidence of a process deeply related to quantum gravity. For a recent review on gravitational waves from inflation we refer the reader to [27].

In Fig. 2 we show the best constraints in the (r, n_s) plane, and the implications on some of the most interesting and popular slow-roll models of inflation. The constraints have been obtained from the full mission temperature data of the *Planck* satellite in combination with a first release of large angular scales polarization data. In particular, summarizing some of the most important results, in turns out that the *Planck* data (including lensing ones) measure the scalar spectral index of the curvature perturbations [16]

$$n_s = 0.968 \pm 0.006 \quad (68\%\ \mathrm{CL}), \tag{20}$$

and strongly constrain their scale dependence,

---

[2] For the importance of B-mode detection for inflation see, e.g. the two recent papers [25, 26].

$$\frac{dn_s}{d\ln k} = -0.003 \pm 0.007 \ \ (68\% \ \text{CL}). \tag{21}$$

As far as the amplitude of the primordial gravitational waves from inflation is concerned, the upper bound on the tensor-to-scalar-ratio (at the pivot scale of $k_* = 0.002$ Mpc^{-1}) is

$$r_{0.002} < 0.11 \ \ (95\% \ \text{CL}), \tag{22}$$

a limit which is independent of, and compatible with, the one obtained by the joint analysis of the BICEP2/Keck Array and *Planck* measurements [28], which yields $r < 0.12, \ \ (95\% \ \text{CL})$. At present the best constraints on the amplitude of the inflationary gravitational waves come from the latest BICEP2/Keck array data, which, using their measurements of the CMB B-mode polarization, yields [29]

$$r_{0.05} < 0.09 \ \ (95\% \ \text{CL}). \tag{23}$$

These constraints are somewhat more robust than those obtained from the temperature power spectra, since the scalar density perturbations cannot generate B-modes polarization. They are also less model dependent, since they do not assume the consistency relation (17) for single-field models of inflation. Th BICEP2/Keck array team find an even stronger constraint, when adding *Planck* temperature (and other) data, finding [29] $r_{0.05} < 0.07 \ \ (95\% \ \text{CL})$.

In terms of the slow-roll parameters, the main constraints are [16]

$$\epsilon_V < \ 0.012 \qquad (95\,\% \ \text{CL}), \tag{24}$$
$$\eta_V = -0.0080^{+0.0088}_{-0.0146} \quad (68\,\% \ \text{CL}), \tag{25}$$
$$\xi_V = \ 0.0070^{+0.0045}_{-0.0069} \quad (68\,\% \ \text{CL}). \tag{26}$$

What are the main implications for single-field models of slow-roll inflation? As one can see from Fig. 2 (see the detailed analysis in [16]), some of the large field models, typical of the so called chaotic inflation [30] with a potential of the form

$$V(\phi) = \lambda \left(\frac{\phi}{\mu}\right)^n, \tag{27}$$

are (strongly) disfavoured by the data. This is the case of, e.g. $n = 4, 3$, while the simple quadratic potential starts to be under pressure. However, the cases with a fractional power of n corresponding to the original models of axion monodromy [31] are compatible with *Planck* data. At the other "extreme" we find the so called R^2-models of inflation, the first model of inflation to be proposed in [2], originally motivated by a modification of Einstein gravity due to quantum corrections. This model is based on the action

$$S = \int d^4x\sqrt{-g}\frac{M_{\rm pl}^2}{2}\left(R + \frac{R^2}{6M^2}\right). \tag{28}$$

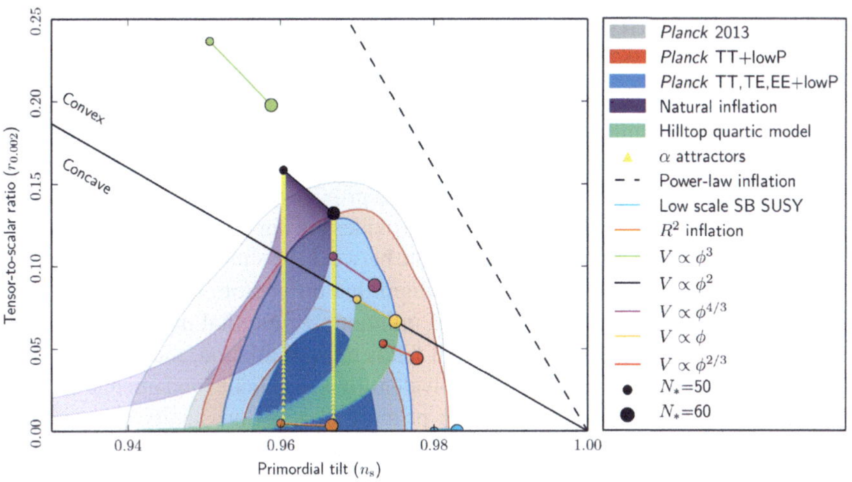

Fig. 2 68 and 95% CL marginalized joint regions for n_s and $r_{0.002}$ from *Planck* in combination with Baryon Acoustic Oscillation (BAO) data sets. The figure shows also the the theoretical predictions of some single-field slow-roll models of inflation. Figure taken from [16]

It is well known that this model can be equivalently described to a slow-roll model with a potential $V(\phi) = \Lambda^4 \left(1 - e^{-\sqrt{2/3}\phi/M_{\rm pl}}\right)^2$ in the Einstein frame. In particular, one of the prediction of this model is a very low level of primordial gravitational waves, with $r \simeq 12/N^2$, which therefore corresponds to $r \simeq 4 \times 10^{-3}$. The fact that this model is the preferred one w.r.t. other models (predicting a larger tensor-to-scalar ratio) definitely revived the interest in such a model and also spurred various theoretical investigations about its possible fundamental motivation. In particular, it was shown how a potential similar to (28) can arise within supergravity [32–34]. For example, the superconformal α-attractors described by the potential [35]

$$V(\phi) = \Lambda^4 \tanh^{2m}\left(\frac{\phi}{\sqrt{6\alpha}M_{\rm pl}}\right). \tag{29}$$

These models, via the two parameters α and m, are able to span a large region in the parameter space of the $(r - n_s)$ plane, from the large field models $V(\phi) \propto \phi^m$ to the R^2 models for $\alpha \gg 1$ or $\ll 1$ respectively. Finally, notice that models of natural inflation [36] with a potential

$$V(\phi) = \Lambda^4 \left[1 + \cos\left(\frac{\phi}{f}\right)\right], \tag{30}$$

or the so-called hilltop models of inflation [37] with $V(\phi) \approx \Lambda^4 \left(1 - \frac{\phi^p}{\mu^p} + \ldots\right)$, although still compatible with the data, are disfavoured w.r.t to the R^2 models [16].

Tests of Gaussianity

As we have pointed out in the list of section "Basic Predictions of Inflation", models of single-field slow-roll inflation predict *almost* Gaussian perturbations. The deviations from a pure Gaussian distribution are predicted to be tiny, of the order of the slow-roll parameters [38–41]. This fact, on one hand provides a clear test for the simplest models of inflation, and, on the other hand, provides the opportunity to probe specific observational signatures that can discriminate between the simplest models and many other realizations of the inflationary paradigm that differ from the simplest picture (e.g. single-feld vs. multi-field models of inflation). Primordial non-Gaussianity is a measure of the interactions of quantum fields during inflation, therefore learning about primordial non-Gaussianity is equivalent to learn about fundamental physics at extremely high energies. Looking for primordial non-Gaussianity means going beyond the power spectrum analyses and to look for 3- and 4-point correlation functions (called, respectively, the bispectrum and the trispectrum in Fourier space) of the primordial curvature perturbations. This allows to possibly exploit a wealth of information which is compressed in the power spectrum alone, and to discriminate among competing models of inflation which would remain otherwise degenerate on the basis of the power spectrum alone. To give the idea of the discriminatory power of primordial non-Gaussian signatures, let us stress that an observation of deviations from a Gaussian distribution above the theoretical threshold set by the simplest models of inflation would rule out *tout court all models* of single-field slow-roll inflation. These are just some of the reasons why searching for observational signatures of primordial non-Gaussianity has emerged in the last 10 years or so as one of the most crucial observable in data analysis related to Inflation models.[3] They also explain why this issue has driven so much attention from a theoretical point of view, leading to the proposal of new inflationary mechanisms to generate the primordial seeds for structure formation. In the following, we briefly summarize some of the main features about models of primordial non-Gaussianity and the current status of the observational constraints.

Models of Primordial Non-Gaussianity

One of the most powerful way to search for primordial non-Gaussian signature is to consider the primordial bispectrum

$$\langle \Phi(\mathbf{k}_1)\Phi(\mathbf{k}_2)\Phi(\mathbf{k}_3)\rangle = (2\pi)^3\delta^{(3)}(\mathbf{k}_1+\mathbf{k}_2+\mathbf{k}_3) f_{NL}\, F(k_1,k_2,k_3), \tag{31}$$

where, following some historical conventions, we have used the perturbation Φ which is related on super-horizon scales to the comoving curvature perturbation $\mathcal{R}$ by $\Phi = -(3/5)\mathcal{R}$ (and reduces to the Baardeen's gauge-invariant gravitational potential in the matter dominated epoch and on super-horizon scales). Here, the

[3] For some specific reviews on this topic see, e.g. [42–46].

so-called 'nonlinearity parameter' f_{NL} [39, 47–49], is a properly normalized amplitude that measures the non-Gaussianity level in the bispectrum. The so-called 'shape' (and scale dependence) of the bispectrum is instead encoded in the dependence of the function $F(k_1, k_2, k_3)$ on the type of triangle (i.e. the configuration) formed by the three wavevectors [49]. Different non-Gaussianity shapes are linked to distinctive physical mechanisms generating the primordial curvature (density) perturbations in the early Universe. The primordial bispectrum is simply related to the (angular) CMB bispectrum by *linear* physical processes, which are encoded in the CMB radiation transfer functions describing the post-inflationary physics and the formation of CMB anisotropies from the initial inflationary seeds. Therefore if primordial non-Gaussian signatures are generated during inflation these should be present also in the CMB anisotropies that we observe.

One of the most important theoretical achievement in this field has been the recognition that the violation of the minimal conditions which are at the base of the simplest models of inflation (discussed in the previous sections) can lead to different *classes* of inflationary models, each of which characterized by specific (and potentially measurable) non-Gaussian signatures. The conditions we are referring to are that (i) inflation is driven by a single scalar field which (ii) generates the primordial curvature perturbations, (iii) slowly rolling along a smooth (featureless) potential via a (iv) canonical kinetic term, (v) lying in an initial Bunch-Davies vacuum state. Each corresponding class leads to a distinctive shape for the bispectrum (and trispectrum), which reflects the underlying physical mechanism that generates the non-Gaussian signal. These different types of primordial non-Gaussianity are in fact generated by different types of interactions of the inflaton fields and/or non-linear evolution during (or immediately after) the inflationary phase. Let us summarize them in the following (for more details see [50, 51]).

- **Local non-Gaussianity** It is characterized by a signal that peaks for the so-called 'squeezed configurations', i.e. for triangles where $k_1 \ll k_2 \simeq k_3$ (or permutations). The bispectrum is given by

$$
\begin{aligned}
B_{\Phi}^{local}(k_1, k_2, k_3) &= 2f_{NL}^{local}\Big[P_{\Phi}(k_1)P_{\Phi}(k_2) + P_{\Phi}(k_1)P_{\Phi}(k_3) + P_{\Phi}(k_2)P_{\Phi}(k_3)\Big] \\
&= 2A^2 f_{NL}^{local}\left[\frac{1}{k_1^{4-n_s}k_2^{4-n_s}} + cycl.\right]. \qquad (32)
\end{aligned}
$$

where $P_{\Phi}(k) = A\,k^{4-n_s}$ is the primordial power spectrum. There is a simple reason why the bispectrum is maximal for squeezed configurations: in this case the primordial non-Gaussianity in the curvature perturbation is generated, in general, on super-horizon scales (on scales large than the Hubble radius). This makes the form of the non-Gaussianity (non-linearities) *local in real space* and therefore, in Fourier space, the bispectrum correlates large and small Fourier perturbation modes. For example a typical form of the primordial gravitational potential leading to a bispectrum like the one in Eq. (32) is the local form that has been used as a first parameterization in the first studies of primordial non-Gaussianity (see, e.g. [38, 39, 47, 48, 52])

$$\Phi(\mathbf{x}) = \Phi_L(\mathbf{x}) + f_{NL}^{local}(\Phi_L^2(\mathbf{x}) - \langle\Phi_L^2(\mathbf{x})\rangle), \quad (33)$$

where $\Phi_L(\mathbf{x})$ is the linear Gaussian gravitational potential. The bispectrum (32) typically arises in the context of multi-field models of inflation, where light scalar fields (which can be different from the inflaton) play an important role in the generation of the curvature perturbation (and in some cases also in driving the accelerated expansion as well). In this scenario typically the non-Gaussian signal is generated by two sources of non-linearities: the development of the field fluctuations and the transfer of these fluctuations into the (inflaton) adiabatic curvature perturbation sector [53]. The main contribution from these non-linearities come from the dynamics on superhorizon scales, leading therefore to a local bispectrum of the form (32). This is the case, e.g. of the curvaton models [54–58], where the (isocurvature) fluctuations of a scalar field different from the inflaton generates the adiabatic curvature perturbations when it decays after inflation, or to the case of multiple inflation, where inflation is driven by more that a single-scalar field [59]. Notice, in particular, that the curvaton scenario naturally predicts a lower bound of order unity for the non-linearity parameter [60], so that a local non-Gaussianity smaller than this value would rule these models out.

- **Equilateral and orthogonal non-Gaussianity**
'Equilateral' non-Gaussianity arises when the non-Gaussian signal correlates fluctuation modes that are of comparable wavelengths. Therefore, the bispectrum peaks for equilateral configurations, triangles in Fourier space with $k_1 \simeq k_2 \simeq k_3$. The so-called equilateral template usually adopted for data analysis is given by

$$B_{\Phi}^{equil}(k_1, k_2, k_3) = 6A^2 f_{NL}^{equil} \left\{ -\frac{1}{k_1^{4-n_s} k_2^{4-n_s}} - \frac{1}{k_2^{4-n_s} k_3^{4-n_s}} - \frac{1}{k_3^{4-n_s} k_1^{4-n_s}} \right.$$
$$\left. - \frac{2}{(k_1 k_2 k_3)^{2(4-n_s)/3}} + \left[\frac{1}{k_1^{(4-n_s)/3} k_2^{2(4-n_s)/3} k_3^{4-n_s}} + (5 \text{ permutations}) \right] \right\},$$

Such a kind of bispectrum is a typical prediction of the so-called 'general single-field' models of inflation [61]. Here, the inflaton is a single scalar field but it is characterized by derivative interactions, e.g. of the type $(\dot{\delta\phi})^3$ and $\dot{\delta\phi}(\nabla\delta\phi)^2$. Therefore such interactions are suppressed on super-horizon scales, being the gradients negligible in this regime, and the correlations between three Fourier modes is maximal when all of them cross the Hubble radius approximately at the same time, which corresponds to an equilateral configuration. Typical models of inflation of this kind are inflation models where the inflaton has a non-canonical kinetic term (a Lagrangian which is a generic function of the inflaton field and its kinetic term $X = g^{\mu\nu}\partial_\mu\phi\partial_\nu\phi$, $\mathcal{L}(\mathcal{X}, \phi)$). For example, the DBI action [62, 63] proposed in the string theory context is of this form. A characteristic of such class

of models is that the inflaton fluctuations can propagate with a sound speed $c_s < 1$, and this is directly related to a potential high signal from this kind of bispectra, since it can raise the amplitude of the bispectrum as $f_{NL} \simeq c_s^{-2}$.
Here we do not go into the details of the so-called 'orthogonal' non-Gaussianity. We just mention that it is an independent form (template) of the bispectrum that turns out to be 'orthogonal' to the equilateral bispectrum, fully describing the parameter space of a theory like, e.g. $\mathcal{L}(\mathcal{X}, \phi)$, characterized by the two derivative interactions mentioned above. As it has been shown by exploiting the methods of an effective field theory approach to inflationary perturbations [64], in general, the equilateral and the orthogonal templates are able to fully describe the parameter space of non-Gaussianity for very general models of single-field inflation [65]. Examples include k-inflation models [66] (where inflation is driven by the kinetic term of the inflaton field), models characterized by more general higher derivative interactions of the inflaton field, such as ghost inflation [67], and more generally effective field theories of inflation [64], or Galileon inflation models [51, 68].

- **Folded non-Gaussianity** This kind of non-Gaussianity is characterized by a correlation among the perturbation modes that is maximal for the so-called 'folded' (or 'flattened') triangles in Fourier space, that is to say for triangles with $k_1 + k_2 \simeq k_3$. Typically these shapes are a clear signature of a deviation of the initial quantum state of the inflaton field from the standard Bunch-Davies vacuum state (defined as that state at asymptotically early times and short distances in which the inflaton fluctuations reduce to the one of a free scalar field in flat space) [44, 69]. This kind of non-Gaussianity can also arise in single-field models of inflation with higher derivative interactions which are related to Galileon models of inflation [68, 70, 71].

The search for primordial non-Gaussianity performed during the last years by the *Planck* team is by far more extensive than what briefly summarized above and it is not limited only to the above types and models of primordial non-Gaussianity. The ones we have briefly discussed are what are now considered as the standard shapes of primordial non-Gaussianity. A detailed analysis has been performed [51], in search for other kinds of primordial bispectra which carry the peculiar imprints of quite distinctive deviations from the models of single-field slow-roll inflation. We are referring, in particular, to bispectra which are strongly scale-dependent, arising, e.g. in the context of inflationary models where there are sharp features, e.g. in the inflaton potential (see, e.g. [47, 72]), or in models where the inflaton potential has periodic modulations (the so-called resonant non-Gaussianity) (see, e.g. [73]), which, e.g. can arise from the small breaking of a shift-symmetry in a a string theory context with axions (the axion monodromy inflation [74]). In these cases various oscillatory features both in the power spectrum and bispectrum can arise, with definite predictions about their correlations. Other interesting non-Gaussian signatures can arise in inflation models where primordial vector fields play a non-negligible role, leading to statistical anisotropic and direction dependent bispectra [75, 76]. Let us finally mention the fact that with the latest *Planck* data analysis, bispectra characterized by parity violation generated in the tensor (primordial gravitational waves) (see, e.g. [77–80]) have started to be analyzed [51].

Going to Higher Order: Trispectra from Inflation

The trispectrum (the Fourier counterpart of the 4-point correlation function) can be characterized by two amplitudes: g_{NL} and τ_{NL}, which, in terms of field interactions, are related to a contact and a scalar-exchange interaction term, respectively. Standard single-field models of inflation predict trispectrum amplitudes which is very low, of the order of the slow-roll parameters. The simplest example is that of the 'local' trispectrum

$$\begin{aligned}\langle \Phi(\mathbf{k}_1)\Phi(\mathbf{k}_2)\Phi(\mathbf{k}_3)\Phi(\mathbf{k}_4)\rangle &= (2\pi)^3\delta^{(3)}(\mathbf{k}_1+\mathbf{k}_2+\mathbf{k}_3+\mathbf{k}_4)\\ &\times \left\{\frac{25}{9}\tau_{NL}\left[P_\Phi(k_1)P_\Phi(k_2)P_\Phi(k_{13})+(11\,perm.)\right]\right.\\ &\left.+\,6g_{NL}\left[P_\Phi(k_1)P_\Phi(k_2)P_\Phi(k_3)+(3\,perm.)\right]\right\}, \qquad (34)\end{aligned}$$

where $k_{ij} \equiv |\mathbf{k}_i + \mathbf{k}_j|$. Along with the bispectrum, intense analyses of non-Gaussian signatures in terms of the trispectrum have recently started using CMB data [51]. These analyses consider for the first time also shapes of the trispectra beyond the g_{NL} and τ_{NL} model of Eq. (34), as those arising from derivative interactions as $\delta\dot{\sigma}^4$ and $(\nabla\sigma)^4$ [81–83]. The theoretical and observational motivations of such an effort are manifold. Improvements in the trispectrum constraints can be crucial, contributing to further discriminate among different models of inflation. For example, the trispectra (34), peaking in 'squeezed' configurations (with a similar meaning as the local bispectrum) can arise only within multi-field models of inflation. Also, there are inflationary models that predict the trispectrum as the first relevant higher order correlator, having a signal-to-noise ratio higher than the bispectrum. Examples include specific curvaton models [84, 85] and some multiple field models of inflation [59, 86]; and there exist technically natural models where the bispectrum remains naturally suppressed with respect to a larger signal-to-ratio in the trispectrum [81, 87, 88] (see also [51] and Refs. therein). For these cases a detection of a trispectrum along with a null detection of the bispectrum would be a clear smoking gun. To give further the idea of the discriminatory power of an observational analysis of inflationary trispectra let us mention that if there is a non-vanishing squeezed bispectrum there must be indeed a trispectrum with a τ_{NL} amplitude that should obey to the constraint $\tau_{NL} \geq (6f_{NL}^{local}/5)^2$. Therefore if, from an observational point of view we would be able to falsify such a relation (the so called Suyama-Yamaguchi relation [89]) we would be in a position to rule out also models inflation that involve more than a single scalar field.

Present Constraints on Primordial Non-Gaussianity

The most stringent constraints on primordial non-Gaussianity come from the analysis of the CMB data of the ESA *Planck* satellite. Some of the main results are summarized in Table 1, where the actual constraints on the standard scale-invariant shapes of primordial non-Gaussianity are reported. We refer the reader to the corresponding *Planck* paper [51], in particular, for the details of the data analysis procedure. Let us just mention here that the results have been obtained via three classes of estimators (the so-called separable template-fitting (KSW), the binned estimator and the modal estimator). All these estimators have been cross-validated using Gaussian and non-Gaussian simulations. The results have been checked to be stable against using different component separation techniques and pass a series of validation tests, including the impact of residual foregrounds. The third column in Table 1 shows the constraints obtained after the subtraction of the so called ISW-lensing effect, which is the main secondary (not primordial in origin) contaminant to the primordial signals. This contamination is due to a cross-correlation between the Integrated-Sachs-Wolfe effect, redshifting the CMB photons as they pass through late time potential wells, and the lensing effect affecting the same photons at low redshift. Such a cross-correlation couples short with large wavelengths perturbation modes, thus being mainly a contaminant for the local type bispectrum. Some short comments are worth to be made here. First, such constraints are, in terms of CMB temperature data, the best constraints one can ever get. Essentially, they are extremely close to the ones that one could obtain from a cosmic variance limited experiment. Let us stress that having a value of f_{NL} of order unity means the following: we are testing deviations from a pure Gaussian distribution of the order of one part to 10^5 (this is clear if one refers, e.g. to Eq. (33). This can be well considered as one of the most precise measurements in

Table 1 2015 *Planck* constraints (68% CL) on the scale-invariant f_{NL} parameters of the primordial local, equilateral and orthogonal shapes. On the second column independent single-shape results are shown. On the third column results with the ISW-lensing bias subtracted are reported (and the equilateral and orthogonal shapes are analyzed jointly since they span the typical bispectrum parameter space of general single-field models of inflation). All the results are obtained via the KSW estimator from the SMICA foreground-cleaned map (see [51] for details). Both temperature (T) and preliminary constraints using temperature and polarization data (T + E) are shown. Table adapted from [51]

Shape (T)	f_{NL} (independent)	f_{NL} (ISW-lensing subtracted)
Local	10 ± 5.7	2.5 ± 5.7
Equilateral	-13 ± 70	-16 ± 70
Orthogonal	-56 ± 33	-34 ± 33
Shape (T + E)		
Local	6.5 ± 5.0	0.8 ± 5.0
Equilateral	3 ± 43	-4 ± 43
Orthogonal	-36 ± 21	-26 ± 21

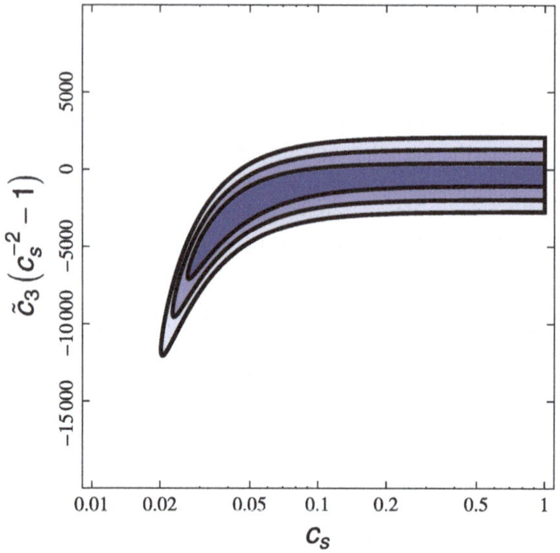

Fig. 3 Bispectrum constraints on a typical general single-field model of inflation (characterized by a non-canonical kinetic term) using an effective field theory approach to inflationary perturbations. The bispectrum parameter space corresponds to equilateral and orthogonal bispectra, see, e.g. Eq. (34). Their amplitudes depend on the inflaton sound speed c_s and a second parameter $\tilde{c}_3$. The figure shows the 68, 95 and 99.7% confidence regions in the single-field inflation parameter space $(c_s, \tilde{c}_3(c_s^{-2} - 1))$ obtained using the constraints in Table 1. Figure taken from [16]

cosmology. As such it can be also considered as one of the most stringent tests of the standard models of single-field slow-roll inflation. In fact all the analysis performed in [51] are consistent with a Gaussian nature of primordial fluctuations. On the other hand such strong limits allow to dramatically reduce in various cases the parameter space of various inflationary models that go beyond the simplest models. An example is shown in Fig. 3. This figure shows, in terms of an effective field theory approach to the inflaton fluctuations, the constraints on the amplitudes of the two interactions terms $(\dot{\delta\phi})^3$ and $\dot{\delta\phi}(\nabla\delta\phi)^2$, that usually characterize general single field models of inflation with a non-canonical kinetic term. In particular this plot answers to a simple and quite natural question: what is the sound speed at which fluctuations propagate in the very early universe? In fact one obtain the limit

$$c_s \geq 0.02 \qquad 95\% \text{ CL} \tag{35}$$

Of course this is just an example and many more analyses on the implications of the non-Gaussianity constraints have been performed and thoroughly described in [51]. As another example, let us mention, e.g. the fact that the *Planck* results on the bispectrum has been able to rule out some models (the so called k-inflation) where inflation is driven by the kinetic term of the inflaton field [50].

Table 2 *Planck* constraints on different primordial trispectra (see [50, 51] for details). If not specified error bars are 68% CL.

Shape (T)	Trispectrum amplitude
Local	$\tau_{NL} < 2800$ (95% CL)
Local	$g_{NL} = (-9.0 \pm 7.7) \times 10^4$
$\dot{\sigma}^4$	$g_{NL} = (-0.2 \pm 1.7) \times 10^6$
$(\partial\sigma)^4$	$g_{NL} = (-0.1 \pm 3.8) \times 10^5$

Table 2 summarizes the most recent constraints on some inflationary trispectra. Along with the limit on g_{NL} and τ_{NL} for a local model of the type described in Eq. (34), other new trispectra shapes have been considered in the last *Planck* data analysis. In particular an interaction term of the type $(\partial\sigma)^4$ is distinctive signature of multifield models of inflation (in the sense of an interaction that is technically natural). These new trispectra has been considered with some minimal requirements on the inflationary Lagrangian in the context of the effective field theory approach to inflationary perturbations, by focusing on quartic interaction terms in the inflationary action with the lowest possible number of derivatives satisfying the inflationary symmetries [83]. Figure 4 shows the constraints in the ($g_{NL}^{(\partial\sigma)^4}$-$g_{NL}^{(\dot{\sigma})^4}$) in a multifield model of inflation, where σ is a spectator field whose quantum fluctuations are converted to adiabatic curvature perturbations. As an example of the efficiency and versatility of the statistical estimators employed in the *Planck* data analysis, let us stress that, in addition to constraining specific bispectra and measuring the f_{NL} values for specific shapes, the different *Planck* pipelines can also obtain a smoothed reconstruction of the full CMB angular bispectrum. Figure 5 shows in fact the CMB bispectrum as 'seen' by the *Planck* satellite. In particular this figure shows the reconstruction of the full three-dimensional CMB bispectrum, using the *Planck* 2015 temperature data, obtained via the so called modal reconstruction technique (see, e.g. [90, 91]).

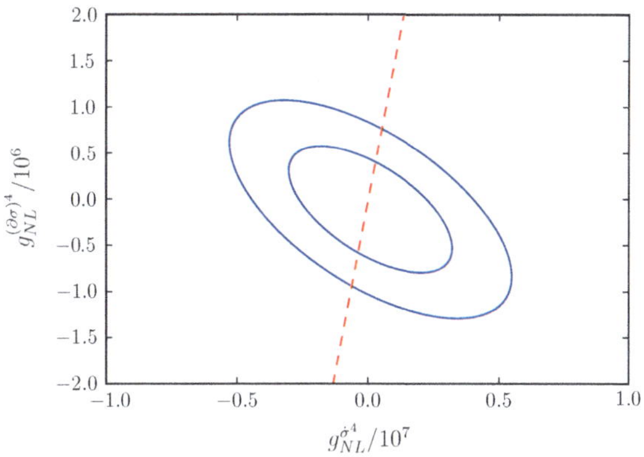

Fig. 4 68 and 95% confidence regions in the ($g_{NL}^{(\partial\sigma)^4}$-$g_{NL}^{(\dot{\sigma})^4}$) plane obtained from the *Planck* trispectrum constraints. Figure taken from [16]

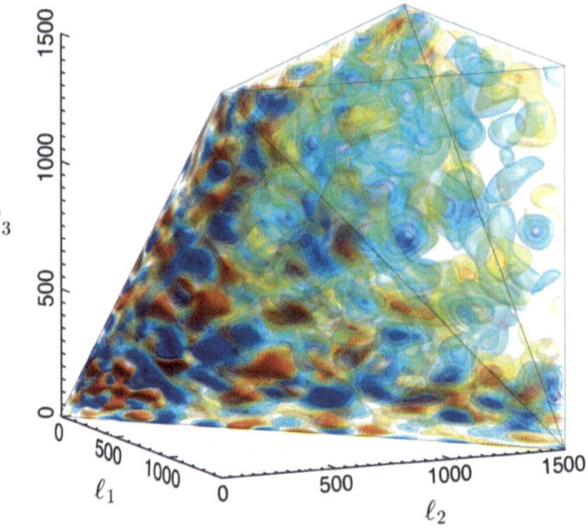

Fig. 5 Full three-dimensional modal bispectrum reconstruction for *Planck* 2015 temperature data. Notice the oscillations in the squeezed limit, e.g. for the configurations with $\ell_3 \ll \ell_2 \sim \ell_3$, that correspond to the ISW-lensing signal. Figure taken from [16]

A Free Gaze into the Future of Inflation: Open Issues, Forthcoming and Future Experiments

It is definitely fair to say that the inflationary scenario in the last years, the years of *precision* Cosmology, has gathered a lot of observational successes. This has made inflation an essential building block of the standard cosmological model (the ΛCDM model + initial conditions provided by inflation). However, many interesting issues remain open.

Despite the simplicity (and appealing features) of the inflationary scenario, and despite the many observational successes, the *precise* inflationary model has still to be clearly identified from an observational point of view. In fact there are still various pathways to strongly improve our knowledge of the fundamental mechanisms underlying inflation and of the finest details of the inflationary dynamics. Related to this point, from the theoretical effort point of view, a complete satisfactory theory of inflation, in the sense of an UV complete theory, is still missing, even if various proposals in that direction continue to be developed (see, e.g. [62, 92–97]). It is clear, from what we said before, that the measurements of the B-mode polarization pattern in the CMB, or the ones related to primordial non-Gaussianity, play a crucial role from this point of view. On one hand the primordial B-modes of CMB polarization can set, quite unambigously, the energy scale at which inflation takes place, thus establishing for the first time a direct link to various particle physics contexts within which one should search for a *theory* of inflation. On the other hand both the level of B-mode in the CMB polarization, and signatures of primordial non-Gaussianity, can provide information about possible symmetries underlying the quantum field theory structure of inflation (and about gravity) given the sensitivity of inflation to High Energy Physics (for more details see the arguments in [22]). Moreover primordial non-Gaussianity can act as a very efficient observable to select (or rule out) among different models of inflation which go beyond the standard-single field models of slow-roll inflation. For example, it could be very efficient in clearly discriminating between single-field models of inflation and multi-field models of inflation. Other important questions to keep in mind are the following: are there clear and well motivated theoretical thresholds which help to identify a clear goal for future experiments aiming at pinning down the correct model of inflation? What are the observables and the forthcoming and next generation measurements that could allow to make a further breakthrough in our knowledge of inflation? Finally, it is fair to say, from a more general point of view, that inflation has been successful over the years also because Early Universe models alternative to inflation have been ruled out over the years (such as, e.g. cosmic strings or textures as a mechanism for structure formation) or because sound and well understood alternatives are lacking. In any case it is interesting to consider such alternatives and to test them observationally to further strengthen the assessment inflation. Let us consider (some of) the above points in more detail. We will also consider the prospects for forthcoming and next generation experiments to test inflationary models, keeping in mind that the measurements of the *Planck* satellite have essentially extracted all the information about the primordial fluctuations as far as CMB temperature anisotropies are concerned.

- **Spectral index**
  As we mentioned above, there is now evidence of a departure from exact scale invariance $n_s = 1$ (the so called Harrison-Zeldovich spectrum) at over 5σ (see Eq. (20)). Also this result can be considered as a strong indication that inflation took place, since it reflects that a dynamical process must have taken place in the early universe, leading to a small deviation from exact scale invariance (possibly with a small breaking of symmetries protecting n_s to deviate from 1). In fact almost all the models of slow-roll inflation predict small deviations from scale invariance, of the order of the slow-roll parameters (and for various models of the order of $1/N$ where $N \simeq 50$–70 is the number of e-folds to the end of inflation). For this observable we could say that the well motivated threshold of $|n_s - 1| \sim 10^{-2}$ has been successfully reached by the present observations. However, it is important to further reduce the error bars on such a parameter, which would allow to further discriminate between different inflationary models still allowed by the data).There is plenty of room to further improve our constraints on n_s exploiting CMB polarization. For example, including E-mode polarization data the *Planck* satellite should be able to improve by approximately 30%. On the other hand future space missions should be able achieve a large improvement, by combining temperature and polarization data (considering TT, TE and BB power spectra). For example, with a proposed satellite mission such as CORE+ (see, e.g. [98]) we could expect an improvement of a factor of three w.r.t. to *Planck* [99], while a satellite like *PRISM* is expected to yield a factor of five improvement w.r.t. *Planck* [100].
- **Running of the spectral index**
  At the moment there is no evidence for a running of the spectral index (see Eq. (21)). A running of the spectral index is a generic prediction of inflation models. In the

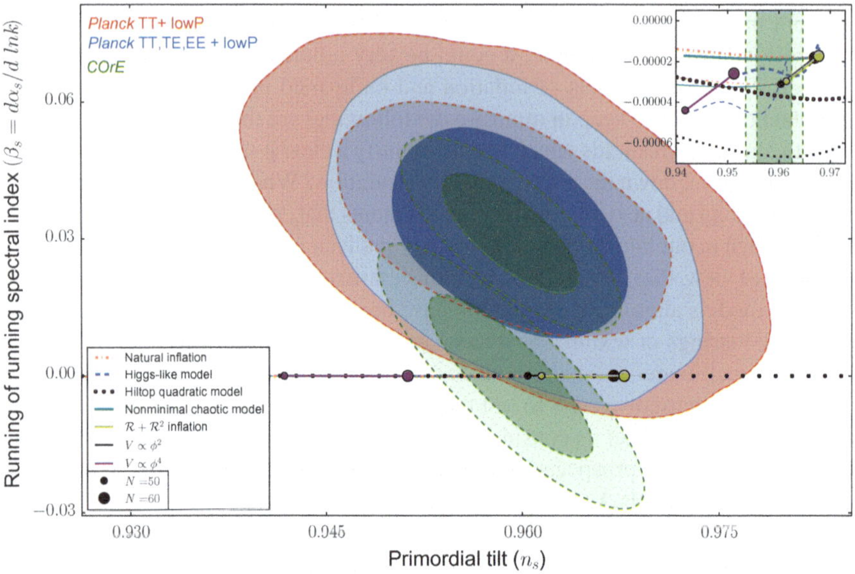

Fig. 6 68 and 95% predicted contours from the CMB mission COrE (considering a 1% foreground removal in power). Also shown are the 68 and 95% allowed contours from *Planck* data. Figure taken from [103]

case of single-field models of slow-roll inflation it turns out to be second-order in the slow-roll parameters (and it would depend also on (minus) the third derivative of the inflaton potential which is expected to be the dominant term, given the present constraints in the slow-roll parameters ϵ_V and η_V). A large negative running of the spectral index (larger than 10^{-2}) would in general be difficult to reconcile with the standard single-field models of slow-roll inflation (see, e.g. [101]), even if this is not impossible (see e.g [102] which considers models with periodic modulations in the inflaton potential, that can arise in the context of axion monodromy inflation). Just to give an example of how well it would be possible to improve in this respect, an experiment like *CORE+* or *PRSIM* could reach an improvement of a factor of three [99, 100] (see Fig. 8). Relevant improvements on the running of the running of the spectral index can also be achieved [103] (see Fig. 6).

- **Tensor-to-scalar ratio *r* and amplitude of primordial gravitational waves from CMB B-mode polarization**
  Given the huge consequences that the detection of primordial gravitational waves from inflation would have on Cosmology (and Fundamental Physics in general), it is worth to ask whether there is a motivated theoretical threshold that should be considered as a clear goal for future experiments. At the moment we are probing values of r of the order of 0.1. As argued, e.g. in Ref. [104] a value of $r \sim 2 \times 10^{-3}$ could be considered as a reasonable reference value to consider for future missions in the case we do not detect primordial gravitational waves with larger amplitude.

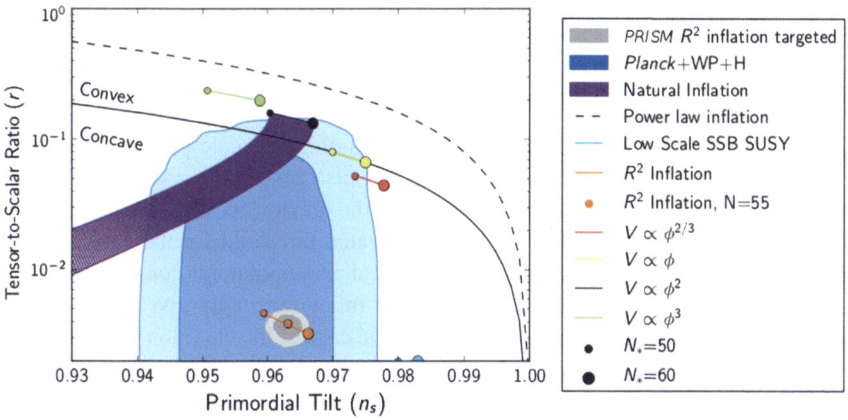

Fig. 7 Constraints on single-field models of slow-roll inflation from *Planck* 2013 temperature data and the forecasted constraints from *PRISM* for a fiducial value of $r = 5 \times 10^{-2}$ (and not assuming de-lensing). Figure taken from [100] (and adapted from [154])

For example, this value corresponds to the benchmark between small-field and large-field models of inflation (characterized, respectively, by an excursion of the inflation field which is sub/super planckian). Moreover, it also approximately corresponds to the value expected in the R^2 models of inflation, which at the moment shows the best compatibility with the data (see Fig. 2). This is actually the sensitivity that future satellite missions specifically designed to measure the CMB B-mode polarizations aim to reach (for example, next generation satellites such as COrE [105], *PRISM* [100], LiteBIRDE [106] and PIXIE [107]). As an example we show what a satellite like *PRISM* could achieve (see Fig. 7) in the $(r - n_s)$ plane. Moreover let us keep in mind that almost all of the Early Universe models proposed as alternatives to inflation predict a very small (undetectable) tensor-to-scalar ratio (this is the case, e.g. of the ekpyrotic models (see, e.g. [108]). Therefore detecting the tensor fluctuations generated by inflation via the B-modes polarization pattern in the CMB would be a clear smoking gun of inflation.

- **Primordial non-Gaussianity**
  The inclusion of the E-mode polarization of the CMB can allow the *Planck* satellite measurements to improve the constraints by 20% up to 45% depending on the kind of non-Gaussianity that is tested (a preliminary analysis including the full mission polarization data as already been performed in the latest *Planck* data release, see [51]). Proposed, future missions like, e.g. *PRISM* [100], to be considered as cosmic variance limited in (E-mode) polarization up to $l_{max} \simeq 3000$ could improve by a factor of almost three our constraints on primordial non-Gaussianity. It is worth stressing that the use of polarization data can be particularly efficient in reduce systematic effects. Moreover it has been demonstrated that CMB measurements that are cosmic variance limited in polarization could reach a sensitivity to primordial non-Gaussianity via the bispectrum channel $\langle EEE \rangle$ that is

larger than the sensitivity that could be reached separately through CMB temperature data alone [100, 105, 109]. This shows the important cross-check and gain of information that can be achieved in maximally exploiting CMB polarization data. The complementarity is a direct consequence of different distinctive signatures of CMB polarization as a function of the cosmological scale probed w.r.t to the CMB temperature anisotropries, and it holds also for the analysis of the power spectra parameters that we have discussed in the previous points (for a discussion see, e.g. [16]). A figure of merit for a motivated threshold for the non-linearity parameter f_{NL} would be $f_{NL} \sim 1$ (both for local and equilateral non-Gaussianity). In fact, it turns out that multifield models of inflation usually give rise to such a level of primordial non-Gaussianity (as for the case of the curvaton model, which has a lower bound on f_{NL} of order unity [110]). For general single-field models of inflation (characterized by higher derivative interactions) such a threshold would correspond to an inflaton that is very weakly interacting, and whose derivative interactions do not interfere with the standard slow-roll dynamics (for a discussion see, e.g. [111]). Of course this figure of merit must be taken with care since there might be models that evade this conditions. It is clear that such a threshold is right at reach of the present CMB constraints for the forecasted sensitivity for future CMB experiments. Notice also that reaching such a threshold could allow to explore in a robust way all those contributions to CMB non-Gaussianity that are due to the non-linear evolution of cosmological perturbations (at second-order in the fluctuations) after inflation, thus providing a new test of (weak) non-linearities in General Relativity (see, e.g. [112–127]).

Of course one would like to push down the sensitivity further in order to reach the level of primordial non-Gaussianity predicted by the standard single-field models of sow-roll inflation which is $f_{NL} \sim \mathcal{O}(\epsilon, \eta) \simeq 10^{-2}$ [40, 128], since there are still two (or more) orders of magnitude between the $f_{NL} \simeq 1$ region and the prediction of standard inflation. As we mentioned in section "Neutrinos in Cosmology" the first robust measurements of the CMB trispectrum of different shapes have been performed in [51]. From this point of view there is large room for improvements, both in reducing the errors bars and in terms of the types of primordial trispectra that can be analyzed (corresponding to different classes of inflationary models).

- **Future/futuristic experiments and new observables to test inflation**
There is an increasing interest towards new directions that in the future could open up new windows into the study of the physics of inflation. This has spurred in the most recent years a vivid discussion about new future (or futuristic) experiments and observables that can be exploited to reduce the errors bars on various inflationary parameters, especially on (but not limited to) different shapes on primordial non-Gaussianity. It is in fact quite clear that, apart from the reasons given above, this is a particularly pressing question as far as non-Gaussianity shapes different from the local one are concerned. Let us give here a couple of examples, mentioning some potential (huge) improvement that could be reached in principle. We focus on CMB spectral distortions and 21 cm radiation. Our examples are not exhaustive, and must be considered as a proof of principle, because they are based on somewhat ideal forecasts for the observables discussed, which neglect

some crucial observational aspects such as the impact of foregrounds and possible systematic effects. Nevertheless, the breakthrough that these observations could in principle reach clearly make it worth studying them in more details.

- CMB spectral distortions
  The frequency CMB spectrum is extremely close to a perfect blackbody, with deviations (the so-called 'CMB spectral distortions') that are constrained at present to be $\Delta I_\nu / I_\nu \leq 10^{-4}$, as constrained by the COBE/FIRAS instrument [129, 130]. However within the current ΛCDM model we *do* expect tiny distortions to be present. The reason is that primordial density perturbations at very small scales are expected to be damped by various mechanisms (such as, e.g. the Silk-damping process), thus injecting some energy into the baryon-photon plasma (including processes at very high redshifts) (see, e.g. [131, 132] and references therein). It is this injection that produces the spectral distortions. The effect depends on the primordial power spectrum predicted by inflation. Therefore such an observable could probe the inflationary fluctuations on very small scales, from *kpc* to *Mpc* scales, corresponding to phases of the inflationary evolution which are completely unexplored at present. This has stimulated the discussion about the results that can be obtained with present and future proposed satellites (such as PIXIE [107] or *PRISM* [100]). For example, spectral distortions can be very sensitive to the running of the spectral index [133] exactly because of the long leverage arm that they can reach in terms of scales probed. For example, recently it has been argued [134] that an experiment with a sensitivity 3 times better than the proposed satellite PIXIE could guarantee a clear signal: either the measurement of the (monopole) of the μ-type (chemical potential) spectral distortions (which is expect to be of 1.9×10^{-8}) or the exclusion of a positive running spectral index. An experiment like *PRISM*, with a sensitivity 10 times better than *PIXIE* should reach an error of about 0.002 on the running of the spectral index (see Fig. 8).

  Another interesting point, related to CMB spectral distortions and the primordial universe, that could realize a breakthrough in the search for primordial non-Gaussianity, is the $T\mu$ cross-correlation between CMB temperature anisotropies and chemical CMB spectral distortions, as originally proposed by [135]. Such a cross-correlation is particularly sensitive to primordial non-Gaussianities that correlate long and short wavelengths of the perturbation modes, in primis local non-Gaussianity and other primordial non-Gaussianity peaking in the squeezed configurations. In this case a cosmic variance limited experiment can reach values of $f_{NL}^{loc} \ll 1$ and at the level of those predicted by the standard single-field models of slow-roll inflation. Moreover the auto-power spectrum $\mu\mu$ could be sensitive to very low values of the τ_{NL} amplitude of the trispectrum [135]. The same observable can be also very sensitive to other models or scenarios, like, e.g. running of primordial non-Gaussianities or signature of violation of statistical homogeneity and isotropy [136–138]. Recently, it has been pointed out [139] that also the bispectrum $TT\mu$ can be another new observable able to measure values of the primordial local trispectrum g_{NL} of the order of unity.

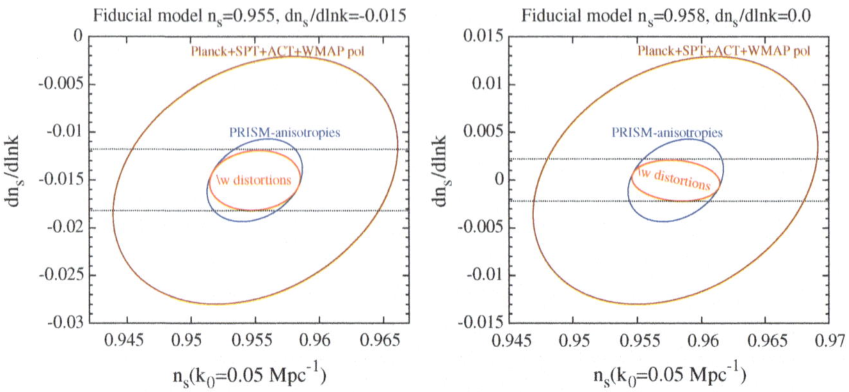

Fig. 8 Fisher forecasts for constraints on the primordial power spectrum for PRSIM [100]. On the *left panel* a power spectrum with running is considered. On the *right panel* a power spectrum without running is analyzed. A combination CMB anisotropies and CMB spectral distortions will allow to reach an error bar ±0.002 for the running, significantly improving the constraints from current CMB experiments. Figure taken from [100]. See also [133]

This also would represent a huge improvement (of 3–4 orders of magnitude) w.r.t to present CMB constraints.

– Large Scale Structure and 21 cm observations
It is well known that forthcoming or future large galaxy surveys (such as EUCLID, LSST or SKA) offer another opportunity to test some of the main inflationary parameters. To give an example, let us focus on the improvement that could be reached on the constraints on primordial non-Gaussianity. Two of the most investigated observables in this case are the (small scale) bispectrum of galaxies and, much more recently, the so-called large-scale 'scale-dependent' halo bias of dark matter halos [140, 141]. In the first case, for local primordial non-Gaussianities, a value of f_{NL}^{loc} as low as a few could be reached (see for details [142]). Scale-dependent halo bias consists in a scale-dependent feature induced in the bias between galaxies and the underlying dark matter fluctuations induced by the presence of large-scale fluctuations that modulate the small density fluctuations of dark matter. Such a modulation arises if the underlying primordial curvature perturbation field is non-Gaussian and couples long with short wavelengths, as it happens in the local primordial non-Gaussianity scenario. The modulation then implies that dark matter density perturbations overcome the threshold for the formation of dark mater halos (within which galaxies form) in a scale-dependent pattern. Therefore scale-dependent halo bias is particularly sensitive to local (or other squeezed) forms of primordial non-Gaussianities. The sensitivity to f_{NL}^{loc} could be three times better than those of the galaxy bispectrum (see, e.g. [143] and discussions in [111, 144]), the

advantage being that, since the effect is enhanced on very large scales, such an observable is not so influenced by the non-linear evolution of gravity as the small scale galaxy bispectrum is. On the other hand the bispectrum of galaxies can in principle test also shapes of primordial non-Gaussianity that are different from the local one, a possibility that in any case has been discussed very recently also in the case of scale-dependent halo bias (see, e.g. [145–147]). To give some example, let us mention that a survey like Euclid should reach a sensitivity to $f_{NL}^{loc} \sim 2$. Therefore with future surveys value of f_{NL}^{loc} of order unity could be reached. Such an observational threshold is very important, since it could lead to a further strong indication in favour of single-field models versus multifield models of inflation. It is definitely fair to say that, especially systematic (instrumental and observational) effects should be kept under exquisite control in order to be able to realize such an improvement w.r.t. to CMB measurements. The LSS survey that we have mentioned are spectroscopy and/or photometric surveys. Recently the exploitation of large radio surveys (such as SKA or EMU-ASKAP) has been lively discussed (see, e.g. [148]). An advantage in this case would be a large volume at disposal from the radio surveys. For example, as discussed in [147], values of $f_{NL}^{loc} \sim 10$ can be reached for local primordial non-Gaussianity (see also the points made in [149]), and interesting improvements on different primordial non-Gaussianity shapes can be reached as well (as, e.g. for primordial non-Gaussianity with anisotropic signatures motivated by early universe models where gauge fields play a relevant role during inflation).

Another futuristic observational way that could open up (in a similar way to CMB spectral distortions) the way to a true breakthrough into our understanding of inflation relies on the exploitation of the brightness temperature fluctuations in the 21 cm line of neutral hydrogen. In this case (similarly to the CMB spectral distortions) the number of modes probed is 2–3 orders of magnitude larger than those probed by CMB anisotropies and a very wide window of high redshifts are involved ($z \sim 10$–100). To give an example, still related to primordial non-Gaussianity, various works have shown that in principle (with a cosmic variance dominated experiment) one could reach sensitivities of order $f_{NL}^{loc} \sim 0.03$, for local non-Gaussianity, and, e.g. $f_{NL}^{equil} \sim 0.04$ [150–152]).

Finally let us mention a very recent proposal to exploit the anisotropies of the Cosmic Infrared Background (CIB) as a powerful observable to (local) primordial non-Gaussianity [153]. In this case primordial non-Gaussianity leaves an imprint in the angular power spectrum of the CIB in the form of a scale-dependent halo bias. Given the large volume covered by the CIB, such a signature could potentially allow to measure a primordial non-Gaussian signal with $|f_{NL}^{loc}| < 1$ in future CMB satellite missions.

Some Aspects of Modified Gravity in Cosmology

In this section we will briefly mention some issues related to scenarios relevant for Cosmology where modifications of gravity (w.r.t. to Einstein General Relativity theory) might play a significant (or even a crucial) role. We will not be exhaustive as far as the topic of possible modifications of gravity are concerned, also because other aspects of the same argument are treated in more details in Chap. 5 of this book. In the following, we want just to briefly emphasize some topics of particular interest that have stimulated some recent activity in this field.

As far the the inflationary scenario is concerned, let us remember that if the density (scalar curvature) perturbations do arise from inflation, then the typical (upper bound to the) energy scale $E_{inf} = V^{1/4}$ is given by $E_{inf} \sim \sqrt{\delta} M_{pl}$, where δ is the amplitude of the density perturbations (see Eq. (18)). Since we know from observations that $\delta \sim 10^{-5}$, this means that the energy scale of inflation might be extremely high, very close to the one of Grand Unification, and not too far away from energy scales where quantum gravity effects might be relevant or leave some significant effective imprints. Therefore, although the majority of the inflationary models are based on Einstein gravity, it might be that Einstein gravity is not the correct description at such high energies. In fact, the very first model of inflation to be proposed in [2], Eq. (28), has been motivated by modifications of Einstein gravity from one-loop quantum effects. An approach that can be followed, using an effective field theory approach to single-field models of inflation is to write down a Lagrangian that contains all generally covariant terms up to two derivatives built with the metric and one scalar field, that is assumed to drive inflation [155]:

$$\begin{aligned}\mathcal{L} = \sqrt{-g}\Big[&\frac{1}{2}M_{Pl}^2\Omega(\psi)^2R - \frac{1}{2}h(\psi)g^{\mu\nu}\partial_\mu\psi\partial^\mu\psi - U(\psi) + f_1(\psi)\left(g^{\mu\nu}\partial_\mu\psi\partial_\nu\psi\right)^2 + \\ &+ f_2(\psi)g^{\rho\sigma}\partial_\rho\psi\partial_\sigma\psi\Box\psi + f_3(\psi)\left(\Box\psi\right)^2 + f_4(\psi)R^{\mu\nu}\partial_\mu\psi\partial_\nu\psi + f_5(\psi)R\,g^{\mu\nu}\partial_\mu\psi\partial_\nu\psi \\ &+ f_6(\psi)R\,\Box\psi + f_7(\psi)R^2 + f_8(\psi)R^{\mu\nu}R_{\mu\nu} + f_9(\psi)C^{\mu\nu\rho\sigma}C_{\mu\nu\rho\sigma}\Big] \\ &+ f_{10}(\psi)\epsilon^{\mu\nu\rho\sigma}C_{\mu\nu}{}^{\kappa\lambda}C_{\rho\sigma\kappa\lambda}. \end{aligned} \tag{36}$$

Some terms that modify Einstein gravity appear immediately, as those related to R^2 and $R_{\mu\nu}R^{\mu\nu}$ terms, terms related to the Weyl-squared terms $C^{\mu\nu\rho\sigma}C_{\mu\nu\rho\sigma}$, or non-minimal couplings between the scalar field and gravity, as, e.g. the first two terms in Eq. (36). Some of the terms in Eq. (36) can actually lead to ghosts, i.e. the terms proportional to f_2, f_3, f_6 and f_8, and therefore one might want set to zero these coefficients. Modifications of Einstein gravity relevant during inflation have been considered (in a framework similar to this) in, e.g. [156–160]. This is an incomplete list of works on the subject. Our aim is not to make a thorough review of this kind of models. Rather we want to put in evidence only some recent directions of investigation and only some specific signatures that can arise in the context of models of inflation arising within modified gravity models, which are related to some of the observational features discussed in the previous section. In particular, notice that recently there has been quite an intense study of primordial non-Gaussian features arising from these models. The derivative self-interactions of the inflaton field appearing in Eq. (36), e.g. do give rise to primordial non-Gaussianity of the equilateral type (see Eq. (34)). On the other hand, in [161] it has been shown that, starting from Eq. (36), and assuming a slow-roll dynamics of the inflaton field (such that the various coefficients $f_i(\psi)$ can be treated as constants at leading order in the slow-roll parameters), also significant primordial non-Gaussianity of the (quasi) local type can be generated. The terms that generate such a signal reduce to a lagrangian that, starting from Eq. (36), reads

$$\mathcal{L} = \sqrt{-g}\left[f(R) - \frac{1}{2}g^{\mu\nu}\partial_\mu\psi\partial_\nu\psi - U(\psi)\right], \tag{37}$$

where $f(R)$ is the generalization of the first term in an expansion in powers of the Ricci scalar (i.e., $f(R) = \frac{1}{2}M_{pl}^2R + R^2/12M^2$). This action contains a scalar degree of freedom, described by the $f(R)$ term, additional to the inflaton field ψ. Such a model can be recast in the Einstein frame (also using a Weyl transformation $g_{\mu\nu} \rightarrow e^{-2\omega}g_{\mu\nu}$), where the action appears as a two-field interacting model (see [161] for more details)

$$\tilde{\mathcal{L}} = \sqrt{-g}\left[\frac{1}{2}M_{pl}^2R - \frac{1}{2}g^{\mu\nu}\gamma_{ab}\partial_\mu\varphi^a\partial_\nu\varphi^b - U_1(\varphi_1) - e^{-4\varphi_1/\sqrt{6}M_{pl}}U(\varphi_2)\right], \tag{38}$$

where

$$U_1(\varphi_1) = -e^{-4\varphi_1/\sqrt{6}M_{pl}}\Lambda\left(\phi\left(\omega\left(\varphi_1\right)\right)\right), \tag{39}$$

and the field metric is given by

$$\gamma_{ab} = \begin{pmatrix} 1 & 0 \\ 0 & e^{-2\varphi_1/\sqrt{6}M_{pl}} \end{pmatrix}. \tag{40}$$

In the Einstein frame the inflaton field is $\psi = \phi_2$, while the other dynamical degree of freedom ϕ_1 in this frame would correspond to the 'new' degree of freedom associated to a deviation from Einstein gravity, in this case due to an $f(R)$ modification. In the Einstein frame the non-trivial interactions between the extra field ϕ_1 and the inflaton field can produce a significant level of primordial non-Gaussianity that peak for quasi-local configurations, providing in this context a realization of the so-called 'quasi-single' field models of inflation [162]. The amplitude of the non-Gaussianity produced can be $f_{NL} \sim \mathcal{O}(1)$, which would be at the reach of future LSS surveys via the scale-dependent halo bias effect discussed in section "A Free Gaze into the Future of Inflation: Open Issues, Forthcoming and Future Experiments".

A related interesting direction that has been developed quite recently is the investigation of primordial non-Gaussianity arising in the tensor perturbations, i.e. in the primordial gravitational waves generated during inflation from terms that go beyond the ones already analyzed for the standard-single field models of slow-roll inflation in [41]. In particular, in Refs. [163–165] the tensor bispectrum arising from terms that are cubic in the Weyl tensor has been considered

$$\Delta\mathcal{L} = f(\phi)\left[\sqrt{-g}C^3 + \tilde{C}C^2\right], \tag{41}$$

Here C is the Weyl tensor, $C^3 = C^{\alpha\beta}{}_{\gamma\delta}C^{\gamma\delta}{}_{\sigma\rho}C^{\sigma\rho}{}_{\alpha\beta}$, $\tilde{C}C^2 = \epsilon^{\alpha\beta\mu\nu}C_{\mu\nu\gamma\delta}C^{\gamma\delta}{}_{\sigma\rho}C^{\sigma\rho}{}_{\alpha\beta}$, $\epsilon^{\alpha\beta\mu\nu}$ is the $4D$ Levi-Civita tensor, and λ is a scale that sets the value of the higher derivative corrections. The first term in general gives rise to a small undetectable (at present) amplitude of primordial non-Gaussianity. The second interaction in Eq. (41) can generate interesting signatures related to parity violation in the bispectrum of gravitational waves (with relevant amplitudes) in the case in which f is not a constant (otherwise the bispectrum vanishes [164, 165]). The latter signatures are, at the level of the bispectrum, the counterpart of other parity violating features that could arise in the power spectrum from the last term in Eq. (36). The predictions and the observational signatures for the power spectrum in this case have been thoroughly studied, e.g. in [166–168]. One of the main observational signatures is a cross-correlation between the E- and B-mode of CMB polarization that can be induced only in the presence of parity breaking [166].

There exist many other approaches different from the effective action written in Eq. (36). One can consider models derived within specific high-energy theories or derived under some symmetry criteria. An example is given by the models of modified gravity arising within the so-called Galileon symmetry [169]. Moreover the Galileon field can arise within fundamental physics scenarios (e.g. [170, 171]).

The Galileon model is based on the action [172, 173]

$$S = \int d^4x\sqrt{-g}\left(\frac{1}{2}R + \sum_{n=0}^{3}\mathcal{L}_n\right), \tag{42}$$

with

$$\begin{aligned}
\mathcal{L}_0 &= c_2 X\,,\\
\mathcal{L}_1 &= -2(c_3/\Lambda^3)X\Box\phi,\\
\mathcal{L}_2 &= 2(c_4/\Lambda^6)X\left[(\Box\phi)^2 - \left(\nabla_\mu\nabla_\nu\phi\right)^2\right] + (c_4/\Lambda^6)X^2R,\\
\mathcal{L}_3 &= -2(c_5/\Lambda^9)X\left[(\Box\phi)^3 - 3\Box\phi\left(\nabla_\mu\nabla_\nu\phi\right)^2 + 2\left(\nabla_\mu\nabla_\nu\phi\right)^3\right]\\
&\quad + 6(c_5/\Lambda^9)X^2G_{\mu\nu}\nabla^\mu\nabla^\nu\phi\,.
\end{aligned} \tag{43}$$

Here X is the kinetic term, $X = -\nabla_\mu\phi\nabla^\mu\phi/2$, $(\nabla_\mu\nabla_\nu\phi)^2 = \nabla_\mu\nabla_\nu\phi\nabla^\mu\nabla^\nu\phi$, and $(\nabla_\mu\nabla_\nu\phi)^3 = \nabla_\mu\nabla_\nu\phi\nabla^\mu\nabla^\rho\phi\nabla^\nu\nabla_\rho\phi$, while the c_i are dimensionless coupling coefficients and Λ is the cut-off of the theory. An interesting signature of these models is a modified consistency relation

$$r = \epsilon_V \bar{c}_s, \tag{44}$$

where $\bar{c}_s$ is given by a particular combination of the coupling constants c_i of the Galileon field and it is related to the sound speed of the Galileon field (see [16, 51] for more details). This is just an example of the fact that, usually, in models of inflation arising within modified gravity scenarios the standard consistency relation (17) can receive significant corrections (for other examples see the review on gravitational waves from inflation [27]). In particular, in these kind of models a blue spectrum of primordial gravitational waves can be produced (which is not possible within the standard models of slow-roll inflation discussed in section "Inflation in the Early Universe"). Notice that for the models of Galileon inflation analyzed in [16, 51] there is an interesting interplay between the parameters determining the power spectrum observables and the bispectrum arising in this kind of models (which is of the equilateral and orthogonal type). Such an interplay allows to break some degeneracies which are usually present among the different parameters determining the power spectra, once the constraints on primordial non-Gaussianity are considered.

Let us mention that the Galileon models described in Eq. (42) have been introduced first of all as possible models to explain the *present* acceleration of the universe, in terms of a modification of Einstein gravity.

In this context a much studied generalization is given by the Horndesky theory [174, 175], which has been also constrained by the *Planck* team in [176]. The Horndesky theory represents he most general class of 4D local scalar-tensor theories that contain second-order equations of motion. It is clear that building the most general healthy scalar tensor-theory of gravity is particularly important for building models of inflation and dark energy. There is an intense research activity at the present time devoted to explore even more general healthy theories that go beyond Horndesky, as those studied in [177, 178] (the so called G^3 theories) and in [179, 180]. For the latter theories there is, in particular, an intense discussion at the moment about their stability properties (like the absence of ghosts), see, e.g. [181–183, 183, 184]. Finally, an interesting issue for all modified gravity models that aim at explaining the present acceleration of the Universe is to reproduce the very accurate tests of General Relativity on small scales, like the tests carried on in the Solar system. For many of the proposed scenarios that we have mentioned above this can be carried on by suppressing the deviations from General Relativity via some 'screening' mechanism (for a review see, e.g. [185]).

Neutrinos in Cosmology

Cosmological data offer the exciting opportunity to obtain tight and accurate constraints on some neutrino properties, most notably their number and the total neutrino mass. These constraints are extremely competitive and provide an independent and complementary approach to laboratory ground-based experiments. In this section we will review the main effects that neutrino induce at a linear level, the non-linear modelling, the constraints that can be obtained from data sets, possible small-scale extension of the standard cosmological model (warm dark matter), and finally the forecasts that could be obtained using future data sets.

Linear Theory of Cosmological Neutrinos

The hot Big Bang model predicts the existence of a relic cosmic neutrino background (CNB), which has not been directly detected yet, but whose properties have been partly measured in an indirect way by using cosmic microwave background, Large Scale Structure (LSS) data and light abundances of the elements as predicted by primordial nucleosynthesis. Neutrinos are produced at high temperature by weak interactions that keep them in equilibrium till expansion makes this process not effective. In these early stages, neutrinos are coupled to the plasma with a Fermi-Dirac distribution with a temperature T_ν and a negligible chemical potential. When the weak interaction rate becomes smaller than the expansion rate, neutrinos decouple; a rough and simple estimate of the decoupling temperature can be obtained from $\Gamma_\nu = \langle\sigma_\nu n_\nu\rangle \sim H = (8\pi\rho/3M_P^2)^{1/2}$, with n_ν the neutrino number density, σ_ν the cross section of the electron-neutrino process and $\rho \sim T^4$ the total energy density dominated by radiation and M_P the Planck mass, which results in a temperature of 1 MeV. It has been shown that neutrino decoupling can be approximated to a reasonable degree of accuracy as an instantaneous process that preserves the Fermi-Dirac spectrum due to the fact that both momenta and temperature scale as a^{-1}.

Neutrino number density per flavour can easily be calculated from the Fermi-Dirac distribution once T_ν is known, while neutrino energy density should be computed numerically and presents two well-defined analytical limits for which $\rho_\nu \propto T_\gamma^4$ (like radiation) if $m_\nu \ll T_\nu$ and $\rho_\nu = m_\nu n_\nu$ if $m_\nu \gg T_\nu$ (like matter).

In Fig. 9 (left panel) we show the evolution of the densities after neutrino decoupling for a standard (spatially flat and with a cosmological constant) model with with $h = 0.7$ and $\Omega_\Lambda = 0.70$, $\Omega_b = 0.05$, $\Omega_\nu = 0.0013$ and $\Omega_{cdm} = 1 - \Omega_\Lambda - \Omega_b - \Omega_\nu$; in this case the masses are distributed according to the normal hierarchy with $m_1 = 0$, $m_2 = 0.009$ eV and $m_3 = 0.05$ eV (the density evolutions are scaled to the power 1/4 in order to better appreciate the different behaviours). In the right panel, we show the evolution of densities w.r.t. the critical density as a function of redshift and temperature: once they become non-relativistic, in the matter dominated era, ν_2 and ν_3 experience an enhancement of their densities. From the figure it is easy to appreciate one unique feature of neutrino particles: they become non-relativistic late in time (at least one of them, as we will see later) in the structure formation era. Their contribution to the energy density, in the non-relativistic limit is mass dependent and we have the following expression: $\Omega_\nu = \frac{\Sigma_i m_i}{93.14h^2\mathrm{eV}}$, which could be of order unity for eV masses. If we rely on the cosmological constraint that $\Omega_m \sim 0.3$, we can easily obtain a conservative upper limit $M_\nu = \Sigma_i m_i = 15$ eV on the total neutrino masses.

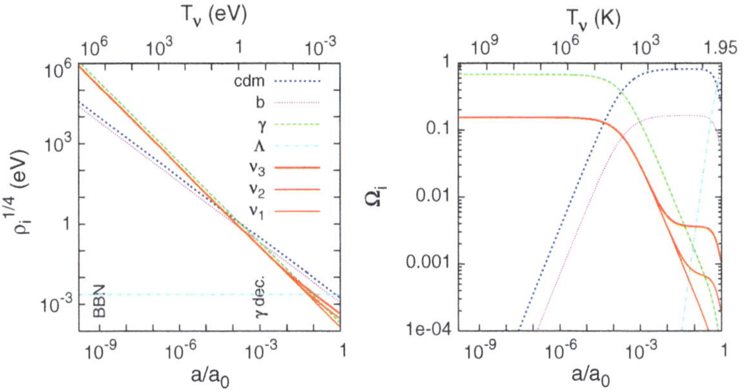

Fig. 9 Evolution of the background densities as a function of redshift and neutrino temperature (*top axis*) from neutrino decoupling until now for a standard flat cosmological model with neutrino masses distributed according to the normal hierarchy. In the *left plot* densities to the power 1/4 (in eV units) while in the *right plot* densities normalized by the critical density are reported. The density of the neutrino mass states ν_2 and ν_3 is clearly enhanced once they become non-relativistic. Taken from Ref. [187]

In the relativistic limit, after neutrino decoupling, and during radiation era their contribution to the energy density of the Universe is usually parameterized in terms of the effective neutrino number N_{eff} as

$$\rho_r = \rho_\gamma + \rho_\nu = \left[1 + \frac{7}{8}\left(\frac{4}{11}\right)^{4/3} N_{eff}\right]\rho_\gamma \tag{45}$$

From the analysis made at Large Electron-Positron Collider (LEP) we know that the number of light neutrinos sensitive to weak interactions is $N_\nu = 2.9840 \pm 0.0082$, while an accurate calculation of neutrino decoupling gives $N_{eff} = 3.046$. Any departure from the latter value should indicate non-standard neutrino physics or contribution from other relativistic particles. The value of N_{eff} is constrained by Big Bang Nucleosynthesis (BBN), since ρ_r in Eq. 45 enters in the Hubble expansion rate. In particular a different value of the expansion rate impacts on light element abundances. At present BBN is providing at limit of $\Delta N_{eff} \leq 1$ at 2σ C.L. [186], while a tighter limit is provided by the CMB and will be discussed later.

The LSS of the Universe is mainly investigated by using the matter power spectrum, which is affected by neutrinos. After decoupling, non-relativistic neutrinos are described by a collisionless fluid where the particles have a thermal velocity $v_{th} \sim 158(1+z)$ (1 eV/m) km/s, this in turns allows to define a neutrino free-streaming wavenumber as a typical distance over which neutrino particles travel between an initial time and t. This latter is defined as:

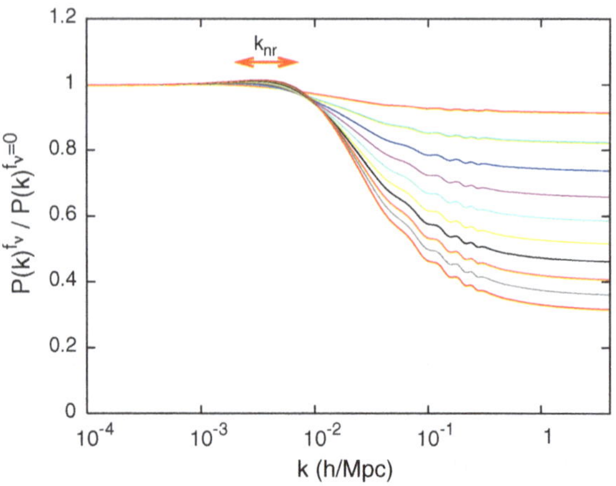

Fig. 10 Ratio of the matter power spectrum including three degenerate massive neutrinos with density fraction f_ν to that with three massless neutrinos. The parameters $(\omega_m, \Omega_\Lambda) = (0.147, 0.70)$ are kept fixed, and from top to bottom the curves correspond to $f_\nu = 0.01, 0.02, 0.03, \ldots, 0.10$. The individual masses m_ν range from 0.046 to 0.46 eV, and the scale k_{nr} from 2.1×10^{-3} h Mpc^{-1} to 6.7×10^{-3} h Mpc^{-1} as shown on the *top* of the figure. k_{eq} is approximately equal to 1.5×10^{-2}h Mpc^{-1}. Taken from Ref. [187]

$$k_{FS}(t) = \left(\frac{4\pi G\bar{\rho}a^2(t)}{v_{th}^2(t)} \right)^2, \tag{46}$$

and is the key quantity able to describe the damping of small scale power. It is easy to see that in the matter domination era the comoving free-streaming length decreases (unlike the physical free-streaming length that increases), as a consequence, for neutrino particles that become non-relativistic in this era the comoving free streaming wave number passes through a minimum $k_{nr} \sim 0.018\Omega_m^{1/2}$ (m/1 ev) h/Mpc at the time of the transition at the time of the non-relativistic transition when $m = \langle p \rangle = 3.15T_\nu$. The physical effect is a suppression of power of the small scales due to the fact that neutrinos cannot be confined to regions smaller than their free-streaming scale since their thermal velocities are larger than escape velocities from the potential wells (on those scales). Modes with $k < k_{nr}$ are never affected by neutrino free-streaming and evolve without experiencing any suppression. In Fig. 10 the neutrino free streaming effect is shown for different values of neutrino masses.

On large scales $k < k_{nr}$ the matter power spectrum can be shown to depend only on the primordial power and on the total matter density Ω_m, while the small scales are reduced for two main reasons: the lack of neutrino perturbations on the total matter power and the slower growth rate of the CDM and baryon components. By using Boltzmann codes, one can see that the *linear* plateau-like suppression at small scale which saturates at ~1 h/Mpc is well described by $\Delta P(k)/P(k) \sim -8f_\nu$ with $f_\nu = \Omega_\nu/\Omega_m$.

For masses that are below 0.6 eV (as the current bounds suggest), the neutrinos are still relativistic at the time of photon decoupling thereby they cannot impact on the evolution of CMB perturbations. However, they have two effects: in terms of background evolution they shift the redshift of matter radiation equality (main effect); in terms of secondary anisotropies they play a role for, e.g. Integrated Sachs-Wolfe effect (ISW), weak lensing etc. Neutrinos with sub eV mass are counted as radiation at the time of equality and as matter now; thereby when M_ν is varied there should either be a change in the non-relativistic matter density $\omega_m = \Omega_m h^2$ today or in z_{eq} A shift in the radiation matter equality affects both amplitude and positions of the peaks; the presence of a different amount of ordinary matter change the angular diameter distance to last scattering (that impacts over the whole CMB power spectrum) and on the tail at small scales due to secondary anisotropies. Another smaller effect induced by neutrinos is the early ISW due to the fact that when they are non relativistic they reduce the time variation of the gravitational potential. The CMB power spectrum is thus sensitive to background effects of the total neutrino masses that present different degeneracies compared to other LSS probes.

In this section we have described the two main effects induced by neutrinos, one at the background level and the second at the level of density perturbations. These effects can be simply parameterized by two numbers N_{eff} and M_ν. We will now refine the particle physics estimates that can be obtained on these two numbers by using LSS data. The broad and conservative range that is our starting point is given by neutrino-less double beta decay and tritium beta decay and by the results on solar and atmospheric neutrinos that give: 0.06 (0.1) eV $\leq M_\nu \leq$ 6 eV (where 0.06 or 0.1 is determined by the hierarchy) (for a recent review see [188, 189]). We will also present limits in terms of N_{eff} even if standard neutrino particles are fixed to 3 from the LEP experiment.

Non-linear Modelling of Cosmological Neutrinos

Here we will extend linear theory by precisely quantifying the neutrino induced effects on the matter power spectrum, in the *non-linear* regime. The first approach is numerical: implementing the neutrino fluid in an N-body code in order to fully address the small scale behaviour. Since the work of [190], it is clear that in order to properly simulate non-linear neutrino clustering it is not sufficient to implement a total linear matter power spectrum in the initial conditions that is calculated in the presence of massive neutrinos: their thermal velocities (which will not be accounted for using this approach) will impact dramatically on their small scale clustering properties. In [191, 192], using an approach similar to that of [190], neutrinos are implemented in the public available code GADGET using two different methods: (i) a 'grid-based' approach for which the neutrino clustering is obtained at a linear level using a Boltzmann code the gravitational potential is calculated at the mesh points of the particle mesh grid (used for calculating long-range forces) and the neutrino contribution is added when forces are calculated by differentiating this potential; (ii)

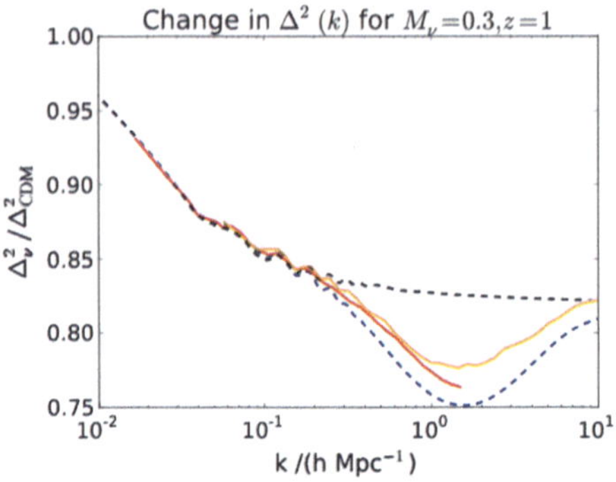

Fig. 11 Neutrino induced suppression as obtained from N-body neutrino simulations. *Dashed black curve* is the linear theory prediction, while the *blue dashed* is a naive (wrong) extension of a simple halo model. Results of simulations are represented by the *orange* and *red curves* for a total neutrino mass of 0.3 eV at $z = 1$. Taken from Ref. [193]. Note that this suppression is plotted versus the CDM power spectrum

a 'particle-base' approach for which neutrinos are implemented as an independent set of particles and evolved using the same method for standard N-body collision-less particles. Technical aspects of neutrino implementations and convergence tests are discussed in [191].

In Fig. 11 we show the neutrino induced suppression, by plotting the ratio of a massive neutrino model w.r.t. CDM power at $z = 1$ for at total neutrino mass of 0.3 eV as found in [193]. The linear theory is shown as a dashed black curve (where it can be appreciated the flat suppression at small scales), the blue dashed curve is a simple extension of the halo model obtained by using a total matter power that include neutrinos, the simulations are represented by the red and orange curves and display a characteristic spoon-shaped feature. This feature is induced by the different evolution and coupling of small scale non-linearities w.r.t. the evolution of linear modes in a massive neutrino cosmology. Here we can thus appreciate one key aspect of non-linear neutrino clustering: a scale and redshift dependent feature at small scales.

The probability distribution function of neutrino peculiar velocities can also be computed from N-body neutrino simulations, as obtained in [194] using the particle based implementation and compared to the one of cold dark matter. The peculiar velocity field offers another opportunity to investigate neutrino clustering in the non-linear regime by estimating the departure from the Fermi-Dirac unperturbed distribution at the same redshift. These results have non-trivial consequences in terms of redshift space distortions and can offer also quantitative predictions for neutrino densities and velocities at the Earth position for direct detection experiments.

We can safely summarize these findings in the following way: suitable modifications of N-body codes allow to control the behaviour of the neutrino fluid simulated in terms of collision-less particles in momentum space at the percent level [190–192].

A different approach to neutrino non-linearities is offered by the neutrino halo model developed in [195] which is able to capture the non-linear behaviour of matter clustering with a ~20% accuracy up to very non-linear scales of $k = 10$ h/Mpc (which would be affected by baryon physics). The largest discrepancies, in this case, are present in the range $k = 0.5$–1 h/Mpc where the 1-halo and 2-halo terms are comparable and are present also in a massless neutrino cosmology. By modelling the neutrino non-linear density field as a sum of a linear and clustered component one can also predict the neutrino power spectrum and the cold dark matter-neutrino cross-power spectrum up to $k = 1$ h/Mpc with ~30% accuracy. For masses below 0.15 eV in [195] it is found the neutrino halo model captures the neutrino induced suppression, casted in terms of matter power ratios between massive and massless scenarios, with a 2% agreement with the results of N-body/neutrino simulations. This is a remarkable results since it allows to compute the neutrino and matter clustering without running computationally expensive simulations and in extremely accurate way.

An interest opportunity to test the non-linear behaviour of neutrinos is offered also by the large scales, like those usually probed by Baryonic Acoustic Oscillations (BAOs). Non linearities at the BAO scale are quantified in [196] where it is shown that the evolution with redshift of the matter correlation function and its dependence on the neutrino masses is well reproduced in a simplified version of the Zeldóvich approximation. These scales are particularly important for the following reason: while in linear theory the BAO peak decreases for increasing M_ν values, the effect of nonlinear structure formation has an opposite tren, since the peak broadening by large scale flows is less effective. As a result of this combined effect, the BAO peak decreases by 0.6% for $M_\nu = 0.15$ eV and increases by ~1.2% for $M_\nu = 0.3$ eV, with respect to a massless neutrino cosmology with equal value of the other cosmological parameters. By extending the analysis to redshift space and to halos, an analytic prescription is provided and neutrino induced non-linearities could be tested by future precision clustering data at the BAO scales.

Another interesting and unexpected result derived from the analysis of N-body neutrino simulations [194, 197–199] is the fact that theoretical expectations of the mass function of dark matter haloes for massive neutrino cosmologies are in better agreement with numerical results when the linear cold dark matter power spectrum is used instead of the total matter power spectrum. Basically, both the universality of the mass function and a scale independent bias is recovered only when using the cold dark matter power spectrum as computed in the corresponding massive neutrino model. This is witnessing the fact that neutrinos participate to the clustering at a 'background' level and has important implications in terms of constrained derived from cluster number counts.

Neutrino Constraints

In this section, we review the neutrino mass bounds that can be derived from current cosmological data. Note that the quoted confidence limits are all based on the Bayesian inference method, and are given at the 95% confidence level after marginalization over all free cosmological parameters. Assuming the minimal 6-parameter ΛCDM model as a baseline model and adding just M_ν as an extra parameter the constraints that can be obtained from the latest CMB only data of Planck [200] range from $M_\nu < 0.49$–0.71 eV depending on the the fact that we add the polarization and CMB lensing data to the TT power spectrum. If we add external data sets, more specifically BAO, Supernovae Ia (JLA compilation) and a prior on the Hubble parameter this limit shrinks to $M_\nu < 0.194$ eV Adding polarization data at small scales produce only a modest improvement, while adding the CMB lensing information has the effect of pulling slightly the likelihood towards larger neutrino masses (but still consistent with zero) as shown in Fig. 30 of [200].[4] Note that in [201] the chosen prior on the Hubble constant is different (higher value) compared to the one used by [200], this also contributes to tighten the neutrino bounds.

Referring to the most recent years, two different positions emerge: on one side, we find claims that cosmology provides us a hint for non-zero neutrino masses. On the other, we have very tight upper limits on M_ν.

In the former case, it has been suggested [204, 205] that a total non zero neutrino mass around 0.3 eV could alleviate some tensions present between cluster number counts (selected both in X-ray and by Sunyaev-Zeldovich effect) and weak lensing data.

However, evidence for non-zero neutrino masses either in the active or sterile sectors seems to be claimed in order to fix the significant tensions between different data sets: CMB and baryonic acoustic oscillations BAOs on one side and weak lensing, cluster number counts and high values of the Hubble parameter on the other.

In the latter case, the limit on $\Sigma = M_\nu$ is so stringent, that it better agrees with the normal hierarchy spectrum, rather than with inverted hierarchy one. The tightest experimental limits on Σ are usually obtained by combining CMB data with the ones probing smaller scales. In this way, their combination allows a more effective investigation of the neutrino induced suppression in terms of matter power spectrum, both in scale and redshift. Recently, a very stringent limit, $\Sigma < 146$ meV (2σ C.L.), was set by [206]. This limit has then been further refined with the latest CMB data and is now 0.12 eV [202]. New tight limits were also presented after the data release by the Planck Collaboration in 2015 [200]. Some of the most significant results are reported in Table 3. The bounds on Σ indicated by these post-Planck studies are quite small, but they are still larger than the final sensitivities expected, especially thanks to the inclusion of other cosmological data sets probing smaller scales Therefore, these small values cannot be considered surprising and, conversely, margins of further progress are present. Figure 12 shows future constraints on neutrino masses that can

---

[4]http://wiki.cosmos.esa.int/planckpla2015/images/0/07/Params_table_2015_limit95.pdfonpage311.

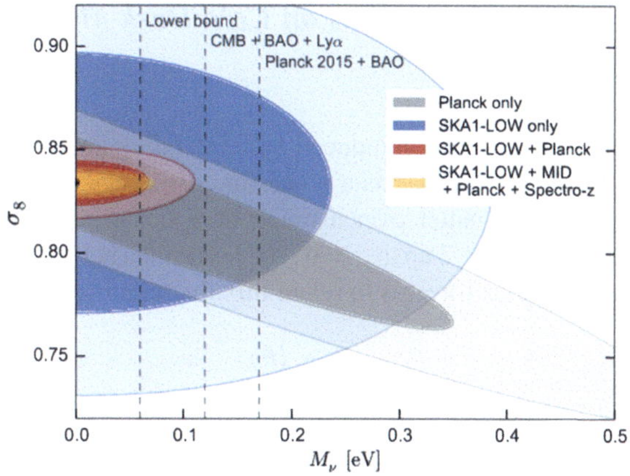

Fig. 12 Forecast marginal 1- and 2σ constraints on M_ν and σ_8, for SKA1-LOW and Planck, and for the combination of all experiments considered here. The combined constraint (*red*) is much better than the individual constraints because multiple parameter degeneracies are broken by combining the datasets. The lower bound, CMB + BAO + Lyα, and Planck 2015 95% limits are shown as vertical dashed lines from left to right respectively. Taken from Ref. [215]. Note the difference degeneracies exploited by the data combination

Table 3 Summary of the constraints obtained on total neutrino mass. All values are 2σ confidence level obtained in a Bayesian analysis

References	M_ν (eV)	Data	Notes
[200]	<0.49–0.7	CMB	1 param. ext. (with and without lensing)
[200]	<0.32	CMB + BAO	2 param. ext. (+N_{eff}) including lens. and pol. at large scales
[200]	<0.21	CMB + BAO	1 param. ext. including lens. and pol. at large scales
[200]	<0.194	CMB + BAO + SNIa + H0	1 param. ext. including lens. and pol.
[200]	<0.17	CMB + BAO	1 param. ext. including lens. and pol.
[200]	<0.153	CMB + BAO + H0	1 param. ext. including lens. and pol. including large scales and bispectra
[202]	<0.12	CMB + BAO + Lyalpha	1 param. ext. including CMB pol.
[203]	<0.13	CMB + BAO + gal. clustering	1 param. ext. including CMB pol.
[201]	<0.0993	CMB + BAO + clusters + H0 + τ	1 param. ext. including CMB pol., clusters, H0 and strong prior on τ

be achieved by combining the intensity mapping data with current CMB and BAO observations.

The Lyman-α Forest: Constraints on Cold Dark Matter Coldness

The Lyman-α forest, the absorption induced by intervening neutral hydrogen along the line-of-sight to a distant quasar, is a powerful cosmological tool ideally suited to probe the clustering of matter over a range of scales from below the Mpc to hundreds of Mpc and from $z = 2$ up to $z = 6$ [207]. The idea behind the cosmological applications of the Lyman-α forest is to relate flux fluctuations:

$$\delta_F = \frac{F - \langle F \rangle}{\langle F \rangle}, \tag{47}$$

to matter fluctuations. This can be done in several ways and the simplest is to make use of the so-called Fluctuating Gunn-Peterson approximation[5]:

$$\tau_{GP} = \frac{\pi e^2}{m_e c} f_\alpha \lambda_\alpha H^{-1}(z)\, n_{HI} \tag{48}$$

with n_{HI} being the neutral hydrogen density, that relates the optical depth to the underlying density of neutral hydrogen, f_α being the oscillator strength and $\lambda_\alpha = 1215.67$ Åbeing the Lyman-α absorption wavelength. The assumption that the gas producing the absorption is in photoionisation equilibrium implies that $n_{HI} \propto \rho^2 T^{-0.7}/\Gamma$, where Γ is the photoionisation rate. Furthermore, if one assumes that the gas temperature scales as $T = T_0(\rho/\langle\rho\rangle)^\gamma$, which is set by the balance between photo-heating and adiabatic cooling due to the expansion of the universe and has been found to be a good approximation of the gas thermal state at low-densities, one obtains:

$$\tau \propto A(z) \left(\frac{\rho}{\langle\rho\rangle}\right)^\beta \tag{49}$$

where the redshift dependent A factor $\beta = 2 - 0.7\gamma$ and will depend also on cosmological parameters, atomic physics and on the photoionisation rate. The observed quantity is the transmitted flux $F = \exp(-\tau)$ and at first order it can be easily seen that flux fluctuations are related to the linear density contrast as $\delta_F \propto -A\beta\delta_{lin}$. Non-linearities in the density fields and those induced by peculiar velocities complicate the picture above and simple analytical insights or semi-analytical models must be replaced by more reliable and accurate hydrodynamic simulations of intergalactic medium structures performed either with smoothed-particle hydrodynamics, Eulerian or adaptive mesh refinements codes.

The use of Lyman-α forest data to probe matter clustering has been pioneered by Croft and co-workers at the end of the nineties: a measurement of the linear matter power spectrum at small scales and high redshift has been presented, by using high and medium resolution quasar spectra together with the so-called 'effective

[5]This approximation neglects non-linearities induced by the fact that the signal is in redshift space.

bias' method, $P_F(k) = b^2_{eff}(k, z) \times P_{lin}(k)$, that allowed an inversion of the one-dimensional flux power to infer the underlying matter power spectrum. After that, other groups used a set of about 30 high-resolution high signal-to-noise quasar spectra taken with the UVES/VLT spectrograph and a suite of full hydrodynamic simulation, that explored several thermal histories, to derive the matter clustering at $z \sim 2.1$ and $z \sim 2.7$. These data have been combined in a series of paper with WMAP data in order to probe the long-lever arm of the matter power spectrum and get constraints on the running of the spectral index and inflationary parameters [209]. A new era in the Lyman-α forest field has entered with the advent of the SDSS survey that has allowed to obtain the 1D flux power spectrum from a set of 3000 low-resolution quasars in the range $z = 2.2$–4.2 over two decades of wavenumbers [210] and to infer the linear matter power spectrum amplitude, slope and curvature at $z = 3$ and at a comoving scale of about ~8 Mpc/h with unprecedented precision, by means of approximate hydro simulations. Again the SDSS data have been combined with other large scale structure probes to get very tight constraints in terms of neutrino mass fractions and cosmological parameters like running of the spectral index and inflation. More recently, BOSS/SDSS-III has measured the 3D clustering of the flux by exploiting the signal in the transverse direction from a set of 50000 quasar spectra: this new data set has allowed to measure at high significance the presence of BAO (Baryonic Acoustic Oscillations) at $z \sim 2.2$ and a new measurement of the 1D flux power has also been recently provided [206].

Lyman-α forest data are currently providing the tightest constraints in terms of WDM properties and there are two main reasons for this. First of all, the one-dimensional power spectrum is a projection of the 3D one and contains information down to very small scales:

$$P_{1D,F} = \frac{1}{2\pi} \int_k^{\infty} P_{3D,F}(y) y \, dy \tag{50}$$

and thereby is sensitive to the cutoff induced by WDM. Secondly, Lyman-α forest data span high redshift where the WDM cutoff in terms of matter power is more pronounced and much closer to the linear behaviour, in fact the Lyman-α forest flux power is particularly sensitive at environments around the mean density, closer to the linear regime and this is especially true at high redshift, due to the strong evolution of the mean flux level. There is also another reason that plays an important albeit minor role: the thermal broadening depends on the temperature, which becomes colder at high redshift, and is a fixed number in velocity space while the free-streaming length scales as $\sqrt{1+z}$, making the thermal contribution to a possible WDM cut-off less prominent at high redshift.

In Table 4 we present a summary of the constraints, in terms of the mass of a thermal relic, that have been obtained by using Lyman-α forest data.

The first constraint was obtained by [208]: by using N-body simulations only and a set of eight high resolution spectra, the authors looked also at the flux probability distribution function and not only at the flux power and obtain a lower limit of 0.75 keV. The main limitations of this work were due to the fact that no hydro simulations were used and a proper marginalization over nuisance parameters was not done. Reference [209] used instead an effective bias method to model the flux-matter relation and a set of full hydro simulations to explore the bias in WDM scenarios relying on high-resolution UVES spectra at $z = 2.1, 2.7$. In this case, the authors found a 2σ lower limit of 0.55 keV for a thermal relic and the nuisance parameters were accounted for (and marginalized over) by allowing an extra normalization error on the data. In this paper the authors also quote a 2σ lower limit of 2 keV for a sterile neutrino in the so-called Dodelson-Widrow scenario and an upper limit for a gravitino of 16 eV 2σ C.L. in a model for which this particle is not the total amount of DM. Subsequently, [210] exploited the unique capabilities of the SDSS flux power spectrum (about 3000 low-resolution low signal-to-noise QSO spectra spanning the redshift range $z = 2.2$–4.2) and showed that the constraints derived from this data set were much tighter due to the wide redshift range probed that allowed to break the degeneracies between cosmological and thermal/nuisance parameters. They obtained a limit of $m_{WDM} > 2.5(14)$ keV for a thermal relic (sterile neutrino) at the 2σ C.L. The analysis made was based on a set of approximate hydro simulations that however explored fully the multi-dimensional likelihood space. The numbers derived above found confirmation in an independent analysis of the SDSS data made by [211] in which a suite of full hydrodynamical simulations were used at the expenses of a relative poor scanning of the multi-dimensional likelihood space obtained with a Taylor expansion of the flux power. In this work, the limits found were: $m_{WDM} > 2(10)$ keV for a thermal relic (sterile neutrino) at the 2σ C.L., in good agreement with the analysis of [210] (Fig. 13).

After this [212], explored the very high redshift regime by using 55 high resolution Keck spectra at $z = 2$–6.4 and obtained the limits $m_{WDM} > 1.2(5.6)$ keV for a thermal relic (sterile neutrino) at the 2σ C.L. However, these limits greatly improved by adding the SDSS data that allowed to break the degeneracies between thermal

Table 4 Summary of the constraints obtained on the mass of a WDM relic by using Lyman-α forest data. Apart from [208] all the other quoted values are 2σ confidence level obtained in a Bayesian analysis

References	m_{WDM} (keV)	Data	Notes
[208]	>0.75	8 high-res.	Not marginalized, N-body only
[209]	>0.55	30 UVES spectra	Eff. bias, hydro sims.
[210]	>2.5	~3000 SDSS spectra	Approx. hydrod., full likelihood expl.
[211]	>2	~3000 SDSS spectra	Full hydro, approx. likelihood expl.
[212]	>4.5	~3000 SDSS, ~60 Keck sp.	Full hydro, approx. likelihood expl.
[213]	>3.3	28 high-z MIKE+HIRES sp.	Full hydro, good likelihood expl.

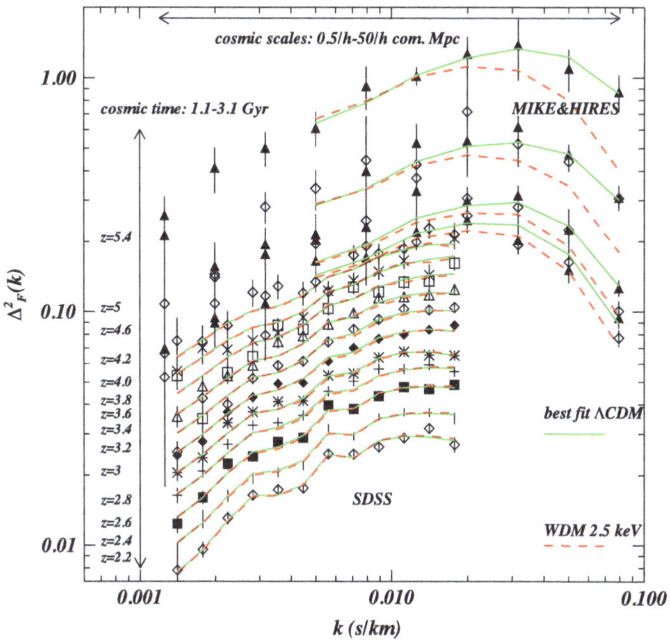

Fig. 13 One-dimensional flux power spectrum in dimensionless units ($\Delta^2(k) = P_F(k) \times k/(2\pi)$) for the low res. and high res. quasar spectra as compiled in [213]. These data points span $z =$ 2.2–5.4, a period of about 2 Gyr and about two decades in wavenumber space. The best fit ΛCDM model is shown as the *blue line*, while the *orange dashed curves* are for a WDM model with a $m_{WDM} = 2.5$ keV which is excluded by the data at very high significance (note that in this case the other parameters have been kept to their best fit values and only m_{WDM} is changed)

and WDM cut-offs. A colder (hotter) IGM will result in an increase (suppression) of the flux power due to the thermal broadening of the lines, which is different: this 'thermal' effect could either erase or boost the WDM induced suppression. The advantage of having a wide redshift range allows to appreciate the different redshift evolution of the WDM and thermal cut-offs and to lift or break their mutual degeneracies. Thanks to their constraining power, these data allowed to obtain $m_{WDM} >$ 4.5 (28) keV for a thermal relic (sterile neutrino) at the 2σ C.L. In this analysis a second order Taylor expansion of the flux was used, but again the parameter space was not explored fully and large numerical corrections were made to the flux power in the highest redshift bins.

In [213] these numbers have been revised by using a very comprehensive grid of hydro simulations that embrace a conservative range of different thermal history. At these high redshift it is also likely that galactic feedback and astrophysical effects have a much weaker impact in terms of flux power, as galaxy formation models seems to suggest. In this case the marginalization over nuisance parameters has been made fully in the most relevant parameter space and hydro simulations at higher resolutions have been employed. The data used were the highest redshift Keck spectra complemented by an equal number of MIKE (Magellan spectrograph) at poorer resolution. The flux power spectrum has been measured at $z = 4.2, 4.6, 5, 5.4$ down to the scales of $k \sim 0.1$ s/km, roughly corresponding to (very non-linear) scales $\lambda = 50$ Mpc/h. The final results, that also allow for a conservative extra error on the data side of about 30% and is not sensitive to continuum fitting uncertainties, give $m_{WDM} > 3.3$ keV for a thermal relic at the 2σ C.L., after having marginalized over nuisance, ultra-violet fluctuations, instrumental resolution, noise of the spectrograph. From this data set and analysis the authors concluded that thermal relics of masses 1 keV, 2 keV and 2.5 keV are disfavoured by the data at about the 9σ, 4σ and 3σ C.L., respectively. WDM models where there is a suppression in the linear matter power spectrum at (non-linear) scales corresponding to $k = 10$ h/Mpc which deviates more than 10% from a ΛCDM model are disfavoured by the data. Given this limit, the corresponding 'free-streaming mass' below which the mass function may be suppressed is $\sim 2 \times 10^8\,\mathrm{h}^{-1}\,M_\odot$. There appears thus to be very little room for a contribution of the free-streaming of WDM to the solution of what has been termed the small-scale crisis of CDM.

These models have been refined further by accounting for the case of a mixed C+WDM model, where an analysis of the SDSS and UVES data was presented in [214]. In this work the main results were expressed in terms of a non-resonantly produced sterile neutrino and give $m_{NRP} > 8$ keV (frequentist 99.7% confidence limit) or $m_{NRP} > 12.1$ keV (Bayesian 95% credible interval) in a pure WDM model. For the mixed model, they obtained limits on the mass as a function of the WDM fraction (percentage) to be smaller than 60% for any value of the WDM particle mass (frequentist 99.7% confidence limit); while the Bayesian joint probability allows any value of the mass (for $m_{NRP} > 5$ keV) at the 95% confidence level, provided that the fraction of WDM is below 35%, for any value of the WDM particle mass. This limit can be roughly translated into a thermal relic mass and implies that fractions of WDM below 35% can be accommodated only for masses above $m_{WDM} > 1.1$ keV.

In [214] a mechanism of resonantly produced sterile neutrino, that occurs in the framework of the νMSM (the extension of the Standard Model with three right-handed neutrinos), is analyzed and it was shown that their cosmological signature can be approximated by that of mixed C+WDM and for each mass greater than or equal to 2 keV, there exists at least one model of sterile neutrino accounting for the totality of dark matter, and consistent with Lyman-α and other cosmological data. However, the transfer function for such candidates is quite different from the one of the thermal relic and no direct comparison with thermal masses can be made.

These lower limits seem to be conflicting with the upper limits obtained on the masses of such particles coming from the observations of the cosmic X-ray background and are: $m_{s\nu} < 1.8$ keV at 95%. In fact, in addition to the dominant decay mode into three active neutrinos, the light sterile neutrino can decay into an active one and a photon with the energy $E_s = m_s/2$. Thus, there exists a possibility of direct detection of neutrino decay emission line from the sources with big concentration of DM, e.g. from the galaxy clusters or other astrophysical objects. Similarly, the signal from radiative sterile neutrino decays accumulated over the history of the Universe could be seen as a feature in the diffuse extragalactic background light spectrum. However, the constrains above assume a very simple model for sterile neutrino production and can be circumvented by considering other models that are more physically motivated [214].

Overall, Lyman-α offers a unique probe of the matter power spectrum down to very small scales and the tightest constraints in terms of CDM coldness. The most recent constraint $m_{WDM} > 3.3$ keV is suggesting that the cosmic web as probed by the Lyman-α forest data is quite cold and the values of WDM masses (0.5–1.5 keV) that are typically used in order to solve the missing satellite, the cusp-core and the 'too-big-to-fail' problems for the dynamical properties of the most massive dwarf galaxies at low redshift are in strong tension with the limits above.

Future Constraints

In [216] we make use of the Monte Carlo Markov Chains (MCMC) technique and assume two sets of theoretical errors. The first error is meant to account for uncertainties in the modelling of the effect of neutrinos on the non-linear galaxy power spectrum and we assume this error to be fully correlated in Fourier space. The second error is meant to parametrize the overall residual uncertainties in modelling the non-linear galaxy power spectrum at small scales, and is conservatively assumed to be uncorrelated and to increase with the ratio of a given scale to the scale of non-linearity. It hence increases with wavenumber and decreases with redshift. With these two assumptions for the errors and assuming further conservatively that the uncorrelated error rises above 2% at $k = 0.4$ h/Mpc and $z = 0.5$, we find that a future Euclid-like cosmic shear/galaxy survey achieves a 1-sigma error on the total neutrino mass close to 32–25 meV, sufficient for detecting the total neutrino mass with good significance. If the residual uncorrelated errors indeed rises rapidly towards smaller scales in the non-linear regime as we have assumed here then the data on non-linear scales does not increase the sensitivity to the total neutrino mass. Assuming instead a ten times smaller theoretical error with the same scale dependence, the error on the total neutrino mass decreases moderately from $\sigma(M_\nu) = 18$–14 meV when mildly non-linear scales with 0.1 h/Mpc $< k <$ 0.6 h/Mpc are included in the analysis of the galaxy survey data. Similar bounds are provided by exploiting the DESI instrument and could result in error bars that are 20 meV for the total neutrino mass. In terms of cold dark matter coldness at small scales new results will be obtained in the next few years using medium resolution spectrographs like WEAVE and X-Shooter that could possibly allow to reach sensitivities to thermal relics around 4 keV.

Conclusions

We have reviewed the recent progress made on several hot cosmological topics. The main conclusions can be summarized as follows.

- **Inflation**. Inflation has been successful in explaining the primordial seeds for structure formation, and all the observations we have at our disposal are fully consistent with its main predictions. In particular there is now a clear and robust evidence of a power spectrum of primordial fluctuations that deviates from scale-invariance. Still, the precise mechanism leading to the generation of the initial conditions in our universe is not clearly identified. Improving our sensitivity on a number of inflationary parameters can allow to reach such an outstanding goal and to understand the finest details of the inflationary dynamics. In particular constraining the level of primordial gravitational waves would be of paramount importance, allowing, among other things, to firmly establish the energy scale of inflation. Up to now we have solid and robust upper bounds on the tensor-to-scalar ratio r (with $r < 0.09$ at 95% CL). Planned future satellite missions, coadiuvated by several efforts from balloon an ground based experiments, can reach some threshold values that can firmly advance our knowledge on inflationary models, such as values as low as $r \sim 10^{-3}$. Improvements in other inflationary parameters, such as the running of the spectral index, are also foreseen that will allow to probe other crucial and complementary aspects of inflation.
- **Primordial non-Gaussianity**. The present constraints on primordial non-Gaussianity provided by the *Planck* satellite represent one of the tightest tests of the simplest models of inflation that has been passed successfully. Measuring primordial non-Gaussianity has become a high-precision measurement of modern cosmology and it allows to test the fundamental interactions among the quantum fields present during inflation and among such fields and gravity. Still, a clear distinction among different models, e.g. single-field versus multi-field, can be achieved only by further improving the sensitivity to primordial non-Gaussianity. Various improvements can be implemented, starting from the full exploitation of CMB polarization data to measurements of non-Gaussian signatures in the LSS (which will become available with, e.g. the Euclid satellite mission), to CMB spectral distortions and more futuristic directions, like high redshift 21 cm fluctuations. Several efforts are undergoing in all these directions (including the very recent proposal to exploit the Cosmic Infrared Background anisotropies) to firmly reach a sensitivity to the non-linearity parameter $f_{NL} \sim 1$, and to eventually reach in the future the level $f_{NL} \sim 10^{-2}$ predicted by the standard models of single-field slow-roll inflation.
- **Neutrinos**. The effect that neutrinos induce can be quantified both at the background level both in their anisotropies, while the first has been clearly detected in CMB data, the second one has not been found yet and would show a 5–10% suppression of power in the total matter power spectrum. This measurement is of primary importance since it will provide the first detection of the cosmological neutrino background in the structure formation era, however the fact that neutrino signatures are evident and more pronounced at medium/small scales poses challenges in the modelling of neutrino induced non-linearities in the biased tracers used to constrain their properties. Present constraints are in the range $M_\nu < 0.12$–0.2 eV, while future missions in the next decade are likely to detect a non-zero neutrino mass since forecasted 1σ sensitivities are in the range 15–20 meV.

- **Cold dark matter at small scales**. The small scale regime is particularly interesting since it can provide a direct link to the most fundamental properties of the dark matter particles (i.e. their thermal velocities, or collisional nature). The standard cosmological model is facing challenges at small scales since the dynamical properties of small dwarf galaxies are not in agreement with the theoretical expectations (this is the main tension, but there are others which are linked to this one). Intergalactic medium data provide the tightest constrains on cold dark matter coldness ruling out thermal relics of masses <3.3 keV at the 2σ C.L. This limit is showing that in this case warm dark matter will closely resemble cold dark matter in terms of structure formation. Forecasts with future instruments and satellite could allow to reach the 4–5 keV limit, in particular by exploiting the high redshift regime by means of 21 cm intensity mapping experiments.

Overall, the next decade will witness important cosmological efforts (DESI, Euclid, CMB-IV experiments) and is likely to provide new discoveries in terms of fundamental physical properties of the Universe.

Acknowledgements The authors are supported by the Astroparticle INFN I.S. "INDARK". Matteo Viel is supported by the ERC-StG "cosmoIGM". The work of Nicola Bartolo and Sabino Matarrese was supported in part by ASI/INAF Agreement I/072/09/0 for the Planck LFI Activity of Phase E2. We thank Fabio Finelli, Michele Liguori, Francisco Villaescusa-Navarro for useful discussions.

References

1. R. Brout, F. Englert, E. Gunzig, Ann. Phys. **115**, 78 (1978). https://doi.org/10.1016/0003-4916(78)90176-8
2. A.A. Starobinsky, Phys. Lett. B **91**, 99 (1980). https://doi.org/10.1016/0370-2693(80)90670-X
3. D. Kazanas, Astrophys. J. **241**, L59 (1980). https://doi.org/10.1086/183361
4. K. Sato, Mon. Not. R. Astron. Soc. **195**, 467 (1981)
5. A.H. Guth, Phys. Rev. D **23**, 347 (1981). https://doi.org/10.1103/PhysRevD.23.347
6. A.D. Linde, Phys. Lett. B **108**, 389 (1982). https://doi.org/10.1016/0370-2693(82)91219-9
7. A. Albrecht, P.J. Steinhardt, M.S. Turner, F. Wilczek, Phys. Rev. Lett. **48**, 1437 (1982). https://doi.org/10.1103/PhysRevLett.48.1437
8. V.F. Mukhanov, G.V. Chibisov, JETP Lett. **33**, 532 (1981). [Pisma Zh. Eksp. Teor. Fiz. **33**, 549 (1981)]
9. S.W. Hawking, Phys. Lett. B **115**, 295 (1982). https://doi.org/10.1016/0370-2693(82)90373-2
10. A.H. Guth, S.Y. Pi, Phys. Rev. Lett. **49**, 1110 (1982). https://doi.org/10.1103/PhysRevLett.49.1110
11. A.A. Starobinsky, Phys. Lett. B **117**, 175 (1982). https://doi.org/10.1016/0370-2693(82)90541-X
12. J.M. Bardeen, P.J. Steinhardt, M.S. Turner, Phys. Rev. D **28**, 679 (1983). https://doi.org/10.1103/PhysRevD.28.679
13. L.F. Abbott, M.B. Wise, Nucl. Phys. B **244**, 541 (1984). https://doi.org/10.1016/0550-3213(84)90329-8
14. V.F. Mukhanov, JETP Lett. **41**, 493 (1985). [Pisma Zh. Eksp. Teor. Fiz. **41**, 402 (1985)]
15. P.A.R. Ade et al., arXiv:1502.01589 (2015)

16. P.A.R. Ade et al., Astron. Astrophys. arXiv:1502.02114
17. H.V. Peiris et al., Astrophys. J. Suppl. **148**, 213 (2003). https://doi.org/10.1086/377228
18. A. Kogut et al., Astrophys. J. Suppl. **148**, 161 (2003). https://doi.org/10.1086/377219
19. M. Sasaki, Prog. Theoret. Phys. **76**, 1036 (1986). https://doi.org/10.1143/PTP.76.1036
20. V.F. Mukhanov, Sov. Phys. JETP **67**, 1297 (1988). [Zh. Eksp. Teor. Fiz. **94N7**, 1 (1988)]
21. D.H. Lyth, Phys. Rev. Lett. **78**, 1861 (1997). https://doi.org/10.1103/PhysRevLett.78.1861
22. D. Baumann et al., AIP Conf. Proc. **1141**, 10 (2009). https://doi.org/10.1063/1.3160885
23. M. Zaldarriaga, U. Seljak, Phys. Rev. D **55**, 1830 (1997). https://doi.org/10.1103/PhysRevD.55.1830
24. M. Kamionkowski, A. Kosowsky, A. Stebbins, Phys. Rev. D **55**, 7368 (1997). https://doi.org/10.1103/PhysRevD.55.7368
25. S. Dodelson, Phys. Rev. Lett. **112**, 191301 (2014). https://doi.org/10.1103/PhysRevLett.112.191301
26. M. Kamionkowski, E.D. Kovetz, 1510.06042
27. M.C. Guzzetti, N. Bartolo, M. Liguori, S. Matarrese, arXiv:1605.01615 (2016)
28. P. Ade et al., Phys. Rev. Lett. **114**, 101301 (2015). https://doi.org/10.1103/PhysRevLett.114.101301
29. P.A.R. Ade et al., Phys. Rev. Lett. **116**, 031302 (2016). https://doi.org/10.1103/PhysRevLett.116.031302
30. A.D. Linde, Phys. Lett. B **129**, 177 (1983). https://doi.org/10.1016/0370-2693(83)90837-7
31. E. Silverstein, A. Westphal, Phys. Rev. D **78**, 106003 (2008). https://doi.org/10.1103/PhysRevD.78.106003
32. J. Ellis, D.V. Nanopoulos, K.A. Olive, Phys. Rev. Lett. **111**(12), 111301 (2013). doi:https://doi.org/10.1103/PhysRevLett.111.129902, https://doi.org/10.1103/PhysRevLett.111.111301
33. F. Farakos, A. Kehagias, A. Riotto, Nucl. Phys. B **876**, 187 (2013). https://doi.org/10.1016/j.nuclphysb.2013.08.005
34. S. Ferrara, R. Kallosh, A. Van Proeyen, JHEP **1311**, 134 (2013). https://doi.org/10.1007/JHEP11(2013)134
35. R. Kallosh, A. Linde, D. Roest, JHEP **1311**, 198 (2013). https://doi.org/10.1007/JHEP11(2013)198
36. K. Freese, J.A. Frieman, A.V. Olinto, Phys. Rev. Lett. **65**, 3233 (1990). https://doi.org/10.1103/PhysRevLett.65.3233
37. L. Boubekeur, D. Lyth, JCAP **0507**, 010 (2005). https://doi.org/10.1088/1475-7516/2005/07/010
38. D.S. Salopek, J.R. Bond, Phys. Rev. D **42**, 3936 (1990). https://doi.org/10.1103/PhysRevD.42.3936
39. A. Gangui, F. Lucchin, S. Matarrese, S. Mollerach, Astrophys. J. **430**, 447 (1994). https://doi.org/10.1086/174421
40. V. Acquaviva, N. Bartolo, S. Matarrese, A. Riotto, Nucl. Phys. B **667**, 119 (2003). https://doi.org/10.1016/S0550-3213(03)00550-9
41. J. Maldacena, JHEP **5**, 13 (2003). https://doi.org/10.1088/1126-6708/2003/05/013
42. N. Bartolo, E. Komatsu, S. Matarrese, A. Riotto, Phys, Rep. **402**, 103 (2004). doi:https://doi.org/10.1016/j.physrep.2004.08.022
43. M. Liguori, E. Sefusatti, J.R. Fergusson, E.P.S. Shellard, Adv. Astron. **2010**, 980523 (2010). https://doi.org/10.1155/2010/980523
44. X. Chen, Adv. Astron. **2010**, 638979 (2010). https://doi.org/10.1155/2010/638979
45. E. Komatsu, Class. Quantum Gravity **27**(12), 124010 (2010). https://doi.org/10.1088/0264-9381/27/12/124010
46. A.P.S. Yadav, B.D. Wandelt, Adv. Astron. **2010**, 565248 (2010). https://doi.org/10.1155/2010/565248
47. L.M. Wang, M. Kamionkowski, Phys. Rev. D **61**, 063504 (2000). https://doi.org/10.1103/PhysRevD.61.063504
48. E. Komatsu, D.N. Spergel, Phys. Rev. D **63**, 063002 (2001). https://doi.org/10.1103/PhysRevD.63.063002

49. D. Babich, P. Creminelli, M. Zaldarriaga, JCAP **8**, 9 (2004). https://doi.org/10.1088/1475-7516/2004/08/009
50. P.A.R. Ade et al., Astron. Astrophys. **571**, A24 (2014). https://doi.org/10.1051/0004-6361/201321554
51. P.A.R. Ade, et al., Astron. Astrophys, arXiv:1502.01592
52. L. Verde, L. Wang, A.F. Heavens, M. Kamionkowski, Mon. Not. R. Astron. Soc. **313**, 141 (2000). https://doi.org/10.1046/j.1365-8711.2000.03191.x
53. N. Bartolo, S. Matarrese, A. Riotto, Phys. Rev. D **65**(10), 103505 (2002). https://doi.org/10.1103/PhysRevD.65.103505
54. S. Mollerach, Phys. Rev. D **42**, 313 (1990). https://doi.org/10.1103/PhysRevD.42.313
55. A. Linde, V. Mukhanov, Phys. Rev. D **56**, 535 (1997). https://doi.org/10.1103/PhysRevD.56.R535
56. K. Enqvist, M.S. Sloth, Nucl. Phys. B **626**, 395 (2002). https://doi.org/10.1016/S0550-3213(02)00043-3
57. D.H. Lyth, D. Wands, Phys. Lett. B **524**, 5 (2002). https://doi.org/10.1016/S0370-2693(01)01366-1
58. T. Moroi, T. Takahashi, Phys. Lett. B **522**, 215 (2001). https://doi.org/10.1016/S0370-2693(01)01295-3
59. C.T. Byrnes, K.Y. Choi, Adv. Astron. **2010**, 724525 (2010). https://doi.org/10.1155/2010/724525
60. N. Bartolo, S. Matarrese, A. Riotto, Phys. Rev. D **69**(4), 043503 (2004). https://doi.org/10.1103/PhysRevD.69.043503
61. X. Chen, M.X. Huang, S. Kachru, G. Shiu, JCAP **1**, 2 (2007). https://doi.org/10.1088/1475-7516/2007/01/002
62. E. Silverstein, D. Tong, Phys. Rev. D **70**(10), 103505 (2004). https://doi.org/10.1103/PhysRevD.70.103505
63. M. Alishahiha, E. Silverstein, D. Tong, Phys. Rev. D **70**(12), 123505 (2004). https://doi.org/10.1103/PhysRevD.70.123505
64. C. Cheung, A.L. Fitzpatrick, J. Kaplan, L. Senatore, P. Creminelli, JHEP **3**, 14 (2008). https://doi.org/10.1088/1126-6708/2008/03/014
65. L. Senatore, K.M. Smith, M. Zaldarriaga, JCAP **1**, 28 (2010). https://doi.org/10.1088/1475-7516/2010/01/028
66. C. Armendariz-Picon, T. Damour, V. Mukhanov, Phys. Lett. B **458**, 209 (1999). https://doi.org/10.1016/S0370-2693(99)00603-6
67. N. Arkani-Hamed, P. Creminelli, S. Mukohyama, M. Zaldarriaga, JCAP **4**, 1 (2004). https://doi.org/10.1088/1475-7516/2004/04/001
68. C. Burrage, C. de Rham, D. Seery, A.J. Tolley, JCAP **1**, 014 (2011). https://doi.org/10.1088/1475-7516/2011/01/014
69. R. Holman, A.J. Tolley, JCAP **5**, 1 (2008). https://doi.org/10.1088/1475-7516/2008/05/001
70. N. Bartolo, M. Fasiello, S. Matarrese, A. Riotto, JCAP **8**, 8 (2010). https://doi.org/10.1088/1475-7516/2010/08/008
71. P. Creminelli, G. D'Amico, M. Musso, J. Noreña, E. Trincherini, JCAP **2**, 6 (2011). https://doi.org/10.1088/1475-7516/2011/02/006
72. X. Chen, R. Easther, E.A. Lim, JCAP **6**, 23 (2007). https://doi.org/10.1088/1475-7516/2007/06/023
73. X. Chen, R. Easther, E.A. Lim, JCAP **4**, 10 (2008). https://doi.org/10.1088/1475-7516/2008/04/010
74. R. Flauger, L. McAllister, E. Pajer, A. Westphal, G. Xu, JCAP **6**, 9 (2010). https://doi.org/10.1088/1475-7516/2010/06/009
75. N. Bartolo, S. Matarrese, M. Peloso, A. Ricciardone, Phys. Rev. D **87**(2), 023504 (2013). https://doi.org/10.1103/PhysRevD.87.023504
76. M. Shiraishi, E. Komatsu, M. Peloso, N. Barnaby, JCAP **5**, 002 (2013). https://doi.org/10.1088/1475-7516/2013/05/002

77. J.L. Cook, L. Sorbo, Phys. Rev. D **85**, 023534 (2012). https://doi.org/10.1103/PhysRevD.86.069901, https://doi.org/10.1103/PhysRevD.85.023534. [Erratum: Phys. Rev. **D86**, 069901 (2012)]
78. N. Barnaby, J. Moxon, R. Namba, M. Peloso, G. Shiu, P. Zhou, Phys. Rev. D **86**, 103508 (2012). https://doi.org/10.1103/PhysRevD.86.103508
79. R.Z. Ferreira, M.S. Sloth, JHEP **12**, 139 (2014). https://doi.org/10.1007/JHEP12(2014)139
80. M. Shiraishi, A. Ricciardone, S. Saga, JCAP **1311**, 051 (2013). https://doi.org/10.1088/1475-7516/2013/11/051
81. N. Bartolo, M. Fasiello, S. Matarrese, A. Riotto, JCAP **1009**, 035 (2010). https://doi.org/10.1088/1475-7516/2010/09/035
82. L. Senatore, M. Zaldarriaga, JHEP **04**, 024 (2012). https://doi.org/10.1007/JHEP04(2012)024
83. K.M. Smith, L. Senatore, M. Zaldarriaga, arXiv:1502.00635 (2015)
84. M. Sasaki, J. Valiviita, D. Wands, Phys. Rev. D **74**, 103003 (2006). https://doi.org/10.1103/PhysRevD.74.103003
85. C.T. Byrnes, M. Sasaki, D. Wands, Phys. Rev. D **74**, 123519 (2006). https://doi.org/10.1103/PhysRevD.74.123519
86. K. Ichikawa, T. Suyama, T. Takahashi, M. Yamaguchi, Phys. Rev. D **78**, 063545 (2008). https://doi.org/10.1103/PhysRevD.78.063545
87. L. Senatore, M. Zaldarriaga, JCAP **1101**, 003 (2011). https://doi.org/10.1088/1475-7516/2011/01/003
88. D. Baumann, D. Green, Phys. Rev. D **85**, 103520 (2012). https://doi.org/10.1103/PhysRevD.85.103520
89. M. Kawasaki, K. Nakayama, T. Sekiguchi, T. Suyama, F. Takahashi, JCAP **11**, 019 (2008). https://doi.org/10.1088/1475-7516/2008/11/019
90. J.R. Fergusson, M. Liguori, E.P.S. Shellard, JCAP **1212**, 032 (2012). https://doi.org/10.1088/1475-7516/2012/12/032
91. J.R. Fergusson, M. Liguori, E.P.S. Shellard, Phys. Rev. D **82**, 023502 (2010). https://doi.org/10.1103/PhysRevD.82.023502
92. S. Kachru, R. Kallosh, A.D. Linde, J.M. Maldacena, L.P. McAllister, S.P. Trivedi, JCAP **0310**, 013 (2003). https://doi.org/10.1088/1475-7516/2003/10/013
93. X. Chen, Phys. Rev. D **71**(6), 063506 (2005). https://doi.org/10.1103/PhysRevD.71.063506
94. J.M. Cline, in *Les Houches Summer School—Session 86: Particle Physics and Cosmology: The Fabric of Spacetime Les Houches*, France, July 31–August 25, 2006, https://inspirehep.net/record/734399/files/arXiv:hep-th_0612129.pdf
95. R. Kallosh, Lect. Notes Phys. **738**, 119 (2008). https://doi.org/10.1007/978-3-540-74353-8_4
96. C.P. Burgess, L. McAllister, Class. Quantum Gravity **28**, 204002 (2011). https://doi.org/10.1088/0264-9381/28/20/204002
97. D. Baumann, L. McAllister, *Inflation and String Theory* (Cambridge University Press, 2015), https://inspirehep.net/record/1289899/files/arXiv:1404.2601.pdf
98. http://conservancy.umn.edu/handle/11299/169642
99. J. Errard, S.M. Feeney, H.V. Peiris, A.H. Jaffe, JCAP **1603**(03), 052 (2016). https://doi.org/10.1088/1475-7516/2016/03/052
100. P. Andrè et al., JCAP **1402**, 006 (2014). https://doi.org/10.1088/1475-7516/2014/02/006
101. R. Easther, H. Peiris, JCAP **0609**, 010 (2006). https://doi.org/10.1088/1475-7516/2006/09/010
102. T. Kobayashi, F. Takahashi, JCAP **1101**, 026 (2011). https://doi.org/10.1088/1475-7516/2011/01/026
103. M. Escudero, H. Ramrez, L. Boubekeur, E. Giusarma, O. Mena, JCAP **1602**(02), 020 (2016). https://doi.org/10.1088/1475-7516/2016/02/020
104. P. Creminelli, D.L. Lpez Nacir, M. Simonovi?, G. Trevisan, M. Zaldarriaga, JCAP **1511**(11), 031 (2015). https://doi.org/10.1088/1475-7516/2015/11/031
105. F.R. Bouchet, et al., arXiv:1102.2181

106. T. Matsumura et al., Mission design of LiteBIRD (2013). https://doi.org/10.1007/s10909-013-0996-1. 1311.2847 [astro-ph.IM] [J. Low. Temp. Phys. **176**, 733 (2014)]
107. A. Kogut et al., JCAP **1107**, 025 (2011). https://doi.org/10.1088/1475-7516/2011/07/025
108. J.L. Lehners, Adv. Astron. **2010**, 903907 (2010). https://doi.org/10.1155/2010/903907
109. E. Sefusatti, M. Liguori, A.P.S. Yadav, M.G. Jackson, E. Pajer, JCAP **0912**, 022 (2009). https://doi.org/10.1088/1475-7516/2009/12/022
110. N. Bartolo, S. Matarrese, A. Riotto, A. Vaihkonen, Phys. Rev. D **76**, 061302 (2007). https://doi.org/10.1103/PhysRevD.76.061302
111. M. Alvarez, et al. (2014)
112. D. Nitta, E. Komatsu, N. Bartolo, S. Matarrese, A. Riotto, JCAP **0905**, 014 (2009). https://doi.org/10.1088/1475-7516/2009/05/014
113. N. Bartolo, S. Matarrese, A. Riotto, JCAP **0701**, 019 (2007). https://doi.org/10.1088/1475-7516/2007/01/019
114. N. Bartolo, S. Matarrese, A. Riotto, JCAP **0606**, 024 (2006). https://doi.org/10.1088/1475-7516/2006/06/024
115. N. Bartolo, S. Matarrese, A. Riotto, JCAP **0605**, 010 (2006). https://doi.org/10.1088/1475-7516/2006/05/010
116. N. Bartolo, A. Riotto, JCAP **0903**, 017 (2009). https://doi.org/10.1088/1475-7516/2009/03/017
117. N. Bartolo, S. Matarrese, A. Riotto, Adv. Astron. **2010**, 157079 (2010). https://doi.org/10.1155/2010/157079
118. N. Bartolo, S. Matarrese, A. Riotto, JCAP **1202**, 017 (2012). https://doi.org/10.1088/1475-7516/2012/02/017
119. A. Lewis, JCAP **1206**, 023 (2012). https://doi.org/10.1088/1475-7516/2012/06/023
120. P. Creminelli, C. Pitrou, F. Vernizzi, JCAP **1111**, 025 (2011). https://doi.org/10.1088/1475-7516/2011/11/025
121. Z. Huang, F. Vernizzi, Phys. Rev. Lett. **110**(10), 101303 (2013). https://doi.org/10.1103/PhysRevLett.110.101303
122. S.C. Su, E.A. Lim, E.P.S. Shellard (2012)
123. G.W. Pettinari, C. Fidler, R. Crittenden, K. Koyama, A. Lewis, D. Wands, Phys. Rev. D **90**(10), 103010 (2014). https://doi.org/10.1103/PhysRevD.90.103010
124. G.W. Pettinari, C. Fidler, R. Crittenden, K. Koyama, D. Wands, JCAP **1304**, 003 (2013). https://doi.org/10.1088/1475-7516/2013/04/003
125. A. Naruko, C. Pitrou, K. Koyama, M. Sasaki, Class. Quantum Gravity **30**, 165008 (2013). https://doi.org/10.1088/0264-9381/30/16/165008
126. C. Fidler, K. Koyama, G.W. Pettinari, JCAP **1504**(04), 037 (2015). https://doi.org/10.1088/1475-7516/2015/04/037
127. R. Saito, A. Naruko, T. Hiramatsu, M. Sasaki, JCAP **1410**(10), 051 (2014). https://doi.org/10.1088/1475-7516/2014/10/051
128. J.M. Maldacena, JHEP **05**, 013 (2003). https://doi.org/10.1088/1126-6708/2003/05/013
129. D.J. Fixsen, E.S. Cheng, J.M. Gales, J.C. Mather, R.A. Shafer, E.L. Wright, Astrophys. J. **473**, 576 (1996). https://doi.org/10.1086/178173
130. J.C. Mather et al., Astrophys. J. **420**, 439 (1994). https://doi.org/10.1086/173574
131. J. Chluba, R.A. Sunyaev, Mon. Not. R. Astron. Soc. **419**, 1294 (2012). https://doi.org/10.1111/j.1365-2966.2011.19786.x
132. J. Chluba, in *Proceedings, 49th Rencontres de Moriond on Cosmology* (2014), pp. 327–334, https://inspirehep.net/record/1298250/files/arXiv:1405.6938.pdf
133. R. Khatri, R.A. Sunyaev, JCAP **1306**, 026 (2013). https://doi.org/10.1088/1475-7516/2013/06/026
134. G. Cabass, A. Melchiorri, E. Pajer, arXiv:1602.05578
135. E. Pajer, M. Zaldarriaga, Phys. Rev. Lett. **109**, 021302 (2012). https://doi.org/10.1103/PhysRevLett.109.021302
136. M. Biagetti, H. Perrier, A. Riotto, V. Desjacques, Phys. Rev. D **87**, 063521 (2013). https://doi.org/10.1103/PhysRevD.87.063521

137. R. Emami, E. Dimastrogiovanni, J. Chluba, M. Kamionkowski, Phys. Rev. D **91**(12), 123531 (2015). https://doi.org/10.1103/PhysRevD.91.123531
138. M. Shiraishi, M. Liguori, N. Bartolo, S. Matarrese, Phys. Rev. D **92**, 083502 (2015). https://doi.org/10.1103/PhysRevD.92.083502
139. N. Bartolo, M. Liguori, M. Shiraishi, JCAP **1603**(03), 029 (2016). https://doi.org/10.1088/1475-7516/2016/03/029
140. N. Dalal, O. Dore, D. Huterer, A. Shirokov, Phys. Rev. D **77**, 123514 (2008). https://doi.org/10.1103/PhysRevD.77.123514
141. S. Matarrese, L. Verde, Astrophys. J. **677**, L77 (2008). https://doi.org/10.1086/587840
142. E. Sefusatti, E. Komatsu, Phys. Rev. D **76**, 083004 (2007). https://doi.org/10.1103/PhysRevD.76.083004
143. C. Carbone, L. Verde, S. Matarrese, Astrophys. J. **684**, L1 (2008). https://doi.org/10.1086/592020
144. K.N. Abazajian et al., Astropart. Phys. **63**, 55 (2015). https://doi.org/10.1016/j.astropartphys.2014.05.013
145. J. Byun, R. Bean, JCAP **1503**(03), 019 (2015). https://doi.org/10.1088/1475-7516/2015/03/019
146. T. Giannantonio, C. Porciani, J. Carron, A. Amara, A. Pillepich, Mon. Not. R. Astron. Soc. **422**, 2854 (2012). https://doi.org/10.1111/j.1365-2966.2012.20604.x
147. A. Raccanelli, M. Shiraishi, N. Bartolo, D. Bertacca, M. Liguori, S. Matarrese, R.P. Norris, D. Parkinson, arXiv:1507.05903
148. S. Camera, et al., 1501.03851
149. T. Baldauf, M. Mirbabayi, M. Simonovi?, M. Zaldarriaga, arXiv:602.00674 (2016)
150. A. Cooray, Phys. Rev. Lett. **97**, 261301 (2006). https://doi.org/10.1103/PhysRevLett.97.261301
151. A. Pillepich, C. Porciani, S. Matarrese, Astrophys. J. **662**, 1 (2007). https://doi.org/10.1086/517963
152. J.B. Muoz, Y. Ali-Hamoud, M. Kamionkowski, Phys. Rev. D **92**(8), 083508 (2015). https://doi.org/10.1103/PhysRevD.92.083508
153. M. Tucci, V. Desjacques and M. Kunz, arXiv:1606.02323 [astro-ph.CO]
154. P.A.R. Ade et al., Astron. Astrophys. **571**, A22 (2014). https://doi.org/10.1051/0004-6361/201321569
155. S. Weinberg, Phys. Rev. D **77**, 123541 (2008). https://doi.org/10.1103/PhysRevD.77.123541
156. A.L. Berkin, K.I. Maeda, Phys. Rev. D **44**, 1691 (1991). https://doi.org/10.1103/PhysRevD.44.1691
157. S. Mollerach, S. Matarrese, Phys. Rev. D **45**, 1961 (1992). https://doi.org/10.1103/PhysRevD.45.1961
158. A.A. Starobinsky, S. Tsujikawa, J. Yokoyama, Nucl. Phys. B **610**, 383 (2001). https://doi.org/10.1016/S0550-3213(01)00322-4
159. F. Di Marco, F. Finelli, R. Brandenberger, Phys. Rev. D **67**, 063512 (2003). https://doi.org/10.1103/PhysRevD.67.063512
160. T. Qiu, J.Q. Xia, Phys. Lett. B **744**, 273 (2015). https://doi.org/10.1016/j.physletb.2015.03.061
161. N. Bartolo, D. Cannone, R. Jimenez, S. Matarrese, L. Verde, Phys. Rev. Lett. **113**(16), 161303 (2014). https://doi.org/10.1103/PhysRevLett.113.161303
162. X. Chen, Y. Wang, Phys. Rev. D **81**, 063511 (2010). https://doi.org/10.1103/PhysRevD.81.063511
163. J.M. Maldacena, G.L. Pimentel, JHEP **09**, 045 (2011). https://doi.org/10.1007/JHEP09(2011)045
164. J. Soda, H. Kodama, M. Nozawa, JHEP **08**, 067 (2011). https://doi.org/10.1007/JHEP08(2011)067
165. M. Shiraishi, D. Nitta, S. Yokoyama, Prog. Theoret. Phys. **126**, 937 (2011). https://doi.org/10.1143/PTP.126.937

166. A. Lue, L.M. Wang, M. Kamionkowski, Phys. Rev. Lett. **83**, 1506 (1999). https://doi.org/10.1103/PhysRevLett.83.1506
167. S. Alexander, J. Martin, Phys. Rev. D **71**, 063526 (2005). https://doi.org/10.1103/PhysRevD.71.063526
168. M. Satoh, JCAP **1011**, 024 (2010). https://doi.org/10.1088/1475-7516/2010/11/024
169. A. Nicolis, R. Rattazzi, E. Trincherini, Phys. Rev. D **79**, 064036 (2009). https://doi.org/10.1103/PhysRevD.79.064036
170. C. de Rham, G. Gabadadze, Phys. Lett. B **693**, 334 (2010). https://doi.org/10.1016/j.physletb.2010.08.043
171. C. de Rham, G. Gabadadze, Phys. Rev. D **82**, 044020 (2010). https://doi.org/10.1103/PhysRevD.82.044020
172. C. Deffayet, G. Esposito-Farese, A. Vikman, Phys. Rev. D **79**, 084003 (2009). https://doi.org/10.1103/PhysRevD.79.084003
173. C. Deffayet, S. Deser, G. Esposito-Farese, Phys. Rev. D **80**, 064015 (2009). https://doi.org/10.1103/PhysRevD.80.064015
174. G.W. Horndeski, Int. J. Theoret. Phys. **10**, 363 (1974). https://doi.org/10.1007/BF01807638
175. C. Deffayet, X. Gao, D.A. Steer, G. Zahariade, Phys. Rev. D **84**, 064039 (2011). https://doi.org/10.1103/PhysRevD.84.064039
176. P.A.R. Ade, et al. (2015)
177. J. Gleyzes, D. Langlois, F. Piazza, F. Vernizzi, JCAP **1502**, 018 (2015). https://doi.org/10.1088/1475-7516/2015/02/018
178. J. Gleyzes, D. Langlois, F. Piazza, F. Vernizzi, Phys. Rev. Lett. **114**(21), 211101 (2015). https://doi.org/10.1103/PhysRevLett.114.211101
179. X. Gao, Phys. Rev. D **90**, 081501 (2014). https://doi.org/10.1103/PhysRevD.90.081501
180. M. Crisostomi, M. Hull, K. Koyama, G. Tasinato, JCAP **1603**(03), 038 (2016). https://doi.org/10.1088/1475-7516/2016/03/038
181. D. Langlois, K. Noui, JCAP **1602**(02), 034 (2016). https://doi.org/10.1088/1475-7516/2016/02/034
182. D. Langlois, K. Noui (2015)
183. J.B. Achour, D. Langlois, K. Noui (2016)
184. M. Crisostomi, K. Koyama, G. Tasinato, JCAP **1604**(04), 044 (2016). https://doi.org/10.1088/1475-7516/2016/04/044
185. P. Brax, Class. Quantum Gravity **30**, 214005 (2013). https://doi.org/10.1088/0264-9381/30/21/214005
186. G. Mangano, P.D. Serpico, Phys. Lett. B. **701**, 296 (2011)
187. J. Lesgourgues, S. Pastor, Phys. Rep. **429**, 6 (2006)
188. S. Dell'Oro, S. Marcocci, M. Viel, F. Vissani, J. Cosmol. Astropart. Phys. **12**, 023 (2015)
189. S. Dell'Oro, S. Marcocci, M. Viel, F. Vissani, arXiv: eprint 1601.07512 (2016)
190. J. Brandbyge, S. Hannestad, T. Haugblle, B. Thomsen, J. Cosmol. Astropart. Phys. **08**, 020 (2008)
191. M. Viel, M.G. Haehnelt, V. Springel, J. Cosmol. Astropart. Phys. **06**, 015 (2010)
192. C. Wagner, L. Verde, R. Jimenez, Astrophys. J. Lett. **752**, 02 (2012)
193. S. Bird, M. Viel, M.G. Haehnelt, MNRAS, **420**, 2551 (2012)
194. F. Villaescusa-Navarro, S. Bird, C. Pea-Garay, M. Viel, J. Cosmol. Astropart. Phys. **03**, 019 (2013)
195. E. Massara, F. Villaescusa-Navarro, M. Viel, J. Cosmol. Astropart. Phys. **12**, 053 (2014)
196. M. Peloso, M. Pietroni, M. Viel, F. Villaescusa-Navarro, J. Cosmol. Astropart. Phys. **07**, 001 (2015)
197. E. Castorina, E. Sefusatti, R.K. Sheth, F. Villaescusa-Navarro, M. Viel, J. Cosmol. Astropart. Phys. **02**, 049 (2014)
198. M. Costanzi, B. Sartoris, M. Viel, S. Borgani, J. Cosmol. Astropart. Phys. **10**, 081 (2014)
199. F. Villaescusa-Navarro, F. Marulli, M. Viel, E. Branchini, E. Castorina, E. Sefusatti, S. Saito, J. Cosmol. Astropart. Phys. **03**, 011 (2014)

200. Planck Collaboration: Planck 2015 results. XIII. Cosmological parameters, arXiv:1502.01589 (2015)
201. E. Di Valentino, E. Giusarma, O. Mena, A. Melchiorri, J. Silk, arXiv:1511.00975
202. N. Palanque-Delabrouille, C. Yeche, J. Lesgourgues, G. Rossi, A. Borde, M. Viel, E. Aubourg, D. Kirkby, J.-M. LeGoff, J. Rich, N. Roe, N.P. Ross, D.P. Schneider, D. Weinberg, J. Cosmol. Astropart. Phys. **02**, 045 (2015)
203. A.J. Cuesta, V. Niro, L. Verde, arXiv:1511.05983
204. R.A. Battye, A. Moss, Phys. Rev. Lett. **112**, 5, id. 051303 (2014)
205. M. Wyman, D.H. Rudd, R.A. Vanderveld, W. Hu, Phys. Rev. Lett. **112**, 051302 (2014)
206. N. Palanque-Delabrouille, C. Yeche, J. Baur, C. Magneville, G. Rossi, J. Lesgourgues, A. Borde, E. Burtin, J.-M. LeGoff, J. Rich, M. Viel, D. Weinberg, J. Cosmol. Astropart. Phys. **11**, 011 (2015)
207. A. Meiksin, Rev. Modern Phys. **81**, 4 (2009)
208. V.K. Narayanan, D.N. Spergel, R. Dave, C.P. Ma, ApJ **543**, 103 (2000)
209. M. Viel, J. Lesgourgues, M.G. Haehnelt, S. Matarrese, A. Riotto, Phys. Rev. D **71**(6), id. 063534 (2005)
210. U. Seljak, A. Makarov, P. McDonald, Trac. Phys. Rev. Lett. **97**, 191303 (2006)
211. M. Viel, J. Lesgourgues, M.G. Haehnelt, S. Matarrese, A. Riotto, **97**, 071301 (2006)
212. M. Viel et al., Phys. Rev. Lett. **100**, 041304 (2008)
213. M. Viel, G.D. Becker, J.S. Bolton, M.G. Haehnelt, Phys. Rev. D **88**(4), id. 043502 (2013)
214. A. Boyarsky, J. Lesgourgues, O. Ruchayskiy, M. Viel, J. Cosmol. Astropart. Phys. **05**, 012 (2009)
215. F. Villaescusa-Navarro, P. Bull, M. Viel, Astrophys. J. **814**(2), 146 (2015)
216. B. Audren, J. Lesgourgues, S. Bird, M.G. Haehnelt, M. Viel, J. Cosmol. Astropart. Phys. **01**, 026 (2013)

The manufacturer's authorised representative in the EU is Springer Nature Customer Service Centre GmbH, Europaplatz 3, 69115 Heidelberg, Germany. If you have any concerns regarding our products, please contact ProductSafety@springernature.com

Printed and bound by CPI Group (UK) Ltd, Croydon, CR0 4YY
05/10/2026
02228139-0018